The Nature of the Environment

Andrew Goudie

School of Geography and the Environment, University of Oxford

The Nature of the Environment

Fourth Edition

First published 1984
Reprinted 1985, 1987, 1988
Second edition 1989
Reprinted 1990, 1992
Third edition 1993
Reprinted 1994, 1996, 1999, 2000

Fourth edition published 2001

2 4 6 8 10 9 7 5 3 1

Blackwell Publishers Ltd
108 Cowley Road
Oxford OX4 1JF
UK

Blackwell Publishers Inc.
350 Main Street
Malden, Massachusetts 02148
USA

British Library Cataloguing in Publication Data

A CIP catalogue record for this book is available from the British Library.

Library of Congress Cataloging-in-Publication Data

Goudie, Andrew.
The nature of the environment / Andrew Goudie.— 4th ed.
p. cm.
Includes bibliographical references and index.
ISBN 0–631–22463–7 (acid-free paper) — ISBN 0–631–20069–X (pbk. : acid-free paper)
1. Physical geography. I. Title.
GB54.5 .G68 2000
910′.02—dc21

00–045446

Typeset in 10 on 11.5 pt Sabon
by Ace Filmsetting Ltd, Frome, Somerset
Printed in Great Britain by
Biddles Ltd
www.biddles.co.uk

This book is printed on acid-free paper.

Frontispiece A montane forest ecosystem on the slopes of a volcano in the Mount Bosavi National Park, Papua New Guinea

Contents

Windows

Preface

The content of courses in physical geography has changed markedly in recent years for three main reasons. First, there has been a desire to make physical geography more relevant to human affairs, to integrate it more closely with human geography, to become more concerned with natural hazards, to investigate environmental problems and to assess the human impact on environmental change. Second, geography as a whole has become involved with process, measurement and numerical analysis. Third, and most important, near-revolutionary developments have taken place in the subject matter of physical geography – notably in ecology, hydrology, plate tectonics and our knowledge of the Pleistocene.

In chapters 1 and 2 we cover developments in such matters as plate tectonics and climatic change, as well as background information on global patterns of natural phenomena. One aim of the book is to stress the way in which the components of the environment are integrated at different scales. Accordingly, Part II deals with the four main world zones (polar, mid-latitude, desert and tropical), describing their features, showing how the different components of the environment are interrelated and explaining their problems, hazards and some of the modifications wrought by man. Part III deals with two rather special environments that occur in any one of the four main zones – mountains and coasts – but the same approach is followed as in the previous four chapters. Part IV moves from major environments to a consideration of some major classes of phenomena (e.g. rivers), and once again interrelationships are explored and human implications assessed.

The overall purpose of the book, therefore, is to impart modern information on the human environment at various scales from the global to the local; to integrate the study of geomorphology, climatology, hydrology, pedology and biogeography; and to consider the ways in which we ourselves both mould and are moulded by our landscape and environment.

In this fourth edition I have used the framework of earlier editions, but have added an extra chapter, expanded and updated the guides to further reading, added a list of key concepts at the end of each chapter, suggested some points for review, increased the number of windows quite substantially, updated many of the examples used, and stressed still further the importance of hazards, natural environmental changes, and human impacts. There are also many new plates, diagrams and tables. I hope that this fourth edition will be even more useful and accessible to its readers than its predecessors.

To avoid the disturbance that results when the text is broken up by detailed referencing, I have refrained from providing such documentation. This means that I cannot give the acknowledgement I would like to the work and ideas of other authors, to whom I am greatly in debt.

Acknowledgements

I am most grateful to Jane Battersby and Jan Burke for their tremendous assistance in preparing this volume. They have assisted me with rare skill and efficiency. I am also grateful to Ailsa Allen and David Sansom for drawing the new figures.

The author and publisher gratefully acknowledge permission to reproduce copyright material in this book.

Figure 1.12: R. S. Dietz, 'Geosynclines, mountains and continent building', *Scientific American*, 226 (3), 1972, p. 37, copyright © Ikuyo Tagawa Garber, executrix to the Estate of Bunji Tagawa; figure 1.14: H. Brown and M. V. Lomolino, *Biogeography*, 2nd edn (2 figures) (Sinauer Associates, Inc., Sunderland, MA, 1998); figure 1.17: Whitmore et al., *The Earth as Transformed by Human Action*, eds B. L. Turner et al. (Cambridge University Press, Cambridge, 1998), fig. 2.1; window 1.2: Allegre, *The Behaviour of the Earth* (Cambridge University Press, Cambridge, 1998), fig. 29; window 1.3: I. E. Whitehouse, *Zeitschrift für Geomorphologie*, suppl. 69 (Gebruder Borntraeger Verlagsbuchhandlung, 1988); window 1.5: D. G. Howell, *Tectonics of Suspect Terrain* (Chapman and Hall, London, 1989), with kind permission of Kluwer Academic Publishers, Dordrecht; figure 2.21 (a and b): J. T. Houghton et al. (eds), *Climate Change: The IPCC Scientific Assessment* (Cambridge University Press, Cambridge, 1996: copyright © Intergovernmental Panel on Climate Change, Geneva); J. T. Houghton, *Global Warming: The Complete Briefing* (Lion Books, Oxford, 1994); figure 2.22: Shukla et al., *Science*, 247, 1990, p. 1322: Copyright © 1990 American Association for the Advancement of Science; window 2.4: Broecker and Denton, reprinted from *Quaternary Science Review*, 9, 1990, pp. 305–41: copyright © 1990 with permission of Elsevier Science; window 2.5: C. Y. Jim, 'The forest fires in Indonesia 1997–1998: possible cause and pervasive consequences', *Geography*, 84, 1999, pp. 251–60, reproduced courtesy of The Geographical Association; window 2.8: J. C. Ritchie, E. H. Eyles and C. V. Haynes, reprinted by permission of *Nature*, 314, 1985, pp. 352–5, copyright © 1985 Macmillan Magazines Ltd; figure 3.2: J. H. Brown and M. V. Lomolino, *Biogeography*, 2nd edn (Sinauer, Sunderland, MA, 1998); window 3.2: R. Cochrane, 'The changing state of the vegetation cover in New Zealand', from A. G. Anderson (ed.), *New Zealand in Maps* (Hodder and Stoughton Educational, London, 1974); figure 4.4: J. Nye, 'The mechanics of glacier flow', reprinted from the *Journal of Glaciology*, 2, 1952, fig. 8, with permission of the International Glaciological Society; figure 4.5: Larry W. Price, *Mountains and Man: A Study of Process and Environment* (University of California Press, Berkeley, 1981), fig. 5.15; figure 4.14: R. J. Rice, *Fundamentals of Geomorphology* (Longman, Harlow, 1977), figs 13.5 and 13.6, copyright © R. J. Rice 1977; figure 4.19: J. R. Mackay, *Canadian Geotechnical Journal*, 7, 1970, fig. 6.6, reproduced by permission of NRC Research Press, Ottawa; window 4.5: H. M. French, *The Periglacial Envi-*

ronment, 2 edn (Longman, Harlow, 1996), fig. 17.5, and D. A. Anisimov, *Physical Geography*, 10, 1989, pp. 282–93 (V. H. Winston and Son, Columbia, 1989); window 4.6: C. Warren, *Geography Review*, 10 (4), 1997, pp. 2–6, adapted by permission of the American Geographical Society, New York; figure 5.1: M. D. Newson and J. Hanwell, *Systematic Physical Geography* (Macmillan, London, 1982, fig. 2.10; figure 5.5: I. D. White, D. N. Mottershead and S. J. Harrison, *Environmental Systems: An Introductory Text* (Allen and Unwin, London, 1984), fig. 8.10; figure 5.8: E. M. Bridges, *World Soils* (Cambridge University Press, Cambridge, 1970), fig. 5.6; figure 5.9c: T. Marsh and R. Monkhouse, *Weather*, 45, 1990, reproduced courtesy of the Royal Meteorological Society, Reading; figure 5.10: A. H. Perry, *Environmental Hazards in the British Isles* (George Allen and Unwin, London, 1981, courtesy of Routledge), fig. 3.5; figure 5.12: A. H. Perry, *Environmental Hazards in the British Isles* (George Allen and Unwin, London, 1981, courtesy of Routledge) fig. 5.2, and J. Glasspoole and H. Rowsell, 'Absolute droughts and partial droughts over the British Isles 1906–1940', *Meteorological Magazine*, 76, 1947; window 5.2: C. Zabinski and M. B. Davis, *The Potential Effects of Global Climate Change in the United States*, ed. J. B. Smith and D. Tirpak (US Environmental Protection Agency, Washington, DC, 1989), Appendix D, 5.1–5.19, and P. Kauppi and M. Posch, *The Impact of Climatic Variations on Agriculture*, vol. 1, ed. M. L. Carter et al., reproduced by permission of Kluwer Academic Publishers, Dordrecht, 1988; window 5.3: J. Rose et al. from *Geomorphology and Soils*, ed. K. S. Richards, R. R. Arnett and S. Ellis (Allen and Unwin, London, 1985), fig. 18.2; window 5.4: K. Pye, *Progress in Physical Geography*, 8 (Hodder and Stoughton, London, 1984); window 5.6: M. Parry and D. K. C. Jones, 'Future climatic change and land use in the UK', *Geographical Journal*, 159, 1993, fig. 11; figure 6.4: G. N. Louw and M. Seely, *Ecology of Desert Organisms* (Longman, Harlow, 1982), figs 3.2, 3.8 and 4.11, and J. E. W. Dixon and G. N. Louw, *Madoqua*, 11, 1978, figs 5 and 4; window 6.1: N. Lancaster (ed.), *The Namib Sand Sea: Dune forms, Processes and Sediments*, 90 6191 697 6, 1989, 23 cm, 192pp, EUR 93.50/US$110.00/GBP66, A. A. Balkema, PO Box 1675, Rotterdam, Netherlands; figure 7.3: S. Nieuwolt, *Tropical Climatology: An Introduction to the Climates of Low Latitudes* (John Wiley and Sons, New York, 1977) figs 6.4 and 6.5, copyright © S. Nieuwolt, 1977; figure 7.6: D. R. Harris (ed.), *Human Ecology in Savanna Environments* (Academic Press, London, 1980), fig. 2; figure 7.9: Peter Haggett (ed.), *Geography: A Modern Synthesis*, 3rd edn (Harper and Row Publishers, Inc., New York, 1979), fig. 8.4, copyright © Peter Haggett, 1979; figure 7.13: D. R. Harris (ed.), *Human Ecology in Savanna Environments* (Academic Press, London, 1980), fig. 4; figure 7.14: J. Davies, *Geographical Variation in Coastal Development* (Longman, Harlow, 1980), figs 43, 45 and 46, and P. A. Furley and W. W. Newey, *Geography of the Biosphere* (Butterworth, London, 1983), fig. 11.11, reprinted by permission of Butterworth Heinemann Publishers, a division of Reed Educational and Professional Publishing Ltd; figure 7.15: Stokes, Judson and Picard, *Introduction to Geology: Physical and Historical* (Prentice-Hall, Englewood Cliffs, NJ, 1978), p. 311, reprinted by permission of Prentice-Hall Inc., Upper Saddle River, NJ; window 7.1: D. Barker and D. Miller, 'Hurricane Gilbert: anthropomorphising a national disaster', figs 1 and 2, from Area 22, 1990, The Institute of British Geographers; figure 8.5b: H. Walter, *Vegetation of the Earth* (Springer Verlag, 1973: Heidelberg Science Library, vol. 15, 1973); figure 8.6: L. W. Saan from *Arctic and Alpine Environments*, ed. H. E. Wright and W. H. Osburn (Indiana University Press, Bloomington, Ind., 1967); figure 8.7: J. R. Flenley, *The Equatorial Rainforest: A Geological History* (Butterworth, London, 1979), reprinted by permission of Butterworth Heinemann Publishers, a division of Reed Educational and Professional Publishing Ltd; figure 9.4: R. J. Small, *The Study of Landforms* (Cambridge University Press, Cambridge, 1970), fig. 184; figure 9.13: R. H. Meade and S. W. Trimble, from 'Changes in sediment loads in rivers of the Atlantic drainage of the United States since 1900', publication of the International Association of Hydrological Science, 113, 1974; figure 10.1: I. G. Simmons, *Biogeographical Processes* (Allen and Unwin, London, 1982), fig. 2.3; figure 10.2: A. S. Boughey, *Ecology of Populations* (Macmillan, London, 1986), p. 102, copyright ©

1968 by A. S. Boughey; figure 10.3: adapted from Eugene P. Odum, *Fundamentals of Ecology*, 3rd edn (Holt, Rinehart and Winston, CBS College Publishing, n.d.), copyright © W. B. Saunders College; figure 10.5: R. J. Whittaker, *Island Biogeography* (Oxford University Press, Oxford, 1998); figure 10.6: modified from M. Gorman, *Island Ecology* (Chapman and Hall, London, 1979), with kind permission from Kluwer Academic Publishers, Dordrecht; figure 10.9: C. Elton, *Ecology of Invasions by Animals and Plants* (Methuen, London, 1958), p. 22; figure 10.10: Nature Conservancy Council, *Summary of Objectives and Strategy* (1984), p. 7; figure 10.12: World Conservation Monitoring Centre, *Global Biodiversity* (Chapman and Hall, London, 1992), fig. 16.5, copyright © World Conservation Monitoring Centre; window 10.2: C. Bronmark and L. A. Hansson, *The Biology of Lakes and Ponds* (Oxford University Press, Oxford, 1998), fig. 6.7; figure 11.4: D. Kelletat, *Zeitschrift für Geomorphologie*, suppl. 81, figs 4 and 5; figure 11.7: Peter W. Birkeland and Edwin E. Larson, *Putnam's Geology*, 3rd edn (Oxford University Press, Oxford, n.d.); figure 12.1: M. A. Carson and M. Kirkby, *Hillslope Form and Process* (Cambridge University Press, Cambridge, 1972), fig. 5.2; figure 12.6: M. J. Selby, *Hillslope Materials and Processes* (Oxford University Press, Oxford, 1982), fig. 7.29; figure 13.1: I. Fenwick and B. J. Knapp, *Soil Processes and Response* (Duckworth, London, 1982), figs 6.28 and 6.29, reprinted by permission of Gerald Duckworth & Co. Ltd; figure 13.4: modified from D. I. Smith and T. C. Atkinson, 'Process, landforms and climate in limestone regions', from *Geomorphology and Climate*, ed. E. Derbyshire (John Wiley and Sons, New York, 1976); figure 13.5: M. Meybeck, *American Journal of Science*, 287, 1987, pp. 401–28; window 13.1: Song Lin Hoa, *Progress in Physical Geography*, 5 (4) (Hodder and Stoughton, London, 1981); figure 14.1: Frederick K. Lutgens and Edward J. Tarbuck, *The Atmosphere: An Introduction to Meteorology*, 7th edn (Prentice-Hall, Englewood Cliffs, NJ, 1989), reprinted by permission of Prentice-Hall, Inc., Upper Saddle River, NJ; figure 14.2: A. Henderson-Sellers and P. J. Robinson, *Contemporary Climatology* (Longman, Harlow, 1986), fig. 3.20; figure 14.7: K. Gilman and M. D. Newson, *Soil Pipes and Pipeflow: A Hydrological Study in Upland Wales* (Geo Books, 1980), fig. 17; figure 14.10: modified from J. C. Rodda, 'The flood hydrograph' in *Water, Earth and Man*, ed. R. J. Chorley (Methuen, London, 1967), fig. 9.12; figure 14.13: H. C. Pereira, *Land Use and Water Resources* (Cambridge University Press, Cambridge, 1973), fig. 8; window 14.4: James T. Teller, in *North America and Adjacent Oceans during the Last Deglacialisation*, ed. W. F. Ruddiman and H. E. Wright Jr (Geological Society of America, 1987), fig. 3.2; figure 15.2: K. S. Richards, *Rivers, Forms and Processes in Alluvial Channels* (Methuen, London, 1982), fig. 2.3; figure 15.9: F. Press and R. Siever, *Earth*, 3rd edn (W. H. Freeman and Co., San Francisco, 1982), figs 7.43 and 7.44; figure 15.13: I. Statham, *Earth Surface Sediment Transport* (Oxford University Press, Oxford, 1977), fig. 6.14, and D. I. Smith and P. Stopp, *The River Basin* (Cambridge University Press, Cambridge, 1978), fig. 14.10; figure 15.14: D. Walling and A. H. A. Kleo, 'Sediment yields of rivers in areas of low precipitation: a global view', from *The Hydrology of Areas of Low Precipitation*, Proceedings of Canberra Symposium, December 1979, pp. 479–93, courtesy of the International Association of Hydrological Sciences; window 15.1: *New Scientist* 14 April 1990; figure 16.2: P. Brimblecombe, reprinted from *Atmospheric Environment*, vol. 11, fig. 1, copyright © 1977 with permission from Elsevier Science; figure 16.6: K. S. Richards and Wood, in *River Channel Changes*, ed. K. J. Gregory (John Wiley and Sons, New York, 1977), fig. 24.6; figure 16.7: A. Gameson and A. Wheeler, in *Recovery and Restoration of Damaged Ecosystems*, eds J. Carins, K. L. Dickson and E. E. Hendricks (The University Press of Virginia, Charlottesville, 1977), fig. 4, reprinted with permission of the University Press of Virginia.

Table 3.2: A. Grainger, *Controlling Tropical Deforestation* (1992), reproduced with the kind permission of Kogan Page Ltd; table 5.4: A. H. Perry, *Environmental Hazards in the British Isles* (Allen and Unwin, London, 1981), courtesy of Routledge; table 8.1: R. Geiger, *The Climate near the Ground* (Harvard University Press, Cambridge, Mass., 1965), p. 444, copyright © Friedrich Vieureg &

Sohn, Weisbaden; table 8.2: K. Hewitt, 'Risk and disasters in mountain lands', in *Mountains of the World: A Global Priority*, ed. B. Messerli and J. D. Ives (Parthenon Publishing, Carnforth, Lancs, 1997), pp. 371–406, copyright © B. Messerli; table 12.2: D. K. Keefer, 'Earthquake induced landslides and their effect on alluvial fans', *Journal of Sedimentary Research*, 68, 1999, pp. 84–104; table 13.4: R. P. C. Morgan, *Soil Erosion and Conservation*, 2nd edn (Longman, Harlow, 1995), table 1.1; window 15.2: Illinois State Water Survey Miscellaneous Publication, 151 (1994), reproduced by permission of Illinois State Water Survey, Champaign, Ill.

Plates

Frontispiece WWF/Dr Tomas Schultze-Westrum; Part II title Planet Earth Pictures/E.C.G. Lemon; Part III title WWF; Part IV title WWF/Charles Vaucher; Part V title Greenpeace Photo Library/Midgley; 1.2 Grant Heilman/William Felger; 1.4 Eros Data; 1.5 window Steve Haynes; 2.1 University of Dundee; 3.4 National Portrait Gallery; 3.5 Australian Information Service; 3.5 bottom, 4.6, 9.3, 11.3, 11.7, 14.2 Popperfoto; 4.2, 4.8 Boston Museum of Science/Bradford Washburn; 4.4 J. Allan Cash; 4.10 A. L. Washburn; 4.11, 4.13 Colin Monteath; 4.2 window HM the Queen in Right of Canada, reproduced from the Collection of the National Air Photo Library with permission of Energy, Mines & Resources, Canada; 5.7 Roger Tutt; 5.8 Wessex Water; 6.1 Central advocate; 6.15 Nigel Press; 6.2 window National Archives Photo 114-SD-5089 (American Image 33, Soil Conservation Service); 6.3 window Hutchison Library/Bernard Regent; 7.3 Oxfam/Mike Goldwater; 7.4 Geoslides; 7.6 South American Pictures; 7.8 Katz/Mansell/Timepix; 7.1 window Popperfoto/Reuters; 8.2, 8.4, 12.8 Mountain Camera/John Cleare; 8.3, 11.10 Grant Heilman; 9.2, 9.5 Cambridge University Library; 9.4 Nature Conservancy Council; 9.8 Aerofilms; 10.5 Museum of Oxford; 10.4 window Panos Pictures/Jiri Polacek; 11.4 ICCE/Daisy Blow; 11.5, 11.11, 11.12, 11.13, 15.3 Eric Kay; 11.6 Contact/Colorific/Douglas Kirkland; 12.6, 12.9 Professor Geoffrey J. Martin; 12.7 Royal Geographic Society; 13.8 Jerry Wooldridge; 13.1 window Panos Pictures/Sean Sprague; 14.1 window K. Shone; 14.4 Hutchison Library/Maurice Harvey; 16.1 Malcolm Newdick; 16.2 Panos Pictures/Barbara Klass.

Colour Plates

1 Edward Derbyshire; 2 Soil Survey & Land Research Centre; 5 Hutchison Library/S. Errington; 7 Hutchison Library/Bernard Regent; 12, 13 Nature Images Inc./Helen Longest-Slaughter; 15 Louisiana State University/Dr L. Rouse.

Part I

The Global Framework

1 Global Geological Background

1.1 The Ancient Earth

Our solar system consists of planets such as the earth, moons, asteroids, comets, meteorites, dust and gas – and a central star – the sun. We do not know exactly how the solar system formed, though the most favoured theory is that condensation took place from a great revolving gas cloud in space, which gradually contracted under gravity. This made the central mass hot enough for thermonuclear reactions to set in, and so a new star was formed – the sun.

Although we may not be sure *how* the solar system originated, we are rather more certain about *when* it originated. Dates for the oldest earth rocks, the oldest moon rocks and some stony meteorites all suggest that the solar system is about 4600 million years old. This is an almost inconceivable time span, especially when we think that our own species has inhabited the earth for only a minute fraction of that time – probably around 2–3 million years.

The world we know today thus has an enormous and complicated history, and to understand the present configuration of the major elements of our planet – the oceans and the continents – we need to look at this history.

Opposite The weathering and erosion of closely bedded sedimentary rocks has produced the distinctive 'pancake rock' formations on the west coast of the South Island of New Zealand.

1.2 Core, Mantle and Crust

When it first formed, the earth must have been a molten mass. As it gradually cooled it became dif-

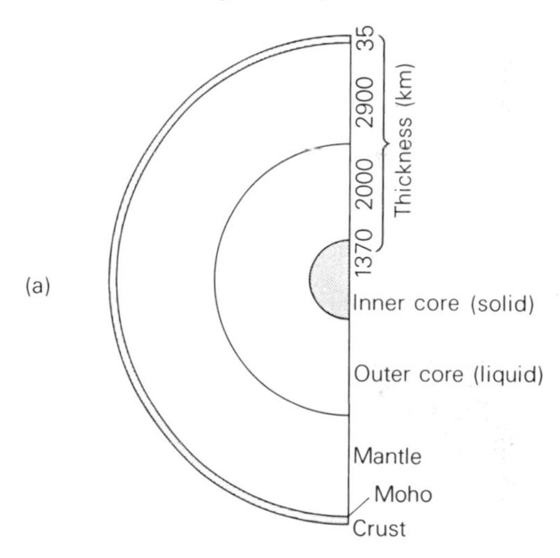

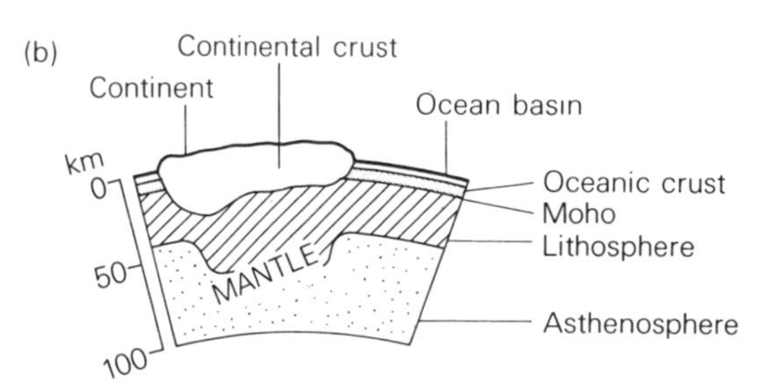

Figure 1.1 The structure of the earth: (a) the major zones; (b) the outer layers.

ferentiated into a series of concentric layers. In the interior of the earth (figure 1.1) we have a layer called the *core*, the outer parts of which have the properties of a liquid and the inner parts of which have the properties of a solid. We know that it is very dense, and that it is probably composed largely of iron, with lesser amounts of other elements such as nickel. We also know that it has very high temperatures – probably of the order of 5500 °C – and very high pressures.

The middle layer of the earth is called the *mantle*. This is thick (*c*.2800 km) and composed of material in the solid state. It consists of 'heavy rock' including peridotite (composed largely of the silicate minerals olivine and pyroxene), dunite (pure olivine) and eclogite (a dense form of basalt). Temperatures within the mantle range from 5000 °C near the core to 1300 °C just below the crust. There are two main parts of the mantle. One part is so hot that it is semi-molten and deformable; this is called the *asthenosphere*. It is overlain by a more rigid layer called the *lithosphere*.

The outer layer of the earth, called the *crust*, is very much thinner, generally being between 6 and 70 km thick, and very much less dense than the underlying mantle, from which it has been derived by complex processes operating over many millions of years. The surface of separation between crust and mantle is called the *Mohorovičić Discontinuity* (after a Yugoslavian seismologist), but is often simplified to *Moho*. The crust occurs as two types: continental and oceanic. The continental crust (known as *sial*) is not very dense (averaging 2.7 g cm^{-3} compared with 3.0–3.3 for oceanic crust and 3.4 for the upper mantle), but it tends to be relatively thicker (averaging 35–40 km, and reaching 60–70 km under high mountain chains) than the oceanic crust (called *sima*), which for its part averages only about 5–6 km in thickness. The continental crust has a varied and complicated composition, though it tends to be granitic above and basaltic beneath. The crust of the ocean basins consists simply of basaltic rocks – the granite is absent.

One particularly fascinating and important discovery of recent years has been that the ocean floors (and underlying oceanic crust) are relatively young. Some parts of the continental crust in Greenland and southern Africa are older than 3500 million years, whereas the oceanic crust is nowhere older than 250 million years.

1.3 The Nature of the Ocean Floors

Viewed from space, one of the most striking features of the earth is how large a proportion of it is covered by the waters of the oceans. Only about 29 per cent of the earth's surface is composed of dry land – the rest is ocean. If we could remove all the water from the ocean basins some other remarkable facts would be revealed: first, the ocean floors are extremely complicated in their relief; second, that relief is characterised by some extraordinary mountains and enormous valleys or trenches.

Seaward from the coast there is generally a gen-

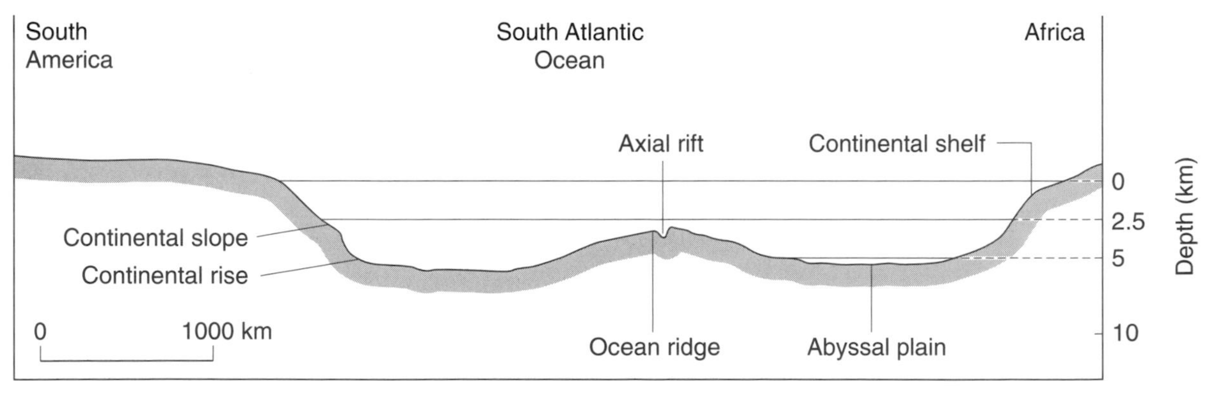

Figure 1.2 A diagrammatic cross-section of the South Atlantic to illustrate some of the main features of the geomorphology of the oceans.

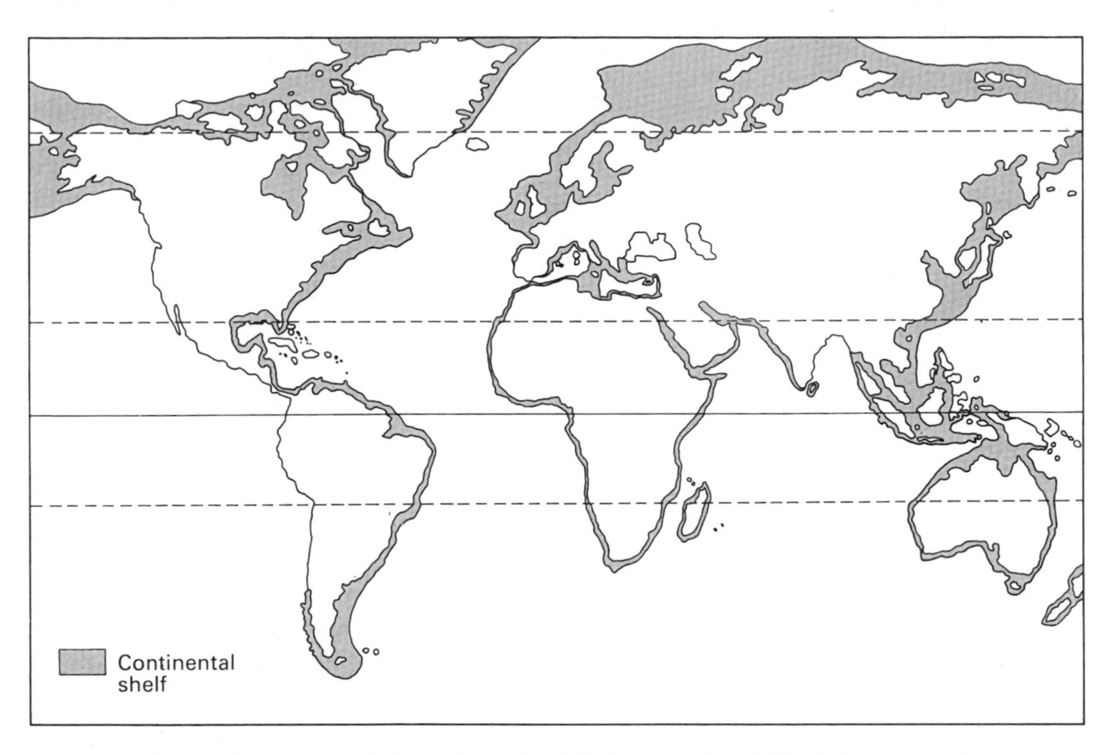

Figure 1.3 The distribution of continental shelves. Note their relatively great extent in high latitudes of the Northern Hemisphere including around the British Isles. Large expanses of the continental shelves became dry land when world sea levels were depressed during the glacial phases of the last Ice Age (Pleistocene).

tly sloping platform that is called the *continental shelf* (figure 1.2), with a water depth of no more than about 180 m. In some parts of the world this shelf is very extensive, notably off China, Canada, northern Australia and western Europe (figure 1.3). Elsewhere, for example off the western coast of South America, it is very narrow, and deep water is reached very quickly as one moves off shore. The seaward edge of the continental shelf is marked by the *continental slope*; this adjoins the *continental rise*, which runs down to the *abyssal plains*. Continental slopes and rises throughout the world are cut by deep valleys called *submarine canyons*. These are generally a few hundred metres deep and several kilometres wide, usually with a V-shaped cross-section. They are probably created by the erosional effects of *turbidity currents* – powerful sediment-laden currents that sweep off the continental slope – though in some areas of powerful downwarping they may have originated as fluvial valleys.

The abyssal plains lie at an average depth of about 5 km, but they are broken by a series of mountains. Some of these crests may be tall enough to rise above the ocean level as islands; the ones that do not break the surface are either drowned peaks (*seamounts*) or have had their tops planed off (*guyots*). Whatever their form, they all have a volcanic origin.

There is often a central ridge structure in an ocean (figure 1.4) that divides the basin into two major compartments. This is especially clearly demonstrated in the Atlantic; in the Pacific the East Pacific Rise is off-centre. These *mid-oceanic ridges* are 2–4 km in height, up to 4000 km wide, and form a nearly continuous submarine mountain range over 40 000 km in length. They sometimes break the ocean surface to give islands and groups of islands like Iceland and Tristan da Cunha. Precisely in the centre of the ridge, at its highest point, there may be a trenchlike feature called the *axial rift*, the form of which (figure 1.2) tempts one to imagine that

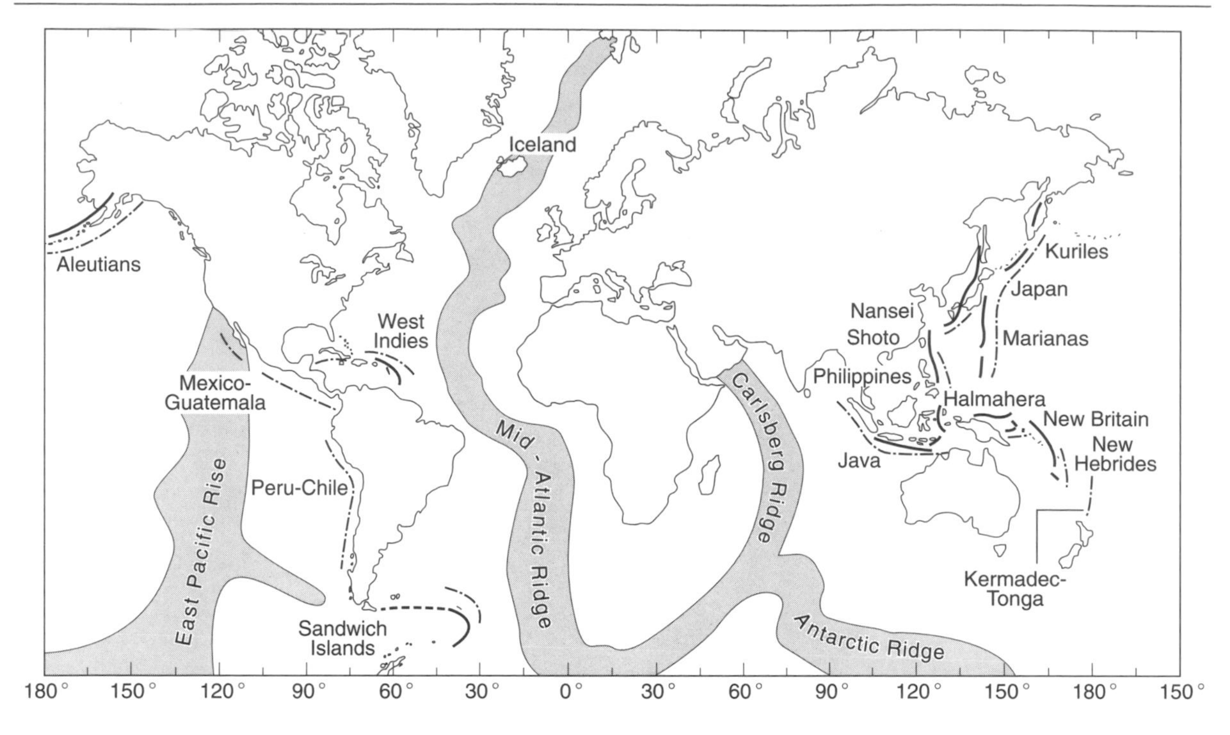

Figure 1.4 The distribution of ocean trenches (solid line), island arcs (dashed line), and mid-ocean ridges (shaded). The ridges are submarine mountains formed by basalt.

perhaps the crust is being pulled apart at this point. (Whether it is or not is something we shall shortly turn to.) The ridges are broken and repeatedly offset by innumerable fractures called *faults*.

Undoubtedly, these oceanic ridges are the most imposing relief features on the face of the earth. One such ridge runs all the way down the Atlantic Ocean, turns east and enters the Indian Ocean, where one branch of it penetrates Africa and the other continues eastwards between Australia and Antarctica before swinging across the South Pacific, eventually penetrating North America at the head of the Gulf of California.

Finally, the ocean floors, and especially the floor of the Pacific Ocean, are characterised by deep, furrow-like *trenches*, hundreds of kilometres long, tens of kilometres wide and reaching to depths in excess of 7 km. They are often flanked by volcanic islands, and may sometimes lie close to major continental mountain chains with volcanic activity (figure 1.4). The deepest place on earth lies in one of these trenches; it is the Nero Deep in the Pacific's Mariana Trench, where the greatest depth below sea level so far determined is 11 033 m. A trench always runs parallel to the edge of a continent or a line of islands, and it tends to have an asymmetric V-shaped cross-section. The steep side of the V always borders the continent or island arc, and the gentler slope leads out to the open sea.

1.4 The Surface of the Land

Basically, the land surface of the earth is characterised by two main categories of surface: active belts of mountain-building; and more inactive regions of old rocks. The former are formed either by *volcanism* or by the breaking and bending of the earth's crust by the forces of *tectonic activity*. Volcanism involves the accumulation of volcanic rock by the extrusion of molten material from within the earth, which is termed *magma*. Many high mountain ranges consist wholly or in part of chains of volcanoes built by lava or ash. In many cases tectonism and volcanism have combined to produce a particular mountain range.

The zones where active mountain-building takes place are rather narrow, and most of them lie along continental margins. They include major chains such as the European Alps and the world's highest range, the Himalayas. These chains are linked in arc-shaped groups to form the two main mountain belts. One of these, the Eurasian–Melanesian belt, begins in the Atlas Mountains of North Africa and runs through southern Europe to Turkey, Iran, the Himalayas and on into south-east Asia. In Indonesia it joins the second belt – the circum-Pacific belt – and runs up into the western Pacific. Here the mountain arcs lie well offshore from the continent of Asia and take the form of *island arcs* (figure 1.4), running through the Philippines, Japan, the Kuriles and the Aleutians. In North and South America, in contrast, this mountain belt is largely on the continents, and includes the Cordilleran ranges in North America and the Andes in South America.

The zones of active mountain-building are therefore very localised in their distribution. The remainder of the continental crust is composed of less active regions where the basement rocks are much older. These *continental shields* are quite extensive (figure 1.5). For the most part they comprise areas of low hills and plateaux beneath which lie igneous and metamorphic rocks. The shields may contain granitic rocks called *cratons*, which have been essentially undisturbed for at least 2400 million years. The old rocks may be covered by younger sedimentary layers, which accumulated when the shields

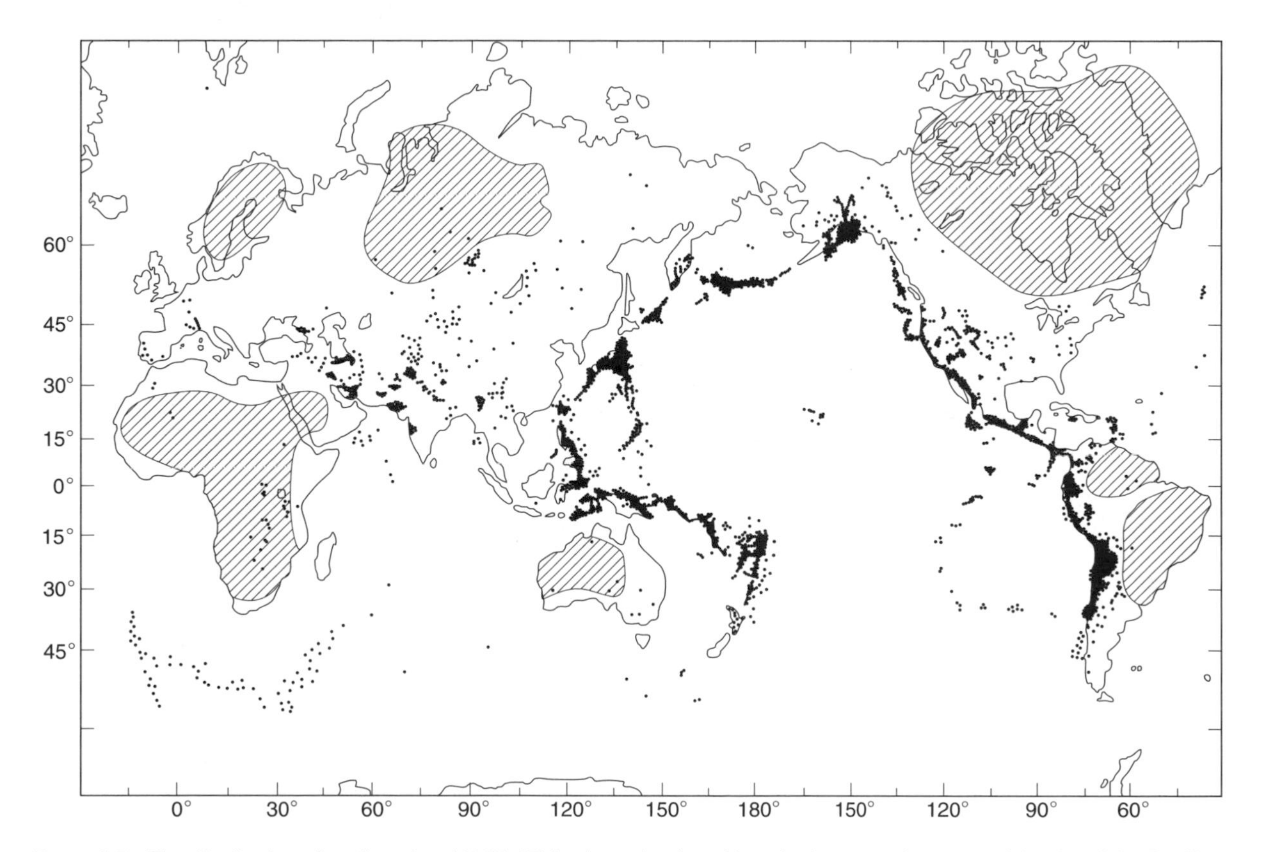

Figure 1.5 The distribution of earthquakes (1957–67) is shown by dots. Note the large number around the rim of the Pacific Ocean. The hatched areas are by contrast much more stable and represent materials cooled from the earliest molten surface of the planet. They are called shields.

subsided and were inundated by shallow seas.

The ancient shields are very extensive in Greenland, north-eastern Canada, eastern South America, much of Africa, the peninsula of India, western Australia and parts of Antarctica.

Another characteristic of these inactive areas is old mountain roots. These are formed largely of sedimentary rocks that were intensely deformed by mountain-building episodes in the past, and frequently were changed by great heat and pressure into metamorphic rocks. Such is the antiquity of these mountain-building phases that extensive erosion has since occurred, so that only the roots of the mountains remain, as chains of long, narrow ridges rising less than a few thousand metres above sea level.

1.5 Earthquakes

We have already seen that there are certain large-scale features that characterise the earth's surface – notably, the ocean ridges and the mountain chains. When we look at the pattern of earthquakes (figure 1.5) we see that they too are highly localised. A major group of earthquakes has occurred along a line formed by the mid-ocean ridges, while another group followed the world's two major mountain chains. As earthquakes represent zones in which the earth's crust is moving and is under stress (see section 11.4 for a further discussion), they give an indication of those parts of the earth where crustal activity is intense. The big question is: why does the distribution of earthquakes, mountain ranges and mid-ocean ridges show the distinctive pattern that it does? Before answering that, however, we should consider another important aspect – the shape that the continents take.

1.6 The Shape of the Continents

If one looks at a map or globe of the world, one of the most striking features is the remarkable way in

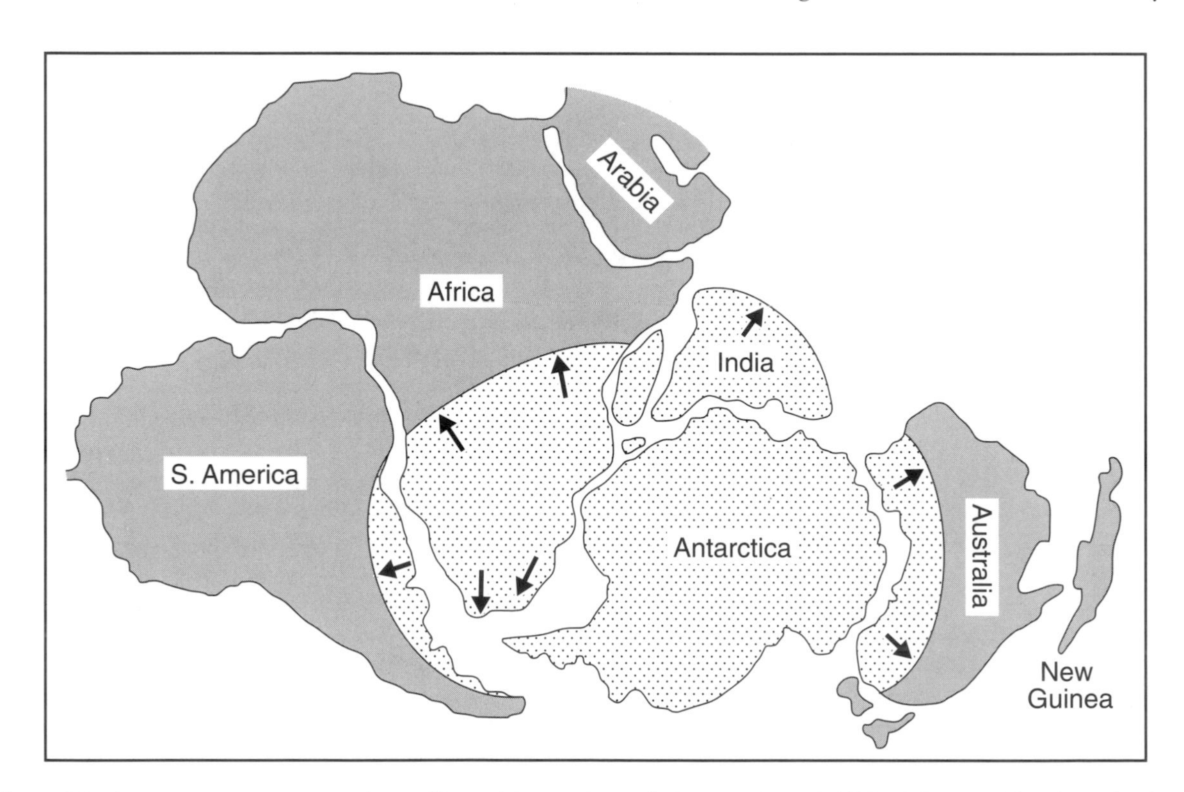

Figure 1.6 An attempt to reconstruct the positions of the continents before continental drift had disrupted Gondwanaland. The dotted area indicates the extent of glaciation on Gondwanaland, with arrows indicating the approximate direction of ice movement.

which, if they were to be pushed up against each other, South America and Africa would fit. This operation can be taken a step further; without too much difficulty, southern Australia can be made to fit on to Antarctica, and so forth, until one giant continent can be assembled like a jigsaw (figure 1.6). Furthermore, geologists have investigated the rocks that occur in, for example, Brazil and south-west Africa, and find that they are much the same in both age and type. In both regions there are shield areas in which the rocks are older than 2 billion years, separated by sharp boundaries from 550 million-year-old rocks. When the two continents are placed in their positions prior to drift, both the rocks and their boundaries correlate well. The structural grain of the continent also matches. Furthermore, from about 550 million to about 100 million years ago, when the two continents were together, there were the same sequences of erosion, glaciation, flooding, sedimentation, coal formation and volcanic activity. The past distribution of certain plants and animals indicates the same thing. For instance, *Mesosaurus* is a small reptile of the Permian Period (*c.*250 million years ago) which is known only from South Africa and southern Brazil; and *Glossopteris* is a fossil leaf abundant in many deposits of the same age in South Africa, South America, Madagascar, India and Australia.

There is also what is sometimes called 'wrong latitude' evidence. Various rocks and fossils that were thought to require rather special climatic conditions for their development were found at latitudes for which such conditions seemed impossible. For example, coal was found in rocks under the icy wastes of Antarctica; yet coal is believed to form in warm subtropical conditions like those of the present-day Florida Everglades. Conversely, evidence was found in central Africa and in lowland India of huge ice sheets which today are restricted to high latitudes. Clearly, it was not easy to reconcile observations of this kind with the present pattern of climatic belts, and so one solution was to suggest that the continents had moved, and had occupied different latitudes at different times.

These considerations have suggested that some areas that are now close together, such as North Africa and southern Europe, had once been far apart, while continents now separated by a wide ocean, such as Australia and Antarctica or Africa and America, were at one time united.

Thus we must conclude that the continents as we know them today have not been immovable, but that their shapes and positions have evolved over time. They are the drifting fragments derived from the break-up of an ancient supercontinent, called Pangaea. (The word comes from the Greek, meaning 'all land'.) The process that caused this break-up is termed *continental drift*.

Pangaea began to disintegrate about 200 million years ago, and initially split into two super-continents, Laurasia in the north and Gondwanaland in the south. By about 180 million years ago Gondwanaland had begun to break up into South America–Africa, Australia–Antarctica, and India. The initial opening up of the South Atlantic between South America and Africa began about 135 million years ago (window 1.1) and India began a rapid journey towards Asia, with which it was to collide about 45 million years ago. Australia and Antarctica began to separate about 45 million years ago, while Europe and North America broke away from each other some 5–10 million years later.

In recent years evidence to support the ancient theory of continental drift has become much more substantial. In particular, geologists have found support for the idea by studies of the past magnetism of the earth (*palaeomagnetism*) and by investigating the ocean floors.

The palaeomagnetic evidence can be explained as follows. The movement of material in the liquid outer portion of the earth's core produces a magnetic field that encompasses the earth. Molten lava, liquid igneous rocks and some water-lain sedimentary rocks all contain substances, especially iron, that respond to magnetism. As they solidify, the grains of susceptible material in the rocks orient themselves towards the magnetic pole that existed at the time of their deposition. The earth's magnetic field sometimes reverses itself, so that the north and south magnetic poles exchange places. Geologically speaking, the reversal takes place quickly and then remains stable for the order of about half a million years. The direction of the stable field is imprinted into new rocks that are formed. In the early 1960s geologists mapped the magnetic directions implanted on rocks in the ocean floor off Ice-

Window 1.1 *The opening of the South Atlantic and its impact*

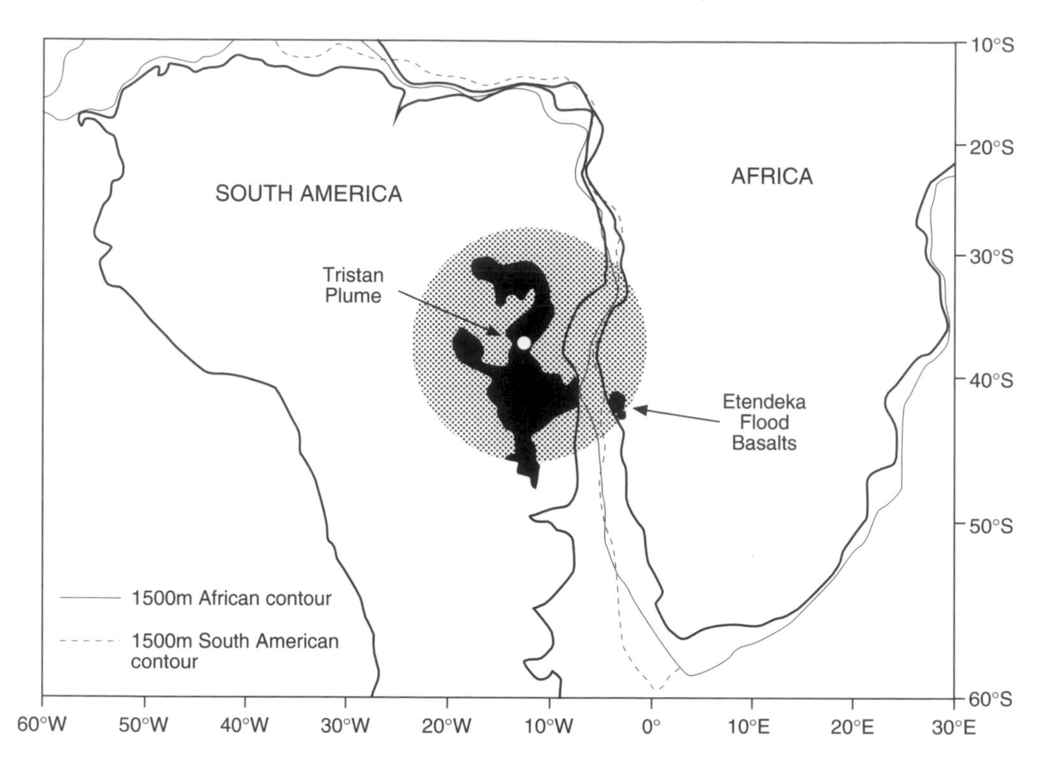

The production of the great flood basalts with sea-floor spreading, the opening of the South Atlantic and the presence of the Tristan Plume.

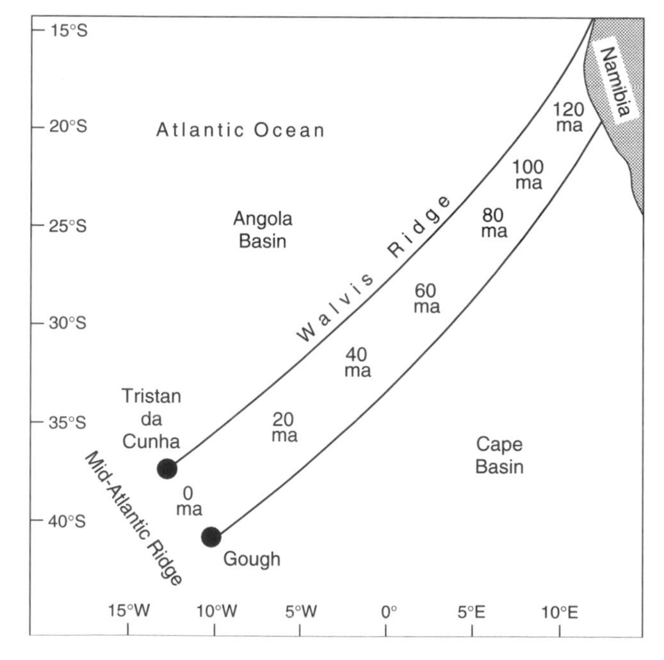

The ages of volcanic rocks in millions of years along the Walvis Ridge.

In late Jurassic to early Cretaceous times continental break-up along the line of the proto-South Atlantic Ocean was initiated. At about 150 million years ago, rifting started at the latitude of the southernmost tip of South America and progressed northwards. The rifting caused the eruption of a large body of volcanic lava (continental flood basalts). These formerly formed one large mass, called the Etendeka lavas in Namibia and the Paraná lavas in South America, but they have now been ruptured into two separate masses on either side of the ocean.

The Walvis Ridge between Namibia and the islands of Tristan da Cunha and Gough marks the trail of the spreading crust across a hot spot or mantle plume. The volcanic rocks of the plume become progressively younger towards the present areas of volcanic activity on the Mid-Atlantic Ridge.

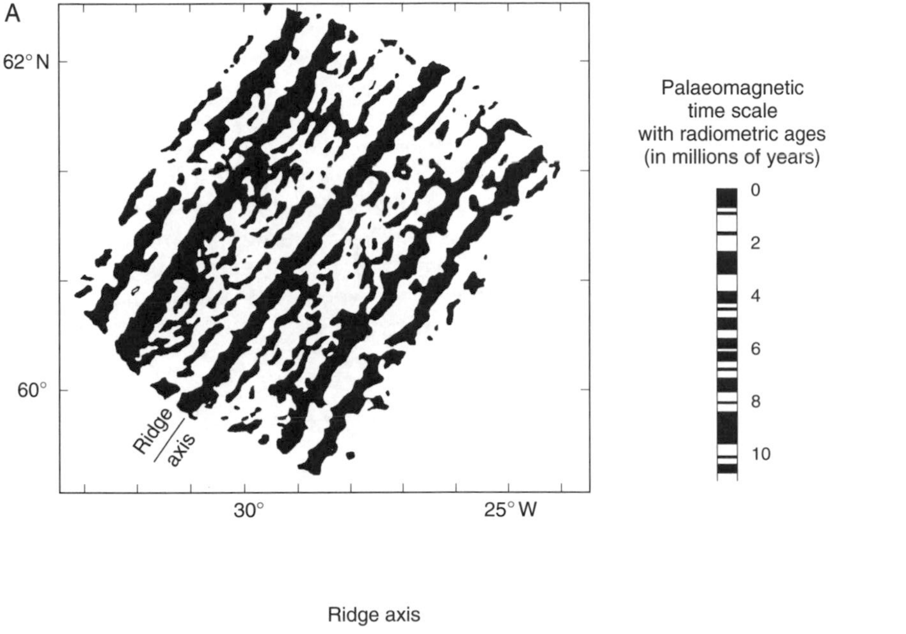

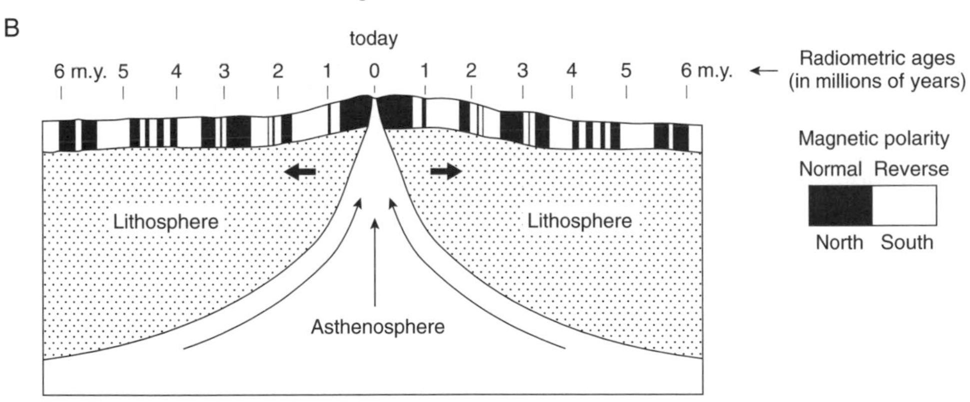

Figure 1.7 Magnetism of the sea floor of the Atlantic, south-west of Iceland. (A) Map of the magnetic stripes showing the near mirror images of the patterns on either side of the volcanic ridge from which spreading is taking place. Black areas are magnetised pointing to a north pole and white areas to a south pole. (B) A cross-section, which also shows the near mirror image of the magnetic pattern.

land and found a remarkable zebra-like pattern, with some rocks being magnetised in one direction and some in the other (figure 1.7).

The explanation for this fossil pattern of magnetisation seems to be that, as lava that has emerged through fissures in the mid-ocean ridge in the vicinity of the axial rift solidifies on the sea floor, it is magnetised in the direction, north or south, of the earth's prevailing magnetic field. The crust retains, like a tape-recording, a magnetic pattern identical

Window 1.2 *The age of the oceans*

The oceans of the world are remarkably young. In geological terms the mean age of the oceanic crust is only about 55 million years, and very little of it is older than Jurassic. The crust gets progressively younger towards the spreading centres or oceanic ridges.

By dating the new lithospheric material formed at the central oceanic ridge, and relating its position to that of the ridge from which it originated, it is possible to estimate with some reliability the long-term rates of sea-floor spreading. The prime dating method that has been employed is magnetostratigraphy, underpinned by potassium argon dating of the lavas that hold the magnetic signal. Rates of spreading range from a low of 6 mm a^{-1} in the northern part of the North Atlantic to rates of over 60 mm a^{-1} in parts of the Pacific Ocean.

The faster-moving plates (Pacific, Nazca, Cocos and Indian) have the common feature that a large fraction of their perimeters is being subducted, whereas, in contrast, the more slowly moving plates (American, African, Eurasian and Antarctic) have large continents embedded in them and do not have significant attachments of downgoing slabs. Press and Siever (1986: 504) put forward a hypothesis to explain this which 'Associates rapid plate motions with the "pull" exerted by large-scale downgoing slabs, and slow plate motions with the "drag" associated with embedded continents'.

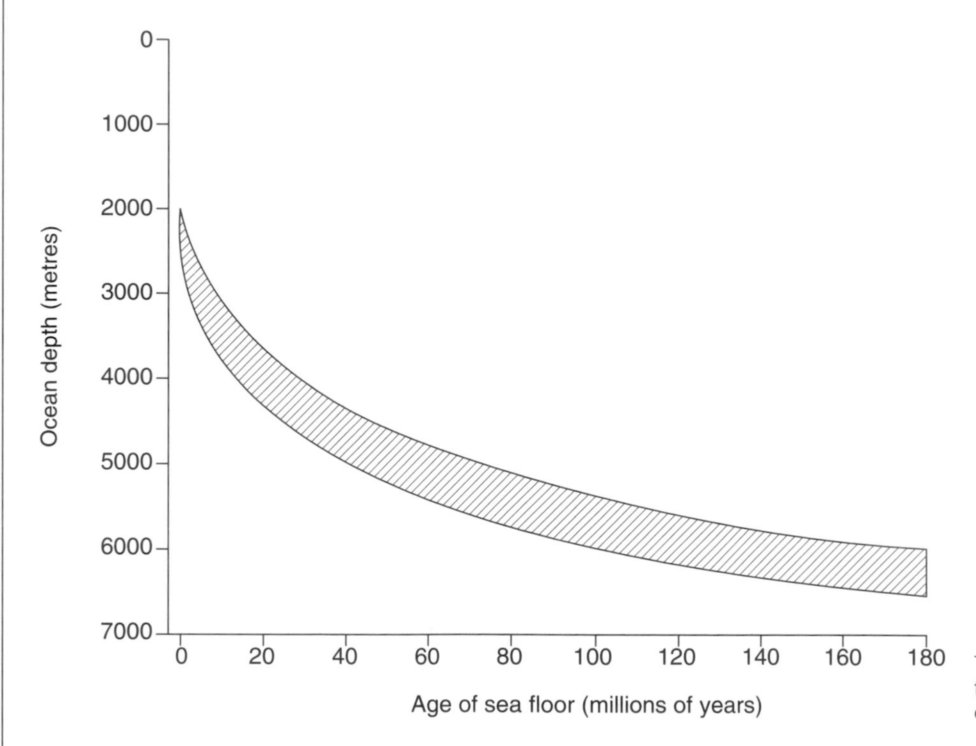

The age of the sea floor in relation to ocean depth.

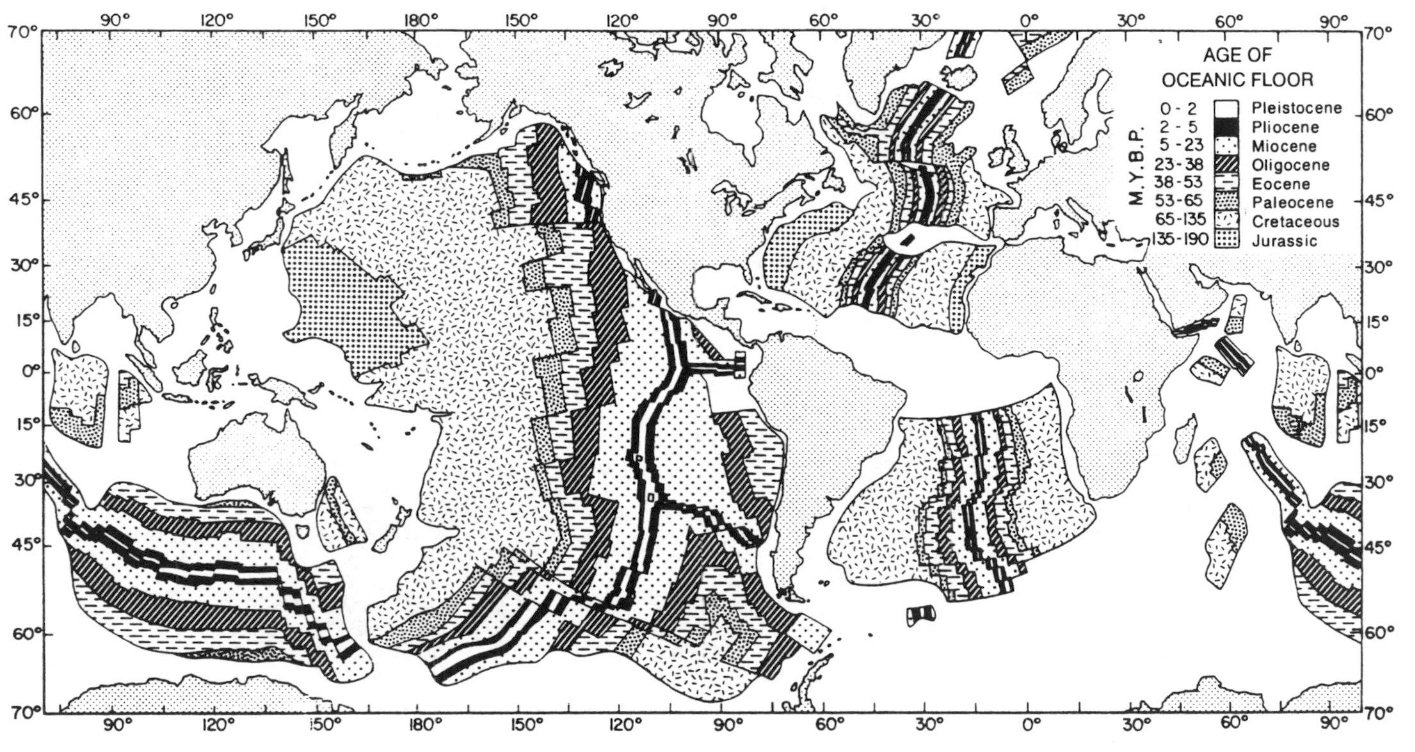

The age of the sea floor, indicating that there is no sea floor older than the Jurassic.

on either side of the ridge. In other words, this demonstrates that material rising from the earth's interior at the mid-ocean ridges slowly spreads across the sea floor.

This was the first real evidence for *sea-floor spreading*, though in the 1950s geophysicists had used palaeomagnetic techniques to determine palaeo-latitudes and thereby to add strength to arguments in favour of continental drift. The speed with which sea-floor spreading takes place can be gauged by palaeomagnetism.

More evidence was to follow, for as it became possible to drill holes into the ocean floor so it became possible to date sub-oceanic rocks by radioactive means. These dates revealed some further interesting patterns that complemented those provided by palaeomagnetism. First, it was clear that most of the ocean basins were covered by comparatively young rocks – rocks less than 200 million years old (window 1.2). Second, as samples were studied from cross-sections of the oceans, the rocks became younger as the mid-ocean ridges were approached. This led to the suggestion that the earth's crust is produced by upwelling magma that is extruded along the mid-ocean ridges, and then moves laterally from the ridge, to be replaced by new liquid magma. This is why the zebra-stripe pattern is produced, why the ocean rocks are so young, and why they become younger towards the ridges.

1.7 The Question of Plates

In the 1960s a new concept of the outer layers of the earth developed that was to tie in with the idea of sea-floor spreading. It was postulated that the earth's crust and the upper part of the mantle (the rigid part called the lithosphere) consist of a set of rigid plates (figure 1.8), which rest on the weaker

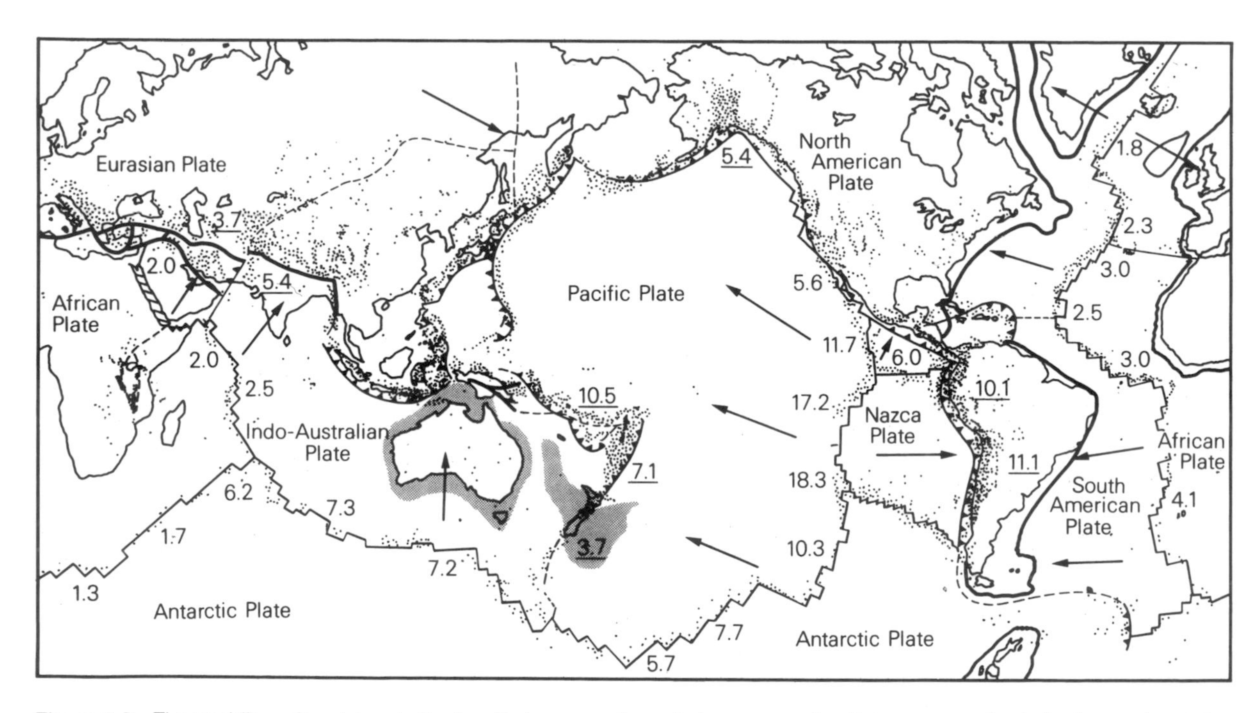

Figure 1.8 The world's major plates, indicating their names, boundaries, zones of collision, zones of subduction and certain major geomorphological phenomena. The rates of sea-floor spreading and of plate convergence (both in cm yr^{-1}) are given. Note the very high rate of spreading in the Pacific.

and deformable asthenosphere and underlie oceans, continents or a combination of the two. Seven major plates and four or five minor plates are recognised, together with eight or nine 'platelets'. They appear to average about 100 km in thickness though they vary from 60 to 300 km, and the larger ones have areas of 65 million km^2! These plates move, and *plate tectonics* is the name given to the study of the way in which these plates move and interact.

Many of the large-scale features of the earth's surface, such as most volcanoes and mountain ranges, are associated with the boundaries between the plates (figure 1.9) (window 1.3). Some plates spread apart along *divergent junctions*, typified by the axial ridge rift at the crest of the Mid-Atlantic Ridge (figure 1.10a). This particular feature results from the contact between the American plates on the one hand and the Eurasian and African plates on the other. Sea-floor spreading takes place along such a boundary as the void between the receding

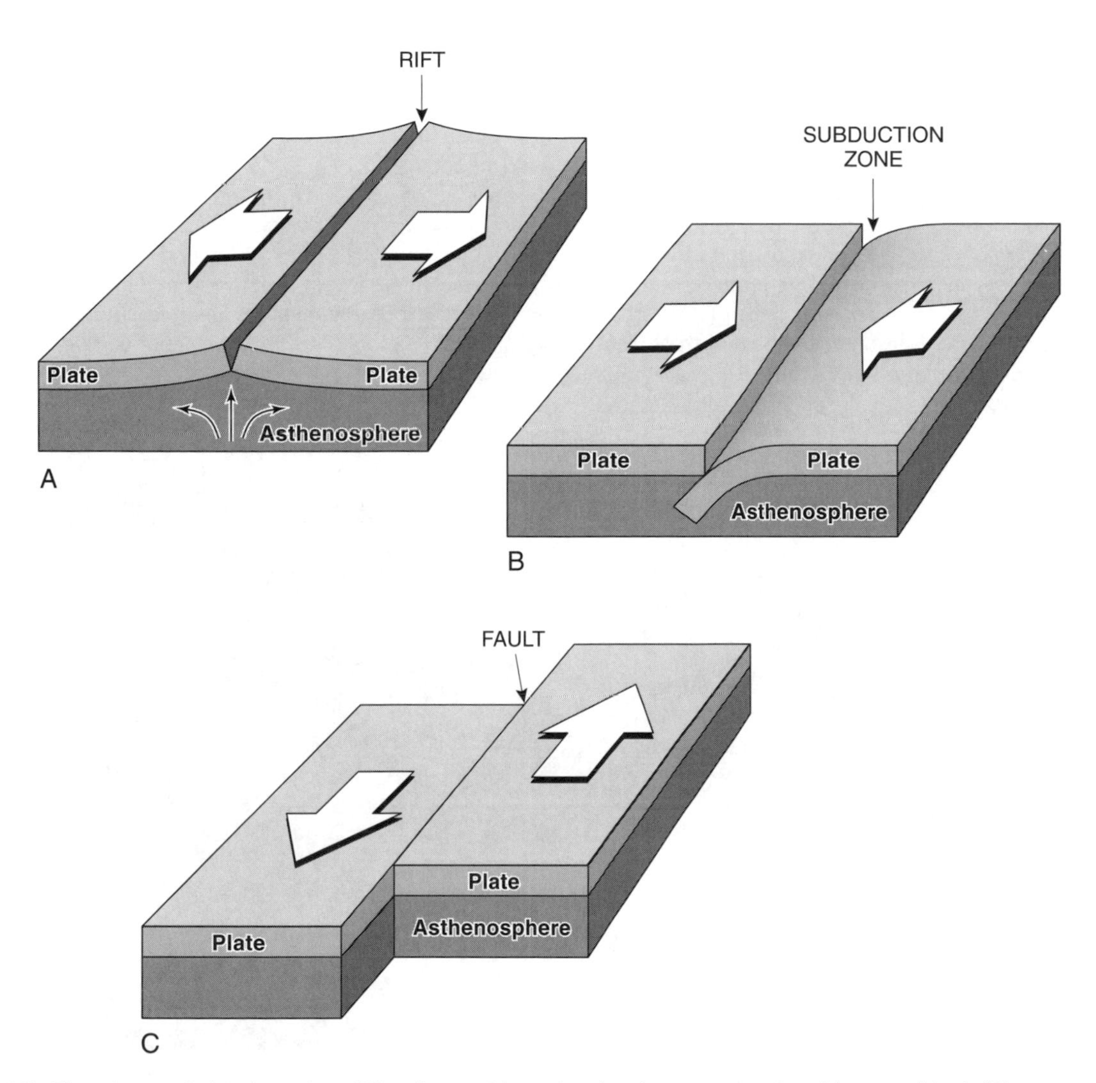

Figure 1.9 Three types of plate boundary: (A) a divergent boundary (as, for example, at a mid-ocean ridge); (B) a convergent boundary (as, for example, on the western side of South America); (C) a transform fault boundary (as, for example, the Dead Sea Rift or the San Andreas Fault).

Window 1.3 The mountains of New Zealand

New Zealand is a mecca for geomorphologists, and possesses a whole range of spectacular landforms. These include alpine mountains, enormous fault lines, large suites of gravel terraces and numerous volcanic features.

As with the Himalayas (see window 1.5), the fundamental reason for the development of such grand scenery is the presence of colliding lithospheric plates. New Zealand is the exposed portion of a fragment of the old super-continent of Gondwanaland. Separation from Gondwanaland may have been complete by about 80 million years ago. During the Cenozoic era the boundary between the Pacific and the Indian/Australian plates developed through that continental fragment. This boundary is marked by a subduction zone (most notably the Kermadec–Hikurangi Trench), which extends southwards from the islands of Tonga into the centre of the North Island of New Zealand. Further south it is marked by a great transform fault – the Alpine Fault – which runs south-westwards through South Island. There is a great deal of compression across this fault, and this compression creates the Southern Alps, which at Mount Cook reach an altitude of 3764 m. These mountains have been rising very rapidly, possibly at as much as 20 mm per year, and have been created since the early Pliocene. They are asymmetric in cross-profile, rising steeply in the west from a narrow coastal plain, but decreasing more gradually through a series of subsidiary ranges and basins to an eastern coastal plain.

The Southern Alps are also shaped by the processes of glaciation, weathering and erosion that are manifestations of the climate. Their crest traps moist air that comes from the Tasman Sea, so that in exposed areas there are massive annual precipitation amounts, which may add up to as much as 12 m. This means that exceptionally large glaciers have formed. In the Pleistocene these glaciers were greatly expanded in volume and carved the troughs in which the great fiords, which serrate the southern coast of New Zealand's west side, developed. Glacial erosion, severe frost weathering, outwash, rockfalls and tectonic uplift combine to make the environment one of intense geomorphological activity.

Milford Sound is a deep fjord on the west coast of New Zealand's South Island. In this area, the Southern Alps, a highly active tectonic zone, were subject to intense glacial erosion during the Pleistocene era.

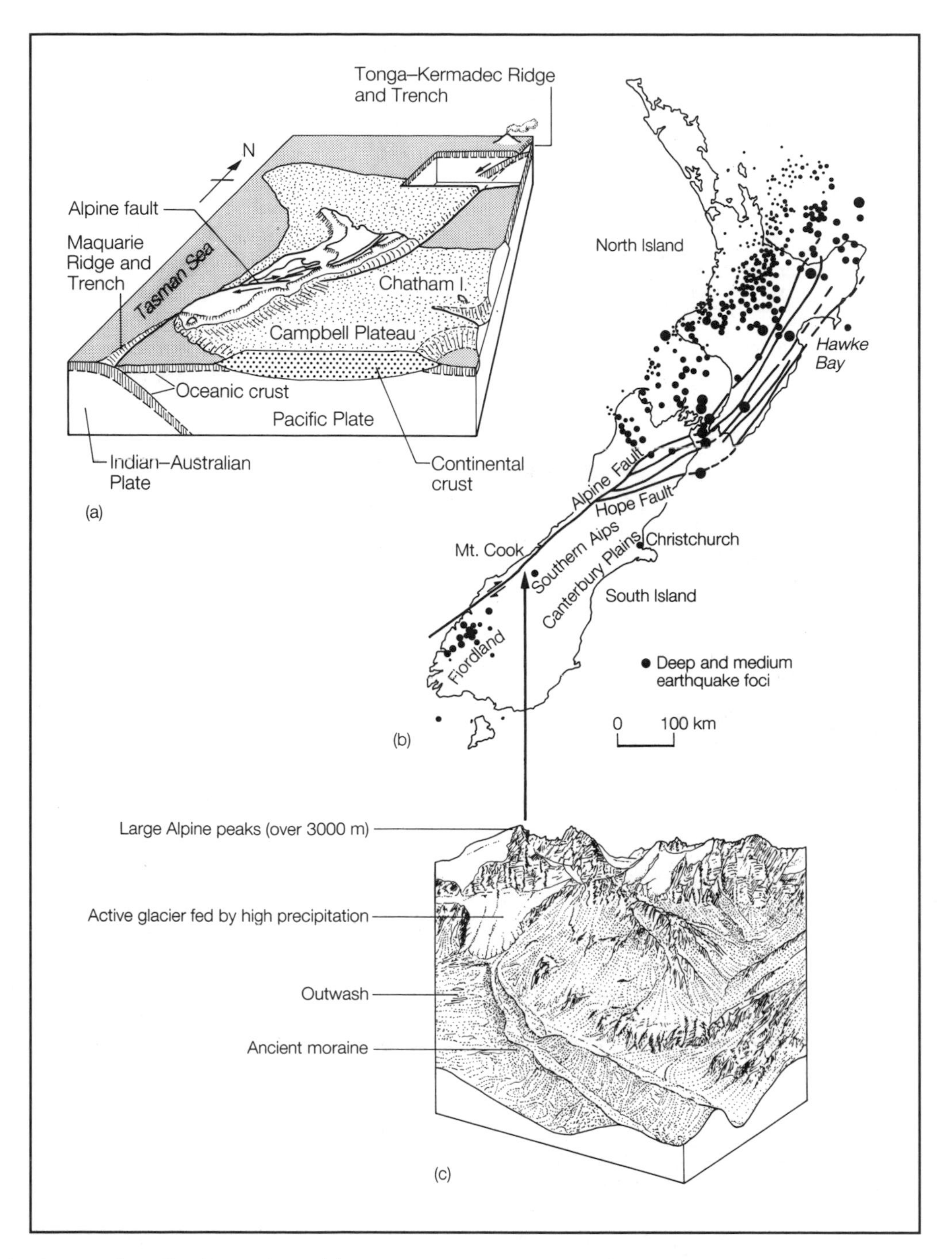

The landscape of the New Zealand Alps (c) in relation to tectonic position (a) and major structures and earthquake activity (b).

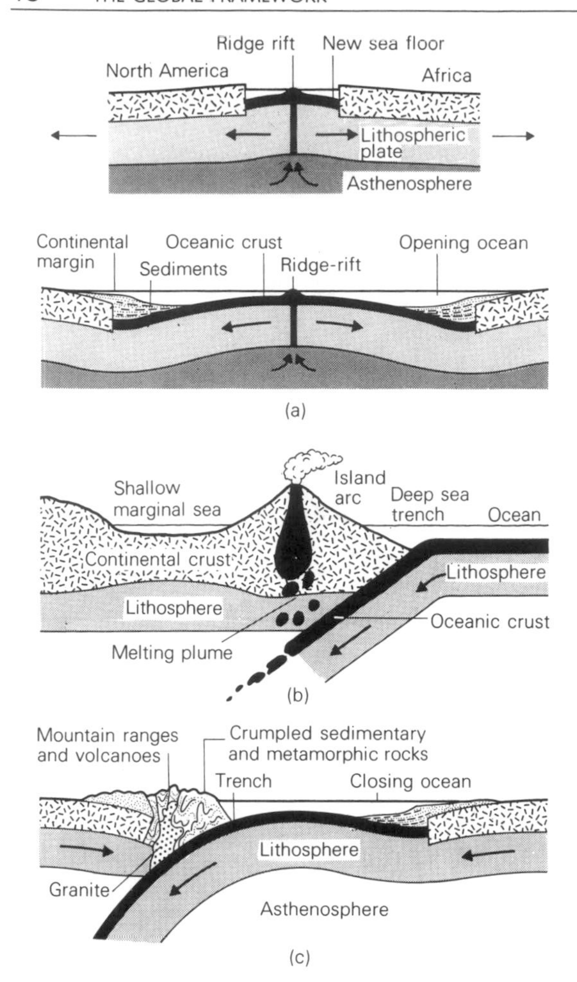

Figure 1.10 The oceans and their development. (a) The oceans developed through mid-ocean rifting. The trailing edges of the receding continents received a great deal of sediment from the land. (b) When plates collide, the thinner, denser oceanic plate tends to sink under the thicker, lighter continental plate, forming an island arc of volcanoes and a deep oceanic trench. (c) As the process continues, orogenesis (mountain-building) takes place, rocks from the ocean floor are uplifted and crumpled, and granite batholiths are intruded.

plates is filled by melted, mobile material that rises from the interior of the earth. The material solidifies in the crack, and the plates grow as they separate. The Atlantic Ocean did not exist 150 million years ago, and it seems to have opened out by this process, leading to the disruption of the ancient super-continent of Pangaea.

The process of divergence is most prevalent in the ocean basins, but there are two major zones of spreading among the continents, one in Africa (plate 1.1) and the other in Asia. The best developed of these is the African zone (figure 1.11), which extends from the Red Sea as a series of rifts through the East African highlands. These rifts are the site of some major lake basins, including Lake Malawi, Lake Tanganyika and Lake Kivu.

In the Asian continent the site of spreading is in Siberia, where there is a rift system of which the deeper parts are occupied by Lake Baikal, and where there has been some Pleistocene volcanic activity.

These continental rifts are major features of the earth's surface; they are many tens of kilometres long, 10–50 km wide and with a central strip of crust downthrown by 1–5 km. They have undergone a gradual extension of a few kilometres during the past 10–30 million years, and may be considered to be incipient plate boundaries (window 1.4).

The rate of divergence either side of the oceanic axial rift does much to explain the nature and distribution of islands and seamounts. Along slowly spreading zones, such as those typical of the Atlantic, volcanic material tends to accumulate relatively close to the locus of spreading, so that there is a well-defined mid-oceanic ridge, with relatively massive mountainous islands like the Azores. However, where, as in the case of the Pacific, the rate of spreading is quite rapid, growing volcanic structures are moved large distances from the zone of origin, where they eventually sink and become dormant. The reason for such sinking is that oceanic crust contracts and becomes more dense as it cools. Thus, if one looks at a map of the floor of the Pacific one notes that there is a very large number of seamounts, and that some of them – guyots – have flat tops because they were planed off by wave action before they sank. Both the seamounts and the guyots have provided platforms on which the many coral reefs, so characteristic of the warmer parts of the Pacific, have grown.

Plainly, however, if plates separate in one place, unless the earth is expanding, they must converge somewhere else. This they do at *convergent junctions*, and the zone of contact is one that is dominated by crumpled mountain ranges, volcanoes,

Plate 1.1 The splitting of plates creates tensions in the earth's crust which can lead to the development of rift valleys as shown here at Baringo in Kenya. The rift is bounded by a lava cliff, Lake Baringo is the floor of the rift, and small volcanoes form islands in the lake.

earthquakes and ocean trenches. At the collision point one plate normally plunges beneath the other, and the material of the submerging plate is gradually reincorporated into the upper mantle and crust. The area where the material is lost is called the *subduction zone*. As the oceanic plate plunges beneath the continental plate with which it has collided, parts of it begin to melt and form magma (figure 1.10b). Some of this reaches the surface as lava erupting from volcanic vents. Furthermore, as an oceanic plate is subducted under its neighbour, plunging down at an angle of between 30° and 60° into the interior of the earth (figure 1.10c), the surface of the crust is drawn down to produce one of the ocean trenches, and the movement is in a series of jerks, which produce earthquakes.

The third type of boundary is called a *transform* or *transcurrent boundary*. This exists where two plates do not come directly into collision but slide past each other along a fault. This type of situation is termed a *conservative zone*, in that plate movement is mainly parallel to the boundary, and therefore crust is neither created nor destroyed. The most famous example of such a boundary is the San Andreas Fault in California (plate 1.2), which separates the northward-moving Pacific Plate from the North American Plate. This is a zone of intense earthquake activity, for the movement along the faults is irregular rather than a smooth process of gradual creep.

The process whereby many mountains are built can be interpreted in terms of plate movement. It starts with the formation of an ocean by sea-floor spreading. Thick muds and silts eroded from the land accumulate on the continental slopes alongside thinner limestones and sands on the continental shelves. After perhaps 100 or 200 million years, a trench forms on one side of the ocean and the ocean crust is subducted under the continent. This may be followed by the consumption of the sink-

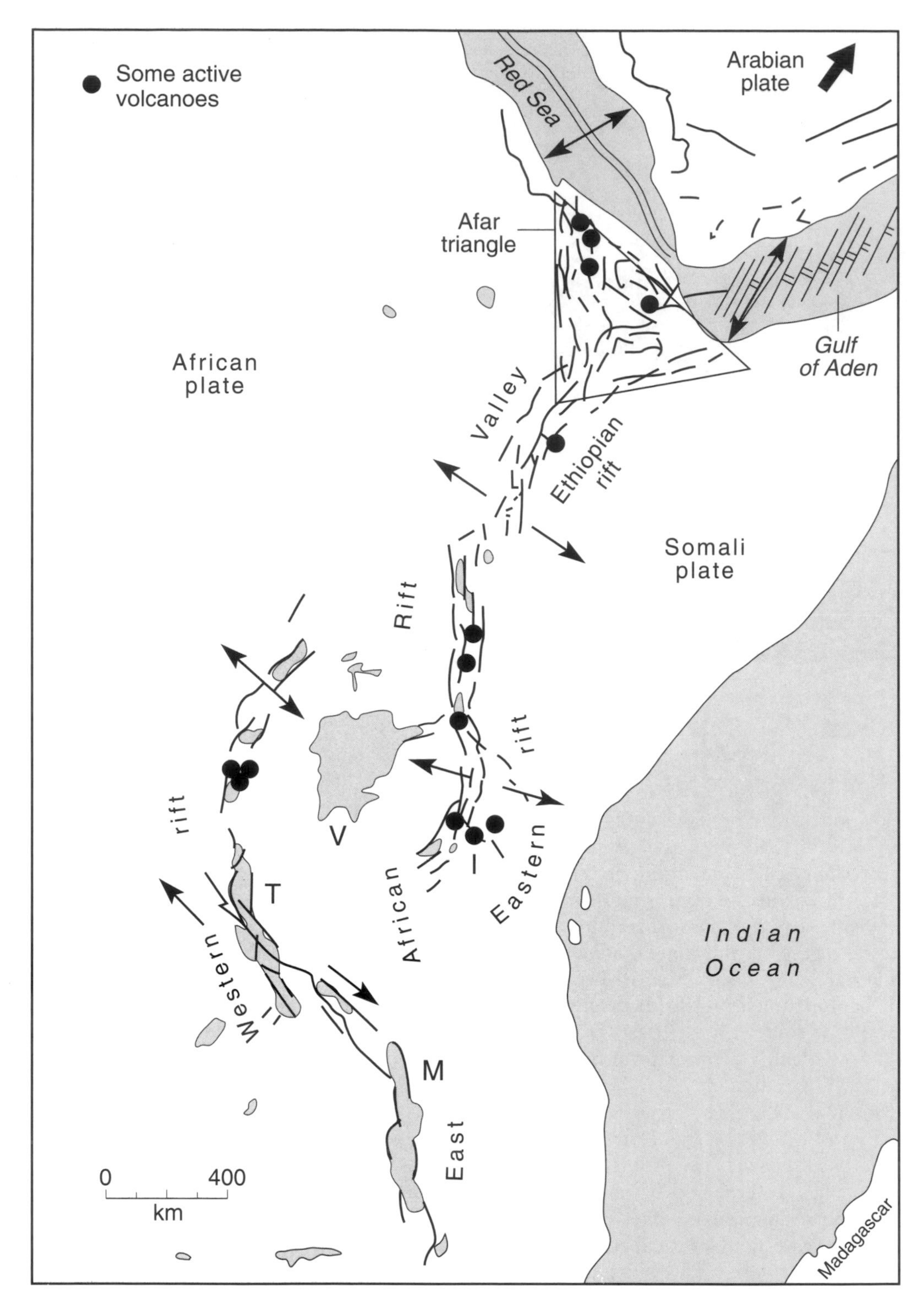

Figure 1.11 The great rift valleys of East Africa comprise a zone where divergence is taking place in the earth's crust and where there has been active volcanic activity. Where the Ethiopian part of the rift joins the Red Sea and the Gulf of Aden, this divergence is taking place about a triple junction – the Afar Triangle.

Window 1.4 Continental disintegration

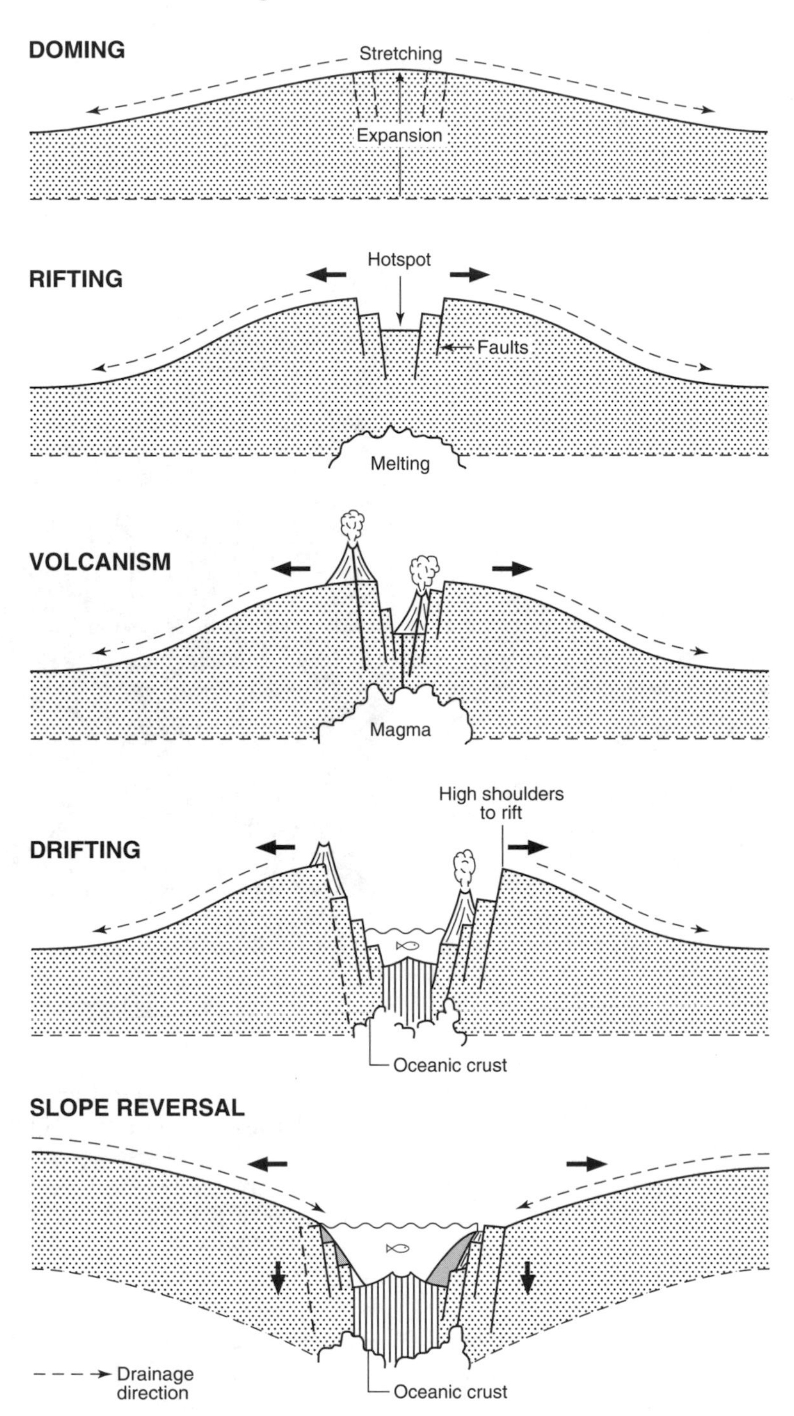

A model of continental disintegration leading to ocean formation.

It is probable that the disintegration of a continent begins with local heating by a hot spot beneath the lithosphere. This causes expansion, which leads to a surface bulge or dome which then cracks to create a rift. The rift is lined by volcanoes, and great spreads of lava may be deposited (as, for example, along the margins of the Red Sea). At this stage in its evolution the rift valley is high above sea level and is bounded by high relief along its sides. As further stretching occurs, the base of the rift sinks below sea level and ocean waters enter the system. Oceanic lavas from the mantle begin to create new ocean crust. The margins of the rift cool as the distance to the new mid-ocean ridge increases, slope reversal takes place, and drainage enters the sea rather than flowing from it.

Plate 1.2 Where two plates move parallel to each other, major faults like the San Andreas Fault in California cut across the landscape.

Plate 1.3 Mount Rakaposhi, one of the highest mountains in the world, is formed of oceanic crust material that was thrust up when the Indian Plate collided with the Eurasian Plate. Intense uplift has created 'the steepest place on earth'.

ing crust, leading to the formation of a chain of volcanoes called an *island arc* (figure 1.10b). After a further period the island arc itself gets swept into the continents and finally the ocean becomes closed and the continents collide. When this happens the sediments of the island arc, together with those of the continental margin, are caught as if in a vice, and so folding and overthrusting takes place in a zone of crumpling. Furthermore, molten igneous rock, generated as the oceanic crust melts in the depths, rises into the crust and so promotes *uplift*. In this way some of the great fold mountains are formed (figure 1.10c).

The formation of island arcs is thus a frequent part of the story of mountain-building. These arcs at present are largely restricted to the Pacific, where they occur around the periphery in a zone known as 'the ring of fire'. They are composed of silica-rich lavas called *andesites*, and include some of the world's most destructive volcanoes. Elsewhere, island arcs may have become incorporated into the continents; for example, the great Karakoram Mountains in northern Pakistan are composed of silica-rich lava, and are thought to be the remains of an island arc that was squeezed into the Eurasian Plate as the Indian Plate moved northwards.

This is the explanation for the fact that the summit of Mount Rakaposhi (7788 m) (plate 1.3), one of the highest mountains in the world, is formed of oceanic crust material (window 1.5).

Some of the old mountain roots that characterise those portions of the earth's surface that are not currently very active, such as the Aravalli Mountains of India, the Urals of Siberia or the Appalachians of the eastern United States, may represent earlier phases of plate movement than do the present zones of active young fold mountains. As we have already noted, much of the continental crust is old,

Window 1.5 *The origin of the Himalayas*

The Himalayas, and associated ranges like the Karakoram, possess peaks of over 8000 m. The summit of Mount Everest is the highest point on the earth's surface. These great mountains make up the world's highest and youngest major mountain belt, and form a great swathe of country, generally 250–350 km wide, which stretches some 3000 km between the borders of Afghanistan and Burma.

The Himalayas have formed in response to the tectonic collision of India with Eurasia. As India broke away from Gondwanaland, it drifted northwards, and as this northward movement progressed, India shifted from a position about 20–40° S of the Equator, at 70 million years ago, to a position between 10° N and 10° S some 40 million years ago. As it did so, the Tethys Oceanic Plate, which lay between India and Eurasia, was subducted beneath the Eurasian Plate, a process which ceased *c.*40–50 million years ago. It was at this time that India collided with Asia for the first time. Prior to that it had been moving northwards at a rate of about 10 cm or more per year. It has kept moving at this reduced, but still appreciable, speed up to the present.

What is interesting is that the process of 'docking' with Asia did not mean the end of northward movement.

The Himalayas are the most impressive mountain chain in the world, and have been created by the collision of the Indian and Eurasian plates. Even the Siwalik foothills shown here have a very marked degree of relief and have been deformed by substantial tectonic processes.

Indeed, since the collision occurred, India has moved some 2000 km further into Asia. This example of what geologists have come to call 'indentation tectonics' has led to extensive thrusting and faulting (with effects which can be detected well into China and south-east Asia), seismic activity, and great uplift of the Himalayas. The rate of uplift has been considerable. The Nanga Parbat region of Pakistan, for instance, has experienced some 10km of uplift in less than 10 million years. Forces of erosion, though great, have not kept pace with this rate of uplift. None the less, the Himalayas are fronted by great fans of fluvial sediment, called *molasse*, along their southern flank: these form the Siwalik foothills.

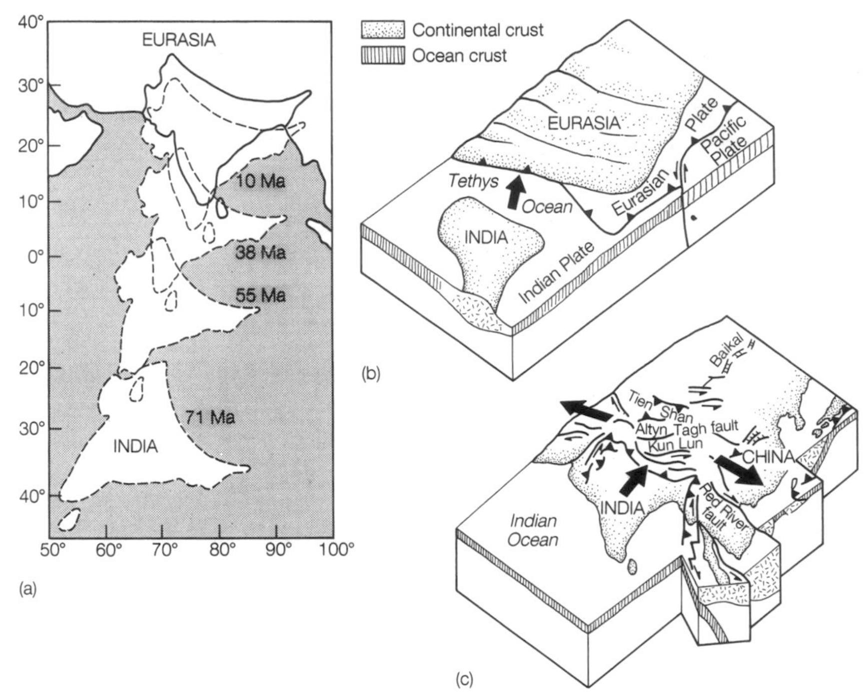

The development of the Himalayas: (a) the northward movement of India with respect to Asia during the Cenozoic era; (b) the structural situation *c.* 60 million years BP, prior to collision; (c) the present-day structural situation, showing how India has not only collided with Eurasia, but has also caused substantial indentation and associated eastward dispersion of continental crust.

Window 1.6 *Wilson cycles*

In the 1960s the geologist J. T. Wilson proposed a hypothesis that continental splitting occurs, that this is then followed by ocean formation and sea-floor spreading, but that eventually closure takes place. Today we find that the Atlantic, Red Sea and the Gulf of California are opening, while the Mediterranean and the Pacific are closing. New continental blocks formed by closure themselves eventually split along the same lines as before to produce yet another ocean in approximately the same location as the preceding one. These complex cycles of ocean basin openings and closings are now called Wilson cycles. This hypothesis was based on the history of the Atlantic Ocean, the present version of which appears to have formed in broadly the same place as its predecessor – the so-called Iapetus Ocean – which had preceded the collision between North America–Greenland and Europe–Africa which had given rise to the Caledonian mountains.

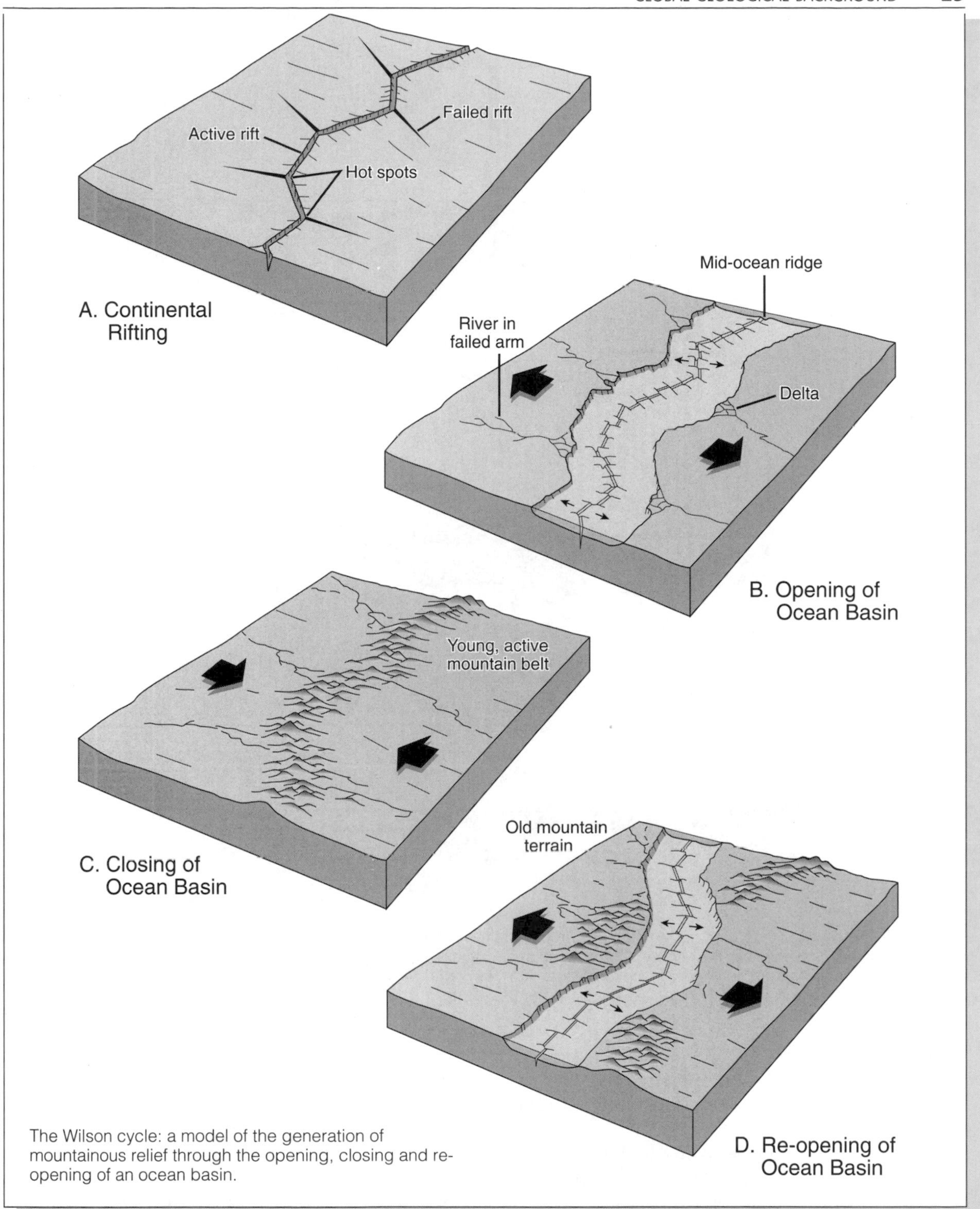

The Wilson cycle: a model of the generation of mountainous relief through the opening, closing and re-opening of an ocean basin.

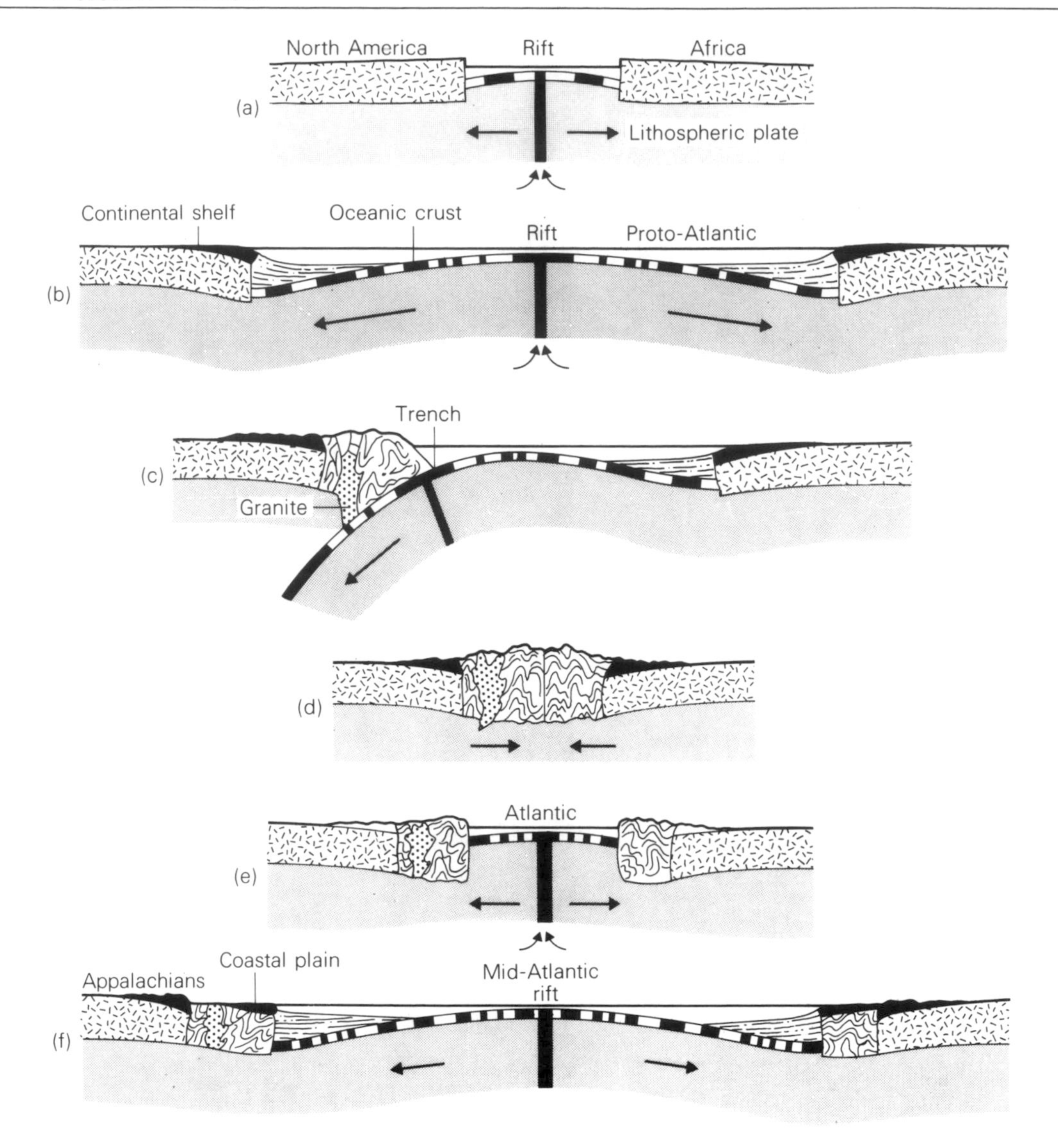

Figure 1.12 (a) The development of the Appalachian Mountains and related systems in the United States by a process of rifting of a lithospheric plate into two plates, North America and Africa; (b) the development of geosynclinal and continental shelf sediments after a period of rifting has taken place; (c, d) the period of igneous intrusion and deformation producing the structure seen today in the Appalachians; (e) the development of the mid-Atlantic rift; (f) continued rifting producing the modern Atlantic basin.

and may therefore have been subjected to more than one phase of plate collision or plate divergence (window 1.6).

This is illustrated for the Appalachians in figure 1.12. They may have formed by a period of rifting (a) and (b), which gradually led on to a period of collision and subduction (c) and to the formation of the mountains (d). Another cycle of rifting followed (e), which split up the mountain zone and led to the opening of the Atlantic Ocean (f). Meanwhile, considerable erosion had led to severe truncation of the continental terrain, producing the

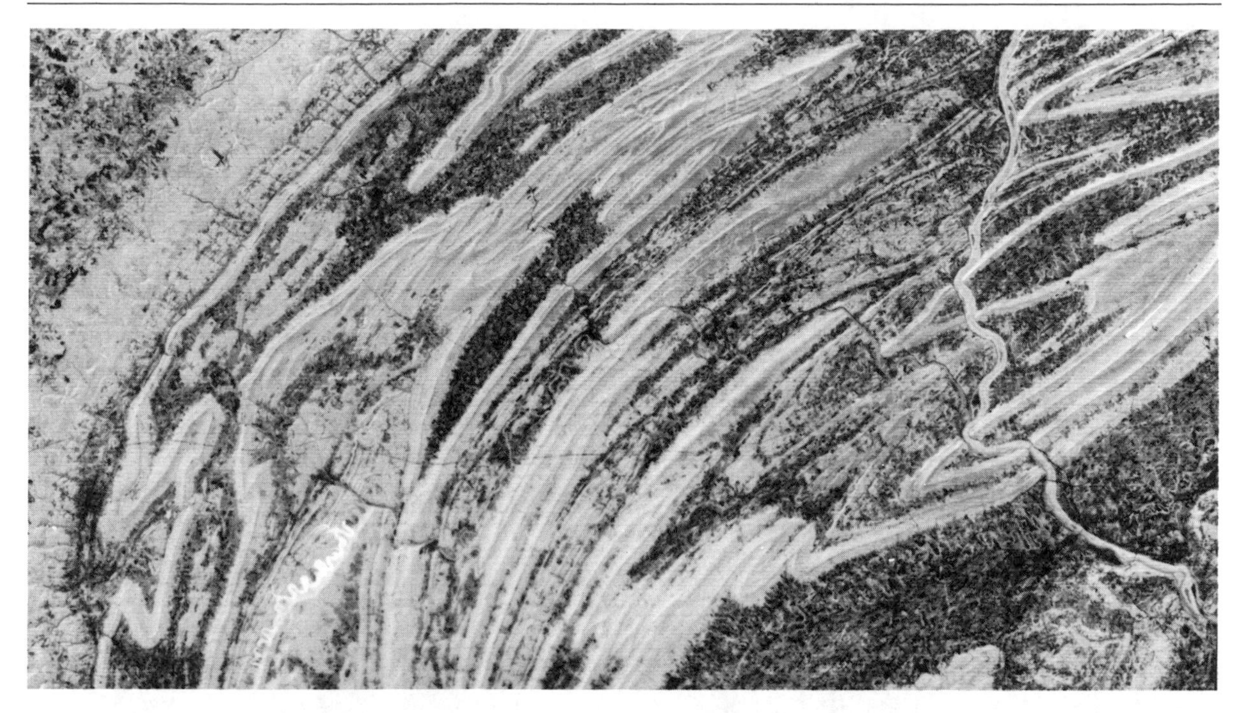

Plate 1.4 A Landsat satellite image shows the great fold structures that underlie the old mountain chain of the Appalachians in the eastern United States. Their relief has been greatly planed off by erosion.

Appalachians we know today (plate 1.4).

The rocks and geological structures of the British Isles (window 1.7) show evidence that such cycles of orogenic activity may have occurred more than once. Indeed, the gradual addition of new orogenic belts of varying rock type, produced by varying intensities of mountain-building, give Britain its enormous diversity of scenery. The Precambrian rocks of the Outer Hebrides and adjacent parts of north-western Scotland represent the edge of an ancient continent, and in the *Caledonian orogenic cycle* (*c*.400 million years ago) very considerable folding and faulting occurred, producing, for example, the faultbounded Scottish Central Valley and initiating the Great Glen Fault further north. In a later cycle, the *Hercynian* (*c*.250 million years ago), igneous activity produced the granite batholith that underlies many of the moorlands of the south-west peninsula of England in Devon and Cornwall; volcanic activity occurred in the Central Valley of Scotland; and a sill – the Great Whin Sill – was introduced into the domed-up northern Pennines. During the *Alpine* cycle, which followed the collision between Africa and Europe, the Alps themselves were formed. This produced ripples in southern England, producing, for example, the dome of rocks that form the backbone of the Weald. Since the Alpine orogeny, the opening up of the Atlantic produced some rifting that created volcanic activity in north-western Scotland and in Antrim. At the coast the lavas form prominent cliffs, and the hexagonal columns produced by the cooling of the basalt are striking features of the Isle of Staffa and of the Giant's Causeway in Northern Ireland.

World coastlines can also be classified and explained in plate tectonic terms. On coasts where two plates converge – *collision coasts* – structural lineations tend to be parallel to the shore. They are relatively straight with high, tectonically mobile hinterlands, and they are fronted by relatively narrow continental shelves. They suffer from earthquakes and volcanic activity, and the earthquakes sometimes generate catastrophic waves called *tsunamis*. The steep, tectonically active hinterland may

Window 1.7 *The opening of the North Atlantic and the landscape of the British Isles*

Prior to 200 million years ago (late Triassic) the North Atlantic did not exist as an ocean, and the continental plates of western Europe, Greenland and North America formed one single plate. During the Middle Jurassic an embryonic plate margin began to develop between Europe and North America, generating new oceanic crust over the entire length of the southern North Atlantic, and causing rifting to occur notably in the North

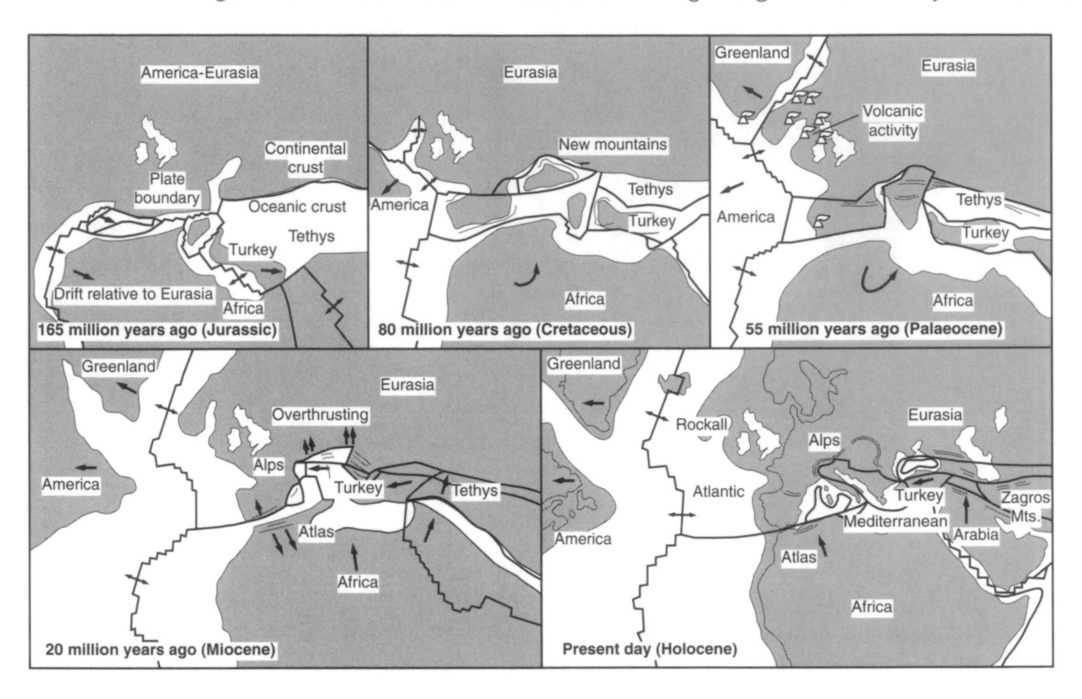

The plate tectonic evolution of the Mediterranean basin and the North Atlantic region since the Jurassic. Note the relatively late opening of the North Atlantic and the associated volcanic activity in the Palaeocene era.

The magnificent Isle of Staffa on the west coast of Scotland is composed of basaltic lava which was produced as rifting accompanied the opening of the Atlantic Ocean.

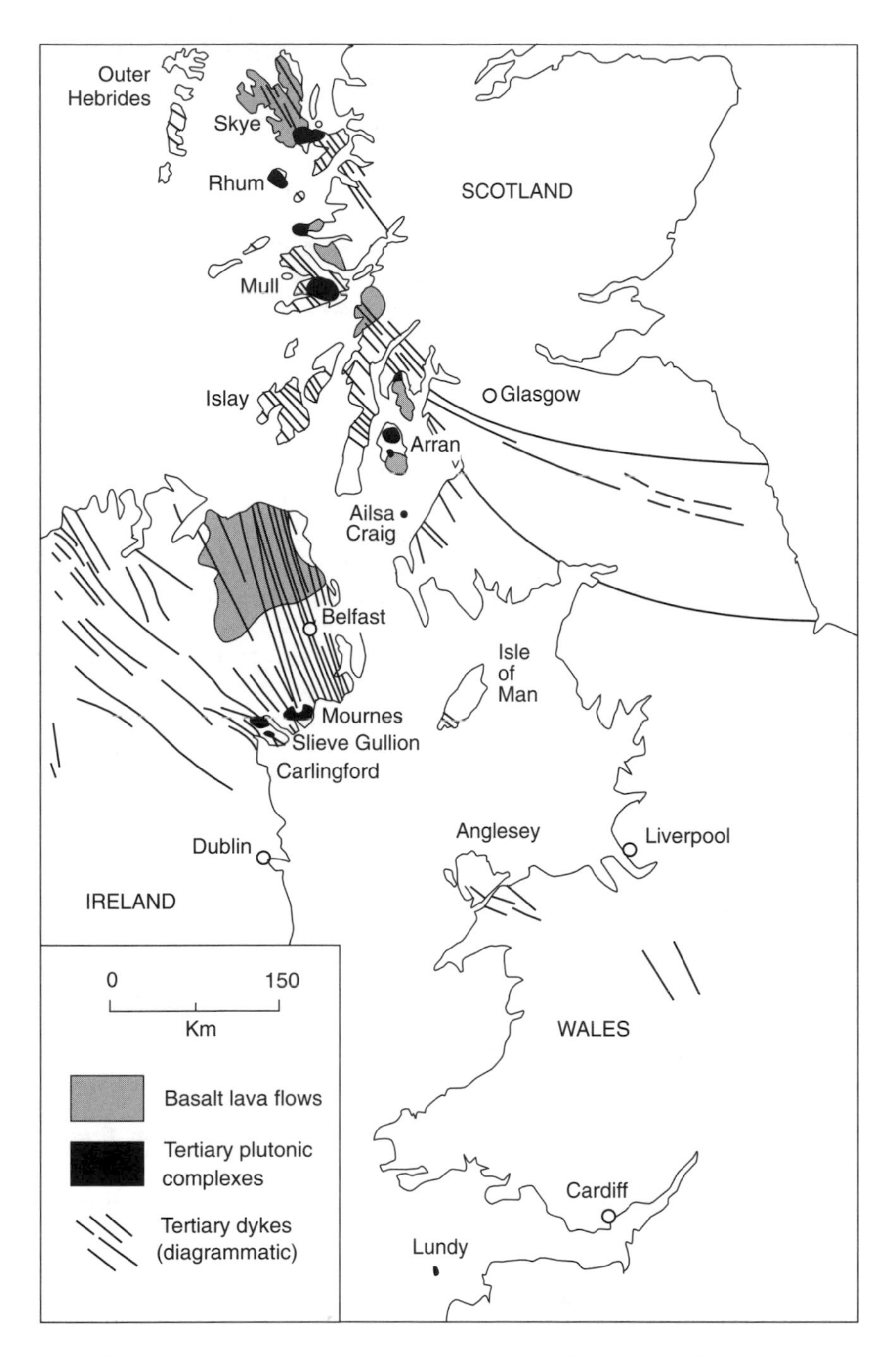

The opening of the North Atlantic produced a major spasm of igneous activity in the British Isles leading to the intrusion of plutons (including the Isle of Lundy), the injection of swarms of dykes and the extrusion of expansive basalt lava flows.

Sea Basin. By the start of the Palaeocene sea-floor spreading had advanced sufficiently that an opening was appearing between Ireland and Greenland.

This phase of plate separation was accompanied by a period of marked crustal extension over much of Britain and Ireland which caused a severe spasm of intrusion and extrusion of igneous material, which lasted from 61 to 52 million years ago. Flood basalts were erupted from Hawaiian-type volcanoes to give lava sheets, notably on Mull (where they are some 2000 m in thickness) and in Antrim. Igneous intrusions developed, including the Lundy granites of the Bristol Channel, which have been dated to 52 million years ago. North-west to south-east trending dyke swarms were also intruded, and extended as far south as Lundy, Snowdonia and the north-west Midlands. Most of this activity had ceased by the early Eocene, for as the Atlantic opened still further igneous activity shifted westwards and became centred upon the location that was eventually to become Iceland.

well be a zone of intense erosion, in which case there will be a considerable amount of sediment deposited along the coast. The existence of mountain-building may mean that there are a series of fossil beaches at high altitudes above the present ones.

The second type of coastline is the *passive margin* or *trailing edge* coast, and it can be subdivided into two categories: Afro- and Amero-. This distinction is fundamental because of its effect on global patterns of fluvial sedimentation. Where the coast on one side of the continent is a collision coast, as in South America, the high, tectonically active rim (in that particular case the Andes) yields a large sediment load to rivers flowing towards the opposite, trailing-edge, coast. Where both coasts are the trailing-edge type this factor does not operate; clear examples of this are provided by the coasts of Africa, where rivers like the Congo and Niger carry relatively small loads in comparison with the Amazon and Mississippi. A major consequence of this is that Amero-trailing-edge coasts are lower-lying, have more extensive sedimentary plains, and possess wider continental shelves. Large deltas may form, the weight of which may cause local subsidence of the earth's crust. Generally, however, these coastlines are tectonically very stable.

We can now see why the continents have the arrangement and shapes they have, why the oceans are dominated by their ridges, why the locations of volcanoes and earthquakes are as they are; and we can also understand how mountain ranges form. The theory of plate tectonics offers an all-embracing explanation, and it has revolutionised thinking about large-scale geomorphology. There are, however, some apparent exceptions. For example, the Hawaiian Islands are great volcanoes, and yet they occur not at the boundary of a plate, as do most volcanoes, but in the *middle* of a plate. In spite of this, they may still help to explain the theory of crustal motion. Indeed, there are some 10 000 central vent volcanoes within the Pacific, some of which have risen above sea level to form volcanic islands but most having remained submerged as seamounts. Many are members of linear chains of oceanic volcanoes that increase in age from one end to the other. The most dramatic example is the Hawaiian–Emperor chain, which consists of 107 islands and seamounts that stretch for about 6000 km across the ocean. The members of this chain get progressively younger as the currently active volcanic island at one end, the island of Hawaii itself, is approached. One theory for this is that for some unknown reason there could be a local 'hot spot' beneath the Pacific Ocean, a point at which magma from the asthenosphere or lower mantle is 'burning' its way up through the lithosphere. It is postulated that the crust is moving from an ocean ridge south-east of Hawaii towards a trench that lies north-westwards. The south-eastern chain of the islands is currently over the hot spot and is the scene of volcanic activity and earthquakes. The formation of the islands to the north-west occurred long ago when that part of the crust was over the hot spot. They have since moved away from the spot and volcanic activity has ceased (figure 1.13). Still further to the north-west is a chain of underwater seamounts – the rem-

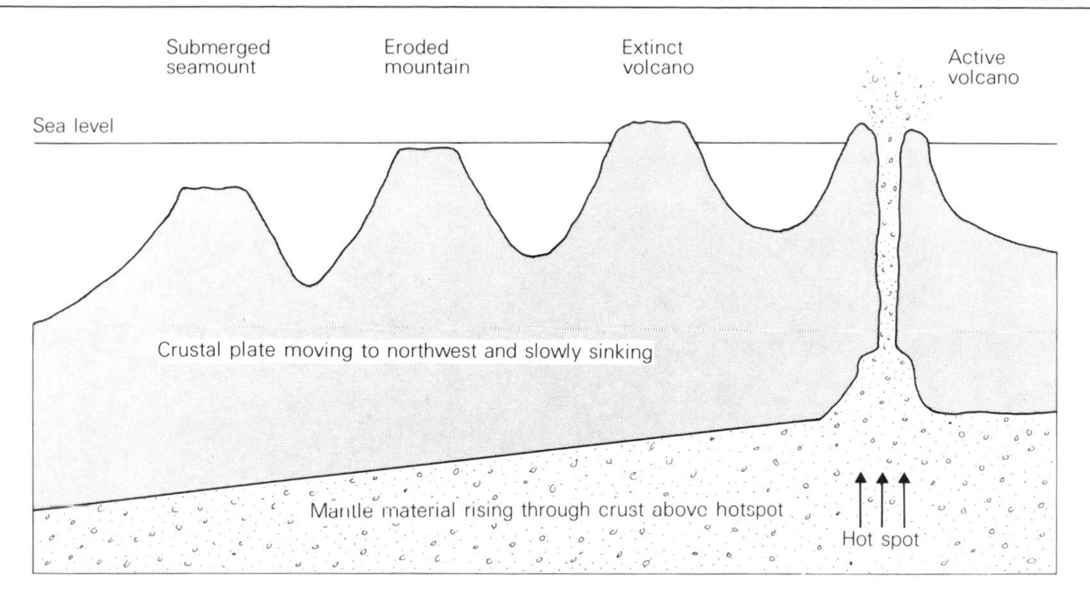

Figure 1.13 Some volcanic islands develop above hot spots in the earth's crust. As the plate moves laterally over the hot spot, so the position of active volcanic activity changes. Away from the spot the volcanoes become less active, more eroded (sometimes to give flat-topped guyots), and eventually suffer subsidence to become submarine mountains or submerged seamounts.

nants of volcanic activity that took place even further back in time.

1.8 Microplates and Exotic Terranes

If one examines the rocks along the west side of much of Alaska, Canada and the United States (figure 1.14) for a width of several hundreds of kilometres, they appear to comprise a mosaic of hundreds of very distinctive segments, the boundaries of which are quite sharp. Adjoining segments may also contain distinct fossil types, indicating that they were probably formed at different latitudes. Some of these so-called *exotic terranes* are fragments of continent and other relicts of island arcs; still others are oceanic plateaux, pieces of oceanic ridges or of sedimentary platforms. They have been termed geological flotsam and jetsam.

Recent palaeomagnetic studies have shown that some of these exotic terranes have travelled some thousands of kilometres, and the conclusion has now been reached that these various terranes had once been scattered throughout much of the Pacific Ocean and had been driven towards North America by sea-floor spreading. Once they reached North America they were compressed, uplifted, and tilted as part of the normal processes of deformation and mountain-building at convergent plate boundaries. These segments, which range from small outcrops of a few kilometres square to larger areas with an extent of many thousands of square kilometres, are also called microplates.

1.9 The Earth's Rock Types

Now that we have considered the major relief patterns both on land and under the oceans, we can turn to a consideration of the materials that make up the earth's surface. Once again we can relate the formation of the major types to the theory of plate tectonics.

Rocks are normally divided into three main types or groups: igneous, metamorphic and sedimentary (figure 1.15).

Igneous rocks

Igneous rocks are composed of mineral crystals that

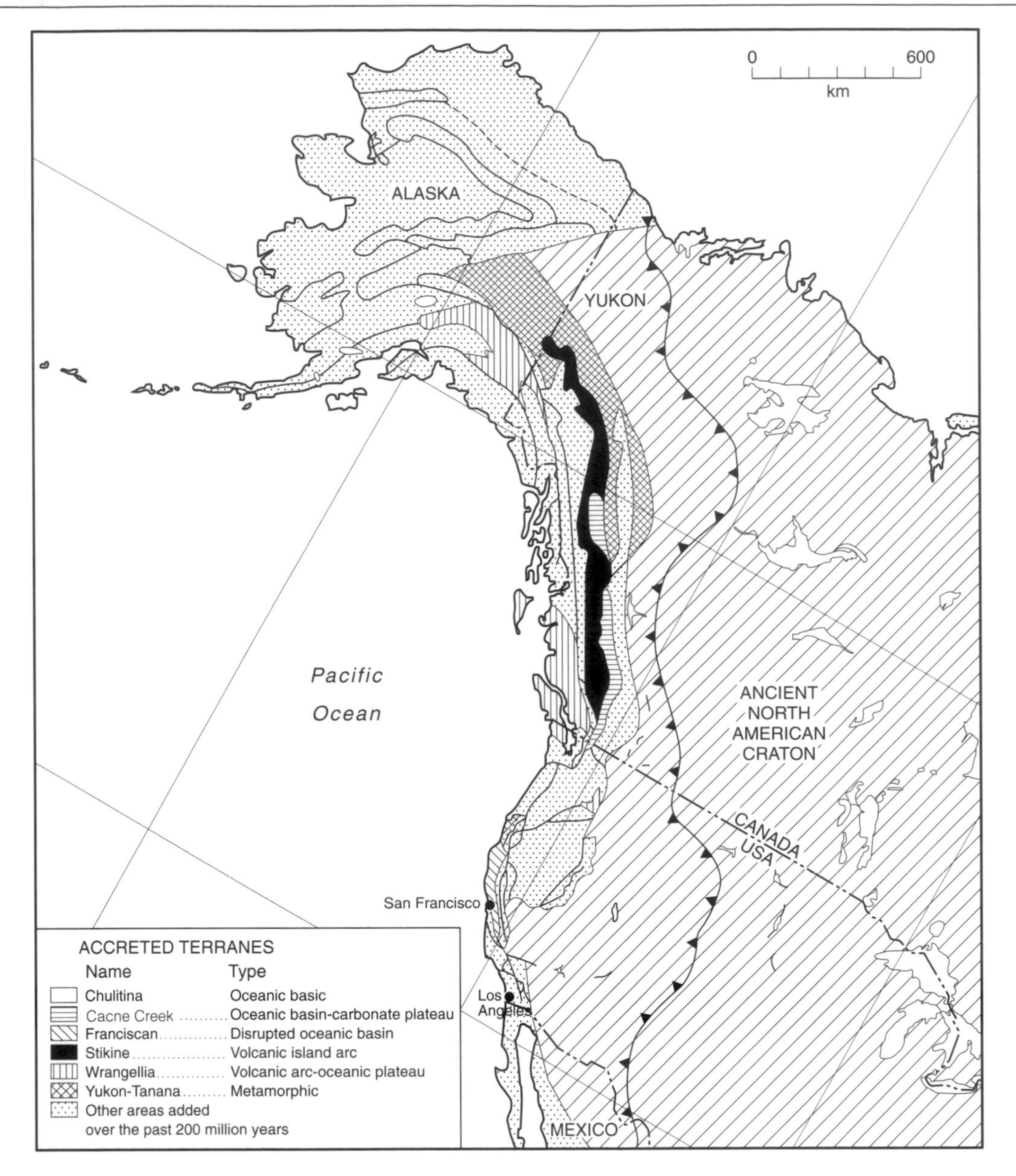

Figure 1.14 Terranes mark the locations of earlier subduction zones along the west coast of North America. Over the past 200 million years, oceanic plates slipped beneath the relatively buoyant continental plates, but islands, seamounts and other superficial features were scraped off and added to the continental plate.

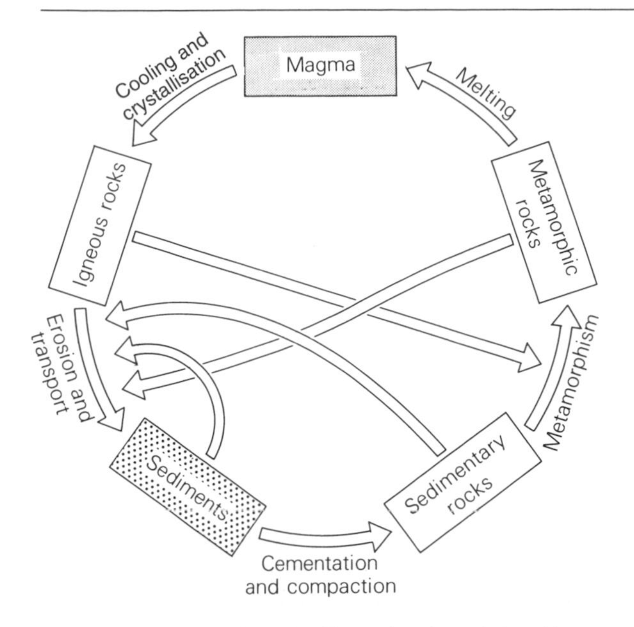

Figure 1.15 The rock cycle, illustrating the relationship between the three main rock types: igneous, metamorphic and sedimentary.

Figure 1.16 Some of the main types of igneous rock mass, showing the difference between extrusive and intrusive types. The batholiths may result from the emplacement of magma during mountain-building caused by plate collision.

Plate 1.5 An isolated mountain (inselberg) in Zimbabwe. It is formed of massive granite, which in this case has very few joints. The granite is composed of coarse crystals produced by slow cooling of magma.

form during the cooling of molten magma. The cooling may take place underground, in which case the products are described as *intrusive* or plutonic; or it may take place at the surface to give *extrusive*, or volcanic, rocks. If the cooling takes place rapidly, as it might at the surface, glassy and fine-grained rocks form; however, slower cooling, which tends to take place at depth, allows bigger crystals to form. Moreover, different crystals may separate out at different temperatures. In these ways a great variety of igneous rock types are formed.

The intrusive types occur in different types of rock body (figure 1.16). When continental crust is deeply melted on a plate margin, it arises in the course of time as enormous dome-like masses called *batholiths*. These are composed of granite (plate 1.5). Some of the magma may be intruded into steep fractures as thin sheets called *dykes*, from which magma may be spread laterally along the *strata* or layers of the rock into which it is being intruded to give flat-lying *sills* and blister-like *laccoliths*.

Magma that succeeds in reaching the surface as an extrusive rock is called *lava*. The most common form is *basalt* (plate 1.6). This is a rock that flows freely and so forms extensive, nearly horizontal, sheets. Volcanoes with such runny basalt lava are produced chiefly in areas where the earth's crust is splitting apart, as for example along the mid-ocean ridges (plate 1.7).

With such fluid magma, gas is easily released and eruptions are non-explosive *effusions*, sometimes of great volume and extent. More viscous types of magma impede the release of dissolved gas, which may therefore sometimes cause an explosion which throws out clouds of *tephra*. A higher silica content in a magma increases its viscosity, and thus the *andesitic* volcanoes typical of destructive plate margins are characteristically explosive in their activity.

Plate 1.6 The rupturing of the earth's crust can create enormous spreads of plateau flood basalts. The many layers of basalt shown here are in the Deccan Plateau of western India and were erupted around 65 million years ago in association with the breaking away of India from Gondwanaland.

Plate 1.7 A field of rough lava on the island of Lanzarote, Canary Islands. The lava was spewed out in a great eruption in 1730–6. Note the volcanic cone in the background.

Metamorphic rocks

Rocks that have had their textures, structures and mineral composition changed by intense heat and/or pressure are called metamorphic rocks. They are a highly important component of the great continental shields and of the basement rocks beneath stable platforms.

They can form in a variety of ways. One group forms in the roots of ancient orogenic belts under the intense forces that are created when tectonic plates collide. This process is termed *regional metamorphism*. Other types are produced when magma is intruded into a host rock, for the temperature of the magma will be greatly in excess of that of the rock through which it is intruded, or upon which it is extruded. Changes may be brought about by partial melting (*contact metamorphism*), by superheating of water and by increasing pressures (*dynamic metamorphism*).

In some cases the degree of change is relatively modest, and the original rocks may do little more than become compact. This is called *low-grade metamorphism*. In *high-grade metamorphism*, on the other hand, the original characteristics of the rock may be totally transformed so that such phenomena as bedding planes, fossils and so on are completely obliterated. Minerals may segregate, forming alternating layers of light and dark minerals, as with the metamorphism of old igneous rocks, like granite, which are thereby converted to gneiss. Often, however, it is sedimentary rocks that are changed: limestone becomes marble and mudstone becomes slate. A good example of this type of situation is provided by the granite batholith of the Dartmoor area in south-west England. The batholith is encircled by a band, often called an *aureole*, of metamorphosed sedimentary rocks.

Sedimentary rocks

These rocks are formed from the destruction of earlier rocks. Weathering, erosion, transport and deposition carry material from one part of the world to another and lay it down as sediments: *marine sedimentary rocks* are deposited under the sea, and *terrestrial* ones on land.

Most sedimentary rocks (plate 1.8) are described

Window 1.8 *An ancient reef in north-western Australia*

About 350 million years ago in the Devonian era, a great reef grew around the land that now forms the north-west corner of Australia. A tropical sea covered what is today the flat desert of the Canning Basin, and the land of that time was a mountainous sandstone. Today, parts of the reef stand up from the floor of the ancient sea, looking almost as if the water has just drained away. These remnants of the reef are the limestone Napier and Oscar ranges of the Kimberley, which in places rise to some 200 metres.

To the south of the reef, flat lands lead to the Great Sandy Desert. To the north is what might be one of the oldest continuously exposed landscapes in the world, the King Leopold Ranges and the Kimberley Plateau. The rocks of this region are at least 700 million years old.

The reef, which has been described by the Australian geologist, P. G. Playford, has a similar structure to a modern reef, with a main reef, a steeply dipping fore reef that once sloped into the sea, and a gently sloping back reef that dipped into the lagoon separating the reef from the continent. Most of the organisms that built the reef were stromatoporoids, an extinct group vaguely reminiscent of, but unrelated to, corals. Other reef-builders were sponges and true corals and stromatolites, a group of cyanobacteria (blue-green algae) that formed mats in tropical seas, and secreted calcium carbonate. Today, where rivers have cut through the layers, as in the Windjana Gorge, you can see the beds of limestone running in different directions depending on which part of the reef you are looking at.

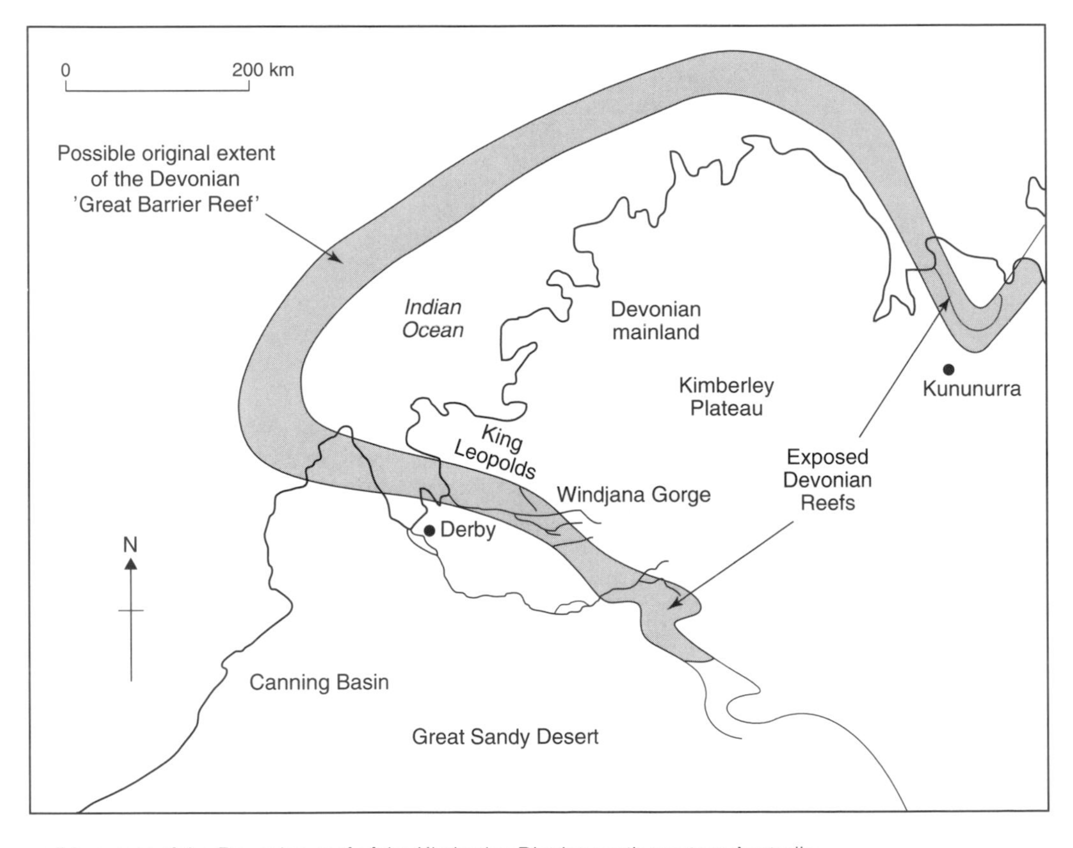

The possible extent of the Devonian reef of the Kimberley District, north-western Australia.

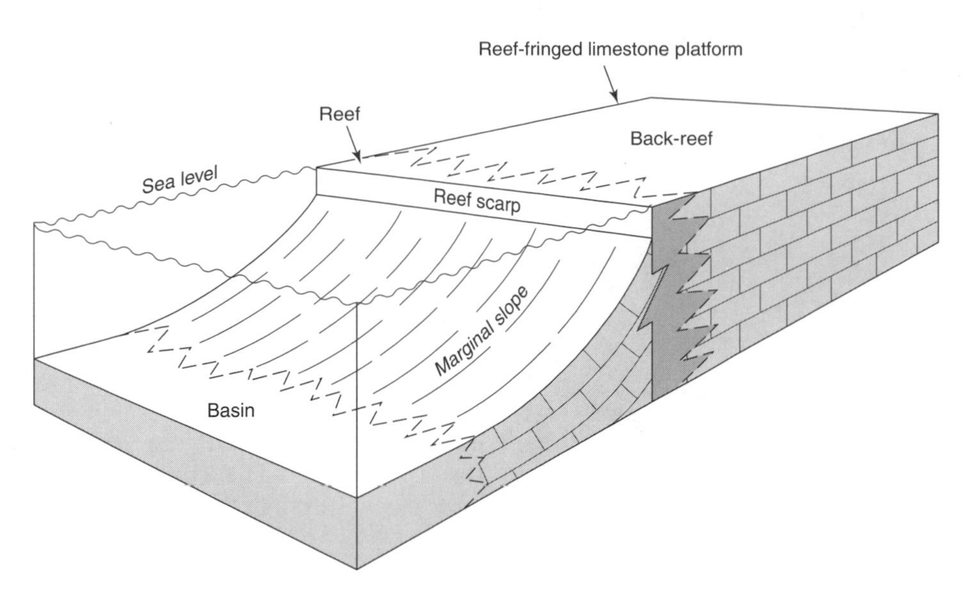

A model of the structure of the Devonian reefs.

The classic face in Windjana Gorge shows the sequence of beds illustrated in the box diagram.

Plate 1.8 Flat-bedded Cambrian–Ordovician sandstones form great cliffs in the Wadi Rum, south-east Jordan. Note also the major near-vertical joints.

as fragmental or *clastic*, and are formed of particles derived from earlier rocks, which have become converted into a harder substance by the process of *lithification*. Thus, gravels become *conglomerates*; angular scree fragments become *breccia*; sands become sandstones; and muds are transformed into mudstones or shales. Other sedimentary rocks are chemical precipitates, such as the limestones that develop on warm continental shelves (window 1.8); while others, like coal, are organic, made out of the fossil remains of animals or plants.

The conversion of a sediment into a sedimentary rock is achieved by a variety of processes. Of great importance is compaction, for as layers and layers of sediment accumulate, they exert pressures on the layers below. This compression squeezes out air and water from between the individual constituent particles, and distorts and flattens the particles themselves so that they interlock with each other. Another important process is cementation. Ground-

Table 1.1 The history of the earth, with subdivisions of time and the main events

Era	*Period*		*Millions of years BP*	*Main events*
Cenozoic	Quaternary	Holocene	0.01	Early civilisations
		Pleistocene	3	Emergence of man
				Start of Ice Age
	Tertiary	Pliocene	7	
		Miocene	26	
		Oligocene	38	Start of main Himalayan folding
		Eocene	54	
		Palaeocene	65	Extinction of dinosaurs
Mesozoic	Cretaceous		136	Main fragmentation of Pangaea; transgression of sea over land
	Jurassic		190	Start of fragmentation of Pangaea
	Triassic		225	Worldwide regression of sea from land
Palaeozoic	Permian		280	Formation of Pangaea
	Carboniferous	Pennsylvanian	315	
		Mississippian	345	
	Devonian		395	Animal life takes to the land
	Silurian		440	First land plants
	Ordovician		500	Vertebrates appear
	Cambrian		570	Major transgression of sea over continents
Precambrian	Proterozoic		2500	First multicellular organisms
	Archaean		?3000	Free oxygen in atmosphere
			?3500	First unicellular organisms
			3780	Age of oldest known terrestrial rocks
			(4600)	Formation of earth

water, seeping through the sediment, may deposit the minerals that it carries in solution, such as calcite, and may in due course unite the whole into a solid mass.

Altogether, sedimentary rocks cover more than two-thirds of the earth's surface and provide the main record of past environments, shifting coastlines and changing climates. Because they often contain fossils, they provide a record of changes in the nature and pattern of plant and animal life, and they have also supplied the principal means of establishing the divisions of geological time.

1.10 The Subdivisions of the Earth's History

Geologists divide up the 4600 million years of the earth's geological history into major eras and periods, as shown in table 1.1. The earliest era is called the Precambrian – a long, complex and poorly understood phase. During it there were periods of mountain-building and at least one ice age. It was a time when the lithosphere, atmosphere and hydrosphere first developed, and the first organisms appeared. Life was, however, essentially primitive, consisting of simple plants like algae. Many of the rocks of the world's great shield areas date back to this time.

In the Mesozoic era life was dominated by the dinosaurs and the gymnosperms (naked seed plants), the climate was basically warm and equable, and the progressive break-up of Pangaea occurred. At the boundary between the Mesozoic and the Cenozoic (that is, between the Cretaceous and the Tertiary) a great spasm of animal extinction occurred that led to the demise of the dinosaurs (window 1.9). In the Cenozoic era the climate deteriorated, sea-floor spreading continued, and there was the Ice Age of Quaternary times.

Human life probably first appeared on earth during the early part of the Ice Age, some 3 million years ago. The oldest human remains have been found in eastern (plate 1.9) and southern Africa. For a very long time the number of humans on the planet was small, and even as recently as 10 000 years ago the global population was probably only about one-thousandth of its size today. Also, for much of that time, humans had only modest technology and limited capacity to harness energy. These factors combined to keep the impact of humans on the environment relatively small. None the less, early humans were not totally powerless. Their stone, bone and wooden tool technology developed through time, improving their efficiency as hunters. They may have caused marked changes in the numbers of some species of animals and in some cases even their extinction. No less important was the deliberate use of fire, a technological development that may have been acquired some 1.4 million years ago. Fire may have enabled even small human groups to change the pattern of vegetation over large areas.

There are at least three interpretations of global population trends over the past 3 million years. The first, described as the 'arithmetic-exponential' view, sees the history of global population as a two-stage phenomenon: the first stage is one of slow growth,

Window 1.9 *The Cretaceous–Tertiary (K–T) extinctions*

About 65 million years ago, a giant meteorite impacted on the earth's surface and at about the same time an estimated 70 per cent of the flora and fauna on earth became extinct. Among the faunal extinctions were those of the dinosaurs. The impact probably occurred at Chicxulut in Yucatan, Mexico. It is probable that the impact of the meteorite caused the ejection of large amounts of fine material into the air and caused enormous wildfires, thereby causing a catastrophic climate change and severe atmospheric pollution (including acid rain). Other scientists believe, however, that volcanic eruptions at much the same time may have been the cause of the extinctions. Thick and extensive flood basalts – the Deccan traps of India (see plate 1.6) – were erupted at the same time and this phase of volcanic activity could have caused cooling of the atmosphere by sulphate aerosols and pollution by acid rain.

Plate 1.9 Some of the earliest fossils of humans have been found in such famous locations as Olduvai Gorge in East Africa.

while the second stage, related to the industrial revolution, displays a staggering acceleration in growth rates. The second view, described as 'logarithmic-logistic', sees the past million or so years in terms of three revolutions – the tool, agricultural and industrial revolutions. In this view, humans have increased the carrying capacity of the earth at least three times. There is also a third view, described as 'arithmetic-logistic', which sees the global population history over the past 12 000 years as a set of three cycles: the 'primary cycle', the 'medieval cycle' and the 'modernization cycle'. These three alternative models are presented graphically in figure 1.17.

Until the beginning of the Holocene, about 10 000 years ago, humans were primarily hunters and gatherers. After that time, in various parts of the world, increasing numbers of them started to keep animals and grow plants. Domestication caused genetic changes in plants and animals as people tried to breed more useful, better-tasting types. Domestication also meant that human populations could produce more reliable supplies of food from a much smaller area than hunter-gatherers. This in turn created a more solid and secure foundation for cultural advance, and allowed a great increase in population density. This phase of development is often called the first agricultural revolution.

As the Holocene progressed, many other technological developments occurred with increasing rapidity. All of them served to increase the power of humans to modify the surface of the earth. One highly important development, with rapid and early effects on environment, was irrigation. This was introduced in the Nile Valley and Middle East over 5000 years ago. At around the same time the plough was first used, disturbing the soil as never before. Animals were used increasingly to pull ploughs and carts, to lift water and to carry produce. Altogether the introduction of intensive cultivation and intensive pastoralism (the use of land for keeping animals) had a profound effect on many environments in many parts of the world. A further significant development in human cultural and technological

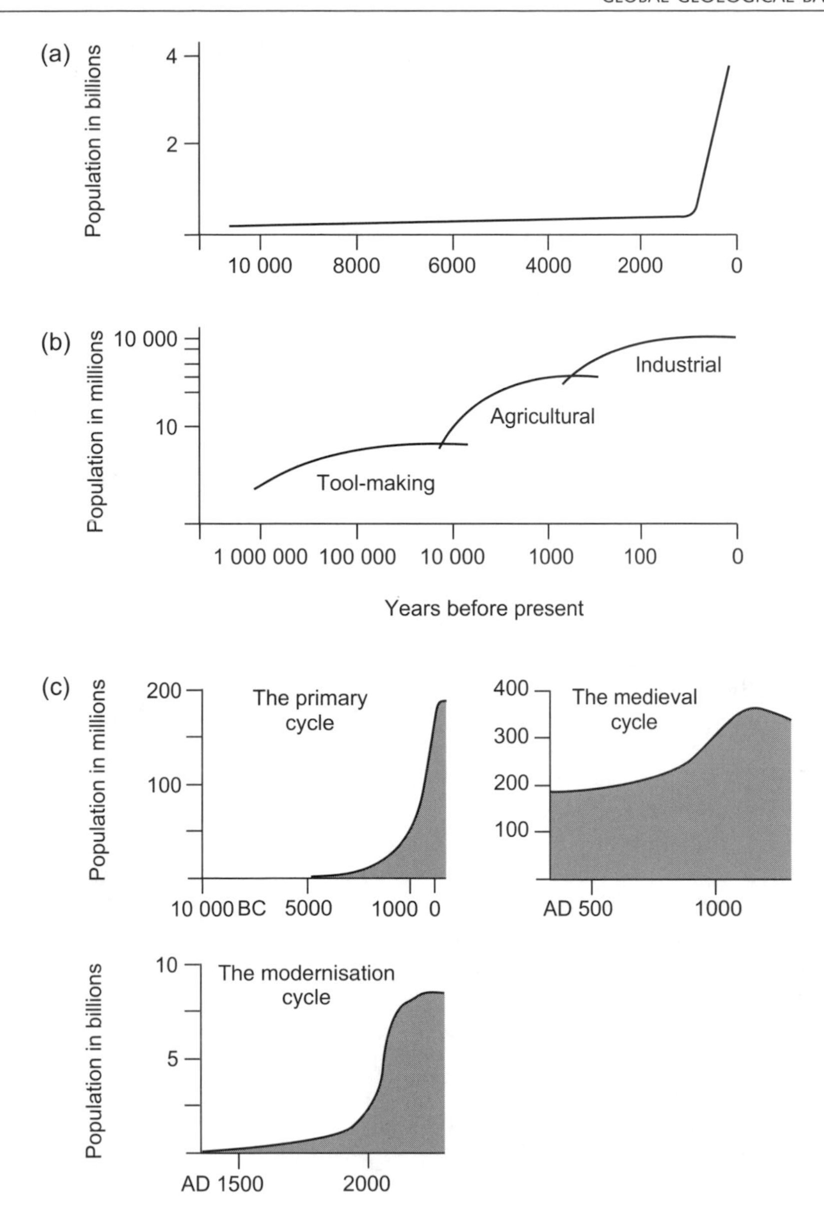

Figure 1.17 Three interpretations of global population trends over the millennia: (a) the arithmetic-exponential; (b) the logarithmic-logistic; and (c) the arithmetic-logistic.

life was the mining of ores and the smelting of metals, begun around 6000 years ago. Metal objects, such as axes and ploughs, gave humans greater power to alter the environment, and the smelting process required large quantities of wood which caused local deforestation.

The processes of urbanisation and industrialisation are two other fundamental developments that have major environmental implications. Even in ancient times, some cities evolved with considerable populations. Nineveh (the Assyrian capital) may have had a population of 700 000, ancient

Rome at its peak may have had a population of around 1 million, and Carthage (on the North African coast), at its fall in 146 BC, had 700 000 inhabitants. Such cities would have exercised a considerable influence on their environs, but this influence was never as extensive as that of cities in the past few centuries. The modern era, especially since the late seventeenth century, has witnessed the transformation of culture and technology through the development of major industries. This 'industrial revolution', like the agricultural revolution, has reduced the space required to sustain each individual and has seen resources utilised more intensively.

Part of this industrial and economic transformation was the development of successful ocean-going ships in the sixteenth and seventeenth centuries. As a result, during this time very different parts of the world became increasingly interconnected. Among other things, this gave humans the power to introduce plants and animals to parts of the world where they had not previously been. The steam engine was invented in the late eighteenth century and the internal combustion engine in the late nineteenth century: both these innovations massively increased the human need for and access to energy, and lessened dependence on animals, wind and water.

Modern science and modern medicine have compounded the effects of the urban and industrial revolutions, leading to accelerating population increase even in non-industrial societies. Urbanisation has gone on speedily, and it is now recognised that large cities have their own environmental problems, and produce a multitude of environmental effects. If present trends continue, many cities in the less-developed countries will become unimaginably large and crowded. For instance, in the year 2000 Mexico City has more than 30 million people – roughly three times the present population of the New York metropolitan area. Calcutta, Greater Bombay (Mumbai), Greater Cairo, Jakarta and Seoul are each in the 15–20 million range. In all, around 400 cities passed the million mark at the end of the twentieth century, and UN estimates indicate that by then over 3000 million people lived in cities, compared with around 1400 million people in 1970. The environmental effects of cities are discussed in chapter 16.

Modern science, technology and industry have also been applied to agriculture. In recent decades some spectacular progress has been made. Examples include the use of fertilisers and the selective breeding of plants and animals. Biotechnology, has, however, immense potential to cause environmental change.

Above all, as a result of the huge expansion of environmental transformation it is now possible to talk about global environmental change. There are two aspects of this: *systemic* global change and *cumulative* global change. Systemic global change refers to changes operating at the global scale and includes, for example, global changes in climate brought about by atmospheric pollution, e.g. the greenhouse effect. Cumulative global change refers to the snowballing effect of local changes, which add up to produce change on a worldwide scale, or change which affects a significant part of a specific global resource, e.g. acid rain or soil erosion. The two types of change are closely linked. For example, the burning of vegetation can lead to systemic global change through processes such as carbon dioxide release and modification of the earth's reflectivity (albedo), and to cumulative global change through its impact on soil erosion and the diversity of life.

Figure 1.18 shows how the human impact on

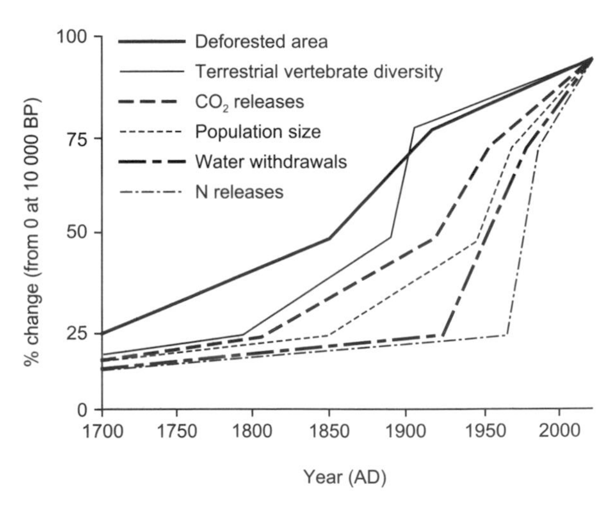

Figure 1.18 Percentage change (from assumed zero impact at 10 000 BP) of selected human impacts on the environment.

six 'component indicators of the biosphere' has increased over time. For each component indicator the total net change clearly induced by humans is put at 0 per cent for 10 000 years ago (before the present = BP) and 100 per cent for 1985. One can then estimate the dates by which each component had reached successive quartiles (that is, 25, 50 and 75 per cent) of its total change at 1985. About half of the components have changed more in the single generation since 1950 than in the whole of human history before that date. Human activities are now causing environmental transformation on the local, regional, continental and planetary scales.

1.11 Conclusions

Some of the major characteristics of the face of the earth can only be understood through an analysis of the present nature of its geological character, and by appreciating how this character has evolved over many millions of years. The other great influence on global patterns is the energy that we receive from the sun. This is the theme of the next chapter.

Key Terms and Concepts

andesite
continental shelf
cratons
extrusive rocks
flood basalts
Gondwanaland
hot spots
igneous rocks
intrusive rocks
island arcs
K–T extinction
magma
metamorphism
ocean ridges
ocean trenches
palaeomagnetism
passive margin
plate tectonics
plumes
rift valleys
rock cycle
sea-floor spreading
seamounts
sedimentary rocks
subduction zones
terranes
trailing edge
transform boundaries
Wilson cycles

Points for Review

- What can the ocean floors tell us about the history of the earth?
- Account for the global patterns of mountains, volcanoes and earthquakes.
- How do we know that sea-floor spreading and continental drift have taken place?
- What landscapes would you expect to find at different types of plate boundary?
- Put the main types of igneous rocks into their plate tectonic settings.
- How have human population numbers developed over the past three million years?
- What are the main ways in which humans are transforming the earth?
- Why in geological history have there been spasms of animal extinctions?
- Compare and contrast the plate tectonic settings of the British Isles and New Zealand.

FURTHER READING

Emiliani, C. (1992) *Planet Earth* (Cambridge: Cambridge University Press). A beautiful survey of the evolution of life, land and the environment.

Goudie, A. S. and Viles, H. (1997) *The Earth Transformed* (Oxford: Blackwell). An introduction to the ways in which humans have modified the earth over the past few millions of years.

Hallam, A. and Wignall, P. B. (1997) *Mass Extinctions and their Aftermath* (Oxford: Oxford University Press). An account of the causes of extinctions that have occurred through earth's history.

Kearey, P. and Vine, F. J. (1996) *Global Tectonics*, 2nd edn (Oxford: Blackwell Scientific). A successful and wide-ranging survey of current views on tectonics.

Lamb, S. and Sington D. (1998) *Earth Story* (London: BBC Books). A beautifully illustrated survey of earth's history.

Press, F. and Siever, R. (1986) *Earth*, 4th edn (San Francisco: Freeman). An attractive American text in physical geology.

Scarth, A. (1994) *Volcanoes* (London: UCL Press). A very readable account of the nature and history of volcanoes.

Selby, M. J. (1985) *Earth's Changing Surface: An Introduction to Geomorphology* (Oxford: Oxford University Press). A substantial overview of the geomorphology of the earth.

Summerfield, M. A. (1991) *Global Geomorphology* (Harlow: Longman Scientific and Technical). A global survey of landforms and processes which is strong on tectonics.

Van Andel, T. H. (1994) *New Views on an Old Planet: A History of Global Change*, 2nd edn (Cambridge: Cambridge University Press). A beautiful survey of the significance of continental drift.

2 Global Climatic Background

2.1 A Vertical Profile through the Atmosphere

Just as we started our consideration of global geology with a vertical section through the earth's interior, so we shall start this consideration of global climate by taking a vertical section through the earth's atmosphere (figure 2.1). The atmosphere consists of a mixture of gases, largely nitrogen and oxygen, that surround the earth to a height of many kilometres. Held to the earth by gravitational attraction, this envelope of air is densest at sea level and thins rapidly upward. This is because air is compressible; therefore near the earth's surface it is much more dense than in higher layers. At sea level the pressure averages about 1013 mb, while at 5000 m it is only 550 mb. This general trend is demonstrated in figure 8.1(b).

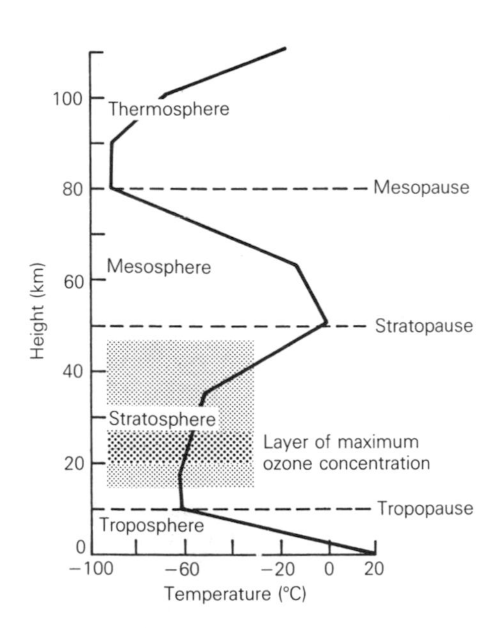

Figure 2.1 The vertical structure of the atmosphere in relation to temperature.

As pressure changes with altitude, so also does temperature, and the atmosphere is divided into layers on this basis. The lowermost layer, in which most of our weather is produced, is the *troposphere*. As we ascend through it the temperature falls at a rate of about 6.4 °C per 1000 m – a rate that is called the *environmental temperature lapse rate*. However, this rate changes abruptly once a certain threshold is reached, and the level at which this change occurs is called the *tropopause*. The altitude of the tropopause varies according to season, but tends to be highest over the equator (*c.*17 km) and lowest over the poles (*c.*10 km). Above the tropopause is a zone called the *stratosphere*, in which the temperature gradually rises, attaining a value of about 0 °C at about 50 km. This zone contains a layer of ozone (O_2) which acts as a filter for some of the burning ultraviolet radiation from the sun. As we ascend still further a temperature decrease sets in, and the point at which this change takes place is called the *stratopause*. Temperature declines to as little as −80° C in the next zone, the *mesosphere*, until a further reversal in trend takes place at the *mesopause* (*c.*80 km above the surface), and brings us into the heat of the *thermosphere*.

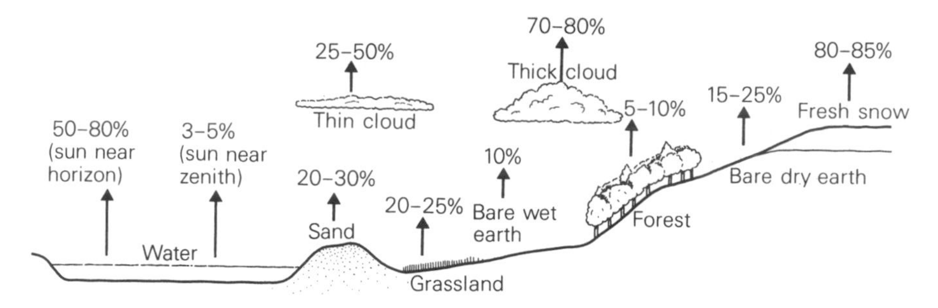

Figure 2.2a The albedo of the earth's surface. The fraction of the total radiation from the sun that is reflected by a surface is called its *albedo*. The albedo for the earth as a whole, called the *planetary albedo*, is about 35 per cent. The albedo varies for different surface types. Note also that the angle at which the sun's rays strike a water surface greatly affects the albedo value.

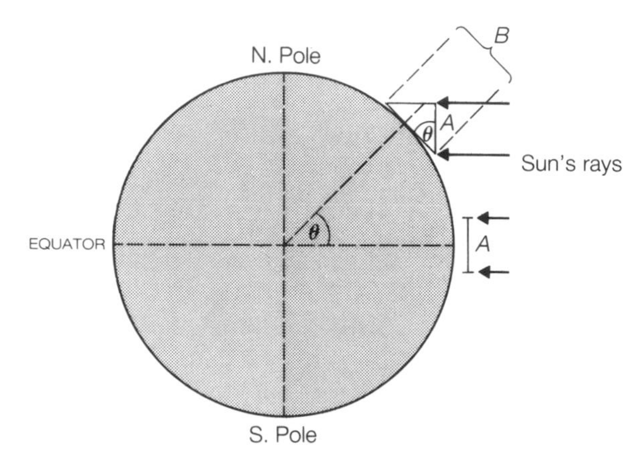

Figure 2.2b The angle of the sun's rays determines the intensity of insolation on the ground. In this example, we can see the effect of latitude on the amount of insolation received per unit area at the Equinox (i.e. when the sun is over the Equator). At high latitudes the same radiation received by area *A* is spread over the larger area *B*. $B = \theta$, where θ is the latitude in degrees.

Of these layers, the one that is of the greatest importance, in that it is the scene of most climate and weather phenomena, is the troposphere.

2.2 Global Climate

Along with geology, climate is the prime control of the physical environment, affecting soils, vegetation, animals (including man) and the operation of geomorphological processes such as ice and wind. The prime control of climate is the heat given out by the sun. Some of this heat creates the energy for the *atmospheric heat engine* which controls the nature of pressure, wind and climatic belts. However, some of the incoming energy never reaches the earth, being absorbed by gases like carbon dioxide (CO_2), dust and water droplets in the atmosphere. Some is reflected back by clouds, and some from the surface of the earth. In all, about 30 per cent is reflected, and the degree of reflection is called the *albedo*. Albedo values range from over 80 per cent for fresh snow to less than 10 per cent in some forested areas (figure 2.2a). The sun's heat reaches the earth by radiation, conduction and convection. By *radiation* solar energy travels through the vacuum of space from the sun to the earth in the form of electromagnetic waves and heats the earth's surface. By *conduction*, air in the thin layer of the atmosphere in contact with the earth is heated directly by the earth's hot surface. By *convection* the heated surface air expands and, being lighter than the air above, rises to be replaced by cooler air.

In the absence of cloud cover, the amount of energy that actually reaches the surface is dependent on the angle of the sun and the duration of daylight. These in turn, of course, depend on latitude. This means that the sun's heat reaches the low latitudes nearer the Equator in excess of that reaching high latitudes, largely because of the higher angle of the incident rays in low latitudes (figure 2.2b). The elevation of the sun is never great in high latitudes, even in mid-summer, while in the polar regions beyond the Arctic and Antarctic Circles, the sun in winter lies below the horizon for considerable periods. However, within low latitudes cloud cover is an important control of the actual amount of solar energy received at the earth's surface; for the heavy cloud of the equatorial zone, especially over land masses, reduces the total annual receipt of solar radiation by as much as 50 per cent. In contrast, the great subtropical deserts experience very little reduction in solar radiation because they are cloudless for much of the year.

If our planet did not rotate and were uniformly covered with one material, the atmosphere would circulate in response to the latitudinal differences in the heat generated from the incoming solar radiation. The fundamental force that causes air to move is the difference in air pressure from place to place, and this in turn is due to differences in temperature. Hot air is less dense than cold air, and so it tends to rise, thereby causing a reduction in pressure. Descending air, on the other hand, increases the pressure. Air will move from areas of high pressure to areas of low pressure, with the speed of movement being related to the pressure gradient. The steeper the pressure gradient, as represented by the closeness of the isobars, the higher will be the velocity.

Relating this to a rotation-free earth, it can be seen that the equatorial regions are zones of intense heating, rising air and low pressures. In the absence of rotation high pressure would develop over the

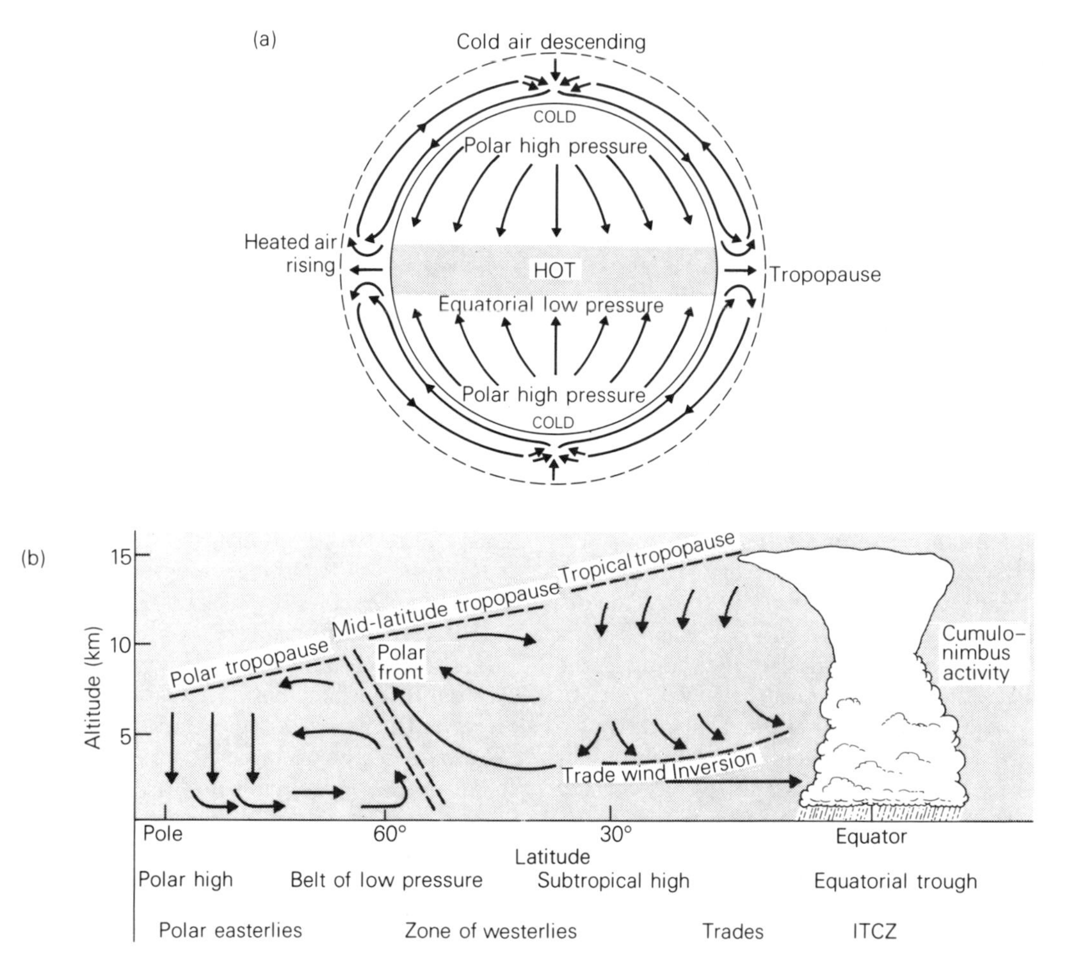

Figure 2.3 The general circulation of the atmosphere: (a) the atmospheric circulation pattern that would develop on a non-rotating planet; (b) the true atmospheric circulation shown schematically in a vertical section from polar regions to the Equator.

Poles, where heating is weakest. There would thus be a pressure gradient between the Equator and the Poles, and a gigantic convectional system would form (figure 2.3a), in which the equatorial low would be fed with air from the polar highs. Thus at the surface air would flow towards the Equator, while aloft it would flow towards the Poles.

In reality, our earth *does* rotate, and furthermore it is not uniformly covered with one material. This introduces a degree of complexity. If we first look at the low latitudes, we find that the global air circulation does indeed behave much like the simple *convection cell model*. Heated air rises near the Equator and begins its journey towards the Poles. However, its progress is hindered by three obstacles. First, as it rises, it no longer has a thick blanket of other air above it, and so it begins to lose much of its heat to space. At the same time, it is further from its principal source of heat energy, the warm earth surface; it gradually loses heat and becomes more dense. Second, because of the shape of the earth, air will converge as it moves poleward; at the Equator the earth has a diameter of about 40 000 km, but its east–west measurement shrinks as the Poles are approached. Third, because of an influence called the *Coriolis Effect*, which is caused by the rotation of the earth, air travelling poleward is deflected to the east.

The Coriolis Effect takes the form of an apparent deflection of a freely moving object or fluid to the right in the Northern Hemisphere and to the left in the Southern Hemisphere. For convenience, scientists have introduced the concept of a fictitious force to relate, mathematically, the size of the deflection to latitude. This Coriolis Force can best be understood by imagining a person standing at the centre of a rotating disc (figure 2.4) facing an object at the rim (part a). When he or she throws a ball directly at the object, the ball travels in a straight line and misses it because the object has rotated further to the left. But to the thrower, who is rotating with the object, the ball appears to have moved in a curved path *away from* the object (part b). Similarly, because of the earth's rotation, winds flowing from high to low pressure are always deflected.

The combination of cooling, compression/convergence and deflection means that, by the time the poleward-moving air has reached about 30° from

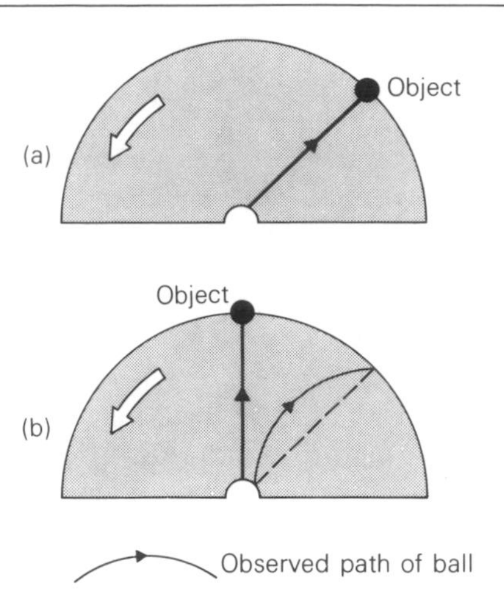

Figure 2.4 Diagrammatic illustration of the Coriolis Force. For explanation see text.

the Equator, it has already subsided. Therefore, a convectional cell does exist, called the Hadley cell, but it does not have the same latitudinal extent that one would have anticipated from the simple model developed on a non-rotating earth. The result of this is that there is a zone of high pressure characterised by subsiding air at about 30° N and S (figure 2.3b). This zone dominates the most important and persistent global wind systems: the trade winds, which blow on the equatorward side, and the westerly winds, which blow on the polar side. The trade winds cover in all about one-half of the earth's surface and converge towards the equatorial low-pressure zone, called the *Intertropical Convergence Zone* (ITCZ).

Between the subtropical high-pressure belt and about 60° latitude is the belt of prevailing westerly winds. In the Northern Hemisphere they are to a certain degree disrupted by land masses; but in the Southern Hemisphere between 40° and 60° S there is an almost unbroken belt of ocean, which enables the westerlies to gather great strength and persistence. The westerlies circulate around a zone of lower pressure called the *polar low*, and involve the entire depth of the troposphere. Especially in the Northern Hemisphere, the uniform flow of the upper-air westerlies is disturbed by the formation of large undulations called *Rossby waves*. These develop along a zone called the *polar front*, which

Plate 2.1 A NOAA-7 satellite image of western Europe and the eastern Atlantic. This large white whirl is a rain-bearing depression which approached Britain on 17 September 1983.

marks the contact between cold polar air, forming the troposphere on the poleward side, and warm tropical air which surrounds the globe on the equatorward side. It is a zone of great instability, along which many atmospheric disturbances are generated (plate 2.1). Associated with the Rossby waves and the polar front is a narrow zone of very high wind speed (on average *c.*125 km h^{-1} in winter) called a *jet stream*, which lies at the level of the tropopause. The westerlies, the Rossby waves and the polar front jet streams (window 2.1) are responsible for a great deal of exchange of heat and energy between high latitudes and low.

The mid-latitude jet stream – the polar front jet – sometimes splits into two and is not continuous round the globe. Several other jet streams are also known to exist. For instance, an almost permanent jet exists over the subtropics centred at latitude 25° N. This westerly jet is located about 13 km above the surface in the region where the Hadley cell and the middle latitude (or Ferrel) cell meet. Another jet flows from west to east at about 15 km over northern India in the summer months when the ITCZ has shifted northwards. Another jet circles the Arctic during the cold polar night at 25–30 km above the surface.

The Rossby waves in the westerlies change their shape to give different patterns (figure 2.5), and once a particular pattern is established it may be persistent. Especially in the Northern Hemisphere, the pattern is related in part to the distribution of land

Window 2.1 *Jet streams and blocking anticyclones*

Jet streams do not flow in a straight line around the world, but in a series of three to six horizontal waves. The number of these waves, and their positions, depend on many factors, including the locations of mountain ranges and the presence of large expanses of cold and warm ocean water. They are also affected by zones of high pressure – blocking anticyclones – that force the jets to divert around them.

The weather of North America is greatly affected by the path of the jet stream. Normal flow of the jet (A) brings mild winters to the United States and leads to rainfall on the west coast. However, if the jet has larger amplitude waves (B), this can cause drought spells in California, bitterly cold and dry conditions east of the Rocky Mountains (as air is brought down from the Arctic) and snowy winters in the east as moisture is picked up in the Gulf of Mexico.

In Europe the jet stream brings a stream of depressions (low-pressure cells) and, with them, rainfall. If a blocking anticyclone becomes established (H) this pattern is changed and drought may result.

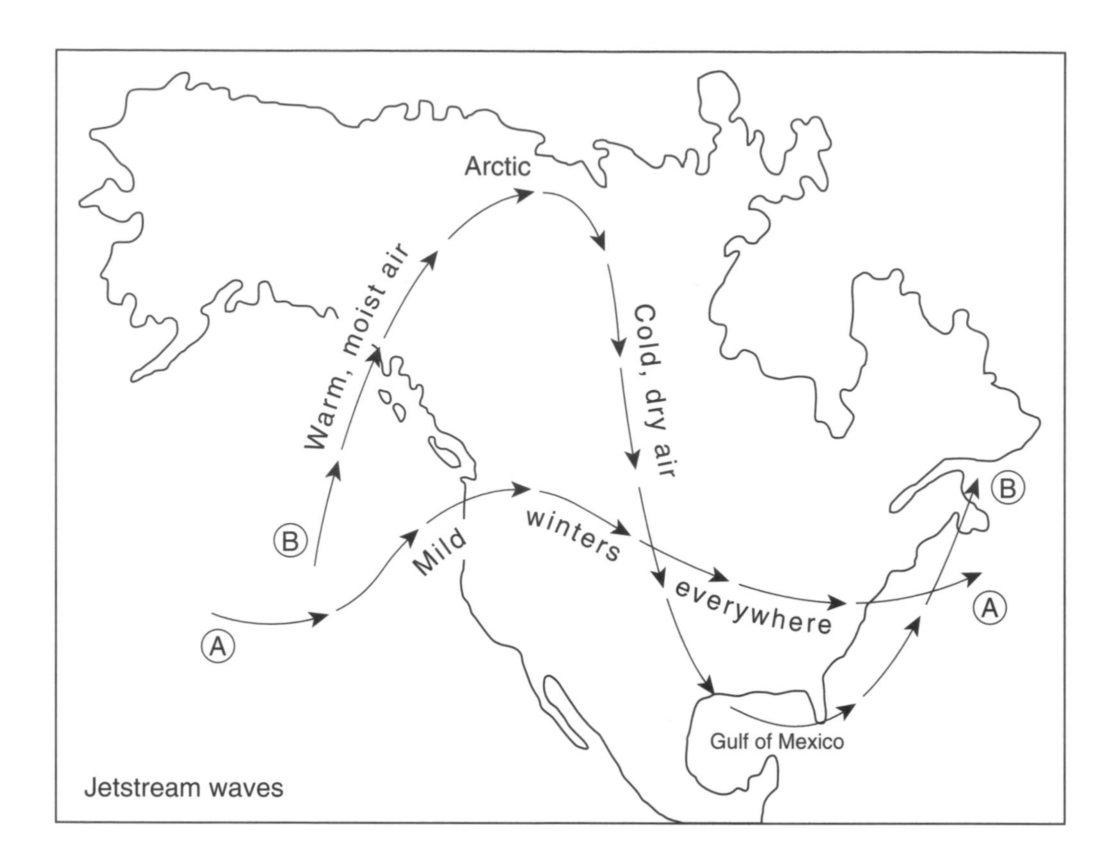

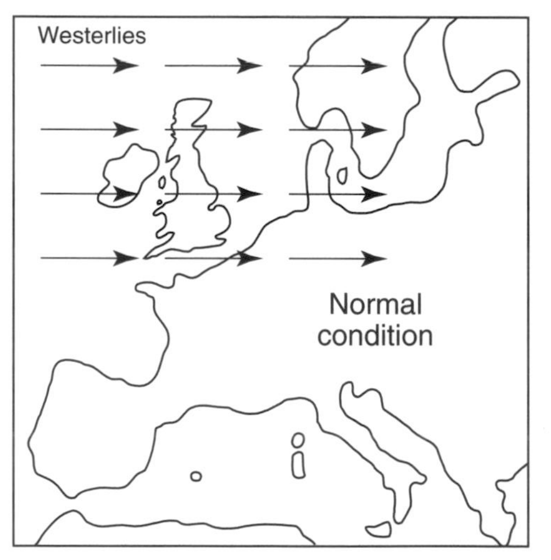

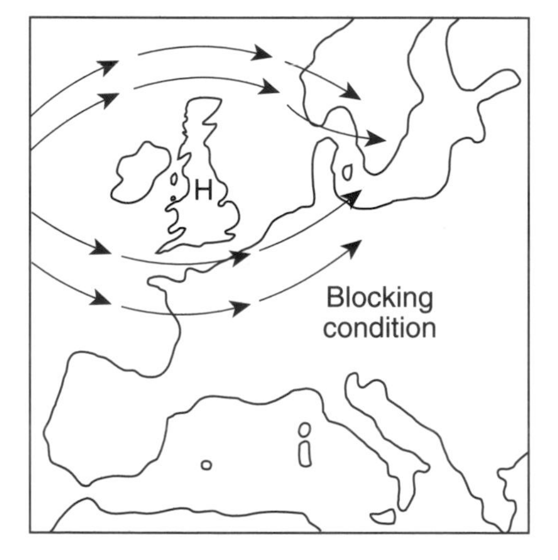

Jet stream: normal conditions and blocking conditions.

and sea, and to the position of high land. A ridge tends to develop over eastern North America and over the Eurasian continent, particularly in winter. The main stationary waves are located at 70° W and 150° E, related to the positions of the Rocky Mountains and the Tibetan Plateau, respectively. In all, there are normally between three and six of the waves around the middle latitudes. The waves sometimes cover only a small latitudinal zone, with the bulk of the movement of air being from the west. This situation is called a *high zonal index*, and is associated with strong and persistent winds that produce mild, wet winters in western Europe. A *low zonal index* develops when the simple pattern of waves and troughs breaks down to give a series of cells. This can disrupt the passage of the rain-bearing depressions and lead to blocking anticyclones, which can be very persistent and can result in long spells of settled or unsettled weather. The great British drought of 1976 and the hot summer

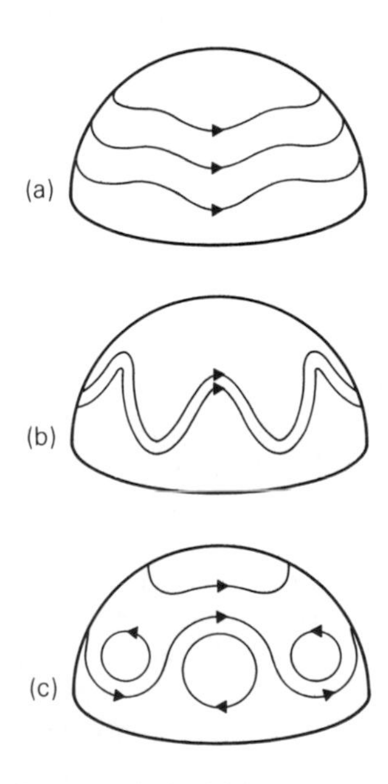

Figure 2.5 The waves in the high-level westerlies of the Northern Hemisphere take on different configurations which are extremely important in terms of surface weather conditions. In the high-index case (a) they have little north–south movement and give changeable weather. The low-index case produces lows (b), and may develop further (c) until the waves break up into a stationary 'blocking' cellular pattern which can produce more extended spells of either good or bad weather.

of 1983 were the result of a blocking anticyclone that lasted a long time.

We can thus summarise the nature of the atmosphere's general circulation. Briefly, it consists of three distinct but collaborative cells. In low latitudes is the so-called *Hadley cell*, with heat near the Equator being transferred from low to high altitudes by great convective activity. Polewards to approximately 30° of latitude is a zone of subsidence throughout all but the lowest kilometre or two of the troposphere. At the surface, trade winds complete the cell, flowing from the north-east in the Northern Hemisphere and from the south-east in the Southern Hemisphere. The trades converge along the ITCZ. In middle latitudes there is a *Ferrel cell*, with westerlies prevailing at all levels in the troposphere, and with weather dominated by transitory depressions and anticyclones. In polar regions, radiative cooling causes air in contact with the earth's surface to contract, which creates subsidence of air in the lower troposphere and an anticyclonic tendency. Surface air that flows away from the Poles is deflected by the Coriolis Effect to become easterlies, but the features associated with the *Polar cell* are rather indistinct and intermittent.

2.3 World Patterns of Precipitation

The nature of the general circulation and the arrangement of the oceans and the continents lead to certain general patterns in the distribution of precipitation (rain, snow, etc.) over the face of the earth (figures 2.6a and 2.6b). In the equatorial belt rainfall levels are high, often exceeding 2000 mm per annum. The prevailingly warm temperatures and high moisture content of the air lead to abundant convectional rainfall over areas such as the Amazon Basin of South America, the Congo Basin of Africa and the archipelago of south-east Asia. Narrow coastal belts of high rainfall extend further polewards along the eastern sides of continents as far as 25–30° from the Equator. These are the coasts where the trade winds blow off the oceans and, encountering coastal hill ranges, produce heavy rainfall. By contrast, in the vast belts of deserts that lie approximately along the tropics the rainfall may be as little as 20–30 mm, and will occur in the vicinity of the subtropical high-pressure cells, where

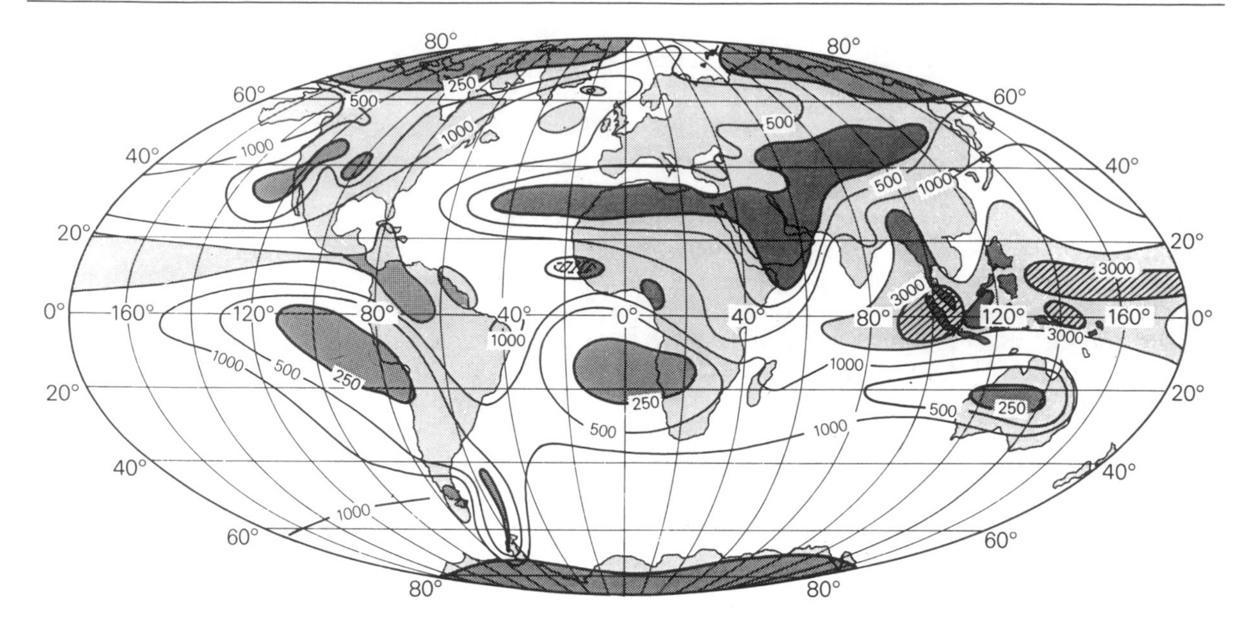

Figure 2.6a World patterns of precipitation. The distribution of precipitation throughout the world in mm per year: dark shading shows precipitation below 250 mm, light shading shows areas with precipitation above 2000 mm, and the diagonal hatching shows areas above 3000 mm.

subsiding rather than rising air is dominant. They are especially well developed on the west coasts of the continents, where the presence of cold ocean currents favours desert development (see section 6.2). Further northwards, in the continental interiors of Asia and North America, dryness results largely from remoteness from ocean sources of moisture.

In the mid-latitudes, especially on west coasts between the latitudes of 35° and 65°, there is another belt of high precipitation in the region of prevailing westerly winds. Where the coasts are mountainous and the air is forced to rise, precipitation can exceed 2000 mm, as in the South Island of New Zealand, Norway, southern Chile and British Columbia. In the Arctic and polar regions the air is so cold that it cannot contain much moisture, so low precipitation conditions return. Except on the west coast belts, annual precipitation poleward of 60° is under 300 mm.

From the point of view of humans, however, it is not only rainfall quantity that is important; one also has to consider the reliability or variability of the rainfall from year to year (figure 2.6c). Once again, it is possible to see clear global patterns. The most striking feature is the very high variability of rainfall in the world's driest areas. Note particularly the great belt of variable rainfall that runs from north-west Africa to the Middle East and central Asia, and the smaller zones in south-western Africa (the Namib), central Australia, north-east Brazil, western South America, the south-west United States and northern Mexico. In the central Sahara the variability may exceed 100 per cent. By contrast, other parts of the world are fortunate in having more reliable and consistent levels of precipitation, which saves them from the worst effects of runs of dry years and droughts. In western Europe the variability may be as little as 10 per cent, and another zone of markedly consistent rainfall is to the east of the Great Lakes in North America. In general, the humid tropics also have low rainfall variability.

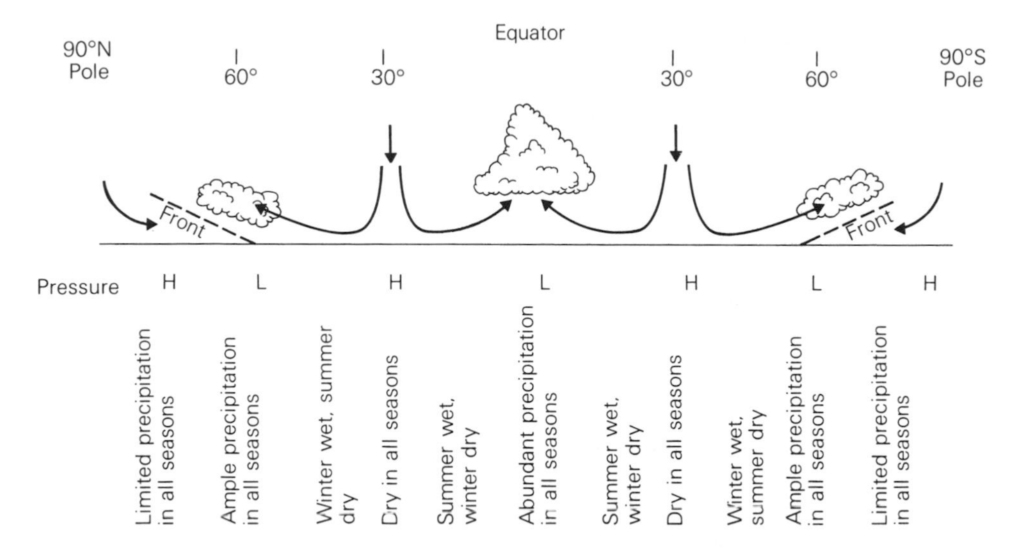

Figure 2.6b Schematic zonation of rainfall belts in a vertical cross-section from Pole to Pole.

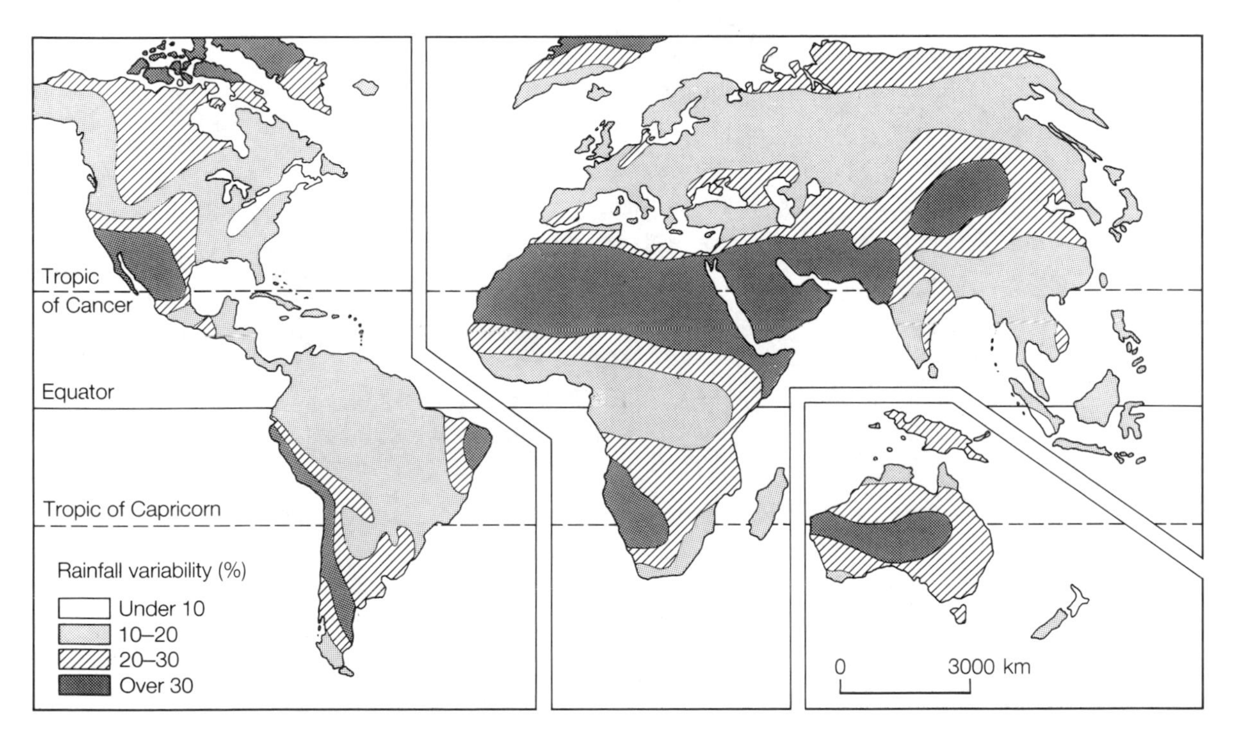

Figure 2.6c World precipitation variability expressed as the percentage departure from the normal. Note the high variability in desert regions.

2.4 World Patterns of Temperature

Because of the way in which radiation receipts from the sun vary with latitude, temperatures tend to decrease from the Equator towards the Poles (figure 2.7a). Moreover, in the equatorial regions the receipt of insolation is fairly uniform throughout the year so that temperatures are uniformly high. Because of their proximity to oceans, which act as a moderating influence, coastal areas also tend to have rather limited temperature ranges over the year. The annual range (figure 2.7b) is moderately large on land areas in the zone of the subtropical high-pressure cells, and because of the frequent absence of cloud cover, maximum temperatures in summer can be very high – higher even than those in the equatorial zone. Large land masses in the sub-Arctic and Arctic zones develop areas of extremely low temperatures in winter but in the summer temperatures can rise to quite high levels, so that over eastern Siberia and northern Canada the annual temperature range may exceed 45 °C. Areas of perpetual ice and snow such as the great ice caps of Greenland and the Antarctic are always intensely cold. Temperatures vary greatly with season in the mid-latitude and sub-Arctic zones because of the annual march of the sun. Thus, in North America the 15 °C *isotherm* (a line drawn to connect all points having the same temperature) lies over Florida in January and near Hudson Bay in July.

The lands and the oceans are very different in the way in which they take up and give out heat. Heat moves into the land by conduction, and as a result only the upper few metres of the land exchange much heat with the atmosphere on an annual basis. Moreover, land materials generally have low heat capacities. For these reasons the heat taken in by the soil in the warm summer is quickly lost in the ensuing cool season. By contrast, heat moves into the oceans by convection and is thus carried to much greater depths than on land. Moreover, water has a high heat capacity, and hence tends to give up its heat much more slowly than the land does. Oceans therefore tend to have a moderating influence on temperatures, so that land areas in winter are typically much colder than the oceans, whereas in the summer they are much warmer. Thus in the interior of continents temperature ranges over the year tend to be higher than they are near oceans.

These differences in heat storage result in pressure differences in the atmosphere. The large land

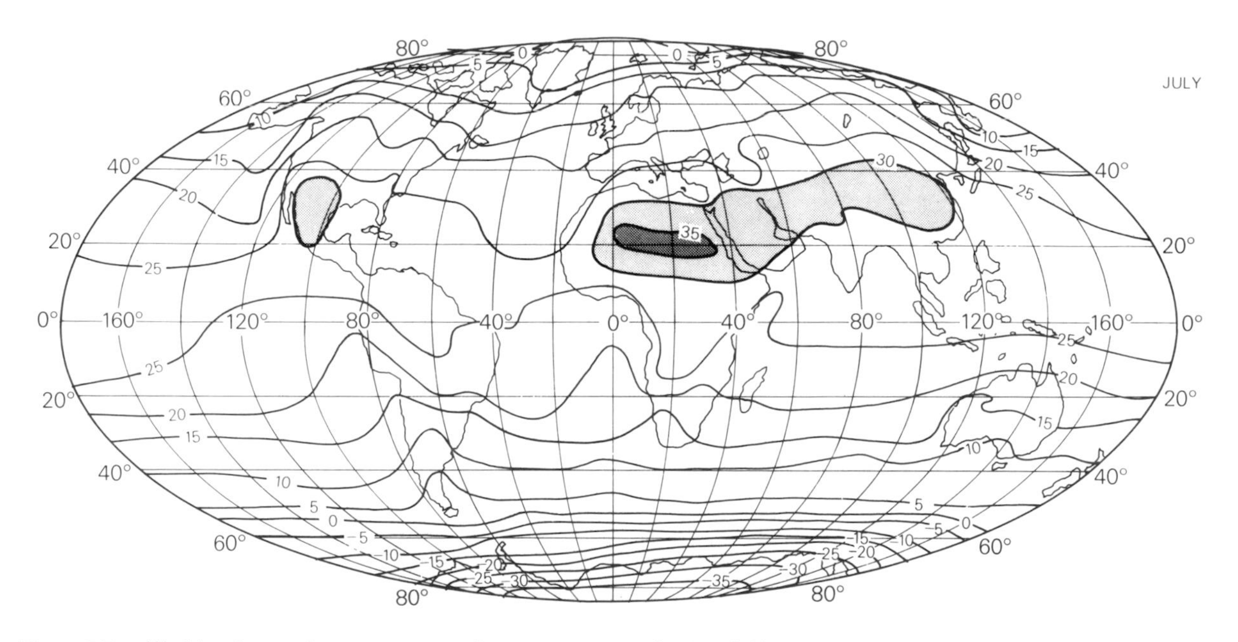

Figure 2.7a World patterns of temperature: surface temperatures for July (°C), expressed as those at sea level.

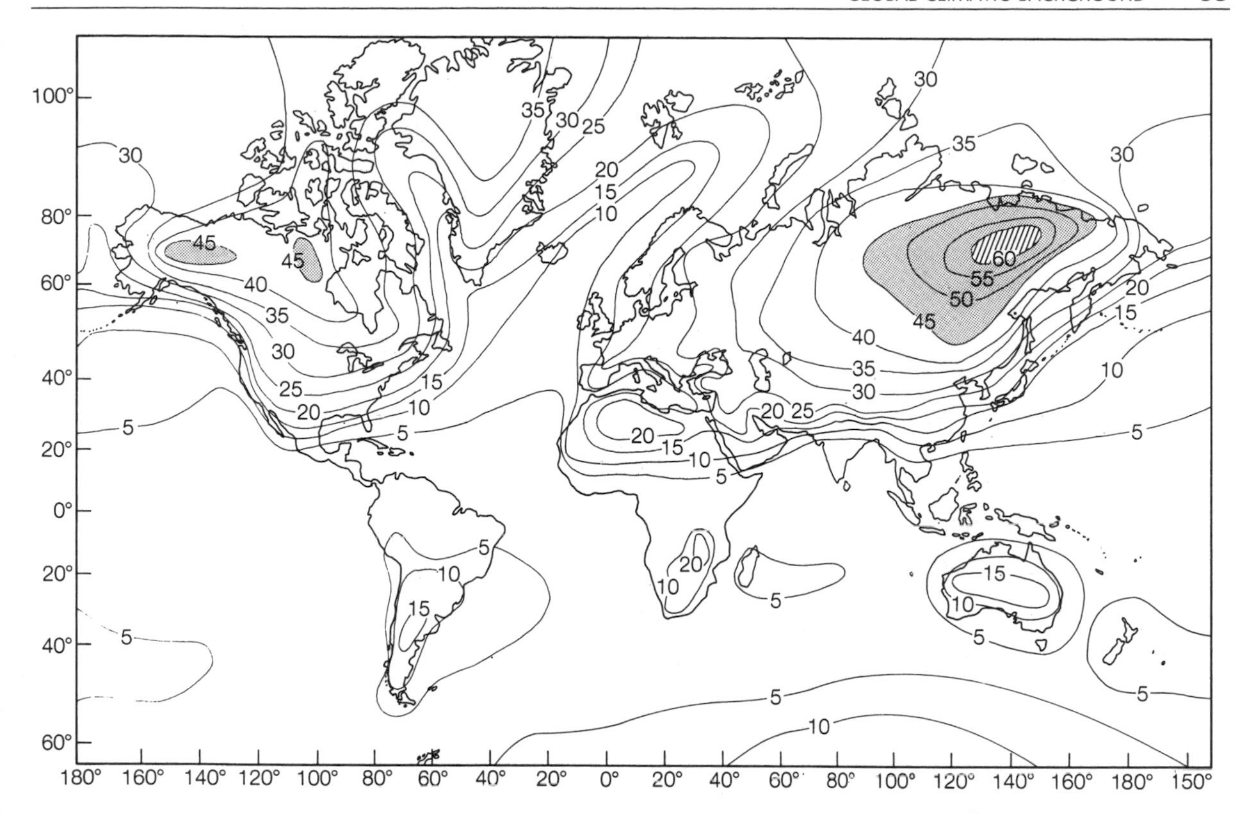

Figure 2.7b World patterns of temperature: annual range of air temperatures (°C) showing differences between the means for January and July. Note the particularly high differences in the interior of eastern Asia.

masses, such as Eurasia, tend to have very high pressures in winter and low pressures in summer. This affects the direction of the major wind belts and contributes, for example, to the wind reversal that is such a feature of the monsoonal lands of Asia (window 2.2) (see section 7.4).

2.5 The Major Climatic Zones

By combining the patterns of temperature, precipitation and evapotranspiration, one can construct a series of major world climatic zones – zones that will form the basis of Part II below, and are important for understanding the world patterns of such phenomena as soils and vegetation.

The *humid tropical zone* surrounds the Equator and extends discontinuously about 10° N and S of it. Mean daily temperature approximates 25–27 °C, and the diurnal variation exceeds the modest annual variation. Precipitation totals vary greatly according to such factors as relief, but are frequently in excess of 2000 mm per annum. One of the outstanding features of this zone is the relative lack of seasonality.

A *semi-arid tropical zone* extends from the humid tropics, and the further one moves away from the Equator, the more seasonal becomes the climate because of the migration of the overhead sun. Rainfall and temperature reach their maximums in the warm wet season when the sun is at its zenith; a dry season then prevails for some months when the sun moves into the opposite hemisphere. Polewards from this zone, and extending to as much as 30° N and S, is the *arid tropical zone*, containing the great hot deserts such as the Sahara. This is a zone of subsiding air, which has sporadic rainfall in time and space, high diurnal temperature ranges and limited cloud cover and high radiation levels. Evapo-

Window 2.2 *The monsoonal lands*

Monsoonal circulations dominate the climates of large continents and their adjacent oceans. In winter, cold, dense, dry air flows outwards towards the warmer oceans, while in summer, intense heating of the land mass results in rising air and a landward flow of moist, oceanic air. The Coriolis Force modifies the radial outward and inward flow directions, leading in the case of Asia to north-east (winter) and south-west (summer) monsoons. Note the particularly large latitudinal movement of the Intertropical Convergence Zone (ITCZ) in the area under the influence of the Asian monsoon.

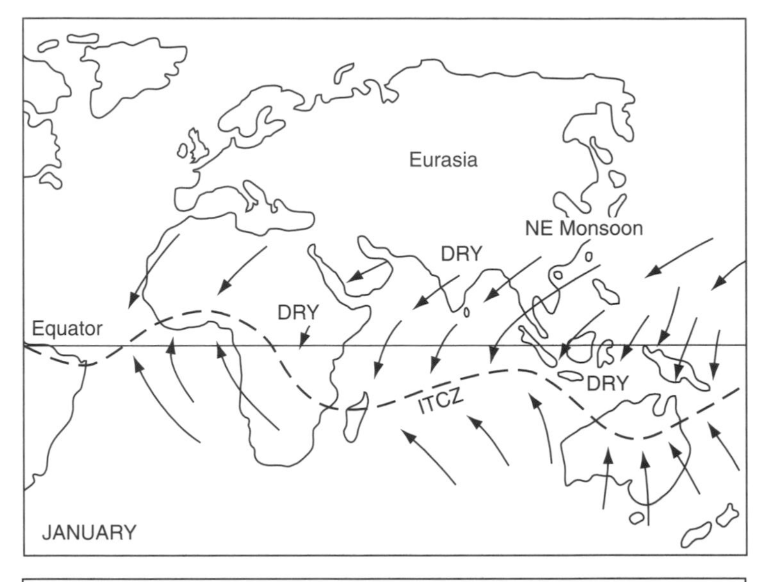

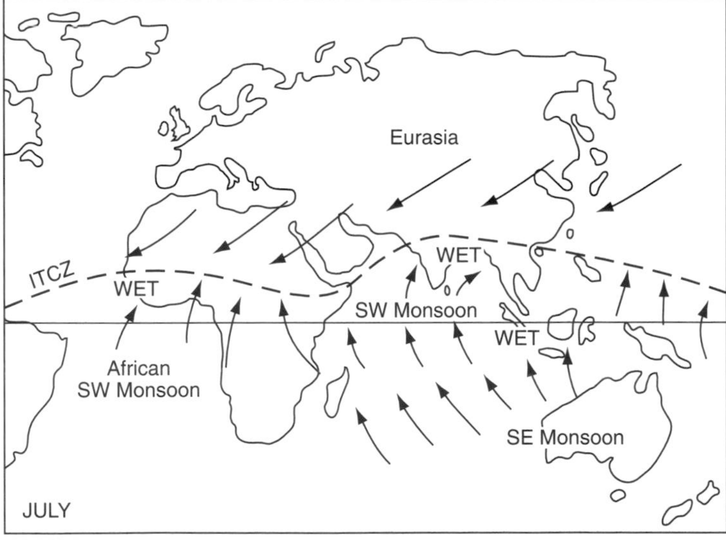

The movement of the Intertropical Convergence Zone (ITCZ) with the seasons.

transpiration greatly exceeds precipitation so that there is a severe water deficit.

This zone is gradually replaced by a belt with more humid conditions, called the *semi-arid sub-tropical zone*. The subsiding dry air of the tropics predominates in summer, while in winter low pressure brings rain and lowered temperatures. These areas thus show very marked seasonal contrasts, and are often described as being of 'Mediterranean' type, though they show great variability in character according to distance from the sea.

At about 50° N and 45° S, the impact of the dry, subsiding air is replaced almost totally by the effects of the polar front, the westerlies and various low-pressure cells. In oceanic areas with a marked maritime influence, such as the British Isles, the annual temperature regime is not great – in the case of the Scilly Isles off the south-west peninsula of England it is only 8 °C. This *humid temperate zone* has mild winters, cool summers and year-round rainfall. Away from the maritime influence the precipitation level decreases; there is more snow in winter, and summer thunderstorm activity is frequent. The annual range of temperature increases, and in the far interior may reach as much as 40–50 °C. This zone is generally called the *semi-arid temperate zone*, but in extreme continental situations the precipitation level becomes so small that an *arid steppe* climate develops.

Moving yet further towards the Poles is the *boreal zone*, though because of the lack of land in the Southern Hemisphere in the latitudes 55–60° S it is only well developed to any great degree across northern North America, Scandinavia and Russia. Summers are cool and moist, but very low temperatures occur in the winter. Edmonton in Canada, for example, which can be taken as a representative station, has a July mean of around 15 °C and a January mean of as little as –8 °C.

The most poleward of our zones is the *Arctic zone*, where the effect of the seasons is intense. For three months of the year there may be 20 hours of daylight. Precipitation levels are generally rather small, and the mean annual temperatures may be low enough for the subsoil to be permanently frozen (a state called *permafrost*).

2.6 The Hydrological Cycle

When they talk about the *hydrological cycle* (see chapter 14 for a fuller discussion), scientists refer to the process whereby water comes to the earth from the atmosphere through precipitation and eventually returns to the atmosphere by evaporation. The largest amounts of water in the whole cycle (figure 2.8) are those involved in direct evaporation from the sea to the atmosphere and in precipitation back to the sea. Evaporation from the land surface, and the transpiration of water from plants, combine with precipitation of water on to the land to play a quantitatively smaller (but from the human point of view, important) part in the cycle. The precipitation on to the land that is not lost by evaporation runs off the surface in the form of streams and rivers (window 2.3).

The great bulk of the world's water is in the oceans – around 97 per cent. For this reason, most of the world's water is salty. Of the 3 per cent of fresh water, about 75 per cent is locked up in the ice caps and glaciers, and much of the rest is present below ground in the rocks as *groundwater*. Only a minute proportion is present at any one time in the rivers (0.03 per cent) and in the lakes (0.3 per cent).

On a global basis, certain patterns emerge (figure 2.8c). Three major zones with an excess of moisture generate substantial stream runoff, while there are two zones where evapotranspiration is so greatly in excess of precipitation that there is a severe moisture deficit.

2.7 Ocean Currents

Ocean currents are controlled by, and contribute to, global climatic conditions. On the one hand, they aid in the exchange of heat between low and high latitudes, and modify extremes of climate. On the other, they are set in motion largely by the prevailing surface winds associated with the atmospheric general circulation. The transference of energy from wind to ocean water is achieved by the frictional drag of the air blowing over the water surface. As with winds, the Coriolis Effect affects the direction of water movement. Temperature differences (figure 2.9(a)) also affect currents through

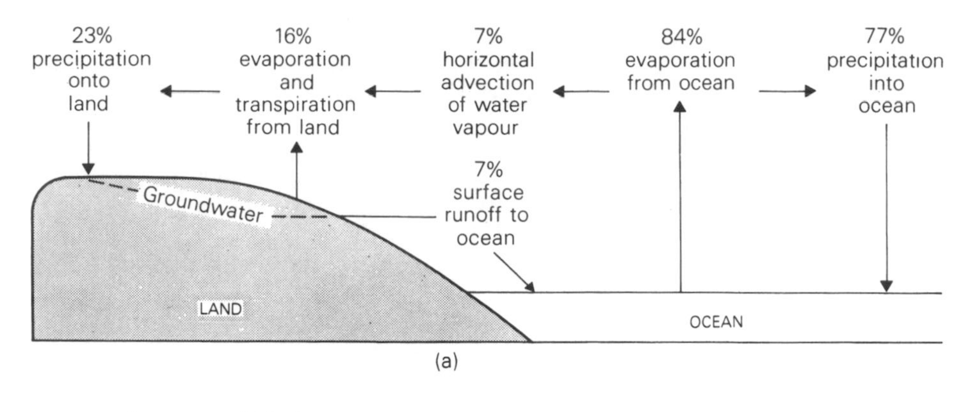

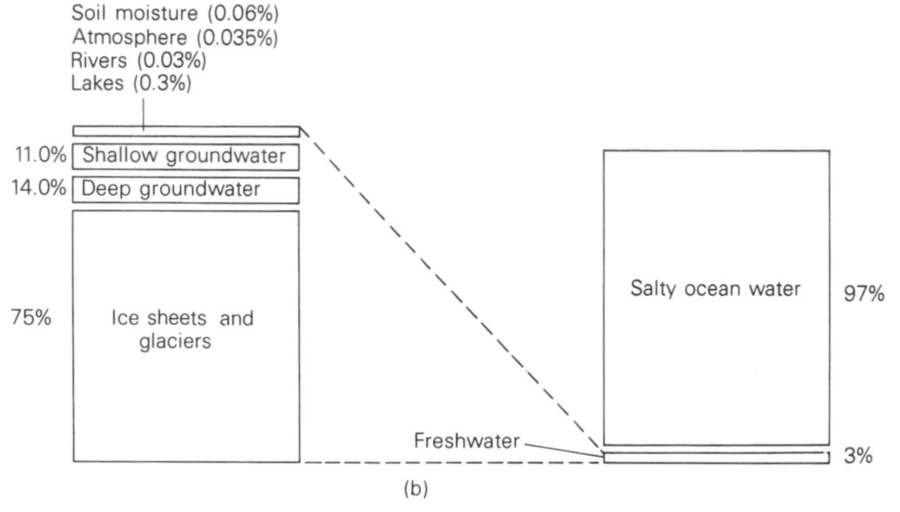

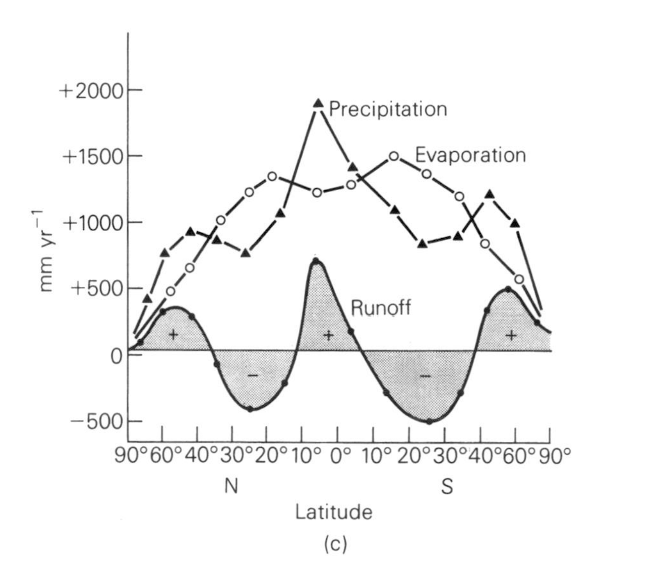

Figure 2.8 The world hydrological cycle: (a) the main components; (b) freshwater and salt proportions; (c) the latitudinal variation in evaporation, precipitation and runoff.

Window 2.3 *Water and sediment in world rivers*

The world's rivers discharge about 35 000 km^3 of water into the oceans each year. The world's ten largest rivers account for about 38 per cent of this total. The Amazon River alone contributes more than 15 per cent of the world total, and more than the combined total of the next seven largest rivers. Indeed, perhaps not surprisingly, it is rivers in tropical areas with heavy rainfall that are the prime contributors of water to the oceans, with about two-thirds of global total coming from southern Asia, Oceania and north-eastern South America. By way of contrast, Africa, because of its generally low rainfall, is a relatively sparse contributor of water to the oceans, with only the Zaire and Niger making any significant input.

The annual average water discharge of the ten biggest (in terms of water discharge) rivers is as follows (in km^3 per year):

River	km^3/yr	River	km^3/yr
Amazon	6300	Mississippi	580
Zaire (Congo)	1250	Yenisei	560
Orinoco	1100	Lena	514
Ganga/Brahmaputra	970	Plata (Plate)	470
Changjiang (Yangtze)	921	Mekong	470

World rivers probably contribute about 13.5 × 10^9 tonnes of material to the oceans each year. This means that the average global yield of sediment per square kilometre of drainage basin is about 150 tonnes per year. Once again there are some very interesting global patterns. For example, the rivers of southern Asia and Oceania contribute about 70 per cent of the total, even though they only account for about 15 per cent of the land area draining into the oceans. North-eastern South America contributes another 11 per cent because of the importance of the Amazon, Orinoco and Magdalena rivers. The very high values for the rivers of Oceania, which encompass those of Taiwan, New Zealand and New Guinea, relate directly to their mountainous terrain, heavy rainfall and relatively small drainage basins, which accumulate and store less sediment than larger rivers.

The average annual suspended load for the ten biggest (in terms of sediment load) rivers is as follows (in 10^6 tonnes per year):

River	10^6 t/yr	River	10^6 t/yr
Ganga/Brahmaputra	1670	Magdalena	220
Huang He (Yellow River)	1080	Mississippi	210
Amazon	900	Orinoco	210
Changjiang (Yangtze)	478	Hunghe (Red River)	160
Irrawaddy	285	Mekong	160

In all, this amounts to about 40 per cent of the world total.

their influence on density. Thus, for example, the chilled surface water of the high-latitude seas will sink to the ocean floor, spreading equatorwards, and displacing upward the less dense, warmer water (window 2.4).

If we look at a map of surface currents (figure 2.9c) certain major features become apparent. Given the importance that we have attached to the subtropical high-pressure zones in the atmosphere, it is perhaps not surprising that in the oceans these are the locations of some major circular movements, called *gyres* (figure 2.9b). To the equatorwards side of each of these great whirls is an *equatorial current* that flows westwards and marks the belt of the trade winds. The equatorial currents are separated by an *equatorial counter-current.*

Poleward of these subtropical gyres is a zone that comes under the influence of the westerlies, and here we have a current type called a *west-wind drift.* This is especially well developed in the Southern

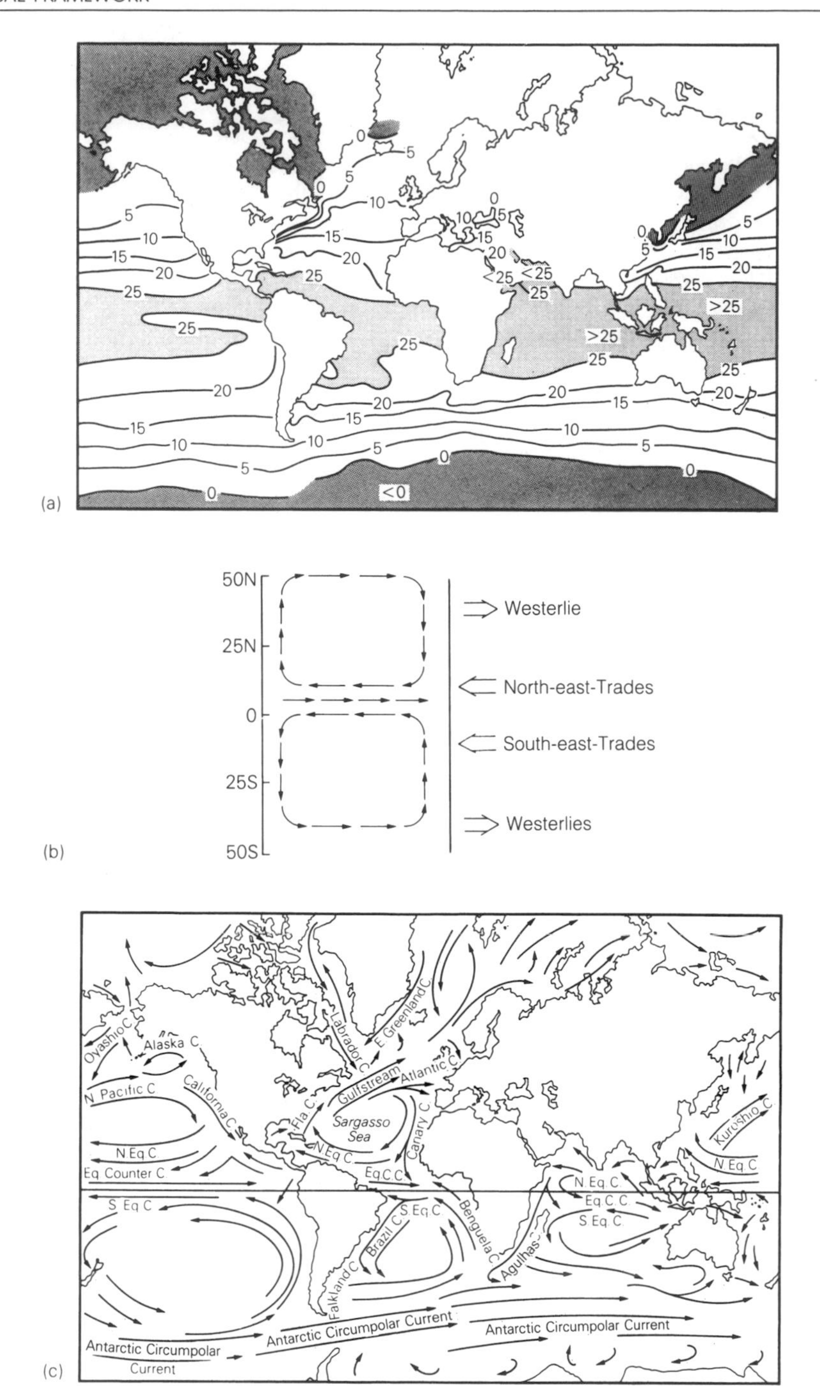

Figure 2.9 Some of the main features of the oceanic circulation: (a) annual mean sea surface temperature over the world (°C); (b) schematic view of the wind-driven circulation of the oceans, with a clockwise gyre in the Northern Hemisphere and a counter-clockwise gyre in the Southern Hemisphere. A counter-current separates the two gyres, and is located between the region of the north-east and south-east trade winds; (c) the actual circulation of the oceans illustrating the main currents.

Window 2.4 *Thermohaline circulation*

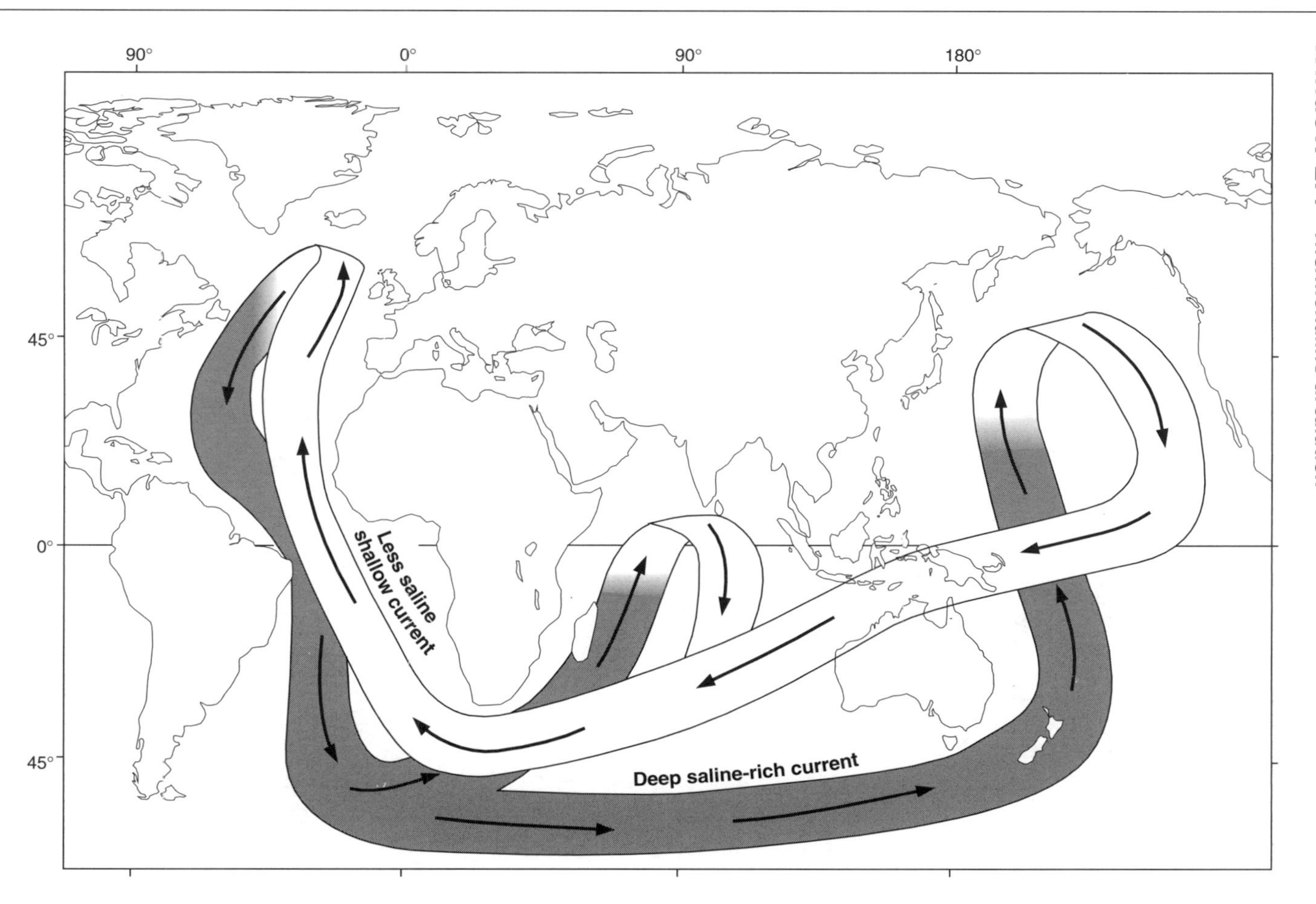

The large-scale salt transport system ('ocean/global conveyor') operating in the present oceans. This compensates for the transport of water (as vapour) through the atmosphere from the Atlantic to the Pacific Ocean. Salt-laden deep water formed in the North Atlantic flows down the length of the Atlantic and eventually northwards into the deep Pacific. Some of this water upwells in the North Pacific, bringing with it the salt left behind in the Atlantic due to vapour transport. Flow of the 'Atlantic conveyor' may have been interrupted during cold episodes.

Climate is closely linked to ocean circulation, so that any change in the pattern of ocean circulation is likely to have major consequences for climate at the regional and even the global scale. Some circulation changes are driven by water temperature and salinity (density) gradients – the so-called thermohaline circulation. Cooler and more saline waters are denser and tend to sink. In modern oceans the densest waters occur in the North Atlantic where the combination of coolness and high salinity leads to large amounts of deep water known as North Atlantic Deep Water (NADW). Within the North Atlantic the thermohaline circulation appears to operate by means of a conveyer system whereby water moves northwards in the upper levels of the ocean, ultimately to sink around latitude 60° N to form the NADW. The return limb of the conveyor at depth transfers this deep water to the southern oceans. At the global scale, it has been suggested that differences in salt concentration between the Atlantic and Pacific Oceans drive a global conveyor through which there is a mass transfer of dense, saline water from the Atlantic to the Pacific at depth, and a compensating counter-current near the surface. If the production of NADW could be varied, as for example by meltwater driving down the salinity of the North Atlantic, this would alter the rate of heat release and so cause regional climatic change.

Hemisphere, extending between latitudes 30° and 70° S.

In low latitudes, the equatorial current turns polewards on encountering the coastlines on the west sides of the oceans, and thus forms a warm current that parallels the coast. Examples of this include the Gulf Stream off North America, the Brazilian Current off Brazil, and the Kuroshio Current off Japan. Such coastlines have higher-than-average water temperatures.

West-wind drifts tend towards the east sides of oceans, and may turn either polewards or equatorwards. The poleward movement is illustrated by the North Atlantic Drift which warms the British Isles, Norway and even the far north-west of Russia, enabling navigation into the Arctic port of Murmansk all the year round. Equatorward flow produces cool currents along the eastern shores of the oceans, such as the Humboldt (Peruvian) Current off South America, the Benguela Current off southern Africa, the Canaries Current off West Africa and the California Current off the western United States. These have an important bearing on the location of some of the world's deserts (see section 6.2).

Finally, in the Northern Hemisphere cold water flows from the largely land-locked polar sea along the west side of the large straits connecting the Arctic Ocean with the Pacific and the Atlantic. Running southwards from the Bering Strait is the cold Kamachatka Current; along the eastern coast of Canada from western Greenland is the Labrador Current; and between Iceland and Greenland is the Greenland Current.

The oceans are basically cold. The water is warm only at the surface, and becomes progressively colder with depth (figure 2.10). Only 8 per cent of the ocean water is warmer than 10 °C, and more than a half is colder than 2.3 °C. This helps to explain the very considerable effect that oceans have on climate. With respect to temperature the ocean, when viewed in cross-section, presents a three-layered structure. At low latitudes throughout the year, and in mid-latitudes in the summer, a warm sur-

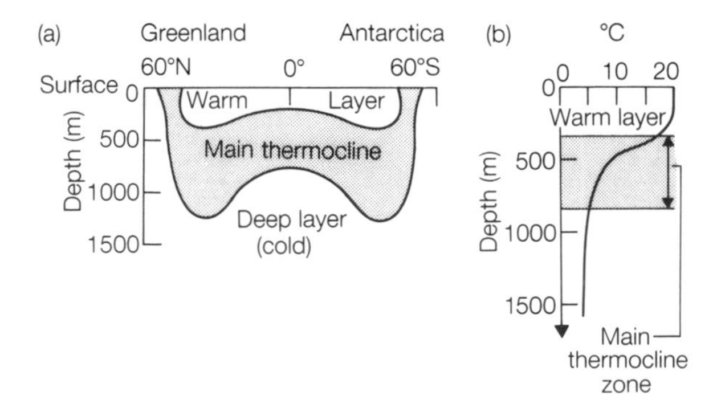

Figure 2.10 The main temperature characteristics of the oceans: (a) a profile of the three main zones of the Atlantic Ocean; (b) the average temperature depth profile for the open ocean in low latitudes.

face layer develops, which because of the mixing effects of wave action may be as much as 500 m thick. In equatorial regions it may have a temperature in excess of 25 °C, but this rapidly declines polewards.

Below this layer temperatures drop rapidly, creating a second layer called the *thermocline*. Beneath this there is cold deep water, temperatures of which range from 0 to 5 °C. In high latitudes the three-layer structure is replaced by a single layer of cold water, which, because it is denser than warm water, 'flows' at depth towards the tropics.

In certain configurations of the coastline, the trend of the currents and the direction of the offshore winds, cold water may move upwards to the surface by a process called *upwelling*. This is a common feature on the west coasts of continents, notably off Peru, Oregon, the western Sahara and south-western Africa. Upwelling creates a narrow strip of ocean water of low temperature that is uncharacteristic of that latitude as a whole. Because the surface waters are being constantly replenished with nutrients from below, such areas are zones of great fertility, thus supporting huge fish populations. By contrast, the Sargasso Sea, which lies in the tropical North Atlantic in the centre of the subtropical high-pressure zone, is a marine desert, in which nutrients cannot be so readily replenished. The warm surface water is markedly less dense than the cooler water beneath and thus forms a stable layer. Because of this, once the nutrients are used up they are not replaced from below (see also section 10.3).

Overall, however, it has to be said that, in comparison with land areas, the oceans are not, in biological terms, very productive. Whereas the *net primary productivity* (NNP – the rate of production of organic matter) of the oceans averages 155 $g\,m^{-2}\,yr^{-1}$, the comparable figure for the continents is 782. This is a topic we shall return to in our consideration of the major world vegetation types (see section 3.1).

2.8 Ocean–atmosphere Interactions

The oceans have a huge heat storage capacity and so have a very fundamental influence on world climates. Should the nature of the ocean temperatures change, then there will be consequential changes of weather patterns on land. Such ocean–atmosphere–interactions have become a major research focus in recent years.

The most interesting illustration of ocean–atmosphere interactions is provided by the *El Niño phenomenon* (window 2.5). Although the term originally referred to a local warm current that runs southward along the coast of Ecuador in the eastern Pacific around the Christmas season (hence 'The Child'), it has now become employed to describe much larger-scale warmings of the eastern equatorial Pacific that last for one or two years and occur at intervals ranging from 2 to 10 years.

The relationship between El Niño and global climate needs to be seen in the context of its atmospheric counterparts, the *Walker Circulation* and the *Southern Oscillation*. Under 'normal' conditions the nature of the Walker Circulation, one of the most important components of the global atmospheric system, is as shown in figure 2.11, with a great longitudinal cell readily observable across the Pacific. Near the coast of South America the winds blow offshore, causing upwelling and the subsequent cold, nutrient rich, offshore waters. However, under El Niño conditions, the circulation pattern is reversed and this causes water temperatures off the coast of South America in the equatorial belt to rise substantially. This see-saw in atmospheric conditions is called the Southern Oscillation.

The oscillation seems to have significance beyond the Pacific coast of equatorial South America, where the presence of warm water may be associated with anomalously high rainfalls over Peru and Ecuador, and with fish mortality in the sea. Recent studies have shown that at times of El Niño numerous persistent climatic anomalies occur elsewhere on earth, such as droughts in Australia, Indonesia and north-east Brazil, severe winters in the USA and Japan, and cyclones in the central Pacific. The oceanic circulation plays the role of a flywheel in the climate system, and is responsible for the extraordinary persistence of the atmospheric anomalies from month to month and even from season to season.

2.9 Climatic Change

As we have already seen in chapter 1, our knowledge of how the world works, and why it is as it is,

Window 2.5 *El Niño, 1997–1998*

El Niño is the term used to describe an extensive warming of the upper ocean in the tropical eastern Pacific lasting up to a year or even more. The negative or cooling phase of El Niño is called *La Niña.* El Niño events are linked with a change in atmospheric pressure known as the Southern Oscillation (SO, see below). Because the SO and El Niño are so closely linked, they are often known collectively as the El Niño/Southern Oscillation or ENSO. The system oscillates between warm to neutral (or cold) conditions every three to four years.

The 1997–8 El Niño was one of the strongest on record, developing more quickly and with higher temperature rises than ever recorded. It developed rapidly throughout the central and eastern tropical Pacific Ocean in April and May 1997. During the second half of the year, it became more intense than the major El Niño of 1982–3, with sea-surface temperature (SST) anomalies across the central and eastern Pacific of 2–5 °C above normal. SSTs exceeded 28 °C across the central and east-central equatorial Pacific beginning in May 1997, as the normal cooling of ocean waters typical of June–October was notably absent. The warming effect of El Niño was a major factor contributing to the record high global temperature in 1997. The estimated global mean surface temperature for land and marine areas averaged 0.44 °C above the 1961–90 base period mean. The previous warmest year was 1995 with an anomaly of +0.38 °C. By mid-January 1998, the volume

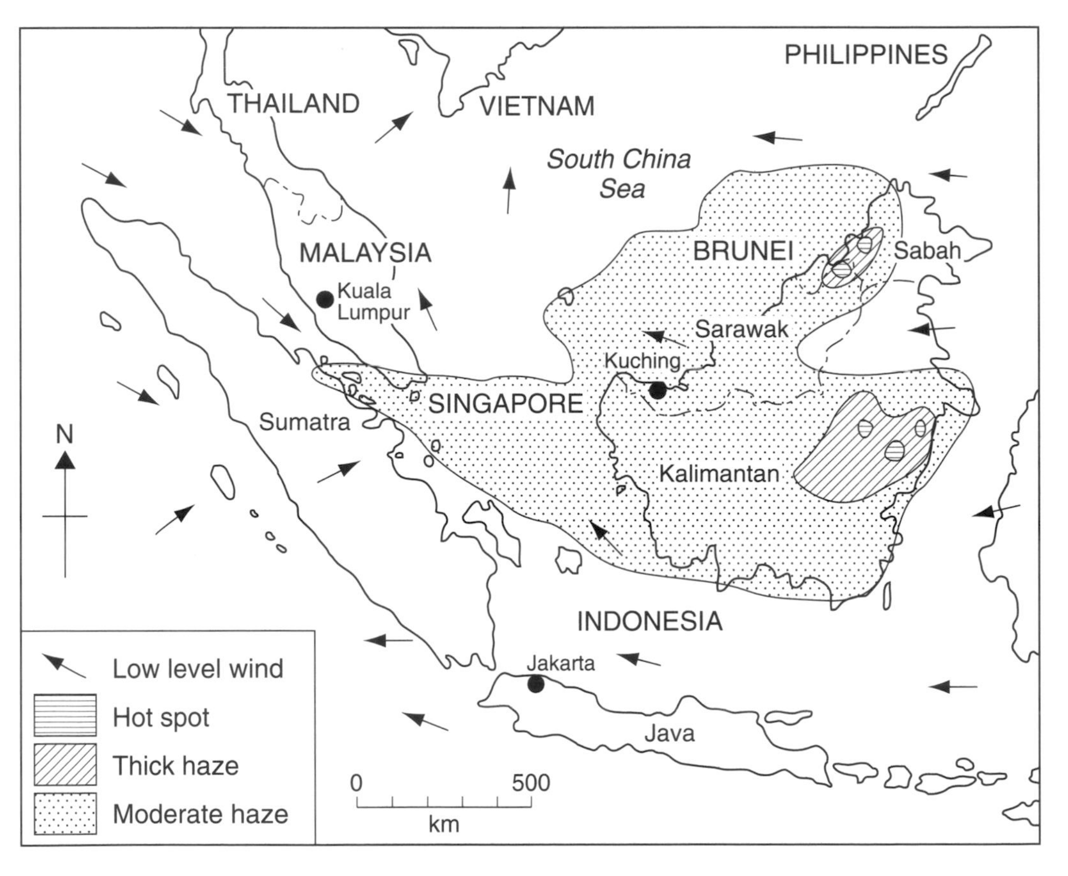

The pattern of forest fires and their haze over south-east Asia in August 1997.

of El Niño's warm water pool had decreased by about 40 per cent from its maximum in early November 1997, but its surface area in the Pacific was still about 1.5 times the size of the continental USA. This warm pool had so much energy that its impact dominated world climate patterns until mid-1998.

Precipitation and temperature anomaly patterns appear to characterise all El Niño warm episodes. These can be summarised as follows:

- The eastward shift of thunderstorm activity from Indonesia to the central Pacific usually results in abnormally dry conditions over northern Australia, Indonesia and the Philippines.
- Drier-than-normal conditions are also usually observed over south-eastern Africa and northern Brazil.
- During the northern summer season, the Indian monsoon rainfall tends to be less than normal, especially in the north-west.
- Wetter-than-normal conditions are usually observed along the west coast of tropical South America, and at subtropical latitudes of North America (the Gulf Coast) and South America (southern Brazil to central Argentina).
- El Niño conditions are thought to suppress the development of tropical storms and hurricanes in the Atlantic but to increase the numbers of tropical storms over the eastern and central Pacific Ocean.

One newsworthy consequence of dry conditions over Indonesia was the occurrence of huge forest fires that polluted the air over large expanses of south-east Asia. In August 1997 a series of forest fires raged in Indonesia and continued until the following June when they were subdued by rainfall. Large tracts of forest were burnt and a great haze of smoke covered extensive areas, especially over Borneo, but extending as far as Sri Lanka. In extreme cases, visibility was reduced to less than 50 m. A warm episode of ENSO brought the worst drought in half a century and set the preconditions for burning. In the western tropical Pacific rainfall totals in 1997 and early 1998 dropped more than 50 per cent below the norm in some months.

Sources

C.Y. Jim (1999) 'The forest fires in Indonesia 1997–1998: possible causes and pervasive consequences', *Geography*, 84: 251–60.
World Meteorological Organisation (1998) *World Climate News*, 13 June.

has been revolutionised in the past few decades by the concept of plate tectonics. Over much the same period another major transformation in ideas has also taken place – again owing much to discoveries on the ocean floors, and again having a bearing on the history of the earth: our ideas about climatic changes over the past few millions of years have been revolutionised.

It has become increasingly apparent that world environments have been subject to frequent and massive changes during the course of the latest period of geological time – the Quaternary. Even in the past 20 000 years the area of the earth covered by glaciers has been reduced to one-third what it was at the glacial maximum; the waters thereby released have raised ocean levels by over a hundred metres; the land, unburdened from the weight of overlying ice, has locally risen by several hundred metres; vegetation belts have swung through the equivalent of tens of degrees of latitude; permanently frozen ground and tundra conditions have retreated from extensive areas of Europe; the rainforests have expanded; desert sand fields have advanced and retreated; inland lakes have flooded and shrunk; and many of the world's finest mammals have become extinct. Thus, on the one hand, the major patterns of climate, ocean currents, fauna, flora, landforms and soils that we have been examining are not very long-lived; and, on the other, they owe much to the happenings of the past few million years.

The changes of climate, and the other environmental changes that they produced, have occurred over a wide range of time scales (figure 2.12). The

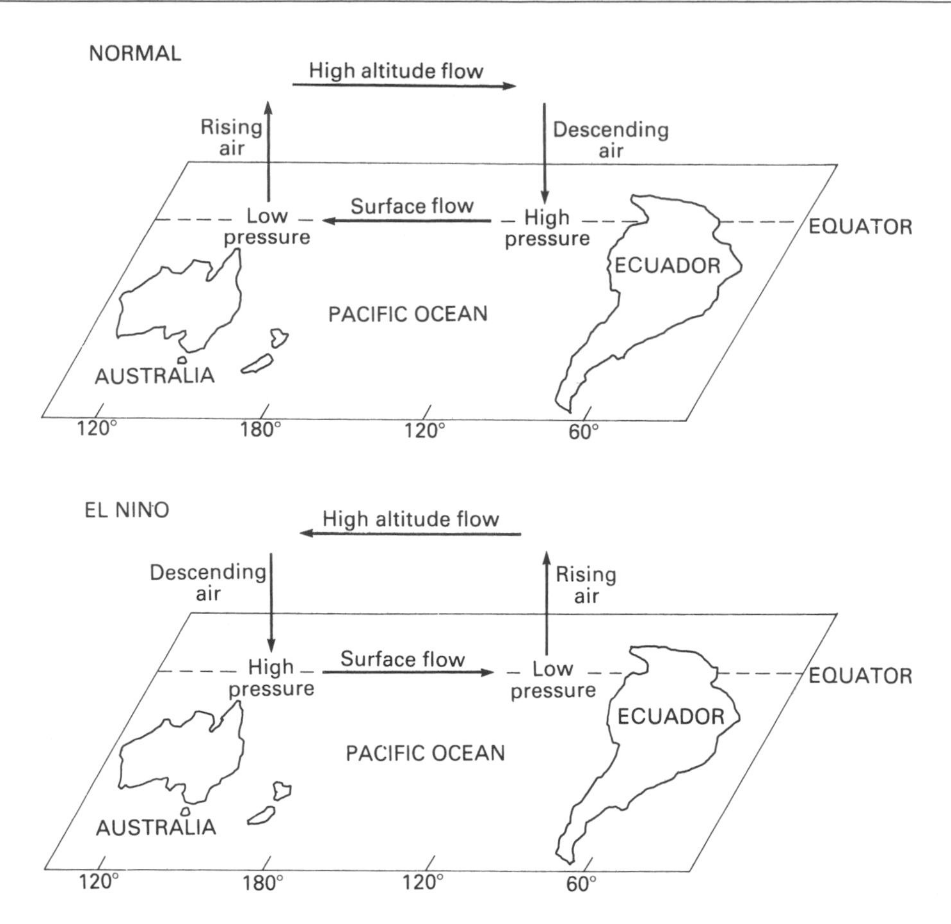

Figure 2.11 The nature of the 'normal' and El Niño conditions in the Pacific Ocean, showing the changing nature of the Walker Circulation.

shorter-term changes include such events as the period of warming that took place in the first decades of the twentieth century, the years of low rainfall and high temperatures which contributed to the formation of the 'Dust Bowl' in the High Plains of the United States in the 1930s, and the run of very dry years since the 1960s that have caused such human misery to the people of the Sahel and Ethiopia in sub-Saharan Africa. The extremely serious climatic deterioration that has afflicted this zone since about 1968 is illustrated by the rainfall data shown in figure 2.13.

But changes lasting hundreds of years were also characteristic of the past 10 000 years (a time often called the Holocene; see table 1.1), including a period of glacial advance between about 1500 and 1850 called the 'Little Ice Age' (window 2.6). This cold event has been identified all over the world, and besides causing glaciers to extend down their valleys, it also caused the retreat of settlement in marginal areas of highland Europe and Iceland, and there seems to have been a period of severe avalanching, landsliding, rockfalls and flooding in countries like Norway and Switzerland. There were other phases of glacial expansion within the Holocene, events to which the name *Neoglacial* is now normally applied. By contrast, there were also certain phases within the Holocene when conditions were rather warmer than they are today. For example, from about AD 750–1300 there was a per-

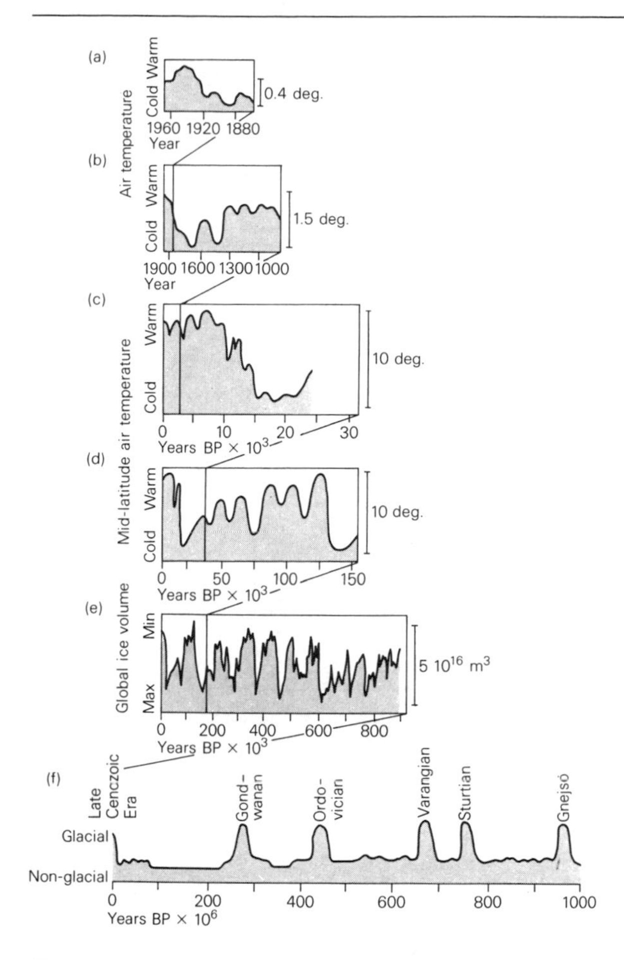

Figure 2.12 The climate of the world has varied by many different time scales over the past billion years: (a) by decade; (b) by century; (c) by millennia; (d) by tens of millennia; (e) by hundreds of millennia; (f) by era.

iod of marked glacial retreat, and this is called the 'Little Optimum' or 'Medieval Warm Epoch'. It was a time when the climate of England was sufficiently mild to permit vine cultivation and wine manufacture as far north as York. A more extended period, the 'Climatic Optimum' or 'Altithermal', witnessed the spread of relatively warm conditions in the earlier millenniums of the Holocene, and temperatures may have been 1–3 °C higher on average than they are today.

The fluctuations within the Pleistocene (the portion of the Quaternary that preceded the Holocene) consisted of major cold phases (called *glacials*) and warm phases (called *interglacials*). The Pleistocene itself consisted of a group of glacial and interglacial events that lasted in total around two or so million years. Such major phases of ice age activity seem to have been separated by around 250 million years.

The cooling that led up to the glacials of the Pleistocene is generally called the *Cenozoic climate decline*. During the Tertiary era, which started at the end of the Cretaceous about 65 million years ago, temperatures showed a general tendency, though not steady or uninterrupted, to fall in many parts of the world. Thus in the North Atlantic region in the early Tertiary conditions favoured a widespread, tropical, moist forest type of vegetation. At the end of the Eocene there was a climatic deterioration so that in the Oligocene the climate of Britain may have been more comparable to that of a region like the south-eastern USA.

The warmth of the first half of the Tertiary (the Palaeogene) in Britain had both local and global causes. At a local scale, Britain was at a lower latitude than today, being 10–12° further south. At a

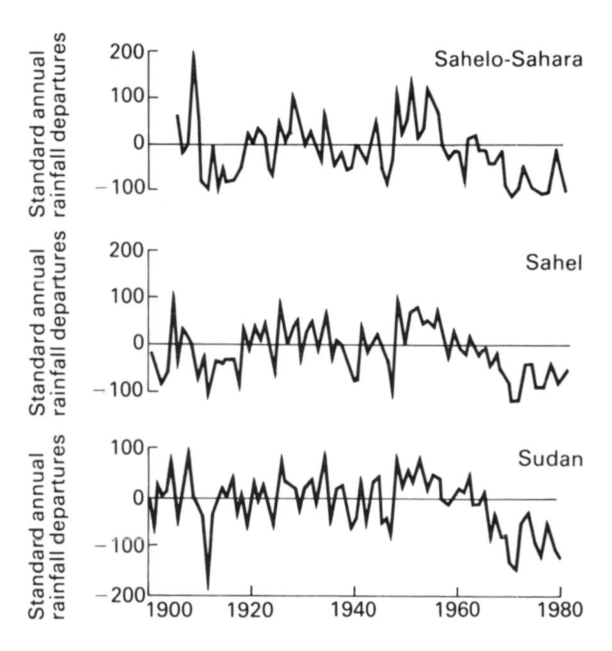

Figure 2.13 Standard annual rainfall departures for three sub-Saharan zones in Africa in the twentieth century. Note the especially dry conditions that have existed since the mid-1960s.

Window 2.6 Glacier retreat since the Little Ice Age

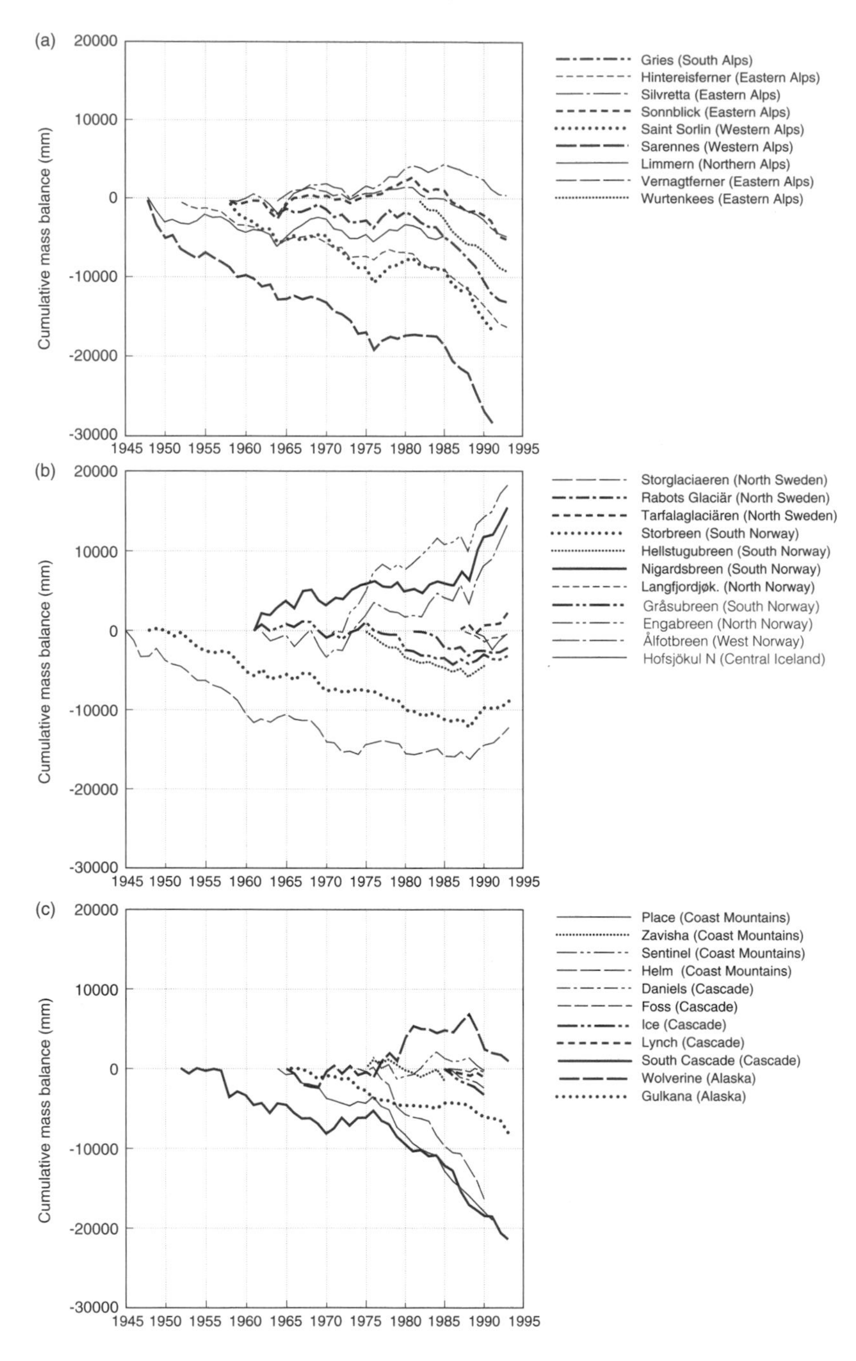

Changing state of glaciers since 1945 in selected regions: (a) European Alps; (b) Scandinavia; (c) North America.

Since the nineteenth century, many of the world's alpine glaciers have retreated up their valleys as a consequence of the climatic changes, especially warming, that have occurred in the past hundred or so years since the ending of the 'Little Ice Age'. Studies in the changes of snout positions obtained from cartographic, photogrammetric and other data therefore permit estimates to be made of the rate at which retreat can occur. The rate has not been constant, nor the process uninterrupted. Indeed, some glaciers have shown a tendency to advance for some of the period. However, if one takes those glaciers that have shown a tendency for a fairly general retreat, it becomes evident that, as with most geomorphological phenomena, there is a wide range of values, the variability of which is probably related to such factors as topography, slope, size, altitude, accumulation rate and ablation rate. It is also evident, however, that rates of retreat can often be very high, being of the order of 20–70 m a^{-1} over extended periods of some decades in the case of the more active examples. It is therefore not unusual to find that over the past hundred or so years alpine glaciers in many areas have managed to retreat by some kilometres. Current glacier tendencies for selected regions are shown in the figure. These are expressed in terms of their mass balance, which is defined as the difference between gains and losses (expressed in terms of water equivalent). In the European Alps (a) a general trend towards mass loss, with some interruptions in the mid-1960s, late 1970s and early 1980s, is observed. In Scandinavia (b), glaciers close to the sea have seen a very strong mass gain since the 1970s, but mass losses have occurred with the more continental glaciers. The mass gain in western Scandinavia could be explained by an increase in precipitation, which more than compensates for an increase in ablation caused by rising temperatures. Western America (c) shows a general mass loss near the coast and in the Cascade Mountains.

The valley of the Mer de Glace, French Alps. The immense boulder in the foreground and the light-coloured moraine plastered on the far side of the valley are products of the Little Ice Age when the glacier was very much thicker than today.

global scale the oceans and continents had a very different form, affecting the patterns of ocean currents and monsoon circulations, but there may also have been much more elevated atmospheric carbon dioxide conditions (creating a greenhouse effect) and a marked reduction in the angle of tilt of the earth's axis which would have affected the amount of incoming solar radiation.

By Pliocene times, the degree of cooling was such that a more temperate flora was present in the North Atlantic region, and at 2.4 million years ago glaciers started to develop in mid-latitude areas and many of the world's deserts came into being.

The realisation that the world had undergone an ice age over the past few millions of years dates back to the 1820s, but it is only in recent decades that the real nature of the changes has become appreciated and that accurate dates have been applied to them. These changes have taken place partly because of the availability of techniques like radiocarbon and potassium–argon dating, developed since the Second World War, and partly because it is now possible to retrieve cores of sediment from the ocean floors. Before these cores were accessible, the evidence for glaciation was obtained from glacial deposits on land, deposits that were repeatedly being destroyed by subsequent glaciation and erosion. By contrast, the oceans have been a more stable environment where the record of deposition has been much less subject to disturbance. Indeed, the problem of estimating the number of glaciations from evidence on the land has been likened in complexity to estimating the number of times that a blackboard has been erased; while using the evi-

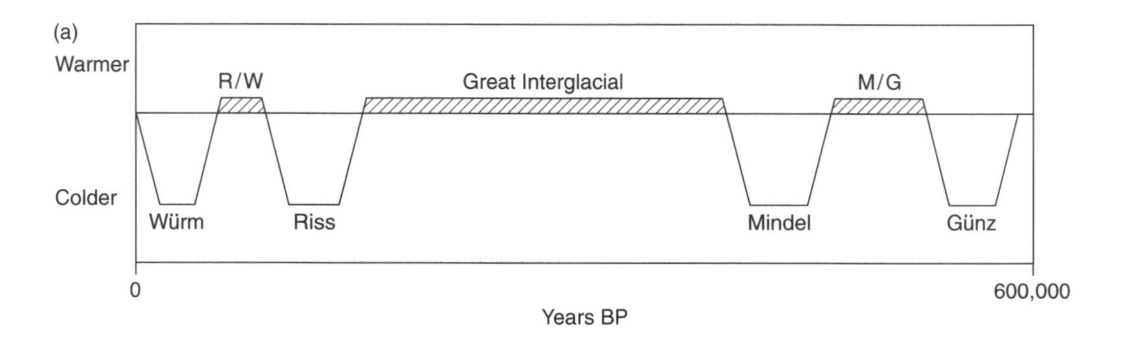

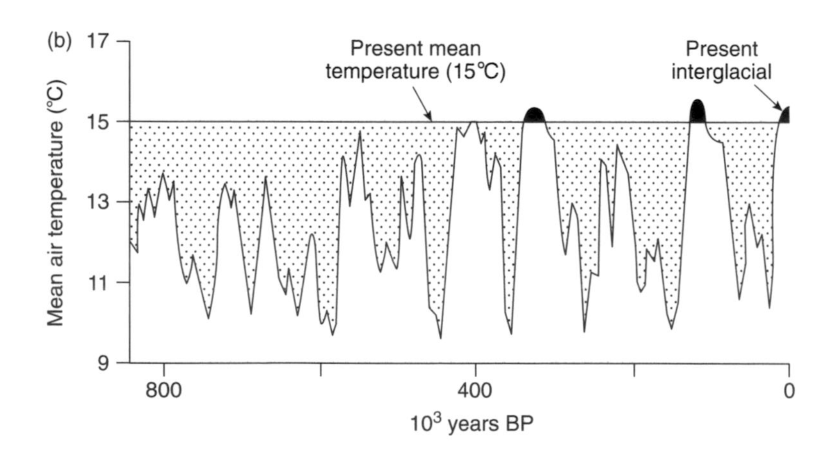

Figure 2.14 (a) The Penck and Brückner model of the Pleistocene; (b) temperature of the earth for the past 850 000 years as inferred from ice volume derived from oxygen isotope measurements of ice cores.

dence provided by layers of sediment in ocean cores has been likened to finding the number of times that a wall has been painted.

In a sense, over the period of the Pleistocene, the ocean floors have been like a dustbin into which layers of sediments and dead organisms have been poured. By looking at these layers it is possible to reconstruct the history of the world's climate in detail. Prior to the availability of such techniques there was wide support for the idea of four major Pleistocene glaciations. This we call the *Penck and Brückner model* (figure 2.14a), after the scientists who propounded it while working in the Bavarian foreland of central Europe (plate 2.2). Now, however, following the work of scientists like Emiliani, since the 1950s we know that the situation has been much more complex.

In the Quaternary period (which comprises the Pleistocene and the Holocene) the gradual and uneven progression towards cooler conditions which had characterised the earth during the Tertiary gave way to extraordinary climatic instability. Temperatures oscillated wildly from values similar to, or slightly higher than, today in interglacials to levels that were sufficiently cold to treble the volume of ice sheets on land during the glacials. Not only was the degree of change remarkable but so also, according to evidence from the sedimentary record retrieved from deep-sea cores, was the frequency of change. In all, there have been about seventeen glacial/interglacial cycles in the past 1.6 million years. The cycles tend to be characterised by a gradual build up of ice volume (over a period of *c.*90 000 years), followed by a dramatic glacial 'termination' in only about 8000 years. Furthermore, over the three or so millions of years during which humans have inhabited the earth, conditions such as those we experience today have been relatively short-lived and atypical of the Quaternary as a whole. Figure 2.14b illustrates the changes that have taken place over the past 850 000 years.

The last glacial cycle reached its peak about 18 000–20 000 years ago, with ice sheets extending over Scandinavia to the north German plain, over most of Britain (except the south) and over North America to 39° N (figure 2.15). To the south of the Scandinavian ice sheet was a tundra steppe underlain by permafrost, and forest was relatively sparse to the north of the Mediterranean (figure 2.16). In low latitudes sand deserts were considerably expanded in comparison with today.

Ice covered nearly one-third of the land area of the earth, but the additional ice-covered area in the last glacial was almost all in the Northern Hemisphere, with no more than about 3 per cent in the Southern. None the less, substantial ice cover developed over Patagonia and New Zealand. The thickness of the now-vanished ice sheets may have exceeded 4 km, with typical depths of 2–3 km. The total ice-covered area at a typical glacial maximum was $40 \times 10^6\,km^2$, compared with the present $15 \times 10^6\,km^2$.

Plate 2.2 A. Penck and E. Brückner were responsible, in the first decade of the twentieth century, for producing one of the most important sequences of glacials and interglacials in the Bavarian valleys: Günz, Mindel, Riss and Würm.

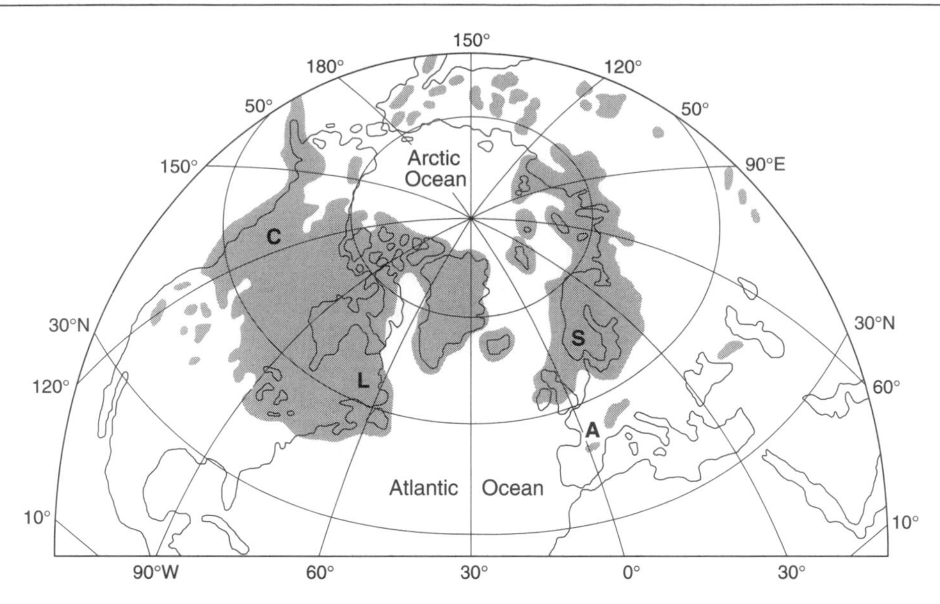

Figure 2.15 The possible maximum extent of glaciation in the Pleistocene era in the Northern Hemisphere (C = Cordilleran ice; L = Laurentide ice; S = Scandinavian ice; A = Alpine ice).

Highly important changes also took place in the state of the oceans. During the present interglacial conditions of the Holocene, the north-eastern Atlantic is at least seasonally ice-free as far north as 78° N in the Norwegian Sea. This condition reflects the bringing of warm water into this region by the Gulf Stream (North Atlantic Current). During the Last Glacial Maximum, however, the oceanic polar front probably lay at about 45° N, and north of this latitude the ocean was mainly covered by sea ice during the winter.

The degree of temperature change that occurred over land was substantial. It was particularly great in the vicinity of the great ice sheets (window 2.7). The presence of permafrost in southern Britain suggests a temperature depression of the order of 15 °C. Mid-latitude areas probably witnessed a lesser decline – perhaps 5–8 °C was the norm – though in areas subject to maritime air masses temperatures were more likely to have been depressed by 4–5 °C.

The cold glacials had a multitude of impacts on the landscape that are still visible today. The ice sheets caused considerable erosion and excavation, producing characteristic landform assemblages with cirques, arêtes, U-shaped valleys, roches moutonnées and other forms. They also transformed drainage patterns as the lacustrine landscapes of the Laurentian Shield (in Canada) and Scandinavia testify. Elsewhere they deposited boulder clay and outwash gravels, some as sheets and some as distinctive landforms (kames, eskers, etc.). Beyond the glacial limit fine particles blown from outwash plains settled to produce great belts of loess in areas like central Europe, Tajikistan, China, New Zealand and the Mississippi valley of the USA. Tundra conditions, with underlying permafrost, created great slope instability and drainage incision, the evidence for which is still very apparent along the escarpments and valleys of southern Britain.

Each glacial cycle had some complexity of form with phases of intense glacial activity and advance (*stadials*), being separated by periods of slightly greater warmth (*interstadials*) when glacial retreat occurred. During the last glacial cycle there were various interstadials, including a particularly marked one during the period 50 000–23 000 BP, and some rather shorter ones nearer the beginning of the cycle.

The most extreme stadial of the last glaciation occurred around 20 000–18 000 years BP, and in Britain is known as the Dimlington stadial. Shortly

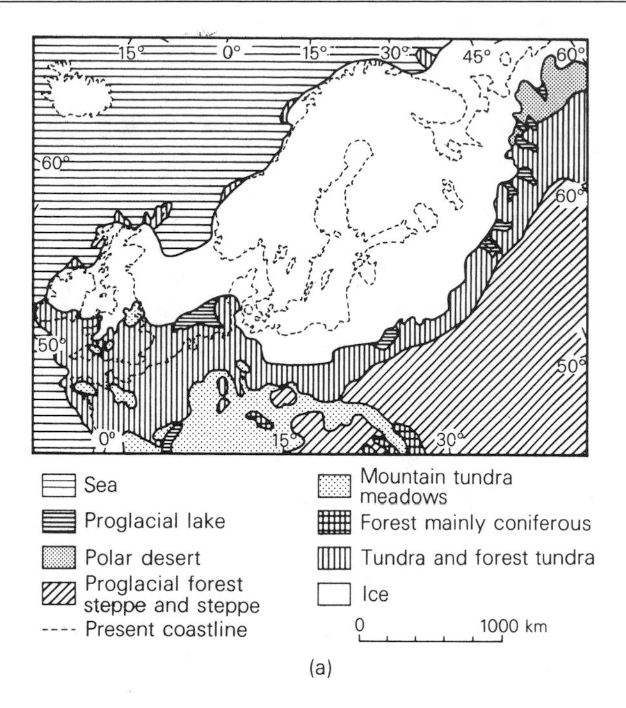

(a)

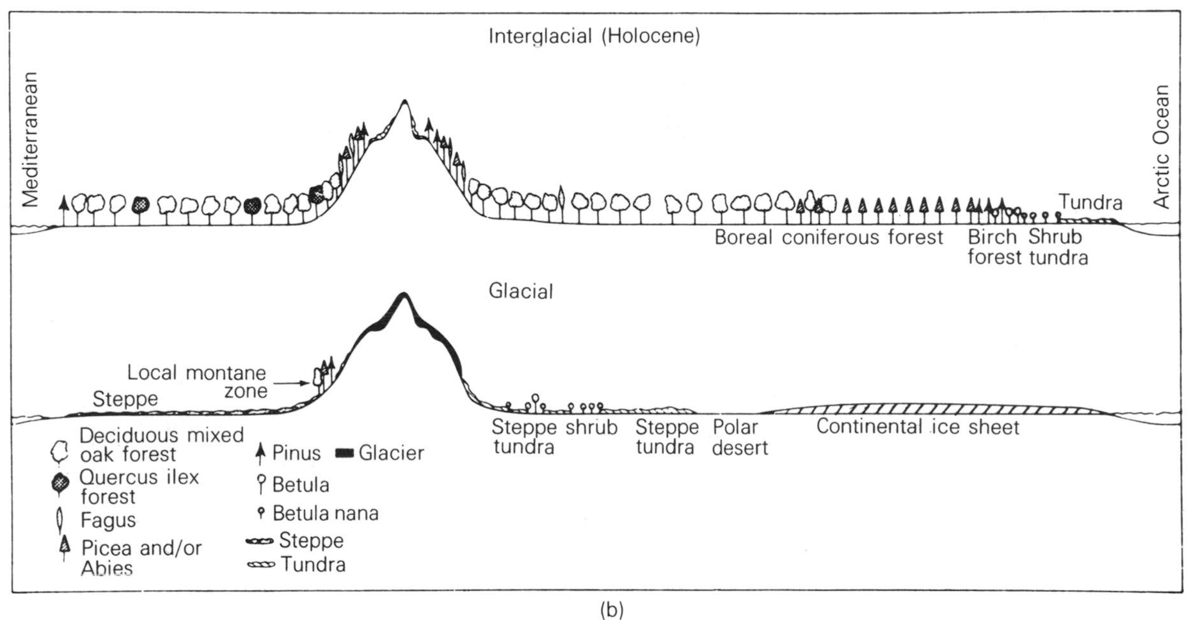

(b)

Figure 2.16 Europe in the Ice Age. (a) A reconstruction of Europe during the maximum of the last glacial phase (*c.*18 000 years BP). Note the great extent of ice cover over Britain and Scandinavia. (b) A comparison of the vegetation conditions of Europe across a north-south transect in an interglacial (top) and a glacial (bottom). Notice the absence of trees during a glacial cold phase.

Window 2.7 *The Ice Age in Britain*

The study of sediment cores from the floor of the North Atlantic on the Rockall Plateau has indicated that substantial quantities of iceberg-rafted debris started to appear around 2.4 million years ago. This probably represents the first time when glacial environments and glaciers appeared in Britain. Since that time there have been repeated alternations of glacials and interglacials, but we are not sure exactly how many. The problem is that later glacial events often remove or bury the evidence of their predecessors.

However, at the maximum extent of glaciation, which may have taken place in a phase called the Anglian, glaciers reached as far south as Bristol, Oxford and north London. This was undoubtedly a very important event in the history of the British Isles, for it radically altered the drainage pattern, eroding the lowland of the Severn Basin and diverting the course of the Thames southwards. The date of this event is uncertain, but may have been about 450 000 years ago.

The last major glacial advance, which peaked only around 18 000 years ago, did not reach so far south, but

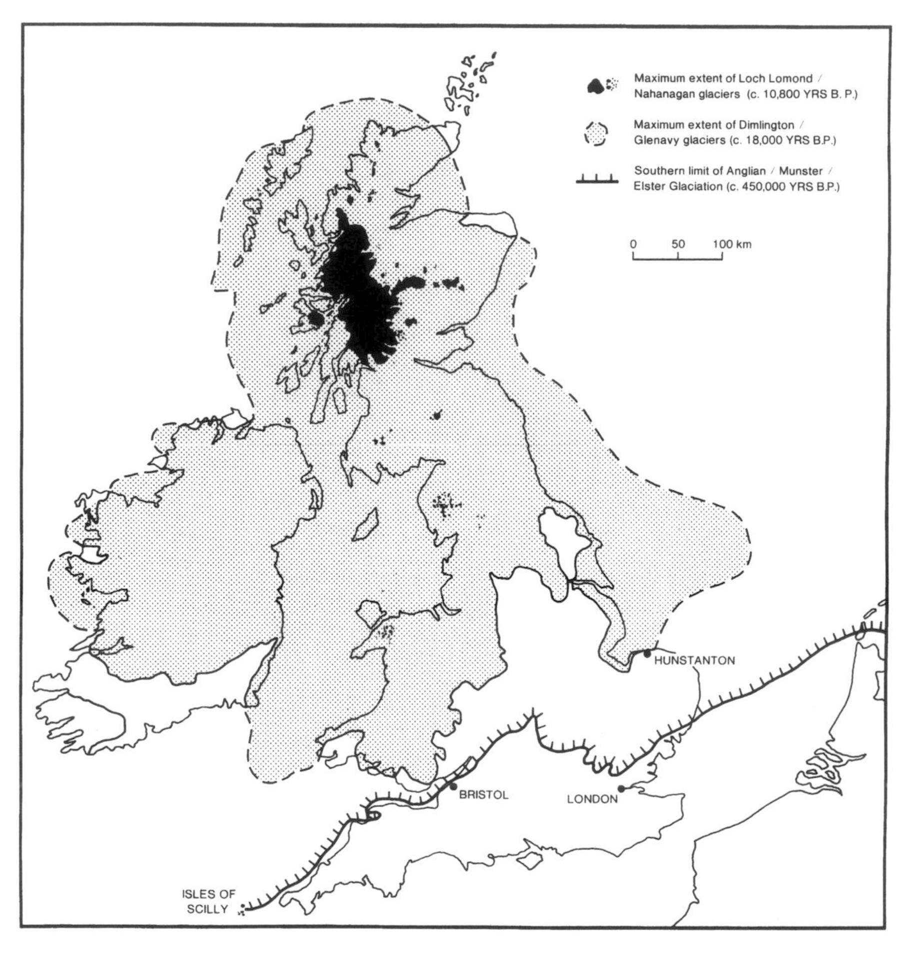

The limits to glaciation in the British Isles at various stages in the Pleistocene.

none the less covered parts of Yorkshire, Norfolk and the Midlands. Ice thicknesses may have been in excess of 1200 m. This ice sheet, named the Devensian, wasted away quickly, and had largely disappeared by 13 000 years ago, leaving most of the Scottish Highlands ice free. However, cold conditions returned to Britain briefly for a short period around 11 000 years ago, in an event called the Loch Lomond Stadial. During this period glaciers reformed in many high-altitude areas (including the English Lake District, Snowdonia and the Brecon Beacons). Their end-moraines are still very evident in the present landscape.

Although this window has highlighted the Anglian and Devensian glaciations and the Loch Lomond Stadial, it is likely that there were many glacial advances and retreats during the Pleistocene. Evidence from other parts of the world, especially from deep-ocean sediments, suggests that there may have been about 17 glacial/interglacial cycles in the past 1.6 million years.

thereafter glaciers began to retreat rapidly only to advance briefly in the Younder Dryas stadial around 11 000 years ago. This event, also called the Loch Lomond Readvance in Scotland, saw the development of cirque glaciers in the British uplands. It ended abruptly around 10 700 years ago, whereupon the world entered the interglacial conditions of the Holocene.

In general terms the Quaternary interglacials were short-lived but appear to have been essentially similar in their climate, fauna, flora and landforms to the Holocene interglacial in which we live today. One of their most important characteristics was that they witnessed the rapid retreat and decay of the great ice sheets and saw the replacement of tundra conditions by forest over the now temperate lands of the Northern Hemisphere. At their peak they may have been a degree or two warmer than now. In recent years considerable information on conditions in the last interglacial (the Eemian) has been obtained from ice cores extracted from the polar ice caps. These ice cores provide a detailed archive of past climatic conditions derived from examination of their chemical, gaseous and particulate contents. In particular their stable isotopic composition provides a means of calculating past atmospheric temperatures. The Greenland Ice Core Project managed to drill through 3029 m of ice under the summit of the Greenland ice sheet, and the core dates back to around 250 000 years ago. During the Eemian there may have been some very rapid, indeed abrupt, climate changes.

For many years it was believed that the climatic changes associated with the glacials and interglacials also affected lower latitudes, and that during glacials wetter conditions existed in the tropics, causing lakes to reach high levels and rivers to flow in areas that are now arid. Such humid phases were called *pluvials*, and the warm dry phases between them were called *interpluvials*. It was widely believed that the humid tropics were little affected by the changes of climate that transformed higher latitudes.

There is indeed much evidence that at certain times in the past there has been more water available in desert areas (window 2.8); huge lakes, for example, filled the now largely dry basins of the south-west United States. However, there is also evidence that in other areas the periods of the glacials were times not of increased humidity, but of reduced precipitation. The most spectacular evidence for this is the great expansion that took place in the distribution of sand dunes in low latitudes. Dunes cannot develop to any great degree in continental interiors unless the vegetation cover is sufficiently sparse to permit sand movement by the wind. If the rainfall is much above 150 mm per annum this is not possible. Studies of air photographs and satellite imagery indicate clearly that degraded ancient dunes, now covered in forest or savanna, are widespread in areas that are now quite moist (perhaps with rainfalls of the order of 750–1500 mm) (see plate 2.3 and colour plate 1). Today about 10 per cent of the land area between 30° N and 30° S is covered by active sand deserts. At or a little after the time of the last great glacial advance, about 18 000 years ago, they possibly characterised almost 50 per cent of the land area in those latitudes (figure 2.17). In between, tropical rainforests and ad-

Window 2.8 *A wet Holocene Sahara*

Of especial importance for vegetation and human activities was the mid-Holocene pluvial, which transformed the Sahara. A good demonstration of this is provided by the pollen analysis undertaken at a site called Oyo in

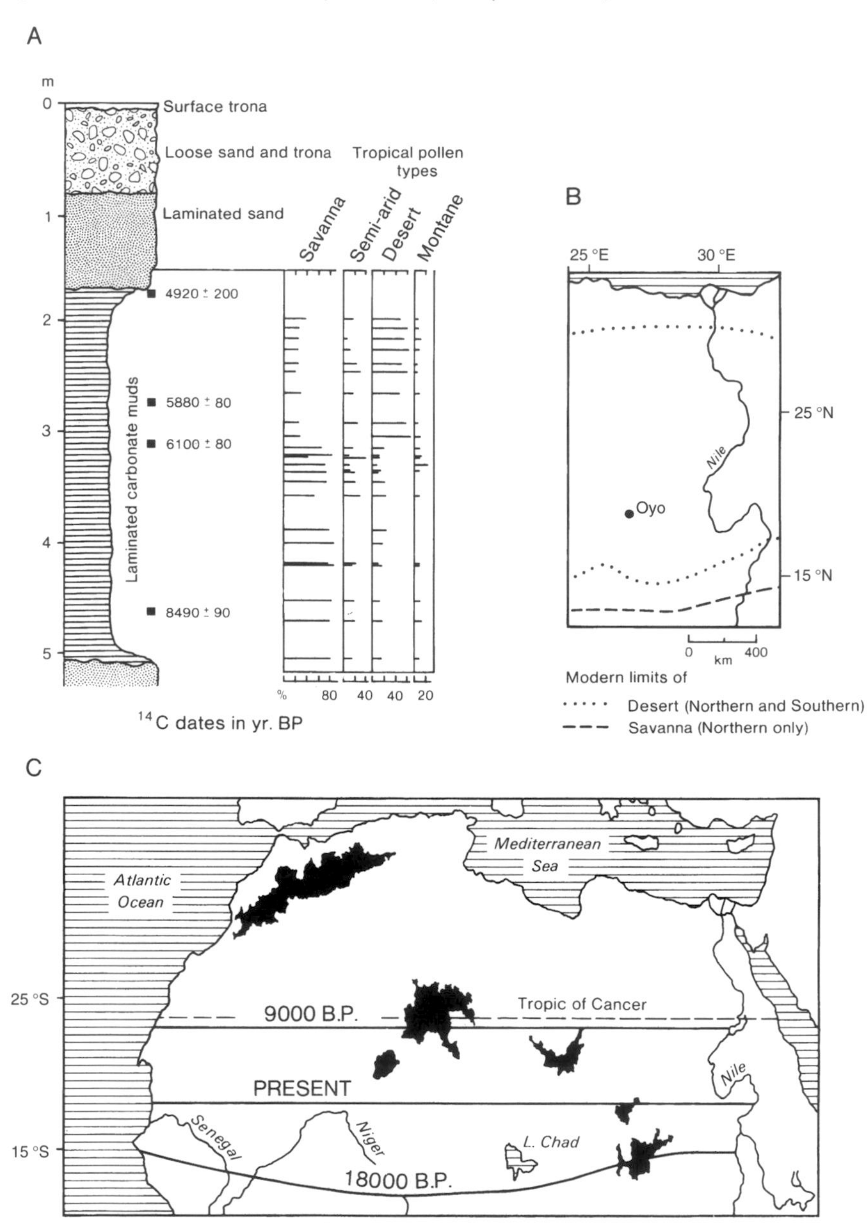

(a) Pollen diagram from Oyo in the eastern Sahara; (b) the location of Oyo; (c) the changing position of the Sahara–Sahel limit.

the eastern Sahara. Pollen spectra at that site, dating from 8500 BP until around 6000 BP, show that there were strong Sudanian savanna elements with tropical affinities in an area which is now hyper-arid. After 6000 BP the lake at Oyo became shallower, and acacia-thorn and then scrub-grassland replaced the sub-humid savanna vegetation. At around 4500 BP the lake appears to have desiccated fully and vegetation disappeared except in a few wadis and oases. Thus, the hyper-arid belt more or less disappeared for one or two millenniums before 7000 BP. The northern limit of the Sahel shifted about 1000 km to the north between 18 000 and 8000 BP and about 600 km to the south between 6000 BP and the present.

jacent savannas were reduced to a narrow corridor, and were much less extensive than they are today.

2.10 Causes of Long-term Climatic Change

As we have already seen, climate has undergone changes over a variety of different time scales and with a varying degree of intensity. An intriguing question is why such changes have taken place. No completely acceptable explanation of climatic change has ever been presented, and no one process acting alone can explain all scales of climatic change.

The complexity becomes evident if we follow the pathway of radiation derived from the ultimate driving force of climate – the sun. First of all, for reasons such as the varying tidal pull being exerted on the sun by the planets, the quality and quantity of outputs of solar radiation may change. It has been recognised that the sun's radiation output changes both in quantity (through association with such familiar phenomena as sunspots, which are dark regions of lower surface temperature on the sun's surface) and in quality (through changes in the ultraviolet range of the solar spectrum). Cycles of solar activity have been established for the short term by many workers, with 11- to 22-year cycles being particularly noted. Sunspot cycles of 80–90 years have also been postulated. The observations of sunspots in historical times have also given a measure of solar activity and one very striking feature of the record is the near absence of sunspots between AD 1640 and 1710, a period sometimes called the Maunder Minimum. It is perhaps significant that this minimum occurred during some of the more extreme years of the rigorous Little Ice Age.

The receipt of such varying radiation at the earth's surface might itself vary because of the presence of fine interstellar matter (nebulae) through which the earth might from time to time pass, or which might interpose itself between the sun and our planet. This would tend to reduce the receipt of solar radiation. Likewise, the passage of the solar system through a dust lane bordering a spiral arm of the Milky Way galaxy might cause a temporary reduction in receipt of the sun's radiation output.

The receipt of incoming radiation will also be affected by the position and configuration of the earth (figure 2.18). Such changes do take place, and there are three main astronomical factors which have been identified as of probable importance, with all three occurring in a cyclical manner. First, the earth's orbit around the sun is not a perfect circle but an ellipse. If the orbit were a perfect circle then the summer and winter parts of the year would be equal in their length. With greater *eccentricity* the length of the seasons will display a greater difference. Over a period of about 96 000 years, the eccentricity of the earth's orbit can 'stretch' by departing much further from a circle and then revert to almost true circularity.

Second, changes take place in the *precession of the equinoxes*, which means that the time of year at which the earth is nearest the sun varies. The reason is that the earth wobbles like a child's top and swivels round its axis. This cycle has a periodicity of about 21 000 years.

Third, changes occur, with a periodicity of about 40 000 years, in the *obliquity of the ecliptic* – the angle between the plane of the earth's orbit and the plane of its rotational equator. This movement has been likened to the roll of a ship with tilt varying from 21° 39′ to 24° 36′. The greater the tilt, the

Plate 2.3 The stripes running from right to left across this Landsat image of Zambia, central Africa, are old fossil dunes that were formed, probably in the Pleistocene, under a formerly more arid climate. Note that in the top right portion of the image some of the tributaries of the Zambezi River show similar alignment.

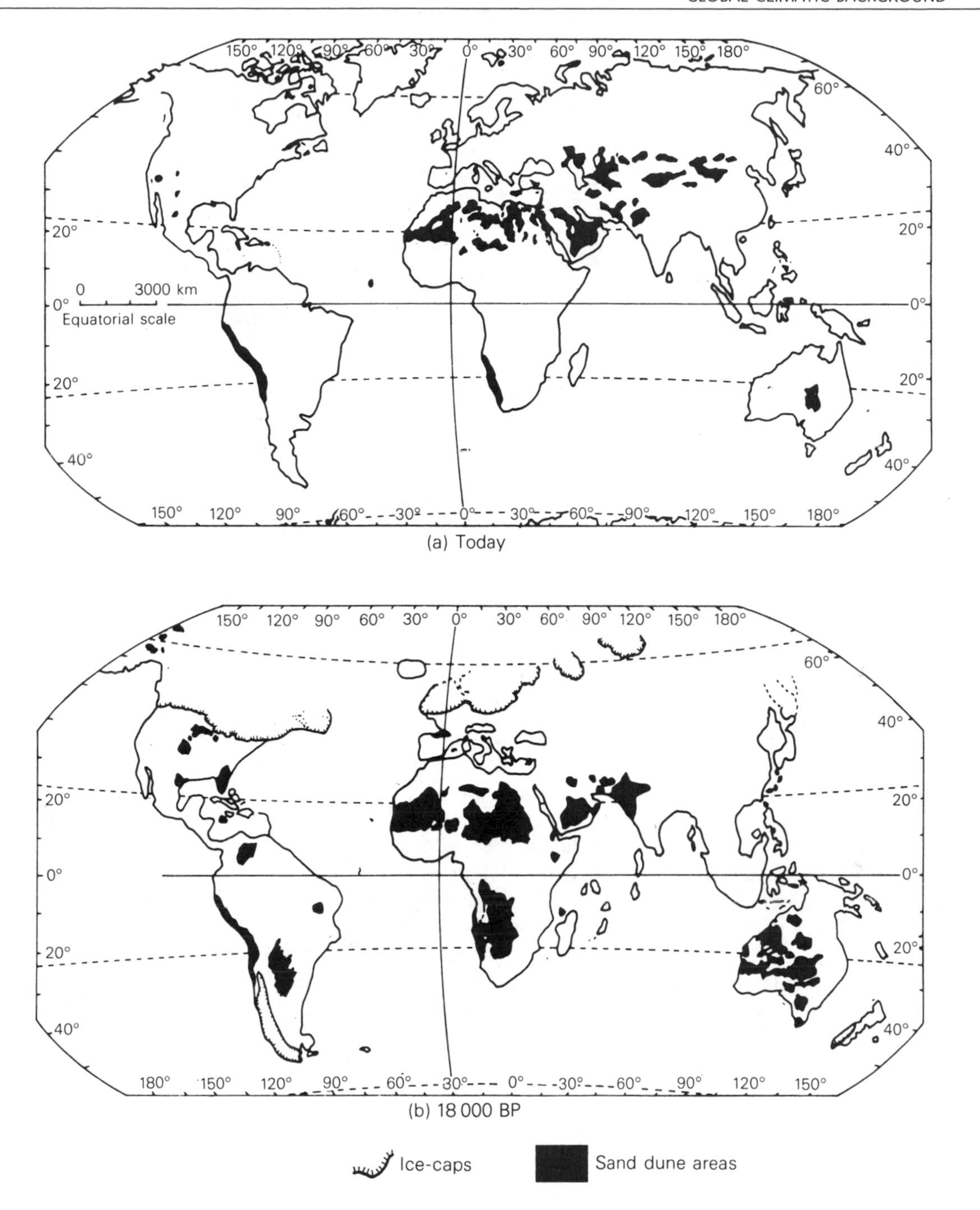

Figure 2.17 Because of greater aridity over large parts of low latitudes at or just after the time of the maximum extent of the last glacial episode, the extent of sand dunes 18 000 years ago (b) was very much greater than it is today (a).

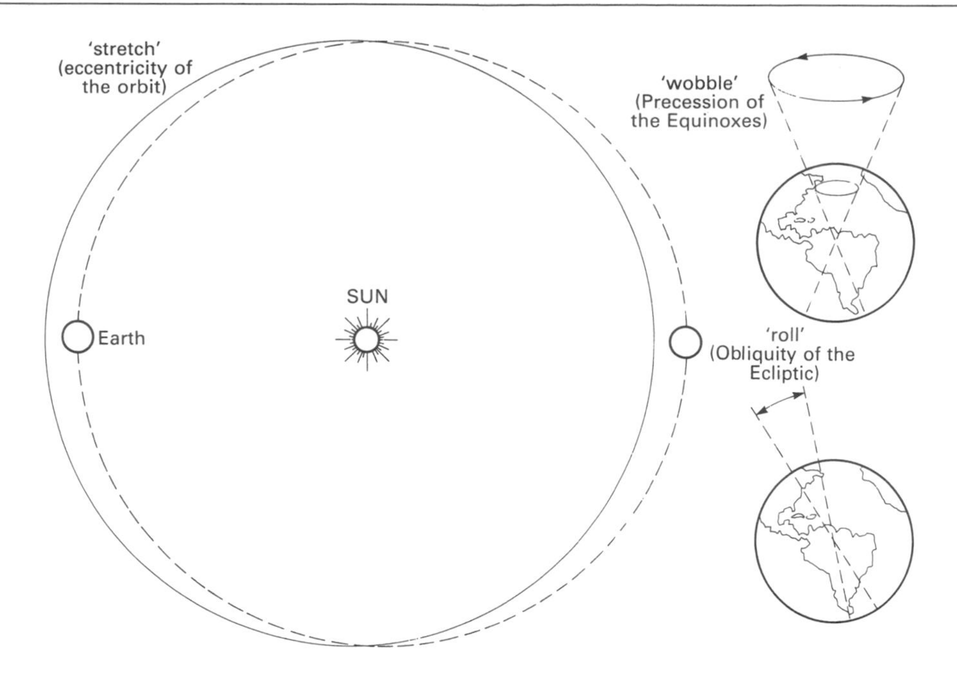

Figure 2.18 The three types of fluctuation in earth geometry involved in the Milankovitch hypothesis.

more pronounced is the difference between winter and summer.

These three cycles comprise what is often called the Milankovitch or Orbital Theory of climatic change. They have a temptingly close similarity in their periodicities to the durations of climatic change associated with the many glacials and interglacials of the past 1.6 million years. Indeed, they have been termed the 'pacemaker of the ice ages'.

Once the incoming solar radiation reaches the atmosphere, its passage to the surface of the earth is controlled by the gases, moisture and particulate matter that are present. Essential importance has been attached to the role of dust clouds emitted from volcanoes. These could increase the backscattering of incoming radiation and thus promote cooling. Volcanic dust veils produced by, for example, the eruption of Krakatoa in the 1880s and by Mount Pinatubo in 1991 (see window 11.4) caused global cooling for a matter of a few years. However, changing levels of volcanic activity are not the only way in which changes in atmosphere transparency might occur. For example, dust can be emplaced into the atmosphere by the wind erosion of fine-grained sediment and soil, and we know from the extensive deposits of wind-laid silts (loess) of glacial age that during the glacial maximums the atmosphere was probably very dusty, contributing to global cooling.

Carbon dioxide, methane, nitrous oxide, sulphur dioxide and water vapour can also modify the receipt of solar radiation. Particular attention has focused in recent years on the role of carbon dioxide (CO_2) in the atmosphere. This gas is virtually transparent to incoming solar radiation but absorbs outgoing terrestrial infrared radiation – radiation that would otherwise escape to space and result in heat loss from the lower atmosphere. In general, through the mechanism of this so-called 'greenhouse effect', low levels of CO_2 in the atmosphere would be expected to produce a 'heat trap'. The same applies to levels of methane and nitrous oxide, which, molecule for molecule, are even more effective greenhouse gases than CO_2. Recently, it has proved possible to retrieve CO_2 from gas bubbles preserved in layers of ice in deep ice cores drilled from the polar regions. Analyses of changes in CO_2 concentrations in these cores have provided truly remarkable results and have demonstrated that CO_2 changes and climatic changes have progressed in approximate synchroneity over the past 160 000 years. Thus the last interglacial around 120 000 years ago was a time of high CO_2 levels, the last glacial maximum around 18 000 years ago of low CO_2 levels. The reasons for the observed natural change in greenhouse gas concentrations are still the subject of active scientific research.

Once incoming radiation from the sun reaches the earth's surface it may be absorbed or reflected according to the nature of the surface, and in particular according to whether it is land or water, covered in dark vegetation or desert, and whether it is mantled by snow.

The effect of the received radiation on climate also depends on the distribution and altitude of land masses and oceans. These too are subject to change in a wide variety of ways – the plates that comprise the earth's crust are ever moving, mountain belts may grow or subside, and oceans and straits open and close. These processes shift areas into new latitudes, transform the world's wind belts and modify the climatically very important ocean currents.

In this discussion of causes it is also crucial to consider feedbacks. Such feedbacks are responses to the original forcing factors that act either to increase and intensify the original forcing (this we call positive feedback) or to decrease or reverse it (negative feedback). Clouds, ice and snow, and water vapour are three of the most important feedback mechanisms. An example of a positive feedback is the role of snow. Under cold conditions this falls rather than rain, it changes the albedo (reflectivity) of the ground surface and causes further cooling of the air above it. Similarly water vapour is a major greenhouse gas and a warmer climate produces more water vapour. This is because the rate of evaporation from the oceans and the water-holding capacity of the air both increase as temperatures rise.

Finally, it may well be that the atmosphere and the oceans possess a degree of internal instability which furnishes a built-in mechanism of change so that some small and random change might, through the operation of positive feedbacks and the passage of thresholds, have extensive and long-term effects. Small triggers might have big consequences.

2.11 Sea-level Changes

The climato-vegetational changes of the Quaternary era were equalled in importance only by the world-wide changes in sea level that took place, though these themselves were caused partially by climatic factors. The sea-level changes are significant because they have affected the configuration of coastlines, the size and existence of islands, the migration of plants, animals and man and the degree of deposition and erosion carried out by rivers in response to a fluctuating base level.

The most important cause of world-wide, or *eustatic*, sea-level change in the Pleistocene was *glacio-eustasy* (see section 4.8). When the ice sheets were three times as voluminous as today, a very large quantity of water was stored in them, and thus there was less water in the oceans. Calculations differ as to the exact quantity of change that would have taken place, but sea levels may well have dropped between 100 and 170 m (figure 2.19), thereby exposing most of the world's continental shelves as dry land. If, on the other hand, the two remaining major polar ice caps were to melt, sea level would probably rise by 66 m above its present position. The ice caps may have been slightly smaller during interglacials than they are today, and this may have caused sea levels to be a few metres higher than now, producing *raised beaches* in some coastal areas (see colour plate 2). In the Holocene, sea level rose very quickly (figure 2.20), especially between 11 000 and 6000 years ago; it flooded the North Sea, broke the land link between Britain and Ireland and flooded many river valleys to give features like the indented coastline of south-west England, with its winding inlets, called *rias*. This transgression is often called the *Flandrian Transgression*.

Other world-wide changes of sea level have been caused by *orogenesis* (mountain-building), associated with plate movement. This, together with seafloor spreading, can change the volume of the oceans, and thus world-wide sea levels. An increase of only 1 per cent in the area of the oceans would lower sea level by around 40 m. The spreading of

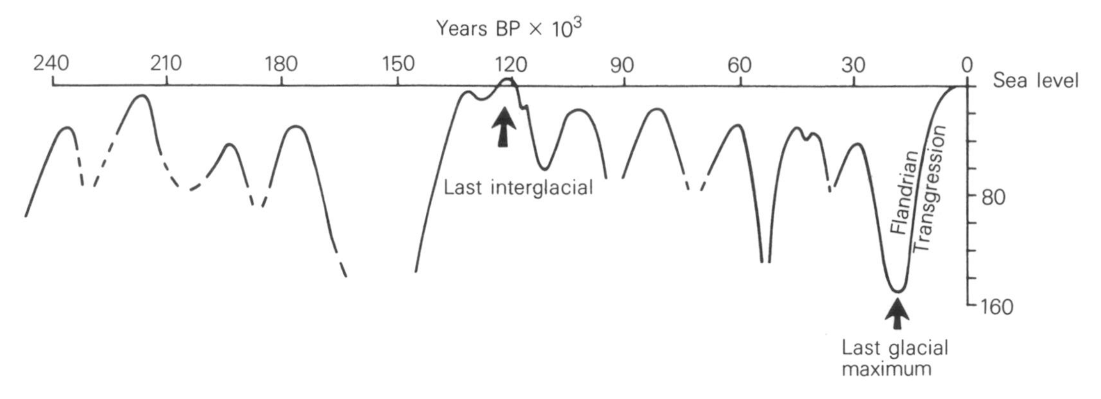

Figure 2.19 The nature of world sea-level change over the past quarter of a million years.

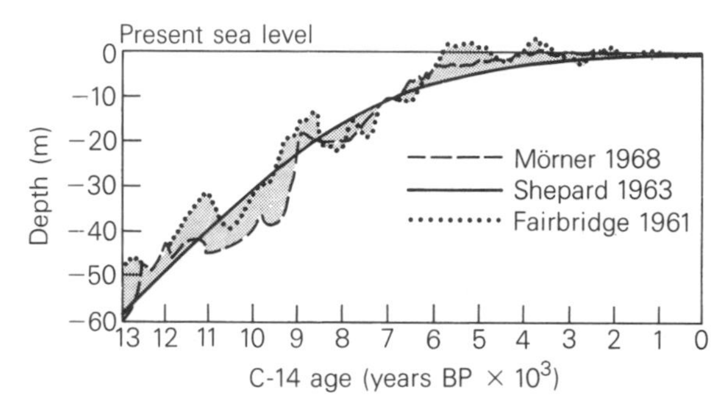

Figure 2.20 The nature of world sea-level change during the Flandrian Transgression of the Holocene.

the ocean basins since the last major stand of high interglacial sea levels some 120 000 years ago may have lowered world sea level by as much as 8 m.

2.12 Future Climates

In recent years several mechanisms have been identified that could cause humans to change the earth's climate.

- *Gas emissions*
  Carbon dioxide
  Methane
  Chlorofluorocarbons
  Nitrous oxide
  } Greenhouse gases
- *Aerosol generation*
  Dust
  Smoke
  Sulphates
- *Thermal pollution*
  Urban heat generation
- *Albedo change*
  Dust addition to ice caps
  Deforestation and afforestation
  Overgrazing
  Extension of irrigation
- *Alteration of water flow in rivers and oceans*
- *Water vapour change*
  Deforestation
  Irrigation

The enhanced greenhouse effect and global warming

Planet earth receives warmth from the sun. Radiation from the sun is partly trapped by the atmosphere. It passes through the atmosphere and heats the earth's surface. The warmed surface radiates energy, but at a longer wavelength than sunshine. Some of this energy is absorbed by the atmosphere, which as a result warms up. The rest of the energy escapes to space. We call this process of warming the 'greenhouse effect' because the atmosphere is perceived to act rather like glass in a greenhouse (figure 2.21). Although the atmosphere consists primarily of nitrogen and oxygen, it is some of the so-called 'trace gases' which absorb most of the heat,

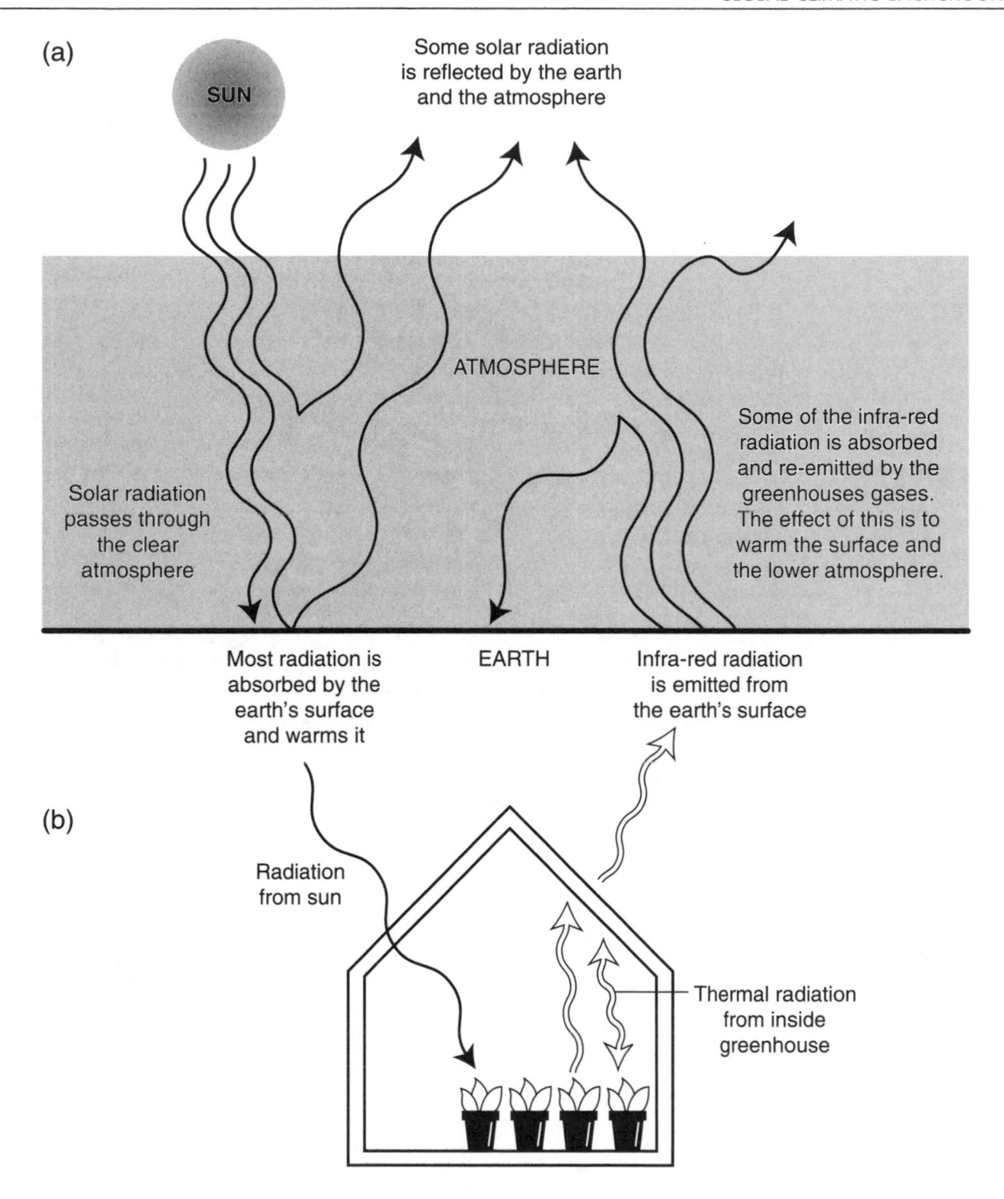

Figure 2.21 (a) The 'greenhouse effect' in the atmosphere; (b) diagram showing how a greenhouse acts as a 'radiation blanket'.

in spite of the fact that they occur in very small concentrations. These are called the 'greenhouse gases'.

Various greenhouse gases occur naturally – water vapour (H_2O), carbon dioxide (CO_2), methane (CH_4), ozone (O_3) and nitrous oxide (N_2O). In recent centuries and decades, however, the quantities of some of these greenhouse gases have started to increase because of human activities. In addition, a new type of greenhouse gas, chlorofluorocarbons (CFCs), has been introduced to the atmosphere in the past 50 years.

Since the start of the industrial revolution two to three hundred years ago humans have been taking stored carbon out of the earth in the form of fossil fuels (coal, oil and natural gas). They burn these fuels, releasing CO_2 in the process. The pre-industrial level of CO_2 in the atmosphere may have been as low as 260–270 parts per million by volume (ppmv). The present level exceeds 360 ppmv and is still rising, as is evident in records of atmospheric composition from various parts of the world. Fossil fuel burning and cement manufacture release over 6 gigatonnes of carbon to the atmosphere as CO_2 each year. Burning of forests and changes in the levels of organic carbon in soils subjected to deforestation and cultivation may also contribute substantially to CO_2 levels in the atmosphere, perhaps by around 2 gigatonnes of carbon each year.

Other gases, as well as CO_2, will probably contribute to the accelerated greenhouse effect. The effect of each on its own may be relatively small, but the effects of all of them combined may be considerable. Moreover, molecule for molecule, some of these other gases may be more effective as greenhouse gases than CO_2. This applies to methane (CH_4) which is 21 times more effective than CO_2, to nitrous oxide (N_2O) which is 206 times more effective, and to the CFCs, which are 12 000–16 000 times more effective.

Where do these other gases come from and why are amounts of them increasing? Concentrations of methane are now over 1600 ppbv, compared to eighteenth-century background levels of 600 ppbv. Methane has increased as a result of the spread of rice cultivation in waterlogged paddy fields, enteric fermentation in the growing numbers of belching and flatulent domestic cattle, and the burning of oil and natural gas. Nitrous oxide levels have increased because of the combustion of hydrocarbon fuels, the use of ammonia-based synthetic fertilisers, deforestation and vegetation burning. The increase in CFCs in the atmosphere (which is also associated with ozone depletion in the stratosphere) results from their use as refrigerants, as foam-makers, as fire control agents, and as propellants in aerosol cans. Use of CFCs is now being restricted by various international agreements.

The earth's climate has become generally warmer over the past century or so, and the 1980s and 1990s saw an unprecedented number of warm years. This has prompted some scientists to propose that global warming, as a result of the accelerated greenhouse effect, has already started. However, the complexity of factors that can cause climatic fluctuations leads many scientists to doubt that the case is yet fully proven. Most, however, believe that if concentrations of effective greenhouse gases continue to rise, and attain double their natural levels by around the middle of the twenty-first century, then temperatures will rise by several degrees over that period. The Intergovernmental Panel on Climate Change (IPCC), which reported in 1990, suggested that global mean temperature might increase during the next century at a rate of 0.3 °C per decade. The IPCC report of 1996 suggested a 'best estimate' of a 2.0 °C increase in temperature by 2100 (with a range of 1–3.5 °C). Cooling effects of aerosols (see below) are taken into account in this prediction. The rise in temperature will not, however, be the same across the globe. In particular, high latitudes (e.g. northern Canada and Eurasia) will show even more pronounced warming, perhaps two to three times the global average.

Such increases in temperature, if they occur, will undoubtedly cause major changes in the general atmospheric circulation. These in turn will cause changes in precipitation patterns. Overall levels of precipitation over the globe will increase as more moisture is released by higher rates of evaporation from the oceans. However, some areas will get wetter while some will get drier. There is still considerable uncertainty about what precise pattern precipitation will take as a result of these changes. The very cold, dry areas of high latitudes may well become moister as a warmer atmosphere will be able to hold more moisture. Some tropical areas may receive more rain as the vigour of the monsoonal circulation and of tropical cyclones is increased. Some mid-latitude areas, like the High Plains of America, may become markedly drier.

There is, however, great uncertainty as to how far the climate may change as a result of the greenhouse effect. The reasons for this uncertainty include:

- doubts about how fast the global economy will grow;

- doubts about what fuels will be used in the future;
- doubts about the speed at which land-use changes are taking place;
- uncertainty regarding how much CO_2 will be absorbed by the oceans and by biota;
- uncertainties about the role of other anthropogenic and natural (e.g. volcanic) causes of climatic change;
- the assumptions that are built into many of our predictive general circulation models (e.g. about the role of clouds);
- the role of possible positive feedbacks and thresholds that may mean changes are more sudden than anticipated, or do not happen at all.

The degree of global warming that is proposed for the coming decades does not at first sight appear enormous. However, it may, over a period that is very short in geological terms, produce warmer conditions than have existed for several million years and set up a series of changes that have important implications both for the environment and for humans. Some of these implications may be benign (e.g. warmer conditions will enable new crops to be grown in Britain) but some of them will be malign (e.g. more frequent and longer droughts in the High Plains of America). Among the *possible* environmental consequences are:

- more intense, widespread and frequent tropical cyclones;
- the melting of alpine glaciers;
- the degradation of permafrost in tundra areas;
- the wholesale displacement of major vegetation belts such as the boreal forests of the Northern Hemisphere;
- rising sea levels and associated flooding of coral reefs, deltas, wetlands, and so on, and accelerated rates of beach erosion;
- decreased flow of water in streams as a result of increased loss of moisture by evapotranspiration;
- reduction in the extent of sea ice in polar waters;
- shifts in the range of certain vector-borne diseases (e.g. malaria).

The role of aerosols

Let us next consider the possible effects of aerosols, a term normally used for smoke, condensation nuclei, freezing nuclei or fog contained within the atmosphere, or other pollutants such as droplets containing sulphur dioxide or nitrogen dioxide. Many atmospheric aerosols (e.g. those derived from volcanoes, sea spray or natural fires) were not placed there by humans. However, humans have become increasingly capable of adding various aerosols into the air. For example, one consequence of the industrial revolution has been the emission of hugely increased quantities of dust or smoke particles into the lower atmosphere from industrial sources. These could influence global or regional temperatures through their impact on the scattering and absorption of solar radiation.

The exact effects of aerosols in the atmosphere are still not clear, however. Whether added aerosols cause heating or cooling of the earth and atmosphere systems depends not only on their intrinsic absorption and backscatter characteristics, but also on their location in the atmosphere with respect to such variables as cloud cover, cloud reflectivity and underlying surface reflectivity. So, for example, over ice caps 'grey' aerosol particles would warm the atmosphere because they would be less reflective than the white snow surfaces beneath. Over a darker surface, on the other hand, they would reflect a greater amount of radiation, leading to cooling. Thus it is difficult to assess precisely the effects of increased aerosol content in the atmosphere.

Uncertainty is heightened because of the two contrasting tendencies of dust: the backscattering effect, producing cooling, and the thermal-blanketing effect, causing warming. In the second of these, dust absorbs some of the earth's thermal radiation that would otherwise escape to space, and then re-radiates a portion of this back to the land surface, raising surface temperatures. Natural dust from volcanic emissions tends to enter the stratosphere (where backscattering and cooling are the main consequences), while anthropogenic dust more frequently occurs in the lower levels of the atmosphere, where it could cause thermal blanketing and warming.

Industrialisation is not the only source of particles in the atmosphere, nor is a change in temperature the only possible consequence. Intensive agricultural exploitation of desert margins can create a dust pall in the atmosphere by exposing larger areas of surface materials to deflation in dust storms. This dust pall can change atmospheric temperature enough to cause a reduction in convection, and thus in rainfall. Observations of dust levels over the Atlantic during the drought years of the late 1960s and early 1970s in the Sahel suggest that the degradation of land surfaces there led to a threefold increase in atmospheric dust at that time. It is thus possible for human-induced desertification to generate dust which in turn increases the degree of desertification by reducing rainfall levels.

Dust storms, generated by deflation from land surfaces with limited vegetation cover, occur frequently in the world's drylands (see section 6.9). They happen naturally when strong winds attack dry and unvegetated sandy and silty surfaces. Their frequency also varies from year to year in response to fluctuations in rainfall and wind conditions. At present, however, in some parts of the world, the dust entering the atmosphere as a result of dust storms is increasing because of the effects of human activity. In particular, processes such as overgrazing strip the protective vegetation cover from the soil's surface. Elsewhere, surfaces may be rendered more susceptible to wind attack because of ploughing or disturbance by wheeled vehicles.

Atmospheric aerosols can be important sources of cloud-condensation nuclei. Over the world's oceans a major source of such aerosols is dimethylsulphide (DMS). This compound is produced by planktonic algae in seawater and then oxidises in the atmosphere to form sulphate aerosols. Because the albedo of clouds (and thus the earth's radiation budget) is sensitive to the density of cloud-condensation nuclei, any factor that has an impact on planktonic algae may have an important impact on climate. The production of such plankton could be affected by water pollution in coastal seas or by global warming. Anthropogenically derived sulphate aerosols could significantly increase planetary albedo, through their direct scattering of short-wavelength solar radiation and their modification of the short-wave reflective properties of clouds. Thus they could exert a cooling influence on the planet.

Land cover changes

Another major possible human-induced cause of climate change is change in the reflectivity (albedo) of the ground surface and the proportion of solar radiation which the surface reflects (see figure 2.2a). Land-use changes create differences in albedo which have important effects on the energy balance of an area. Tall rainforest may have an albedo as low as 9 per cent, while the albedo of a desert may be as high as 37 per cent. There has been growing interest in the possible consequences of deforestation on climate as a result of the associated change in albedo. Ground deprived of vegetation cover as a result of deforestation and overgrazing has a much higher albedo than ground covered in plants. This could affect temperature levels.

Some scientists have argued that the increase in surface albedo resulting from a decrease in plant cover would lead to a reflection outwards of incoming radiation, and an increase in the radiative cooling of the air. Consequently, they argue, the air would sink to maintain thermal equilibrium by adiabatic compression, and cumulus convection and its associated rainfall would be suppressed. A positive feedback mechanism would appear at this stage: namely, the lower rainfall would in turn adversely affect plants and lead to a further decrease in plant cover.

This view has been disputed by other scientists who point to the effect of vegetation on evapotranspiration. They point out that vegetated surfaces are usually cooler than bare ground, since much of the solar energy absorbed is used to evaporate water. They conclude from this that protection from overgrazing and deforestation might be expected to lower surface temperatures and thereby reduce, rather than increase, convection and precipitation.

The models used by some scholars suggest that removal of the humid tropical rainforests could also have direct climatic effects. Deforestation in the Amazon Basin would lead to reductions in both precipitation and evaporation as a result of the changes in surface roughness and albedo. The sur-

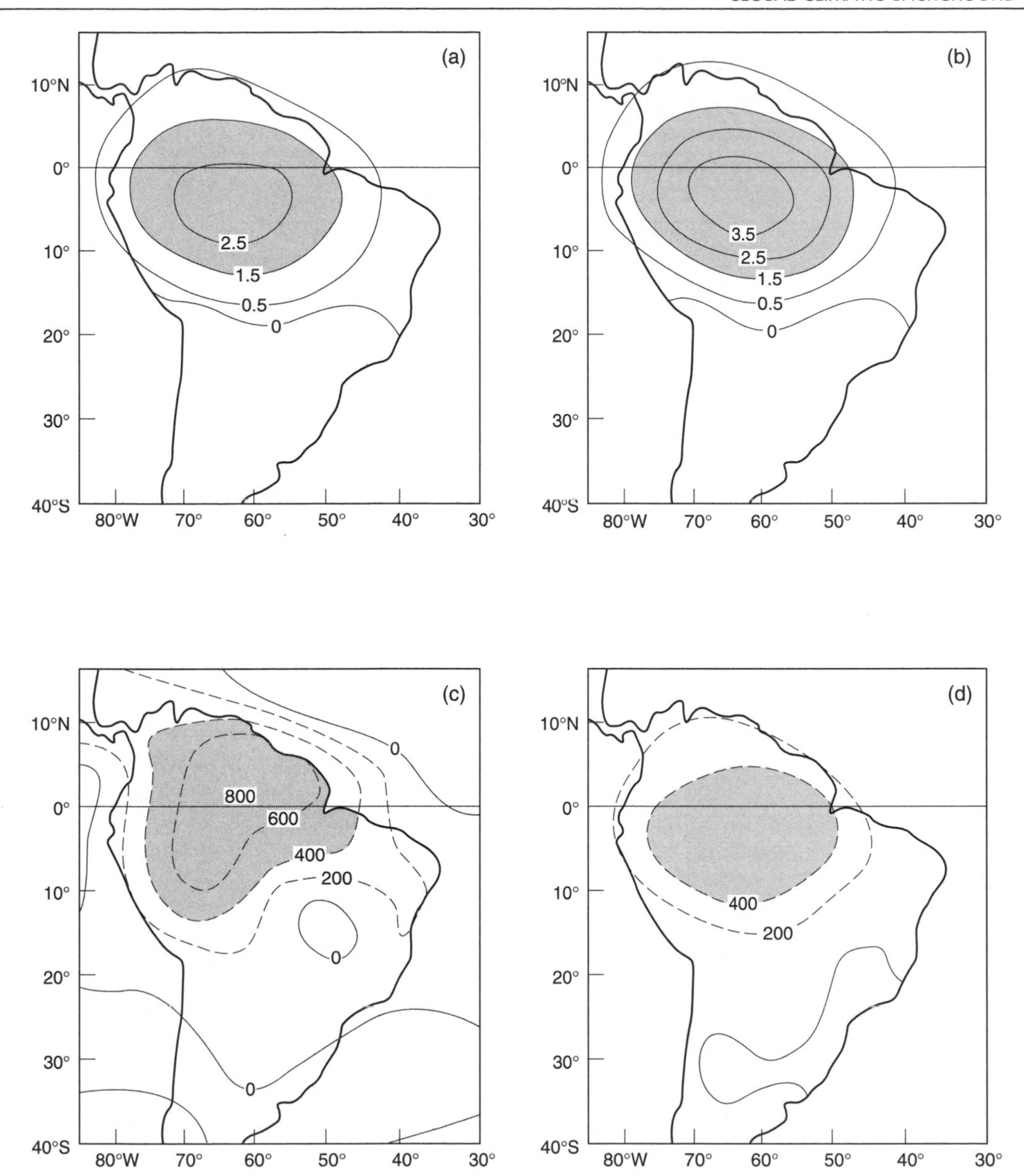

Figure 2.22 Predictions of the change in climate following a conversion of Amazonian rainforest to grassland: (a) temperature increase (°C); (b) evaporation decrease (mm per year); (c) rainfall decrease (mm per year); (d) evapotranspiration decrease (mm per year).

face roughness effect occurs because rainforest has quite a jagged canopy and this in turn affects wind flow.

In addition, the present use of irrigation over about 0.4 per cent of the earth's surface (1.3 per cent of the land surface) is decreasing the albedo of irrigated areas, possibly on average by 10 per cent. The corresponding change in the albedo of the entire earth–atmosphere system would amount to about 0.03 per cent – enough perhaps to maintain the global mean temperature at a level nearly 0.1 °C higher than it would otherwise be.

A change in land use can also lead to a change in the moisture content of the atmosphere. It is possible, for example, that if humid tropical rainforests are cut down, the amount of moisture transpired into the atmosphere above them will be reduced. This would reduce the potential for rain (figure 2.22). The spread of irrigation could have the opposite effect, leading to increased atmospheric humidity levels in the world's drylands. The High Plains of the USA, for example, are normally covered with sparse grasses and have dry soils throughout the summer. Evapotranspiration there is very low. In the past four decades, however, irrigation has been developed throughout large parts of the area. This has greatly increased summer evapotranspiration levels. There is strong statistical evidence that rainfall in the warm season has been increased by the use of irrigation in two parts of this area: one extending through Kansas, Nebraska and Colorado, and a second in the Texas Panhandle. The largest absolute increase was in the latter area. Significantly, it occurred in June, the wettest of the three heavily irrigated months.

2.13 Conclusions

In these first two chapters we have seen that the most important large-scale patterns on the face of the earth are controlled either by geological or by climatic factors. It is evident that certain phenomena, such as soils, vegetation, ocean currents, landforms and hydrology, show global patterns that are related in their distribution to the major climatic zones. Some of these will be discussed in the next chapter, while Part II will involve a consideration of the physical geography of some of the world's most important zones: polar regions, periglacial regions, temperate regions, arid regions and the tropics.

■ *Key Terms and Concepts*

aerosols
albedo
atmospheric heat engine
Cenozoic climate decline
Coriolis Effect
dimethylsulphide (DMS)
El Niño
Flandrian Transgression
general circulation
glacial
glacio-eustasy
greenhouse effect
Hadley cell
hydrological cycle
interglacial
Intertropical Convergence Zone (ITCZ)
jet stream
lapse rate
Little Ice Age
Milankovitch hypothesis
orogenesis
pluvial
precipitation variability
Quaternary
Rossby wave
Southern Oscillation
thermocline
thermohaline circulation
upwelling
Walker circulation
zonal index

■ *Points for Review*

- What are the main features of the general circulation of the atmosphere?
- Account for the distribution of the world's (a) wettest areas and (b) driest areas.
- What effects do the oceans have on the world's climate?
- What do you understand about the El Niño phenomenon and its importance?

- What was the world like at the maximum of the last glaciation?
- What are the natural causes of climate change?
- How may humans achieve climate change?
- Why does sea level change?

FURTHER READING

Arnell, N. (1996) *Global Warming, River Flows and Water Resources* (Chichester: Wiley). A study of potential future hydrological changes.

Barry, R. G. and Chorley, R. J. (1998) *Atmosphere, Weather and Climate*, 7th edn (New York/London: Routledge). The classic introductory text.

Eisma, D. (ed.) (1995) *Climate Change: Impact on Coastal Habitation* (Boca Raton, Florida: Lewis). A study of the effects of climate change and sea-level rise.

Gates, D. M. (1993) *Climate Change and its Biological Consequences* (Sunderland, Mass.: Sinauer). A textbook that considers some of the effects of climate change on the biosphere.

Goudie, A. S. (1992) *Environmental Change* (Oxford: Clarendon Press). An introduction to the profound changes of the past three million years.

Graedel, T. E. and Crutzen, P. J. (1993) *Atmospheric Change: An Earth System Perspective* (San Francisco: Freeman). A well-illustrated study of changes, past, present and future.

Houghton, J. T. (1994) *Global Warming: The Complete Briefing* (Oxford: Lion). An introduction by a leading scientist.

Houghton, J. T., Meira Filho, L. G., Callander, B. A., Harris, N., Kaltenberg, A. and Maskell, K. (eds) (1996) *Climate Change 1995: The Science of the Climate Change* (Cambridge: Cambridge University Press). A report from the Intergovernmental Panel on Climate Change on the causes and trends of climate change.

Jones, J. A. A. (1997) *Global Hydrology* (Harlow: Longman). An authoritative discussion of the links between climate and hydrology at the global scale.

Kemp, D. D. (1994) *Global Environmental Issues: A Climatological Approach*, 2nd edn (London: Routledge). An introductory survey.

Lowe, J. J. and Walker, M. J. C. (1997) *Reconstructing Quaternary Environments*, 2nd edn (Harlow: Longman). A very strong treatment of techniques for environmental reconstruction.

Roberts, N. (1998) *The Holocene*, 2nd edn (Oxford: Blackwell). A well-illustrated story of the past 10,000 years.

Williams, M. A. J. and Balling, R. C. (1996) *Interactions of Desertification and Climate* (London: Arnold). A study of land-cover changes and their effects.

Williams, M. A. J., Dunkerley, D., De Deckker, P., Kershaw, P. and Chappell, J. (1998) *Quaternary Environments*, 2nd edn (London: Arnold). An up-to-date survey, with an Australian perspective, on climatic and other changes in the Quaternary.

3 The Organic World

Having just discussed the global patterns of climate, and how they have changed through time and may change in the future, we now move on to a consideration of some of the global patterns of organic phenomena: plants, animals and soils. These are components of what is often called the *biosphere*.

3.1 Major Vegetation Types

Just as it is possible to classify and map major climatic types, so it is possible to classify and map the main types of vegetation on the face of the earth (window 3.1). At a very gross scale, the distribution of vegetation types is closely related to the distribution of climatic types, and one of the classic ideas of biogeography was that if an area of land – bare of vegetation because of some event like a volcanic eruption – became colonised by vegetation, there would be a gradual *succession* of vegetation changes until the optimum for the conditions, known as the *climatic climax community*, established itself (see chapter 10).

There is no doubt that climate does affect plant growth. In a broad sense, meagre rainfall results in desert vegetation, light rainfall causes grassland and abundant rainfall produces forest. Likewise, temperature plays a large part in accounting for the differences between, by way of example, tropical rainforest and the coniferous forests of higher latitudes. Low temperatures tend to result in lower plant growth and smaller size, and frosts can be lethal to certain species. Many plants cannot make active growth if the temperature is below 6 °C and for most of the species of temperate deciduous forest there must be at least six months with temperatures above such a minimum.

On the other hand, any simple relationship between the climate and vegetation type may be obscured by differences in such factors as soil type, the activity of fires and the actions of man. The problem of mapping the major zones is compounded by the fact that their delimitation will inevitably be arbitrary, for there are very few clear-cut lines in nature. There is also the perennial problem of whether the zones should be small in number, simple in type – and thus probably overgeneralised – or whether one should go for more categories, but then inevitably cause a lack of clarity.

Bearing these problems in mind, one can see that the map of vegetation types (figure 3.1) is remarkably similar to a map of major climatic types, and this is brought out further in table 3.1.

The rate at which vegetative matter is produced, together with the actual amount of vegetative material that exists in a particular area (the *biomass*), also shows broad patterns that are related to climate (figure 3.2). In desert areas and in the tundra, biomass levels are very low; they are intermediate in the boreal forest and mid-latitude deciduous forests, and reach their highest levels in the humid tropical rainforest. These differences in the production of organic matter have great importance in terms of soil development and rock weathering in different parts of the world.

Window 3.1 Mediterranean vegetation

Regions with Mediterranean climates tend to occur on the western coasts of continents between 30° and 40° latitude. Although these regions are separated by thousands of kilometres of oceans, they support superficially rather similar plant life because the plants (though unrelated to each other in terms of their lineage) have independently succeeded in evolving similar adaptations to this distinctive climatic environment.

The Mediterranean type of environment is unique because it has cool, wet winters and dry, hot summers. Plants therefore have had to adapt to severe droughts and to periodic wildfires. Their distinctive shrub communities, many of them now much modified by human activities, are variously called chaparral, mattoral, macchia, maquis or fynbos (see window 3.3).

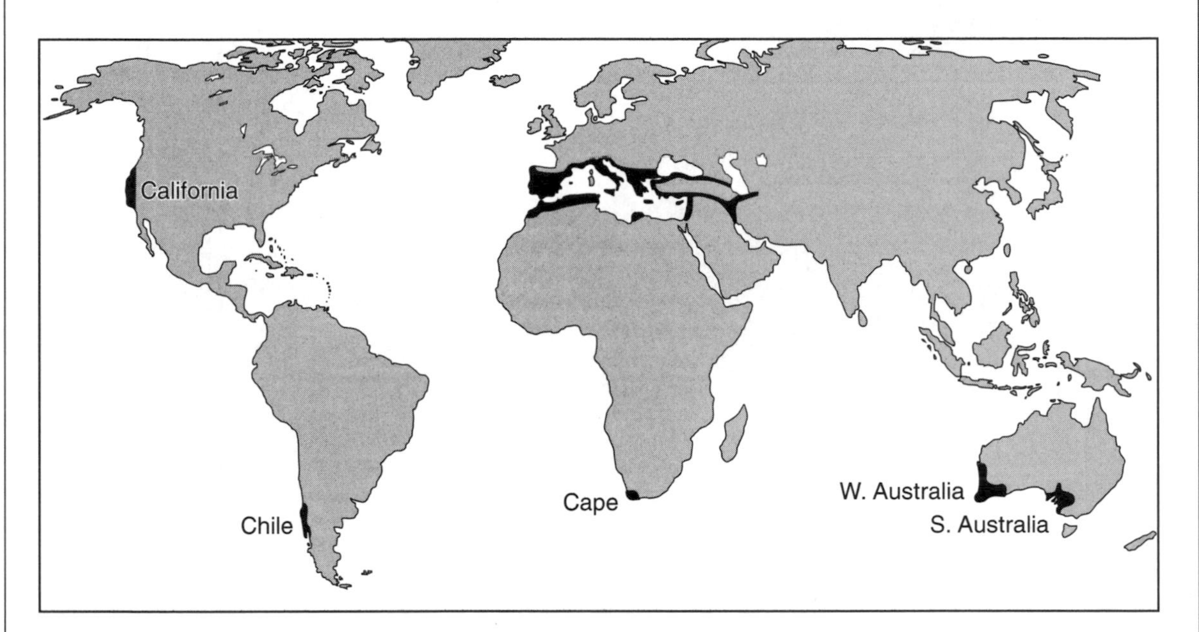

Areas of Mediterranean vegetation.

Table 3.1 The relationship between climatic zone and vegetation type

Climate	*Vegetation*
Humid tropical zone	Rainforests, with mangroves on coasts
Semi-arid seasonal tropics	Savanna
Arid tropics	Desert scrub, or vegetationless
'Mediterranean'	Evergreen woodlands and shrubs
Humid temperate (maritime)	Temperate deciduous forest
Cool temperate (continental)	Temperate grasslands, steppe etc.
Boreal zone	Coniferous and birch forests
Arctic zone	Tundra, shallow-rooted shrubs

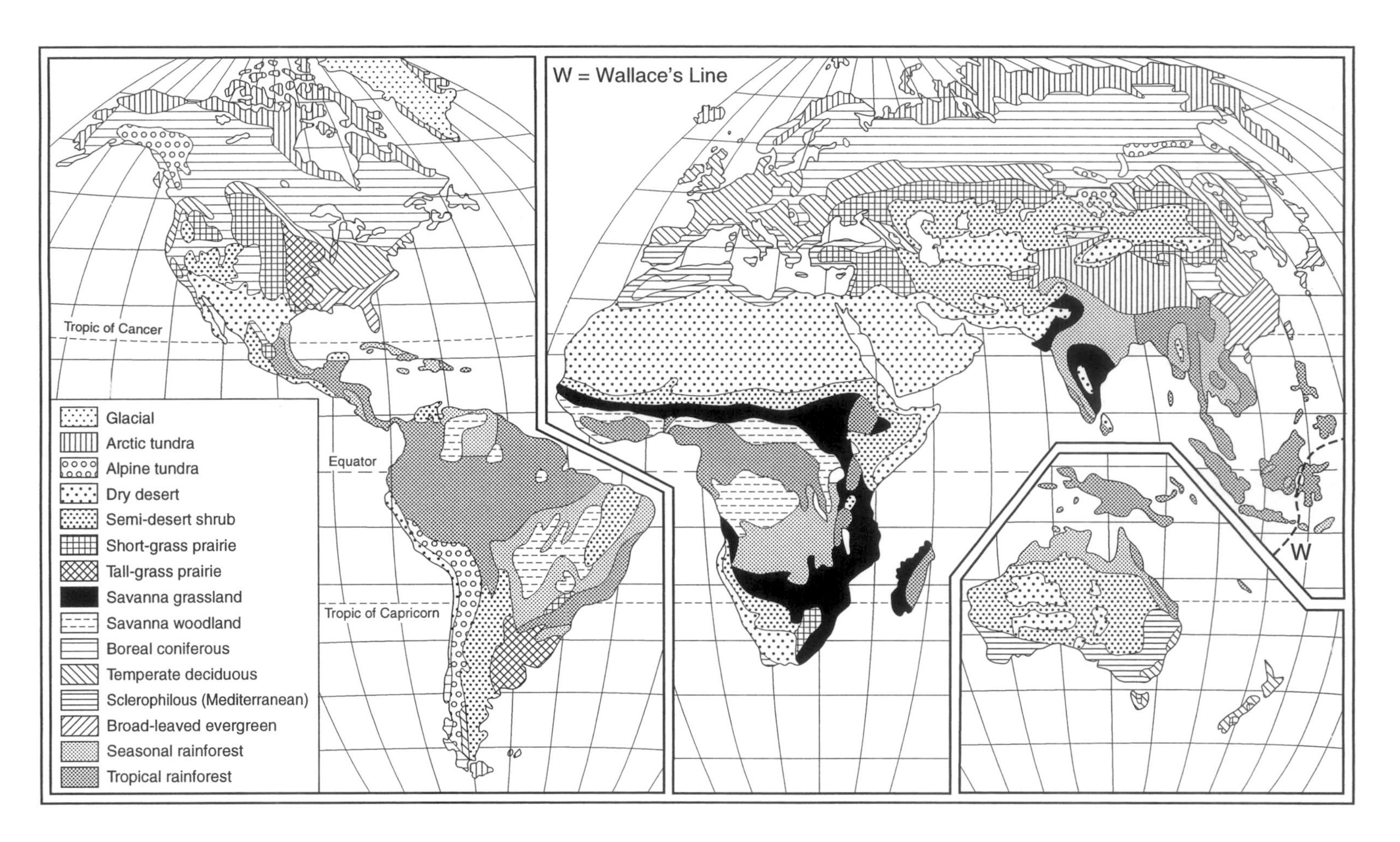

Figure 3.1 Vegetation types of the world.

Altitude modifies the general distribution pattern, and introduces local complexity (figure 3.3). Unless the relief is very slight, each altitudinal zone will afford suitable conditions for the appearance of plants in general characteristic of a zone or zones in higher latitudes; subtropical plants will occur here and there in tropical regions, temperate plants will occur in both tropical and subtropical zones, and so on. The effect of altitude is discussed further in chapter 8.

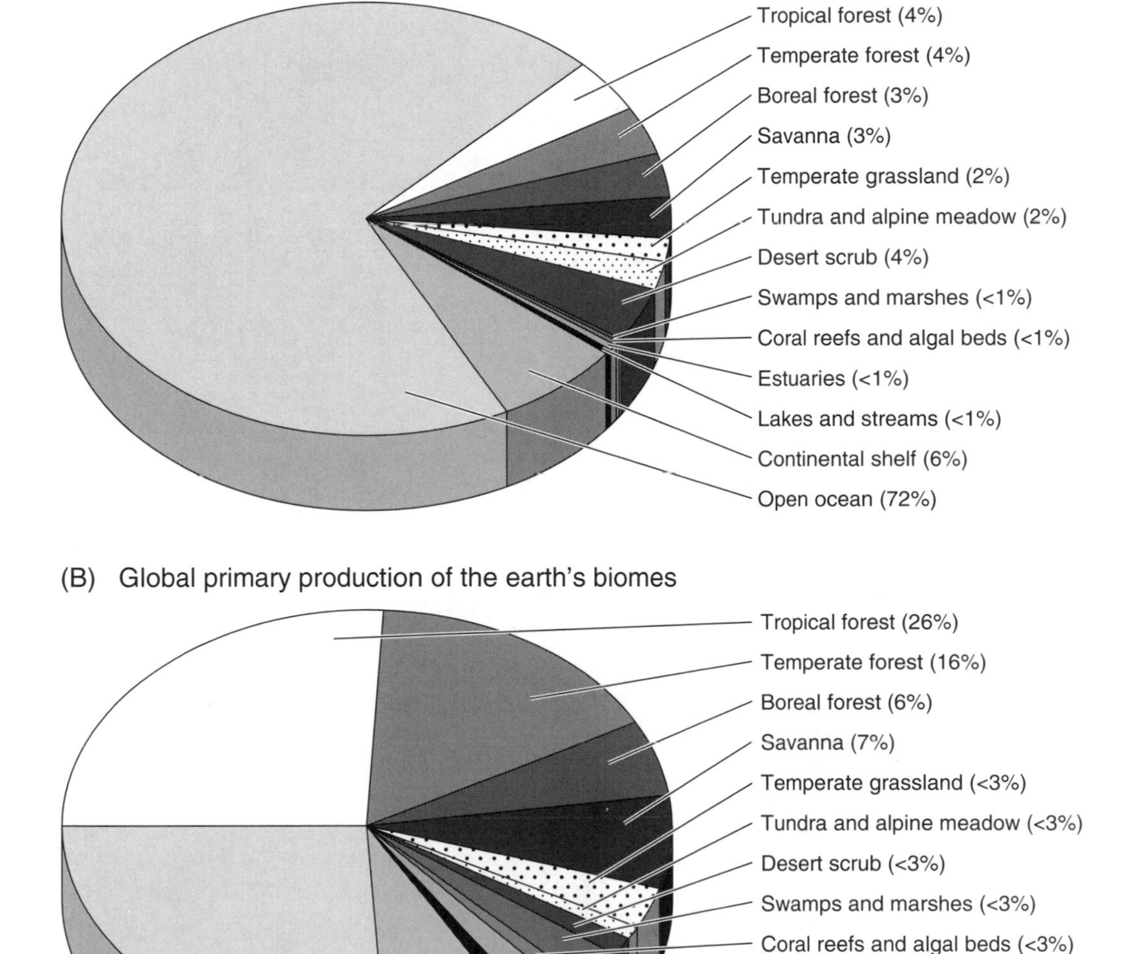

Figure 3.2 (above and overleaf) Characteristics of world biomes: (A) relative areas; (B) global primary production; (C) primary productivity; (D) biomass in relation to area.

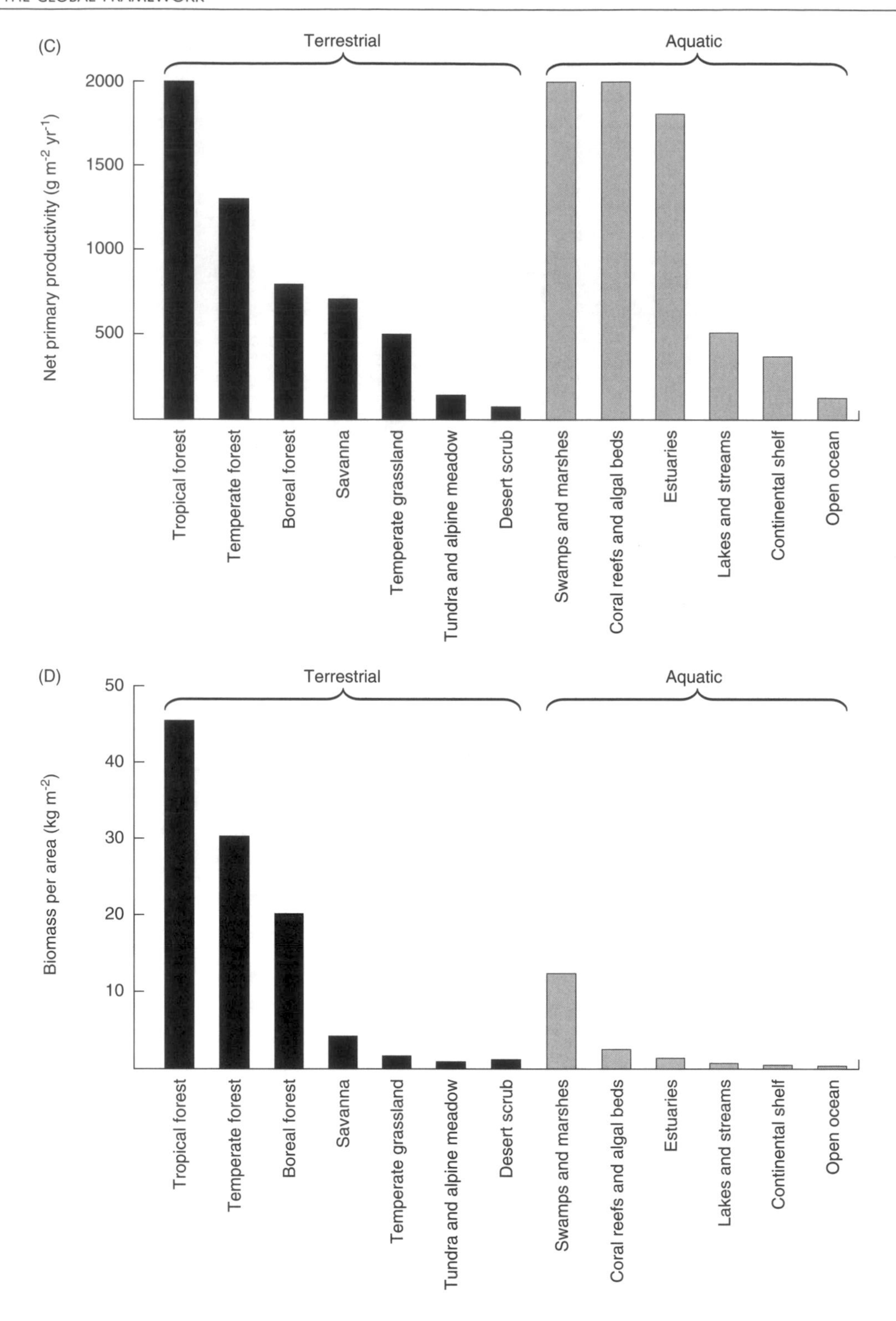
(C)
Terrestrial
Aquatic
Net primary productivity (g m-2 yr-1)
2000
1500
1000
500
Tropical forest
Temperate forest
Boreal forest
Savanna
Temperate grassland
Tundra and alpine meadow
Desert scrub
Swamps and marshes
Coral reefs and algal beds
Estuaries
Lakes and streams
Continental shelf
Open ocean
(D)
Terrestrial
Aquatic
Biomass per area (kg m-2)
50
40
30
20
10
Tropical forest
Temperate forest
Boreal forest
Savanna
Temperate grassland
Tundra and alpine meadow
Desert scrub
Swamps and marshes
Coral reefs and algal beds
Estuaries
Lakes and streams
Continental shelf
Open ocean

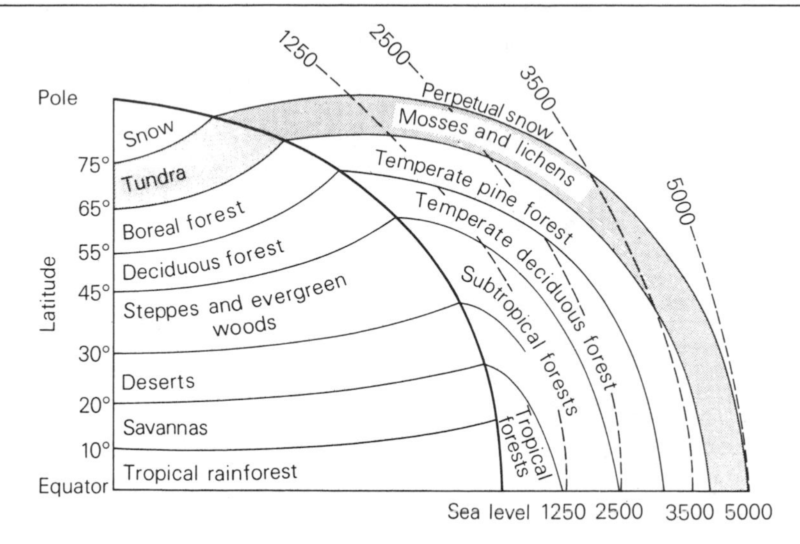

Figure 3.3 The modification of the world's major vegetation zones by altitude.

3.2 Human Modification of Major Vegetation Types

The global patterns and nature of some of the earth's important vegetation assemblages are being substantially changed by human actions, including fire, grazing and deforestation.

The use of fire

There are many good reasons why humans, from our early Stone Age ancestors onwards, have found fire useful:

- to clear forest for agriculture;
- to improve the quality of grazing for game or domestic animals;
- to deprive game of cover or to drive them from cover;
- to kill or drive away predatory animals, insects and other pests;
- to repel or attack human enemies;
- to make travel quicker and easier;
- to provide light and heat;
- to enable them to cook;
- to transmit messages, by smoke signals;
- to break up stone for making tools or pottery, smelting ores, and hardening spears or arrowheads;
- to make charcoal;
- to protect settlements or camps from larger fires by controlled burning;
- to provide spectacle and comfort.

Fire has been central to the life of many groups of hunter-gatherers, pastoralists and farmers (including shifting cultivators in the tropics). It was much used by peoples as different from one another as the Aboriginals of Australia, the cattle-keepers of Africa, the original inhabitants of Tierra del Fuego ('the land of fire') in the far south of South America and the Polynesian inhabitants of New Zealand (window 3.2). It is still much used especially in the tropics, and above all in Africa. Biomass burning appears to be especially significant in the tropical environments of Africa in comparison with other tropical areas. The main reason for this is the great extent of savanna which is subjected to regular burning. As much as 75 per cent of African savanna areas may be burned each year. This is probably an ancient phenomenon in the African landscape which occurred long before people arrived on the scene. Nevertheless, humans have greatly increased the role of fire in the continent, where they may have used it for over 1.4 million years.

Fire is crucial to an understanding of some major biome types, and many biota have become adapted to it. For example, many savanna trees are fire-resistant (plate 3.1). The same applies to the shrub vegetation (*maquis*) of the Mediterranean lands, which contains certain species which thrive after burning by sending up a series of suckers from ground level. Mid-latitude grasslands (e.g. the prairies of North America) were once thought to have developed in response to drought conditions during much of the year. Now, however, some have argued that this is not necessarily the case and that, in the absence of fire, trees could become dominant. The following reasons are given to support this suggestion:

Window 3.2 *The transformation of New Zealand*

New Zealand was only settled very recently, first by Polynesians (around 1200 years ago) and then by Europeans (around 200 years ago). The Polynesians carried out extensive firing of vegetation in pre-European settlement times, and hunters used fire to facilitate travel and to frighten and trap a major food source – the flightless moa (now extinct). The changes in vegetation that resulted were substantial. The forest cover was reduced from about 79 per cent to 53 per cent, and fires were especially effective in the drier forests of central and eastern South Island in the rain shadow of the Southern Alps. The fires continued over a period of about a thousand years up to the period of European settlement.

Islands like New Zealand have been vulnerable to the effects of introduced plants, which spread explosively. Gorse is one pernicious example, and nearly 60 per cent of all plant species in New Zealand are now aliens. It has often been proposed that the introduction of exotic terrestrial mammals has had a profound effect on the flora of New Zealand. Among the reasons that have been put forward for this belief are that the absence of native terrestrial mammalian herbivores permitted the evolution of a flora highly vulnerable to damage from browsing and grazing, and that the populations of wild animals (including deer and opossums) that were introduced in the nineteenth century grew explosively because of the lack of competitors and predators.

The now-extinct moa (*Diornis giganteus*) of New Zealand (1500 years BP).

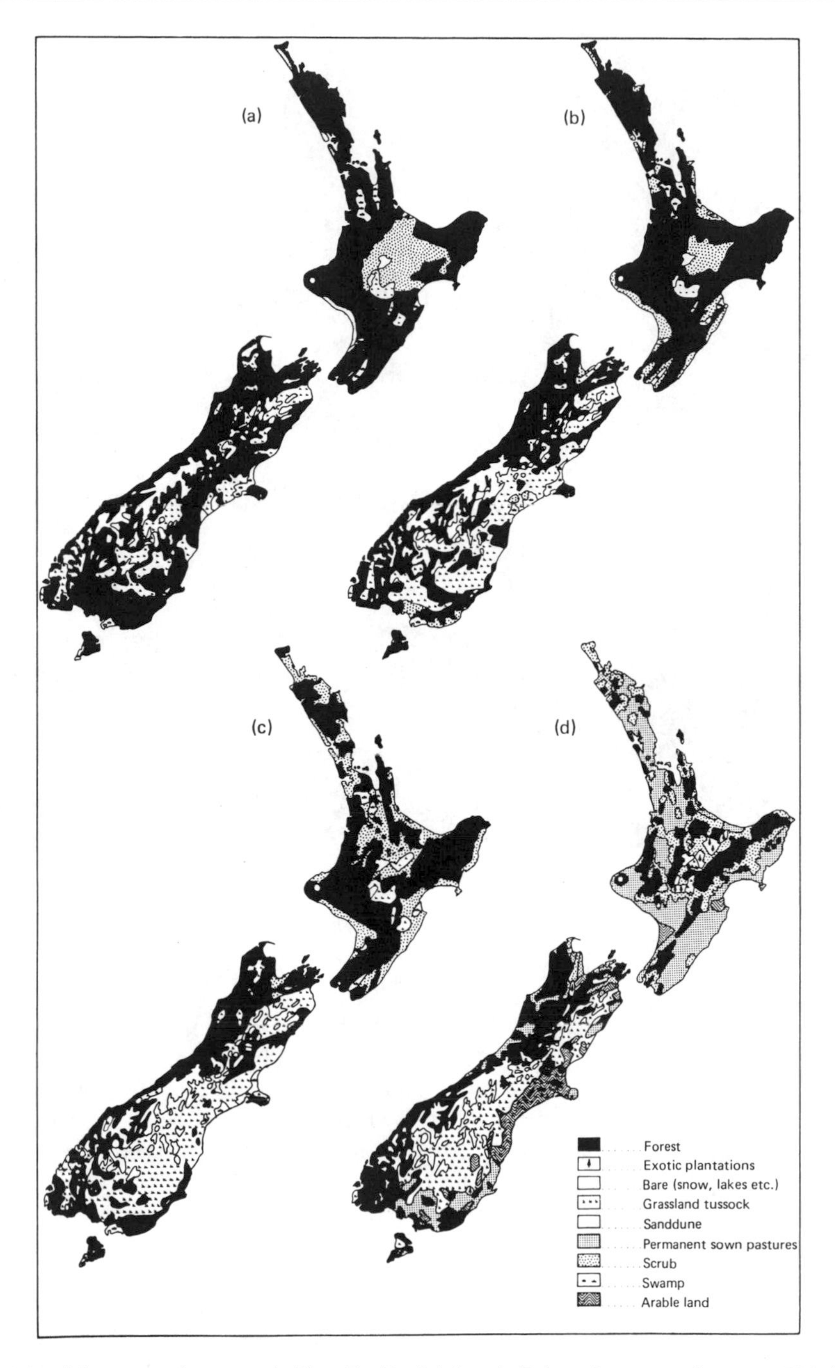

The changing state of the vegetation cover in New Zealand: (a) early Polynesian vegetation, *c.*AD 700; (b) pre-classical Maori vegetation, *c.*AD 1200; (c) pre-European vegetation, *c.*AD 1800; (d) present-day vegetation.
Source
R. Cochrane in A. G. Anderson (ed.) (1977) *New Zealand in Maps*, section 14 (London: Hodder and Stoughton).

Plate 3.1 The 'Pindan' savanna of tropical north-western Australia. This landscape is frequently burnt by the Aboriginal inhabitants and has been for thousands of years. This probably has a profound effect both on the distribution and the characteristics of this wooded grassland.

- planted groves and protected trees seem able to flourish;
- some woodland species, notably junipers, are remarkably drought-resistant;
- trees grow along escarpments and in deep valleys where moisture is concentrated at seeps and in shaded areas, and where fire is least effective: the effects of fire are greatest on flat plains where there are high wind speeds and no interruptions to the course of the fire;
- where fires have been restricted, woodland has spread into grassland.

Fire rapidly alters the amount, form and distribution of plant nutrients in ecosystems, and has been used deliberately to change the properties of the soil. Both the release of nutrients by fire and the value of ash have long been recognised, notably by those involved in slash-and-burn techniques. However, once land has been cultivated, the loss of nutrients by leaching and erosion is very rapid. This is why the shifting cultivators have to move on to new plots after only a few years. Fire quickly releases some nutrients from the soil in a form that plants can absorb. The normal biological decay of plant remains releases nutrients more slowly. The amounts of phosphorus (P), magnesium (Mg), potassium (K) and calcium (Ca) released by burning forest and scrub vegetation are high in relation to both the total and the available quantities of these elements in soils.

Grazing

A second major means of transforming vegetation assemblages is through the grazing and trampling activities of domestic stock. In particular, many of the world's grasslands have long been grazed by wild animals like the bison of North America or the large game of East Africa, but the introduction of pastoral economies also affects their nature and productivity.

Light grazing may increase the productivity of wild pastures. Nibbling, for example, can encourage the vigour and growth of plants, and in some species the removal of coarse, dead stems permits succulent sprouts to shoot. Likewise the seeds of some plant species are spread efficiently by being carried in the guts of cattle, and then placed in favourable seedbeds of dung or trampled into the soil surface. Moreover, the passage of herbage through the gut and out as faeces modifies the nitrogen cycle, so that grazed pastures tend to be richer in nitrogen than ungrazed ones. Also, grazing can increase species diversity by opening out the community and creating more niches.

On the other hand, heavy grazing may be detrimental. Excessive trampling when conditions are dry will reduce the size of soil aggregates and break up plant litter to a point where they are subject to removal by wind. Trampling, by puddling the soil surface, can accelerate soil deterioration and erosion as infiltration capacity is reduced. Heavy grazing can kill plants or lead to a marked reduction in their level of photosynthesis. In addition, when relieved of competition from palatable plants or plants liable to trampling damage, resistant and usually unpalatable species expand their cover.

In general terms it is clear that in many parts of the world the grass family is well equipped to withstand grazing. Many plants have their growing points located on the apex of leaves and shoots, but grasses reproduce the bulk of fresh tissue at the base of their leaves. This part is least likely to be damaged by grazing and allows regrowth to continue at the same time that material is being removed.

Communities severely affected by the treading of animals tend to have certain distinctive characteristics. These include diminutiveness (since the smaller the plant is the more protection it will get from soil-surface irregularities); strong ramification (the plant stems and leaves spread close to the ground); small leaves (which are less easily damaged by treading); tissue firmness (cell-wall strength and thickness to limit mechanical damage); a bending ability; strong vegetative increase and dispersal (for example, by stolons); small hard seeds which can be easily dispersed; and the production of a large number of seeds per plant (which is particularly important because the mortality of seedlings is high under treading and trampling conditions).

Deforestation

Clearing forests is probably the most obvious way in which humans have transformed the face of the earth. Forests provide wood for construction, for shelter and for making tools. They are also a source of fuel, and, when cleared, provide land for food production. For all these reasons they have been used by humans, sometimes to the point of destruction.

Forests, however, are more than an economic resource. They play several key ecological roles. They are repositories of biodiversity; they may affect regional and local climates and air quality; they play a major role in the hydrological cycle; they influence soil quality and rates of soil formation; and prevent or slow down soil erosion.

We do not have a clear view of how fast deforestation is taking place. This is partly because we have no record on a global scale of how much woodland there is today, or how much there was in the past. It is also because there are disagreements about the precise meaning of the word 'deforestation'. For example, shifting cultivators and loggers in the tropics often leave a certain proportion of trees standing. At what point does the proportion of trees left standing permit one to say that deforestation has taken place? Also, in some countries scrub is included as forest while in others it is not.

What we do know is that deforestation has been going on for a very long time. Pollen analysis shows that it started in prehistoric times, in the Mesolithic (around 9000 years ago) and Neolithic (around 5000 years ago). Large tracts of Britain had been deforested before the Romans arrived in the islands in the first century BC. Classical writers refer to the effects of fire, cutting and the destructive nibble of goats in Mediterranean lands. The Phoenicians were exporting cedars from Lebanon to the Pharaohs and to Mesopotamia as early as 4600 years ago. A great wave of deforestation occurred in western and central Europe in medieval times. As the European empires established themselves from the sixteenth and seventeenth centuries onwards, the activities of traders and colonists caused forests to contract in North America, Australia, New Zealand and South Africa, especially in the nineteenth century. Temperate North America, which was wooded from the Atlantic coast as far west as the Mississippi River when the first Europeans arrived, lost more woodland in the following 200 years than Europe had in the previous 2000. At the present time, the humid tropics are undergoing particularly rapid deforestation. Some areas are under particularly serious threat, including south-east Asia, West Africa, Central America, Madagascar and eastern Amazonia (figure 3.4).

Since pre-agricultural times approximately one-fifth of the world's forests has been lost. The highest losses (about a third of the total) have been in temperate areas. However, deforestation is not an unstoppable or irreversible process. For example, a 'rebirth of forest' has taken place in the USA since the 1930s and 1940s. Many forests in developed countries are slowly but steadily expanding as marginal agricultural land is abandoned. This is happening both because of replanting schemes and because of fire suppression and control. Also, in some cases the extent and consequences of deforestation may have been exaggerated.

Views vary as to the present rate of rainforest removal but estimates by the Food and Agriculture Organization put the total annual deforestation in 1990 for 62 countries (representing some 78 per cent of the tropical forest area of the world) at 16.8 million hectares. This figure is significantly higher than the one obtained for these same countries for the period 1976–80 (9.2 million hectares per year).

The loss of moist rainforests in some of the world's humid tropical regions is a very major concern. The consequences are many and serious (table 3.2). The causes are also diverse and include encroaching cultivation and pastoralism (including cattle ranching), mining and hydroelectric schemes, as well as logging operations themselves.

One particular type of tropical forest ecosystem coming under increasing pressure from various

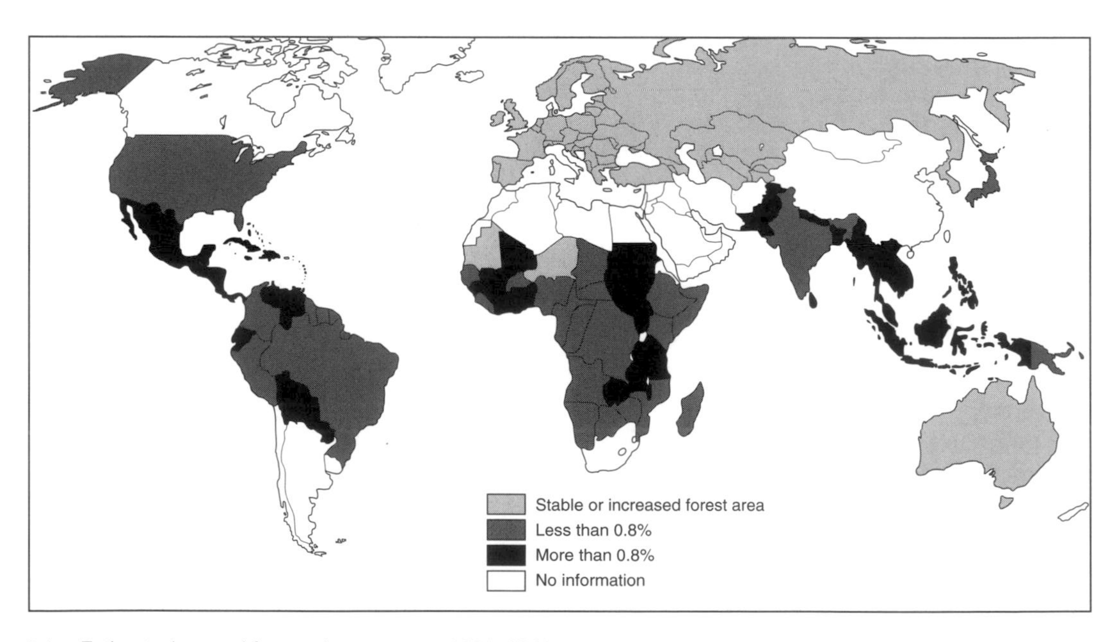

Figure 3.4 Estimated annual forest change rates, 1981–1990.

Table 3.2 The consequences of tropical deforestation

Type of change	*Examples*
Reduced biological diversity	Species extinctions Reduced capacity to breed improved crop varieties Inability to make some plants economic crops Threat to production of minor forest products
Changes in local and regional environments	More soil degradation Changes in water flows from catchments Changes in buffering of water flows by wetland forests Increased sedimentation of rivers, reservoirs etc. Possible changes in rainfall characteristics
Changes in global environments	Reduction in carbon stored in the terrestrial biota Increase in carbon dioxide content of atmosphere Change in global temperature and rainfall patterns through greenhouse effects Other changes in global climate due to changes in land surface processes

Source: A. Grainger (1992) *Controlling Tropical Deforestation* (London: Earthscan)

Plate 3.2 Mangrove forests, which fringe the coastlines of many tropical regions, as here in Mauritius in the Indian Ocean, are an important wetland habitat that is under increasing human pressure.

human activities is the mangrove forest characteristic of inter-tidal zones (plate 3.2). These ecosystems constitute a reservoir, refuge, feeding ground and nursery for many useful and unusual plants and animals. In particular, because they export decomposable plant debris into adjacent coastal waters, they provide an important energy source and nutrient input to many tropical estuaries. In addition they can serve as buffers against the erosion caused by tropical storms – a crucial consideration in low-lying areas like Bangladesh. In spite of these advantages, mangrove forests are being degraded and destroyed on a large scale in many parts of the world, either through exploitation of their wood resources or because of their conversion to single-use systems such as agriculture, aquaculture, salt-evaporation ponds or housing developments. To give two examples: mangrove areas in the Philippines converted to fish ponds have increased from less than 90 000 hectares in the early 1950s to over 244 000 hectares in the early 1980s; while in Indonesia logging operations are claiming 200 000 hectares of mangrove each year.

On a global basis, it has been calculated that since 1700 about 19 per cent of the world's forests and woodlands have been removed. Over the same period the world's cropland area has increased by over four and a half times, and between 1950 and 1980 it amounted to well over 100 000 km^2 per year.

3.3 Floral Realms

From the time in the distant geological past when plants spread rapidly around the world until the present, the plant cover has never ceased to evolve. Mutations have emerged continuously, and their survival and establishment have depended on ecological circumstances at the time of their appearance. Climates have changed; continents have shifted, isolating some plant groups and joining others; new mountain barriers have developed; and sea level has varied, creating or destroying land-bridges between continents. For these reasons the world today does not have a uniform flora, but we can none the less detect areas where there is some coherence in the distribution pattern of particular species. These can be generalised into maps – maps of *flora* and not of vegetation types. The distinction between the two is an important one: the flora of an area is the sum total of all the plant species in it; vegetation is the kind of plant cover in that area. Thus, although two floral regions may be similar in their vegetation, because, for example, they are both areas of tropical rainforest, they do not necessarily have much in common botanically or share more than a few genera.

An example of floral regionalisation is shown in figure 3.5a, and it has certain broad similarities to a map of the faunal realms, a topic we shall turn to shortly. Although there can be many small regions (in this example, 37), they can be lumped into six major groups called *kingdoms*: Holarctic, Palaeotropical, Neotropical, Cape (window 3.3), Australian and Antarctic.

The nature of the floral realms has been modified by human activities, for people are important agents in the spread of plants and other organisms. Some plants are introduced deliberately by humans to new areas: these include crops, ornamental varieties and miscellaneous landscape modifiers (trees for reafforestation, cover plants for erosion control and so on). Indeed, some plants, such as bananas and breadfruit, have become completely dependent on people for reproduction and dispersal, and in some cases they have lost the capacity for producing viable seeds and depend on human-controlled vegetation propagation.

However, some domesticated plants, when left to their own devices, have shown that they are capable of at least ephemeral colonisation, and a small number have successfully naturalised themselves in areas other than their supposed region of origin. Examples of such plants include several umbelliferous annual garden crops (fennel, parsnip and celery) which, though native to Mediterranean Europe, have colonised waste lands in California. The Irish potato, which is native to South America, grows unaided in the mountains of Lesotho. In Paraguay, orange trees (originating in south-east Asia and the East Indies) have demonstrated their ability to survive in direct competition with natural vegetation.

Plants that have been introduced deliberately (plate 3.3) because they have recognised virtues can be usefully divided into an economic group (for example, crops, timber trees etc.) and an ornamen-

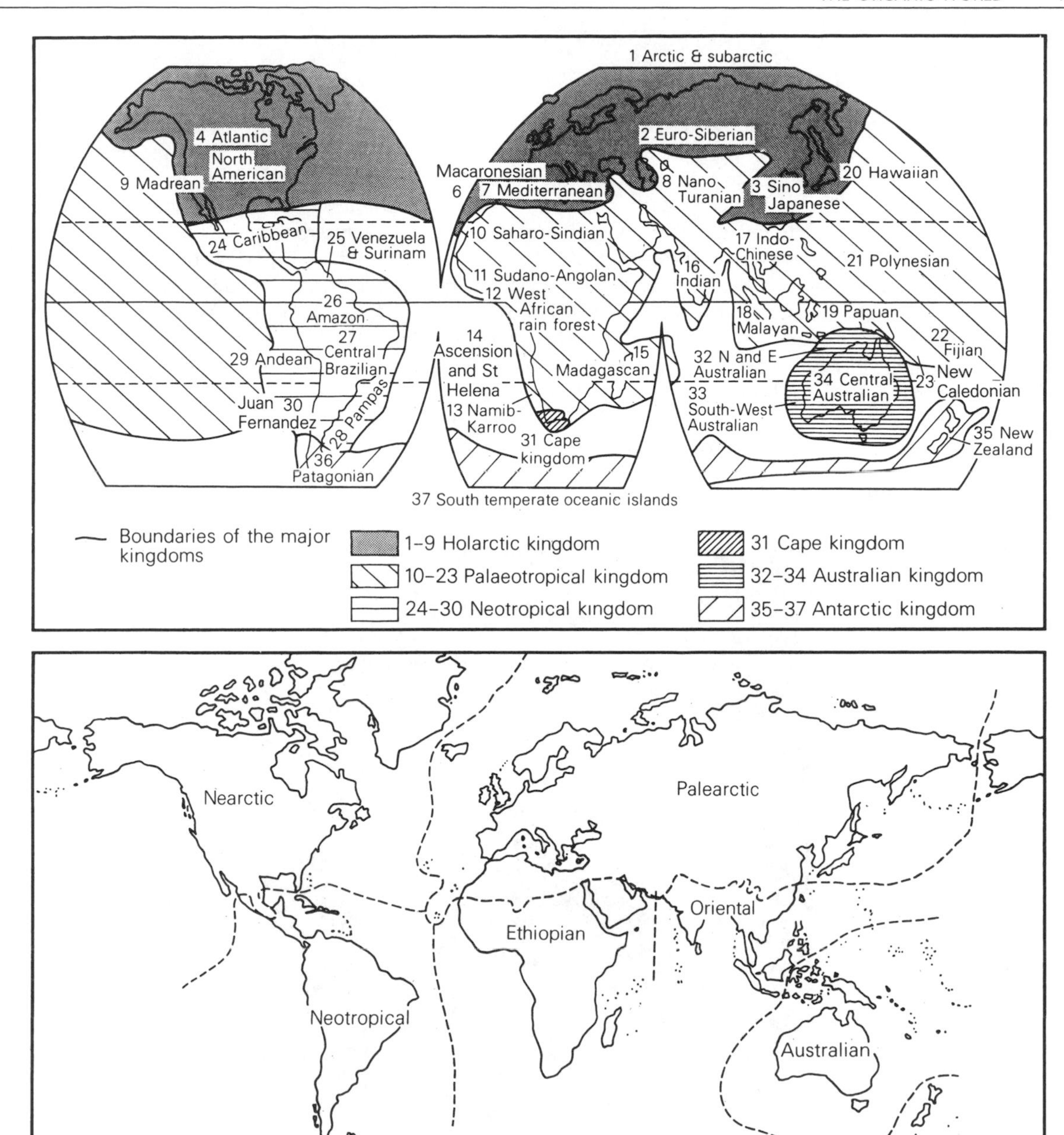

Figure 3.5 The major (a) floral and (b) faunal realms of the world.

Window 3.3 *The Cape floral kingdom*

The world has six floral kingdoms. Five of these cover huge areas (such as Australia or the Northern Hemisphere), but one of them is small and restricted to only the southern tip of Africa. This is the Cape floral kingdom. Although it is small, it is immensely rich in plant species – 1300 per 10 000 km^2. An important component of this flora is called fynbos, which is dominated by woody shrubs, including the famous proteas. Over 7700 plant species are found in fynbos and of these roughly 70 per cent are endemic to the area – that is, they are found nowhere else in the world. There are, for example, 600 different species of ericas or heaths in the fynbos, but only 26 in the rest of the world. The plants have evolved in virtual isolation from the rest of the world over tens of million of years, and have developed adaptations to the area's Mediterranean climate.

The state of the fynbos has been undermined by miscellaneous human activities, including the spread of towns, land clearance for agriculture, and afforestation. However, one of the most severe human impacts is caused by the explosive spread of a variety of introduced exotic plants, including trees or large shrubs belonging to the genera *Acacia*, *Hakea* and *Pinus*. Many of these were introduced from Australia and have overwhelmed the native flora over extensive areas.

Visit this web site:http:/www.botany.uwc.ac.za/fynbos/

Typical fynbos vegetation with proteas in the Cederberg Mountains of the Cape, South Africa.

Plate 3.3 Many plants were moved around the world deliberately to stock botanical gardens. The wonderful gardens at Pamplemousses in Mauritius were established in part to aid the dissemination of useful plants. As the inscription says (in French), 'The gift of a useful plant appears more precious to me than the discovery of a gold mine and is a more durable monument than a pyramid.'

tal or amenity one. In the British Isles, the great bulk of deliberate introductions before the sixteenth century had some sort of economic merit, but only a handful of the species introduced thereafter were brought in because of their utility. Instead, plants were introduced increasingly out of curiosity or for decorative value.

Many plants, however, have been dispersed accidentally as a result of human activity: some by adhesion to moving objects, such as individuals themselves or their vehicles; some among crop seed; some among other plants (like fodder or packing materials); some among minerals (such as ballast or road metal); and some by the carriage of seeds for purposes other than planting (as with drug plants).

The accidental dispersal of such plants and organisms can have serious ecological consequences. In Britain, for instance, many elm trees died in the 1970s because of the accidental introduction of Dutch elm disease fungus which arrived on imported timber at certain ports, notably Avonmouth and the Thames Estuary ports. There are also other examples of the dramatic impact of some introduced plant pathogens. In western Australia the great jarrah forests have been invaded and decimated by a root fungus, *Phytophthora cinnamomi*. This was probably introduced on diseased nursery material from eastern Australia, and the spread of the disease within the forests was facilitated by road building, logging and mining activities that involved movement of soil or gravel containing the fungus. More than 3 000 000 ha of forest have been affected.

Ocean islands have often been particularly vulnerable. The simplicity of their ecosystems inevitably leads to diminished stability, and introduced species often find that the relative lack of competition enables them to broaden their ecological range more easily than on the continents. Moreover, because the natural species inhabiting remote islands have been selected primarily for their dispersal capacity, they have not necessarily been dominant or even highly successful in their original continental setting. Therefore, introduced species may prove more vigorous and effective. There may also be a lack of indigenous species to adapt to conditions such as bare ground caused by humans. Thus introduced weeds may catch on.

There are a number of major threats that invasive plants pose to natural ecosystems:

1 Replacement of diverse systems with single species stands of aliens, leading to a reduction in biodiversity, as for example when Australian *acacias* have invaded the fynbos heathlands of South Africa.
2 Direct threats to native faunas by change of habitat.
3 Alteration of soil chemistry. For example, the African *Mesembryanthemum crystallinum* ac-

cumulates large quantities of salt. In this way it salinizes invaded areas of Australia and may prevent the native vegetation from establishing.

4 Alteration of geomorphological processes, especially rates of sedimentation and movement of mobile landforms (e.g. dunes and salt marshes).
5 Plant extinction by competition.
6 Alteration of fire regime. For example, in Florida, USA the introduction of the Australasian *Melaleuca quinquenervia* has increased the frequency of fires because of its flammability, and has damaged the native vegetation which is less well adapted to fire.
7 Alteration of hydrological conditions (e.g. reduction in groundwater levels caused by some species having high rates of transpiration).

3.4 Faunal Realms

Next, in this study of present global patterns, we must take a look at the distribution patterns of the world's animals, and try to subdivide the world into zoogeographical regions or realms (figure 3.5b). Of the various attempts that have been made to do this, the most famous is that of a contemporary of Charles Darwin, A. R. Wallace (plate 3.4). In the late nineteenth century Wallace established six major regions, to which he gave specific names as they do not correspond very precisely with political or cultural areas. They are still widely accepted as useful, though subdivisions and amalgamations have been made from time to time. Some scientists have considered the Neotropical and Australian regions to be zoologically so different from the rest of the world, and from one another, as to rank as regions equivalent to the remaining four put together. In this classification there are three realms: Neogea (Neotropical), Notogea (Australia) and Arctogea (the rest of the world). Another proposal is that Palearctic and Nearctic do not merit separate regional status, and should thus be combined into one region – the Holarctic. As with all schemes of regionalisation and classification, some people are 'lumpers' and some are 'splitters'.

Plate 3.4 One of the greatest zoogeographers of all time was A. R. Wallace, who divided the world up into a series of faunal realms or regions. He noted the considerable differences between the fauna of Australia and of Asia, and the boundary between these two faunal areas is normally called Wallace's Line.

The important point, however, is that there are major differences in the nature of the species of animals found in different parts of the world. This reflects many factors, including present and past climates, the absence or presence of former land connections between continents, the operation of continental drift and the different operation of evolution in different parts of the world. For example, the fauna of Australia may have been similar at one time to that in other parts of the world because it was part of the great supercontinent of Pangaea or Gondwanaland. However, for some considerable time it has been isolated by a zone of deep water from Asia, and so exchanges have been limited, and evolution has been able to take place in comparative isolation, producing some unique species that are adapted to the particular environmental conditions of that continent. It is for these sorts of reasons that the Australian fauna is so special (plate

Plate 3.5 Australia has many endemic species of fauna, including marsupials. Isolation has enabled the evolution of such strange beasts as the koala bear (bottom, left) and the platypus (top).

3.5). Apart from bats, there are only nine families of mammals, and eight of these are unique. The dominant mammal fauna is marsupial; it is made up of six families, none of which occurs in the New World, where are found the only other living marsupials. The remaining two families of Australian mammals belong to a separate subclass of mammals, Monotremata. They are those bizarre egg-laying beasts, the duckbilled platypus and the spiny ant-eaters.

Since we have developed the means of long-distance travel, particularly across the oceans, we have greatly modified the distribution of animal species. In some cases people have been the unwitting cause of the introduction of foreign species to an area, as for example when cats and rats escaped from ships that visited tropical islands. At other times they have deliberately introduced particular species – for sport, for economic gain or from nostalgia – and as a consequence some species, like the trout, have a vastly greater distribution than they would have had without our assistance.

3.5 The World's Great Soil Orders

When a rock is first exposed at the earth's surface by erosion, it becomes subjected to the action of atmospheric and biological agents. Mechanical weathering by frost and other such processes achieves the first stages in soil formation by fragmenting the rock. Chemical weathering gradually changes the minerals of the rock, and some of the easily soluble components thus released are removed into streams by being *leached* out of the surface layers, while others may be involved in the nourishment of invading micro-organisms. As time passes, these increase in bulk, in complexity of life form and in their effect on the soil mantle, and eventually an organic rich layer may form at the surface. A soil may therefore be defined as an aggregate of many individual physical, chemical and biological processes that can be classified into various types, which gradually lead to the development of distinct layers or horizons by additions, removals, transfers and transformations of materials and energy. Most notable additions are organic matter and gases; removals involve salts and carbonates; transfers are of humus and sesquioxides; and transformations occur of, for example, primary minerals to secondary clays. All these processes take place in various combinations more or less simultaneously, the balance between them governing the nature of the soil profile.

Thus, within the soil numerous processes go on continuously as matter is transferred from one horizon to another, is added to the soil from above and is lost to plant roots from below. Such processes (which are discussed in greater detail in chapter 13) depend in part upon climatic conditions, so that at a gross scale certain patterns of soil development can be identified.

Soils that develop in areas of low precipitation, where rates of evaporation are high, have a water deficit for much of the time and undergo a process called *calcification*. When rain does fall it is sufficient to penetrate the upper soil layer, dissolve some calcium and percolate downwards. However, there is insufficient rainfall to leach the soil effectively, and soon the available water is evaporated or absorbed, leading to the deposition of the calcium carbonate.

In cool, moister climates, soil development tends to be influenced by *podzolisation* (window 3.4). Under such circumstances the water supply is more ample, so that the rainfall is sufficient to leach soluble materials quite thoroughly from the upper horizons, leaving only the rather inert silica behind. The leaching is assisted and accelerated by the presence of organic matter from the upper humus layer, and the minerals are transferred downwards to accumulate in the B horizon as an iron-cemented layer called an ironpan (section 13.2).

The hot wet regions of the tropics undergo *lateritisation*. High rainfall and high temperatures combine to promote intense chemical weathering; high temperatures promote bacterial action and organic matter is rapidly destroyed so that only a very limited humus layer is developed. These soils are often red in colour, and are dominated by large quantities of iron and aluminium sesquioxides.

Processes such as these help to account for the major differences in soils in different major climatic zones. The US Department of Agriculture has identified ten main soil *orders*, and their classification is called the Comprehensive Soil Classification System (CSCS) or, alternatively, *The Seventh Approxi-*

mation. We can appreciate their distribution and character by taking a hypothetical continent from the Northern Hemisphere (figure 3.6).

In the south-east region of this hypothetical continent, hot, moist conditions produce lateritisation processes and lead to *oxisols*. In the dry, hot south-west are the *aridisols*, which are affected by calcification processes. In the cool and cold north pod-

Window 3.4 Podzols

Podzols are characterised by the presence, just below the surface, of an ashy-coloured horizon. It is from this that they derive their Russian name (*pod*, 'under' and *zola*, 'ash'). They are very extensive in a circumpolar belt which extends approximately from the Arctic Circle southwards to the latitude of St Petersburg (in Europe) and the northern shores of the Great Lakes (in North America). They are particularly well developed on permeable sands and gravels and occur on some of the heathlands of Britain. They are frequently associated with coniferous boreal forest. Their horizons are as follows. Below the raw humus layer there is a grey and somewhat structureless Ea horizon from which virtually all free iron has been removed. Beneath this is the B horizon of illuviation which typically includes a humus-enriched layer (Bh) and a strong brown or rusty coloured Bs horizon of iron and aluminium enrichment. High available soil moisture and organic material promote the development of these horizons.

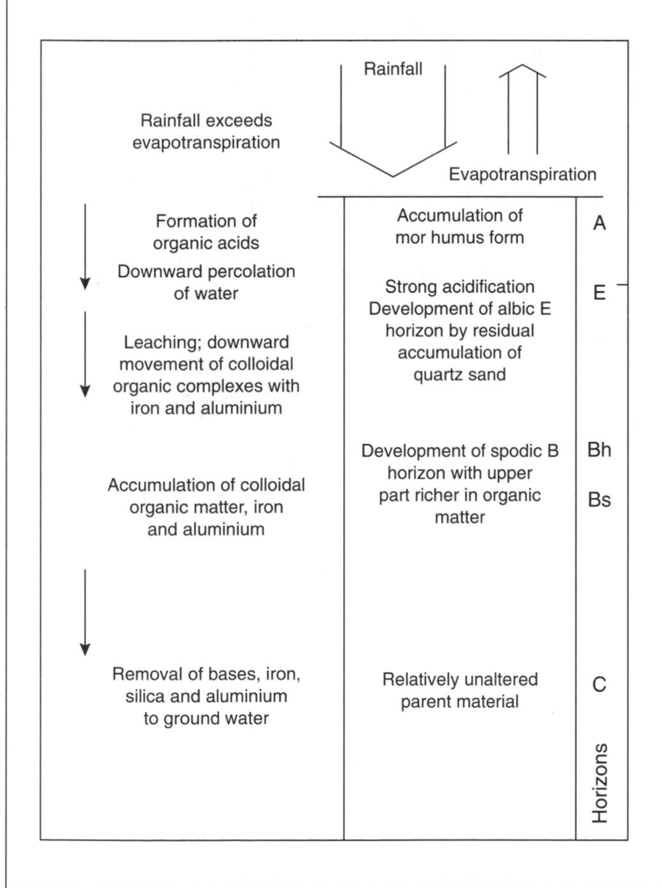

The process of podzolisation.

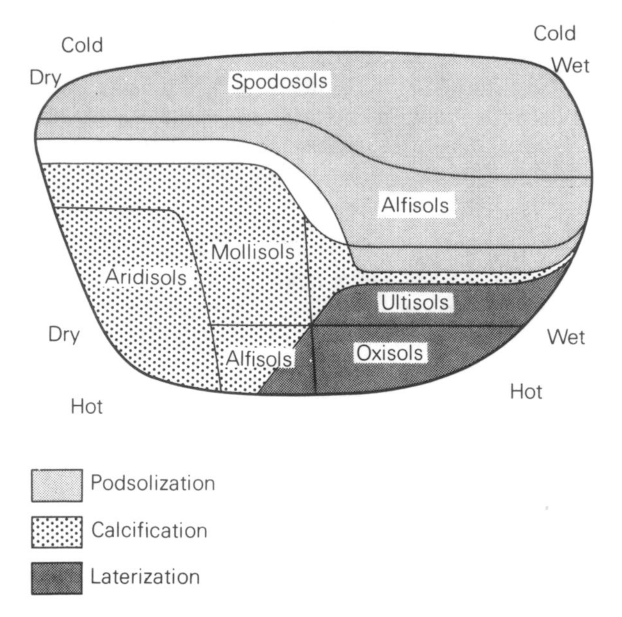

Figure 3.6 Schematic representation of the great soil orders on a hypothetical continent in the Northern Hemisphere.

zolisation occurs, producing *spodosols*. *Mollisols* are located in the intermediate positions between dry and moist climates, in areas of grassland vegetation. They display some calcification, but in addition possess a dark humus-rich upper layer. On the moist side of the mollisols lie the *alfisols*, located between arid and subhumid soils on one side and more humid ultisols on the other. These are grey-brown soils that commonly occur beneath deciduous forest. They are acid and have a lower horizon of clay accumulation. *Ultisols* develop where there is a pronounced summer wet season and a water-deficit dry season. They are quite deeply weathered and are transitional towards oxisols, often displaying the characteristic reddish-yellow coloration in the B horizon owing to the concentration of iron oxides.

The other four main soil orders are less clearly associated with any particular climatic regime. *Entisols* are soils that either have not existed long enough to develop mature horizonation (i.e. they are rec*ent*) or lie on parent materials, such as quartz

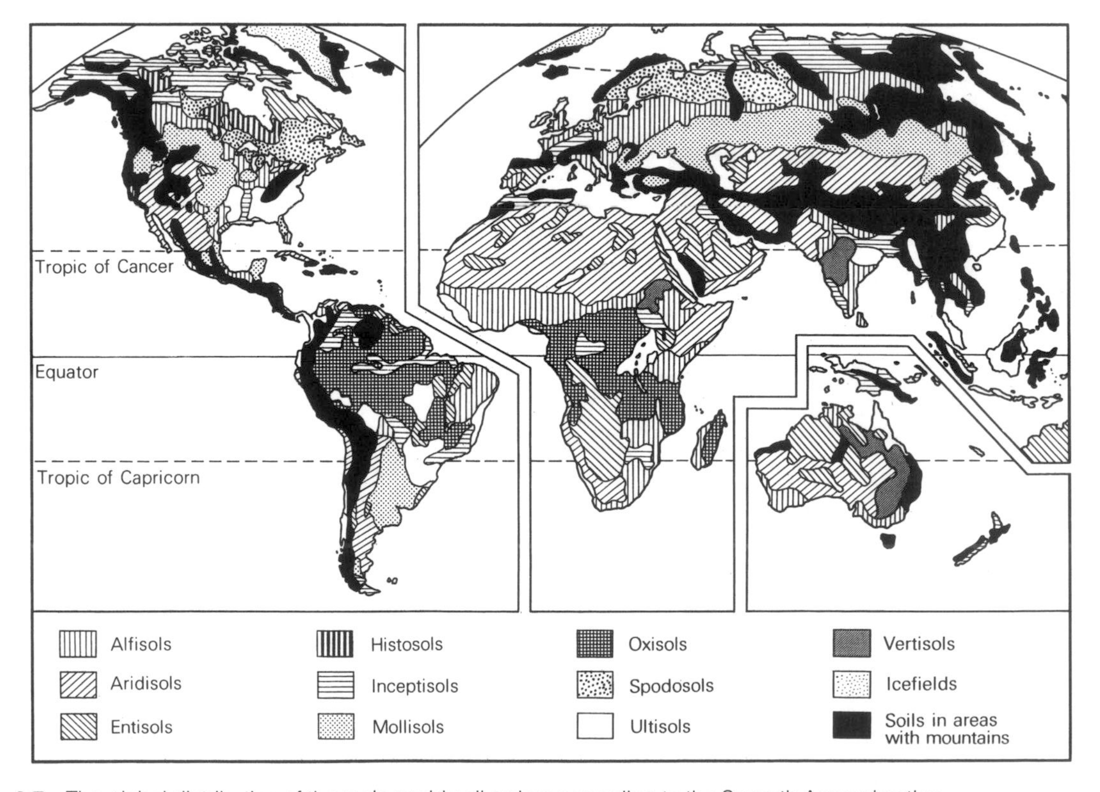

Figure 3.7 The global distribution of the main world soil orders according to the Seventh Approximation.

Table 3.3 US Department of Agriculture soil classification: the Seventh Approximation (1975)

Order	*Suborder*	*Characteristics/environment*
Alfisols (soils with an argillic horizon and moderate to high base content)	Aqualfs	Gleying features
	Boralfs	Others in cold climates
	Udalfs	Others in humid climates (including most leached brown soils)
	Ustalfs	Others in subhumid climates
	Xeralfs	Others in sub-arid climates
Aridisols (desert and semi-desert soils)	Argids	With argillic horizon (i.e. zone of clay accumulation)
	Orthids	Other soils of dry areas
Entisols (immature usually azonal soils)	Aquents	Gleying features
	Arents	Artificially disturbed
	Fluvents	Alluvial deposits
	Psamments	Sandy or loamy sand textures
	Orthents	Other entisols
Histosols	Fibrists	Plant remains very little decomposed
	Folists	Freely draining histosols
	Hemists	Plant remains not recognisable because of decomposition; found in depressions
	Saprists	Plant remains totally decomposed (black)
Inceptisols (moderately developed soils, not in other orders)	Andepts	Volcanic ash
	Aquepts	Gleying features
	Plaggepts	Man-made surface horizon
	Tropepts	Tropical climates
	Umbrepts	Umbric epipedon (i.e. dark-coloured surface horizon of low base status); hills and mountains
	Ochrepts	Other inceptisols (including most brown earths) of mid-high latitudes
Mollisols (soils with a dark A horizon and high base status, e.g. chernozems, rendzinas)	Albolls	With argillic and albic horizons
	Aquolls	Gleying features
	Rendolls	Highly calcareous materials
	Borolls	Others in cold climates
	Udolls	Others in humid climates
	Ustolls	Others in subhumid climates
	Xerolls	Others in sub-arid climates
Oxisols (soils with an oxic horizon or with plinthite near surface)	Aquox	Gleying features
	Humox	With a humose A horizon
	Torrox	Oxisols of arid climates
	Orthox	Others in equatorial climates
	Ustox	Others in subhumid climates
Spodosols (soils with accumulation of free sesquioxides and/or organic carbon e.g. podzols)	Aquods	Gleying features
	Ferrods	Much iron in spodic horizon
	Humods	Little iron in spodic horizon
	Orthods	Both iron and humus accumulation
Ultisols (soils with an argillic horizon, but low base content)	Aquults	Gleying features
	Humults	With a humose A horizon
	Udults	Others in humid climates
	Ustults	Others in subhumid climates
	Xerults	Others in sub-arid climates
Vertisols (cracking clay soils with turbulence in profile)	Torrerts	Usually dry (cracks open for 300 days per year)
	Uderts	Usually moist (cracks open and close several times a year)
	Usterts	Cracks remain open 90 days per year (in monsoon climates)
	Xererts	Cracks remain open 60 days per year

dune sand, that do not readily evolve into horizons. *Inceptisols* include soils formed on the alluvium deposited by major rivers. *Vertisols* are clayey soils characterised by deep, wide cracks in the dry season. These cracks close up in the wet season when the available moisture increases causing the clays to swell, but before it does so a portion of the surface material has washed into the cracks: it becomes in*vert*ed. Finally *histosols* are primarily organic matter rather than mineral soils and occur in bogs, moors or as peat accumulations where waterlogging is prevalent.

Thus the Seventh Approximation has ten soil orders (table 3.3 and figure 3.7), and although their terminology may at first sight appear forbidding and bewildering, it is relatively easy to understand once the principles of its construction have been grasped. The name of each order is based on syllables that are intended to convey the major attributes of that class, and we have already seen the origin of the terms entisol and vertisol. To construct class names at the suborder level we take two formative elements: the first indicating the characteristics of the soil or its environment (such as *aqu*, indicating wetness; see table 3.4), and the second being a suffix derived from the name of the order. Thus we could have the suborder *Aquox*, being an oxisol with gleying features indicative of wetness.

3.6 Human Modifications of Soil

Soils, being thin, heavily exploited and taking a long time to form, are prone to profound modifications in the face of human pressures. Some of the modifications are beneficial (plate 3.6), but others are detrimental as can be seen when we consider the ways in which humans alter some of the key soil-forming factors:

1 *Parent material*
Beneficial: adding mineral fertilisers; accumulating shells and bones; accumulating ash locally; removing excess amounts of substances such as salts.

Table 3.4 Formative elements in names of suborders of Seventh Approximation

Formative element	*Meaning*
alb	Presence of albic horizon (a bleached eluvial horizon)
and	Ando-like (i.e. volcanic ash materials)
aqu	Characteristics associated with wetness
ar	Mixed or cultivated horizon
arg	Presence of argillic horizon (a horizon with illuvial clay)
bor	Of cool climates
ferr	Presence of iron
fibr	Fibrous
fluv	Floodplain
fol	Presence of leaves
hem	Presence of well-decomposed organic matter
hum	Presence of horizon of organic enrichment
ochr	Presence of ochric epipedon (a light-coloured surface horizon)
orth	The common ones
plagg	Presence of a plaggen epipedon (a man-made surface 50 cm thick)
psamm	Sandy texture
rend	Rendzina-like
sapr	Presence of totally humified organic matter
torr	Usually dry
trop	Continually warm
ud	Of humid climates
umbr	Presence of umbric epipedon (a dark-coloured surface horizon)
ust	Of dry climates, usually hot in summer
xer	With annual dry season

Plate 3.6 The micro-plots in Lanzarote, Canary Islands, used for growing vines and other tree crops, are essentially composed of soils that have been created from volcanic deposits by the labours of the islanders.

Detrimental: removing through harvest more plant and animal nutrients than are replaced; adding materials in amounts toxic to plants or animals; altering soil constituents in a way to depress plant growth.

2 *Topography*
Beneficial: checking erosion through surface roughening, land forming and structure building, raising land level by accumulation of material; land levelling.
Detrimental: causing subsidence by drainage of wetlands and mining; accelerating erosion; excavating.

3 *Climate*
Beneficial: adding water by irrigation; rainmaking by seeding clouds; removing water by drainage; diverting winds etc.
Detrimental: subjecting soil to excessive insolation, to extended frost action, to wind etc.

4 *Organisms*
Beneficial: introducing and controlling populations of plants and animals; adding organic matter including 'night-soil', loosening soil by ploughing to admit more oxygen; fallowing; removing pathogenic organisms, e.g. by controlled burning.
Detrimental: removing plants and animals; reducing organic content of soil through burning, ploughing, over-grazing, harvesting etc.; adding or fostering pathogenic organisms; adding radioactive substances.

5 *Time*
Beneficial: rejuvenating the soil by adding fresh parent material or through exposure of local parent material by soil erosion; reclaiming land from under water.
Detrimental: degrading the soil by accelerated removal of nutrients from soil and vegetation cover; burying soil under solid fill or water.

Among the more important human modifications of soil types are salinisation in irrigated areas and soil erosion and degradation by wind and water (see sections 6.14 and 13.6).

3.7 Climatic Geomorphology: The Influence of Climate, Soil and Vegetation

Although we started this part of the book by considering the pattern of major world landforms (shields, ocean ridges etc.) in terms of plate tectonics and global geology, it is also undoubtedly true that the nature of the landforms may, at a global scale, owe much also to the nature of the climate in the area, and to the influence that climate has through its effect on the nature of soil and vegetation. Because of this, attempts have been made by various climatic geomorphologists to delimit *mor-*

Table 3.5 Büdel's morphogenetic zones of the world

Zone	*Present climate*	*Past climate*	*Active processes (fossil ones in brackets)*	*Landforms*
(1) Of glaciers	Glacial	Glacial	Glaciation	Glacial
(2) Of pronounced valley formation	Polar, tundra	Glacial, polar, tundra	Frost, mechanical weathering, stream erosion (glaciation)	Box valleys, patterned ground etc.
(3) Of extra-tropical valley formation	Continental, cool temperate	Polar, tundra continental	Stream erosion (frost processes, glaciation)	Valleys
(4) Of subtropical pediment and valley formation	Subtropical (warm; wet or dry)	Continental, subtropical	Pediment[a] formation (stream erosion)	Planation surfaces and valleys
(5) Of tropical planation surface formation	Tropical (hot, wet or wet–dry)	Subtropical, tropical	Planation, chemical weathering	Planation surfaces and laterite[b]

[a] A pediment is a low-angle, concave rock surface at the base of a high-angle slope.
[b] Laterite is an iron-rich and/or aluminium-rich crust characteristic of a tropical region and caused by rock weathering.

Table 3.6 Wilson's morphogenetic systems of the world

System	*Dominant geomorphological processes*	*Landscape characteristics*
(1) Glacial	Glaciation, snow action (nivation), wind action	Glacial scour, alpine topography, moraines, kames, eskers etc.
(2) Periglacial	Frost action, solifluction	Patterned ground, outwash plains, solifluction, lobes etc.
(3) Arid	Desiccation, wind action, running water	Dunes, salt pans, deflation basins etc.
(4) Semi-arid (sub-arid)	Running water, rapid mass movements, mechanical weathering	Pediments, fans, badlands, angular slopes with coarse debris
(5) Humid temperate	Running water, chemical weathering, creep (and other mass movements)	Smooth slopes, soil-covered, ridges and valleys, extensive stream deposits
(6) Selva	Chemical weathering, mass movements, running water	Steep slopes, knife-edge ridges, deep soils (laterites included), coral reefs

phogenetic regions. The concept behind this is the theory that, under a certain set of climatic conditions, particular geomorphological processes will predominate; these will give to the landscape of a region characteristics that will set it apart from those of other areas developed under different climatic conditions. Because of the frequency and nature of climatic changes, it is necessary to consider the influence not only of present climates, but also of past climates.

One classification attempt has been that of Büdel, a German geomorphologist. His regionalisation scheme is illustrated in table 3.5. For comparison, table 3.6 sets out a more recent scheme, that of Wilson, an American geomorphologist. Whichever scheme is adopted, it is evident that climate does control the distribution of certain important phenomena and processes, including glaciation, permafrost, coral-reef growth, dune formation, frost weathering and wind erosion. These are the topics that we shall consider further in Part II.

Key Terms and Concepts

biomass
biomes
climatic climax community
climatic geomorphology
deforestation
faunal realms
fire
floral kingdoms
floral realms
introductions
lateritisation
leaching
maquis
morphogenetic regions
net primary productivity
podzolisation
Seventh Approximation
soil orders
succession

Points for Review

- Which are the world's most and least productive biomes?
- How does fire modify ecosystems?
- Describe and account for the distribution of the world's grasslands.
- How does grazing modify the ecosystem?
- Describe the location, causes and consequences of deforestation.
- What are plant invasions and why is their study important?
- What do you think are the main soil types in the world and what are their main characteristics?
- Give examples of beneficial and detrimental alteration of soil properties by humans.

FURTHER READING

Bailey, R. C. (1988) *Ecoregions* (New York: Springer). A relatively concise description of the world's main vegetation types.

Birkeland, P. W. (1999) *Soils and Geomorphology*, 3rd edn (New York: Oxford University Press). A classic look at the links between soils and landforms.

Bridges, E. M. (1997) *World Soils*, 3rd edn (Cambridge: Cambridge University Press). An excellent introduction to the world's soil types.

Brown, J. H. and Lomolino, M. V. (1998) *Biogeography*, 2nd edn (Sunderland, Mass.: Sinauer). A treasury of ideas on what true biogeography is.

Crutzen, P. J. and Goldammer, J. G. (1993) *Fire in the Environment* (Chichester: Wiley). A particularly useful study of fire's importance.

Drake, J. A. et al. (eds) (1989) *Biological Invasions: A Global Perspective* (Chichester: Wiley). An edited collection of essays that considers both animal and plant invaders.

Grainger, A. (1992) *Controlling Tropical Deforestation* (London: Earthscan). An up-to-date introduction with a global perspective.

Holzner, W., Werger, M. J. A., Werger, I. and Ikusima, I. (eds) (1983) *Man's Impact on Vegetation* (The Hague: Hunk). A wide-ranging edited work with examples from many parts of the world.

Johnson, D. L. and Lewis, L. A. (1995) *Land Degradation: Creation and Destruction* (Oxford: Blackwell). A broadly based study of intentional and unintentional causes of many aspects of land degradation.

McTainsh, G. and Boughton, W. C. (1993) *Land Degradation Processes in Australia* (Melbourne: Longman Cheshire). An Australian perspective on soil modification.

Meyer, W. B. and Turner, B. L. (eds) (1994) *Changes in Land Use and Land Cover: A Global Perspective* (Cambridge: Cambridge University Press). An excellent edited survey of human transformation of the biosphere.

Morgan, R. P. C. (1995) *Soil Erosion and Conservation*, 2nd edn (Harlow: Longman). A revised edition of a fundamental work.

Paton, T. R., Humphreys, G. S. and Mitchell, P. B. (1995) *Soils: A New Global View* (London: UCL Press). A somewhat controversial attempt to link soils to their situation in the landscape.

Simmons, I. G. (1979) *Biogeography: Natural and Cultural* (London: Arnold). A useful summary of natural and human impacts on ecosystems.

Whittaker, R. J. (1998) *Island Biogeography* (Oxford: Oxford University Press). A survey of the nature, evolution and conservation of island life.

Wilson, E. O. (1992) *The Diversity of Life* (Harvard, Mass.: Belknap Press). A beautifully written and highly readable discussion of biodiversity.

Part II

Major World Zones

4 Cold Environments

4.1 Polar Climates

Enduring cold typifies the climate of high latitudes. For six months at the North and South Poles the sun is out of sight entirely. For another six it is constantly above the horizon, but as it is never very high solar radiation is weak. The sun's rays are too oblique to be genuinely effective. Moreover, much of the solar energy is either reflected by the snow and ice or involved in melting snow and evaporating meltwater, so that neither the land surface nor the air adjacent to it becomes warm. At the Arctic and Antarctic Circles (latitude 66°) the daily period of sunlight varies from 24 hours at the time of the summer solstice to zero hours at the winter solstice.

The equatorward boundary of polar climates is generally taken as the line where the mean temperature of the warmest month is not more than 10 °C. This line broadly corresponds with the poleward limit of tree growth. In the Northern Hemisphere the threshold isotherm swings well poleward of the Arctic Circle over most of Asia and Alaska, coincides reasonably well with that parallel over lowland Europe, and lies to the south of it over much of Greenland and eastern North America. In the Southern Hemisphere the only extensive area of land with polar climates is the ice-covered Antarctic continent. Because its single land mass is centred near the South Pole and surrounded by extensive oceans with relatively uniform temperatures, the Antarctic possesses much greater uniformity and simplicity of climate than the Arctic.

Opposite Scientist and seals amid the snow and ice of the Antarctic.

Both areas are major sources of chill air; this tends at all times to move equatorward, converging in mid-latitudes with warmer air from the subtropical anticyclones and thereby playing a primary role in the generation of mid-latitude disturbances along the zones of contact between cold and warm air.

Polar climates can be subdivided into two types: *ice-cap climates* and *tundra climates*. In the case of the former, the average temperatures of all months are below freezing (0 °C) so that vegetation growth is impossible, and a permanent ice-and-snow cover prevails. At the South Pole the warmest month (December) has a mean temperature of –28 °C and the three coldest (July, August, September) a mean temperature of –59 °C. At Vostok, in the interior of Antarctica and at a height of 3500 m, the average August temperature is no less than –68 °C and the thermometer has been known to drop to nearly –90 °C. Information about precipitation is limited, but it is generally believed to be scanty in amount; for the low temperatures, the low specific humidity and the extreme stability of the air, as indicated by persistent strong temperature inversions, inhibit snowfall. At Eismitte in interior Greenland the annual precipitation is equivalent to only 80–100 mm of water, while the whole Antarctic continent receives on average less than 150 mm of water annually.

The tundra class of climate, which is confined almost exclusively to the Northern Hemisphere, is intermediate between the ice-cap climate of perpetual snow and ice, and the climate of the mid-latitudes. This belt lies between the 0 °C warmest-month isotherm on its polar side and the 10 °C warmest-month isotherm on its equatorward side.

Usually only a few months have average temperatures above freezing, and frosts may occur at any time. The continuous but weak summer sun may free the land of its snow cover for a few months and cause the upper layers of the soil to thaw, but the subsoil remains frozen from year to year as *permafrost* (section 4.9). Precipitation does occur, usually in summer, but is generally low, in the range 75–450 mm per annum. In some maritime regions, such as the Aleutian Islands between Alaska and Siberia, precipitation may be higher, attaining 1500 mm per year, and precipitation levels may also increase as one moves equatorwards.

The extreme cold of high-latitude and high-altitude areas creates problems for humans, but such problems are greatly accentuated when the extreme cold is accompanied by high wind velocities. This produces a phenomenon called *wind chill*. On a cold but calm day, warm human skin heats the air next to it, which in turn passes heat to the next layer and so on until the heat is carried away from the body. This is not a particularly efficient way to lose heat because the conductivity of air is low. Therefore calm air feels relatively warm. A wind, however, *pulls* heated air away from the surface. Cold air replaces skin-warmed air, and the faster the air moves the more heat is taken from the body. This makes it feel much colder than the thermometer reading, as is demonstrated in figure 4.1.

4.2 Vegetation and Wildlife

Because of the climatic definition we employed earlier, polar regions lie beyond the limit of real trees. Indeed, the area and its vegetation are described by the Lappish word *tundra*, meaning ‘barren land’. In reality, however, some plants have adapted to the extreme conditions of the tundra (table 4.1). They belong to five main groups: lichens, mosses, grasses and grass-like herbs, cushion plants and low shrubs. Nearly all of them are perennials, for climatic conditions provide too short a growing season for an annual’s life-cycle to be completed. They spring into life for just a few months in summer, remaining dormant for up to ten months per year. Most manage to achieve only a very low annual growth rate, partly because of low temperatures and partly because of the high winds. Wind chill is just

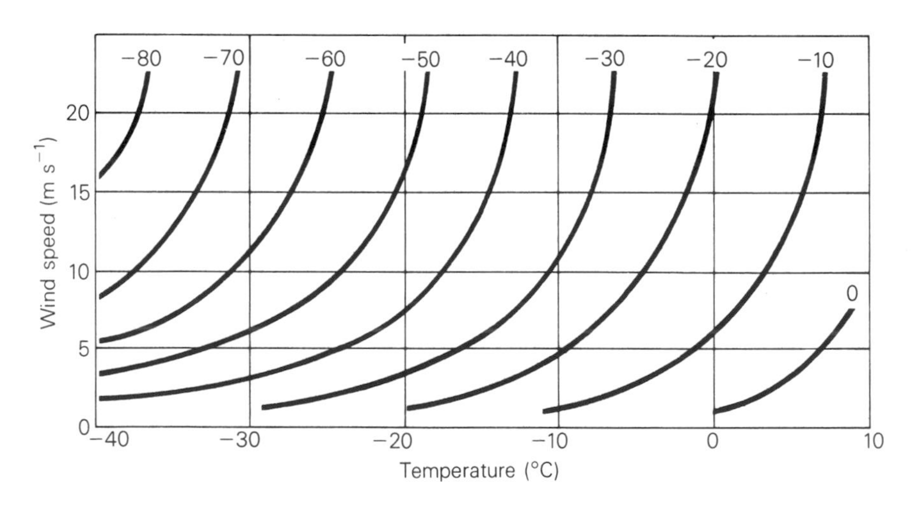

Figure 4.1 Wind chill. The numbers of the slanting lines indicate the effective temperature in °C of air at a given temperature and wind speed.

Table 4.1 Some adaptations of Arctic vegetation to polar climates

Adaptation	*Effect*
Prostrate shrub	Insulation between snow, warmer micro-climate, less wind
Cushion plants	Warm micro-climate and less wind
Annuals rare	Growing season too short for full cycle
Herbaceous perennials common	Large underground root structure, storage of food over winter
Reproduction often by rhizomes, bulbs or layering	Avoids reliance on completing flower-to-seed cycle
Pre-formed flower buds	Maximises time for seed production
Growth at low positive temperatures	Maximises length of growing season
Optimum photosynthesis rate at lower temperatures than most plants	Maximises length of growing season
Frost resistance	True of flowers, fruits and seeds
Longevity	Suitable for 'opportunist' lifestyle, e.g. lichens may live for several thousand years
Drought resistance	Suitable for rock surfaces or arid climates

Source: modified from D. E. Sugden (1982) *Arctic and Antarctic* (Oxford: Blackwell), table 5.1

as discouraging for plants as it is for man – indeed, the reduction of wind speed near the ground is one of the reasons for the widespread development of dwarf vegetation. Plant growth is further hampered by the fact that Arctic soils are also relatively infertile; they tend to lack nitrogen, one of the most important of plant nutrients. Tundra plants also have to face being covered with snow for extended periods; they have to endure very severe frosts; and they have to withstand the disturbance of the soil by permafrost.

Some scientists divide tundra plants into two main types. The first is the *fell field community*, where the vegetation is so scantily developed that the ground is never covered with a complete sward. Rock surfaces have lichens and some cushion plants such as moss campion (*Silene acaulis)* and various saxifrages. The second type consists of the *tundra communities*. In the sward shrub-heath tundra there are creeping willows, dwarf birch and berry-bearing members of the *Vaccinium* family, interspersed with Arctic heather and a carpet of mosses and lichens. In other areas there may be grassland tundra with a great development of sphagnum moss and cotton grass. The grass and sedges often occur as tussocks. Finally, there may be willow and alder thickets, especially where abundant summer moisture is available, as along stream banks and below some snow patches.

In Antarctica, because of the almost universal lack of vegetated ground, there are virtually no truly land animals or birds – the life is very largely marine, with some beasts, notably penguins and seals, venturing on to the land when conditions permit. However, in the northern polar regions there is a much richer land fauna, though it is still a relatively poor range of species compared with more favourable climatic zones. For example, it has been calculated that of 8600 bird species in the world only 70 breed in the Arctic, while of 3200 mammals in the world only 23 occur north of the tree-line. There are no reptiles. Some individual faunal species may, however, occur in large numbers, e.g. mosquitoes, lemmings and caribou.

Polar animals have adapted themselves to extreme cold and fluctuations in temperature (table 4.2). One way of doing this is to grow fur (as, for example, with the polar fox and the musk ox) or feathers. Sometimes the fur may have certain characteristics that add to its insulational properties. For instance, the hair of the caribou is thicker at the tip than at the base, thus entrapping air which assists in insulation; it also contains many air cells within the hair stem. Another form of adaptation is to hibernate during the coldest months, though where ground temperatures are very low even this is not possible. Most of the Arctic insects spend the winter in a larval or even egg form. The third form of adaptation is to migrate during the coldest months to areas where food may be available. Birds are especially prone to do this, though some of the large mammals may also migrate (generally in

Table 4.2 Faunal adaptations to polar climate

Stress	*Consequence*
Severe climate	Low number of species Low mean densities
Low temperature	High-quality fur insulation Increased metabolic rates
Snow	Life below snow patch for smaller animals Large herbivores favour soft/thin snow
Short summer	Birds migrate Breeding cycle compressed Large clutch/litter size

Source: after D. E. Sugden (1982) *Arctic and Antarctic* (Oxford: Blackwell), table 5.3

search of a winter food supply), notably the caribou.

4.3 Glacier Types

The earth is at present in an ice age, and around 10 per cent of its surface is covered by glaciers. Moreover, over the past few million years the extent of such ice has been frequently much more expanded than now, covering around one-third of the earth's surface (see section 2.9). Glaciers are therefore important both because they cover substantial parts of the earth's surface and because they have left their imprint on the land they formerly covered.

The biggest glaciers are called *ice sheets* (window 4.1). These have a flattened dome-like cross-section and are hundreds of kilometres in width. The most famous ice sheets, because of their size, are those of the two polar regions – Antarctica and Greenland. During the cold, glacial phases of the last ice age, which took place during the Pleistocene, there were also two other enormous ice sheets, one over Scandinavia and Britain, and another, called the Laurentide Sheet, over much of North America. Those dome-shaped masses with a smaller area, less than about 50 000 km^2, are called *ice caps*. We can summarise these *non-valley glaciers* as follows:

Ice sheet: more than 50 000 km^2, with a flattened dome, which buries underlying relief.

Window 4.1 The polar ice sheets and ice shelves

The polar ice sheets of Greenland and Antarctica are enormous. Excluding water in the ground, glacier ice represents 80 per cent of the world's freshwater, of which 99 per cent is locked up in these two ice sheets. The Antarctic ice sheet covers a continent that is a third bigger than Europe or Canada and twice as big as Australia. It attains a thickness that can be greater than 4000 m, thereby inundating entire mountain ranges. The Greenland ice sheet only contains 8 per cent of the world's freshwater ice (Antarctica has 91 per cent), but nevertheless covers an area ten times that of the British Isles. The Greenland ice fills a huge basin that is rimmed by ranges of mountains, and has depressed the earth's crust beneath.

The Antarctic ice sheet is bounded over almost half of its extent by ice shelves. These are floating ice sheets nourished by the seaward extensions of the land-based glaciers or ice streams and by the accumulation of snow on their upper surfaces. Ice-shelf thicknesses vary, and the seaward edge may be in the form of an ice cliff up to 50 m above sea level with 100–600 m below. At its landward edge the Ross ice shelf is 1000 m thick. It covers an area greater than that of California.

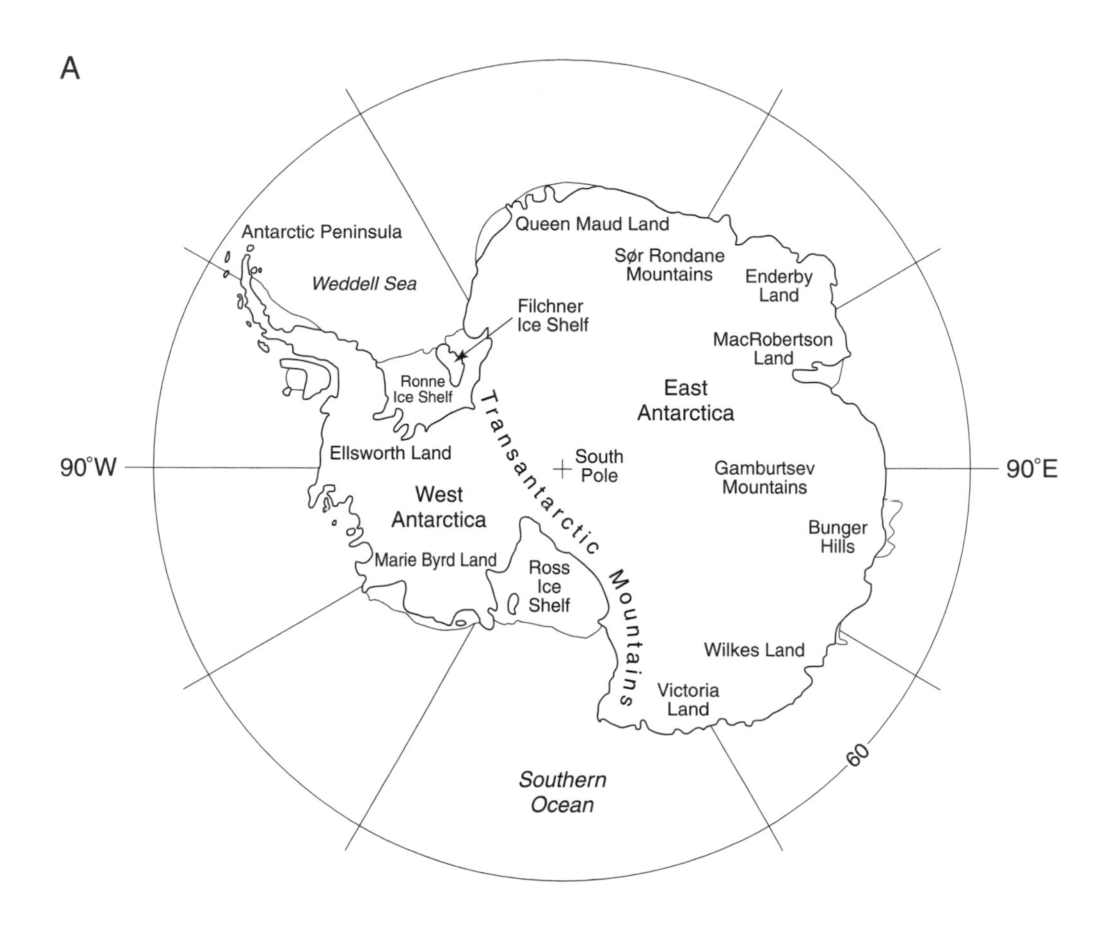

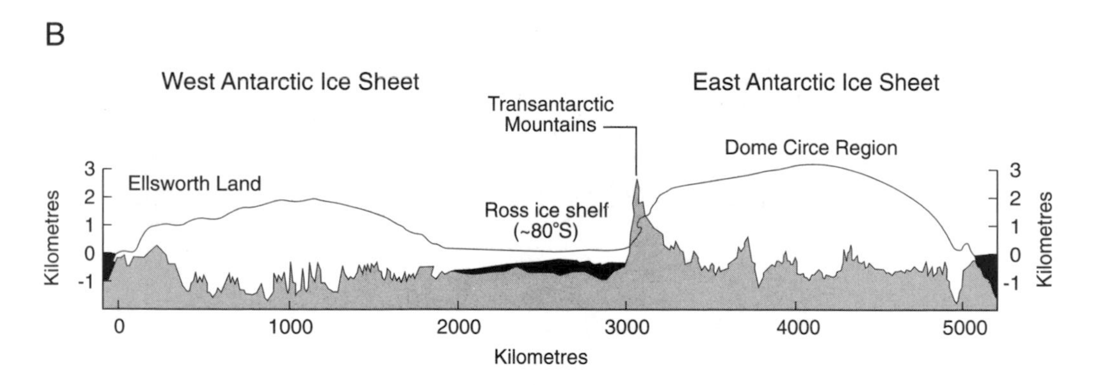

Above and overleaf The Antarctic ice sheet and shelves. (A) Location map of Antarctica. (B) Cross-section through the East and West Antarctic ice sheets, showing the irregular nature of the bedrock surface, ice thickness, and the floating ice shelves. (C) Subglacial relief and sea level. The white areas are below sea level. (D) Surface elevations on the ice sheet in metres.

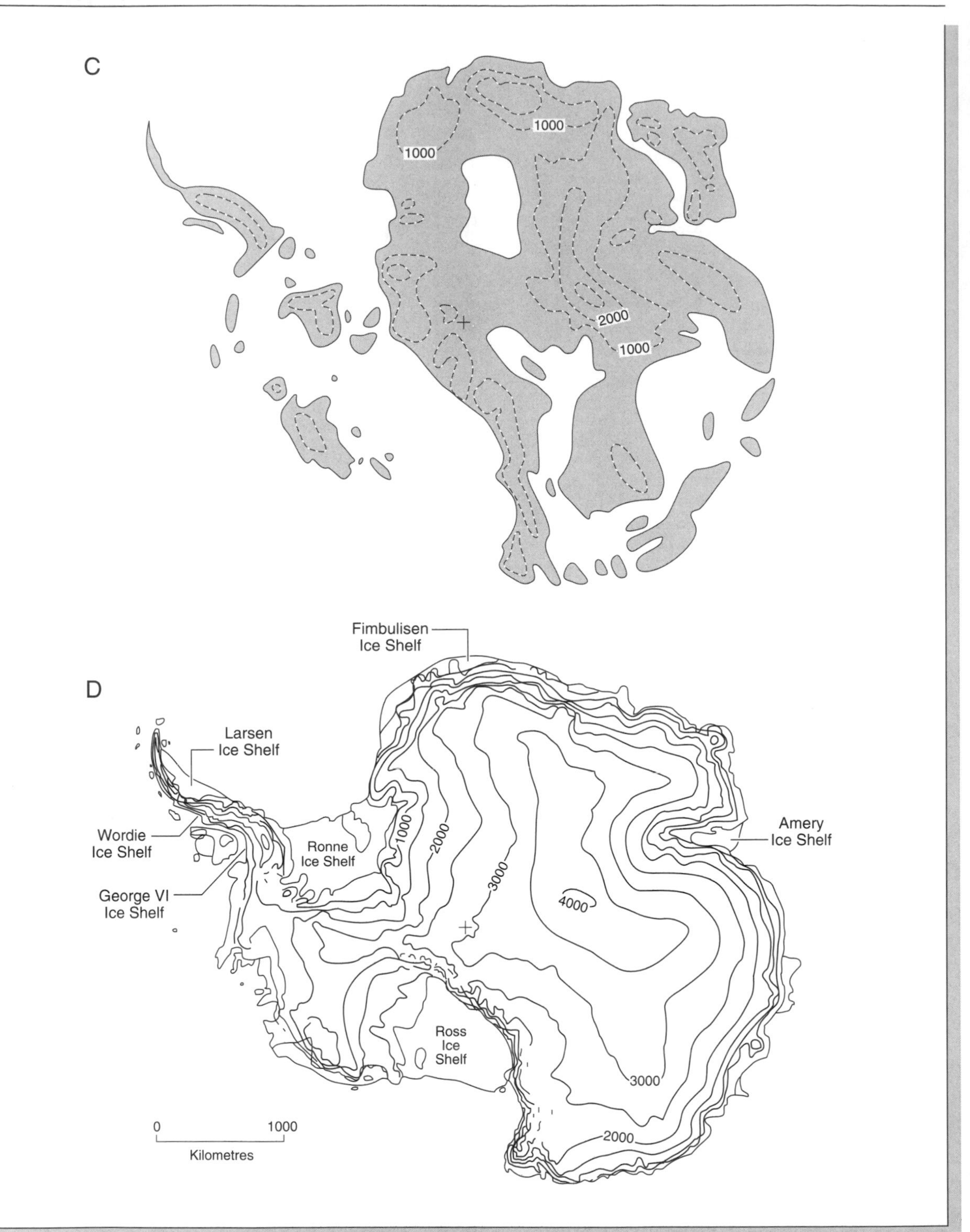
C
1000
1000
2000
1000
D
Fimbulisen Ice Shelf
Larsen Ice Shelf
Wordie Ice Shelf
George VI Ice Shelf
Ronne Ice Shelf
1000
2000
3000
4000
Amery Ice Shelf
Ross Ice Shelf
3000
2000
0
1000
Kilometres

Ice cap: a small ice sheet with an area less than 50 000 km^2, but which still buries the landscape.
Ice dome: the central part of an ice cap or ice sheet.
Outlet glacier or ice: a stream of ice that drains part of an ice sheet or ice cap, and which often passes through confining mountains.
Ice shelf: a thick, floating ice sheet which is attached to a coast.
Ice field: a relatively flat and extensive mass of ice.

The other main class of glacier is the *valley glacier* (plate 4.1). These 'rivers of ice' occupy basins or valleys in upland areas, and can be further subdivided into miscellaneous types as follows:

Valley glacier: a body of ice that moves down a valley under the influence of gravity and is bounded by cliffs.
Cirque glacier: a small ice body that occupies an armchair-shaped hollow in mountains which has been cut into bedrock; the hollows are sometimes also called cwms or corries.
Niche glacier: a small upland ice body resting upon a sloping rock face or in a shallow hollow that the glacier itself has modified only slightly.
Diffluent glacier: a valley glacier that diverges from a trunk glacier and crosses a drainage divide through a diffluence col.
Piedmont glacier: a glacier that leaves its confining rock walls and spreads out to form an expanded foot glacier; formed on a lowland or at a mountain foot.

Large valley glaciers in the Karakoram Mountains in the western Himalayas may be 60 km or more long, while small niche and cirque glaciers may extend only a few hundreds of metres.

4.4 The Formation of Glacier Ice

When snow crystals fall they have an open, feather-like appearance, and therefore a low density. This is why it is generally reckoned that 100 mm of snow is equivalent to only 10 mm of rainfall. However, if the snow crystals are compacted by the weight of overlying snow, or if they are partially melted, they are converted to a mass of partially consolidated ice crystals, with interconnected air spaces between them. Such material is called *firn* or *névé*, and as it develops further it becomes denser and denser until most of the air spaces are eliminated and pure ice develops. In the lower-latitude glaciers this process of transformation may take a few years, whereas in colder areas, such as central Antarctica, the transition may take a few thousands of years. Most glaciers are composed of the ice produced from snow that has been modified in this way, but on some glaciers refrozen meltwater can make up much of the mass. Either way, if the climate and relief of an area are favourable, sufficient ice may accumulate so that, under the influence of gravity, the mass begins to move and thereby becomes a glacier.

The existence and development of a glacier de-

Plate 4.1 The Mer de Glace glacier in the French Alps. The stripe down the middle is a medial moraine composed of till.

pend on the simple fact that in its accumulation area snowfall is greater than thaw. In cold regions a winter snowfall of 1 m may be enough to keep a glacier going, whereas in a warmer area 8–10 m would be needed. The altitude below which glaciers cannot form (*the glaciation level*) depends on the balance between snowfall and thaw and also on temperature. Temperatures in turn largely depend on latitude and altitude above sea level, precipitation, the nature of winds and the distance from the sea. Thus the glaciation level is low in polar regions and high in the tropics, low in coastal areas, and high in continental interiors. For instance, in the Antarctic the glaciation level is at sea level, whereas in the European Alps it is at 2600 m in the west and as much as 3200 m in the east (where it is drier and the accumulation of snow on the glaciers is smaller).

4.5 How Glaciers Move

That glaciers flow is made evident by a consideration of their velocity characteristics. A glacier's surface velocity is highest near the centre line and diminishes towards the sides, where friction against the rock will bring it close to zero. Velocity also tends to decrease with depth, especially in the lower parts near the bed (figure 4.2).

Although at first it is difficult to understand how a seemingly rigid mass of ice can propel itself, glaciers do move, and while they do not progress with anything like the same speed as a river, their movement is none the less easily measured, and is sometimes appreciable. Some glaciers occasionally surge (plate 4.2), and may for short periods reach speeds of 5 m per *hour* or more, but most move at only a fraction of that speed, say 50 m per year.

The movement of the glacier takes place in three main ways: by *sliding* over bedrock; by internal deformation (*creep*) of the ice; and by alternate *compression and extension* of the ice mass in response to changes in the bedrock surface below the ice. The first of these processes – sliding over the base – is relatively easy to envisage. Ice normally forms from water at a temperature of 0 °C, but the temperature at which water freezes is reduced under pressure, and as a glacier moves it will exert pressure and therefore some melting may take place

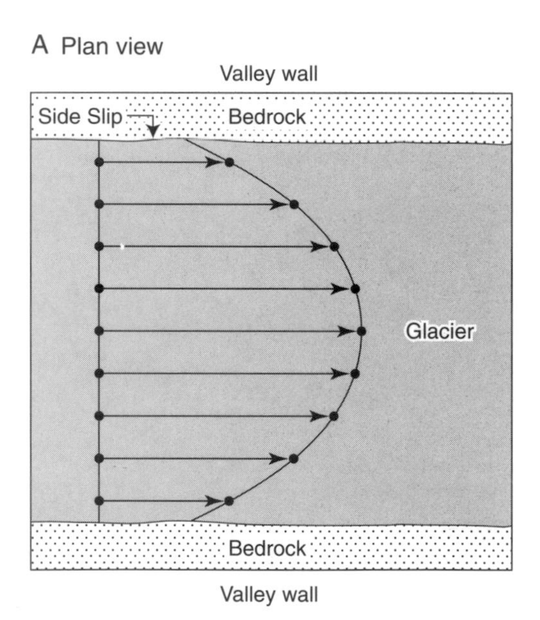

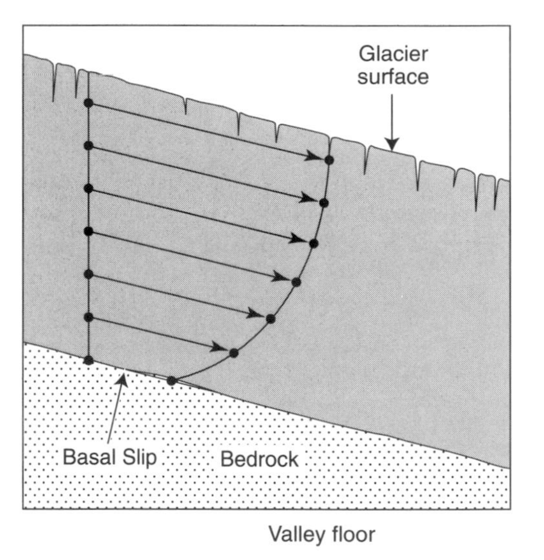

Figure 4.2 The speed of movement of glaciers depends on internal deformation and by slipping and sliding. The arrows in these diagrams show the profiles of movement, with the highest velocities being in the centre (A) and the upper parts (B) of the glacier. Note also the amount of basal slip and side slip.

Plate 4.2 In some mountainous areas, glaciers are prone to surge at exceptional speeds. Glaciers that have this characteristic, such as the Sustina Glacier, Alaska, have highly deformed moraines.

at its base. For this and other reasons a thin film of water may exist between the glacier and the bedrock. This film reduces friction and hence allows the glacier to slide. Plainly, such a film of water is more likely to occur in glaciers whose temperature is close to their melting point (the so-called 'warm' glaciers) than in glaciers of extremely cold areas.

Basal slip is an important cause of movement. Expressed as a percentage of the total movement, the amount of slip in temperate glaciers has been found to vary from 10 to 75 per cent with a mean value of around 50 per cent. The presence of a water film or unfrozen deformable bed sediment at the base of the glacier is an important requisite, and at least a portion of the movement probably takes place by the refreezing (*regelation*) of water around small irregularities in the bedrock floor. This occurs because there is increased pressure on the upstream side of any protuberance, which leads to local melting along that area. The water then flows to the downstream side, where it refreezes under conditions of diminished pressure (figure 4.3).

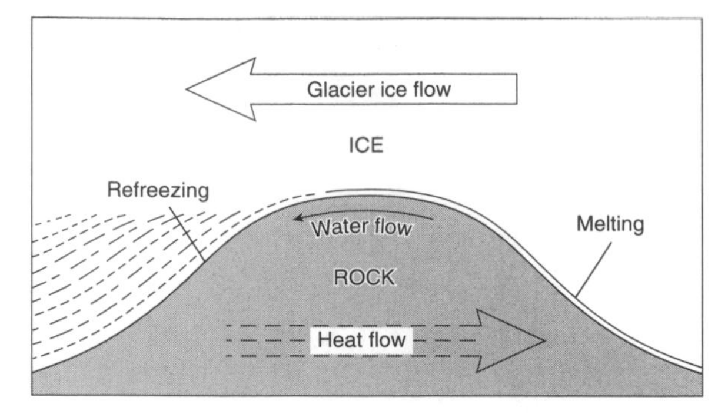

Figure 4.3 The regelation sliding mechanism. Regelation sliding takes place because high pressures occur on the upstream side of a rock obstacle or protruberance. These cause a lowering of the pressure melting point and the melting of ice immediately up-glacier of the obstacle. The resulting meltwater migrates to the down-glacier side of the obstacle. Because pressures are lower there, the water refreezes because the pressure melting point is higher. Thus the ice bypasses an obstacle by temporarily turning to water and back again.

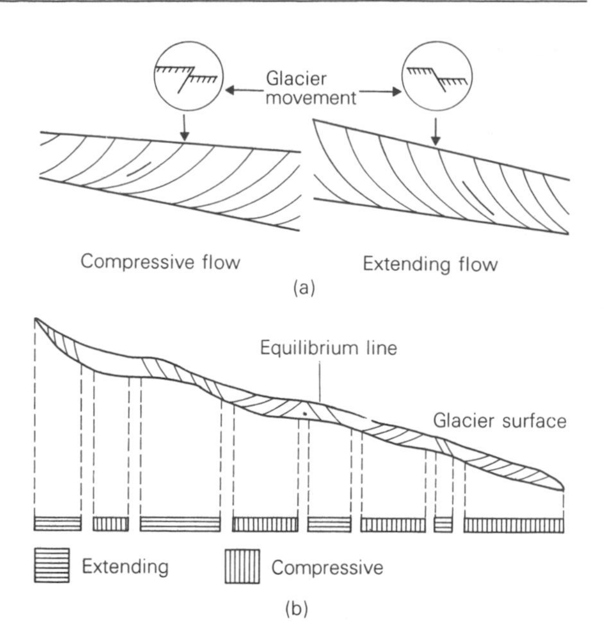

Figure 4.4 Extending and compressive flow of a glacier: (a) the two types of flow and their associated slip lines; (b) the distribution of the two types of flow down the long profile of an idealised glacier. Compressive flow tends to accentuate existing concavities.

Ice is brittle and plastic at the same time. Hit it with a hammer and it will crack like a piece of brittle glass; leave a horizontal bar of ice supported at only one end and it will, given time, deform plastically under its own weight. In the case of glacier ice the deformation, or creep, will take place under the action of gravity, depending on the stresses involved – the weight of the overlying ice column and the slope of the upper ice surface. Thus the rate of plastic deformation will increase with increasing surface slope. Temperature also plays a role, for cold ice moves less readily than ice that is close to melting point, rather in the same way as oil is more viscous at low temperatures. Other things being equal, therefore, a thick steep glacier in the temperate zone will flow faster than a thin, flat ice cap in a polar area.

Under some conditions compression and extension take place because the ice cannot deform sufficiently quickly to the stress within the ice. As a result ice fractures, and movement takes place along a plane. *Tensional fractures,* where the ice on either side of the fracture is separated, are exemplified by crevasses in the upper layers of the glacier. *Shear fractures* occur where thrusting takes place along a slip plane or fault (see figure 4.4 and section 4.6).

Alpine glaciers have generally been recorded as moving at velocities between 20 and 200 m per year, but may accelerate to rates in excess of 1000 m per year down the steeper slopes. Some of the fastest rates are found on outlet glaciers of the polar ice sheets, where velocities as high as 7000 m per year have been recorded. Some glaciers are prone to phases of accelerated flow that are called surges (window 4.2).

4.6 Glacial Erosion

The Victorian writer John Ruskin once remarked that a glacier could no more erode its valley than custard could erode a custard bowl, suggesting that glaciers tend to lie rather impotently in their valleys, that the valleys were there before the glaciers, and that ice is softer than rock just as custard is softer than the sides of a dish. Glaciers do indeed often lie in pre-existing valleys, and ice is indeed softer than rock, but glaciers are by no means impotent agents of erosion. Recent studies of the quan-

Window 4.2 *Surging glaciers*

Some glaciers are affected by periodic surges, in which ice is transmitted down-glacier at speeds which are far greater than the norm. Waves of ice may move down surging glaciers at rates of as much as 350 m per day. Such activity may have a periodicity of between 15 and 100 years. Surges have a variety of consequences:

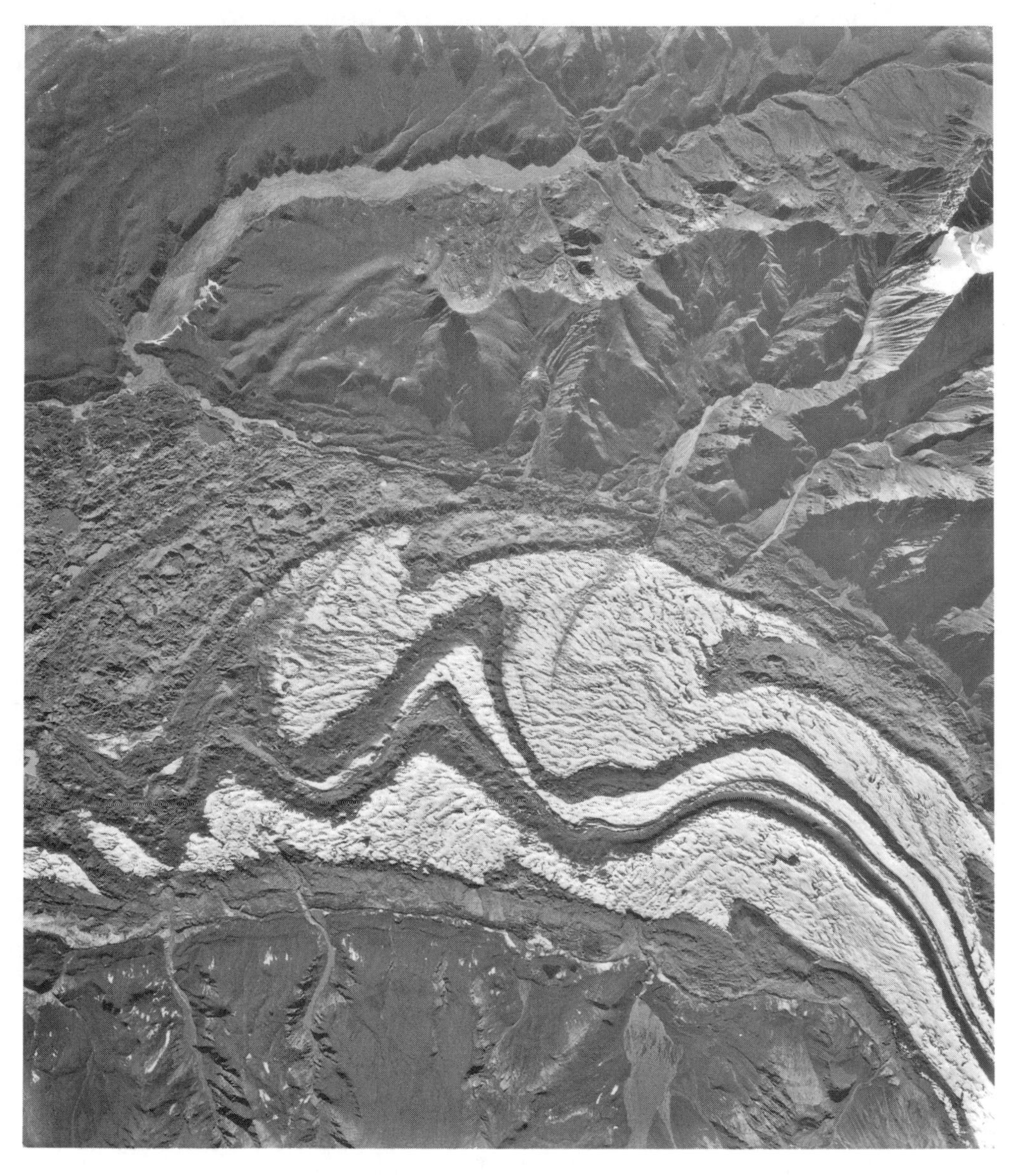

Surging glaciers, including this example from Canada, can be identified on aerial photos by the looped and deformed character of their medial moraines.

1 The ice front may be displaced forward rapidly. Horizontal displacements of up to 11 km have been recorded.
2 A surge reduces the surface slope in the upper parts of the glacier and steepens it in the lower parts:
   - as a result, former lateral moraines may be left high and dry.
   - tributary glaciers may be shorn off and left hanging.
3 Medial and terminal moraines may suffer severe distortion.
4 Icebergs may be shed into bays and fiords in large quantities, threatening shipping.

Why do some glaciers surge? The answer may lie in the glacier's 'plumbing system'. This was explained thus by Richards and Sharp (1988: 4):

> Over a period of decades prior to a surge, a glacier is in a state of gradual 'build up'. This build-up involves the thickening of an upper 'reservoir' area and the thinning of a lower 'receiving' area, with the net effect that the long profile of the glacier becomes progressively steeper over time. In the reservoir area this means that the stresses acting on the ice near the glacier bed become greater year by year, and that the basal ice deforms more and more rapidly. Not only does this result in the gradual speed up of the glacier, but it also means that it becomes harder and harder for the melt-waters to prevent their drainage channels cut into the basal ice from being closed down by ice deformation. Eventually the point is reached where the water pressure required to keep the channels open is so high that the water is forced to flow out of the channels and into the film and cavity system spread across the whole of the glacier bed. Once this happens, the drainage of waters through the glacier becomes much less efficient and a sub-glacial reservoir develops. This acts like a cushion which keeps the glacier separated from the bed and allows very rapid sliding to occur.

Eventually this rapid sliding causes such high stresses that the surge is propagated down-glacier, causing rapid thickening of the glacier and a sharp increase in velocity. Should this activity reach the glacier's snout, a rapid advance may occur, which ploughs into, deforms and overrides moraine and outwash materials.

However, as the glacier advances and the ice thickness in the reservoir area is reduced, the stresses which both drive the flow of the glacier and close down the internal plumbing system of drainage channels are gradually reduced. The drainage channels begin to open again, the subglacial reservoir becomes drained, and the all-important water cushion that permitted rapid sliding is removed. Thus the glacier slows down, permitting the whole cyclical process to start again.

Richards and Sharp (1988) also recognise an alternative hypothesis to explain surging in those glaciers that rest on beds of soft sediment. In this case, meltwater drainage normally occurs via a network of pipes incised into the sediments. As the glacier builds up prior to surge, however, the stresses acting on the sediments increase until they start to deform. Once this happens, the pipes are destroyed and the water is forced to flow through the pore spaces between sediment particles. Drainage via this route is much less efficient than via the pipes and water gets backed up beneath the glacier. As a result, the water pressure rises, reducing the strength of the sediments and increasing the extent of the bed deformation, which is responsible for the high surge velocities. In this case, the surge stops when the spreading of the glacier reduces stresses to levels which are insufficient to deform the sediments, allowing the pipes to become re-established as the major drainage route through the glacier.

Thus some glaciers are potentially unstable environmental features, and engineering structures such as roads and hydroelectric schemes need to be designed with this in mind.

Source

Keith Richards and Martin Sharp (1988) *The Geography Review*, 1(5): 2–10.

tities of material being carried away by glacial meltwater streams indicate that erosion in a glaciated catchment may frequently be equivalent to a ground surface lowering of 2000–3000 mm per thousand years, perhaps ten times the norm for ordinary fluvial catchments without glaciers. How then is this high rate of denudation achieved?

First of all, glaciers can in some respects be likened to a conveyor belt. If a rockfall puts a vast amount of coarse debris on to a glacier surface, for example, or if frost-shattering sends down a mass of angular rock fragments on to the glacier surface, it can then be transported, almost whatever its size, down valley. The range of debris size that can normally be carried by a river is much more limited. Second, beneath glaciers there is often a very considerable flow of meltwater. This may flow under pressure through tunnels in the ice at great speed, and may be charged with coarse debris from the bed of the glacier. Such subglacial streams are highly effective at wearing down the bedrock beneath a glacier. Third, although glacier ice itself might not cause marked erosion of a rock surface by *abrasion,* when it carries coarse debris at its base some abrasion can occur. This grinding process has been observed directly by digging tunnels into glaciers, but there is other evidence for it: rock beneath glaciers may be *striated* or scratched (just as wood may be scratched by sandpaper), and much of the debris in glaciers is ground down to a fine mixture of silt and clay called *rock flour.*

Glaciers also cause erosion by means of *plucking.* If the bedrock beneath the glacier has been weathered in preglacial times, or if the rock is full of joints, the glacier can detach large particles of rock. As this process goes on, moreover, some of the underlying joints in the rock may open up still more as the overburden of dense rock above them is removed by the glacier. This is a process called *pressure release.*

In order to account for the irregular erosion at the base of a glacier by which deep basins and intervening bars may be formed, variation in the nature of glacier flow down valley needs to be considered. Particularly helpful in this context is *the theory of extending and compressing flow* (section 4.5). Figure 4.4 shows the theoretical pattern of slip planes or potential shear surfaces associated with the two types of flow. Where the flow of the glacier is accelerating (extending), a series of planes of weakness is likely to develop within the ice, curving down tangentially to the glacier bed. In contrast, in the *ablation zone,* where melting takes place, and those parts of the glacier where the flow is decelerating (compressing), the planes of weakness will tend to develop curving upward tangentially from the bed in a down-glacier direction. This type of plane becomes pronounced near the snout of the glacier and also at the base of steep slopes. The significance of this for glacial longitudinal profiles, or *long profiles,* is that in the zones of compression the ice may be expected to carry up debris from the glacier floor into the ice – thereby causing erosion. In this way basins at the base of a steep slope or ice fall, once initiated, may become self-perpetuating. Irregularities in the bed of the glacier influence the glacier flow, and in turn the different types of flow can influence the bed. Once irregularities in the glacier bed have appeared, therefore, they will tend to become accentuated. It is this tendency of *positive feedback* that probably accounts for so many of the differences between the long profiles of glacial valleys and those of ordinary river valleys, the former being irregular and the latter being more smoothly concave downstream.

As debris-laden ice grinds and plucks away the surface over which it moves, characteristic landforms are produced which give a distinctive character to glacial landscapes (figure 4.5). Of the features resulting from glacial quarrying, one of the most impressive is the *cirque.* This is a horseshoe-shaped, steep-walled, glaciated valley head. It is called a *cwm* in Wales, and a *corrie* in Scotland. In Europe and North America, where cirques occur they are nearly always orientated between north-west and south-east, and there are two reasons for this. First, north-facing slopes receive less insolation from the sun, and this helps to preserve small glaciers. Second, the main snow-bearing winds are from the west, and eddies would ensure that snow banks were preserved on the lee slopes, i.e. those facing east.

The origin of cirques is still uncertain but they may develop from an initial hollow in which a snow bank accumulated. The presence of the snow bank would increase the amount of diurnal and seasonal

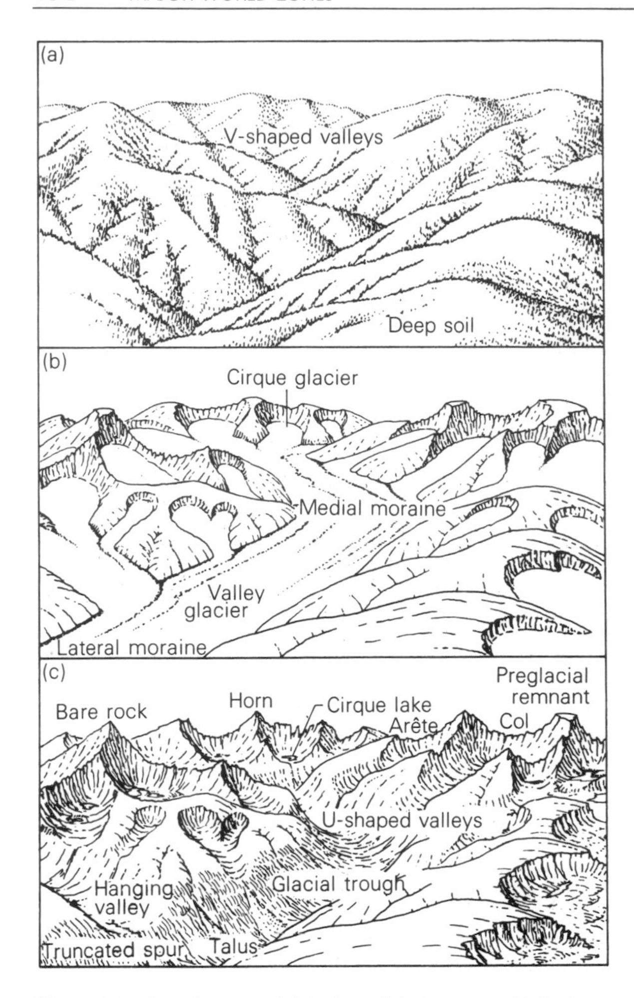

Figure 4.5 Landscapes (a) before, (b) during and (c) after glaciation.

frost weathering caused by meltwater in the hollow, so that it would gradually be enlarged until it was big enough to hold a small glacier. This glacier would then start to cause plucking and quarrying at its head, probably as a result of the downward percolation of water and the frost-shattering of the rocks in the headwall. Another process that seems important is the rotational movement of the ice, which excavates closed basins in the floors of cirques. Should the glacier disappear such basins become the sites of small lakes (figure 4.6), which add immeasurably to the beauty of the landscape.

As cirques evolve they eat back into the hill mass in which they have developed. When several cirques lie close to one another, the divide separating them may become progressively narrowed until it is reduced to a thin, precipitous ridge called an *arête* (figure 4.7). Should the glaciers continue to whittle away at the mountain from all sides, the result is the formation of a pyramidal *horn*, of which the Matterhorn in the European Alps is the most famous example.

Some of the most spectacular erosional effects of glaciation observable today are those of valley glaciation (see colour plate 3). The lower ends of spurs and ridges are blunted or truncated; the valleys assume a U-shaped configuration (plate 4.3); they become more linear; and hollows or troughs are excavated in their floors. Many high-latitude coasts, such as those of Norway (plate 4.4) and western Scotland, are flanked by narrow troughs, called *fiords*, which differ from land-based glacial valleys in that they are submerged by the sea. Some

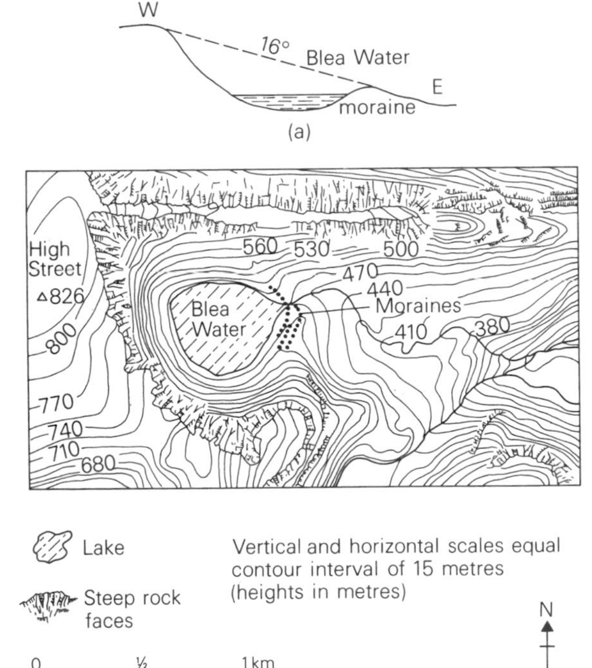

Figure 4.6 Blea Water in the English Lake District – a typical cirque basin. Note the approximately north-easterly orientation of the basin.

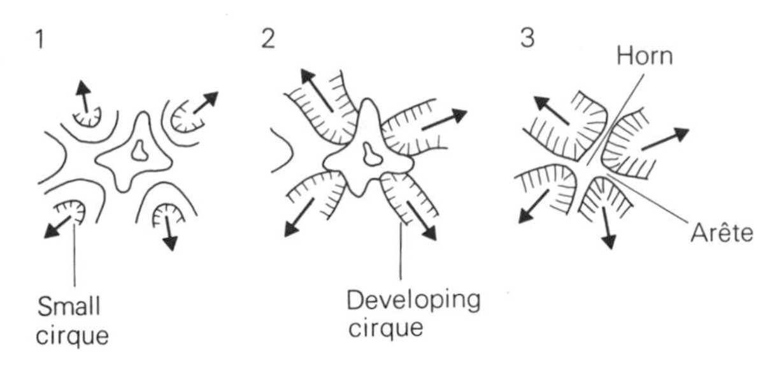

Figure 4.7 Stages in the development of horns and arêtes by backward extension of cirques.

of them are extraordinarily deep, and bear witness to the power of glaciers to excavate to great depths – the maximum known depths of Greenland, Norwegian and Chilean fiords are of the order of 1300–1400 m, and in Antarctica fiords almost 1000 m deeper are known. These great depths are usually separated from the open oceans by a *sill* of solid rock, over which the depth is only around 200 m or so. *Fjards* are related to fiords. They are coastal inlets associated with the glaciation of a lowland coast, and therefore lacking the steep walls characteristic of glacial troughs. A good example of a fjard coast is that of the state of Maine in the United States.

A further erosional effect of valley glaciers is the breaching of watersheds, for when ice cannot get away down a valley fast enough – perhaps because its valley is blocked lower down by other ice or because there is a constriction – it will overflow at the lowest available point, a process known as *glacial diffluence*. The result of this erosion is the creation of a *col*, or a gap in the watershed.

A striking example of present-day diffluence is provided by the Rimu Glacier in the Karakoram Mountains of the Indian sub-continent (figure 4.8). The Rimu consists of three branches fed by numerous minor glaciers draining from the surrounding

Plate 4.3 A classic U-shaped valley with truncated spurs developed in the granite mountains of north-central Portugal.

Plate 4.4 The Trollfjord in the Lofoten Islands in Norway illustrates the great amount of incision that can be created by glacial erosion in areas with heavy precipitation. Some Norwegian fiords are over 1000 m deep.

peaks. Part flows into the Yarkand, which flows into central Asia, and part flows into the Shayok, which eventually flows into the Indus and thence into the Indian Ocean.

The long profile of a glacial trough is characteristically irregular: steps, basins and reversed gradients occur. The factors determining where pronounced deepening of the valley floor takes place have been the subject of many arguments. For example, it has been proposed that where two glaciers join there is a great increase in erosive capacity. Elsewhere it can be shown that the glaciers have plucked preferentially at areas of non-resistant and closely jointed rock. Sometimes the basins may be areas where the rock was deeply weathered by chemical action or by freeze–thaw in preglacial times. Alternatively, if the valley is for some reason laterally constricted, say by a zone of resistant rock, the glacier enlarges its cross-sectional area by eroding its floor more deeply.

Tributary valleys to the main glacial trough have their lower ends cut clean away as the spurs between them are ground back and truncated. Furthermore, the floor of a trunk glacier is deepened more effectively than those of feeders from the side or at the head, so that after a period of prolonged glaciation such valleys are left hanging high above the main trough. Such *hanging valleys* have often become the sites of waterfalls.

The development of an ice sheet tends to scour the landscape. In Canada there are vast expanses of empty territory where the Pleistocene glaciers scoured and cleaned the land surface, removing almost all the soil and superficial deposits and exposing the joint and fracture patterns of the ancient crystalline rocks beneath. Streamlined and moulded rock ridges develop (plate 4.5), including *roches moutonnées* (figure 4.9a). In parts of Scandinavia and New England these may be several kilometres long and have steepened faces of more than 100 m in height. They are interspersed with scoured hollows which may be occupied by small lakes when the ice sheet retreats. In Scotland relief that is dominated by this mixture of rock ridges and small basins is called *knock and lochan* topography.

In areas made of tough limestone, such as the

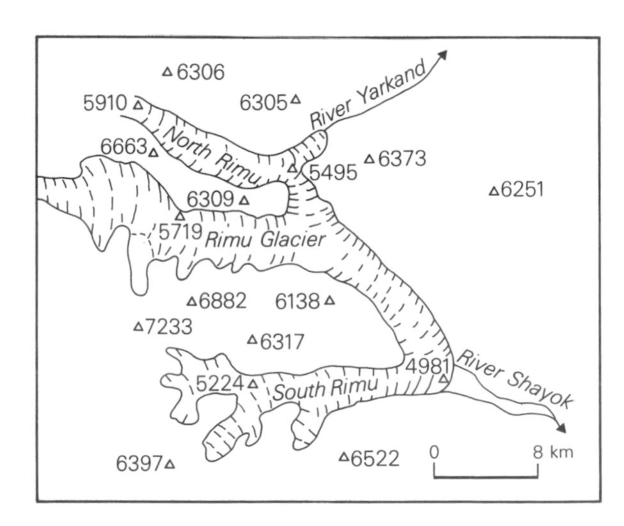

Figure 4.8 An example of glacial diffluence: the Rimu Glacier in the Karakoram Mountains (heights in metres).

Plate 4.5 A glacier moving across metamorphic rocks in the Hunza Valley of northern Pakistan created this example of a *roche moutonnée*.

Carboniferous limestone landscapes of northern England and western Ireland, glacial scour has planed off the limestone, exposing bedding plane surfaces, and stripping off any overlying, superficial materials. The result is the development of bare rock surfaces, called *limestone pavements*, some of which may show distinct signs of glacial moulding. The micro-forms developed on their surfaces are the result of post-glacial solutional modification.

Another feature, known as *crag-and-tail* (figure 4.9b), is the result of the presence of an obstructive mass of rock, the crag, that lies in the path of oncoming ice. This mass protects softer rocks in its lee from the effects of glacial erosion, for the ice appears to have moved over and around the 'crag' leaving a gently sloping 'tail' in its lee. The best known example of this phenomenon is in the city of Edinburgh, where Edinburgh Castle (plate 4.6) rests on a resistant igneous plug (the crag) with its tail of limestone and sandstone sloping eastward along the line of the 'Royal Mile'.

Glacial modification on a major scale can also be achieved in low-lying areas of *scarp and dip* topography. The characteristic form of scarp to the south of the generally accepted limits of glacial advance can be seen in the Chilterns of southern England, where there is a prominent chalk escarpment with a firm crestline and a considerable degree of

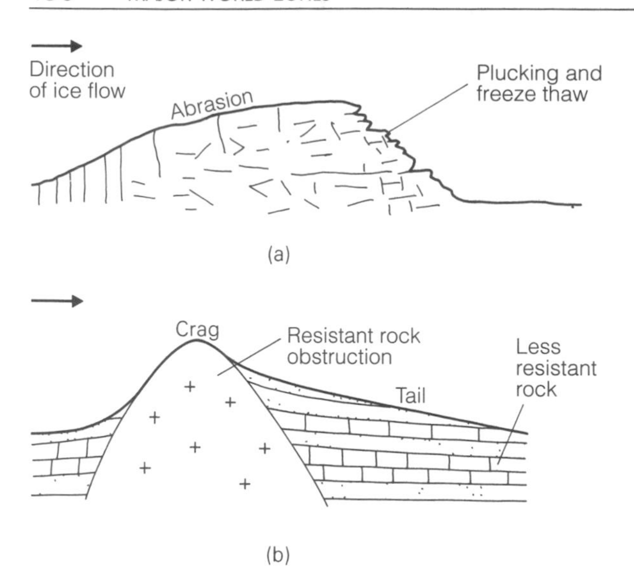

Figure 4.9 Two glacial landforms: (a) roche moutonnée; (b) crag-and-tail.

Plate 4.6 The imposing bulk of Edinburgh Castle rests on a resistant igneous crag that was shaped by glacial erosion.

relief (figure 4.10). To the north of this glaciation line the form of the scarp is radically different: the typical *cuesta* (escarpment) form is replaced by a low and relatively even-crested plateau. Thus, from the Goring Gap to Ivinghoe Beacon the main scarp stands within 1.5 km of the basal outcrop of the chalk; but further north the main escarpment is found progressively farther from the basal outcrop, and in the Breckland of East Anglia it may be removed by as much as 12–14 km. It is also greatly flattened. In effect, the ice has lowered and displaced the line of the escarpment.

Glacial erosion does appear to be highly selective, which leads us to hypothesise that within an ice sheet, which may as a whole be slow-moving, there are local streams of high velocity. In areas of high relief (as, for example, on the west coast of Scotland) one would have suitable conditions for high velocities to occur, and where the pre-existing relief was favourable and rocks were erodible, there would be fast-moving ice streams beneath the glacier that would be capable of great erosive activity.

Glaciers produce a great deal of *meltwater*, some of which flows on the glacier, some within, and some over the bedrock floor. It often flows very swiftly; moreover, the meltwater that moves in tunnels at the base of a glacier may be subject to very great pressure, and under certain conditions it can flow uphill. Some streams may also contain large quantities of debris that may give them further erosive power. In Denmark and Germany great subglacial meltwater rivers flowing southwards towards the wasting margin of the last Scandinavian ice sheet are thought to have cut the so-called *tunnel valleys*, some of which are over 70 km long and in places over 100 m deep. Some such channels have humped sections in their long profiles, indicating that the subglacial streams accomplished much erosion while flowing uphill under great pressure.

Lakes are a widespread feature of many areas that have been glaciated, and they result from a variety of erosional and depositional causes. Some are caused by glacial overdeepening, some are the result of blockage of drainage by moraines, and others may be the result of uneven deposition and melting of till sheets. In some cases a combination of erosion and deposition may explain their development; this is the case with both the lakes at the foot of the Italian Alps (figure 4.11) and the Great Lakes in the United States. During the Ice Age some glacial lakes were huge. The North American ice sheet, around 25 000 years ago, dammed many of the northward-draining rivers of central Canada and north-central United States to form Lake Agassiz,

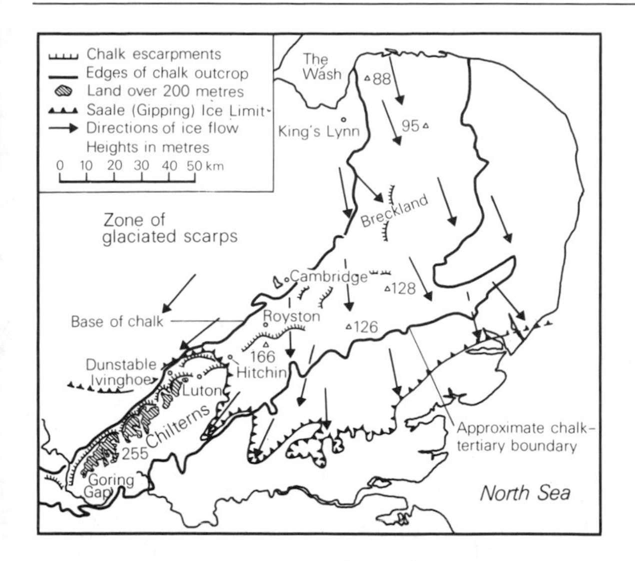

Figure 4.10 During a major Pleistocene ice advance the chalk escarpment of eastern England was greatly modified by erosion. Compare the position of the escarpment, with regard to the position of the base of the chalk, in the zone of glaciated scarpland with that of the scarp to the south of the glacial line.

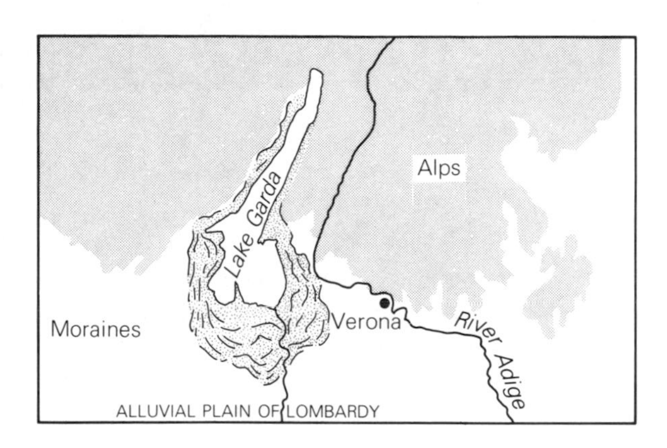

Figure 4.11 Lake Garda at the foot of the Italian Alps, with its bordering lateral and terminal moraines. It owes about 150 m of its depth to the thickness of the moranic barrier, but below that level to a maximum depth of 350 m it occupies an ice-excavated rock basin.

which, at various times, inundated an area of 950 000 km², though the maximum extent at any one time was considerably less.

4.7 Glacial Deposits

Because glaciers are so effective at erosion and transport, they are also associated with large quantities of debris. According to its location with respect to the glacier, such debris transported by an ice mass may be divided into three main categories: *englacial* debris (which occurs within the glacier), *supraglacial* debris (which occurs on the glacier surface), and *subglacial* debris (which occurs at the base of the glacier). Deposition of the transported material is a complex process, but that fraction deposited directly from the ice is called *till*. It consists of a wide range of grain sizes, and is thus often called *boulder clay*. It also possesses very little stratification, frequently contains far travelled erratic material, tends to have *clasts* with edges and corners blunted by abrasion, and often has its larger particles showing a preferred orientation or alignment (see colour plate 4).

Traditionally, two types of till have been recognised: *lodgement till*, which is laid down subglacially when debris is released directly from the sole of the ice, and *ablation till*, which accumulates initially in a supraglacial position and is later lowered to the ground surface by undermelting. Ablation till can be further subdivided into *meltout till* and *flow till*: the former is the direct product of ablation continuing beneath a cover of detritus, while the latter consists of debris that has built up on the ice and after saturation with meltwater becomes so unstable that it flows or slumps into nearby hollows.

Till may be deposited as a series of distinctive landforms. Ridges of glacially deposited material occurring along the terminal margins of glaciers and ice sheets are called *end moraines* or *terminal moraines* (plate 4.7). Across lowland regions they may be traced for hundreds of kilometres, and those laid down by valley glaciers may be over 100 m in height. They are partly the result of debris being deposited in the ablation zone, and partly the consequence of the pushing of debris by the glacier. On the sides of glaciers there may be prominent ridges called *lateral moraines* (plate 4.8).

In some areas, particularly where a glacier opens out on to a plain, the till is deposited as swarms of rounded hummocks, called *drumlins*. These sometimes occur in a regular pattern, and the term 'basket of eggs relief' is often applied to a glacial

Plate 4.7 A large terminal moraine blocking the Esmark Valley in Norway.

landscape of this type. Drumlins are streamlined hills which may vary in size from only a few metres in height to over 50 m. They are commonly between 1 and 2 km in length and about 0.5 km wide. They generally lie with their long axes parallel to the inferred direction of ice movement and have an approximately elipsoid plan.

Closely related to drumlins are a whole series of other streamlined forms described under the broad heading of *fluted ground moraines*. Some of the largest flutes are up to 20 km long, 100 m wide and 25 m high, but they are generally much smaller than this and are found just beyond the glacial front.

There is also a series of landforms that, while not strictly speaking made of till, owe their form and origin to glacial agency. Included in this category are *kames* and *eskers*. These are the product of glacial rivers (figure 4.12). Kames are features produced at the margins of the ice and eskers are primarily forms developed beneath the ice. Kames consist of irregular undulating mounds of bedded sands and gravels that are essentially a group of alluvial cones or deltas deposited unevenly along the front of a stagnant or gradually decaying ice sheet. Some kames occur as *kame-terraces* and were formed along the trough between the glacier and the valley side, forming narrow, flat-topped, terrace-like ridges. *Perforation* or *moulin kames* are formed in hollows and perforations in decaying ice, and if they develop along major crevasses they may have an angular dog-leg form.

Eskers are elongated ridges of stratified gravel,

Plate 4.8 Glaciers are extremely effective as transporters of debris. The Barnard Glacier in Alaska shows a whole suite of medial and lateral moraines. Note also the truncated spurs of the main valley, caused by substantial erosion.

usually thought to be the casts of streams formed either on the ice or beneath it. The largest examples may be 100 m high and extend for tens of kilometres (figure 4.13). Often they are sinuous.

Some attempts have been made to create a model of drift landform zonation. Deposition takes place in two zones. In the zone of active ice, streamlining takes place and lodgement till and drumlins are deposited. As one moves towards the outer edge of the ice sheet, stagnant ice becomes dominant, with eskers and end moraines being deposited.

The meltwater that flows from an ice mass creates large outwash plains which are sometimes called after their Icelandic name of *sandur* (plural *sandar*). These are characterised by multi-thread channels called *braids*, by abundant coarse debris and by marked variations in discharge, including catastrophic floods.

In some cases ice may dam up a river so that a lake forms. If the ice dam persists long enough a

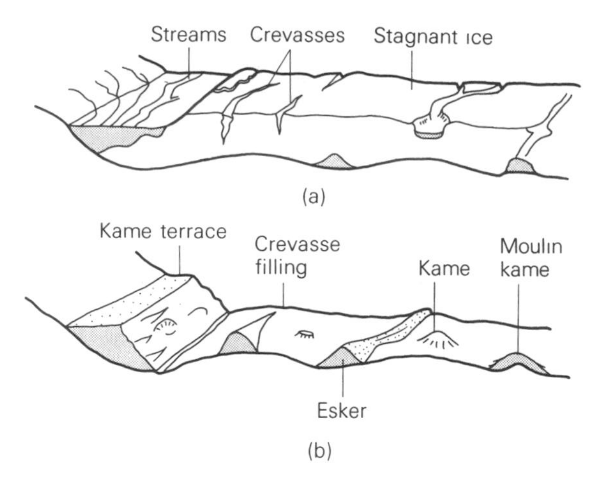

Figure 4.12 The process of formation of some fluvio-glacial landforms: (a) stagnating ice; (b) resulting landforms.

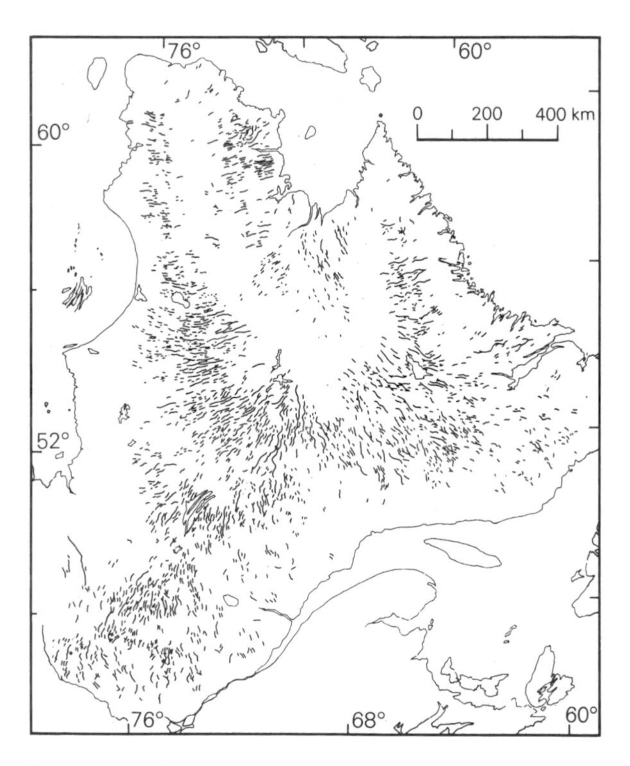

Figure 4.13 The pattern of eskers left behind as the last ice sheet wasted in the Labrador Peninsula, Canada.

shoreline may be created, lacustrine sediments may develop, deltas may grow at the mouths of any inflowing streams and an overflow channel may be cut across some convenient col. The classic illustration in Britain is the parallel roads of Glen Roy in Scotland (figure 4.14a). As the ice gradually retreated, so lower cols were exposed, which permitted the lake to overflow at lower levels. At each lower stage a *strandline* was cut by the lake. The overflow channels may persist as important features of the landscape even when ice has disappeared completely. Similarly, in the English Midlands (figure 4.14b) Welsh ice ponded up the drainage of the Severn–Avon lowlands to form Lake Harrison. Its shoreline can be traced for a distance of 55 km at an elevation of around 125 m – the height of the overflow channel at Fenny Compton by which the lake water overflowed into the Cherwell and thence into the Thames. Another form of drainage modification is the *glacier margin channel* (see figure 4.15). These are extremely widespread in areas of slow deglaciation.

4.8 Glacier Ice and Sea-level Change

If one visits many coasts, the evidence of sea-level change is clear to see. Where there are stranded beach deposits, marine shell beds and platforms backed by steep cliff-like slopes, one has evidence of emerged shorelines. Elsewhere there is often evidence of submerged coastal features such as the drowned mouths of river valleys (*rias*), glaciated valleys (fiords), submerged dune chains, *notches* and *benches* in submarine topography, and remnants of forests or peat layers at or below present sea level.

The fluctuations in degree of glaciation (see section 2.9) play a major role in explaining fluctuations of sea level. The expansion and contraction of ice sheets causes an alternating uptake and release of water into the world's oceans, which in turn causes alternating regressions and transgressions. Such changes are called *glacioeustatic*. During glacial maximums, world sea level fell by the order of 100–170 m, while during interglacials sea levels would be at a level similar to, or slightly higher than, today's. If the two main existing ice caps – Greenland and Antarctica – were to melt totally, sea level would be raised a further 66 m.

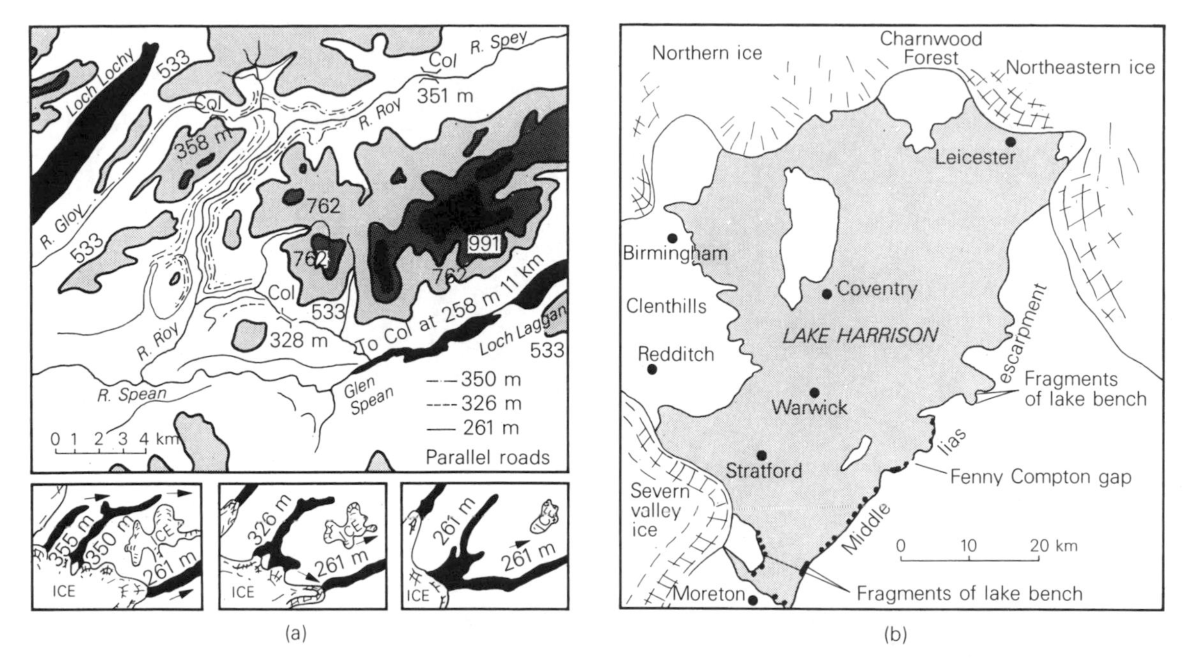

Figure 4.14 Two examples of glacial lakes. (a) The parallel roads of Glen Roy, Scotland. The three small sketches show three stages in the evolution of the lake system as the main ice withdrew towards the south-west. (b) Lake Harrison in the English Midlands at its maximum extent.

Another fundamental way in which ice sheets would affect sea level is through *glacio-isostasy* (window 4.3). The presence of large ice masses causes a depression of the earth's crust, but when such ice masses melt the crust rises by way of compensation. The present polar ice caps are big enough to cause considerable isostatic depression, so that interior Greenland, for instance, has a bedrock surface that is at or below present sea level. If the several thousands of metres of ice above this bedrock were to melt, geophysicists estimate that the bedrock surface would slowly rise to form a plateau about 1000 m above sea level. In the areas that were covered by Pleistocene ice but have since been released from its weight, the degree of isostatic uplift in post-glacial times (about the last 11 000 years only) has been considerable: around 300 m in both North America and Scandinavia.

Figure 4.15 Diagrammatic representation showing the development during deglaciation of glacier margin hillside channels.

4.9 Permafrost

Having considered the importance of glacier ice in the environment, we must now consider a no less important form of ice – *permafrost* – and consider the nature of periglacial landscapes (window 4.4). Permafrost is perennially frozen ground. It underlies approximately one-fifth to one-quarter of the land surface of the earth (table 4.3), with large areas in Siberia, Canada and Alaska and other major occurrences in Greenland, Spitzbergen and north-

Window 4.3 *Glacio-isostasy*

The principle of glacio-isostasy can be summarised thus: during glacial phases, water loads were transferred from the oceanic 70 per cent of the earth's surface to the glaciated 5 per cent. This led to depression of the crust, whilst the release of the weight of the ice resulting from melting leads to uplift.

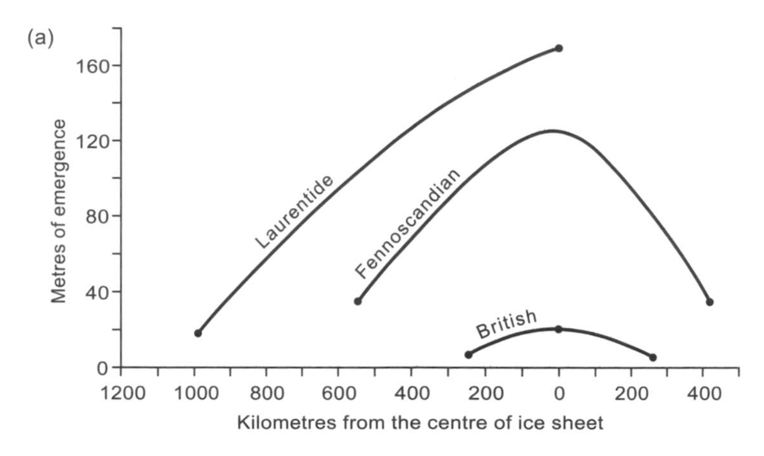

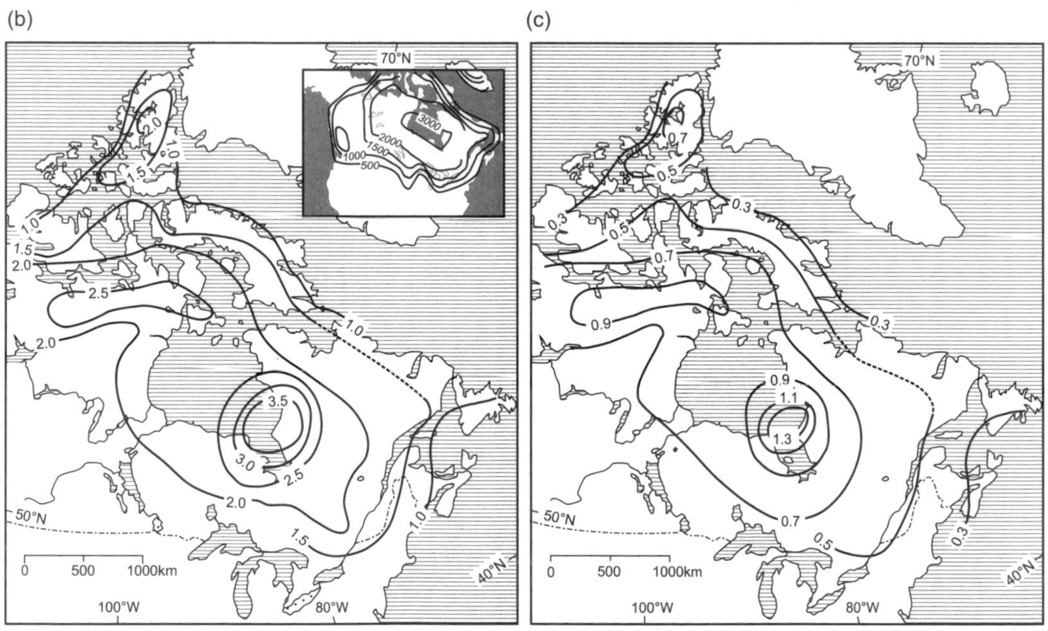

Glacio-isostatic adjustment. (a) Cross-section through the Laurentide, Fennoscandian and British ice sheets showing the amounts of isostatic recovery in the past 7000 years. (b) Average rate of Holocene uplift in metres per 100 years for northern and eastern Canada. The inset shows the thickness of the Laurentide ice sheet at 18 000 BP (contours in metres). (c) Present rate of uplift in metres per 100 years for northern and eastern Canada.

The degree of isostatic change, and the rate at which it occurred, are related to the differing volumes of the various ice caps. This is illustrated when one looks at the amount of isostatic recovery that has taken place in the Holocene in North America, Fennoscandia and the British Isles. It has been greatest in the area vacated by the Laurentide ice cap, the largest of the three ice caps under consideration, and least in the area vacated by the British ice cap, which was the smallest of the three.

The degree of maximum isostatic uplift has been considerable: around 300 m in North America and 307 m in Fennoscandia, but less in Great Britain. The ice caps of Greenland and Antarctica, moreover, are still exerting sufficient weight to create a considerable degree of isostatic depression. Much of the bedrock surface of interior northern Greenland is currently at or below present sea level. Gravity readings and ice-thickness determinations obtained by trans-Greenland expeditions suggest that before the ice sheet formed, the presently low-lying bedrock areas of northern Greenland, some of which extend below sea level, were in the form of a plateau about 1000 m above sea level. If the ice were to be removed, the bedrock surface would slowly rise up to this height once again.

In areas peripheral to the ice sheet, such as parts of the eastern coast of the United States, the Baltic and the North Sea, there are indications that zones not subject to an ice-load bulged up during glacial phases, perhaps because of volumetric displacement of the upper mantle's low-velocity layer, but that they have collapsed peripherally in post-glacial times, giving greater submergence than could be explained by the eustatic Flandrian (Holocene) Transgression alone. This is thought to be the result of some compensatory transference of sub-crustal material. Areas like northern Canada are still rising at rates that can exceed 1 m in a century.

ern Scandinavia (figure 4.16). Permafrost also exists offshore, particularly in the Beaufort Sea of the Western Arctic and in the Laptev and East Siberian Seas, and at high elevations in mid-latitudes, such as the Rocky Mountains of North America and the interior plateaux of central Asia. It occurs not only in the tundra and polar desert environments poleward of the tree-line, but also in extensive areas of the boreal forest and forest–tundra environments.

Above the layer of permanently frozen ground there is usually a layer of soil in which temperature conditions vary seasonally, so that thawing occurs when temperatures rise sufficiently in summer but freeze in winter or on cold nights. This zone of freeze–thaw processes is called the *active layer*. It varies in thickness, ranging from a depth of 5 m where unprotected by vegetation to typical values of 15 cm in peat.

Conventionally, two main belts of permafrost are identified. The first is the zone of *continuous permafrost* (figure 4.16): in this area permafrost is present at all localities except for localised thawed zones, or *taliks*, existing beneath lakes, river channels and other large water bodies which do not freeze to their bottoms in winter. In the *discontinu-*

Window 4.4 *The term 'periglacial'*

What do we mean by the term 'periglacial'? Literally speaking, it means 'near to a glacier'. Unfortunately, however, many cold parts of the world are not near glaciers (e.g. parts of Siberia), while in some parts of the world (e.g. New Zealand) large glaciers may descend into temperate woodland. Nowadays, therefore, the term is used to describe areas where frost and snow (nival) processes are active, and may or may not refer to locations close to glaciers (e.g. the higher parts of the Scottish Highlands). Likewise, periglacial areas are not necessarily underlain by permanently frozen ground (permafrost).

Table 4.3 Some facts about permafrost

Country	*Area (10^6 km²)*	*Percentage of total area of state*
Former USSR	11.0	49.7
Mongolian People's Republic	0.8	
China (without Tibet)	0.4	
Tibet	1.5	
Alaska, USA	1.5	80
Canada	5.7	
Greenland	1.6	
Antarctica	13.5	
Total	36.0	24% of total land area of the world

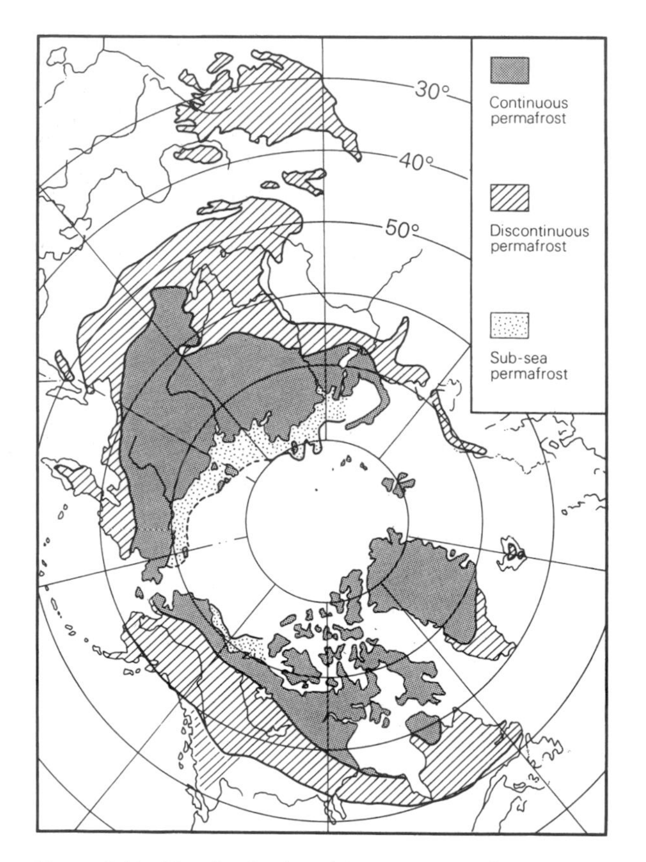

Figure 4.16 The distribution of the main permafrost types in the Northern Hemisphere.

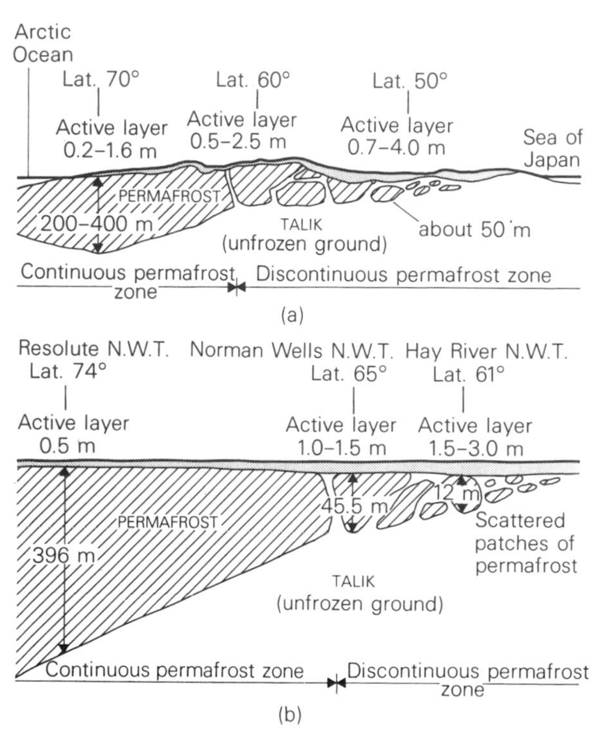

Figure 4.17 Vertical distribution of permafrost and active zones in longitudinal transects through (a) Eurasia and (b) Northern America.

ous permafrost zone small scattered unfrozen areas appear (figure 4.17).

Maximum known depths of permafrost reach 1400–1450 m in northern Russia and 700 m in the north of Canada, regions of intense winter cold, short cool summers, minimal vegetation and limited snowfall. In general, the thickness decreases equatorwards. Sporadic permafrost tends to occur between the –1 °C and –4 °C mean annual air temperature isotherms, while continuous permafrost tends to occur to the north of the –7 °C to –8 °C isotherm.

4.10 Ground Ice

The ice that occurs in the ground in permafrost areas takes on a variety of forms. Sometimes it may occur as *massive ice,* tens of metres thick. Elsewhere it may occur as *ice wedges* (plate 4.9); these tend to occur as polygonal crack patterns, with the wedges being about 1.0–1.5 m across at the surface and extending as tapering wedges 3.0 or more metres into the ground. The average dimensions of the polygons range from 15 to 40 m. They are thought to form because, when soils are cooled to below –15 °C to –20 °C, there is a shrinkage of the volume of ice held within the soil and this leads to the development of cracks and fissures in the frozen ground. In the following spring, moisture collects in these fissures and freezes, preventing the crack from closing when the ground expands as the temperature rises. It is believed that this process is self-perpetuating, since the wedge of ice provides a plane of weakness which is reopened under stress the following year, allowing the deposition of a further layer of ice.

Plate 4.9 An ice segregation developed in the tundra of Arctic Canada.

Plate 4.10 The build-up of ice can create small ice-cored hills called pingos. This particularly large example from northern Canada is starting to collapse.

Another form of ground ice is described by an Eskimo (Inuit) word – *pingo.* Pingos are ice-cored hills that have been domed up from beneath by ice growth (plate 4.10). They vary from a few metres to over 60 m in height and are up to 300 m in diameter. Normally two types have been recognised. *Open-system pingos* occur in areas of thin or discontinuous permafrost where surface water can penetrate into the ground and continue to circulate in the unfrozen sediments beneath the permafrost. When this water rises towards the ground surface, it freezes to form localised bodies of ice which force the overlying sediments upwards. *Closed-system pingos*, on the other hand, are much more charac-

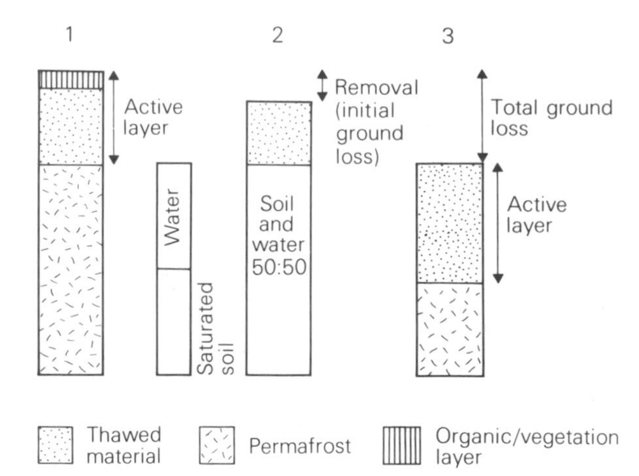

Figure 4.18 Diagram illustrating how disturbance of high ice-content terrain can lead to permanent ground subsidence.

teristic of low-lying areas with continuous permafrost and are brought about by the local downward growth of permafrost into a previously unfrozen zone.

4.11 Thermokarst

The term *thermokarst* is used to describe irregular hummocky terrain often studded with small water-filled depressions created by the melting of ground ice. This terrain appears similar to the sinkhole topography of limestone regions, but despite the apparent similarity of topography it should be made clear that thermokarst is not a variety of *karst*: the latter is a term applicable to limestone areas where the dominant process is a *chemical* one – solution (see section 13.5). Underlying the development of thermokarst is a *physical* process – ground ice melting in permafrost regions.

The development of thermokarst is primarily the result of the disruption of the permafrost and an increase in the depth of the active layer. We can illustrate this by a simple example (figure 4.18). In stage 1 we have an undisturbed tundra soil which has an active layer 45 cm thick. Beneath this is frozen ground which is supersaturated: it yields, on thawing, 50 per cent excess water and 50 per cent saturated soil. In stage 2 some of the upper ground surface is removed (say 15 cm), together with its insulating vegetation, and as a result, under the bare ground conditions, the active layer equilibrium thickness might increase to 60 cm. As only 30 cm of the active layer remains, 60 cm of the permafrost must thaw before the active layer can thicken to 60 cm (since 30 cm of water will be released). Thus in stage 3 the surface subsides 30 cm because of the thermal melting associated with the degrading permafrost, to give an overall depression of 45 cm.

There are many reasons why thermal disequilibrium and subsequent permafrost degradation might take place: climate might change; there might be a forest fire; erosion might remove the upper soil layer; or humans might strip off the vegetation during the building of a road or other structure, or warm the permafrost by building a house or constructing a pipeline. Thus, although some thermokarst is natural, human intervention in a permafrost area can cause subsidence to occur. Such subsidence can be very dangerous for engineering structures. Very major changes to permafrost could occur as a result of future global warming (window 4.5).

4.12 Ice Segregation and Frost Heaving

When the water in a soil starts to freeze, there is a tendency for unfrozen water within the soil to migrate towards where the freezing is taking place. This has two consequences: freezing takes place unevenly in the soil, leading to the formation of *ice segregations*; and freezing of the soil leads to local heaving of the ground surface because of the localised migration of water and its subsequent expansion upon freezing – there is a volume expansion of approximately 9 per cent when water turns into ice.

Like thermokarst development, frost heaving is a major problem that the engineer has to face in the development of permafrost areas. *Differential frost heave* is a particularly serious engineering problem. Structures are frequently built on piles in many of the world's environments, but in periglacial areas the piles are subject to heave by frost pro-

Window 4.5 Permafrost in a warmer world

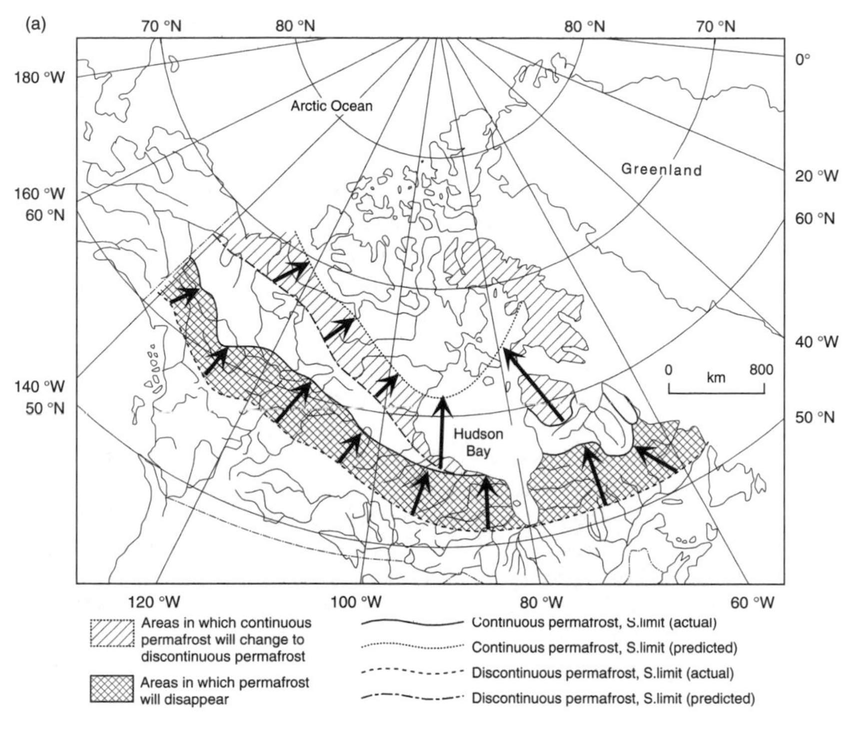

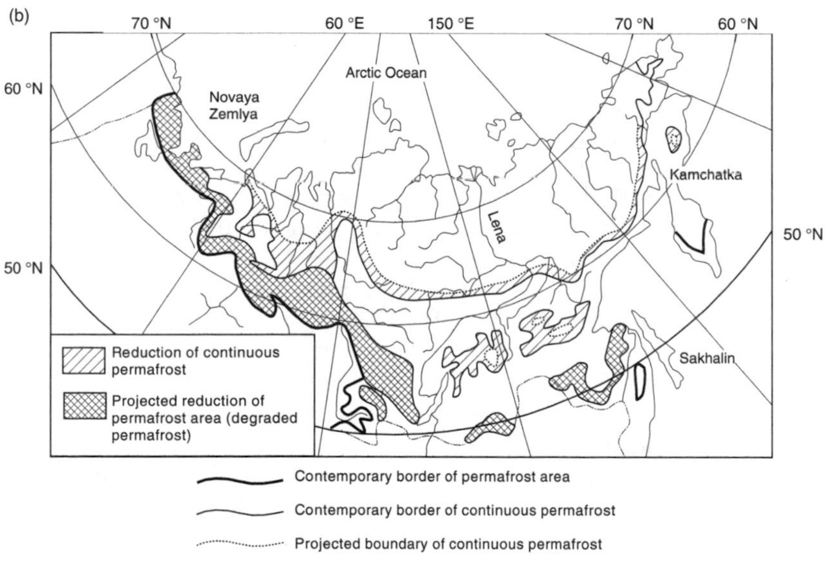

Projection of changes in permafrost with global warming for (a) North America and (b) Siberia.

Under most models of climatic change, the degree of temperature increase will be greatest in high latitudes. With a doubling of greenhouse gas concentrations, a high-latitude temperature rise of perhaps as much as 6–8 °C might be expected. Given that the existence and distribution of permafrost is so closely related to temperature, it is probable that major changes will occur. Average northward displacement of the southern permafrost boundary by 150 ± 50 km would be expected for each 1 °C warming so that a total *maximum* displacement of between 1000 and 1500 km is possible. The Intergovernmental Panel on Climate Change suggests a 15 per cent shrinkage in total permafrost area by 2050. The figure shows predictions of permafrost displacement for Canada and Siberia.

However, major uncertainty surrounds the rate at which permafrost degradation will occur. As it is probably a slow process, permafrost will continue to exist for an extended time in areas of *continuous* permafrost. In areas of discontinuous or sporadic permafrost the rate will vary greatly depending on local material conductivity, snow cover and vegetation. Changes in vegetation type and in snow cover in a warmer world may modify the direct consequences of warmer surface temperatures.

There is historical evidence that permafrost can degrade relatively rapidly. For example, during the warm optimum of the Holocene (*c.*6000 years ago) the southern limit of discontinuous permafrost in the Russian Arctic was up to 600 km north of its present position. Similarly, during the warming phase of recent decades, researches have demonstrated that along the Mackenzie Highway in Canada, between 1962 and 1988, the southern fringe of the discontinuous zone moved north by about 120 km in response to an increase over the same period of 1 °C in mean annual temperature.

Where the permafrost is ice rich, or contains massive ground-ice segregations, subsidence and settling due to thawing will occur, inducing a thermokarst relief which can alter drainage patterns and stream courses. Coastal retreat will also gather momentum as permafrost degrades in coastal lowlands, and large areas may be inundated as depressions are lowered to below current sea levels because of thaw settlement. River banks and lake and reservoir shorelines might also become amenable to faster rates of erosion, while slopes could become less stable and the active zone become thicker.

cesses. This is especially true in those areas of discontinuous permafrost where the active layer is thick. In parts of Alaska many bridges show the effects of heave (figure 4.19). It is not uncommon for a thawed zone to occur beneath the river channel, so that piles inserted in the streambed itself are subjected to minimal frost heave. Likewise, piles inserted into the permafrost on either side of the river are relatively unaffected, since the freezing of the piles to the permafrost provides a resistance to the upward heaving of the seasonally frozen zone. In between these two zones, the piles are located in the zone of seasonal freezing, with the result that up-arching of both ends of the bridge may occur.

Within periglacial areas, the seriousness of frost heave and freeze–thaw processes varies with climatic conditions and material types (figure 4.20). The most hazardous zone is that in which summer temperatures are high enough for thaw to take place but winter temperatures are low enough for freezing to occur to a great depth. In very cold areas where there is very little summer thaw the problem is less, and in relatively mild areas where the degree of winter freezing is reduced the problem is also diminished. Sands and gravels are relatively immune to the worst aspects of frost heave, whereas sediments with a large content of silts and clays are highly susceptible.

One of the prime consequences of frost action in the soil is the development of soil and vegetation patterns. On flat ground, circles, nets and polygons with dimensions several metres across may develop; on moderate slopes, steps; and on steeper slopes, stripes. Six main causes for the development of such patterned ground have been recognised:

1 The upward movement of stones by freeze–thaw activity.

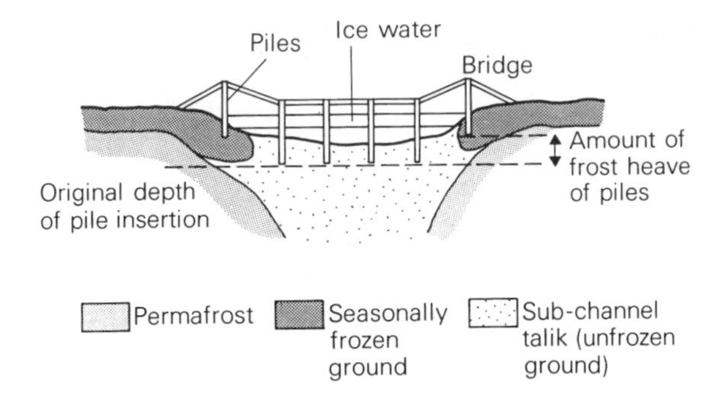

Figure 4.19 Diagrammatic illustration of how frost-heaving of piles inserted in the layer of seasonally frozen ground can result in bridge deformation.

2 Differential heaving produced by the effect of alternate freezing and thawing on sediments of different grain size.
3 Solifluction (a type of soil movement on slopes) (see section 4.14).
4 Pressure effects such as those produced by the down-freezing of the active layer in the autumn.
5 Contraction effects.
6 The effect of peat and vegetational cover on the rapidity of freezing and thawing.

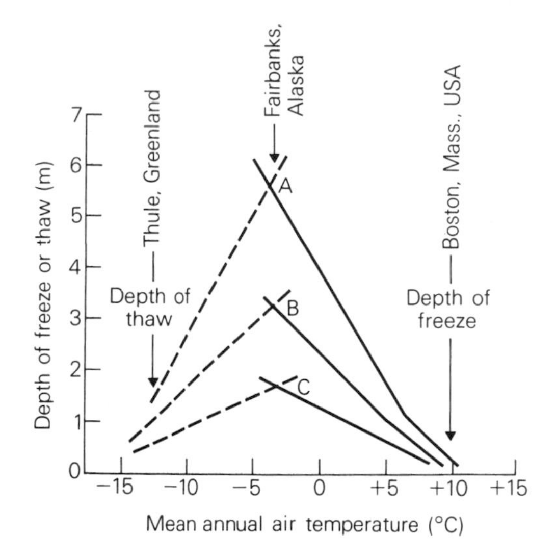

Figure 4.20 The amount of freeze and thaw penetration in relation to mean annual temperature. In very cold areas (e.g. Thule, Greenland), the amount of *thaw* is highly restricted, whereas in warmer areas (e.g. Boston) the amount of *freezing* is limited. It is the zone in between (exemplified by Fairbanks, Alaska) in which the amount of freeze-thaw activity is the greatest, and the associated risks most severe. The risk varies according to the nature of the materials. A is the relationship for highly susceptible materials, B is the relationship for average conditions, and C is the relationship for less susceptible materials.

4.13 Frost Shattering and Soil Formation

In areas where frosts are frequent, alternations of freezing and thawing are an effective cause of rock breakdown so long as moisture is present. The breakdown is not the result of a chemical change in the constituents of the rock, as would be brought about by *chemical weathering*, but is a mechanical breakdown – a type of weathering called *physical weathering*.

Frost leads to mechanical breakdown in two main ways. The first method depends on the fact that when water freezes at 0 °C it expands by about 9 per cent. This creates pressures, calculated to be around 2100 kg cm^{-2}, that are higher than the tensile strength of rock (generally less than 250 kg cm^{-2}). The second method results from the fact that when water freezes in rock or soil the ice attracts small particles of water that have not frozen from the adjoining pores and capillaries. Nuclei of ice crystal growth are thus established, and some workers believe that this mode of ice crystal growth is a more potent shattering agent than the change in volume that occurs when water becomes ice.

The rate at which frost shattering occurs depends on many factors. One of the most important controls is the presence of moisture; laboratory experiments have shown that the amount of disintegration in rocks supplied with abundant moisture is much greater than that in similar rocks containing less moisture. For this reason, dry tundra areas and cold deserts may undergo less extreme frost weathering than moister environments. A second major factor is the frequency and magnitude of temperature changes, although much controversy surrounds this subject. There is still doubt as to whether rapid freezing of moisture in rocks during, say, a daily cycle of freeze–thaw, is more effective than slow freezing (during, say, a seasonal cycle of freeze–thaw), and

Plate 4.11 Avalanches pose a major natural hazard in periglacial areas and occur in a variety of forms. Here we have an example of a powdery dry-snow avalanche in the Himalayas.

whether intense cold during the period of freezing is more potent than temperatures just a few degrees below freezing. In general, laboratory experiments tend to show that the number of freeze–thaw cycles is more important than their intensity. This would tend to suggest that those areas with high moisture availability and many daily cycles of freezing and thawing in the course of a year would be most prone to the effects of frost weathering. This in turn indicates that in extremely cold polar areas there may be less frost shattering than in environments where daily freeze–thaw cycles are more frequent (e.g. Iceland, or low-latitude–high-altitude areas such as the Himalayas).

The nature of the rock type is also a vital factor. Working from both laboratory experimentation and field observation, geologists have ranked rock types in their order of susceptibility. Rocks such as tough quartzites and igneous rocks tend to be most resistant, while shales, sandstones and porous chalk tend to be least resistant.

In many periglacial environments of both past and present, extensive upland surfaces of angular rocks and boulders (called *blockfields*) are a dramatic exemplification of the role of frost. Other associated features include large *screes,* and angular, frost-riven *tors* (see section 5.16).

Although little has so far been said specifically about the soils of tundra areas, in effect many of the main processes that form them, together with many of their characteristics, have already been described. It is obvious that the presence of permafrost, the development of an active layer, the operation of *solifluction,* frost weathering and frost heaving, the existence of low evapotranspiration rates, the development of a limited plant cover and the tendency for summer waterlogging all combine to make the soils of tundra regions highly distinctive. The waterlogging tends to promote *gleying* (section 13.2); the frost heaving tends to disturb what might be called 'normal' soil-layer development; and low rates of organic decomposition may lead to the accumulation of peat.

4.14 Slope Processes, Avalanches and River Regimes

Slope processes of mass wasting occur with great power in the periglacial environment. There are various reasons for this. First, the presence of permafrost provides a barrier to the downward infiltration of water into the ground, thereby inducing high pressures in the soil pores in the surface layers. Second, the layer above the permafrost, the active layer, is extremely moist and highly unstable. Third, the permafrost acts as a lubricated surface over which the materials of the active layer can flow. Fourth, the process of frost heave tends to cause a downhill movement of material under the influence of gravity.

For these reasons, hillside slopes tend to be characterised by instability, and *slope failure* is frequent, posing problems for buildings and other engineering structures. On exposed cliff faces blocks may be detached by frost weathering, while on more gentle slopes there are frequent flows of debris and soil. Indeed, many people regard *solifluction* – the slow flowing from higher to lower ground of masses of waste saturated with water – as one of the most widespread processes of soil movement in periglacial regions. This process can operate over low-angle slopes (as low as 1°) and tends to involve movement of the upper 50 cm of the soil at rates of around 0.5–5.0 cm per year. It tends to create a series of *lobes* on slopes, resulting in a landscape that bears some resemblance to the wrinkled hide of an aged elephant.

Another feature of some periglacial slopes is the development of *cambering.* Where there is a jointed rock, like a limestone, overlying clay, as there is along the Cotswold Escarpment in the English Midlands, joints and bedding planes would have been opened up by frost wedging during the Ice Age. Moreover, thawing of the underlying clays would have made them very mobile, so that they would have been squeezed outwards by the weight of the overlying caprock of limestone, which in turn would sag at the edges and be cracked apart. Large masses would slip down the hillside.

One special form of mass movement that is common in areas of snow and ice is the *avalanche* (plate 4.11). These occur especially in areas of steep slopes – generally above about 22° – but their momentum may carry them on to lower slopes, particularly if they are channelled into a gully (plate 4.12). Northerly and western slopes (in the Northern Hemi-

Plate 4.12 Many mountains in periglacial areas like the high Sierra Nevadas in California are furrowed by chutes which act as preferred routes for avalanches and may be enlarged by them. It is advisable not to locate buildings or other structures at the bases of such chutes.

sphere) are especially dangerous, for the lack of sun causes the snow to stabilise less readily.

There are various different types of avalanche (figure 4.21), but the two most important are the airborne dry-snow avalanches (plate 4.13) and the flowing wet-snow avalanches. In the first of these the devastation is largely the result of the associated shock waves, which are powerful enough to decimate mature forests and collapse buildings. A wave of snow-free air precedes the actual avalanche, and within the avalanche itself there are violent internal gusts that can travel at speeds in excess of 300 km per hour. In the case of the flowing wet-snow avalanches it is the sheer weight of the snow that causes the devastation. One destructive avalanche in the Italian Alps in 1885 comprised no less than 2.5 million tonnes of snow and its energy was equivalent to about 300 million horsepower (a powerful ocean liner, for comparison, generates about 150 000 horsepower). The power of such avalanches may be augmented by some of the boulders and other debris that they pick up in the course of their movement (see also window 8.1).

Rivers in periglacial areas have very distinctive *runoff regimes* that reflect the importance of the short summer melting season (figure 4.22). Snow and ice tend to melt rapidly in the early summer (late June or early July in the case of the Canadian Arctic), and as the summer progresses, and there is less snow to melt, the runoff steadily decreases. This type of regime is described as *nival*. However, in those types of river basin where permanent snow or ice fields occur, melting goes on throughout the summer months whenever it is warm enough. Peak runoff under these conditions is often delayed until late July or early August. Such regimes are often called *proglacial*. In some proglacial streams a potentially hazardous event is the occurrence of a *jökulhlaup* – a large-magnitude, localised flood surge caused by the bursting of ice dams along the margins of the ice (window 4.6).

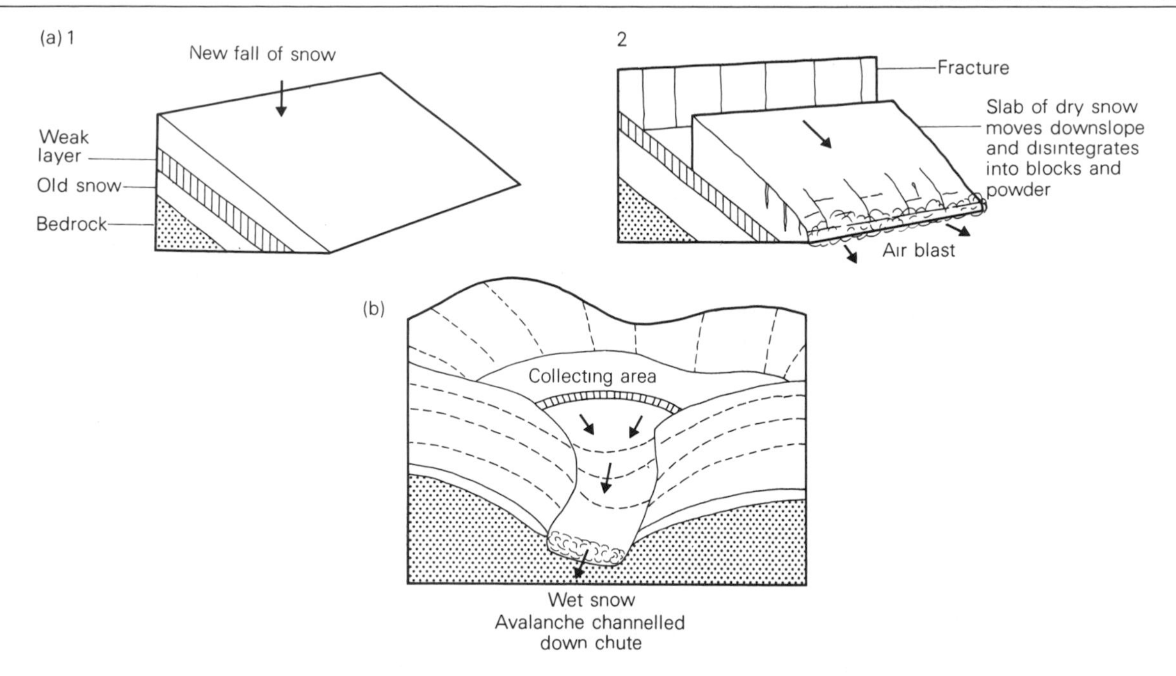

Figure 4.21 Snow avalanches. (a) The dry-snow type. A new fall of snow adds extra weight to the top layer of snow. The extra pressure causes it to start moving over an underlying layer that was deposited in an earlier snow fall. As the slab moves over the earlier layer it fragments under the force of gravity. Some damage will be caused by the air blast that precedes it. (b) The wet-snow type of avalanche. Snow accumulates in a collecting area and when meteorological conditions are suitable it starts to flow down a pre-existing channel or chute.

Plate 4.13 A slab avalanche of dry snow in the Southern Alps of New Zealand.

Window 4.6 *Jökulhlaups*

Some glacial systems are characterised by the periodic or occasional release of large volumes of stored water in catastrophic outburst floods. These outbursts are generally known by their Icelandic name 'jökulhlaup', but are also known as 'débâcles' in mainland Europe.

Jökulhlaups may be triggered by three main mechanisms: the sudden drainage of an ice-dammed lake below or through a barrier of ice; lake water overflow and rapid incision of ice, bedrock or sediment barriers by river erosion; and the growth and collapse of subglacial reservoirs. They not only occur in Iceland, but also in New Zealand, the Americas, Karakorams and mainland Europe. The sudden discharge of massive amounts of water and sediment leads to the construction of outwash plains and can cause profound damage to roads and bridges.

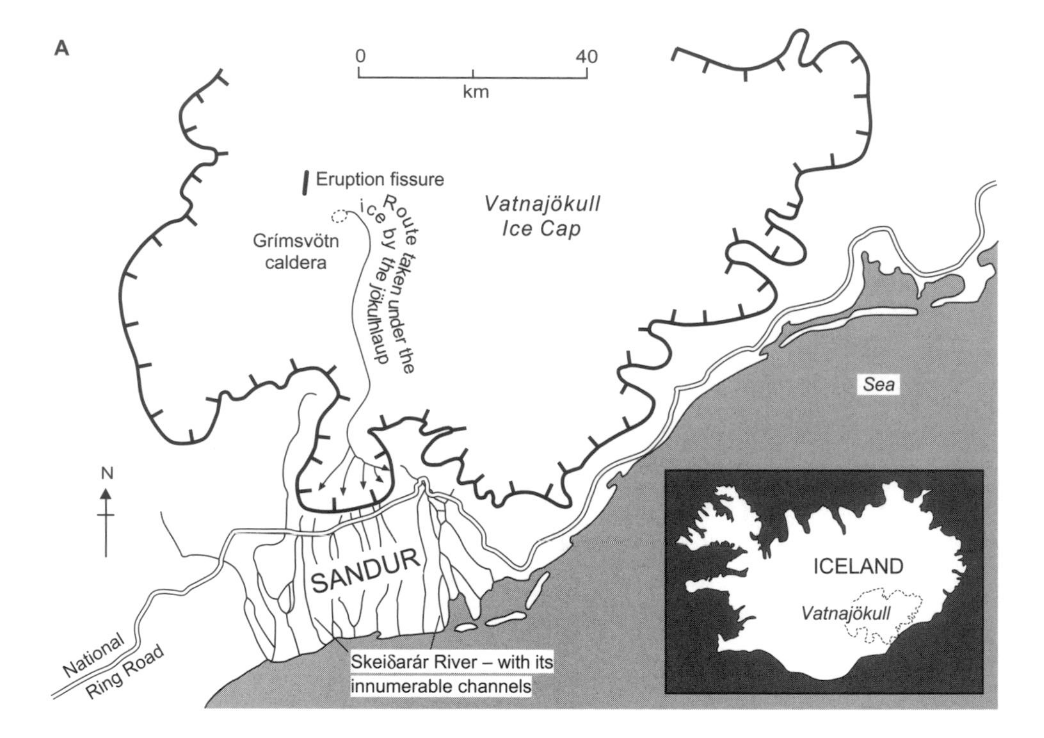

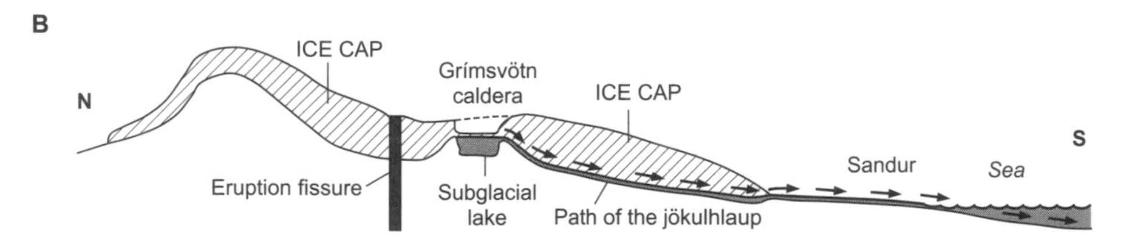

(A) Map of the Vatnajökull ice cap showing the location of the fissure eruption, the Grimsvötn caldera where the subglacial lake built up, the 50 km route taken under the ice cap by the floodwaters, and the broad sandur plain that was inundated. The inset map shows Vatnajökull. (B) Schematic cross-section through Vatnajökull to illustrate the relationship between the volcanic centres, the subglacial lake in the Grimsvötn caldera, and the path of the jökulhlaup.

In Iceland some jökulhlaups can be triggered by volcanic eruptions beneath ice caps (e.g. the 1996 eruption below Vatnajökull). Although present-day outburst floods are awesome in their power, their geographical impacts and discharges are minor in comparison with the floods released from the great Laurentide ice cap of North America during the last glaciation of the Pleistocene.

The Icelandic jökulhlaup of November 1996 was caused by volcanic activity along an eruption fissure beneath the ice cap. This led to accumulation of water in a subglacial lake which suddenly drained under the ice cap, causing massive floods to flow across the outwash plains (sandur) between the ice and the sea. Some 10 km of the national ring road were completely washed away and a further 10 km were severely damaged.

Source

Charles Warren (1997) 'Ice, fire and flood in Iceland', *The Geography Review*, 10 (4): 2–6.

4.15 Environmental Problems and Hazards for Tundra Development

We have already seen that the tundra environment has certain characteristics that have to be dealt with if humans are to undertake successful development of the area, as in Siberia, Alaska and elsewhere. These can be listed as follows:

- avalanches near steep slopes
- slope instability associated with solifluction
- rockfalls caused by frost action
- thermokarst subsidence – both natural and induced by people
- ground heaving (e.g. pingo growth, frost boils)
- spring snow melt floods
- glacier burst and glacier damming floods

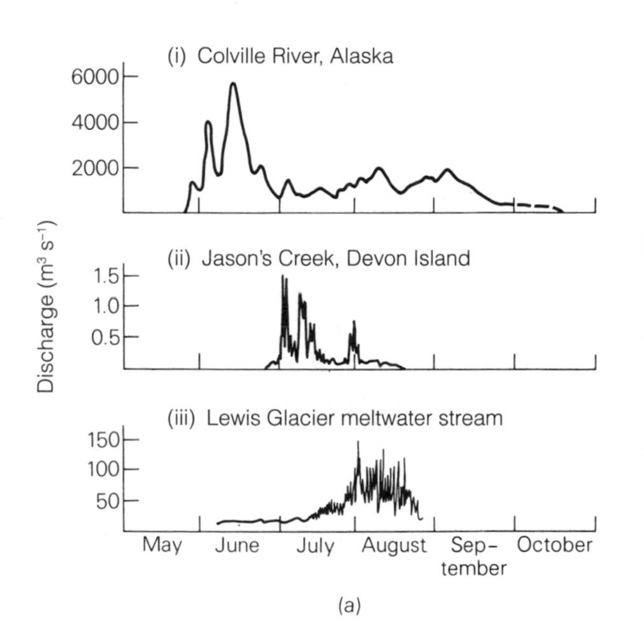

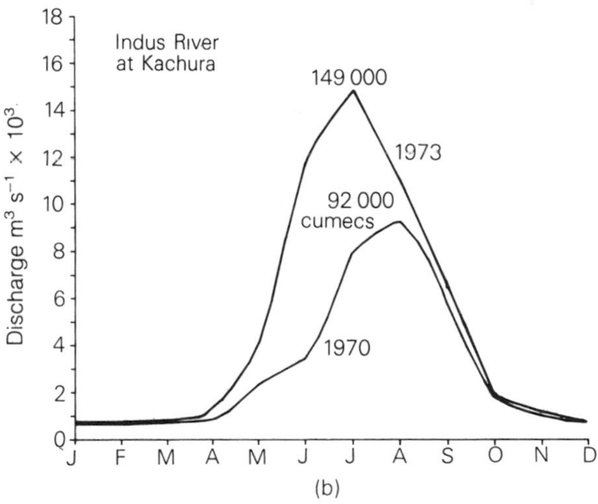

Figure 4.22 (a) The seasonal runoff regime for periglacial rivers in North America; and (b) the seasonal regime for the Indus, which derives the bulk of its flow from mountains of the Karakorams and Himalayas. The melting of snow and glacier ice in the warm summer months produces this highly seasonal regime. Although the size of the peak flow varies from year to year, the peak always occurs between June and August.

- ice jams on rivers
- icings
- frost heave of piles
- freezing of public utilities (e.g. sewers, water-pipes)
- lack of water in winter for effluent disposal
- wind chill
- very short growing season
- lack of sunlight in winter months
- severe snowfalls
- bitterly low temperatures
- icebergs and pack ice offshore
- infertile soils
- bogs etc., creating transport problems

Some of these hazards, such as avalanches and floods, can be avoided by building settlements in locations that are not prone to them. Other problems can be overcome by appropriate engineering solutions. Thus, thermokarst subsidence produced by the heat given out by buildings or pipelines can be lessened by building them on an insulating pad of gravel or on piles above the ground. Alternatively, in the case of thermokarst subsidence produced by vegetation disturbance, great care can be taken to limit this by keeping vehicles to narrow strips rather than allowing them to range widely over the countryside. Public utilities such as sewers and water mains can be placed in specially designed and insulated boxes that run above ground on supports, and are called *utilidors*. *Icings,* which are sheet-like masses of water that form at the surface in winter when water issues from the ground, can be a major hazard to motorways, but they can be evaded by, for example, avoiding the creation of road cuttings, from the walls of which water seepage might occur.

Floating sea ice poses major problems for coastal development, navigation and the exploitation of ocean resources. A distinction can be made between *sea ice,* formed by direct freezing of ocean water at temperatures below –2 °C, and *icebergs* and *ice islands,* which are bodies of land ice broken free from glaciers and ice shelves. Sea ice is relatively thin (generally less than 5 m), and when it completely covers the sea surface is called *pack ice.* However, wind and current action may break the pack ice up into individual *ice floes* separated by narrow strips of open water called *leads.* When ice floes are brought together by winds, they buckle and turn upward into a chaotic landscape of rugged pressure ridges that make travel difficult. Icebergs, formed by the carving of blocks of glacier ice, drift under the influence of ocean currents in a partly submerged form. Large, tabular examples are called ice islands, and these may be enormous – 30 km across and 60 m or more thick.

With care, these fragile and sometimes dangerous tundra environments, with their easily disrupted permafrost and their slow-growing plants, can be developed; but it requires a deep and sympathetic appreciation of their character.

Key Terms and Concepts

basal creep
cirques
englacial
esker
frost heave
frost shattering
glacial diffluence
glaciation level
glacio-isostasy
ground ice
jökulhlaup
kame
moraine
nival regime
patterned ground
permafrost
pingo
solifluction
subglacial
supraglacial
surging glacier
thermokarst
till
tundra
wind chill

Points for Review

- How do plants and animals adapt to polar conditions?
- How do glaciers move?
- Are glaciers effective agents of erosion?
- Why do glaciated landscapes often contain many lakes?
- Attempt a classification of the main types of glacial deposit.
- What impact does the presence of permafrost have on landscape?
- What problems might a road builder encounter in a tundra environment?

FURTHER READING

Benn, D. I. and Evans, D. J. A. (1998) *Glaciers and Glaciation* (London: Arnold). A very large advanced text.

Dixon, J. C. and Abraham A. D. (1992) *Periglacial Geomorphology* (Chichester: Wiley). An edited collection of research papers.

Fogg, G. E. (1998) *The Biology of Polar Habitats* (Oxford: Oxford University Press). An overview of life at the two poles.

French, H. M. (1999) *The Periglacial Environment*, 2nd edn (London: Longman). A predominantly geomorphological treatment, with a strong emphasis on permafrost.

Hambrey, M. and Alean, J. (1992) *Glaciers* (Cambridge: Cambridge University Press). A beautifully illustrated text on glaciers.

Hansom, J. D. and Gordon, J. E. (1998) *Antarctic Environments and Resources: A Geographical Perspective* (Longman: Harlow). A new, wide-ranging study of Antarctica.

Knight, P. G. (1999) *Glaciers* (Cheltenham: Stanley Thornes). A comprehensive and detailed study of knowledge about glaciers.

Sugden, D. E. (1982) *Arctic and Antarctic: A Modern Geographical Synthesis* (Oxford: Blackwell). A survey of all aspects of polar environments.

5 The Mid-latitudes

5.1 The Westerlies

The world's mid-latitudes are dominated by upper winds that blow from the west. They encircle the globe with a series of large waves more than 2000 km across called *Rossby waves*. These waves affect the interface between the colder air masses of high latitudes and the warmer air masses of low latitudes. While they may be accentuated by the disturbing effect of mountain barriers like the Rockies, they are inherent features of a rotating fluid system which has a thermal gradient applied to it. In the Northern Hemisphere, at any one time, the predominant number of Rossby waves is five. Within these westerlies there is a narrow band of high-velocity winds, called the *jet stream,* which flow at speeds of up to 300 km h^{-1} at altitudes between 9000 and 15 000 m.

The character of the Rossby waves is important in affecting mid-latitude weather because surface weather systems, such as depressions (areas of high pressure), are related to them. As figure 5.1a indicates, the soaring and diving action of the upper waves pulls air columns upwards and downwards, producing low-pressure and high-pressure systems respectively.

The shape of the waves can vary considerably (figure 5.1b), for the essentially west–east trajectory can be deformed so that the troughs and ridges become accentuated, ultimately splitting up into a cellular pattern with pronounced meridional (north–south) flow at certain longitudes. The deep, relatively permanent *blocking anticyclones,* which may exist in association with increased meridional flow, act as a barrier to the travelling eddies of low pressure, and so can have an important effect on weather. In England the very cold winter of 1963 and the extraordinarily severe drought of 1976 (see section 5.17) were brought about by blocking anticyclones.

5.2 Cyclones and Anticyclones

Much of the unsettled, windy, cloudy, wet weather of the mid-latitudes is associated with the development of low-pressure systems called *depressions* or *cyclones.* The convergence of air towards the centre of the low-pressure zone is accompanied by uplift and subsequent cooling of air, which in turn produce cloudiness and precipitation. By contrast, much of the settled, fair, sunny weather in these latitudes is associated with anticyclones, in which the air subsides and spreads outwards, causing warming.

These mid-latitude cyclones depend for their development on the coming together of large bodies of air of contrasting physical properties. Such bodies of air are called *air masses.* Air masses differ widely in temperature (from very warm to very cold), and in moisture content (from very dry to very moist). The boundaries between air masses in the mid-latitudes are relatively sharply defined, and are called *fronts.* The character of the air masses depends on where they come from (obviously, equa-

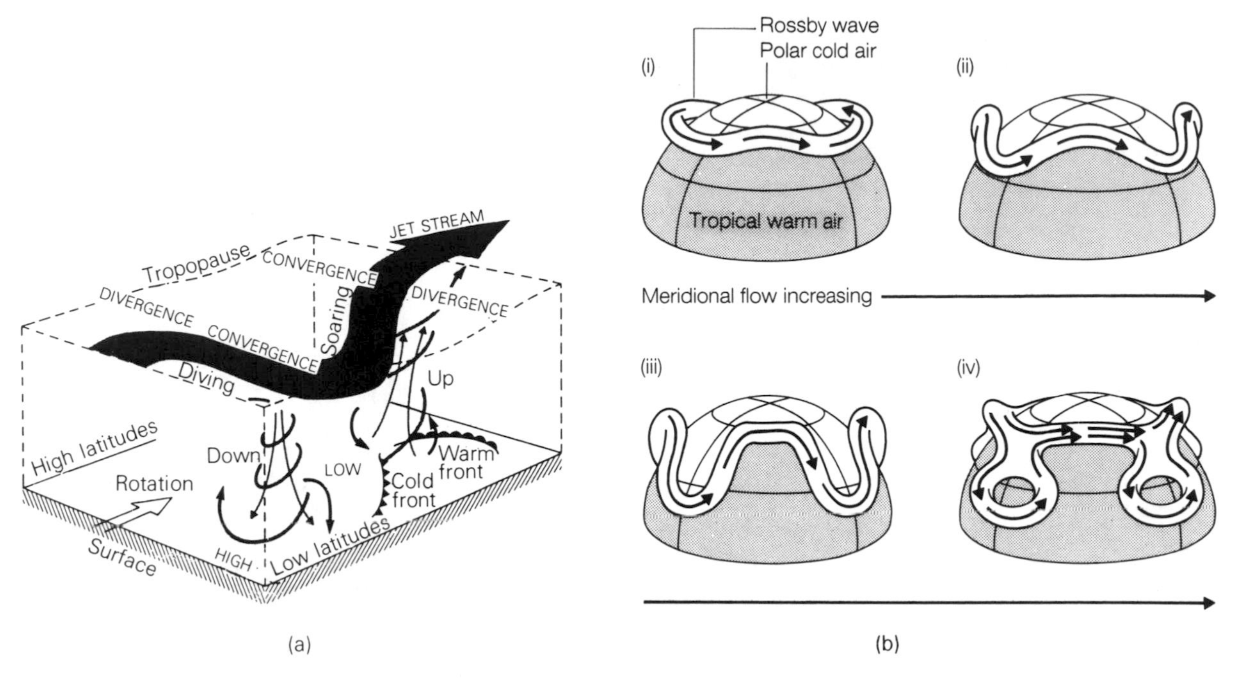

Figure 5.1 Rossby waves and their relationship to surface conditions: (a) the relationship between waves in the upper air and areas of high and low pressure near the surface; (b) the different configurations of the waves as they affect the development of blocking anticyclones, etc. They start off with a gentle undulation (i), which gradually becomes accentuated (ii and iii), until cells of cold and warm air are formed (iv). As development progresses flow becomes less zonal and more meridional.

torial air will be very different from polar air) and the type of surface over which they developed (maritime air will have very different moisture characteristics from continental air).

Many of the day-to-day changes in mid-latitude weather are associated with the formation and movement of fronts. Masses of cold polar air meet warm, moist tropical air along the *polar front*. Instead of mixing and diffusing freely, these two very different air masses remain clearly defined; but they interact along the polar front in great spiralling cyclonic whorls.

We can now follow the stages in the development of such a whorl (figure 5.2). At the start of the cycle (stage A), the polar front is a relatively smooth boundary along which air is moving in two different directions. At stage B a wave begins to form, and cold air is turned in a southerly direction and warm air in a northerly direction, so each invades the domain of the other. In stage C the wave disturbance along the polar front has intensified and deepened; while cold air is now actively pushing southwards along a *cold front*, warm air is actively moving north-eastward along a *warm front*. In stage D the cold front has overtaken the warm front, reducing the zone of warm air to a narrow sector and producing an *occluded front*. Eventually (stage E) the warm air is forced off the ground, isolating it from the source region of warm air to the south. The source of moisture and energy is thus cut off, the cyclonic storm gradually dies out, and the polar front is re-established more or less as it was at the beginning of the cycle in stage A.

The cold front, produced where cold air is invading the warm air zone, is a zone where the colder air mass, being denser, remains in contact with the ground and forces the warmer air mass to rise over it. The warm air tends to rise about 1 km vertically for every 40 km of horizontal distance. Cold fronts are associated with phenomena such as thunder-

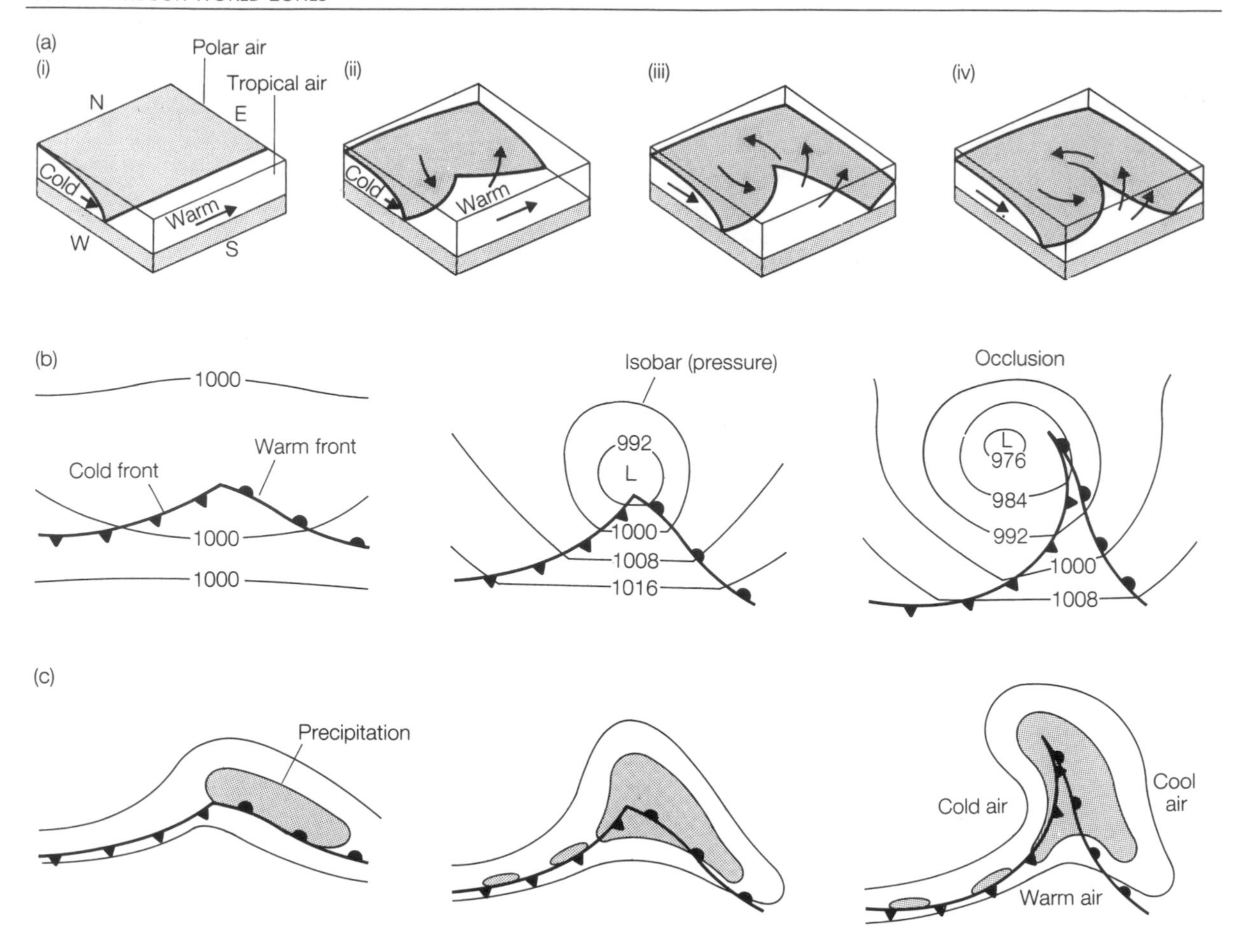

Figure 5.2 Stages in the development of a Northern Hemisphere mid-latitude frontal depression. In (a) the block diagrams show how the wave on the frontal surface gradually occludes over a period of days (i–iv); (b) shows the form of the warm and cold fronts on the ground; while (c) shows the pattern of precipitation.

storms, which are created as unstable warm air is lifted.

In the case of a warm front, warm air is moving into a region of colder air. Once again, the dense cold air remains in contact with the ground, forcing the warm air mass to rise. The slope is generally less steep than for that of a cold front, and is generally between 1 in 80 and 1 in 200. The type of weather associated with the passage of a cyclone and its different types of front is shown in figure 5.3.

5.3 Air Masses

Because the development of fronts is intimately connected with the differences between air masses, it is necessary to consider air masses further at this point. First, a definition: 'Air masses are large bodies of air in which conditions of temperature and humidity are remarkably uniform over distances of about 1000 km.' They form because, if air remains over a part of the earth's surface with little movement for about three to five days, it tends to take on certain characteristics related to the amount of radiation and evaporation there is in the area. These

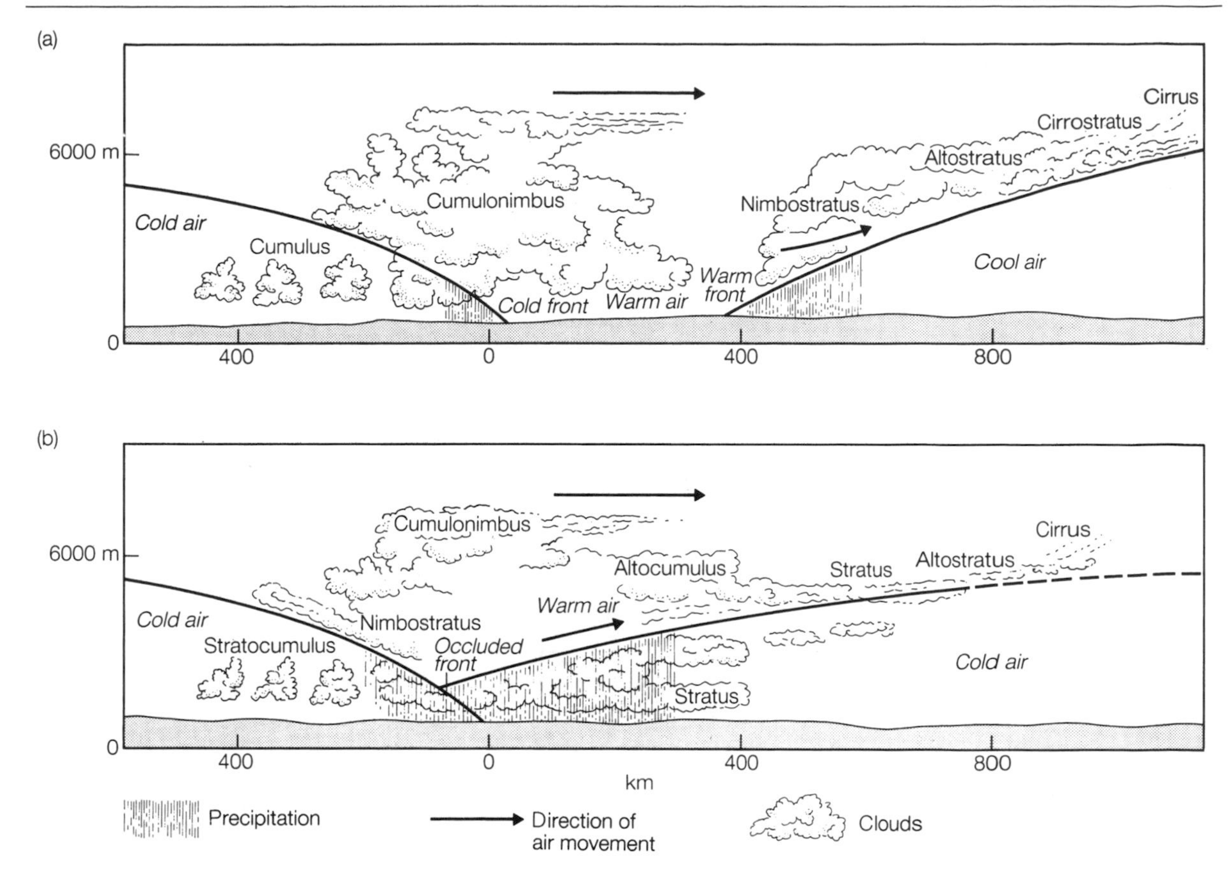

Figure 5.3 Cloud and precipitation conditions associated with the passage of a mid-latitude frontal depression: (a) weather prior to occlusion; (b) weather associated with occluded front. For definitions of the different cloud genera see chapter 14.

in turn affect the air's temperature and humidity. The areas where the mass of air remains stationary for such a period are called *source regions* (figure 5.4), and they are associated closely with the large, slow-moving, high-pressure (anticyclonic) belts of the subtropics and of high latitudes. Thus one can distinguish six main types of source region: those that dominate temperature characteristics (polar air (P) and tropical air (T)) and those that dominate humidity characteristics (continental air (c) and maritime air (m)) (table 5.1).

Britain stands in an area where four source regions are important, and this means that its climate is subjected to a wide range of influences that help to explain its inherent variability over quite short time spans. *Maritime tropical air* (mT) is the major type of air that Britain receives from the tropics. It originates in the oceanic subtropical high-pressure belts, and tends to be warm and humid. In summer, surface heating of this warm, moist air may promote instability and thunderstorm formation. *Continental tropical air* (cT) originates over the great Sahara desert, but it very rarely reaches Britain in its original dry, hot, dusty form, since its high temperature and low relative humidity encourage moisture to be evaporated into it as soon as it crosses the oceans. *Maritime polar air* (mP) reaches Britain after crossing the North Atlantic, and when it travels in a northerly direction it brings heavy clouds and rain. *Continental polar air* (cP) forms over Siberia and northern Scandinavia. In winter it brings intensely cold weather, though it tends to be very

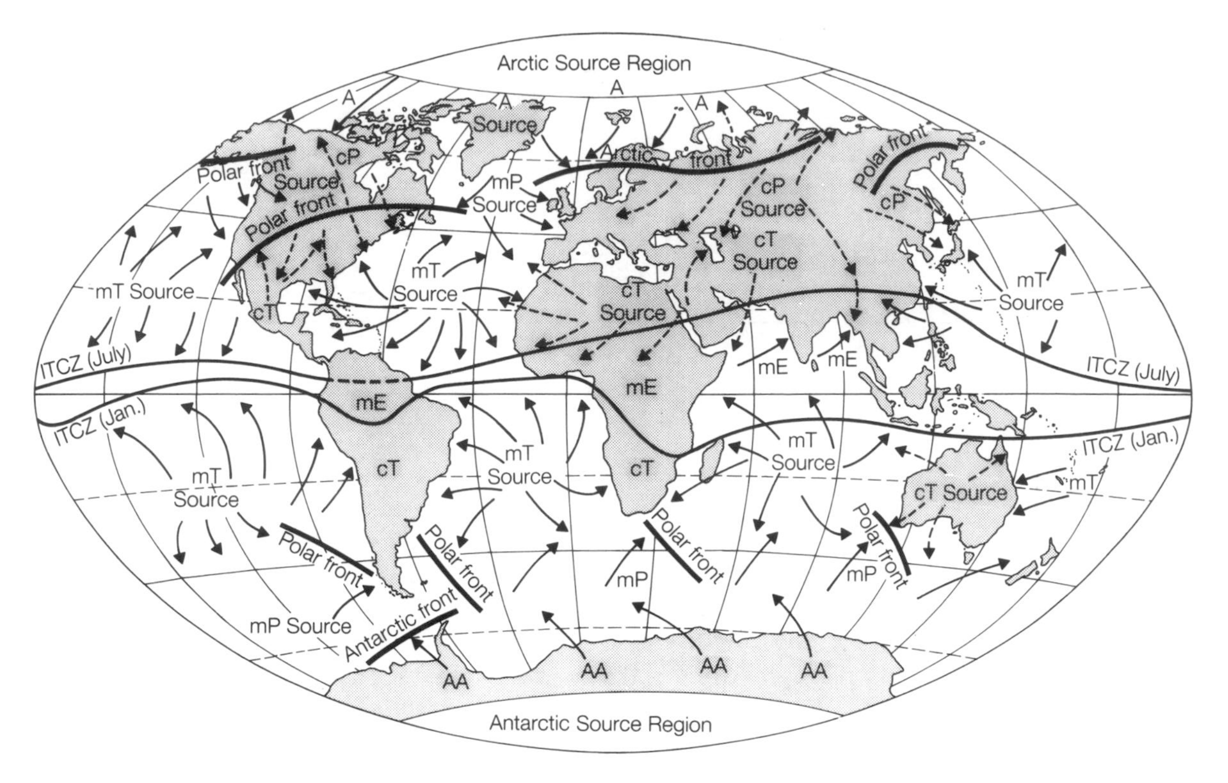

Figure 5.4 The global distribution of air mass source regions. The abbreviations are the same as those explained in table 5.1. The position of the fronts is for July. The dashed arrows represent continental air masses.

dry and stable. When it crosses the North Sea it picks up some moisture, and may bring cloudy and drizzly weather.

In a similar way one can see the impact of various air masses on the USA. Most of the cold air masses are polar continental (cP) in origin, but from time to time arctic air masses cross the north-eastern USA. Polar continental air from source regions in northern Canada and Alaska regularly invades the Great Plains and the Midwest, sometimes stretching further south than the Gulf of Mexico. The polar maritime (mP) air masses have their origin mostly over the North Atlantic and North Pacific; the latter, which are pushed eastwards by the westerly wind systems, have a strong and widespread impact on weather across North America.

Most of the tropical maritime (mT) air masses move northwards from the Gulf of Mexico and the Caribbean, and they affect weather conditions east of the Rocky Mountains (particularly when the warm, moist Gulf air meets the cold, dry polar air coming down from Canada). Tropical continental (cT) air masses rarely pass over the USA because the source area in Mexico and Central America is relatively small. When they do cross the area, however, they bring warm, dry air into the Great Basin and the Southwest.

The air masses that affect Australia and New Zealand are discussed in window 5.1.

5.4 Cool Temperate Climates

The cool temperate climates can be differentiated from the warm temperate climates by one main characteristic: they possess a real cold season that retards or inhibits active plant growth and places a

Table 5.1 Classification of air masses

Major group	*Subgroup*	*Source regions*	*Properties at source*
Polar (P)	Maritime polar (mP)	Oceans poleward of approx. 50°	Cool, rather damp, unstable
	Continental polar (cP)	Continents in vicinity of Arctic Circle; Antarctica	Cold and dry, very stable
	Arctic (A) or Antarctic (AA)	Polar regions	Cold, dry, stable
Tropical (T)	Maritime tropical (mT)	Trade wind belt and subtropical oceans	Moist and warm, stability variable: stable on east side of oceans, rather unstable on west
	Continental tropical (cT)	Low-latitude deserts, chiefly Sahara and Australian deserts	Hot and very dry, unstable
	Maritime equatorial (mE)	Equatorial oceans	Warm, moist, generally slightly stable

Window 5.1 *Air masses in Australia and New Zealand*

In the Southern Hemisphere mid-latitudes, the largest air mass sources are over the great oceans so that most of them are maritime in origin. Three main types dominate the weather of Australia and New Zealand: tropical maritime, tropical continental and polar maritime. The effects of the different types are summarised thus (after Sturman and Tapper, 1996: 121–2):

- *Modified polar maritime (NPm)*. The Southern Ocean is the source region of this cold, moist and unstable air mass. Air originating on the edge of the Antarctic (55–68° S) affects southern parts of Australia and New Zealand, when strong southerly flow follows a cold front, bringing snow and sleet to low levels.
- *Southern maritime (Sm)*. This cool, moist air originates at lower latitudes in the Southern Ocean (35–55° S). It tends to be unstable at low levels and stable aloft, bringing cool, moist and cloudy weather with drizzle to southern Australia throughout the year. Larger quantities of rain are produced with orographic lifting, over New Zealand.
- *Tropical maritime Tasman (tTm)*. Originating in the north Tasman Sea, this air mass is warm, unstable and moist to high levels. With its higher temperatures, the air mass brings warm, cloudy and drizzly weather to the east coast of Australia and north-western New Zealand. Heavier rainfall occurs where there is orographic lifting.
- *Tropical maritime Pacific (pTm)*. This is warmer than the tTm air mass, as it originates further north over the tropical western Pacific, and affects North Queensland for much of the year, bringing heavy rainfall when associated with tropical cyclones. It may also affect parts of northern New Zealand.
- *Tropical maritime Indian (iTm)*. This air-mass type is sourced in the eastern Indian Ocean and is similar to pTm. It affects coastal areas of north-western Australia.
- *Equatorial maritime (Em)*. Very warm, moist and unstable air affects north and north-western Australia during the summer monsoon season. This equatorial maritime air brings very heavy rainfall and high humidity and can affect areas as far south as 30° S.
- *Tropical continental (Tc)*. Central Australia provides the source for this type, being very hot, dry and unstable in summer, but cooler in winter. Cloud and rainfall are limited by both the lack of moisture and the trade-wind inversion, which restricts vertical motion. Much of northern and central Australia is affected by this air-mass type for most of the year, and periodically it brings heatwave conditions to areas further south.
- *Subtropical continental (STc)*. South-central Australia provides the source region for this warm and dry air

mass where it dominates in winter. As with the tropical continental type, subsidence in the subtropical anticyclones of this region has a strong influence on the character of this air mass. The influence of subsidence inversions ensures that surface instability is generally unable to produce deep convective activity.

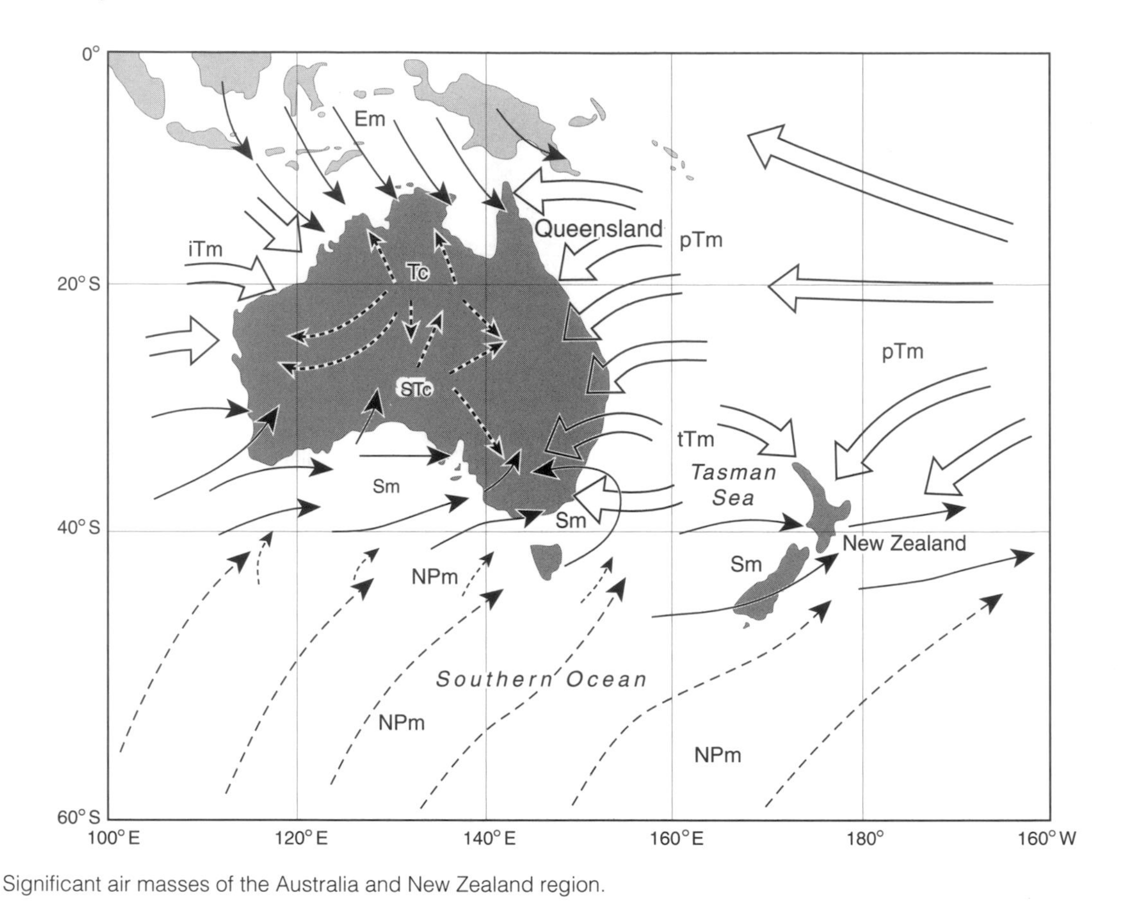

Significant air masses of the Australia and New Zealand region.

check on agricultural activity. The severity of the cold season increases from west to east, for winter warmth and rain come from the west, and this means that the relative position of land and sea are more important than latitude in this zone.

The depressions of the westerlies influence climate throughout the year in this zone rather than primarily in the winter as is the case in the warm temperate regions, though they tend to be repelled by the high pressures that develop over the cold continents in winter. It is therefore useful to make a distinction between maritime and continental types of climate. The areas under maritime or oceanic influence (table 5.2) tend to have a smaller annual range of temperature; a higher humidity and rainfall; a relatively evenly distributed rainfall with a tendency towards a winter maximum; a rather imperceptible gradation of the season with frequent lapses from, for example, spring into winter or autumn into summer; and rapidly fluctuating weather associated with the passage of fronts and anticyclones.

Table 5.2 Representative climatic data: cool temperate climates

	Year	*J*	*F*	*M*	*A*	*M*	*J*	*J*	*A*	*S*	*O*	*N*	*D*
Cool temperate oceanic climate													
Portland, Oregon, USA (45 °N)													
Temperature (°C)	12	4	6	8	11	14	16	19	19	16	12	8	5
Precipitation (mm)	1112	170	140	122	79	59	40	15	15	48	84	165	175
Hokitika, New Zealand (42 °S)													
Temperature (°C)	12	16	16	15	13	10	8	7	8	10	12	13	14
Precipitation (mm)	2951	249	186	247	234	249	247	229	239	234	299	269	269
Cool temperate continental climate													
New York City (41 °N)													
Temperature (°C)	11	−1	−1	4	9	16	21	23	22	19	13	7	1
Precipitation (mm)	1067	84	84	86	84	86	86	105	110	86	86	86	84
Moscow (56 °N)													
Temperature (°C)	4	−11	−9	−5	3	12	17	19	17	11	4	−2	−8
Precipitation (mm)	534	28	25	30	38	48	51	71	74	56	35	40	38

The importance of degree of *continentality* is brought out very clearly in figure 5.5. This diagram shows the annual temperature regimes for five stations all of which are at approximately 52° N. As one moves eastward, away from the moderating influence of the Atlantic Ocean, the severity of the winter cold becomes greater, as does the intensity of the summer heat. The equability of the climate enjoyed in the western parts of Ireland contrasts starkly with the extremes of climate endured in the eastern parts of Russia.

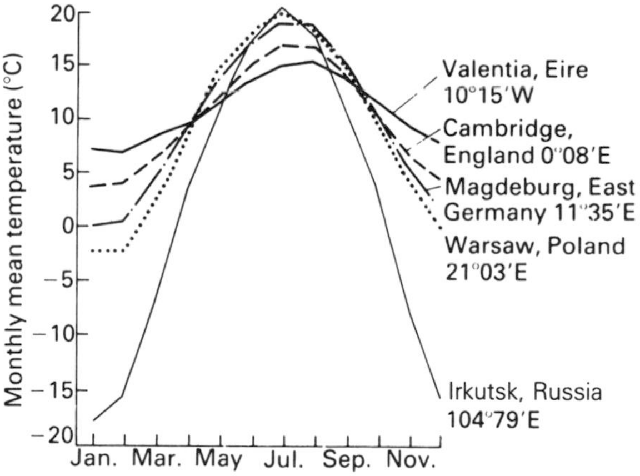

Figure 5.6 shows the average annual temperature range at the earth's surface, this being the difference in mean surface air temperature between the warmest and coldest months. This demonstrates very clearly just how extreme the climate conditions are in the interiors of North America and Eurasia.

Figure 5.5 Annual temperature regimes for five stations along latitude 52° N with mean monthly temperatures corrected to sea level.

5.5 Western Margin Warm Temperate Climates (the Mediterranean Type)

The area that comes under the influence both of the westerlies and the trade winds, according to season, is a zone of climatic transition. This warm temperate or subtropical zone derives its summer influences from the east and its winter influences from the west; summer is therefore continental on western margins and maritime on the eastern; winter, conversely, is maritime on western margins and continental on eastern. Clearly, therefore, eastern or western marginal situation must be a fundamental criterion for the subdivision of warm temperate climates.

The western margin type of climate is often termed 'Mediterranean', although it occurs in locations other than the Mediterranean basin, notably California, Chile, the Cape area of South Africa

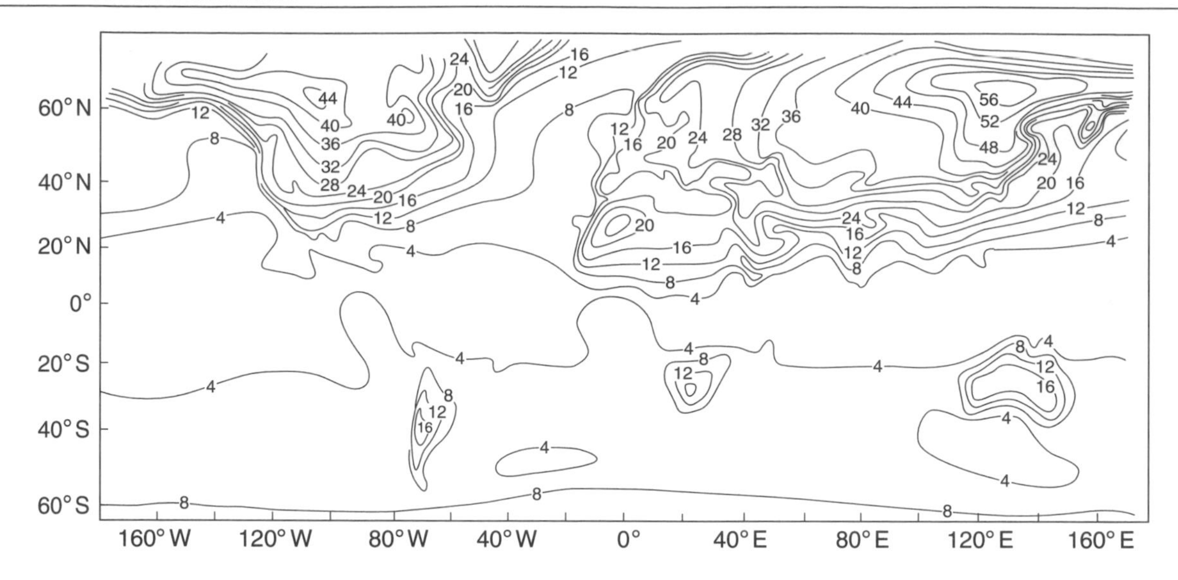

Figure 5.6 Average annual temperature range at the earth's surface (°C). The annual temperature range is the difference in mean surface air temperature between the warmest and coldest months.

and western Australia. While there are differences both between and within these different areas, this climatic type has certain essential ingredients: a winter incidence of rainfall and a more or less complete summer drought; hot summers and mild winters; and a high amount of sunshine, especially in summer.

The annual temperature range is greater here than in the tropics, especially when one moves away from maritime influences, and the daily range of temperature is also considerable, especially in the dry summer months (table 5.3). Rainfall levels (produced mainly by the passage of westerly depressions in the winter months) tend to be between 350

Table 5.3 Representative climatic data: warm temperate climates

	Year	*J*	*F*	*M*	*A*	*M*	*J*	*J*	*A*	*S*	*O*	*N*	*D*
Warm temperate western margin (Mediterranean type)													
Haifa, Israel (33 °N)													
Temperature (°C)	21	14	14	16	19	23	25	28	28	27	24	21	16
Precipitation (mm)	621	180	145	23	18	3	0	0	0	0	13	69	170
Perth, Australia (32 °S)													
Temperature (°C)	18	23	23	22	19	16	14	13	13	14	16	18	22
Precipitation (mm)	862	8	13	18	40	125	175	165	145	84	54	20	15
Warm temperate eastern margin													
Charleston, South Carolina (33°N)													
Temperature (°C)	19	10	11	14	18	23	26	28	27	25	20	14	11
Precipitation (mm)	1207	76	79	84	61	84	130	157	165	132	94	64	81
Sydney, Australia (33 °S)													
Temperature (°C)	17	22	22	21	18	15	12	11	13	15	17	19	21
Precipitation (mm)	1210	91	112	125	137	130	122	127	76	74	74	71	71

and 900 mm per year. Westward-facing shores backed by mountain ranges may have rather high totals. Some of the mountains of Dalmatia to the east of the Adriatic Sea receive over 4500 mm, making them one of the wettest spots in Europe. The length of the wet and dry seasons tends to vary with latitude. Polewards, the rainy season lengthens at each end until the dry season can no longer be said to exist, and one enters the zone of the cool temperate climate. Tunis, on the south side of the Mediterranean, has five months with less than 25 mm of rain, while Genoa, just to the south of the Alps in Italy, has none. Equatorwards, the winter rains begin later and later and cease earlier and earlier until one is in effect in the zone of the trade-wind deserts (see chapter 6).

One of the most interesting climatic features of the Mediterranean lands is the occurrence of a number of local winds of distinctive character. These are associated with the movement of depressions. A moving depression involves air masses originating both on its poleward and its equatorward side, and therefore both cold and warm winds may result. Polar air masses associated with depressions may cause cold winds to burst into normally milder areas, especially if topographical conditions are favourable. In the Mediterranean the *bora* of Croatia and the *mistral* of the south of France blow from the winter high-pressure areas over continental Europe, moving with great ferocity when channelled between uplands. The mistral, for example, is often channelled down the Rhône Valley between the Alps and Massif Central. Comparable winds elsewhere include the *norte* and *papagayo* in Mexico, the *pampero* in the Argentine, and the southerly *burster* in New South Wales, Australia.

A depression moving eastwards along the Mediterranean basin may sometimes lead to an outbreak of hot air from the Sahara. The *gibli* in Tunisia and Libya, the *leveche* in Spain, the *scirocco* in Italy and the *khamsin* in Egypt are notable examples of such winds. They are often hot, dusty and excessively dry, though sometimes, if they have crossed the sea, they may become very humid and unpleasantly sticky. Both the very cold winds and the very hot winds may have most unfortunate effects on crops in the region.

5.6 Eastern Margin Warm Temperate Climates

On the eastern sides of the continents the transition between the trade winds and westerly circulations is represented by a type of climate that, while generally sharing the mild winters and hot summers of the Mediterranean climate, differs fundamentally in the amount and distribution of precipitation. Because they are on shore, the trade winds that bring summer drought to western margins are here rain-bearing; the westerlies that bring cyclonic rain to western margins are here continental; and depressions are less vigorous. Winter rain is thus relatively less than in the zones with a Mediterranean climate, but summer rains are relatively generous, and tropical cyclones may bring severe storms. Areas with this type of climate include Argentina, Uruguay and south Brazil, the south-east coast of South Africa, the Gulf-Atlantic states of the United States and south China. Rainfall is reasonably well distributed (table 5.3), winters are mild, and summers, because of high temperatures and high humidity, can be rather oppressive.

Plate 5.1 In high latitudes or high altitudes in mid-latitudes one of the major vegetation types is the boreal forest, dominated by evergreen conifers. This example comes from the Front Range of the Rocky Mountains in Colorado, USA.

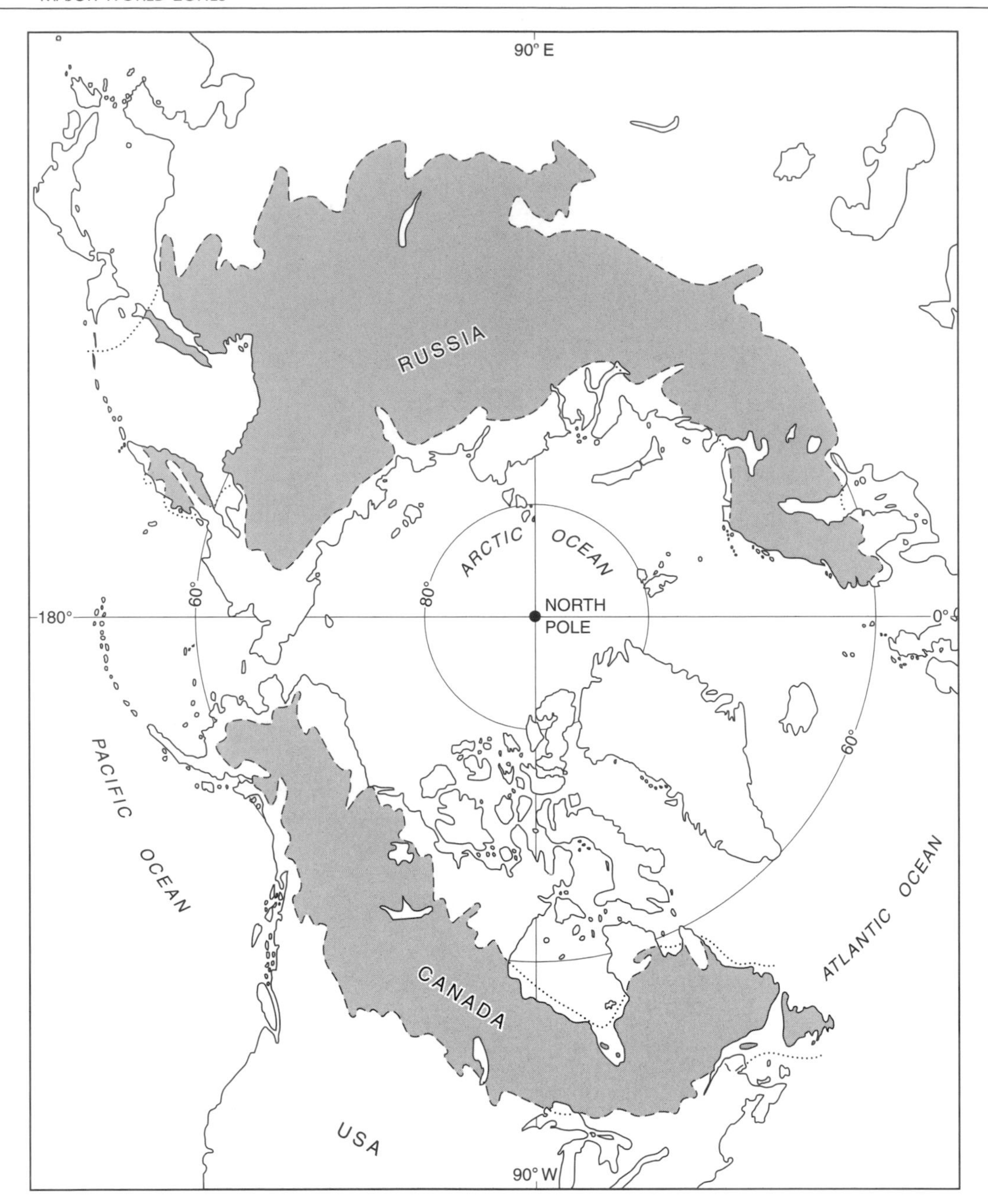

Figure 5.7 The boreal forest, shaded, is one of the world's greatest biomes and covers huge areas in Canada, Scandinavia and Russia. It is composed of hardy species of pine, spruce, larch and fir, though there may be some deciduous hardwoods such as birches, poplars, willows and alders.

5.7 Boreal Forest

To the equatorward side of the tundra zone, summer temperatures gradually rise and trees begin to become an important component of the flora (plate 5.1 and figure 5.7). Evergreen conifers start to thrive, and they have certain adaptations suited for survival: provided that water is available, they can photosynthesise all the year round; the needles are a type of leaf that helps the trees to resist drought either when water is locked up in the form of ice or when strong winds increase transpiration rates; and the shape of the crowns helps them to shed snow so that branches are not snapped off by the weight.

Window 5.2 *Changing mid-latitude vegetation in a warmer world*

Increasing temperatures created by the enhanced greenhouse effect may have profound effects on the vegetation of mid-latitude vegetation belts. For example, the figure shows predicted changes in the northern and southern boundaries of boreal forest in response to climate warming associated with a doubling of atmospheric carbon-dioxide levels. Note how the southern boundary moves from the southern tip of Scandinavia to the northernmost portion.

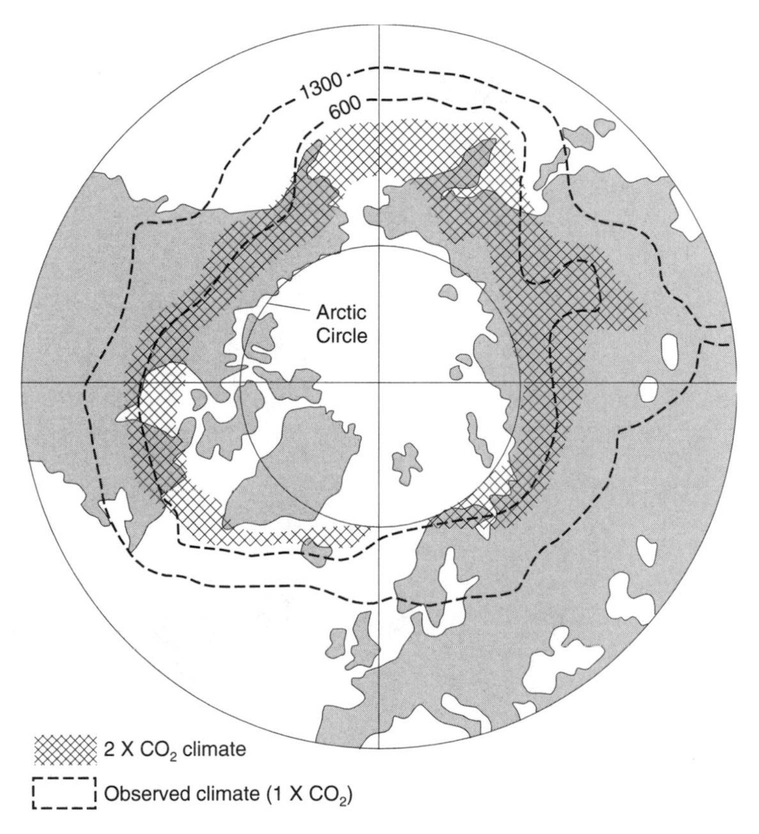

The northern and southern boundaries of the boreal forest are approximately defined by the 600 and 1300 growing degree-day isopleths. These are shown in their current positions and in the positions they would occupy under a warming associated with a doubling of CO_2 levels.

It has also proved possible to model the potential changes in the range of certain tree species in eastern North America for a doubling of atmospheric carbon-dioxide levels. The differences between their present ranges and their predicted ranges in the middle of the twenty-first century is very large.

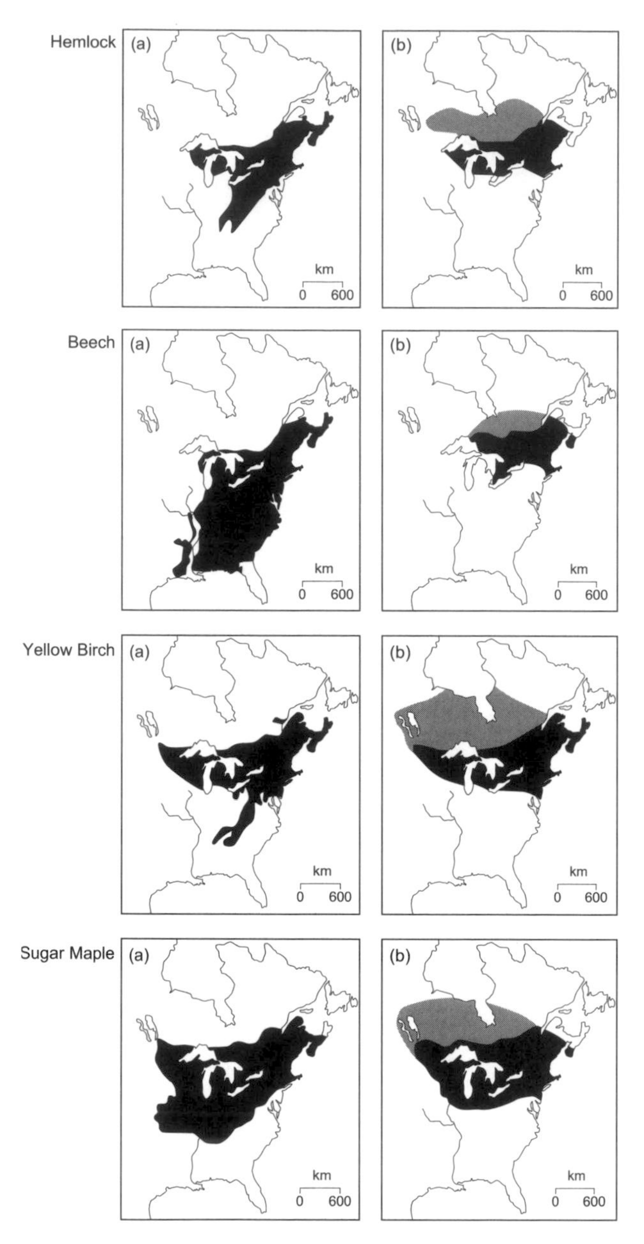

Present and future ranges of four tree species in eastern North America: (a) present range; (b) range in 2090 AD under the GISS GCM 2 × CO_2 scenario. The black area is the projected occupied range considering a rate of migration of 100 km per century. The grey area is the potential projected range with climate change.

Large trees up to 40 m high form a fairly continuous *stand* that permits relatively little light to penetrate to the ground, so that a lower layer of trees is uncommon. None the less, there is usually a fairly continuous ground cover made up of lichens, mosses, low shrubs such as crowberry, and bog-moss like *Sphagnum*. Which of these occurs depends on local conditions of drainage and light.

Spruces (*Picea*) are the dominant tree, but firs (*Abies*), pines (*Pinus*) and larches (*Larix*) may also be important. On the boundary with the tundra, birch scrub may form an important belt of vegetation, while in the Rocky Mountains the deciduous aspen (*Populus tremuloides*) occurs at the tree-line between the alpine grass and herb communities and the evergreens.

The boreal forest is quite productive (see section 3.1), generally producing about $800\,g\,m^{-2}\,yr^{-1}$ of vegetable matter (compared with up to 1200 for the deciduous forest of mid-latitudes). Because of the generally low temperatures, organic matter decays rather slowly, so that litter tends to accumulate as a thick layer on the forest floor. From time to time this litter may be consumed by, and provide fuel for, forest fires. Potential global warming may in the future cause major modification of the boreal forest (window 5.2).

5.8 Deciduous Forest

Equatorwards of the boreal forest comes a forest type in which growth is seasonal; for the trees are deciduous, losing their leaves in the winter. The deciduous habit of the main trees can be seen as a form of dormancy, which is a response to the low winter temperatures and the freezing of water. The dominant trees are 40–50 m in height, and their leaves tend to be broad. They are sometimes hosts for climbers such as ivy (*Hedera helix*), and epiphytes such as mosses, lichens and algae grow on the trunks. Dominance by two or three species is common, and single-species stands are often found. The dominant trees are typically oaks (*Quercus*), elms (*Ulmus*), limes (*Tilia*), beeches (*Fagus*), chestnuts (*Castanea*), maples (*Acer*) and hickories (*Carya*). The species density of trees never reaches the levels found in the tropics, the maximum being about eight species per acre in Europe and 40 in North America. Where light penetrates the canopy formed by these dominants a shrub layer may form, and its species may include hazel (*Corylus*), holly (*Ilex*) and members of the *Rosaceae* (such as hawthorn).

5.9 Mid-latitude Grassland

In the continental interiors of the mid-latitudes, especially where a rainfall total of 300–500 mm occurs, there are some extensive areas of grassland. They have many names (*steppe* in Eurasia, *prairie* in North America, *pampas* in Argentina, *veld* in South Africa etc.). Much of this grassland has in the past two centuries been heavily modified by human activity – ploughed up for wheat or grazed with cattle and sheep – but it is still far from clear whether these grasslands are by origin man-made or are a natural response of plants to certain environmental pressures in this particular environment. It used to be fairly generally thought that prairies were essentially a climatic phenomenon, and botanists tended to believe that, under the prevailing conditions of soil and climate, the invasion and establishment of trees was greatly hindered by the presence of a dense sod. High evapotranspiration levels were thought to give a competitive advantage to herbaceous plants with shallow, densely ramifying root systems, capable of completing their life-cycles rapidly.

However, various lines of evidence have been used to suggest that trees can survive, and indeed flourish, in these areas. Trees planted in plantations or groves have been successful even during drought years. Furthermore, along escarpments and other abrupt breaks in topography, trees are often found in an otherwise treeless area. One possible explanation for this is that fire, whether natural or man-made, is more effective on flat, level surfaces, where there are higher wind speeds and no barriers to its movement, than along escarpments and the like. What may well have been underestimated by some earlier workers is the ability of early and primitive man to use fire – to clear land for agriculture, to improve grazing by removing dead grass and encouraging new growth, to drive wild animals and so on. Many additional fires may have been accidental. Natural fires caused by lightning may also have played a role in denuding flat areas of tree cover, as may intense grazing by wild animals such as bison.

5.10 Mediterranean Evergreen Woodland

The warm temperate Mediterranean climate belt, at least in the moister areas, was probably originally covered mainly by open woodlands consisting primarily of various types of evergreen oaks, though there were also some conifer forests, with pines, firs, cypresses and cedars.

However, where rainfall conditions are less favourable, and/or where human activity has degraded the natural forest through grazing, burning and cutting, the forest appears to degenerate into a low scrub (the *mallee* of Australia, the *garrigue* and *maquis* of southern Europe and the *chaparral* of California). The dominants in this form of vegetation are low trees and shrubs, usually about 3–4 m high. In the Mediterranean proper they include the wild olive (*Olea europœa*), pines, a strawberry tree (*Arbutus unedo*) and some heathy shrubs such as *Erica, Ulex* and *Genista*.

5.11 Soils of Cool Temperate Climates

In areas with cool temperate climates there are four main soil types that have a wide distribution: podzols, brown earths, leached soils and grey soils. There are, of course, many divergences from these zonal types, depending on the interplay of parent material, vegetation type, soil drainage, maturity and other such factors; but what they have in common is the fact that in such cool climates there is sufficient precipitation to maintain an overall downward leaching of any soluble soil constituents. In very general terms, the podzols are typical of the boreal forest; brown earths and grey-brown podzolic soils are normally associated with the deciduous forest; and the grey soils are representative of the transition zone between forest and grassland (steppe).

Podzols (figure 5.8a(i)) are characterised by the presence, just below the surface, of an ashy-coloured horizon. In the Northern Hemisphere they are generally regarded as being found in a circumpolar belt approximately from the Arctic Circle southwards to the latitude of the Great Lakes in North America and to about 50° N in Europe. They develop best on permeable gravels and sands. Podzols are soils in which a redistribution of the soil constituents has taken place through the action of downward-percolating rainwater, which contains decomposition products derived from the thick surface organic layer. This transports iron downwards, and it may accumulate in the B horizon as an iron-enriched layer called an *ironpan* or *hardpan*.

Brown earths and related soils derive much of their character from their association with the deciduous forests. With the annual leaf fall of the deciduous trees and the contribution from the smaller shrubs, herbs and grasses, the litter produced is more varied than that of the coniferous forest. It also has a higher nutritive status and is more easily digested by the many soil-living fauna. Thus there is an efficient breakdown of litter, and the humus is incorporated into the soil by the action of earthworms. The brown earth profile frequently lacks the superficial horizons of organic material associated with the podzols, but leaching is still important and the upper horizons are normally neutral to acid, carbonates having been largely leached away (figure 5.8a(v)).

The *grey soils* (figure 5.8a(ii)) tend to form in continental environments and are found between the podzols and the chernozems. Rainfall levels tend to be modest so that leaching is not as complete as it is in the brown earth zone, and some calcification occurs in the B horizon. A major feature of these soils is the accumulation of clay, derived from the upper horizons, in the B horizon.

5.12 Soils of Warm Temperate Climates

In the Mediterranean region of climate the length of the summer drought is the most important zonal factor in soil formation. Increasing length of drought produces a sequence of soils ranging from brown earths (figure 5.8a(vi)), developing in a leaching environment with less than one month of summer drought, through red and brown Mediterranean soils (figure 5.8a(iii)) to cinnamon soils (figure 5.8a(vii)), where five or six months of drought lead to some soil calcification.

Brown Mediterranean soils developed under the

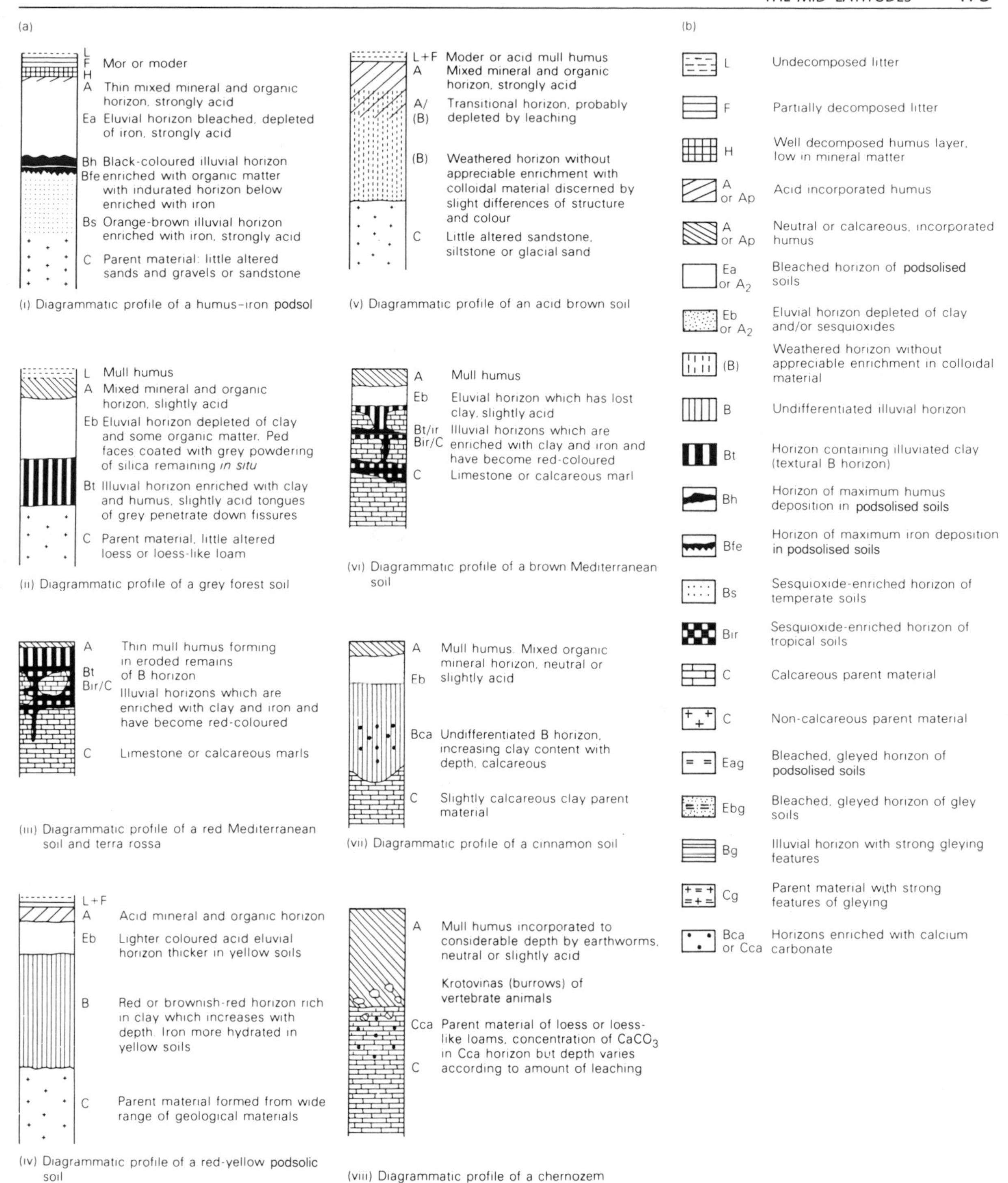

Figure 5.8 (a) Some characteristic soil profiles from mid-latitudes; (b) some of the abbreviations used to describe different characteristics.

original mixed Mediterranean forest dominated by oaks and pines. They are characterised by a brown colour, a friable humus-rich A horizon and a denser and less friable B horizon. In the lower horizons of these soils, clays and complexes of iron and silica are deposited to give the characteristic rich red colour. The *red Mediterranean soils* have resulted in some cases from the development of soils on the eroded remains of the brown Mediterranean soils. Other red soils are formed upon the relic clays resulting from limestone weathering. Where a longer summer drought is experienced, the characteristics of the soil gradually become more typical of semi-arid areas: a light, yellowish brown soil results, called a *cinnamon soil*. The A horizon is moderately rich in organic matter, calcium carbonate concretions occur below about 30 cm in drier localities, and clay content increases from the lower part of the A horizon downwards through the profile.

The soils of the warm temperate east margin climates are typically *red and yellow podzolic soils* (figure 5.8a(iv)), and as they occur between the major areas of podzolisation (poleward) and ferrallitisation (equatorward), evidence of both processes is present. The moist climate throughout the year is conducive to leaching, so that the surface horizons tend to be acid. The clay content increases with depth, partly by translocation of clay down the profile and partly because of clay mineral formation *in situ*. The red and yellow colours of these soils indicate different degrees of iron oxide hydration, with the red soils developing under drier conditions and the yellow developing where moister conditions prevail.

In the warm temperate continental interiors occurs one of the world's most famous zonal soil types: the *chernozem*, or black earths (figure 5.8a(viii)). They are a feature of the Russian steppes and the prairies of the United States and Canada. Many of them have developed on great expanses of aeolian silts, called *loess*, which were laid down on the margins of the greatly expanded Pleistocene ice sheets by winds transporting fine material from areas of moraine and outwash (see section 5.14). The chernozem soils are associated with grassland vegetation. Under the prevailing climatic conditions of this zone, grass growth is vigorous in late summer and the frosts of the winter arrest the process of decomposition. Consequently, losses of organic material are minimised. The loess parent material also provides a source of calcium carbonate. There is a rich soil fauna, including miscellaneous burrowing animals, that cause some mixing in the soil profile. There is a deep, humus-enriched A horizon, while lower down there is a concentration of calcium carbonate caused by mild leaching from above followed by re-evaporation in the warm summers. In drier areas the soils may be less rich in organic matter and are called *chestnut soils*.

5.13 The Impact of Climatic Change on the Landscape

As we saw in chapter 2, the past couple of million years have been marked by very frequent and extreme changes in climate. For much of this time, the world has had a climatic zonation that is very different from what we see today, and this is especially true when we look at the mid-latitudes, for during the colder phases of the Pleistocene they were greatly affected by the expansion of both permafrost and glaciers. Many mid-latitude landscapes, sediments and soils still show, often to a marked degree, the imprint of these changes.

In North America a great sheet of ice was more or less continuous from Atlantic to Pacific, and was composed of two main bodies: the Cordilleran glaciers associated with the coastal ranges and the Rockies, and the great Laurentide ice sheet. At its maximum, the ice extended to approximately the present positions of St Louis, Missouri, and Kansas City. In the British Isles (window 5.3) there were ice sheets that varied in their extent during different glacials, but merged with those of Scandinavia. At its maximum the ice extended as far south as Essex, Oxfordshire and the northern coast of the south-west peninsula. In Europe the alpine glaciers formed a great sheet that may have been over 1500 m thick in places, and reached down to altitudes as low as 500 m on the north side of the Alps and 1000 m on the south side. There was also a great sheet that covered Scandinavia and reached into the North European Plain. This sheet may have been over 3000 m thick. Another extended ice sheet occurred in Siberia. Other major Pleistocene ice sheets developed in Argentina, Tasmania and New Zealand.

Window 5.3 Glacial and periglacial conditions in the British Isles in the Pleistocene

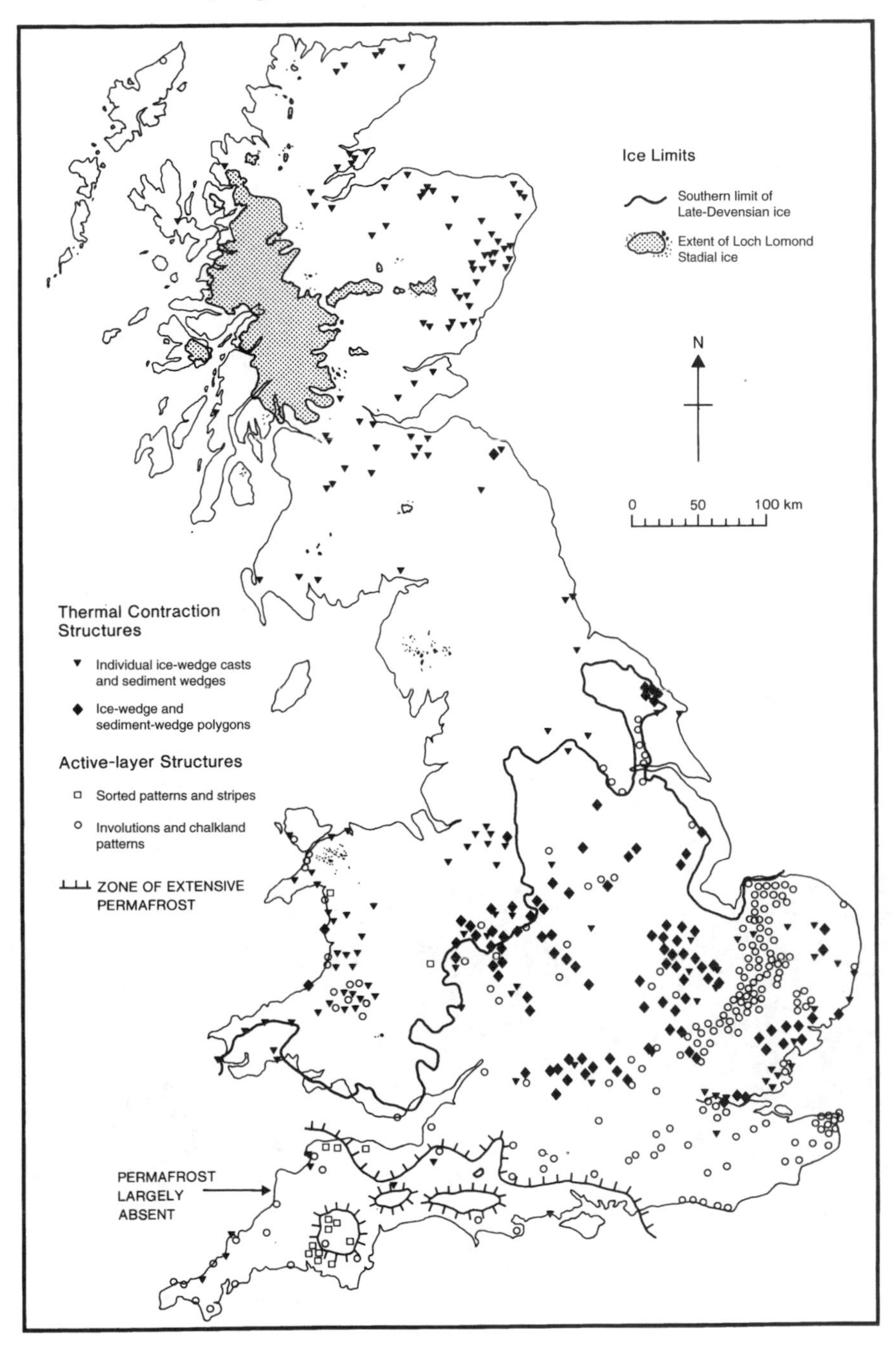

Periglacial phenomena in Britain.

The maximum extent of glaciation in the British Isles is the subject of some debate but it appears that around 450 000 years ago (in the Anglian glaciation) ice sheets extended as far south as the Isles of Scilly, Bristol, Oxford and Essex.

In the last glaciation, which reached its peak around 20 000–18 000 years ago (the Dimlington stadial), ice was less extensive and large expanses of central and eastern England were not covered. The ice retreated after that time but reappeared briefly around 11 000 years ago in highland areas (the Younger Dryas or Loch Lomond stadial).

Areas that were not glaciated were subjected to severe periglacial conditions, and periglacial structures are particularly widespread in areas that were to the south of the limit of the last full glaciation.

As a consequence of these great glacial expansions, many mid-latitude areas show all the imprints that glaciers can create in a landscape (see section 4.6). The mountains of upland Britain show many of the classic features of glacial erosion (U-shaped valleys, amphitheatrical cirques, lakes, hanging valleys, roches moutonnées, drainage diversions etc.), while British lowland areas like East Anglia show a mantle of till and outwash deposits.

Beyond the limits of the great Pleistocene ice sheets, particularly in Europe, there were great areas of open tundra. These areas were frequently underlain by permafrost, and we can find evidence for its presence in southern Britain in the form of fossil ice-wedge polygons and pingos. In East Anglia (plates 5.2 and 5.3), Kent, the Severn and Avon Valleys and parts of Devon there are wide expanses of patterned ground of various types that result from tundra conditions and probably only the extreme tip of the south-west peninsula was unaffected. Per-

Plate 5.2 In the Breckland of East Anglia, near Thetford, stripes of heather (*Calluna*) are one type of pattered ground that is a relict feature of extreme periglacial conditions during the late Pleistocene. Active examples of such features may be found in Alaska today.

1 Late Pleistocene linear sand dunes, now wooded, passing under the Holocene alluvium of King Sound, near Derby, north-western Australia. The dunes formed under drier conditions than those of the present day and at a time when sea levels were low. The rise of sea level in the Holocene – the Flandrian Transgression – flooded them.

2 A raised beach near Portland Bill, Dorset, southern England. At the base is a sloping rock-cut platform. On top of that is a layer of coarse boulders that represent an old storm beach. Above the boulders is an accumulation of beach shingle. This in turn is capped by yellowish sediments that may be of periglacial origin.

3 A typical glaciated landscape in the French Alps near Chamonix. Notice the frost shattered pinnacles (*aiguilles*) and the great degree of glacial incision.

4 The Arolla Glacier in the Swiss Alps. The line of debris running along the glacier is a medial moraine.

5 Loess deposits, such as these from the Tajik Republic in central Asia, can reach great thicknesses. However, their stratigraphy is complex, for at times of reduced accumulation and of land surface stability soils may have developed. This example shows several ancient soils. They may provide very important information on changing environmental conditions in the Pleistocene.

6 A mobile barchan in the Western Desert of Egypt near Kharga. It has submerged a telephone line.

7 The Namib Desert in southern Africa is one of the driest places on the face of the earth, and large dunes some 100 metres high and tens of kilometres long have developed. Sand movement on these linear dunes is facilitated by the almost total absence of vegetation in this hyper-arid environment.

8 The former course of the main coastal railway line in the desert state of Namibia, south-western Africa. The line was invaded by dunes and has now been relocated to another course. Shifting dunes, especially barchans, cause many problems for engineering structures, and there are various techniques that can be used to try to reduce their mobility.

9 Monsoon flooding in Bangladesh is caused both by storm surges coming from the Bay of Bengal and by flooding rivers coming from the mountains of High Asia. It is possible that in a warmer world, resulting from the greenhouse effect, such floods may become more common and severe, as cyclones become more frequent and sea-level rise takes place.

10 Tropical rainforest on the steep slopes behind Rio de Janeiro, Brazil.

11 Along many tropical coastlines mangrove swamps are a major coastal environmental type. They are diverse and productive ecosystems, important wildlife habitats, effective barriers against coastal erosion, and – if managed with care – a valuable source of wood, fish and other economic resources. This example is from Mahe in the Seychelles. In many areas mangroves are being destroyed by human activities.

12 A small coral island from the Great Barrier Reef, Queensland, Australia.

13 A landscape of striking granitic inselbergs near Petropolis, Brazil.

14 Geikie Gorge, on the Fitzroy River in north-western Australia, shows the importance of high flow events resulting from the intense rainstorms of the tropical summer months. The white tide-line is produced when the river rises as much as 18 metres above its winter level.

15 The Hunza Valley in the Karakoram Mountains of north-west Pakistan. This is a very hazardous area because of the large number of landslides, shifts in glacier courses, and the flooding and shifting of the sediment-laden river.

16 The chalk coastline at Durdle Door, Dorset, southern England, demonstrates two important geomorphological phenomena: the role of rockfalls in cliff evolution, and the development of a basal shore platform. Although wave action may be a major process in platform formation, other processes including various types of weathering may also play a major role.

Plate 5.3 During the late Pleistocene the Stiperstones of Shropshire were just on the margins of a great ice sheet. Severe frost weathering caused quartzite rocks to be shattered. The debris formed blockstreams, some of which are sorted into periglacial patterns of stripes and polygons.

mafrost was also important in Europe; although there is still some dispute as to its precise maximum extent, it appears that only the south and central Balkans, peninsular Italy, the Iberian Peninsula and south-west France were unaffected. The degree of change has been substantiated by the study of pollen remains in old peat deposits. Much of the area, of course, was covered by cold steppe and tundra vegetation, rather than by the forests that are the natural vegetation of Europe during interglacial conditions.

5.14 Loess Sheets

One of the great features of much of the mid-latitudes is the development of major sheets of silty sediment, of primarily windblown origin, called *loess* (window 5.4). For vast areas (at least $1.6 \times 10^6\,km^2$ in North America and $1.8 \times 10^6\,km^2$ in Europe), loess blankets pre-existing relief, and in China it has been recorded as reaching a thickness in excess of 330 m. In the Missouri Valley of Kansas in the United States the loess may be 30 m thick, in Argentina it reaches 100 m, in New Zealand it reaches 18 m, in the Rhine Valley 30 m and in Tajikistan it approaches 200 m (see colour plate 5). The maximum depth of loess in Britain is only a couple of metres.

The distribution of loess throughout the world suggests that much of it was derived by deflation (removal by wind) of fine material from the great outwash plains that lay equatorward of the great Pleistocene ice sheets, though in some areas it may have been derived from deflation from desert basins – indeed, there is considerable controversy as to which source is the most important. In any case, there is no doubt that loess has been a most important parent material for soils, especially chernozems; that it has provided light soils that could be readily ploughed by early farmers in Europe; and that, if subjected to erosion by water, it can form some dramatic badlands types of scenery and contribute large amounts of sediment to rivers like the Yellow River in China.

5.15 Dry Valleys and Mis-fit Streams

One of the most striking features of the geomorphology of much of lowland England, especially on limestone, chalk and sandstone, is the presence of dry valleys (plate 5.4) and mis-fit streams. *Mis-fit streams* are streams that are smaller (often by a ratio of about 10 : 1) than the valleys in which they flow, the meanders of the present streams being very much smaller than the meanders made by the valleys as a whole. Some of the streams of the Cotswold dip-slope have very small discharges at present, for example, but they occur in valleys with extremely large meanders, which appear to be quite out of proportion to the streams they contain. They

Window 5.4 The loess of China

Loess is a widespread surface material type consisting largely of windblown silt, but the greatest expanse of this material that is known is in central China, in the provinces of Shanxi, Shaanxi, Gansu and Ningxia. This loess plateau has an area of 317 600 km^2, two and a half times larger than England, and near the city of Lanzhou is the thickest known exposure in the world (over 335 m). The loess is primarily made of quartz silt, and has been accumulating at least since the beginning of the Pleistocene (i.e. over the past 1.6 million years). It tends to become finer in a south-easterly direction.

Two main views have been put forward about the origin of the loess material, though both agree that it has been transported to its present location by winds blowing from the interior. One school of thought believes that the quartz silt that makes up the loess originated as fine material that was produced by the weathering of rocks (by salt, ice crystal growth and other processes) in the deserts upwind. Another school of thought sees the loess silt as having originally been produced by glacial grinding of sediments in the Pleistocene, when glaciers and ice caps may have been more extensive in the high mountains and plateaux of Asia. The former view currently enjoys greater support.

The Loess Plateau is downwind from the world's greatest mid-latitude desert area. Winds blowing from the deserts carried dust which settled out in the more humid lands to the south-east to form the loess.

The loess of China has many human implications. It is a very fertile parent material for soil, and is easily worked, but it is easily eroded, suffers from some severe types of mass movements and slope failures, and delivers large amounts of sediment that choke some of China's greatest rivers.

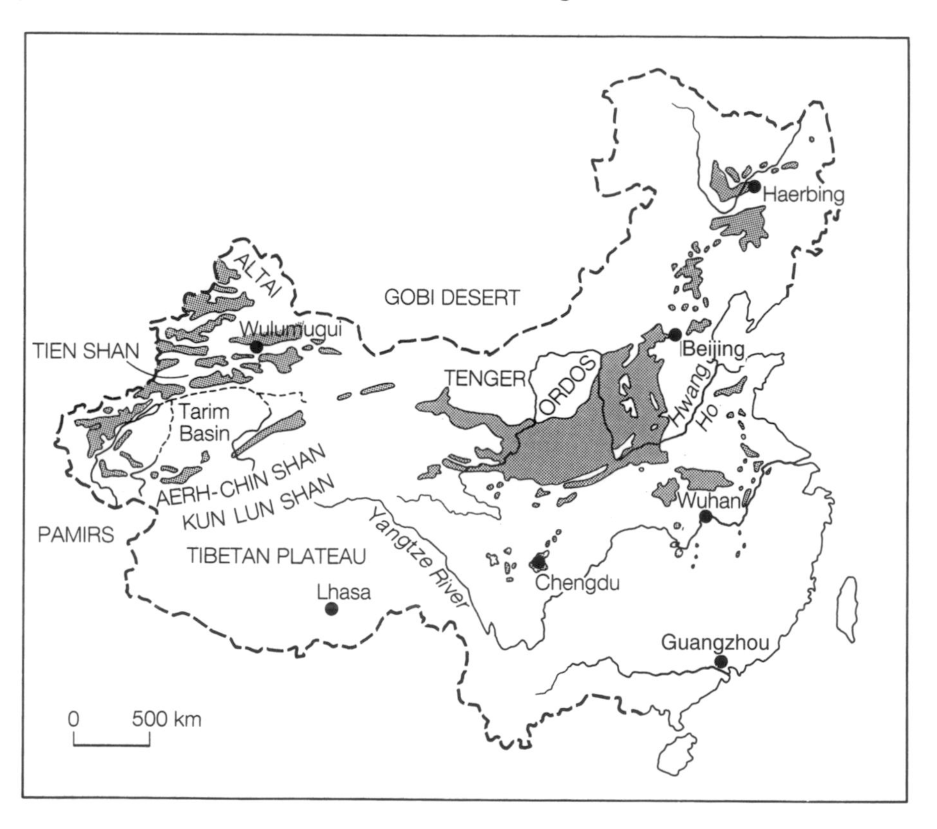

The distribution of loess in China.

The loess lands of China are heavily cultivated. This picture shows the greatest thickness that the loess can attain, the way in which attempts are made to farm and conserve steep slopes by terracing, the development of major landslides (see middle ground) and the formation of depressions by subsurface removal of silty material.

are in a sense dry valleys that have not quite dried up.

There is no common consensus as to the causes of valley meander, mis-fit stream and dry valley development. Some of the main hypotheses that have been postulated are listed below. They can be categorised into uniformitarian hypotheses (those requiring no major changes of climate or base level, but merely the operation of normal processes through time), and palaeoclimatic hypotheses (associated primarily with the major climatic changes of the Pleistocene):

Plate 5.4 The chalklands of England, including the White Horse hill on the Berkshire Downs near Uffington, contain many dry valleys. This photograph shows a valley called the Manger. The furrows on the left side of the valley have an uncertain origin, but may have been cut by Ice Age avalanches.

Uniformitarian

Superimposition from a cover of impermeable rock or sediments
Joint enlargement by solution over time
Cutting down of major through-flowing streams
Reduction of precipitation in catchment area and fall of groundwater levels arising from scarp retreat
Cavern collapse
River capture
Rare events of extreme magnitude

Marine

Failure of systems to adjust to a falling Pleistocene sea level and associated fall of groundwater levels
Tidal scour in association with former estuarine conditions

Palaeoclimatic

Overflow from proglacial lakes
Glacial scour
Erosion by glacial meltwater
Reduced evaporation caused by lower temperatures
Spring snow-melt under periglacial conditions
Runoff from impermeable permafrost

The uniformitarian hypotheses cannot be dismissed, though many workers claim that most dry valleys are the products of cold phases of the Pleistocene. Some dry valleys do appear to head back into areas of impermeable rocks that overlie rocks like limestone, and in such cases it is entirely possible that the former were superimposed from a cover of clay, grit or some such impermeable stratum that has since been removed by erosion. Likewise, it is apparent that in a limestone area, as time goes on, the joints in the limestone will be progressively widened by solution so that the rock mass becomes more and more permeable, thereby reducing the volume of surface stream flow. In the same way, with the passage of time a master stream may cut down preferentially to the others and thereby abstract some of the water from their basins, thus making them go dry. The retreat of an escarpment through time is another factor that could reduce the catchment area of a stream, and it might also cause a reduction in groundwater levels. The collapse of underground caverns or the capture of one river by another (see section 11.7) could cause localised examples of dry valleys, though it would scarcely account for a widespread pattern. We must remember that occasional extreme rainfall can sometimes cause so-called dry valleys to flow, as was the case with the great storm on the Mendips in Britain in 1968 which caused substantial flow down Cheddar Gorge (plate 5.5).

None of the theories so far mentioned is based on an assumption of a change in sea level or a change of climate. However, most workers invoke some such major change to account for the development of dry valleys and mis-fit streams. Some geomorphologists, for example, have maintained that in the early Pleistocene there was a high stand of sea level at around 180 m in parts of south-east England, and that at that time groundwater levels would have been high, so that streams might have flowed on the chalk. As the sea retreated, however, the groundwater level might also have fallen, causing some of the valleys to go dry. Tidal scour may have played a role in widening some of these valleys.

The most popular theories, however, have involved changes of climate in the Pleistocene, for as we have seen, many parts of the mid-latitudes were subjected to a rigorous glacial or periglacial climate in the Pleistocene glacial phases. Large lakes were ponded up in certain areas by the glaciers, and overflow from such lakes could have carved some channels that are now dry. Some of the valleys might even have been scoured by glaciers themselves, or by meltwater running from them. Most importantly, however, cold climatic conditions would have caused low evaporation rates so that runoff could be a greater proportion of rainfall; much of the precipitation would have fallen in the winter months as snow, and when this melted in the summer thaw period rapid runoff would have occurred; and the presence of an impermeable permafrost layer would have enabled surface flow on rocks like limestone which are normally too pervious to permit it. Frost weathering might have broken some of the rocks up and rendered them more susceptible to fluvial erosion.

Some field evidence supports the idea that periglacial conditions may have been responsible, possibly in combination with some of the other hypotheses, for dry valley formation in south-east England. Many of the dry valleys are fronted by a fan of debris, called *coombe rock* or head, which is thought to be frost-shattered debris moved by solifluction. In some dry valleys on the chalk downs in Wiltshire and Dorset the coombe rock may contain extremely large boulders of Tertiary sandstones (called *sarsens*) which occur as block-streams, and once again solifluction is thought to have been the cause of their movement. Moreover, many of the dry valleys in the Chilterns do not seem to be related very strongly in their direction to the joints in the chalk, and this may be because they were fashioned by periglacial torrents working on rock that was impregnated with permafrost. Some of them also display marked slope asymmetry. Under periglacial conditions, the aspect of a slope may have a considerable influence on its form and development, for some slopes exposed to greater sunlight will be more subject to freeze–thaw and solifluction processes, and this will influence the steepness of the valley side slopes.

5.16 Tors

Another landform that has often been attributed to

Plate 5.5 The great Cheddar Gorge of the Mendip Hills in England is a dry valley cut into Carboniferous limestone. It was probably shaped under periglacial conditions, though its lower course may be flooded by extreme storms like that of 1968.

periglacial times is the *tor,* an upstanding mass of rock or boulders that rises above more gentle slopes around (plate 5.6). Such features are not restricted to periglacial areas – they are, for example, often developed on sandstone and granite outcrops in low latitudes – but frost action may be one of the many processes that favour their development. Certainly, many workers believe that the granite tors of Dartmoor, the gritstone tors of the Pennines or the quartzite tors of the Stiperstones in Shropshire are relict features that are inactive today.

One school of thought interprets mid-latitude tors as essentially angular landforms produced by frost shattering, with the weathered debris being transported away from the remaining, more resistant masses by solifluction. These tors tend to be known as *palaeo-arctic tors*. Under this hypothesis, the tors are the remnants of a frost-riven bedrock outcrop surrounded by a low-angle terrace (called an altiplanation terrace) across which the frost-shattered material is removed by solifluction and other mass-movement processes. On Dartmoor, the presence of great spreads of angular debris (called *clitter*), arranged in stripe and polygonal patterns and extending very long distances over quite gentle slopes, lends support to this idea.

However, a second school of thought argues that tors are prepared by deep weathering along joints to produce rounded subsurface core-stones (*woolsacks)* under warm, humid conditions in the Tertiary. It is maintained that these are exhumed by the stripping of the weathered bedrock or *regolith*

Plate 5.6 There are many examples of tors in Britain, and they have developed on a wide range of rock types. The Stiperstones in Shropshire (left) are composed of quartzite, the Pennine tors (top right) are often formed of gritstone, and the Dartmoor tors (bottom right) of granite. Frost weathering and solifluction may have contributed to their development.

(called 'growan' on Dartmoor) by solifluction. Such tors are regarded as being *palaeo-tropical forms*, analogous, but on a smaller scale, to the inselbergs and kopjes of tropical regions (see section 7.11).

5.17 Natural Hazards in Western Europe

Having considered the main characteristics of the present and past environments and landscapes of mid-latitudes, let us now move to a consideration of the natural hazards that afflict people within one mid-latitude environment, western Europe:

storms
gales
convective storms
hail
thunderstorms
tornadoes
snow, frost and cold
floods
  inland
  coastal
droughts
fog
air pollution
water pollution
land pollution
heavy metals
noise
geomorphological hazards
seismicity
mass movements
subsidence

coast erosion
biological hazards
human health hazards
plant and animal disease (e.g. Dutch elm; foot and mouth)
botanical infestations

As the above list shows, although this area is normally thought to be favoured with a relatively equable climate and relatively limited tectonic activity, there is none the less a wide range of natural events (sometimes exacerbated by man) that prove hazardous.

The first class of hazard is the storm. Because western Europe lies within the belt of westerly depressions and from time to time comes under the influence of disturbances with steep pressure gradients, gales can be strong and frequent, especially in the coastal fringes and at higher altitudes. A gale (in which the wind attains a speed of 34 knots or more over any period of ten consecutive minutes during the day) occurs on over 50 days in the year in parts of the Shetlands off north-east Scotland but only on 0.1 days in Kew (in the suburbs of London in south-east England). These gales cause considerable damage during the course of the year. In addition, crops are damaged, transport services delayed and sailors drowned.

An extremely severe wind storm was one which pounded Europe on 15–16 October 1987 (plate 5.7). Wind speeds exceeded 119 knots at Quimper in Brittany and the bell-tower of the medieval abbey at Caen in northern France was blown down. The storm crossed into south-east England, killing 19 people, destroying 15 million trees and depriving hundreds of thousands of homes of their electricity and telephones. The total cost to the insurance industry at 1988 prices was £1000 million. The storm also caused problems in Jutland (Denmark), in the Skaggerak and in south Norway.

Plate 5.7 On 16 October 1987, large areas of south-east England were afflicted by a ferocious storm, which caused much damage to buildings and trees. At Brasted Chart a majority of the trees on the estate were felled.

Western Europe is also influenced by various types of convective storm. Hailstorms do not reach the intensity or frequency of those in some other parts of the world, though in 1846 large hailstones are reported to have smashed 7000 panes of glass in the House of Commons in London! Of more significance is lightning associated with thunderstorms, which kills people and may disrupt electricity supplies. Thunderstorms and severe precipitation events may be increased in incidence through the building of cities (see section 16.1). This has been established for London, The Netherlands and several areas of Germany, including Hamburg and the Ruhr.

A tornado (plate 5.8) is a relatively narrow vortex (a few metres to hundreds of metres in diameter) of rotating winds, which spiral inwards and upwards to the base of a convective thunderstorm cloud. They are typically short-lived (often lasting only a few minutes) and of small area extent (track lengths being typically a kilometre or less, and widths 50–100 m). However, they cause great damage as a result of their swirling winds (velocities may exceed 200 knots), and suction and shearing effects. Larger tornadoes are related to the passage of well-defined fronts or troughs associated with rapidly deepening or very deep depressions. Although they can occur in any season, the greatest frequency is in autumn and early winter, when the presence of relatively warm seas surrounding Europe enhances instability as cold air sweeps from the east. By contrast, the lowest frequencies occur in the period from February to May, when the seas are normally coldest.

Plate 5.8 A tornado in Dade County, Florida, USA in 1925.

Droughts are another type of climatic hazard (plate 5.9). The impact of the severe drought of 1975–6 is sufficiently recent to show that western Europe is far from immune from the effects of occasional damaging drought. The heavy impact of drought over that period arose from the fact that it was particularly long sustained, with few relieving rainy spells. Over much of southern England (figure 5.9), northern France, Belgium and The Netherlands, rainfall between December 1975 and July 1976 was only 40–50 per cent of the long-term normal. The occurrence of heat-wave conditions in high summer aggravated the deficit of soil moisture. The potato crop was severely reduced in France, Denmark, England, Belgium and Germany, and milk yields were also down. In Britain a particular problem was the subsidence of houses that took place on severely desiccated clay soils.

Such periods of drought are associated with the presence of one of two types of persistent anticyclonic system:

(a) north-eastward extensions of the Azores high across southern England;
(b) persistent or repetitive blocking anticyclones centred near to or over Britain: the normal westerly flow with its travelling depressions is diverted north, and also sometimes south, of the high-pressure cell.

Why such anticyclonic patterns should become established in certain areas, and for such long periods, is still a matter of discussion, but it may be that unusual seawater temperatures over the Atlantic play a role.

Other climatic hazards are those associated with snow, frost and cold (figures 5.10 and 5.11). Among

Plate 5.9 Severe droughts can occur in the UK. In the most severe drought on record, in the summer of 1976, this reservoir at Sutton Bingham almost dried up. High temperatures and low levels of rainfall, caused by the existence of a persistent blocking anticyclone, brought on the drought.

the consequences of severe winter weather are high fuel costs, the laying off of outdoor workers, damage to crops and flocks, damage to plumbing and disruption of transport services. In upland areas such as the European Alps, avalanches can be one cause of severe loss of life (see window 8.1).

A further type of hazard is that associated with flooding (see section 15.12), which can be subdivided into floods occurring on the coast (window 5.5) and those occurring inland. Inland floods occur when meteorological conditions produce intense precipitation, prolonged precipitation or rapid snow-melt, acting either singly or in combination. Although the lower evaporation rates of the winter months cause heavier saturation of river catchments after heavy rainfall, much of the most severe flooding in non-alpine regions occurs in the summer months because of extremely localised, short-duration convective precipitation, often associated with thunderstorms. The effects may be magnified by *orographic lifting* (section 14.2) or by the presence of a major urban area (see section 16.1). A wide variety of weather conditions, or synoptic situations, can produce the deep convection and the unstable atmosphere necessary to allow heavy, thundery summer rainfall, but frequently, in the case of Britain, it is small, rather shallow depressions moving from the Bay of Biscay up the English Channel or across southern England that are responsible (figure 5.12). Snow-melt floods occur after severe winters when substantial quantities of snow are rapidly melted by the influx of warm moist air.

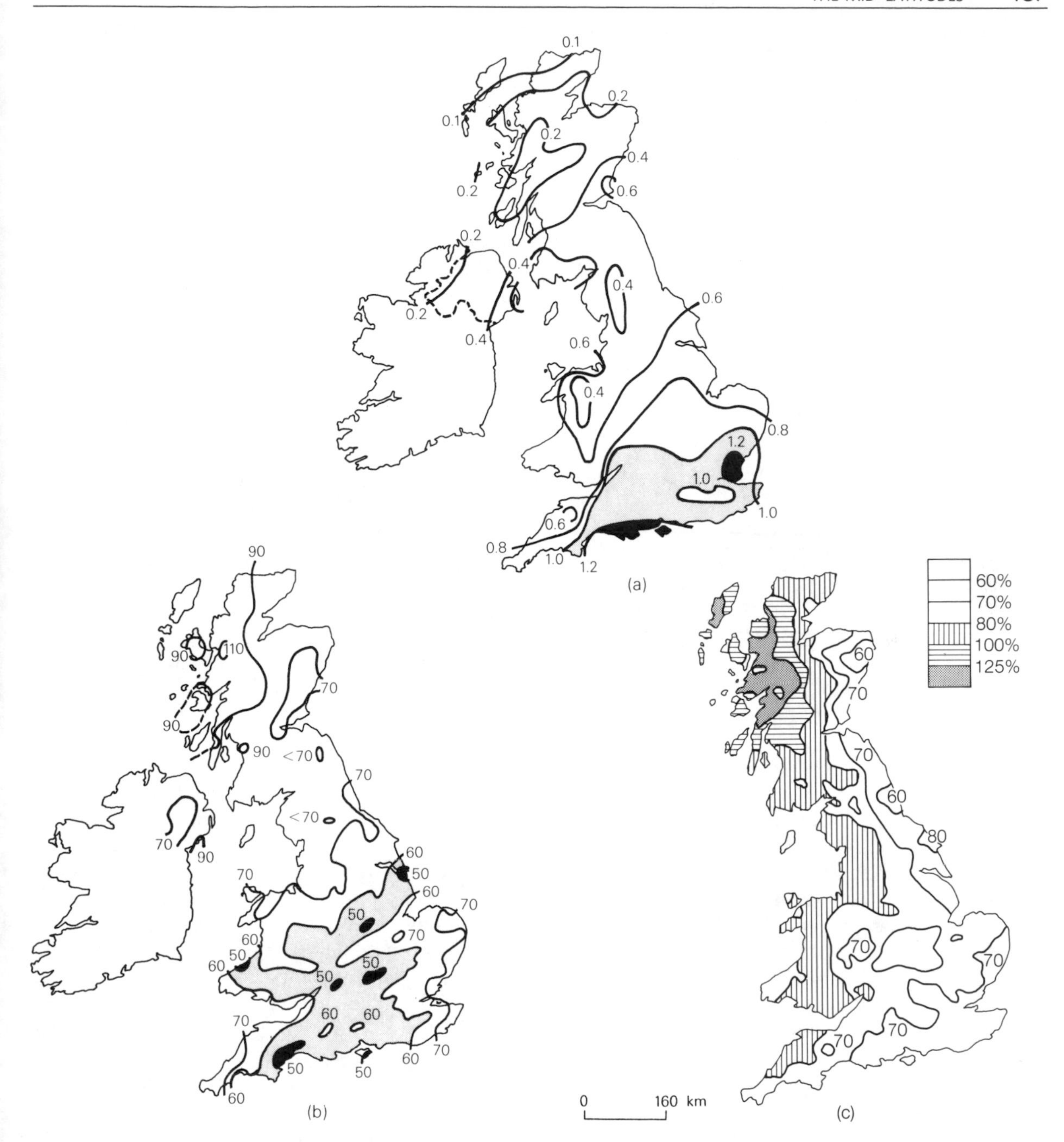

Figure 5.9 Drought conditions in Britain: (a) the regional distribution of mean annual absolute drought; (b) the distribution of rainfall, May 1975–August 1976 (the great drought) as a percentage of the 1916–50 mean; (c) rainfall for the period November 1988–November 1989, another major drought, as a percentage of the corresponding 1941–70 average, showing a rather similar pattern to that experienced in 1976, with normal or above normal values in the north-west of Scotland, but with values less than 60 per cent of average in the east of the country.

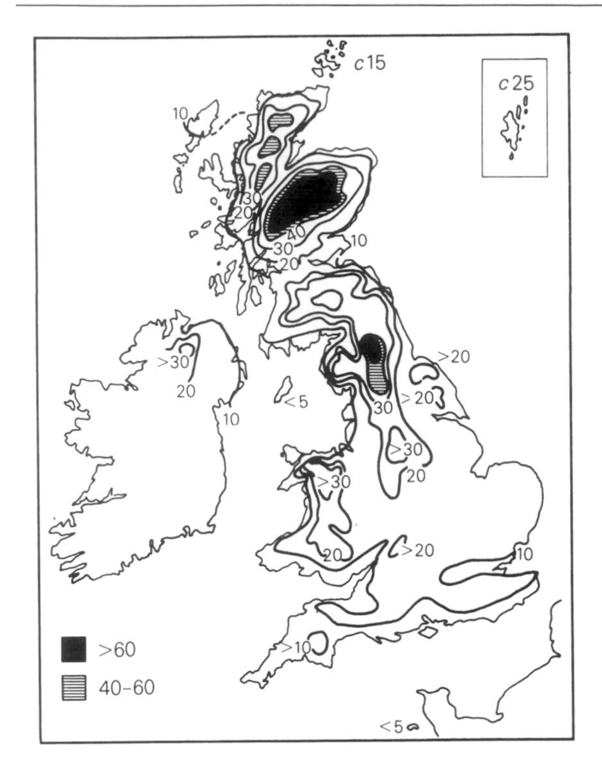

Figure 5.10 Mean number of days with snow lying at 0900 hrs each winter over the period 1941–70.

Coastal flooding is a serious problem, particularly along the low-lying areas bordering the southern North Sea. Cyclonic storms produced cataclysmic tempests and floods that caused great loss of life in the Middle Ages. Four storms along the Dutch and German coasts in the thirteenth century are thought to have killed 100 000 people each. It was North Sea storms that reduced the island of Heligoland from a length of 60 km around the year 800 AD to 25 km by 1300 and only 1.5 km by the twentieth century. The worst coastal flooding event in recent history was the storm of 31 January to 1 February 1953, which killed some 1600 people in The Netherlands and 350 in England. It was caused by a North Sea storm surge, which occurred when strong north-westerly and northerly winds drove water southwards into the increasingly narrow confines of the southern North Sea. The strong winds were associated with a large and intense depression, and mean wind speeds of 70 knots with gusts of 100 knots were observed along the east coast of Scotland. In addition, river levels were high and were further ponded up by high tides pushing up the estuaries. Because south-east England and the coast of the Low Countries are subsiding and world sea levels are rising, the coastal flood risk is becoming more serious.

The final type of meteorological hazard to be considered is fog. This occurs with meteorological conditions that are conducive to the cooling of air below its dew point (table 5.4). *Radiation fog* tends to occur during anticyclonic spells when winds are light. Visibility is worst around dawn, when temperatures are at their lowest, and there is intense cooling under clear skies. This produces low-level inversions of temperatures. The temperature of the air increases with height up to several hundred metres above the ground, so that the lower atmosphere is very stable and little mixing occurs. In areas of industrial pollution, and in areas where topographical conditions favour the development of a frost hollow, the fog is not easily dispersed. *Advection fog* occurs when warm air is cooled, either by its passage over a cool sea or by passage over a cold (often snow-covered) land surface.

Fog is one type of hazard that may be rendered more serious by human activity. A fog caused by a combination of favourable meteorological conditions and particulate emissions by industry may therefore be called a *quasi-natural hazard*. Such fogs are a type of pollution, and together with the other forms they may be injurious to human health.

The seismic hazard of earthquake activity is not one of major significance in western Europe, though the building of sensitive structures such as nuclear power stations requires that it be accorded some attention. A far more important geomorphological hazard is that involving various types of mass movement (plate 5.10), such as landslides and debris flows. These can be accelerated by such processes as deforestation or the creation of unstable spoil heaps in industrial areas.

Another form of geomorphological hazard, partly induced by human activity, is land subsidence. For example, coal mining, salt extraction and the draining of peaty areas have caused the ground surface to be lowered, sometimes catastrophically. Soil erosion is a hazard on light sandy soils, and the pres-

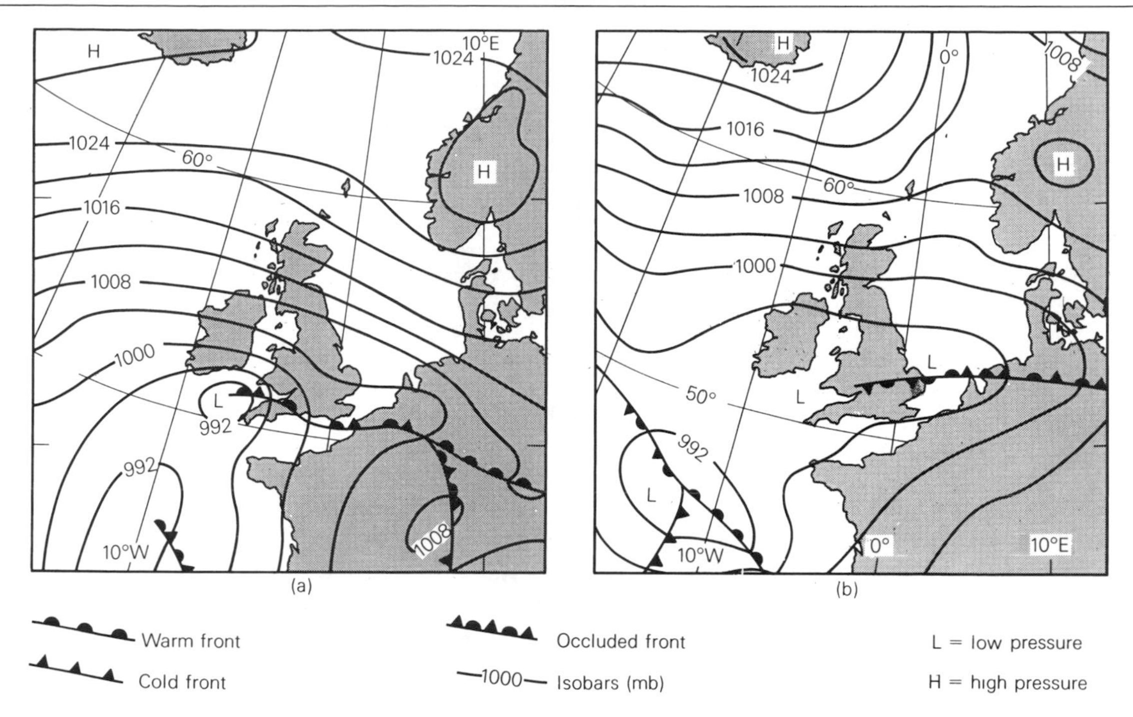

Figure 5.11 Meteorological conditions giving heavy snowfalls over Britain in 1962–3. (a) On 30 December a wave depression had moved northwards from Biscay to a position south of Ireland, having become partially occluded with the warm occlusion moving very slowly northwards in the Channel region. A cold easterly airstream covered most of the British Isles. The front remained quasi-stationary in the Channel region before (b) moving northwards to northern England by 4 January.

Window 5.5 *Floods in north-west and central Europe*

In the last week of January 1995, catastrophic flooding occurred over a large part of north-west Europe. This was caused by heavy rain during that month combined with snow-melt caused by mild temperatures in the Alps and the Ardennes. Some 40000 homes were destroyed in France and many major German cities suffered inundation. The centre of Cologne was submerged under 2 m of water for several days. In The Netherlands about 250000 people had to leave their homes – the largest civilian evacuation there in more than 40 years. One million cattle were also evacuated from flood-prone areas.

Another recent European flood is that which affected eastern Europe in July 1997. This summer flood was caused by heavy rainfall that was deposited by slow-moving depressions. The flooding took place in the Czech Republic, Poland and eastern Germany, and was a characteristic of the Odda (Oder) and Vistula Rivers. The flood was thought to be a 1 in 200–500-year event. About 100 people were killed in Poland and the Czech Republic, and at the height of the flooding Poland had 150 000 evacuees. Parts of Wrocław, an ancient and beautiful university city in southern Poland, were under as much as 4 m of water. Rainfall totals for July 1997 were 351 mm in Ostrava (260 mm more than the norm), 225 mm in Brno (161 mm more than the norm) and 238 mm in Wrocław (154 mm more than the norm).

Source

N. J. Middleton (1996) 'The 1995 floods in northwest Europe', *The Geography Review*, 9 (5): 25–6.

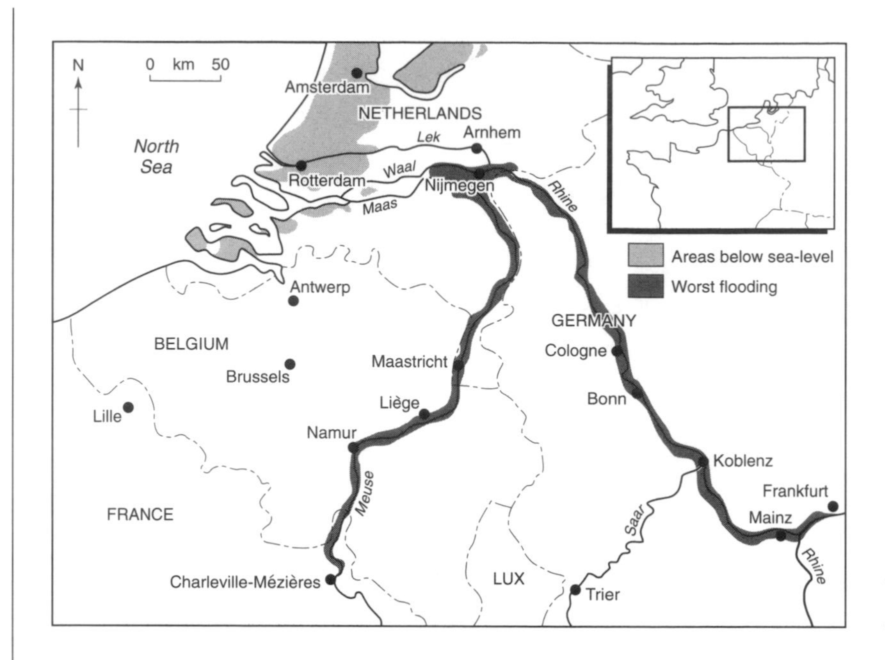

Areas worst affected by floods in north-west Europe in late January/early February 1995.

Floods in Poland, July 1997.

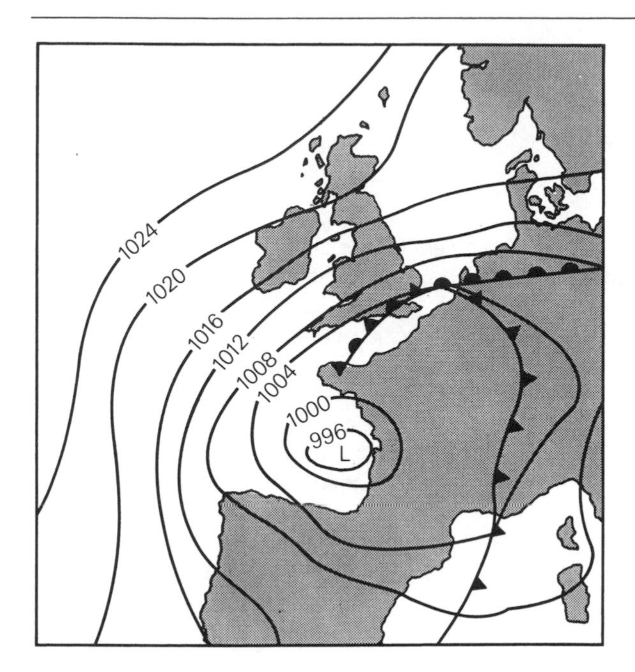

Figure 5.12 A synoptic chart for conditions of very heavy rainfall in England on 15 September 1968.

Plate 5.10 Even in England slope instability can occasionally cause serious problems for transport. In this example, the main road from Manchester eastwards through the Peak District to Sheffield has been disrupted by landslipping at Mam Tor, 'the shivering mountain'.

Table 5.4 Summary of fog characteristics in Britain

Type of fog	*Season*	*Areas affected*	*Mode of formation*	*Mode of dispersal*
Radiation fog	October–March	Inland areas, especially low-lying, moist ground	Cooling due to radiation from the ground on clear nights when the wind is light	Dispersed by the sun's radiation or by increased wind
Advection fog				
(a) Over land	Winter or spring	Often widespread inland	Cooling of warm air by passage over cold ground	Dispersed by a change in air mass or by gradual warming of the ground
(b) Over sea and coastline	Spring and early summer	Sea and coasts, may penetrate a few miles	Cooling of warm air by passage over cold sea	Dispersed by a change in air mass and may be cleared on coast by the sun's heating
Frontal fog	All seasons	High ground	Lowering of the cloud base along the line of the front	Dispersed as the front moves and brings a change of air mass
Smoke fog (smog)	Winter	Near industrial areas and large conurbations	Similar to radiation fog	Dispersed by wind increase or by convection

Source: A. H. Perry (1981) *Environmental Hazards in the British Isles* (London: Allen & Unwin)

Window 5.6 *The impact of global warming on the UK*

If global warming takes place as a result of the build-up of greenhouse gases in the atmosphere, it will cause some significant changes to the environment of the UK. A recent report suggests that the following are some of the main changes in climate.

- Annual rainfall to increase by 5 per cent by the 2020s and by 10 per cent by the 2050s.
- Winter temperatures to increase by the 2050s by between 0.8 °C in the north-west and 2 °C in the south-east.
- Summer temperatures to increase by the 2050s by between 1.2 °C in the north-west and 1.8 °C in the south-east. Assuming an unchanging variability of summer temperature, the probability of a summer being as warm as that of 1995 increases from 0.013 to 0.33 by the 2050s.
- Whilst summer rainfall is expected to increase in northern areas, a decrease in summer rainfall in south-east England of about 10 per cent by the 2050s corresponds to an increase in potential evapotranspiration (PE) of as much as 40 per cent. By contrast, PE decreases in north-west Scotland in both summer and winter.
- Winter rainfall is expected to increase in all areas, but the increase may be largest in southern England.
- The frequency of frost is expected to drop by about 50 per cent by the 2050s, and the frequency of days with temperatures exceeding 25 °C could double.

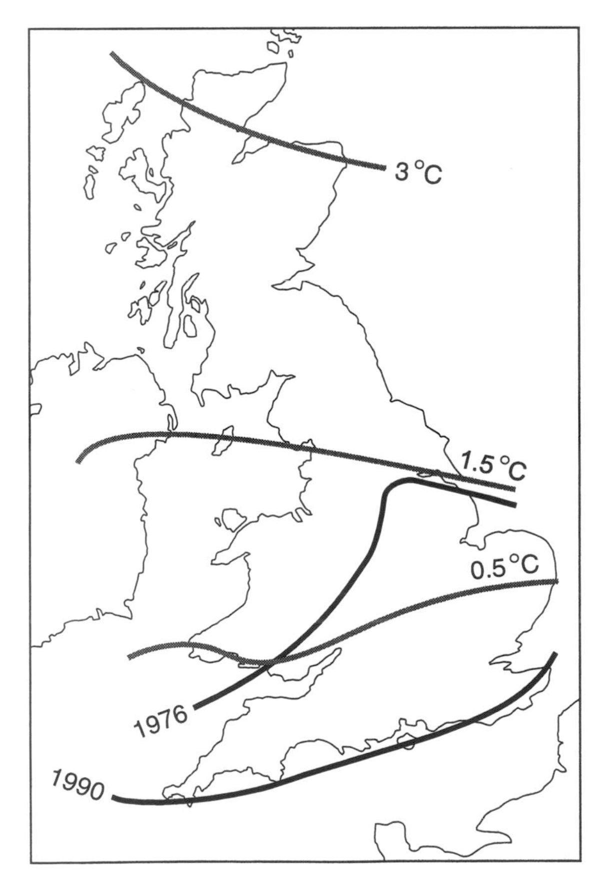

Potential distribution of grain maize in the UK under different warming scenarios.

It is likely that such changes would have an impact on various aspects of the economy, including agriculture. The thermal limits of agriculture would be likely to shift by around 300 km of latitude and 200 m of altitude per degree Celsius. Several crop species, such as wheat, maize and sunflowers, have their contemporary northern limits in the UK. An increase of temperature could, therefore, assuming that soil conditions were suitable, lead to a substantial northward shift of cropping zones. This could transform the British agricultural landscape. British fields and rural areas might come to resemble those currently found further south in mainland Europe. For example, the northern limit of grain maize, which currently lies in the extreme south of England, could be shifted across central England by a 0.5 °C increase in temperature, across northern England by a 1.5 °C increase and into the north of Scotland by an increase of 3 °C.

A rise in temperature, apart from transforming the range over which particular crop types could be grown, could be significant for the agricultural sector in other ways. For example, higher temperatures and more frequent summer droughts might reduce crop yields. The occurrence of certain plant pests and diseases could change, for better or worse.

Source

D. Wheeler and J. Mayes (eds) (1997) *Regional Climates of the British Isles* (London: Routledge), ch. 12.

ence of groynes and sea walls along much of the lowland coastline is a testimony to the seriousness of coast erosion on the borders of the North Sea. Once again, the role of man in causing and intensifying the nature of the hazard is important.

The final natural hazards that we need to consider are those that can be termed 'biological hazards'. These can include the health hazards posed for people by certain components of the environment. Many recent studies have shown that certain diseases are not uniformly distributed over western Europe; for example, it is becoming apparent that some types of cancer may be associated with the presence of especially high concentrations of certain rare metals in the ground, and that the incidence of some types of cardiovascular disease may be increased in areas where the water supplies are hard (i.e. rich in salts such as calcium carbonate). Also coming under the heading of biological hazards are diseases that affect plants and animals. For example, since the 1960s an aggressive strain of Dutch elm disease has killed more than 11 million trees in Britain, transforming the rural landscape. Our increasingly mobile society means that biological hazards involving the introduction of various pathogens may become an increasingly serious problem.

It is possible that future climatic changes will cause substantial changes in mid-latitude environments (window 5.2). Window 5.6 discusses some of the impacts that global warming could have on the British Isles.

■ *Key Terms and Concepts*

air masses
blocking anticyclones
boreal
chernozem
continentality
depressions
drought
dry valley
flooding
fog
fronts
jet stream
loess
Mediterranean climates
natural hazard
podzols
Rossby waves
tor
tornadoes
westerlies

Points for Review

- What are Rossby waves and why are they important?
- What do you understand by the term 'air mass'? Illustrate how different air masses give different weather conditions in a particular area of your choice.
- What is continentality?
- What are the main characteristics of climates in the Mediterranean zone?
- Account for the nature and distribution of either boreal forest or mid-latitude grasslands.
- What evidence is there for past climatic change in a mid-latitude area of your choice?
- Describe the causes and consequences of a recent natural catastrophe in western Europe.

FURTHER READING

Ballantyne, C. K. and Harris, C. (1994) *The Periglaciation of Great Britain* (Cambridge: Cambridge University Press). A discussion of the effect of cold climates on the landscape of Britain.

Blondel, J. and Aronson, J. (1999) *Biology and Wildlife of the Mediterranean Region* (Oxford: Oxford University Press). A book with a wider scope than the title suggests.

Bluestein, H. B. (1999) *Tornado Alley* (New York: Oxford University Press). A lavish account of tornadoes in the USA.

Goudie, A. S. (1990) *The Landforms of England and Wales* (Oxford: Blackwell). A survey of the geomorphology of England and Wales.

Goudie, A. S. and Brunsden, D. (1994) *The Environment of the British Isles: An Atlas* (Oxford: Oxford University Press). Some 120 maps of the British Isles, with explanatory text.

Hulme, M. and Barrow, E. (eds) (1997) *Climates of the British Isles* (London: Routledge). A systematic edited work looking at the past, the present and the future.

Perry, A. H. (1981) *Environmental Hazards in the British Isles* (London: Allen and Unwin). A wide-ranging review of mid-latitude hazards in the British context.

Sturman, A. P. and Tapper, N. J. (1996) *The Weather and Climate of Australia and New Zealand* (Melbourne: Oxford University Press). An excellent account of mid-latitude weather from a southern hemisphere perspective.

Wheeler, D. and Mayes, J. (eds) (1997) *Regional Climates of the British Isles* (London: Routledge). This book not only looks at the climate of particular regions of the British Isles, but also discusses the climatic background and the climatic future.

6 Deserts

6.1 Introduction

The deserts of the world (figure 6.1) are areas in which there is a great deficit of water, predominantly because they receive only meagre amounts of precipitation. Whereas even the driest parts of Britain have around 500 mm of rainfall per year, there are many desert weather stations that normally record less than one-tenth of that figure. In some years they may even record no rainfall at all.

It is this shortage of moisture, often worsened by high temperatures, that determines many of the characteristics of the soils, the vegetation, the animals, the landforms and the human activities of such an area. Thus modern systems for defining aridity tend to be based on the concept of the *water balance* – the relationship that exists in a given area between the input of water in the form of precipitation (P), the losses arising from evaporation and transpiration by plants (*evapotranspiration*) (E_t), and any changes in storage (soil moisture, groundwater etc.). By definition, in arid regions there is an overall deficit in water balance over a year, and the size of that deficit determines the degree of aridity. The actual amount of evapotranspiration (AE_t) that occurs will vary according to whether there is any water to evaporate, so climatologists have devised the concept of *potential evapotranspiration* (PE_t), which is a measure of the evapotranspiration that would take place from a standardised surface never short of water. The volume of PE_t will vary according to four climatic factors: radiation, humidity, temperature and wind. C. W. Thornthwaite developed a general moisture index based on PE_t:

When $P = PE_t$ throughout the year, the index is 0.
When $P = 0$ throughout the year, the index is –60.
When P greatly exceeds PE_t throughout the year, the index is +100.

Under this system, areas with values below –40 are regarded as arid, those between –20 and –40 as semi-arid, and those between 0 and –20 as subhumid. The arid category can be further subdivided into arid and extreme arid, with *extreme aridity* being defined as the condition experienced in any locality in which at least 12 consecutive months without any rainfall have been recorded, and in which there is not a regular seasonal rhythm of rainfall.

Extremely arid areas cover about 4 per cent of the earth's land surface, arid about 15 per cent, and semi-arid about 14.6 per cent. Combined, these amount to almost exactly one-third of the earth's land surface area. The deserts occur in five great provinces separated by either oceans or equatorial forests. The largest of these by far includes the Sahara and a series of other deserts extending eastwards through Arabia to central Asia. The southern African province consists of the coastal Namib Desert (window 6.1) and the Karroo and Kalahari inland dry zones. The South American dry zone is confined to two strips – the Atacama along the west coast and the Patagonian Desert along the east coast.

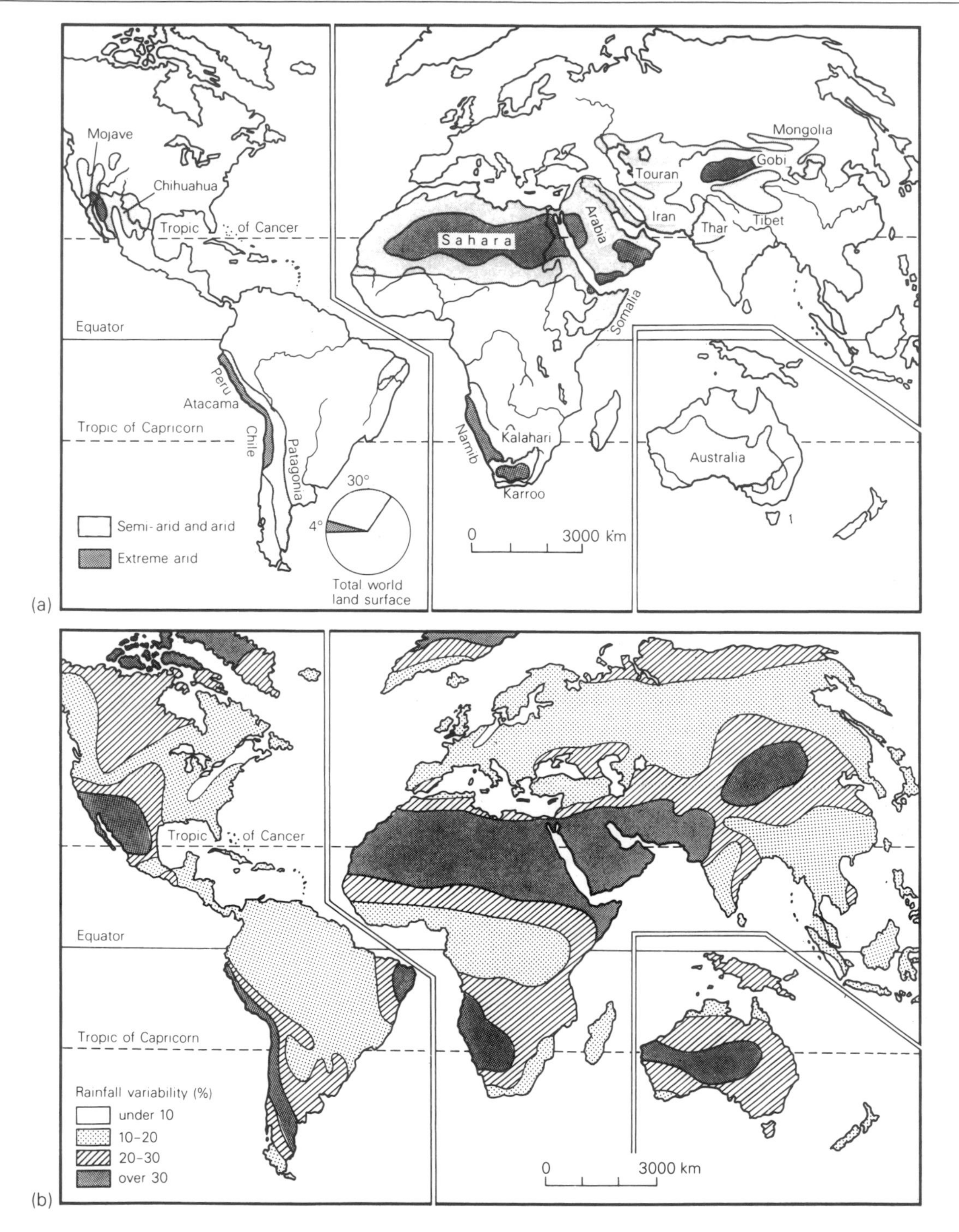

Figure 6.1 (a) The distribution of world deserts; (b) world map of rainfall variability. Note the way in which the major desert belts depicted in (a) show variability that is often in excess of 30 per cent.

Window 6.1 *The Namib Desert*

The Namib Desert, which occupies portions of South Africa, Namibia and Angola, lies on the Atlantic coast of southern Africa, and extends over a distance of more than 2000 km. It is one of the world's driest deserts – some areas receive less than 20 mm of rainfall per year – and owes its aridity in part to its position on the western side of the continent and to the presence offshore of a very cold ocean current, the Benguela Current. The desert probably originated over 20 million years ago, when Antarctica had drifted to its present position near the South Pole, and started to feed cold water into the Southern Ocean.

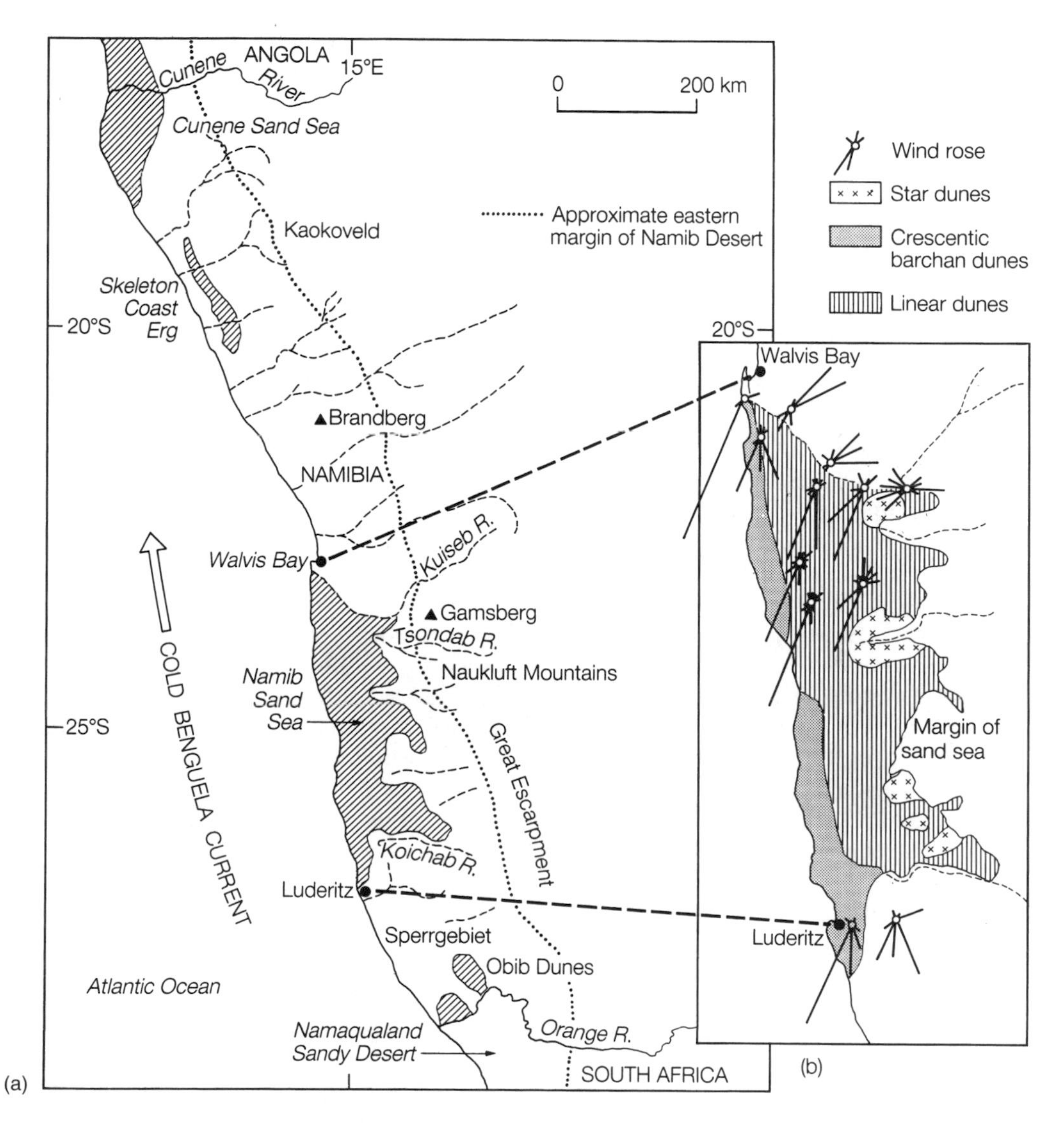

The Namib Desert of southern Africa: (a) location map; (b) main dune types in relation to wind directions.

The Namib displays a variety of desert landscapes, and can be divided into four main areas. In South Africa, south of the Orange River, is the Namaqualand Sandy Namib, which is not as dry as areas to the north, receiving as much as 150 mm of winter rainfall, which is sufficient to maintain a tolerably dense cover of succulent vegetation. This vegetation serves to stabilise extensive areas of superficial sands, many of which were deposited by rivers draining to the coastal plain from the interior, and deriving much of their sediment from areas of granite and Table Mountain sandstone. To the north of the Orange River is a rocky and sand-covered plain, called the *Sperrgebiet*, which extends as far north as Luderitz. It shows the development of yardangs and other forms of wind fluting.

The second zone of the Namib is the great Namib Sand Sea, which lies between Luderitz and the Kuiseb River (near Walvis Bay). Here there are some 34 000 km^2 of sand dunes, with a range of morphologies. In the coastal strip there are barchan dunes shaped by south-westerly winds. In the interior there are large linear dunes, which run more or less from south to north and are shaped by southerly winds with a bi-modal regime. These have a spacing that varies between 1200 and 2800 m, and attain heights of 25–170 m. On the eastern margins of the sand sea there are large star dunes, reaching heights of 200–350 m, created by complex wind regimes.

To the north of the Kuiseb lies the Central Namib Plains, which are largely composed of bare rock, for the Kuiseb River largely cuts off any sand movement from the dunes to the south. The plains are studded by impressive clumps of inselbergs, composed of granitoid rocks. The most impressive of these is the great Brandberg Mountain.

The fourth zone of the Namib lies to the north and consists of a coastal sand sea, known as the Skeleton Coast Erg, and the rocky plains of the Kaokoveld.

The North American desert province occupies much of Mexico and the south-western United States, including the Mojave and Sonoran Deserts. The fifth and final province is in Australia.

Although the water balance approach is one of the most satisfactory ways in which to subdivide desert environments, it must be remembered that there are many other factors that affect the nature of any particular desert area. Thus it is useful to draw a distinction between warm deserts and those deserts that, because of either high latitude or high altitude, have winter frosts. Likewise, coastal deserts such as the Atacama and the Namib will have very different temperature regimes and humidity characteristics from the deserts of continental interiors. There are also many different landscape types, or *relief units*, in deserts; for just as the temperate regions of the world comprise such features as mountains and plains, rivers and deltas, lakes and so on, so deserts have corresponding topographical features.

On a large scale, one can divide these various relief units on the basis of their geological history into shield deserts and mountain and basin deserts. The *shield deserts*, which occur in India, Africa, Arabia and Australia and were once part of the Gondwanaland mass (see section 1.6) have much less relief than the mountain and basin deserts; because they do not suffer from extensive tectonic activity at present, ancient land surfaces have often survived over wide areas. In contrast, the *mountain and basin deserts*, such as those of the south-west USA or Iran, have much steeper relief, and because they are often undergoing present-day mountain-building they tend to have sharp fault junctions between mountains and plains. The block-faulted topography of the American south-west is an excellent example of this type. In California's Death Valley, one of the hottest and driest places on earth, high mountains rising to over 3000 m are in close juxtaposition to shimmering white salt flats that lie below sea level. The high relative relief is a major control of the types of geomorphological processes that operate.

Within desert regions of these two structural types, some areas are dominated by wind, some by water, some by erosion and some by deposition. Various major settings can be identified, each of which has distinctive landforms, surface materials and plant life. According to the classification of an Australian geomorphologist, J. A. Mabbutt, there are the following main types of desert area:

Desert uplands, where geological controls of relief are important, bedrock is exposed, and relief is high.

Desert piedmonts, which are zones of transition separated from the uplands by a break of gradient but which none the less receive runoff and sediments from the uplands, and which have both depositional (e.g. alluvial fans) and erosional forms (e.g. pediments).

Stony deserts, which consist of stony plains and structural plateaux, and may have a cover of stone pavement.

Desert rivers and floodplains (features of *desert lowlands*).

Desert lake basins, which are sumps to which the disorganised drainage progresses, and which are often salty.

Sand deserts, which tend to be beyond the limits of active fluvial activity but often derive their materials by wind action removing erodible material from floodplains or lake basins; they are characterised by dunes.

6.2 What Causes Aridity?

Most of the world's deserts, whatever their form, derive their character from the fact that they receive little rainfall. This is because they tend to occur in zones where there is subsiding air (section 2.3), relative atmospheric stability and divergent air flows at low altitudes, associated with the presence of great high-pressure cells around latitude 30°. Such areas, as we learnt in chapter 2, are but seldom subjected to incursions of precipitation-bearing disturbances and depressions either from the Intertropical Convergence Zone of the tropics or from the belt of mid-latitude depressions associated with the circumpolar westerlies. The trade winds that blow across these arid zones are evaporating winds, and because of the trade-wind inversion they tend to be areas of subsidence and stability.

The subtropical highs are the major cause of aridity, though the desert zones are not always continuous around the earth at 30° N – for instance, the Indian monsoon gives substantial quantities of rain over northern India, and elsewhere the high-pressure cells are disrupted into a series of local cells, notably over the oceans, where air moving clockwise around the equatorial side of the cell brings moisture-laden air to the eastern margins of the continents in the Caribbean, Brazil and Queensland.

The rather general global tendencies produced by the subtropical highs are often reinforced by more local factors. Of these, distance from the sea (*continentality*) is a dominant one and plays a large part in the location and characteristics of the deserts of areas like central Asia. The *rain-shadow effect* produced by great mountains can create arid areas in their lee, as in Patagonia, where the southern Andes play this role. Other deserts are associated with the presence of cold ocean currents offshore. In the case of the Namib and the Atacama, for example, any winds that do blow onshore tend to do so across cold currents produced by the movement of water from high latitudes to low and associated with zones of upwelling of cold waters from the depths of the oceans. Such cool winds are stable because they are cooled from beneath. They also have a relatively small moisture-bearing capacity. In other words, they reinforce the stability produced by the dominance of subsiding air in desert areas.

6.3 Desert Rainfall

The main characteristics of deserts are caused by the very low levels of rainfall, and extremely low levels of precipitation are a particular feature of some coastal deserts. For example, mean annual totals at Callao in Peru are only 30 mm, at Swakopmund in Namibia only 15 mm, and at Port Etienne in Mauritania only 35 mm. In Egypt there are stations where the mean annual precipitation only amounts to 0.5 mm. Years may go by in such areas of extreme aridity when no rain falls at all.

Another highly important characteristic of desert rainfall is that it is highly variable over time (figure 6.1b). This inter-annual *variability* (*V*) can be expressed as a simple index:

$$V\,(\%) = \frac{\text{the mean deviation from the average}}{\text{the average}} \times 100$$

Plate 6.1 Although generally dry, deserts can suffer from floods caused by a combination of abnormal rainfall and impermeable surfaces. In January 1984 severe disruption occurred in the arid heart of Australia, and at Marla Bore the railway was washed away.

European humid temperate stations may have a variability of less than 20 per cent, whereas variability in the Sahara ranges from 80 to 150 per cent. This implies that from time to time, although mean rainfall levels are so low, there can still be individual storms of surprising size (plate 6.1). Indeed, maximum falls in 24 hours may exceed the long-term annual precipitation values. For example, at Chicama in Peru, where the mean annual precipitation over previous years had been a paltry 4 mm, in 1925 394 mm fell in one storm. Similarly, at El Djem in Tunisia (mean annual precipitation 275 mm), 319 mm fell in three days in September 1969, causing severe flooding and creating great geomorphological changes.

It would, however, be wrong to give the impression that all desert rainfall occurs as storms of such ferocity and intensity. Most falls in storms of low intensity. This is brought out clearly when one considers the rainfall statistics for the Jordanian Desert in the Middle East and Death Valley in California (figure 6.2). Both these areas have very low rainfall in terms of mean annual levels (102 and 67.1 mm, respectively), yet on average rain falls on 26 and 17 days respectively, so that the mean rainfall event tends to be only 3–4 mm – which is much the same as for London.

In coastal deserts, with cold currents offshore, the moisture provided by fogs may augment that produced by rain. In the coastal fringes of Namibia the mean annual fog precipitation (35–45 mm) may exceed that from rain, and the fog may occur on up to 200 days in the year and extend over 100 km inland. In Peru the fogs and low cloud provide enough moisture to support a growth of vegetation.

Precipitation in arid zones, in addition to showing temporal variability, also shows considerable spatial variability. For this reason it is often described as being ‘spotty’.

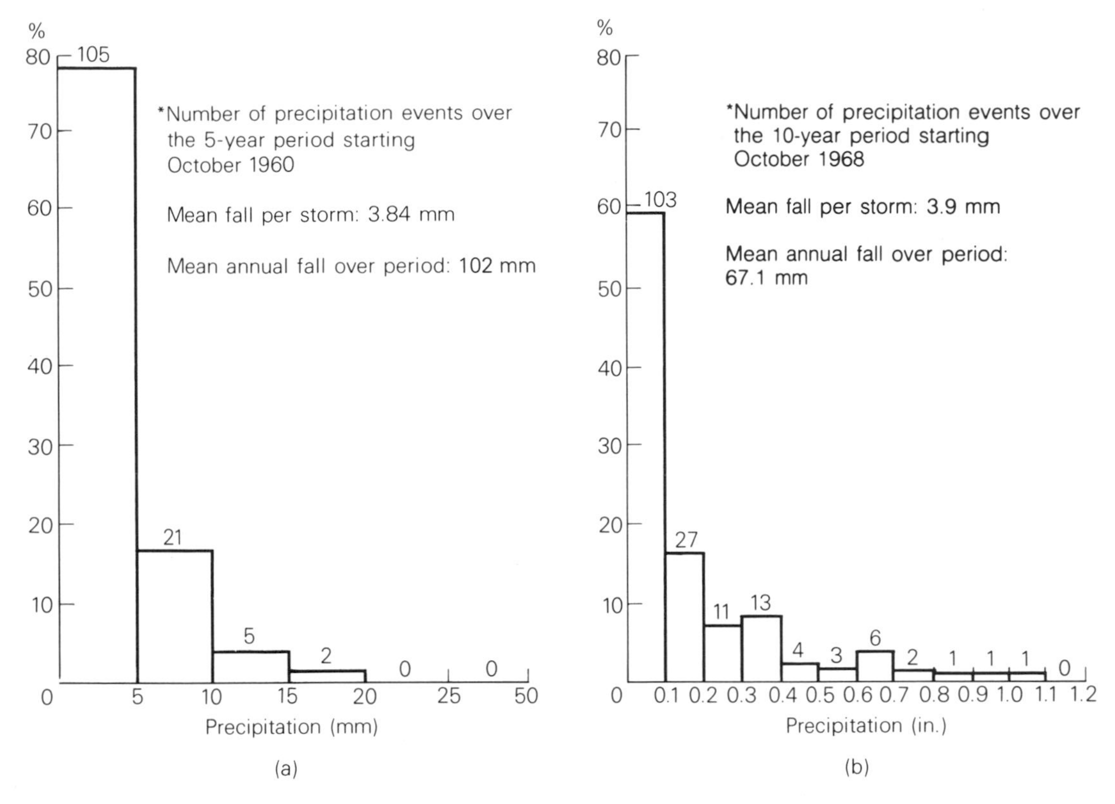

Figure 6.2 Amount of rain falling in each rainy day for two desert locations: (a) H4, Jordan; (b) Death Valley, California. Note that in both cases the average fall per storm is less than 4 mm.

6.4 Desert Temperatures

According to their type, deserts have a wide range of temperature conditions. Interior deserts can be subjected to extremes of temperature, both seasonally and diurnally, that are not equalled in any other climatic region, while coastal deserts tend to have relatively low seasonal and diurnal ranges.

In the case of coastal deserts, the climate is modified and moderated by the presence of cold currents and upwelling. Temperature ranges over the year are low – Callao, in the Peruvian desert, has an annual range of only 5 °C. Daily ranges in such stations are also low, often around 11 °C, and only about half what one would expect in the Sahara. The annual temperature values are also generally moderate (*c.* 19 °C in the Atacama, and 17 °C in the Namib).

By contrast, great extremes of temperature can occur in interior deserts, with maximum shade temperatures exceeding 50 °C. Temperatures in excess of 37 °C may occur for many days on end in the summer months, but because of the clear skies there may be a marked reduction of temperature at night, and daily ranges of 17–22 °C are normal. In the winter months in high-altitude interior deserts, frost can occur frequently.

While some deserts show great daily and seasonal ranges in air temperatures, ground surface temperatures show even greater ranges. Sand, soil and rock have been recorded as reaching temperatures as high as 82 °C, and some Russian workers have reported daily ranges as high as 75 °C! Such extreme values have many implications both for rock weathering and for plant and animal life.

6.5 Past Climates in Deserts

During the Ice Age of the Pleistocene the world was very different from today. In high latitudes the great ice sheets were far more extensive than now, and permafrost extended into areas like southern England. Lower latitudes did not escape these changes; temperature and rainfall conditions fluctuated repeatedly, and many deserts were subjected to increased rainfall: such periods are called *pluvials*, or lacustral phases. Some deserts have also been subjected to even greater aridity than today: such dry phases are called *interpluvials*.

There are several indicators in the present desert environments of higher levels of precipitation in the past: high lake levels marked by ancient shorelines around now dry, salty, closed basins; expanses of fossil soils of humid type, including laterites and other types indicative of very marked chemical changes under conditions of humidity; great spreads of spring-deposited lime, called *tufa*, indicating former higher groundwater levels; vast river systems which are currently inactive and blocked by dune fields; and animal and plant remains, together with evidence of former human habitation, in areas now too dry for people to survive.

The evidence for formerly drier conditions includes the presence of degraded, stable sand dunes in areas that are now too wet for sand movement to occur (see section 2.9).

Some of the pluvial lake basins reached colossal dimensions, especially in the south-west of the United States, where faulting had created a large number of closed basins in which lakes could accumulate during periods of greater humidity. Lake Bonneville (plate 6.2), near Salt Lake City in Utah, now has a water area of 2600–6500 km^2 and is highly saline – in the pluvials of the Pleistocene it had an area of 51 700 km^2 (almost the size of present-day Lake Michigan), and was probably relatively fresh. It was 335 m deeper than it is now.

Likewise, on the margins of the Sahara, Lake Chad underwent major changes in level: it may have been about 120 m deeper than it is now, and it ex-

Plate 6.2 The terraces cut in the hillside behind the school in Utah, USA, were cut by waves in pluvial Lake Bonneville.

tended for hundreds of kilometres north of its present limits. In central Asia there was another colossal lake, in the area of the present Aral and Caspian Seas, that covered over $1.1 \times 10^6 km^2$ and extended 1300 km up the Volga from its present mouth.

The dates for these more humid phases are still the subject of great discussion. In general, it appears that in the American south-west the high lake levels were broadly contemporaneous with the last major expansion of the great ice caps (around 18 000 years BP), but in areas like the margins of the Sahara and in East Africa the last major lake expansion phase (probably one of many) took place just after the start of postglacial times (around 9000 years BP).

Window 6.2 *The 'Dust Bowl' of the 1930s in the USA*

One of the world's greatest agricultural regions is the Great Plains of America. The western part of this area, the High Plains, is an area of dry grasslands that spreads northwards from Texas, and includes parts of the states of Oklahoma, Kansas, Colorado, New Mexico and Nebraska. In the 1930s this area was struck by a major disaster that has been graphically described by Donald Worster in his book *Dust Bowl* (New York: Oxford University Press, 1979):

> Weather bureau stations on the plains reported a few small dust storms throughout 1932, as many as 179 in April 1933, and in November of that year a large one that carried all the way to Georgia and New York. But it was the May 1934 blow that swept in a new dark age. On 9 May, brown earth from Montana and Wyoming swirled up from the ground, was captured by extremely high-level winds, and was blown eastward toward the Dakotas. More dirt was sucked into the airstream, until 350 million tons were riding toward urban America. By late afternoon the storm had reached Dubuque and Madison, and by evening 12 million tons of dust were falling like snow over Chicago – 4 tons for each person in the city. Midday at Buffalo on 10 May was darkened by dust, and the advancing gloom stretched south from there over several states, moving as fast as 100 miles an hour. The dawn of 11 May

During the mid-1930s the High Plains of the United States were hot and dry. Moreover, large areas had also been ploughed up for cereal cultivation. The result of this was that the top soil was exposed to wind erosion, dust-storm generation and sand deposition. This example from South Dakota shows the barren land surface and the way in which sand encroached on farm machinery and buildings.

> found the dust settling over Boston, New York, Washington, Atlanta, and then the storm moved out to sea. Savannah's skies were hazy all day 12 May; it was the last city to report dust conditions. But there were still ships in the Atlantic, some of them 300 miles off the coast, that found dust on their decks during the next day or two.
>
> In the 1930s the Soil Conservation Service compiled a frequency chart of all dust storms of regional extent, when visibility was cut to less than a mile. In 1932 there were 14; in 1933, 38; 1934, 22; 1935, 40; 1936, 68; 1937, 72; 1938, 61 – dropping as the drought relented a bit – 1939, 30; 1940, 17; 1941, 17. Another measure of severity was made by calculating the total number of hours the dust storms lasted during a year. By that criterion 1937 was again the worst: at Guymon, in the panhandle of Oklahoma, the total number of hours that year climbed to 550, mostly concentrated in the first six months of the year. In Amarillo the worst year was 1935, with a total of 908 hours. Seven times, from January to March, the visibility there reached zero – all complete blackouts, one of them lasting eleven hours. A single storm might rage for one hour or three and a half days. Most of the winds came from the southwest, but they also came from the west, north, and northeast, and they could slam against windows and walls with 60 miles-per-hour force. The dirt left behind on the front lawn might be brown, black, yellow, ashy grey, or, more rarely, red, depending upon its source. And each colour had its own peculiar aroma, from a sharp peppery smell that burned the nostrils to a heavy greasiness that nauseated.

The 'black blizzards' of the 'dirty thirties' caused great hardship, and were a manifestation of a great amount of landscape degradation, which today we would call desertification.

What were the causes of this tragedy so movingly captured in John Steinbeck's novel *The Grapes of Wrath* (1939)? Natural causes certainly played a role, for the 1930s were a hot, dry decade in the High Plains, and this reduced the vegetation cover and dried out the soils, making them susceptible to removal by strong winds. However, human activities played a major role. Most notably, large areas of grassland, protected in the absence of ploughing by a dense sod, were ripped open for cereal cultivation. The development of the tractor, the combine and the truck enabled the great plough-up to occur.

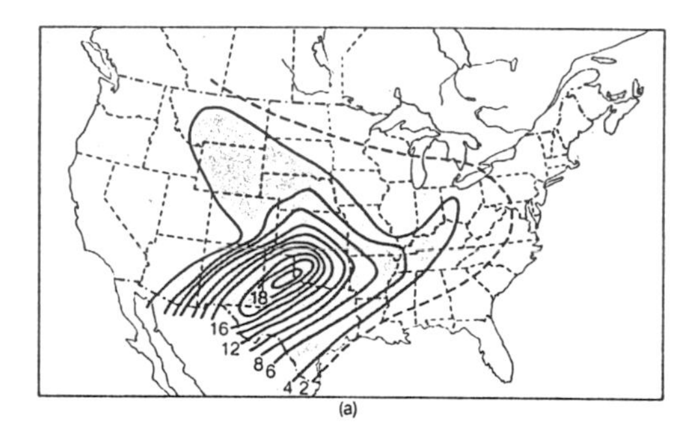

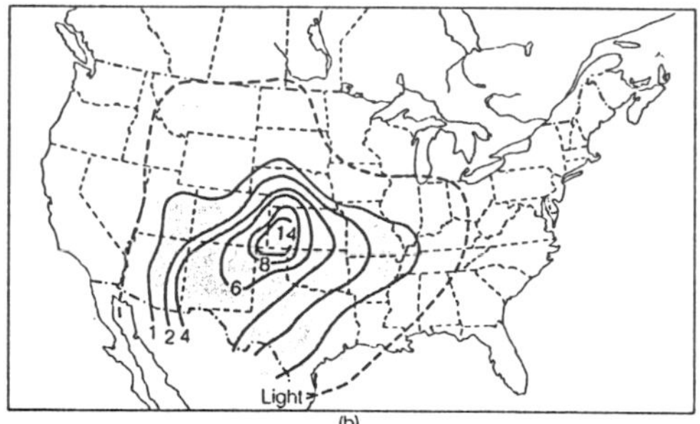

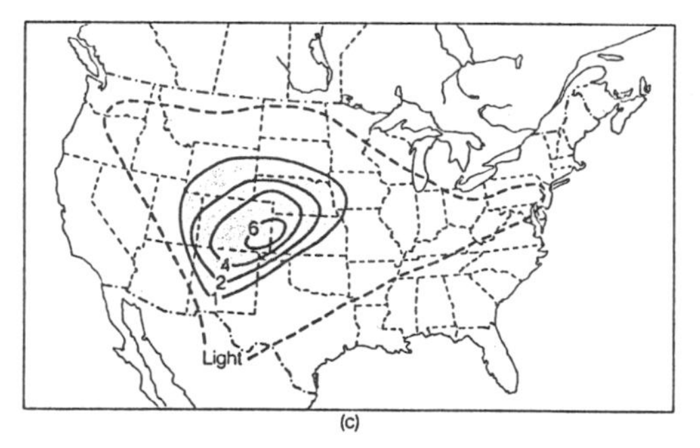

The concentration of dust storms (number of days per month) in the USA in 1936, illustrating the extreme localisation over the High Plains of Texas, Colorado, Oklahoma and Kansas: (a) March; (b) April; (c) May.

Although climatic change of a marked degree appears to have taken place in the Pleistocene and early Holocene, it is apparent from the study of meteorological records, dating back in some cases to the middle of the nineteenth century, that appreciable fluctuations may still take place. Thus, for example, in the 1930s a period of greatly reduced rainfall and higher-than-average temperatures in the United States contributed to the extreme wind erosion and dust blowing of the 'Dust Bowl' years (window 6.2).

Likewise, since the late 1960s a great belt, extending from Mauritania in the west to north-west India in the east, has suffered from a series of prolonged and serious droughts, with rainfall only around two-thirds of the long-term mean. This has caused extensive famines and led to great pressure on the limited vegetation available for grazing animals on the desert margins. Severe land degradation and erosion have ensued.

Plate 6.3 In the hyper-arid Namib Desert, the vegetation cover is generally sparse, but in wet years grasses may grow for a short period.

6.6 Desert Vegetation and Animals

One consequence of the present aridity of deserts is that the vegetation cover is generally low (plate 6.3) – a closed cover is seldom encountered. A useful measure of the degree of vegetation development in an area is its *biomass,* the total amount of living plant material above and below ground. Deserts have a low biomass, often 100 times less than that of an equivalent area of temperate forest. Water is the vital influence on plant growth, of course, and is responsible for this low biomass level. Most plant tissues die if their water content falls too low: the nutrients that feed plants are transmitted by water; water is a raw material in the vital process of photosynthesis; and water regulates the temperature of a plant by its ability to absorb heat and because water vapour lost to the atmosphere during transpiration helps to lower plant temperatures. However, water not only controls the volume of plant matter produced – it also controls the distribution of plants within an area of desert; some areas, because of their soil texture, topographical position or distance from rivers or groundwater, have virtually no water available to plants, whereas others do.

The nature of plant life in deserts is also highly dependent on the fact that they have to adapt to the prevailing aridity. There are two general classes of vegetation: *perennials,* which may be succulent and are often dwarfed and woody; and annuals or *ephemerals,* which have a short life-cycle and may form a fairly dense stand immediately after rain.

The ephemeral plants evade drought. Given a year of favourable precipitation such plants will develop vigorously and produce large numbers of flowers and fruit. This replenishes the seed content of the desert soil. The seeds then lie dormant until the next wet year, when the desert blooms again.

The perennial vegetation adjusts to the aridity by means of various avoidance mechanisms (figure 6.3). Most desert plants are probably best classified as *xerophytes.* They possess drought-resisting adaptations: transpiration is reduced by means of dense hairs covering waxy leaf surfaces, by the closure of stomata to reduce transpiration loss and by the rolling up or shedding of leaves at the beginning of the dry season. Some xerophytes, the *succulents* (including cacti), impound water in their structures. Another way of countering drought is to have a limited amount of mass above ground and to have extensive root networks below ground.

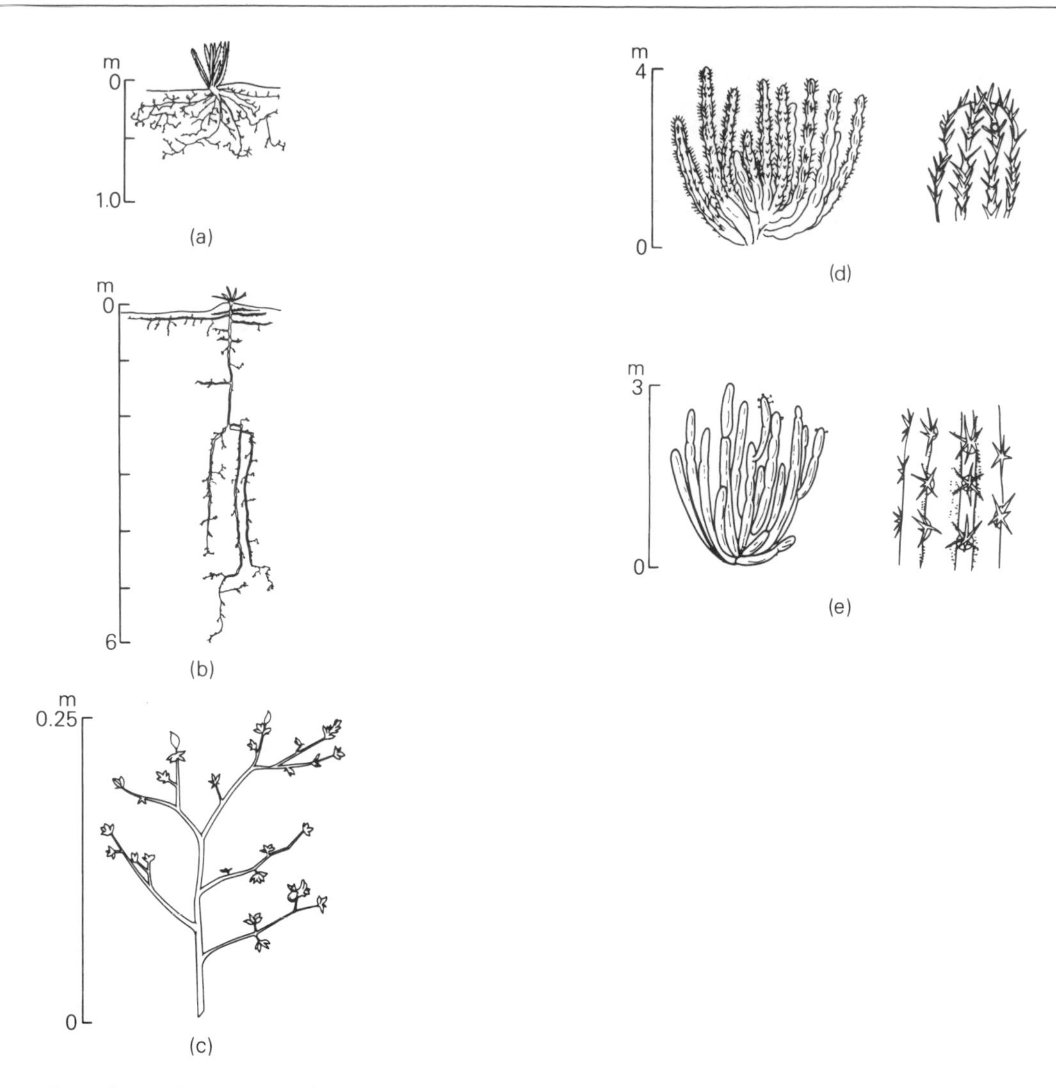

Figure 6.3 Some forms of adaptation of desert plants to the rigours of the environment. (a) Succulent plants, which can store a great deal of water and so can rely on near-surface roots collecting rain from sporadic precipitation events. (b) Acacia bush of a type very characteristic of desert margins, which adopts a dual strategy of having subsurface roots to collect any moisture sporadically available from rainfall, together with a very deep rooting system stretching down to permanently available groundwater in the subsoil and rock. (c) A plant with very small leaves in relation to the size of the branch (*c.*25 cm) thereby limiting evapotranspiration loss. (d) The thorny euphorbia of the Namib Desert in Namibia; and (e) its equivalent in North America, the cactus.

It is not unusual for the roots of some desert perennials to extend downwards more than 10 m. Some plants are woody in type – an adaptation designed to prevent collapse of the plant tissue when water stress produces wilting. Another class of desert plant is the *phreatophyte*. These have adapted to the environment by the development of long tap roots which penetrate downwards until they approach the assured water supply provided by groundwater. Among these plants are the date palm, tamarisk and mesquite. They commonly grow near stream channels, springs or on the margins of lakes. In addition to a shortage of water, many desert plants have to be able to cope with saline soil conditions. Salt-tolerant plants are called *halophytes*.

Animals also have to adapt to desert conditions (figure 6.4), and may do it through two forms of behavioural adaptation: they either escape or retreat. Escape involves such actions as *aestivation*, a condition of prolonged dormancy or torpor during which animals reduce their metabolic rate and body temperature during the hot season or during very dry spells. Seasonal migration is another form of escape, especially for large mammals or birds. The term *retreat* is applied to the short-term escape behaviour of desert animals, and it usually assumes

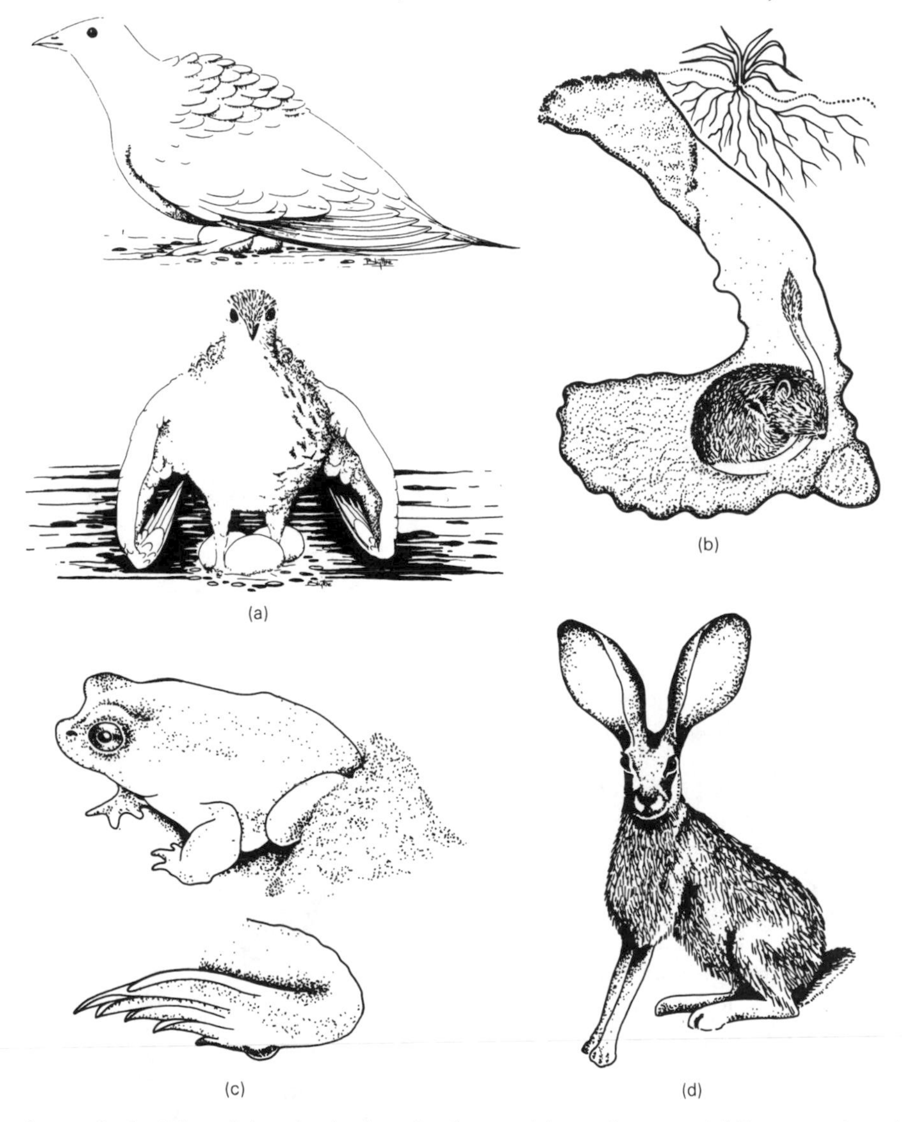

Figure 6.4 Some forms of adaptation of desert animals to the rigours of the environment. (a) The ground-nesting sandgrouse shades its eggs during the hottest hours of the day and reduces the radiation load by erecting the mantle feathers. It also faces into the wind and raises itself above the nest, thereby facilitating convective cooling of the shaded eggs. (b) The kangaroo rat burrows, thereby sheltering from predators, radiation and desiccation. (c) Using its strong feet, which have a horny projection, the spadefoot toad of the Sonoran Desert can escape to depths as great as 90 cm below the surface during unfavourable climatic conditions. (d) The enormous ears of the jack rabbit assist the animal in offloading excessive body heat.

the pattern of a daily rhythm. Birds shelter in nests, rock overhangs, trees and dense shrubs to avoid the hottest hours of the day, while mammals like the kangaroo rat burrow underground.

Some animals have behavioural, physiological and morphological (structural) adaptations that enable them to withstand extreme conditions. For example, the ostrich has plumage that is so constructed that the feathers are long but not too dense. When conditions are hot the ostrich erects them on its back, thus increasing the thickness of the barrier between solar radiation and the skin. The sparse distribution of the feathers, however, also allows considerable lateral air movement over the skin surface, thereby permitting further heat loss by convection. Furthermore, the birds orientate themselves carefully with regard to the sun and gently flap their feathersome wings to increase convective cooling.

6.7 Soils and Surface Materials

Arid environment soils, which are called *aridisols,* have certain general characteristics that are distinctive. Because of the low plant biomass they have a low organic content, and thus are dominantly mineral soils of an immature and skeletal type. They are also little subjected to leaching because of low precipitation levels, so that soluble salts tend to accumulate in the profile at a depth related either to the position of the water table or to the depth of moisture percolation. Such deposition forms one of the few distinct horizons in aridisols, and it occurs nearer to the surface as precipitation diminishes. Another general characteristic is that desert soils have a generally low clay content compared with humid zone soils. In general, clay formation increases with increasing rainfall.

Of these characteristics, salt accumulation is one of the most important. When groundwater is present near the surface, as in the floodplains of through-flowing rivers or in proximity to salt lakes, concentrations of salt may occur at high enough levels to be toxic to plants. Soils of this type with a saline horizon of NaC1 (sodium chloride) are called *solonchaks* (white alkali soils), and those with Na_2CO_3 (sodium carbonate) are called *solonetz* (black alkali soils). High-saline contents affect plant growth in various ways. First, they affect the physical structure of the soils: a high content of sodium salts leads to a dispersal or *deflocculation* of the soil particles or aggregates, causing the soils to lose their structure so that they become relatively impervious and badly aerated. Second, there is the *osmotic effect*, which opposes the entry of water into plant roots, and thus increases total moisture stress. Third, there is a straightforward nutritional effect, whereby toxicity or nutritional imbalance is caused: some salts in large quantities are poisonous.

Although much of the salt ultimately comes from natural sources (from the atmosphere, from rock

Plate 6.4 A typical calcrete hardpan at Barberspan in South Africa. Beneath the book there is a solid crust, and beneath that there is a more friable, nodular zone. Such crusts are a feature of the soils of semi-arid areas.

weathering, from inflowing rivers etc.) and is a normal characteristic of many desert areas, human activity has increased the extent and degree of salinity in many ways. In particular, the development of irrigation for agriculture can lead to a build-up of salt levels in the soil through the mechanism of raising groundwater, so that it is near enough to the surface for capillary rise and subsequent evaporative concentration to occur. Elsewhere over-pumping of fresh groundwater in coastal areas can permit the incursion of more saline water from the sea. The problem of induced salination arising from irrigation is severe; it has been estimated that the percentage of salt-affected and waterlogged soils in the irrigated areas of Iraq and Syria amounts to 50 per cent.

Sometimes the accumulation of soluble materials progresses so far that hard surface or subsurface crusts develop. Warm deserts have a variety of these crusts – sometimes called *duricrusts* – and they are classified according to the nature of the main chemical composition of the cement. *Calcretes,* calcium carbonate crusts, are the most widely distributed, and the numerous different forms that have been described include hardened or indurated horizons some metres thick (plate 6.4). *Silcretes* are crusts formed by the cementation of a matrix by silica in the form of opal, chalcedony or quartz. The most extensive areas of silcrete occur in southern Africa and in Australia, but they are by no means restricted to semi-arid areas, and some develop in wetter zones. *Gypcrete* crusts are composed of calcium sulphate, and while calcretes predominate in areas where precipitation ranges from 200 to 500 mm, gypcretes tend to be found in zones with between 50 and 200 mm. In still drier areas the crusts are often composed of sodium chloride and, exceptionally – notably in the Atacama – sodium nitrate. This distribution pattern, controlled by precipitation levels, would appear to be related to the solubility of the various minerals involved, each mineral having a threshold above which the climate is wet enough to dissolve and leach the near-surface accumulations.

The processes responsible for the development of these duricrusts are as diverse as the crusts them-

Plate 6.5 In desert areas, like the dry valleys of the Karakoram in Pakistan, rock surfaces are often covered by desert varnish. Where the dark varnish has been scratched away, as with these graffiti, the lighter underlying rock is evident.

selves. Some crusts result from salts being deposited in evaporating lakes; some develop *in situ* and involve a relative accumulation of the cementing mineral by the leaching of other minerals; some develop as a result of evaporation of groundwater in areas of high evaporation rates; some develop as a result of the evaporation of laterally moving subsurface and surface waters on features like pediments and alluvial fans; others involve the downward leaching and accumulation of soluble materials that have been brought in as dust.

Another crust, somewhat different from those just described, is the thin patina of iron and manganese oxide that covers many rock surfaces. This is called *desert varnish* (plate 6.5).

Of the various surface types found in deserts, *stone pavements* are one of the most characteristic. These are armoured surfaces composed of rock fragments, either angular or rounded, usually one or two stones thick, set in or on matrices of finer material such as sand, silt and clay. The traditional explanation for these features is that they are a consequence of *deflation*, or the removal by wind, of fine materials from a deposit of initially mixed grain size. This would tend to leave the coarse contents as a lag or residue at the surface (figure 6.5). However, a similar result could occur from water sorting. Coarse materials could remain at the surface after raindrops had dislodged fine materials and they had been flushed away by running water in the form of sheet flow. Vertical processes may also play a role, for both freezing and thawing and wetting and drying can lead to the migration of coarse particles towards the ground surface.

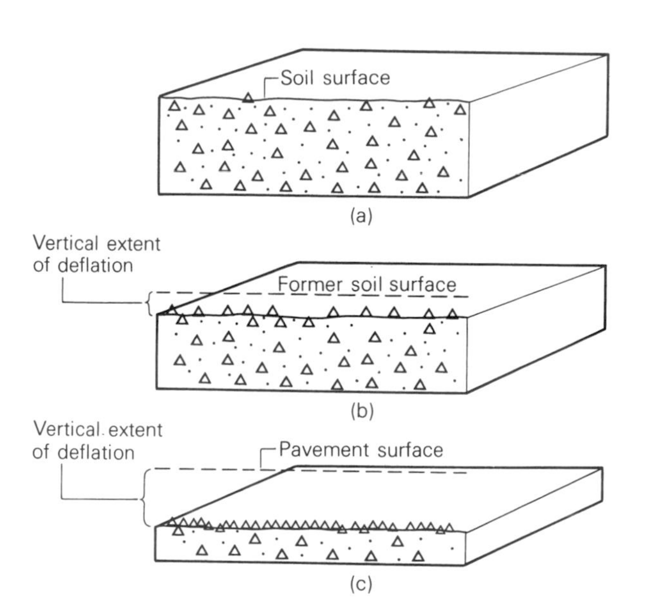

Figure 6.5 The classic deflation model of stone pavement development, with an initial alluvial sediment containing both fine and coarse materials (a) being subjected to deflation (b) until such time as the surface is lowered to such an extent (c) that a coarse gravel lag is left at the surface, the fine material having blown away.

6.8 Insolation and Salt Weathering

Mechanical weathering, involving the disintegration of rock without any chemical change, has often been considered important in the desert environment. There are probably two main types of mechanical weathering: that produced by insolation and that produced by salt.

Insolation weathering is the rupturing of rocks and minerals primarily as a result of large daily temperature changes, which lead to temperature gradients within the rock mass. Areas that are heated expand relative to the cooler portions of the rock and stresses are thereby set up. In igneous rocks, which contain many different types of mineral with different coefficients and directions of expansion, such stresses are enhanced. Moreover, the varying colours of minerals exposed at the surface will cause differential heating and cooling.

Daily temperature cycles under desert conditions may exceed 50 °C, and during the heat of the day the rock surface may exceed a temperature of 80 °C. However, rapid cooling takes place at night, creating, it has been thought, high tensile stresses in the rock. Desert travellers have claimed to hear rocks splitting with sounds like pistol shots in the cool evening air – certainly, split rocks are evident on many desert surfaces.

At first sight the process of insolation weathering seems a compelling and attractive mechanism of rock disintegration. However, in recent years doubt has been cast upon its effectiveness on a variety of grounds. The most persuasive basis for doubting its power was provided by experimental (albeit crude) work in the laboratory by geomorphologists like Blackwelder, Griggs and Tarr. They all found that simulated insolation produced no discernible disintegration of dry rock, but that when water was used in the cooling phase of a weathering cycle disintegration was evident. This highlighted the importance of the presence of water. Likewise, studies of ancient stone buildings and monuments in dry parts of North Africa and Arabia showed very little sign of decay except in areas, for example close to the Nile, where moisture was present. Indeed, there are many situations where there is moisture in deserts: in the Namib coastal desert rocks may be wetted by fog on as many as 200 days in the year; in parts of Israel there may be dews on 150 days in the year; and even in a very arid area like Death Valley in California there are on average 17 days in the year when rain falls. When water combines chemically with the more susceptible minerals in a rock they may swell, producing a sufficient increase in volume to cause the outer layers of rock to be lifted off as concentric shells – a process called *exfoliation*. Thus, some of the weathering that used to be attributed to insolation may now be attributed to chemical changes produced by moisture.

However, the importance of insolation cannot be dismissed entirely. The early experimental work had grave limitations: the blocks used were very small, they were unconfined, and the temperature cycles employed were unlike those encountered in nature. Some recent experiments using a wide range of rock types under more natural temperature cycles have revealed that some micro-cracking can take place.

One agent of mechanical weathering that has come into favour recently is *salt weathering*, for salts of various types are widespread in deserts. Low rainfall levels mean that salts are prone to accumulate rather than being dissolved and carried away in rivers. The salts include common salt (NaC1), sodium carbonate (Na_2CO_3), sodium sulphate

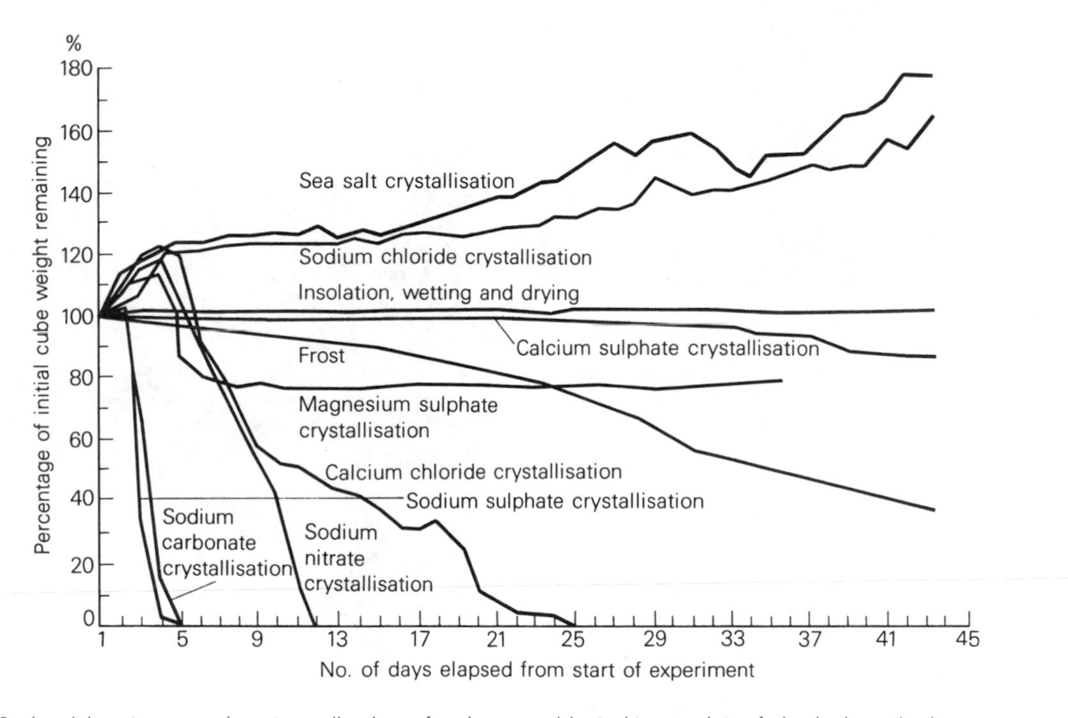

Figure 6.6 In a laboratory experiment, small cubes of rock were subjected to a variety of physical weathering processes, and their weight was recorded after each cycle. Some salts appear highly effective, especially sodium carbonate and sodium sulphate. Some samples increased in weight because of the absorption of salt, but when leached they also showed signs of disintegration.

Plate 6.6 After only 12 years from the date of construction this nurses' home in Bahrain was disintegrating as a result of salt attack. The house was built in a low-lying area, so salty groundwater rose into the foundations and crystallisation occurred.

(Na_2SO_4), magnesium sulphate ($MgSO_4$), gypsum (Ca_2SO_4) and (especially in the Atacama) sodium nitrate (Na_2NO_3). They break up rocks in two main ways. First, when a solution containing salts is either cooled or evaporated, salt crystals will form, and pressures accompanying the *crystallisation* can be great enough to exceed the tensile strength of the rocks in which the solution was contained. Second, salt minerals expand when water is added to their crystal structure. This change of state is termed *hydration*. Some salts exist at atmospheric temperatures and humidities in the non-hydrated form, taking up water of crystallisation and expanding. In the case of Na_2SO_4 and Na_2CO_3 the volume expansion involved may be in excess of 300 per cent. If the salts are in the rock, the pressures generated can break the rock.

There is now plenty of laboratory and field evidence for the power of salt weathering (figure 6.6). Rapid decay of buildings in salty areas has been noted in Bahrain (plate 6.6), Suez and the Indus Valley of Pakistan, while telegraph poles and grave stones in some deserts seem to act like wicks for groundwater solutions and suffer from salt crystallisation as the water evaporates. In Death Valley, California, boulders of resistant igneous rock (20–60 cm in diameter) in alluvial fans can be seen to disintegrate to dust within a few metres of reaching the salt pan (plate 6.7).

Plate 6.7 When fine materials are washed or blown away in deserts, stone pavements may develop, in which a coarse lag of stones mantles the ground surface. These particles in Death Valley, California, have been broken up by salt weathering.

6.9 Wind Action in Deserts

Because deserts are so dry and vegetation cover is so limited, there is little to protect the desert surface against the action of wind, but the importance of wind in shaping desert landscapes has been subject to rapid changes of emphasis. At the end of the nineteenth century many workers who had experience in some of the new colonial territories like South West Africa attributed the formation of isolated hills (*inselbergs*) and the low-angled smooth rock-cut surfaces that surrounded them (*pediments*) to the planing action of sandladen wind, with its supposedly high abrasional ability. This phase has been termed the phase of 'extravagant aeolation'.

In due course, however, there was a reaction against these views, partly because of an increasing body of work in the deserts of North America, with their steep slopes, active tectonics and slightly wetter conditions. Various arguments were marshalled against the proponents of widespread wind planation, and many of them are valid today. First, it was appreciated that rare rainstorms, acting on surfaces with limited vegetation cover and soils with limited infiltration capacities, could cause fluvial processes to be active. Second, the pluvials of the Pleistocene and Holocene were thought to have had an important influence on the moulding of desert surfaces. Third, it was shown that many desert surfaces were protected from the reduction of the surface by wind action, by stone pavements or crusts. Fourth, wind erosion depends on the presence of abrasive sand (which is not present everywhere), and as the sand moves at only a metre or so above the ground surface it operates only over a limited height range. Fifth, many of the features that were attributed to wind erosion either could be explained by other processes (e.g. closed basins, or inselbergs) or were of no very great significance in the landscape (e.g. the curiously shaped rocks called *zeugen*).

None the less, where there are high winds, with plenty of abrasive sand, operating on susceptible surfaces, wind erosion may be important. Indeed, the examination of air photographs and satellite imagery has revealed that extensive areas of bedrock in many deserts have huge bowed, or *arcuate*, grooves with relative relief of more than 100 m; these run for tens of kilometres (plates 6.8 and

Plate 6.8 Yardangs, carved in dark pluvial lake sediments in the Dakhla Oasis, Egypt. These streamlined forms were created by winds coming from the top left-hand direction.

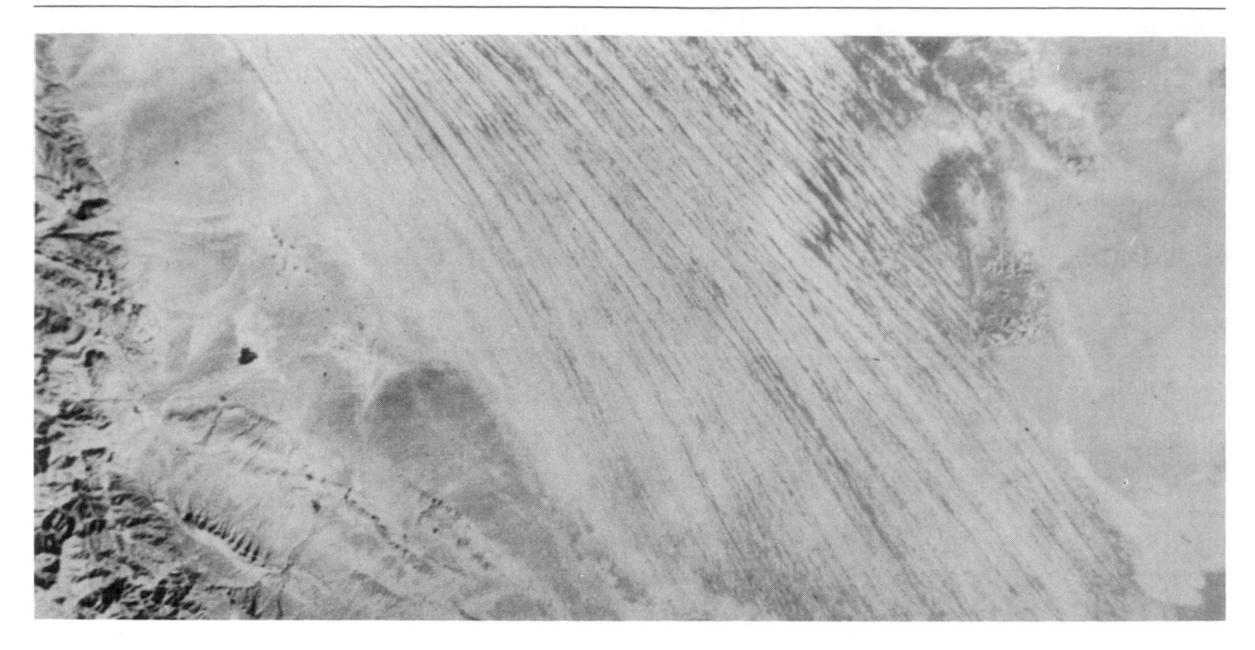

Plate 6.9 In Iran, satellite images show great parallel grooves, called yardangs or kaluts. They are caused by erosion of sediments by strong north-westerly winds. These examples are tens of kilometres long.

6.9), and are aligned with the prevailing winds (see table 6.1). Such areas of groovy ground are called *yardangs*. Many desert depressions (plate 6.10) also appear to have an aeolian origin; for they have a streamlined shape, occur in lines, and have dunes on their lee sides resulting from the deposition of sands excavated by winds from the hollows. Moreover, the general association of so many hollows with deserts implies that 'normal' processes (presumably fluvial action and slope movements) are inadequate to fill them in, and that some abnormal process (presumably wind) is effective at digging

Table 6.1 Yardangs

Location	*Lithology*
Taklimakan, China	Fluvial and lacustrine sediments of Pleistocene age
Lut, Iran	Pleistocene fine-grained horizontally bedded silty clay and limey gypsiferous sand
Khash Desert, Afghanistan	Clay
Sinai	Nubian sandstone
Saudi Arabia	Caliche, limestone
Bahrain	Aeolianite (Pleistocene), dolomite
Egypt	Eocene limestone, lake beds; Nubian sandstone
South-central Algeria	Cretaceous clays, Cambrian claystones
Borkou, Chad	Palaeozoic and lower Mesozoic sandstones and shales
Jaisalmer, India	Eocene limestones
Namib Desert	Precambrian dolomites, granites and gneisses
Rogers Lake, California	Dune sands and lake beds
Northern Peru	Weakly to moderately consolidated upper Eocene to Palaeocene sediments (shales and sandstones)
South-central Peru	Siltstones of upper Oligocene to Miocene age

Plate 6.10 A small closed depression in the arid zone of western Australia. The main processes responsible for its development are probably rock disintegration caused by salt weathering, followed by deflation by wind. Such hollows often have a small dune on their lee sides, called a lunette, which is composed of material deflated from the basin.

them out. It should be remembered, however, that many processes can produce a hole in the ground, and that many desert depressions may be the result of hole-forming processes that inherently have very little to do with aridity: for example, tectonic processes can produce fault basins, solution can create hollows in limestone areas, animals can excavate water holes and so on.

That wind action is an important factor in creating desert landscapes is perhaps best indicated by the occurrence of *dust storms* (plate 6.11) produced by the deflation of fine materials (especially silt) from unvegetated desert surfaces. Such deflation is an insidious form of soil erosion. Dust storms may be of sufficient intensity to reduce visibility to below 1000 m on 20–30 days in the year, and are large enough (with dimensions up to 2500 km) to be seen on satellite imagery. Dust may be transported large distances: on occasions, Saharan dust has reached places as far away as Britain and Florida.

In a dust storm of average violence a cube of air (dimensions 3 m × 3 m × 3 m) might well contain 28 g of dust suspended in it. Thus a storm 500 km × 600 km across might well be sweeping 100 000 000 tonnes of solids along with it!

6.10 Sand Deposition – Dunes

Although the romantic view of deserts envisages landscapes dominated by ever-changing sand dunes, with oases, camels and men in flowing robes, only about one-quarter to one-third of the world's deserts are covered by aeolian sand, so its role in deserts should not be exaggerated. Indeed, in the Ameri-

Plate 6.11 A dust storm in Jazirat Al Hamra, United Arab Emirates, in March 1998.

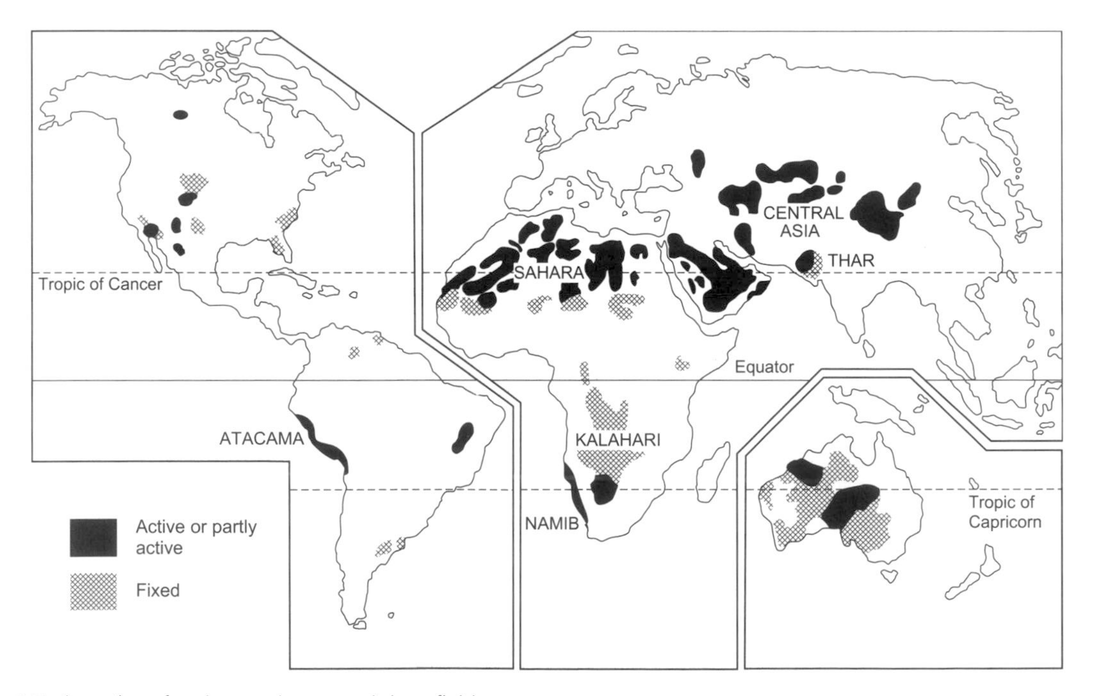

Figure 6.7 Location of major sand seas and dune fields.

Plate 6.12 The movement of sand grains on the crest of a dune in the Thar Desert, India. Most of the grains move along at only a short height above the ground surface by means of a jumping action termed *saltation*.

can deserts sand dunes occupy less than 1 per cent of the surface area. None the less, great *ergs*, or seas of sand, are found nowhere else on earth, and they form some of the most beautiful, repetitious and regular landforms that our planet (and, indeed, Mars) has to offer. Particularly large ergs occur in Arabia and the Sahara (figure 6.7).

The sand that makes up dune fields is a result of wind winnowing grains of an easily transported size from alluvial plains, from lake shores, from sea shores and from weathered rock like sandstone and granite. When wind velocity exceeds the threshold velocity required to initiate sand grain movement (generally this is about 20 km h^{-1}), the grains begin to roll along the ground, but after a short distance this gives rise to a bounding or jumping action called *saltation* (plate 6.12). Grains are taken up a small distance into the airstream (often only a matter of a few centimetres) and then fall back to the ground in a fairly flat trajectory. The descending grains dislodge further particles and thereby the process of saltation is maintained across country. The saltating grains (which generally have a diameter between 0.15 and 0.25 mm) lift larger grains (with diameters generally in the range of 0.25–2.00 mm), which move forward by *surface creep*. The smallest grains (< 0.15 mm) may be carried high up into the air in *suspension*.

Dunes form because saltating grains tend to accumulate preferentially on sand-covered areas rather than on adjoining sand-free surfaces. This seems to result from the check to a strong wind through intensified sand movement over a sand surface and from the lower rate of sand movement where saltating grains 'splash' into loose sand compared with that over firm ground.

The geometric forms of deposition of sand as dunes are very varied (figure 6.8) and depend on

Table 6.2 Relative importance of major dune types in the world's deserts (percentage)

	Thar	*Takla Makan*	*Namib*	*Kalahari*	*Saudi Arabia*	*Ala Shan*	*South Sahara*	*North Sahara*	*North-east Sahara*	*West Sahara*	*Average*
Linear dunes (total)	13.96	22.12	32.84	85.85	49.81	1.44	24.08	22.84	17.01	35.49	30.54
Simple and compound	13.96	18.91	18.50	85.85	26.24	1.44	24.08	5.74	2.41	35.49	23.25
Feathered	—	—	—	—	4.36	—	—	3.56	1.13	—	0.91
With crescentic superimposed	—	3.21	—	—	—	—	—	4.02	7.32	—	1.46
With stars superimposed	—	—	14.34	—	19.21	—	—	9.52	6.15	—	4.92
Crescentic (total)	54.29	36.91	11.80	0.59	14.91	27.01	28.37	33.34	14.53	19.17	24.09
Single barchanoid ridges	8.96	3.21	11.80	—	0.59	8.62	4.08	0.06	—	0.65	3.80
Megabarchans	—	—	—	—	—	—	—	7.18	1.98	—	0.92
Complex barchanoid ridges	16.65	33.70	—	—	14.32	18.39	24.29	26.10	12.55	18.52	16.45
Parabolics	28.68	—	—	0.59	—	—	—	—	—	—	2.93
Star dunes	—	—	9.92	—	5.34	2.87	—	7.92	23.92	—	5.00
Dome dunes	—	7.40	—	—	—	0.86	—	—	0.80	—	0.90
Sheets and streaks	31.75	33.56	45.44	13.56	23.24	67.82	47.54	35.92	39.25	45.34	38.34
Undifferentiated	—	—	—	—	6.71	—	—	—	4.50	—	1.12

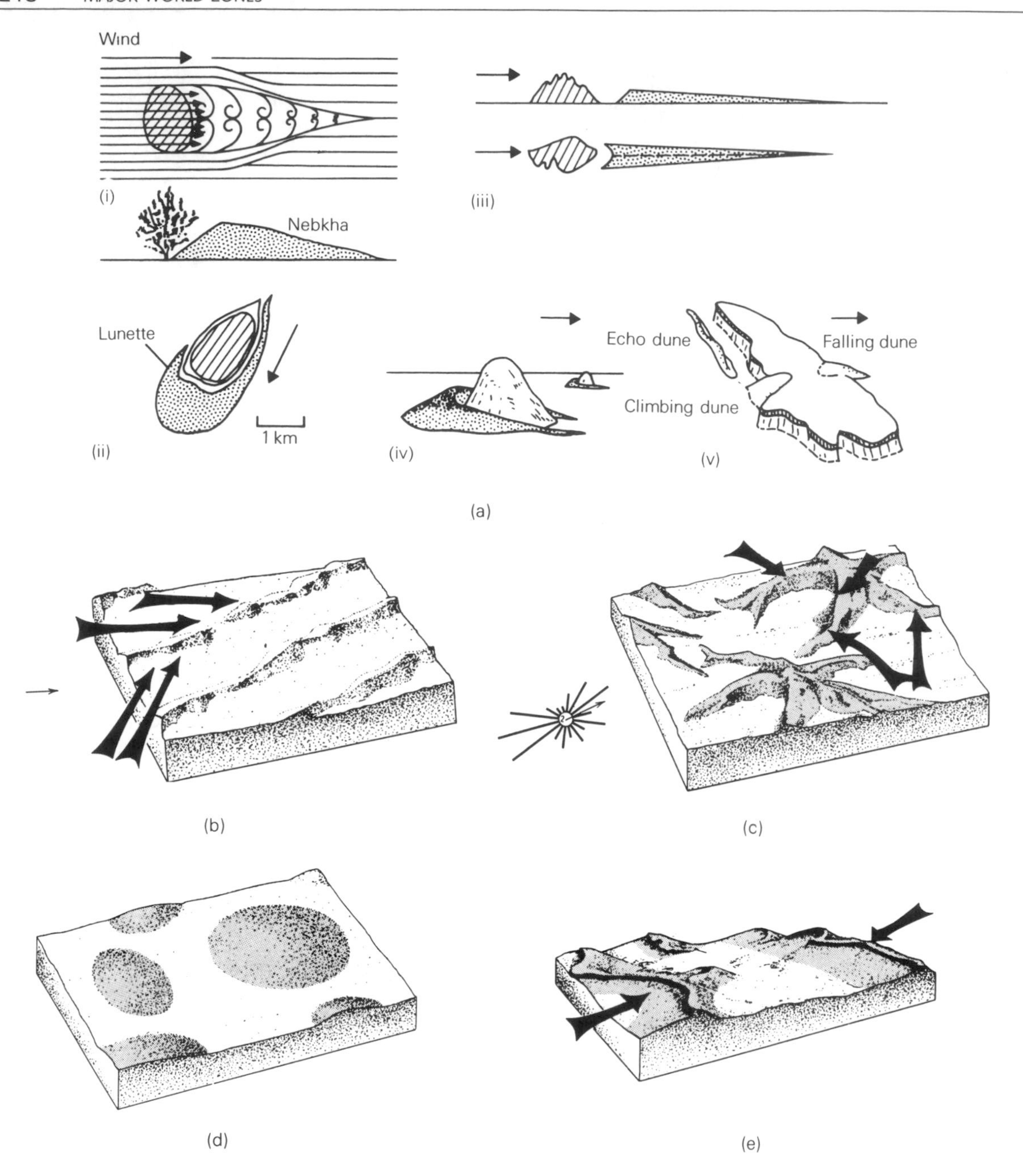

Figure 6.8 Some of the main types of dune form encountered in the world's deserts: (a) obstacle or topographic dunes: (i) a small dune or nebkha in the low-velocity area to the lee of a shrub; (ii) a crescentic lunette formed to the lee of a small desert depression (playa); (iii) wind-shadow dunes formed in the lee of some hills; (iv) a dune formed to the windward of a hill; (v) dune development in the proximity of a plateau; (b) linear dunes, or seifs (the arrows show probable dominant winds); (c) star dunes (the arrows show the effective wind directions); (d) dome dunes; (e) reverse dunes (arrows show the wind directions); (f) parabolic dunes (arrow shows prevailing wind direction); (g) barchan dunes (arrow shows prevailing wind direction); (h) barchanoid ridge (arrow shows prevailing wind direction); (j) transverse dune (arrow shows prevailing wind direction).

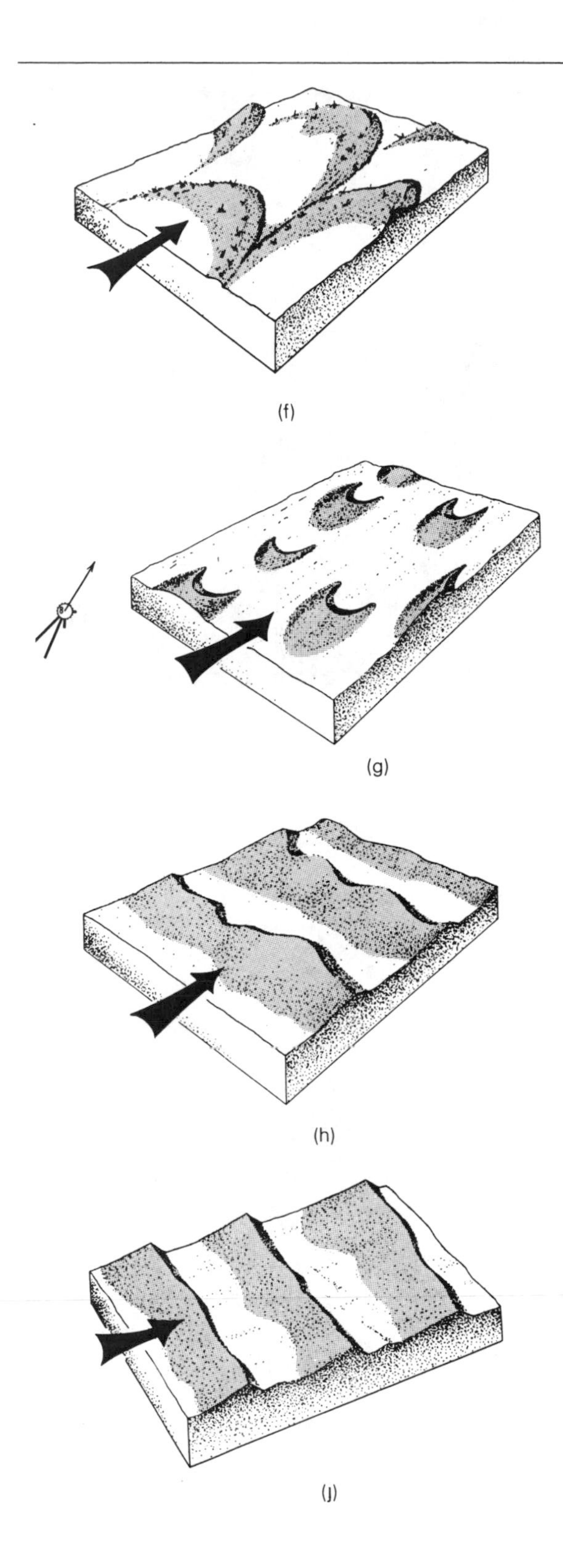

the supply of sand, the nature of the wind regime, the extent of vegetation cover and the shape of the ground surface. As these conditions vary from desert to desert, the relative importance of different dune types also varies (table 6.2).

One class of dune is that formed by the interference of an obstacle, such as a hill, with wind and sand movement. Where the wind velocity is checked by the hill, sand will be deposited sometimes to the lee and sometimes to windward. Large shrubs can have a similar effect on wind flow leading to the formation of *nebkhas* (figure 6.8a(i)). Some obstacle or topographic dunes develop in the lee of desert depressions and such crescentic features are called *lunettes* (figure 6.8a(ii)). Most dunes, however, do not require an obstacle, whether hill, shrub, depression or dead camel. Indeed, the most regular dune forms develop on the most regular surfaces.

Probably the best known and most common basic dune form results from winds having a single dominant direction and the dune being oriented with its axis at right angles to the wind direction (see colour plate 6). Such dunes range from small crescent-shaped types (*barchans*) (figure 6.8g) through parallel rows of *barchanoid ridges* to essentially straight ridges known as *transverse dunes* (figure 6.8j). These dunes are all characterised by slip faces in one direction and represent unidirectional wind movement. Barchan dunes occur where sand supply is limited and transverse dunes where sand is abundant. Dome dunes normally show no obvious slip face and are circular or elliptical in plan view (figure 6.8d). They may result from strong winds that truncate the top and flatten the lee slopes of barchan.

A class of dune in which the form owes much to the presence of a limited vegetation cover or some soil moisture is the *parabolic dune* (figure 6.8f). These are hairpin-shaped with the nose pointing downwind. They may occur in clusters creating rake-like forms.

Linear dunes (plates 6.13 and 6.14), or *seifs*, are straightish ridges with slip faces on both sides that run more or less parallel to the wind (figure 6.8b). They are commonly 5–30 m high and occur as evenly spaced ridges 200–500 m apart. The ridges may extend for tens, or even hundreds, of kilometres and link together in tuning-fork-shaped junc-

Plate 6.13 Sand dunes have a great variety of forms. One common type is the linear dune or seif. These features, which in the case of the Namib examples shown here may be over a hundred metres high, can run parallel for tens of kilometres. The Namib dunes have a more barchanoid form near the coast (on the left of the Landsat satellite image). The northern and eastern boundary is marked by the Kuiseb River.

Plate 6.14 The linear dunes of the inland Kalahari Desert in south-west Africa are smaller than those of the Namib. They are often a few tens of metres high. The road follows a depression between two parallel ridges.

Plate 6.15 This Landsat image of Algeria, North Africa, shows a large field of star dunes. The frame is about 180 km across.

tions that almost invariably point downwind. They occur in areas where there is seasonal or diurnal change in wind direction.

If winds come from many different directions, *star dunes* (or *rhourds*) (plate 6.15), with arms extending radially from a central peak, occur (figure 6.8c). These are up to 150 m high and between 1 and 2 km across. Intermediate in character between the star dune and a transverse ridge is the *reversing dune*, which characteristically forms where two winds from nearly opposite directions are balanced with respect to strength and direction (figure 6.8e).

Although we have discussed the major dune forms individually, in many situations it is apparent how one dune type can grade into or develop from another. The crescentic barchan can, by moving into an area with a different wind regime or sand supply, become deformed into a longitudinal *seif* (see colour plate 7). Likewise, the hairpin-shaped parabolic dune can gradually be elongated until a blow-out occurs so that two longitudinal dunes are developed.

Although wind-direction characteristics have been stressed in the discussion of dune forms, sand supply is also an important control. Barchans tend to occur where there is little sand and almost unidirectional winds; crescentic ridges where sand is abundant and winds are slightly more variable; linear dunes where sand supply is small, but winds are even more variable; and star dunes are found in areas with complex wind regimes and abundant sand supply (figure 6.9).

Even at the greatest scale of investigation – on a continental scale – dunes show a remarkable regularity of pattern. They tend to occur as a *wheel-round* or swirl in an anti-clockwise direction related to dominant continental wind patterns (figure 6.10). In some drylands, dunes are being de-stabilised or reactivated by human activities and this is one facet of the process of desertification (see window 6.3).

Dune reactivation arouses some of the strongest fears among those combating desertification. The increasing population levels of both humans and their domestic animals, brought about by improvements in health and by the provision of boreholes, has led to excessive pressure on the limited vegetation resources. As ground cover has been reduced, so dune instability has increased. The problem is not so much that dunes in the desert cores are relentlessly marching on to moister areas, more that fossil dunes, laid down during the more arid phase peaking around 18 000 years ago, have been re-

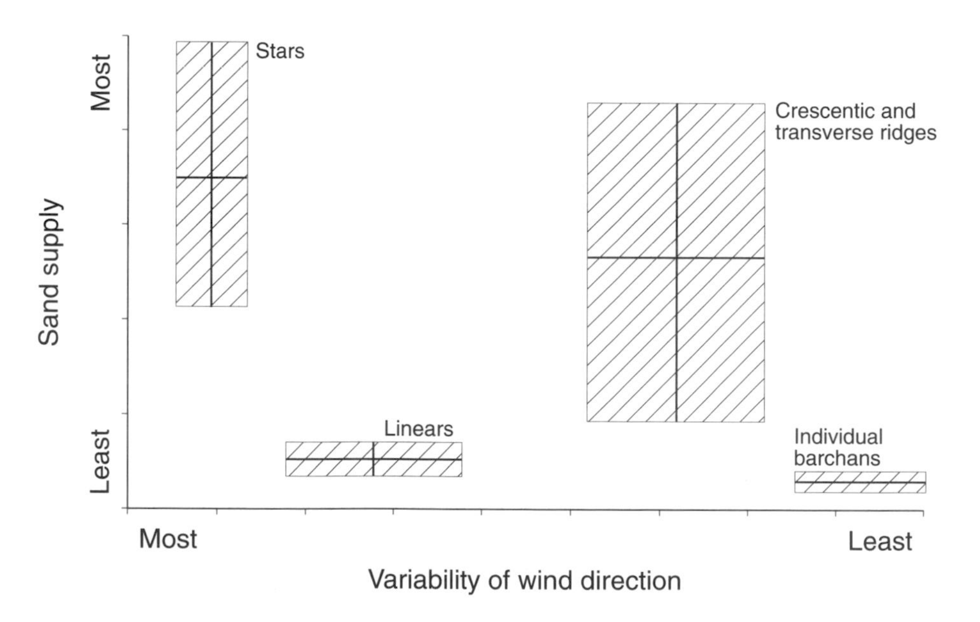

Figure 6.9 A simple model of the relationship between dune form, wind variability and sand supply.

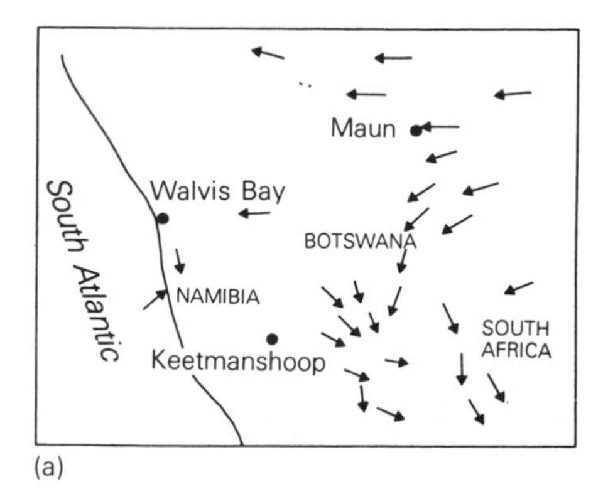

(a)

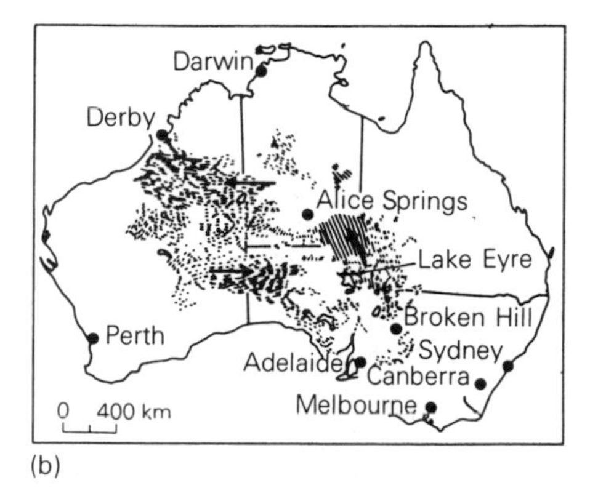

(b)

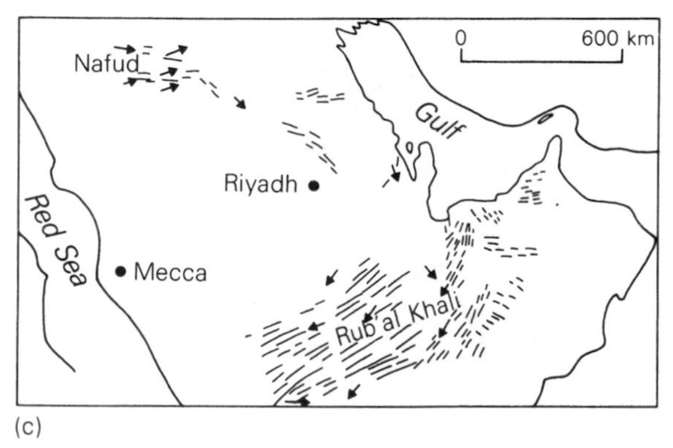

(c)

Figure 6.10 Regional patterns of dune trends to illustrate the characteristic wheel-round form: (a) southern Africa; (b) Australia; (c) Arabia.

activated *in situ* by the removal of stabilizing vegetation.

Many methods are used in the attempt to control drifting sand (see colour plate 8) and moving dunes (table 6.3). In practice, most solutions to the problem of dune instability and sand blowing have involved establishing a vegetation cover. This is not always easy. Plant species used to control sand dunes must be able to endure undermining of their roots, burying, abrasion and often severe deficiencies of soil moisture. Thus the species selected need to have the ability to recover after partial burying, to have deep and spreading roots, to have rapid height growth in the seedlings stages, to promote rapid litter development, and to add nitrogen to the soil through root nodules. During the early stages of growth they may need to be protected by fences (plate 6.16), sand traps and surface mulches. Growth can also be stimulated by the addition of synthetic fertilisers.

6.11 The Work of Rivers

Although we have discussed the role of wind erosion and deposition in the arid environment at some length, present and past river action is also important in moulding desert landscapes (plate 6.17). As we noted earlier, although rainfall quantities may be low overall, substantial falls may occur from time to time. Moreover, many desert surfaces have a number of characteristics that enable them to generate considerable runoff from quite low rainfall intensities. First, the limited vegetation cover provides little organic litter on the surface to absorb water. Second, the sparseness of vegetation means that humus levels in the soil are low, and this, combined with the minimal disturbance by plant roots and a greatly reduced soil fauna, makes the soil dense and compact in texture. Third, as there is virtually no plant cover to intercept the rainfall, rain is able to beat down on the soil surface with maximum force; and fine particles, unbound by vegetation, are redistributed by splash to lodge in pore-spaces and to create a puddled soil surface of reduced permeability. Studies in various areas have shown that, where such crusted and impermeable soils exist, the infiltration rate is only a few millimetres per hour, so that a rainfall rate in excess of this is likely to produce overland flow.

Desert stream channels or valleys are called *wadis,* and while they are normally dry they can be subjected to large flows of water and sediment. Such

Window 6.3 ***Desertification***

One of the biggest environmental issues of the past two decades has been the question of desertification. Maps of the world have been produced which purport to show that huge areas are at risk from varying degrees of land degradation. It is widely reported that desertification affects about 65 million hectares of once-productive agricultural land and threatens the livelihoods of 850 million people.

The term 'desertification' is often confused with 'drought'. However, there is a difference. Drought is a relatively short-term problem, which has acute phases lasting for a few years at a time. By contrast, desertification is a more chronic, long-term problem. Drought does not directly result in desertification unless it is of an extended duration. Normally, when the return of rain heralds the end of drought, vegetation returns. Desertification is also often confused with 'famine', but famine does not directly relate to the state of arid environments but rather to food shortages, some of which may, of course, be caused by droughts or desertification.

The 1977 Nairobi Conference on Desertification depicted four classes of desertification on its maps:

Slight

- little or no deterioration of plan cover or soil

Moderate

- significant increase in undesirable forbs and shrubs; or
- hummocks, small dunes or small gullies formed by accelerated wind or water erosion; or
- soil salinity causing *c.*10–50 per cent reduction in irrigated crop yields

Severe

- undesirable forbs and shrubs dominate the flora; or
- sheet erosion by wind and water has largely denuded the land of vegetation, or large gullies are present; or
- salinity has reduced irrigated crop yields by more than 50 per cent

One of the most severely desertified areas in the United States is the Navajo Lands. One of the most important causes of this phenomenon in that area is overgrazing by domestic stock.

Very severe

- large, shifting, barren sand dunes have formed; or
- large, deep and numerous gullies are present; or
- salt crusts have developed on almost impermeable irrigated soils.

The fundamental causes of most problems of arid-zone degradation relate to twentieth-century explosions in human population levels. These have led to overcultivation, overgrazing, deforestation and salinisation of irrigation systems.

Overcultivation

This is a significant factor in arid-zone degradation, and results from the creep of dry-land agriculture into excessively dry areas. Cropping now takes place in areas receiving as little as 150 mm of annual rainfall in North Africa and the Near East, and 250 mm in the Sahel. Soil left barren after cropping, or after crop failure, is prone to wind and water erosion, further accelerating desertification.

Overgrazing

Overgrazing is widely regarded as a prime cause of desertification. Globally, the population of cattle rose by 38 per cent between 1955 and 1976, and that of sheep and goats by 21 per cent. In many areas the increases in free-ranging livestock populations have exceeded the carrying capacity of the land. Furthermore, in some regions increases in the area under cultivation have reduced available pastureland and intensified pressure on remaining pastures.

The installation of modern boreholes and various types of excavated waterhole has enabled rapid multiplication of livestock numbers, so that severe overgrazing and land degradation frequently occur, primarily in the proximity of the new water sources. Without the rest period that was previously assured by intermittent water supplies, vegetation replaces water as the limiting factor in livestock survival.

Deforestation

The uprooting of woody species is a further fundamental factor in land degradation. The collection of wood

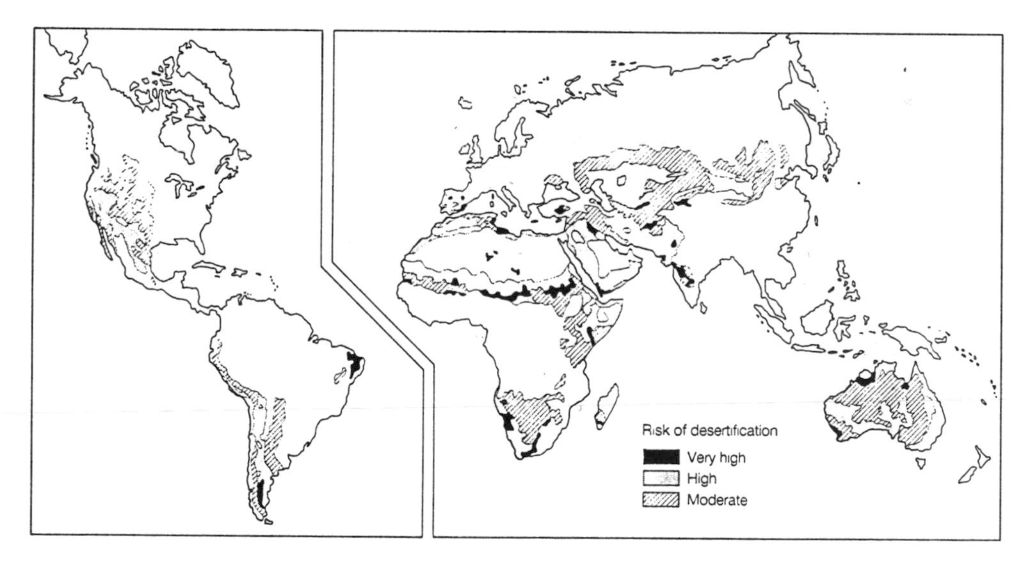

The United Nations Conference on Desertification (Nairobi, 1977) map of areas at high risk of desertification.

for charcoal and firewood is an especially serious problem in the vicinity of large urban areas, where electricity and electrical devices (e.g. stoves) are too expensive for most of the urban poor.

Salinisation
Humanly induced salinisation, resulting from the spread of irrigation, is a sinister and widespread form of land degradation. Soil salinity destroys the soil structure, reduces plant growth and ultimately kills plants.

Plate 6.16 One means of controlling the advance of barchan dunes near Walvis Bay, Namibia – sand fences.

flows, as one might expect from the character of desert rainfall, are sporadic in time and space. In the Sahara there is an average of one flood a year in the semi-arid parts, but in the more arid core some of the wadis may go ten years without carrying water. It is also common for any one stream to carry discharge along only a portion of its course, and floods often peter out because of losses from seepage and evaporation. When these streams flow they carry a considerable sediment load and can pose problems to roads – the flow may erode the road, while the sediment may clog culverts and the like.

A useful measure of geomorphological activity is sediment yield per unit area over a period of time (figure 6.11). This can be determined by such means as recording the quantity of sediment accumulation in reservoirs or measuring the amount of material carried by rivers. Studies suggest that in deserts, as rainfall rises from zero, yields of sediment increase rapidly because more and more runoff becomes available to move sediment; and yet there is still plenty of bare ground susceptible to

Table 6.3 Control techniques for drifting sand and mobile dunes

Problem	*Control methods*
Drifting sand	Enhancement of deposition of sand through creating large ditches, vegetation belts and barriers, and fences (plate 6.16) Enhancement of sand transport by aerodynamic streamlining of surface, or changing surface materials Reduction of sand supply by surface treatment, improved vegetation cover or erection of fences Deflection of moving sand by fences, barriers or vegetation belts
Moving dunes	Removal by mechanical excavation Destruction by reshaping, trenching through dune axis or surface stabilization of barchan arms Immobilization by trimming, surface treatment and fences

Plate 6.17 Even in the depths of Death Valley, California (mean annual rainfall only about 70 mm), the imprint of fluvial processes is clearly visible in this heavily dissected landscape. Some desert areas have exceptionally high drainage densities.

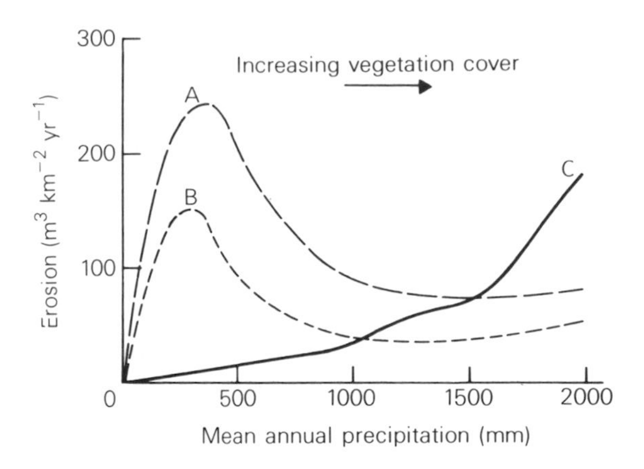

Figure 6.11 Sediment yield in relation to mean annual precipitation. A represents rates of erosion in high-relief situations; B represents rates of erosion in lower-relief situations; C represents rates of limestone solution. Rates of erosion appear to be at a maximum under semi-arid conditions, whereas rates of solution become high only under conditions of much more substantial rainfall.

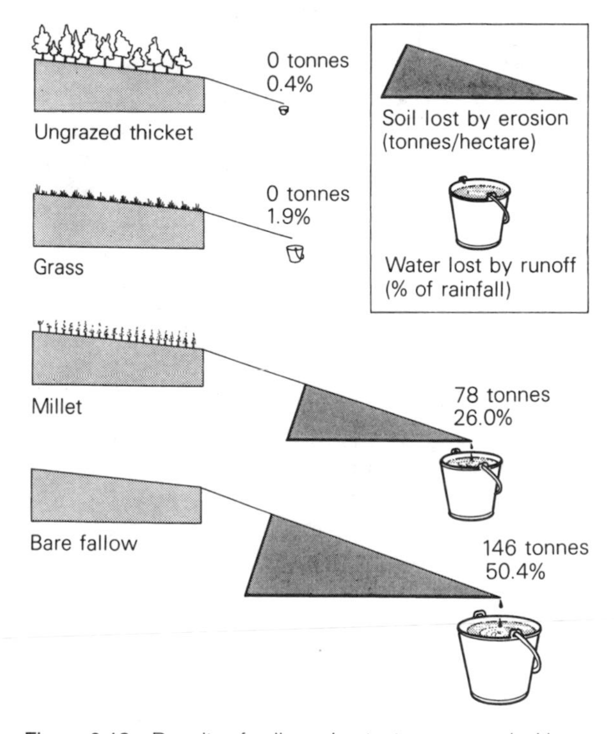

Figure 6.12 Results of soil erosion tests on ground with different vegetation covers in a semi-arid part of Tanzania. The grass-covered plots had little loss of soil and water, whereas rates of soil loss and runoff from bare fallow ground were very high.

rainwash erosion. However, as the amount of vegetation cover increases, sediment yields start to decline. Thus, while in extreme deserts sediment yields are very small, in semi-arid areas they may be among the highest in the world. Moreover, many streams are so full of sediment that there are frequent occurrences of mudflows, in which solid matter may account for between 25 and 75 per cent of the flow. Such large sediment concentrations are important in the formation of alluvial fans, and can lead to rapid sedimentation behind engineering structures such as dams.

On desert margins, disturbance or removal of the vegetation cover by human activity can markedly increase the background natural rates of sediment yield. A study in Tanzania showed that annual soil loss was virtually nil under ungrazed thicket or grass but rose to 78 t ha^{-1} under millet, and to almost 146 t ha^{-1} under bare fallow (figure 6.12). There was a comparable increase in the amount of water lost by runoff, for water infiltration was far less under bare fallow conditions.

Because of the high runoff and sediment yields from some desert surfaces, the imprint of water is often made clear through the development of a considerable degree of dissection or drainage density. On erodible rocks and sediments *badlands* may develop that have *drainage densities* (the total length of drainage line per unit area) between 10 and 20 times the average under humid climates. In parts of arid North America densities may be as high as 350 km km^{-2} compared with 2–8 km km^{-2} over most of Britain. However, in sandy areas in deserts, where surface conditions give a high infiltration rate and hence little runoff, drainage density will be low.

Drainage systems are not always very highly integrated. Much of the drainage, because of high evaporation or high infiltration losses into alluvial fans or aeolian materials, does not reach the sea, and because of the existence of closed basins (see section 6.9), much of the drainage may be *endoreic* (i.e. centripetal, flowing towards the centre). None the less, some desert rivers, including many of those that are of most use to man, such as the Nile, Tigris–Euphrates and Indus, have their sources outside the desert realm, and such *allogenic* or *exogenous* rivers may be through-flowing and perennial.

A characteristic feature of desert drainage systems are *alluvial fans* (see figure 6.13a). These are cones of sediment that occur between the mountain front and a low-lying plain. Their size is variable – small ones may have a radius of only a few tens of metres, while the large ones may be more than 20 km across and 300 m thick at the apex. They form where rivers emerge from their confined upper mountainous courses. At such a point the river can spread out, dissipating energy and decreasing its velocity, so that deposition of sediment occurs. They form most easily where there is a sharp juxtaposition of mountain and lowland, and for this some tectonic activity may be required, as in the Basin and Range Province of the western United States. Moreover, fans derived from easily eroded rocks (such as shales) are larger than those derived from resistant rocks (such as quartzites). The types of flow that deposit the material composing the fans vary from simple streamflows to highly viscous mud

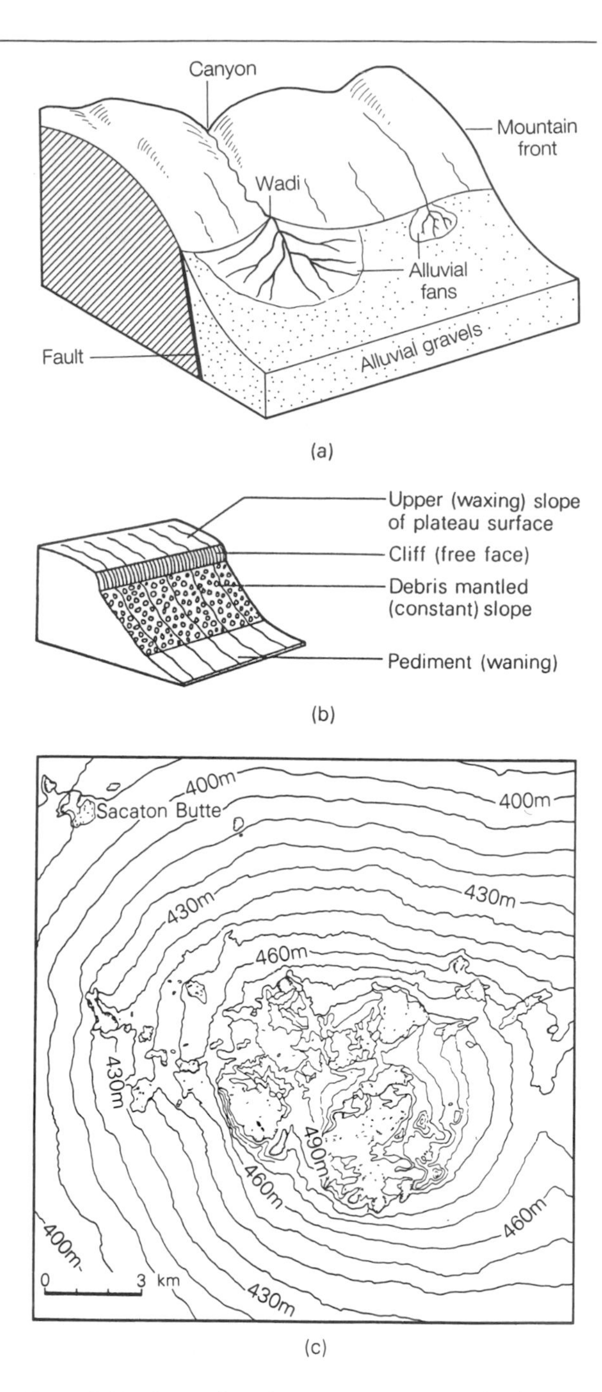

Figure 6.13 Some slope forms encountered in the desert environment: (a) an alluvial fan formed at the mountain front where a river (wadi) leaves a canyon; (b) the four main slope components of some arid profiles; (c) pediments and pediment passes around inselbergs in the Sacaton Mountains, Arizona.

and debris flows. Because of the variability of flow regime and the variable nature of the material composing the fans, they are subject to rapid changes in character, with channels shifting laterally over a wide area and alternately cutting and filling themselves. This presents problems for the establishment of transport links and buildings on fan surfaces.

6.12 Desert Slopes

Slope profiles in arid areas, being little obscured by vegetation, are visually more dramatic than those of most humid regions and many workers have commented on their apparent angularity. Their form can be analysed in terms of an idealised slope profile (see figure 6.13b) with four components: an upper convexity (the waxing slope), a cliff (free face), a straight segment (constant slope) and a basal concavity (pediment). The pediment (see figure 6.13c) and the constant slope are often separated by an abrupt break of slope (plate 6.18). The pediments may coalesce and cover extensive areas (called *pediplains*), and from them may rise isolated residual outliers called *buttes*, which possess one or both of the steep slope components.

Plate 6.18 One of the classic landform assemblages of a desert area such as the Mojave Desert in California is the formation of pediments. There is a very sharp break in slope between them and the mountain front behind. Note the sparse vegetation cover.

The pediments consist of a gently sloping rock-cut surface, generally at an angle of between 1° and 7°, with a few lines of concentrated drainage and a thin veneer of rock detritus. An early explanation for them was that they were formed by sheets of water of high velocity (called *sheetfloods*) which planed down the surface. Such sheetfloods have been described from time to time, but the hypothesis suffers from the perennial problem of the chicken and the egg. Which came, first, the pediment or the sheetflood? A second hypothesis sees pediments as having been shaped by lateral planation; that is, it is envisaged that streams flowing from the mountain front on to the plains would swing from side to side and gradually erode the surface. However, this explanation fails to account for pediments that directly abut the mountain front away from such swinging channels. A third hypothesis involves the role of surface and subsurface weathering. This is likely to be accentuated at the junction between the mountain front and the plain because of the natural concentration of water there through percolation. The weathering will produce fine-grained material that can be removed, in the absence of a vegetation cover, by sheetfloods, wind and other processes.

Rising above the pediments there are often isolated steep-sided hills called *inselbergs* (plate 6.19), formed as a result of gradual slope retreat produced by weathering and erosion at the break of slope. They occur in a wide variety of rock types, including coarse sandstones (e.g. Ayers Rock in Australia) and granites. While not restricted to desert areas, such inselbergs, which may be over 600 m high, are spectacular landscape features in many deserts.

Whether desert slopes develop by parallel retreat to give pediplains and steep-sided residuals or evolve by gradual slope decline through time depends very much on rock type. This is demonstrated very clearly by a study that S. A. Schumm made in semi-arid Dakota (plate 6.20), in the United States. In that area there are two sedimentary rock types occurring under identical climatic conditions, the Brule and Chadron. The Brule had steep (44°) straight

Plate 6.19 A classic inselberg landscape at Gross Spitzkoppie in Namibia. These isolated island hills rise up steeply above a very gently sloping plain – a pediment.

slopes, and erosion had produced outliers with a similar slope profile. By contrast, the Chadron had rounded slopes with broad interfluves, and residuals had lower slope angles than the scarp from which they had become detached by erosion. The explanation for this can be seen in terms of the processes operating on the two different rock types. Because it had a low infiltration capacity, the Brule was characterised by surface wash and so the steep slopes were produced by stream incision. The Chadron produced a surface of clay aggregates with an open texture which had a much higher infiltration capacity, and so the creep of the clay aggregates when saturated with moisture was the main slope-forming process.

6.13 Groundwater

Because of the low rainfalls and the general absence of perennial rivers (with the exception of the great exogenous rivers like the Nile), groundwater is often especially important if arid areas are to be exploited.

Groundwater occurs underground in the rocks and sediments, in zones termed *aquifers*. Such aquifers may have been charged with water either by occasional storms or during pluvials, when arid zones were less extensive than now, or because they extended into presently relatively humid areas. In permeable rocks like the Nubian sandstone of the Sahara, large quantities of water may be involved, and if the arrangement of the strata is favourable artesian flow may occur (see section 14.7). Sand dunes and wadi gravels, both being highly permeable, may also be useful sources of groundwater. Indeed, this is the explanation for some oases in the middle of an erg. When it rains the water infiltrates rapidly into the dune sand and, provided the underlying platform is more or less impermeable, the water accumulates in a dome shape. Lateral seepage from the dunes may be sufficient to water some palm trees. In a similar manner, coastal dunes

Plate 6.20 Steep slopes developed by erosion of the Brule rocks of the Badlands of South Dakota, USA.

along the shores of the Arabian Gulf have been favoured sites for settlement.

However, groundwater resources are neither infinite nor endlessly exploitable without certain problems. First, as some of them are fossil, having been created in pluvial phases, over-pumping may cause their rapid depletion. In parts of the high plains of Texas, water levels have fallen 50 m in 40 years. Second, some of them are too saline to be used for many purposes. Third, over-pumping in coastal areas may cause saltwater from the sea to intrude. The reason for this is that freshwater has a lower density than saltwater, so that a column of seawater can support a column of freshwater; if the freshwater is pumped away too quickly, saltwater rapidly rises to replace it from beneath. Fourth, by pumping out groundwater from the pores of rocks, some subsidence of the ground surface may result, creating engineering problems. In Mexico City 7.5 m of subsidence has occurred (equivalent to a rate of 250–300 mm yr^{-1}) and in the Central Valley of California, 8.5 m.

Groundwater resources are being exploited at an ever-increasing tempo. This is partly because of the availability in the past two or three decades of new means of groundwater exploitation for irrigation purposes. Especially notable in this context is the adoption of a technique called centre-pivot irrigation, which is achieved by pumping water up a borehole and then distributing it across the land by means of a huge revolving boom equipped with sprinklers.

6.14 Dams, Reservoirs and Inter-basin Water Transfer

Besides exploiting groundwater by drilling deep boreholes, man can improve water supplies in arid areas by damming rivers. Indeed, the first recorded dam was built in Egypt some 5000 years ago to enable irrigation to take place. Since then the number and size of dams have increased dramatically: they control floods, reduce the threat of famine and provide electricity.

However, dams have a whole series of environmental consequences (window 6.4) that may or may not have been anticipated (figure 6.14). They trap sediment, which leads to a reduction in the sediment load of the river downstream. This in turn, as is the case with the Aswan Dam on the Nile, may reduce the amount of flood-deposited nutrients on the fields, provide less nutrients for fish in the southeast Mediterranean Sea, and cause coast and riverbed erosion because less sediment is available to cause accretion. The weight of water in the reservoir behind the dam may accelerate earthquake activity by depressing the earth's crust; seepage from the reservoir and from canals may cause build-up of salinity; local climates may change; the diversion of river flow may change salinity in the seas, which in turn may affect their ice cover and affect world climate; water-borne diseases may be more widely transmitted; and the migration of fish may be disturbed.

River flows are now being managed so that major transfers of water are made from one river basin to another. This can have a whole series of consequences, as is made evident when one considers the dire plight of the Aral Sea in central Asia (window 6.5).

Window 6.4 ***The pace and price of irrigation***

The twentieth century was an age of irrigation. World-wide, between 1900 and 1950, the area of land that was irrigated almost doubled, reaching 94 million hectares. By 1990 this figure had surged to 250 million hectares. Approximately one-third of the global food harvest comes from the 17 per cent of the world's cropland that is irrigated. In terms of gross irrigated area, the five largest countries are India, China, the former Soviet Union, the USA and Pakistan.

As the *State of the World Report* (1990) put it, this expansion has not been achieved without an environmental price tag:

> Each year, some 33 000 cubic kilometers of water – six times the annual flow of the Mississippi – are removed from the earth's rivers, streams and underground aquifers to water crops. Practised on such a scale, irrigation has had a profound impact on global water bodies and on the cropland receiving it. Waterlogged and salted lands, declining and contaminated aquifers, shrinking lakes and inland seas, and the destruction of aquatic habitats combine to hang on irrigation a high environmental price.

Of the side-effects of irrigation, the worst damage results from the waterlogging and salinity build-up in irrigated soils. Salinity in soils has a range of undesirable consequences. For example, as irrigation water is concentrated by evapotranspiration, calcium and magnesium components tend to precipitate as carbonates, leaving sodium ions dominant in the soil solution. The sodium ions tend to be absorbed on the colloidal clay particles, deflocculating them and leaving the soil structureless, almost impermeable to water and unfavourable to root development. Poor soil structure and toxicity lead to the death of vegetation in areas of saline patches. This creates bare ground which is vulnerable to erosion by wind and water.

Probably the most serious result of salinisation is its impact on plant growth. This takes place partly through its effect on soil structure, but more significantly through its effects on osmotic pressures and through direct

Waterlogged and salinized fields in the Indus Plain of Pakistan. The spread of irrigation has caused the level of the water table to approach the ground surface, and under conditions of high temperatures and limited rainfall, salt accumulates as a white crust over the ground surface.

toxicity. When a water solution containing large quantities of dissolved salts comes into contact with a plant cell it causes the cell's protoplasmic lining to shrink. This is due to the osmotic movement of the water, which passes out from the cell towards the more concentrated soil solution. The cell collapses and the plant dies.

This toxicity effect varies with different plants and different salts. Sodium carbonate, by creating highly alkaline soil conditions, may damage plants by a direct caustic effect; high nitrate may promote undesirable vegetative growth in grapes or sugar beets at the expense of sugar content. Boron is injurious to many crop plants at solution concentrations of more than 1 or 2 ppm.

There are a variety of reasons why soil salinity is spreading. The most important of these is the growth in the area of irrigated land, which increased from about 8 million hectares in 1800 to 250 million hectares in the 1990s. The extension of irrigation and the use of a wide range of different techniques for water abstraction and application can lead to a build-up of salt levels in the soil. This happens because water abstraction raises the groundwater level so that it is near enough to the ground surface for water to rise to the surface by capillary action. Evaporation then leaves the salts in the soil. In the case of the semi-arid northern plains of Victoria in Australia, for instance, the water table has been rising by around 1.5 m per year so that now, in many areas, it is little more than 1 m below the surface. When groundwater comes within 3 m of the surface in clay soils – less for silty and sandy soils – capillary forces bring moisture to the surface where evaporation takes place, leaving salts behind.

Second, many irrigation schemes spread large quantities of water over the soil surface. This is especially true for rice cultivation. Such surface water is readily evaporated, so that again salinity levels build up.

Third, the construction of large dams and barrages to control water flow and to give a head of water creates large reservoirs from which further evaporation can take place. The water gets saltier. This salty water is then used for irrigation, with the effects described above.

Fourth, water seeps laterally from irrigation canals, especially in highly permeable soils, so that further evaporation takes place. Many distribution channels in a gravity irrigation scheme are located on the elevated areas of a floodplain or riverine plain to make maximum use of gravity. The elevated landforms selected are natural levées, river-bordering dunes and terraces, all of which are composed of silt and sand which may be particularly prone to loss by seepage.

The round field produced by a centre-pivot irrigation system near Port Elizabeth in South Africa. Similar fields, which are often several hundreds of metres in radius, are also a major feature of land use in the western United States. Such irrigation systems can cause a series of environmental problems, including groundwater depletion and soil salinity.

In coastal areas salinity problems are created by seawater incursion brought about by overpumping of fresh groundwater from aquifers. If the aquifer is open to penetration from the sea, salty water tends to replace the freshwater that has been extracted. This is a particularly serious problem along the shores of the Persian Gulf where, because of the dry climate, the freshwater can only slowly be replenished by rainfall.

It has been estimated that salt-affected and waterlogged soils account for 50 per cent of the irrigated area in Iraq, up to 40 per cent of all Pakistan, 50 per cent in the Euphrates Valley of Syria, 30–40 per cent in Egypt and up to 30 per cent in Iran. In Africa, however, where there are fewer great irrigation schemes, less than 10 per cent of salt-affected soils are so affected because of human action. Looking at the problem on a global basis, from 1700 to 1984, the global areas of irrigated land increased from 50 000 to 2 200 000 km^2, while at the same time some 500 000 km^2 were abandoned as a result of secondary salinisation. In the past three centuries irrigation has resulted in 1 million square kilometres of land with diminished productivity due to salinisation.

Given the seriousness of the problem, a range of techniques for the eradication, conversion or control of salinity has been developed. These include the following:

- Provision of adequate subsoil drainage to prevent waterlogging, to keep the water table low enough to reduce the effects of capillary rise and to remove water that is in excess of crop demand.
- Leaching of salts by applying water to the soil surface and allowing it to pass downward through the root zone.
- Treatment of the soil (with additions of calcium, magnesium, organic matter etc.) to maintain soil permeability.
- Planting of crops which do not need much water.
- Planting of crops or crop varieties that will produce satisfactory yields under saline conditions.
- Reduction of seepage losses from canals and ditches by lining them (e.g. with concrete).
- Reduction in the amounts of water applied by irrigation, by using sprinklers and tricklers.
- Storage of heavily salted waste water from fields in evaporation ponds.

Salinisation of irrigated cropland in selected countries.

Country	*% of irrigated lands affected by salinisation*	*Country*	*% of irrigated lands affected by salinisation*
Algeria	10–15	Jordan	16
Australia	15–20	Pakistan	<40
China	15	Peru	12
Colombia	20	Portugal	10–15
Cyprus	25	Senegal	10–15
Egypt	30–40	Sri Lanka	13
Greece	7	Spain	10–15
India	27	Sudan	20
Iran	<30	Syria	30–35
Iraq	50	USA	20–25
Israel	13		

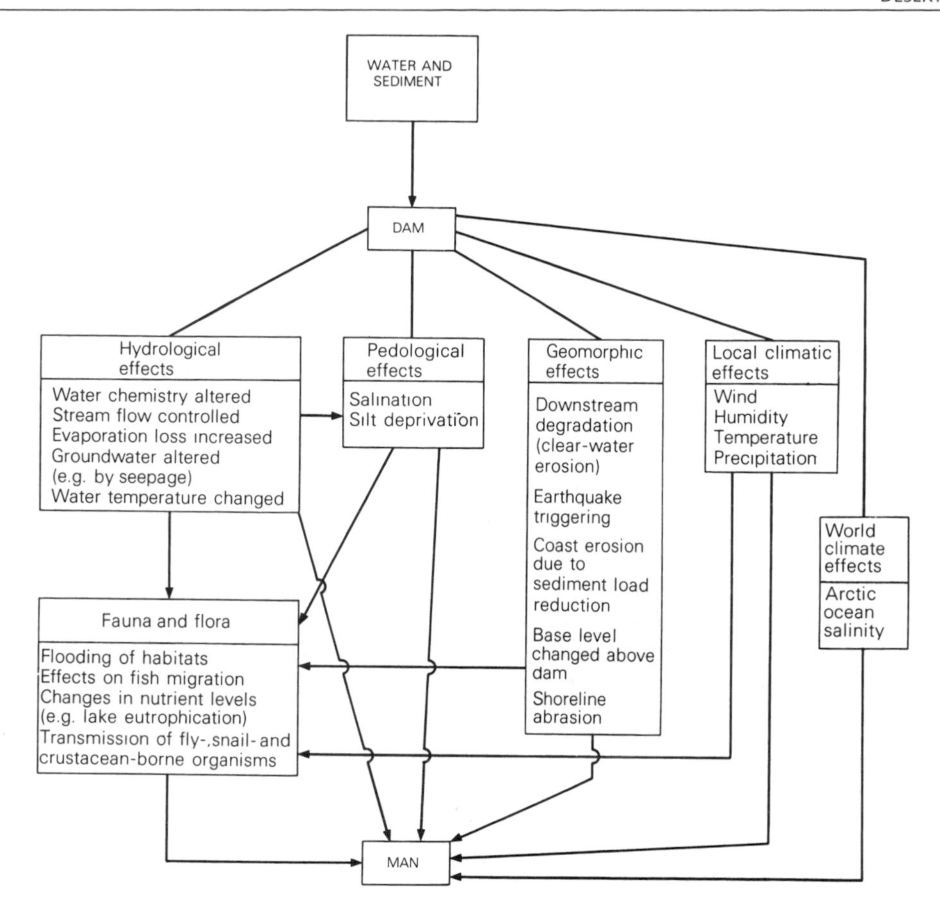

Figure 6.14 Some of the consequences of dam construction on the environment.

Window 6.5 Inter-basin water transfers and the death of the Aral Sea in central Asia

Increasing rates of water consumption and the unequal distribution of water resources from one region to another mean that in many parts of the world long-distance transfers of water are made between river basins. Also, in the world's drylands, large quantities of water are abstracted from rivers to supply irrigation schemes. One of the results of such large-scale modifications of river regimes is that the discharges of some rivers have declined very substantially. This in turn means that the extent and volume of any lakes into which they empty have been reduced.

Perhaps the most severe change to a major inland sea or lake is that taking place in the Aral Sea in the southern part of the former Soviet Union. Until very recently this was the world's fourth largest lake, with a high level of biological activity and a rich and distinctive aquatic fauna and flora. It had considerable commercial fisheries, and was used for transport as well as sporting and recreational activities. It was also a refuge for huge flocks of waterfowl and migratory birds. It may also have exerted a favourable climatic, hydrological and hydrogeological effect on the surrounding area.

However, since the 1960s a dramatic change has taken place. The inflow of water into the lake has decreased markedly and it has now lost more than 40 per cent of its area and about 60 per cent of its water volume. The lake's level has fallen by more than 14 m. Its salinity has increased threefold. Its fauna and flora have been destroyed, so that only a small number of aquatic species has survived. The climate around the lake may also have been affected. The increasing areas of exposed, desiccating and salty lakebed provide an ideal environment for the genesis of dust storms. Such storms now evacuate some tens of millions of tons of salt each year and dump them on agricultural land, reducing crop yields. The human population also seems to be suffering from poorer-quality water supply and from respiratory disorders caused by the blowing salt and dust. It is not surprising, therefore, that the Aral Sea is now regarded as the greatest ecological tragedy of the former Soviet Union.

Why has the inflow of water to the Aral Sea declined so extraordinarily? The main reason was that in the 1950s and early 1960s a decision was taken to expand irrigation in central Asia and Kazakhstan, so that crops like rice and cotton, which consume a great deal of water, could be cultivated in the middle of a desert. Large volumes of fertilizers and herbicides were also used in growing these crops, and these have contributed to the deterioration in water quality. In many cases, too, the irrigation systems themselves were of poor design, construction and operation.

Sources

P. P. Micklin (1988) 'Desiccation of the Aral Sea: a water management disaster in the Soviet Union', *Science*, 241: 1170–5. One of the key papers that drew attention to the situation around the Aral Sea.

P. P. Micklin (1992) 'The Aral crisis: introduction to the special issue', *Post-Soviet Geography*, 33 (5): 269–82. A collection of papers on all aspects of the Aral Sea problem (at http:/www.dfd.dlr.de/app/land/aralsee/chronology.html).

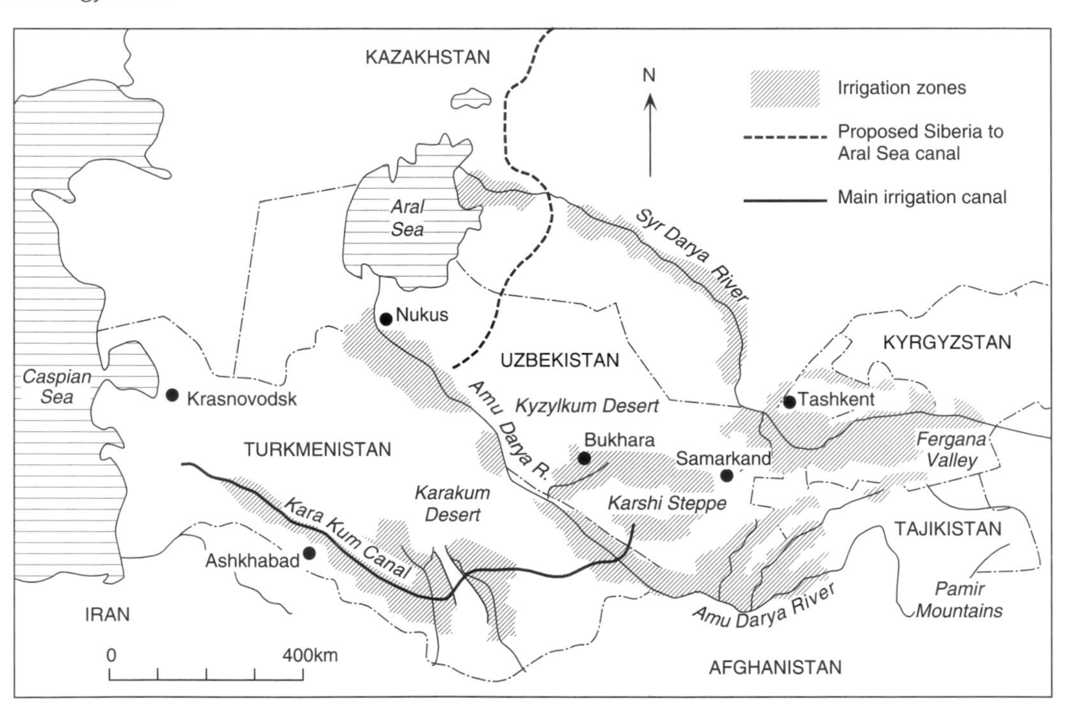

Irrigation and the Aral Sea.

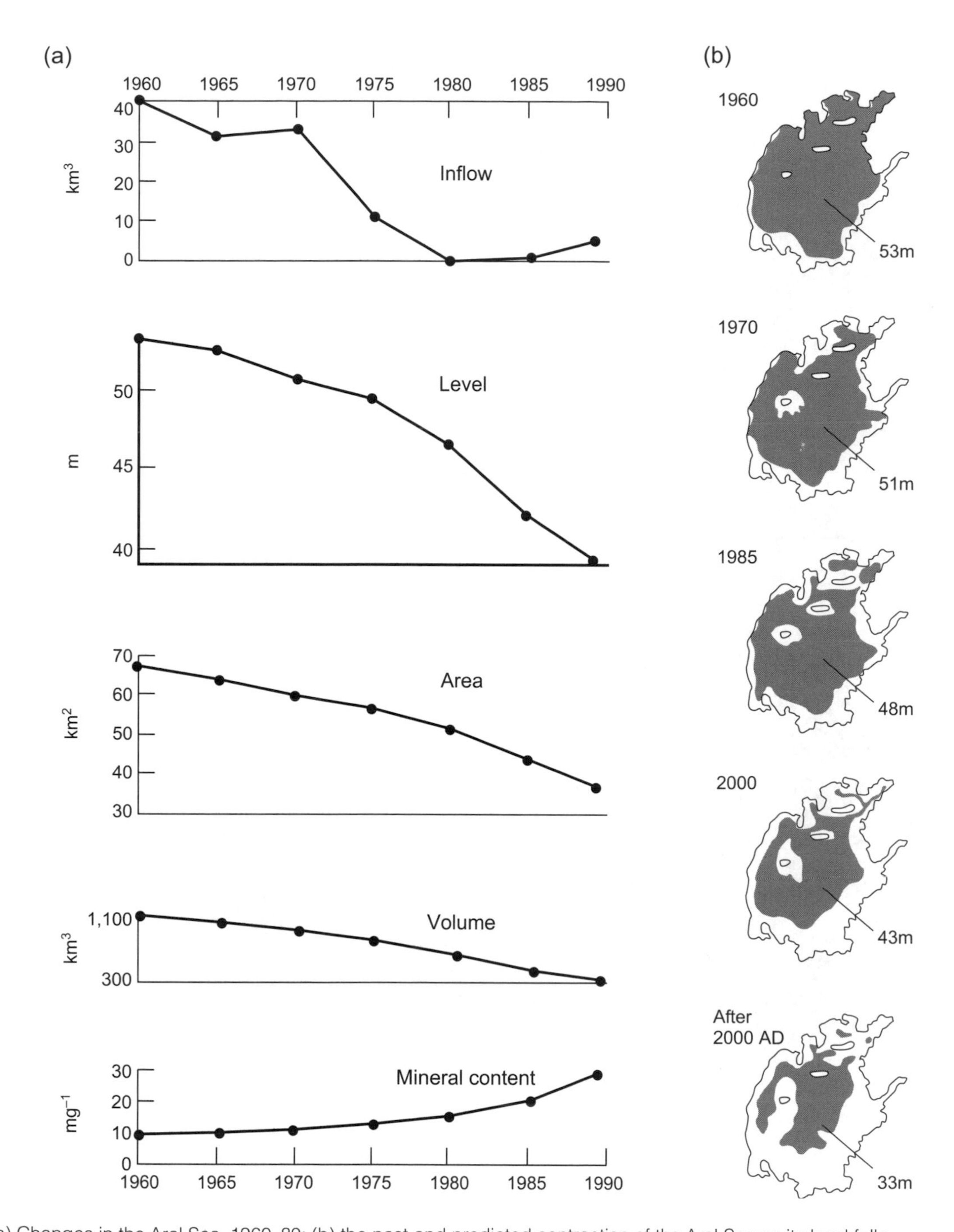

(a) Changes in the Aral Sea, 1960–89; (b) the past and predicted contraction of the Aral Sea as its level falls.

Plate 6.21 Overgrazing in sandy areas, as around a well in southern Botswana, can cause dunes to become active. Virtually all ground cover has been removed by domestic animals.

6.15 Land Degradation

The problems associated with dams and reservoirs provide an indication that desert environments may be very difficult to develop without unfortunate environmental consequences. In recent years the fear has been expressed that human actions may even be leading to the spread of desert-like conditions – a process termed *desertification* or *desertisation*. Not everyone is agreed as to the cause of the spread of desert-like conditions, and the question has been asked whether this process is caused by temporary drought periods of high magnitude, is due to a longer-term climatic change towards aridity, is caused by man-induced climatic change, or is the result of human action through man's degradation of the biological environments in arid zones. Most people now believe that it is produced by a combination of increasing human and animal population levels, which causes the effects of drought years to become progressively more severe, so that the vegetation is placed under severe stress – by over-

Plate 6.22 Overgrazing by cattle and goats near Baringo in Kenya has lowered the ground surface to such an extent that tree roots have been exposed. This type of erosion is one component of the desertification problem.

grazing, by cultivation of marginal areas, by burning shrub vegetation to increase the area of pasture, and by collection of wood for fuel. The removal of vegetation sets in train the operation of such insidious processes as deflation (plate 6.21) and water erosion (plate 6.22). These further reduce the utility of the available land.

6.16 Problems of the Desert Realm

Although deserts cover one-third of the earth's surface, they are generally hostile environments for man, and they are easily degraded. Some of the problems they present are as follows:

sand encroachment on roads etc.
wind erosion of soils
dust storms
salinity
weathering of buildings
reduction of soil quality
restriction of plant growth
floods
debris flows
siltation of dams
silt depletion downstream of dams
salinity build-up
disease transmission
climatic change
groundwater extraction
subsidence
saltwater incursion
smog (produced by sunlight and inversion)
extreme heat
very low humidity

Although spectacular advances in desert utilisation may occur in some of the developed countries and the new oil-rich states of the Middle East (because they have money to invest in long-term research and in projects of dubious short-term economic value), much of the world's desert areas will remain relatively untouched wilderness for decades to come.

■ *Key Terms and Concepts*

alluvial fan
aquifer
aridisols
aridity
biomass
deflation
desertification
duricrust
dust bowl
erg
insolation
inter-basin water transfer
interpluvials
pan
pediment
phreatophyte
pluvials
potential evapotranspiration
salinisation
salt weathering
sheetflood
stone pavement
water balance
xerophyte
yardang

■ *Points for Review*

- What is aridity?
- What causes aridity?
- What evidence is there for climate change in deserts?
- How do plants and animals adapt to desert conditions?
- What are the distinguishing characteristics of desert soils?

- What causes rocks to break down in deserts?
- In what ways does wind mould desert landscapes?
- Describe the main types of dune form and the conditions that give rise to their development.
- Why are rivers effective in moulding desert landforms?
- What happens if groundwater is over-exploited?
- What are the adverse consequences of dam construction?
- What is desertification and how might you seek to control it?

FURTHER READING

Agnew, C. and Anderson, E. (1992) *Water Resources in the Arid Realm* (London: Routledge). A survey of climate, drought, water management and so on.

Beaumont, P. (1989) *Drylands: Environmental Management and Development* (London: Routledge). A particularly useful treatment of the human use of drylands.

Cooke, R. U., Warren, A. and Goudie, A. S. (1993) *Desert Geomorphology* (London: UCL Press). An advanced overview.

Goudie, A. S. (ed.) (1990) *Techniques for Desert Reclamation* (Chichester: Wiley). This edited work looks at some of the solutions that are available for dealing with the problems of desert environments.

Goudie, A. S. and Watson, A. (1990) *Desert Geomorphology*, 2nd edn (Basingstoke: Macmillan). An introduction.

Grainger, A. (1990) *The Threatening Desert: Controlling Desertification* (London: Earthscan). A very readable and wide-ranging review of desertification.

Lancaster, N. (1995) *Geomorphology of Desert Dunes* (London: Routledge). A thorough discussion of dunes.

Livingstone, I. and Warren, A. (1996) *Aeolian Geomorphology: An Introduction* (Harlow: Longman). An accessible treatment of all aspects of the role of wind as a geomorphological agent.

Louw, G. N. and Seely, M. K. (1982) *Ecology of Desert Organisms* (London: Longman). The best treatment of desert plants and animals.

Middleton, N. J. (1991) *Desertification* (Oxford: Oxford University Press). A well-illustrated, simple introduction designed for use in secondary education.

Mortimore, M. (1998) *Roots in the African Dust: Sustaining the Drylands* (Cambridge: Cambridge University Press). A discussion of land degradation in the drylands of Africa.

Thomas, D. S. G. (ed.) (1998) *Arid Zone Geomorphology*, 2nd edn (Chichester: Wiley). A wide-ranging edited compilation.

7 The Tropics

7.1 The General Atmospheric Circulation

The tropics do not constitute one uniform environmental type. Within the zone bounded by the Tropics of Cancer and Capricorn there is a wide range of climatic, vegetational, topographical and pedological characteristics. There are, for instance, rainforests, grasslands, mangroves, corals, mountains and plains, and while some areas teem with human populations others are almost uninhabited. They are all linked, however, by the fact that as a rule they receive much more solar radiation throughout the year than other zones.

Indeed, one of the main functions of the general atmospheric circulation in the tropics is to dissipate to other parts of the globe the surplus heat energy supplied to low latitudes by solar radiation. According to the simple cell model, proposed as early as 1735 by G. Hadley, the heated air from the equatorial regions rises; the air masses are then transported to higher latitudes by upper air currents, sometimes called the *anti-trades*, which gradually slow down at about 20°–30° latitude. At that latitude subsidence of the air takes place, resulting in an area of relatively high pressure – the subtropical highs. The *Hadley cell* is closed by a massive air stream called the trade winds moving at low levels towards the Equator. The two trade-wind systems converge in an area of low pressure called either the *equatorial trough* or the Intertropical Convergence Zone (ITCZ). Although this model has certain deficiencies and is rather oversimplified, it none the less presents a reasonably accurate picture of the main air movements over the tropical oceans (see section 2.2).

During the course of a year, the ITCZ migrates north and south following the sun (figure 7.1), though the latitudinal extent of the migration is

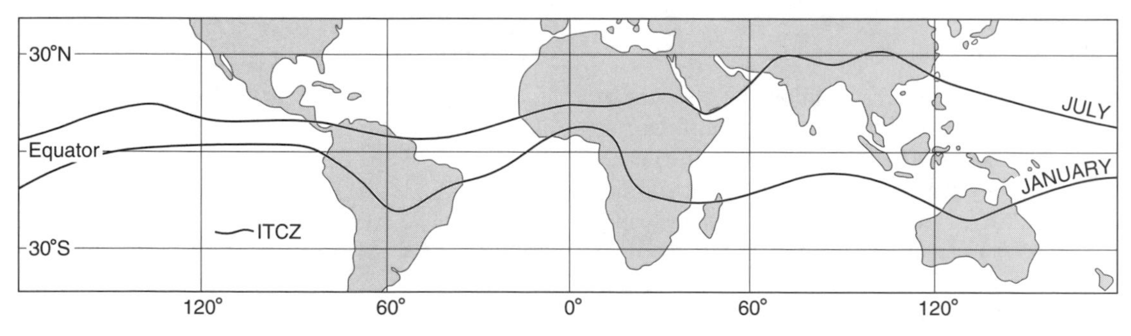

Figure 7.1 The seasonal movement of the Intertropical Convergence Zone (ITCZ).

normally only a few degrees, except for over the Indian Ocean, where it may amount to 30° degrees of latitude – an important consideration in explaining the great Indian monsoon (see section 7.4). The ITCZ is characterised by relatively low surface pressure, rising air movement and a convergence of air masses. Wind speeds are generally low and of variable direction, and are sometimes called the 'doldrums'.

Between the subtropical highs and the ITCZ, the low-level general circulation is dominated by strong and persistent easterly winds called the *trades*. They cover about 20° of latitude in the summer hemisphere and approximately 30° in the winter hemisphere.

Usually three distinct layers are identified in them. The general direction of the *lower trades* is east-north-east (ENE) in the Northern Hemisphere and ESE in the Southern Hemisphere, and they often show a remarkable constancy of direction. They are limited to elevations of about 500 m near the subtropical highs, but near the ITCZ they can reach levels as high as *c*.2500 m. This is therefore a shallow layer, and so the short cumulus clouds rarely produce much precipitation.

Over the lower trades lies the important *trade-wind inversion layer,* in which the temperature increases with elevation at a rate of about 1 °C 100 m^{-1}. The main effect of this inversion is that it acts as an efficient lid, which prevents all upward movements of the air and thus dampens down convectional precipitation. It disappears when convergence prevails, as near the ITCZ. Above the inversion are the *upper trades*, which are low-velocity westerly winds that may reach elevations between 6000 and 10 000 m.

7.2 The Wet Tropics

The wet tropics characterise almost 10 per cent of the earth's land area and form a discontinuous belt that typically extends 5–10° of latitude into each hemisphere. They tend to have a greater latitudinal extent along the eastern side of continents (especially South America) and along some tropical coasts. This is due primarily to the windward position of these areas on the weak western side of the subtropical high, a zone dominated by neutral or stable air – western coasts often have cold offshore currents and zones of upwelling that cause precipitation levels to be low.

The wet tropics have temperatures that usually average 25 °C or more each month, and have both high annual means and small annual ranges. Since these areas lie so close to the Equator, the vertical rays of the sun are always relatively close and changes in the length of daylight throughout the year are slight: hence seasonal temperature variation is minimal. Daily temperature variations greatly exceed seasonal differences, so that while annual temperature ranges in the wet tropics rarely exceed 3 °C, daily temperature ranges are two to five times greater than this. The temperatures tend not to reach the very high absolute levels that may occur in some extra-tropical areas – the highest temperatures recorded in New York (40 °C) are higher than those recorded in equatorial stations like Djakarta in Indonesia and Belém in Brazil.

The wet tropics generally receive from 1750 to 2500 mm of rainfall per year, though topographical differences cause considerable variability. The high rainfall levels are in part the result of thermal convection caused by the extensive heating of the zone, and of converging winds associated with the widespread ascent of warm, humid, unstable air. Completely cloudless days are rare, and cloudiness averages 50–60 per cent. On average, there are 75–100 days with thunderstorms in the course of a year,

Table 7.1 Rainfall intensity, represented as annual mean rainfall per rain-day

	mm
Hong Kong	21.2
Djakarta (Indonesia)	13.5
Rangoon (Burma)	20.9
Calcutta (India)	15.5
Bombay (India)	22.4
Entebbe (Uganda)	12.4
Lagos (Nigeria)	14.4
Accra (Ghana)	13.6
Georgetown (Guyana)	13.3
Quito (Ecuador)	8.5
San Salvador (El Salvador)	16.1
San Juan (Puerto Rico)	10.1
Mean	15.2

Table 7.2 Representative climatic data: tropical climates

	Year	*J*	*F*	*M*	*A*	*M*	*J*	*J*	*A*	*S*	*O*	*N*	*D*
Tropical wet climates													
Singapore (1°18′N)													
Temperature (°C)	27	26	27	27	27	28	27	27	27	27	27	27	27
Precipitation (mm)	2413	252	172	193	188	172	172	170	196	178	208	254	257
Belém, Amazon Valley, Brazil (1°18′S)													
Temperature (°C)	26	25	25	25	25	26	26	26	26	26	26	26	26
Precipitation (mm)	2735	340	406	437	343	287	175	145	127	120	91	89	175
Tropical wet-and-dry climates													
Calcutta, India (23°N)													
Temperature (°C)	26	18	21	26	29	30	29	28	28	28	27	22	18
Precipitation (mm)	1494	10	28	35	51	127	284	308	292	229	110	13	5
Normanton, Australia (18°S)													
Temperature (°C)	27	30	29	29	28	26	23	22	24	27	29	31	31
Precipitation (mm)	952	277	254	155	38	8	10	5	3	3	10	45	142

and the number of rain-days is large (e.g. Belém in Brazil has 243 per year). In some core areas like the western Amazon Basin and the innermost Congo Basin there may be no dry months at all, but more generally there is a brief dry season of one, or at most two, months' duration.

The rainfall in the tropics is of high intensity, which has great implications for landform development and soil erosion, particularly if the protective covers afforded by the canopy of the rainforest are removed (table 7.1). Annual values of rainfall per rain-day for humid temperate stations like London are between 4 and 6 mm; the mean value for the tropical stations listed is 15.2 mm, about three times as much.

7.3 Tropical Seasonal Climates

In between the wet tropics and the great subtropical deserts is a belt of climate where there is usually a smaller total precipitation and one that is much more seasonal than in the wet tropics. This is a zone in which the natural vegetation tends towards savanna rather than rainforest. During the course of a year, with the north–south shifting of the solar radiation belts and the consequent latitudinal displacement of pressure and wind, this belt is alternately affected by the wet ITCZ and equatorial westerlies (in summer) and by the drier parts of the trades and subtropical anticyclones (in winter).

The annual temperature range, while higher than that in the wet tropics, is still rather low and seldom exceeds 8 °C. During the rainy season of the summer months the diurnal range is also low, but in the dry season it is likely to be higher. Frequently the hottest months do not coincide with the time of highest sun but somewhat precede it, for during the rainy season persistent cloud cover and heavier precipitation tend to lower the air temperature (table 7.2).

It is the seasonal distribution of precipitation that makes this zone distinct, and the seasonal contrast is particularly clear in the case of those areas subject to the influence of the monsoons.

7.4 Monsoons

For centuries it has been known that in some low-latitude areas there is a summer and winter reversal of winds accompanied by a distinct seasonal change in rainfall – the wind from one direction being humid, unstable and conducive to rainfall, and that from the other direction being stable and dry. South Asia has the classic example of a monsoon, or seasonal wind reversal, and it is notable for the suddenness with which the onset (*burst*) of the rainy season takes place. However, monsoons also occur elsewhere. They are encountered along the Guinea

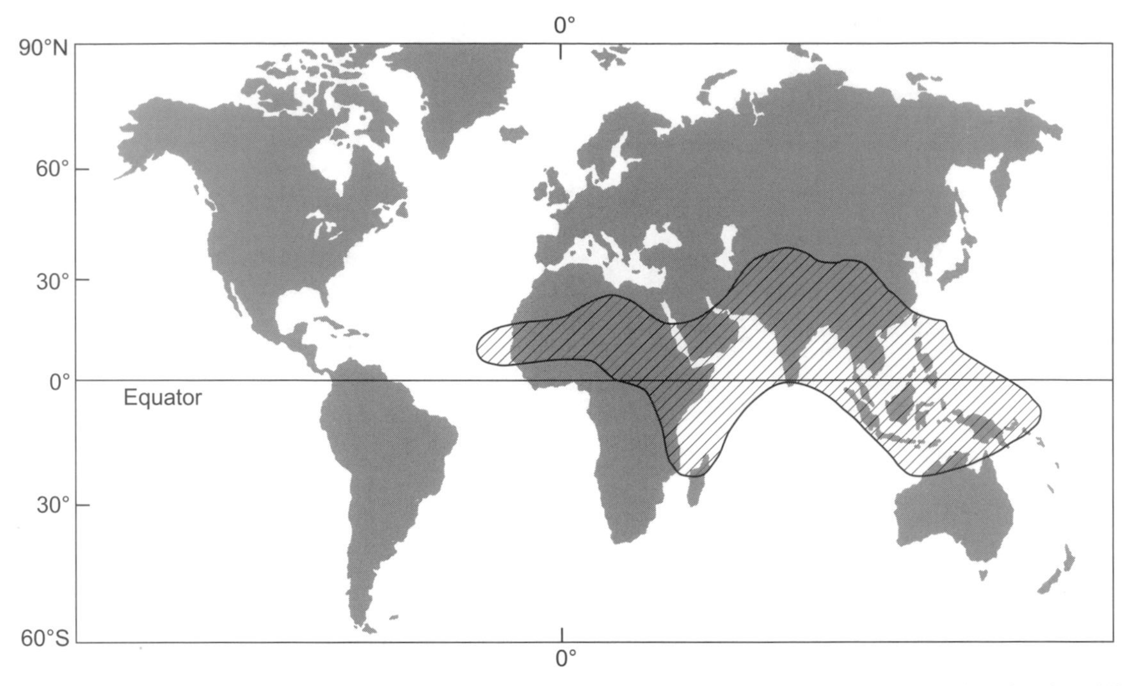

Figure 7.2 The distribution of monsoonal climates. The map is based on four criteria: (i) that prevailing wind direction shifts by at least 120° between January and July; (ii) that the average frequency of prevailing wind directions in January and July exceeds 40 per cent; (iii) that the mean resultant wind in at least one of those months exceeds 3 m s^{-1}; and (iv) that fewer than one cyclone–anticyclone alternation every two years occurs in either January or July in a 5° latitude–longitude rectangle.

coast of West Africa, in East Africa and in northern Australia (figure 7.2).

The traditional explanation for the Indian monsoon is that it is essentially a giant land–sea breeze system caused by seasonal contrasts in temperature between the land masses and the oceans. These temperature differences, which are especially marked in the case of Asia because of its enormous mass, affect the seasonal movement of pressure and winds. Over the oceans the seasonal shifts of heat and pressure zones are rather small, in keeping with the similarly small annual temperature changes. Over land, by contrast, where temperature variations are larger, the movement of the heat and pressure zones is exaggerated. This shows up in the summer and winter movement of the ITCZ, which over India in summer occurs more than 30° away from the Equator. This extreme position is associated with the large temperatures in the continental interior, which create rising air and low pressure, thereby drawing the ITCZ away from the Equator and giving southwesterly winds.

However, this simple thermal explanation of the seasonal wind reversal needs to be supplemented by a consideration of the role of upper air conditions, for these help to explain the sudden burst and the extreme seasonal contrast that are such features of the Indian monsoon.

The breaking of the monsoon is now related to the shifting of the mid-latitude jet stream. Anticyclonic conditions and subsiding air are normally found on the earth's surface a short distance equatorward of the position immediately below the jet. It has been suggested that in winter the jet runs southwards of the great topographical barrier formed by the mountains and high plateaux of Tibet and neighbouring areas. This induces a subtropical anticyclonic cell of dry subsiding air over northern-central India, and divergence of air from this centre gives the oceanward winter monsoon. However, as the winter season recedes, the jet stream slowly moves northwards until it is quickly diverted to the north side of the topographical barrier.

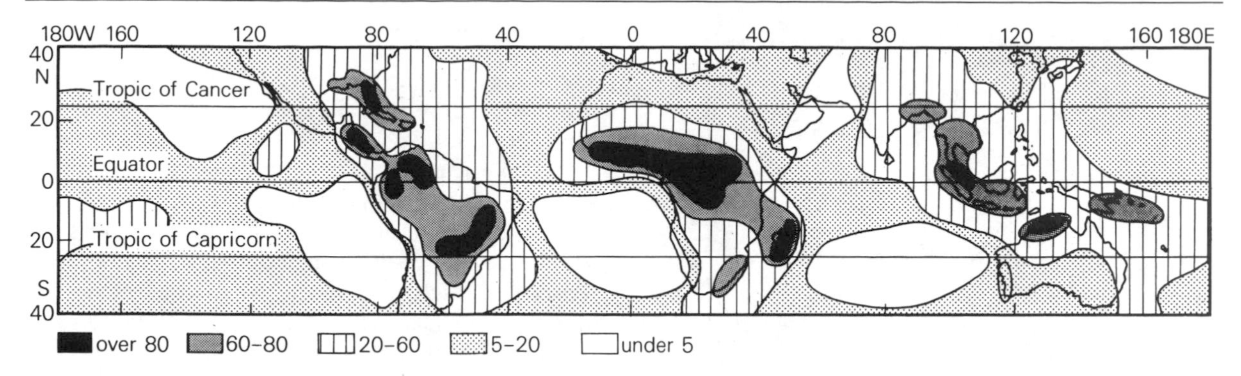

Figure 7.3 Mean number of thunderstorms per year.

7.5 Tropical Disturbances

In comparison with disturbances of the higher latitudes (see section 5.2), those of the tropics show two major differences. First, there is an almost complete absence of the fronts so typical of the mid-latitude depressions, for in the tropics the differences in temperature between two air masses are usually so small that no clear front develops. Second, the Coriolis Force has a lower strength close to the Equator, and this means that most disturbances in very low latitudes are of shorter duration or lower intensity than those of the higher latitudes.

The main prerequisite for the formation of tropical disturbances is the presence of warm and humid air masses in which no inversion exists. This type of air is potentially unstable, and any upward movement will be reinforced by the release of large amounts of latent heat of condensation. Uplift may be caused by heating from the earth's surface (convection), and convectional cells may turn into *thunderstorms*, the smallest and most frequent form of tropical disturbance (figure 7.3). Uplift can also result from the convergence of two air masses, for when this occurs there is no way for the air to go but upward, and so once again thunderstorms may occur, arranged in bands called 'linear systems'. The effects of convergence at low levels in the atmosphere will be accentuated if there is divergence aloft. An upper tropospheric high-pressure area, which creates a general outflow of air, can cause upward movement of air and instability at lower levels. Upward movements of air are also related to the topographical features of the earth's surface – for example, air may be forced upwards over mountains, hills and coastlines.

Some disturbances that occur in the tropics have their origins outside this zone, for depressions from the polar front region of mid-latitudes may occasionally venture far towards the Equator, and then regenerate when meeting warm and humid air masses.

Monsoon depressions

Another type of tropical disturbance is the monsoon depression (figure 7.4a). These are closed isobaric low-pressure systems with a diameter of the order of 500–1000 km. They occur at the rate of about three per month in the Asian summer monsoon, and they develop over the Bay of Bengal and, to a lesser extent, over the Arabian Sea. Convergence in the westerly monsoon current near the surface combined with divergent easterlies aloft initiate an upward air movement, which becomes intensified into a large low-pressure area because of the release of large amounts of latent heat.

Linear systems (squall lines)

Linear systems or squall lines consist of lines or bands of thunderstorms; they can be hundreds of kilometres long and 10–30 km wide. They result from convergence, for instance when local land and sea breezes from different islands meet, as they often do in south-east Asia.

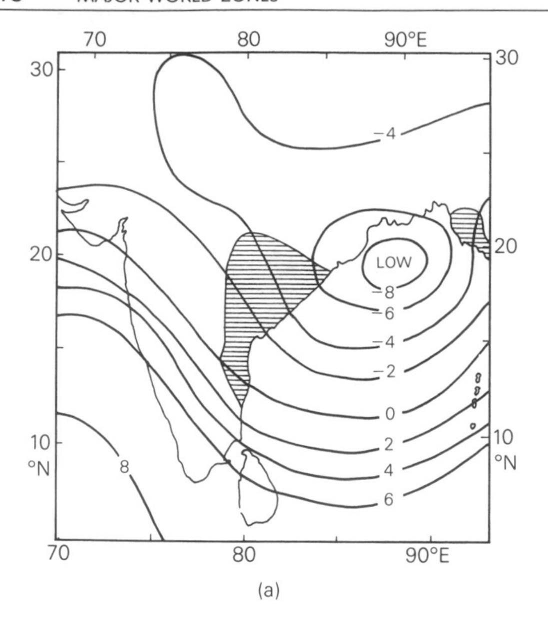

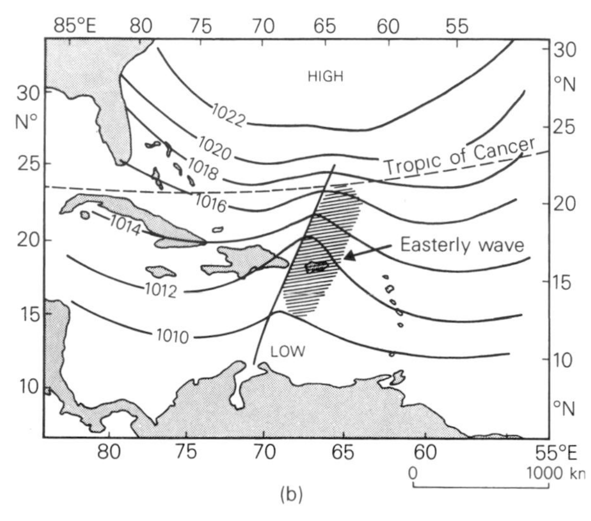

Figure 7.4 Some types of atmospheric perturbation in the tropics: (a) an example of a monsoon depression over the Bay of Bengal, 100 mbr contours in decimetres for 1200 hrs GMT on 20 August 1967; the hatched area represents the zone of continuous rainfall; (b) an easterly wave in the Caribbean area; the hatched area indicates the main rainfall zone.

Wave disturbances

The trades and the equatorial easterlies of the central Pacific develop rain-giving perturbations that are quite unlike the depressions of mid-latitudes. They take the form of wave troughs, which are scarcely detectable in the surface pressure field, and are about half the size of a mid-latitude depression. The air in front of the trough diverges and is characterised by descending drying air, while the air behind the trough converges. This convergence creates instability, so that heavy cumulonimbus and cumulus clouds develop that produce moderate or heavy thundery showers. As these waves move with the easterlies, they are generally called *easterly waves* (figure 7.4b).

The main areas in which these waves occur are the western edges of the large ocean basins, in latitudes between 5° and 20°. They are best known in the Caribbean but also occur in the Pacific. In the former area during the late summer they come as often as every 3–5 days and, in the latter, every 2–3 days. They tend to move at about 15–20 $km\,h^{-1}$ and bring substantial quantities of rain. Some of them occasionally develop into the much more hazardous tropical cyclones.

Tropical cyclones

Tropical cyclones are closed low-pressure systems, generally about 650 km in diameter, which bring violent winds, torrential rainfall and thunderstorms. They usually contain a central region, the *eye*, with a diameter of some tens of kilometres and with light winds and more or less lightly clouded sky. They are given a variety of regional names, including 'hurricane' (West Indies) (window 7.1), 'typhoon' (China Sea), 'willy willy' (Australia) and 'cyclone' (Bay of Bengal).

In order for cyclones to develop, various conditions need to be fulfilled. First, there must be a plentiful supply of moisture, and for this reason distribution of cyclones is closely related to those regions where the highest sea-surface temperatures are found (i.e. the western portions of the tropical oceans in late summer), with temperatures of at least 27 °C. This moisture provides the necessary latent heat to drive the storm and to provide the rainfall.

Window 7.1 *Hurricane Gilbert, Jamaica, 1988*

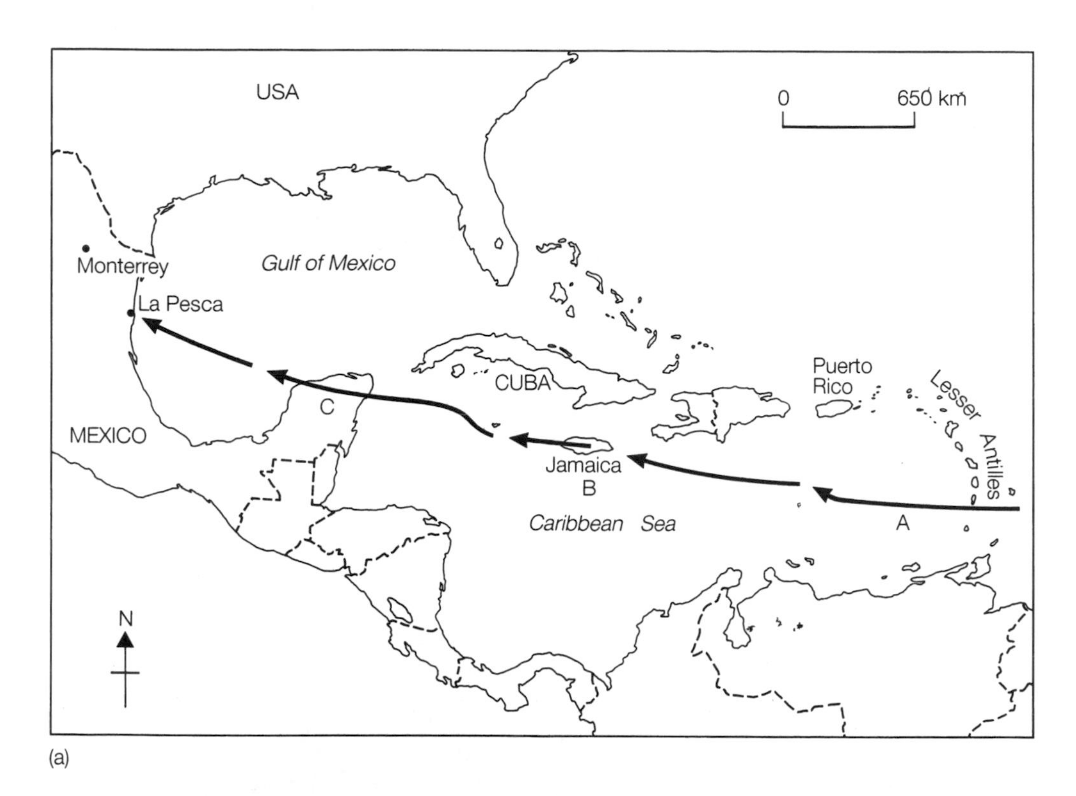

(a)

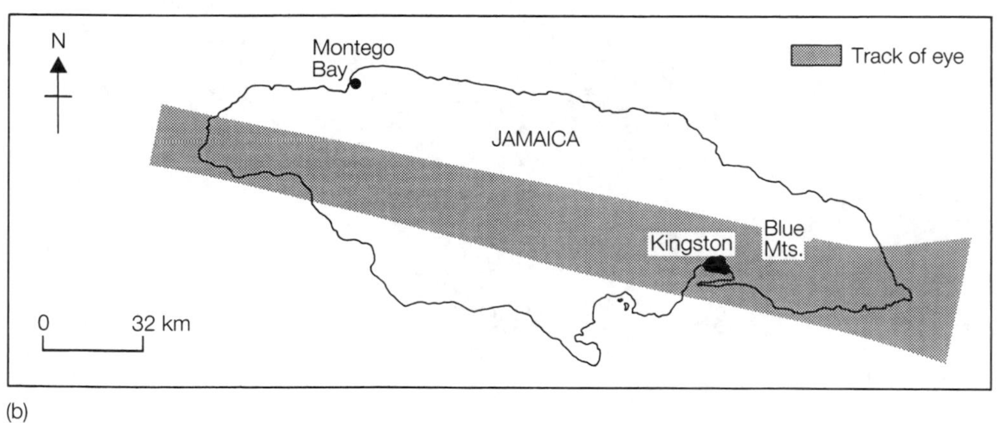

(b)

The track of Hurricane Gilbert: (a) across the Caribbean (A = 10 September, B = 12 September, C = 14 September; (b) across Jamaica.

Hurricane Gilbert was spawned in the Atlantic Ocean some 360 km east of Barbados on 10 September 1988. It tracked across the Caribbean, heading in a westerly direction, reached the island of Jamaica on 12 September, and then made for Mexico, where it lost energy over the Sierra Madre Oriental, south of Monterrey.

Hurricane Gilbert was particularly intense, with a pressure that at one time plummeted to 888 mb, the lowest pressure ever recorded in the Western Hemisphere. Its eye had a particularly large diameter at *c*.40 km.

The effects of this dramatic event on Jamaica were considerable. Prior to its onslaught many of the drainage basins on the island, most notably in the steep and deeply dissected Blue Mountains, had been badly degraded by accelerated erosion, itself the result of generations of poor land use. The hurricane thus acted upon an environment that was already in a fragile state. Many trees and shrubs, including whole plantations, were defoliated or blown down, and the heavy rainfall (up to 900 mm on the higher ground) caused severe landsliding and soil erosion. The coastline was severely battered and eroded by large waves, and the combined surge and swell reached 7.6 m on the east and north coasts of the island.

Damage from this one hurricane probably cost Jamaica between US$800 million and US$1000 million, a value which exceeds the country's annual foreign exchange earnings from exports. Furthermore, the nation's ability to earn foreign exchange was greatly reduced by the damage that occurred to the all important banana and coffee crops: for example, 98.5 per cent of the banana acreage were damaged.

The death rate was relatively modest. Forty-five people died in Jamaica, but in Mexico, where the hurricane ended up, 330 deaths were reported.

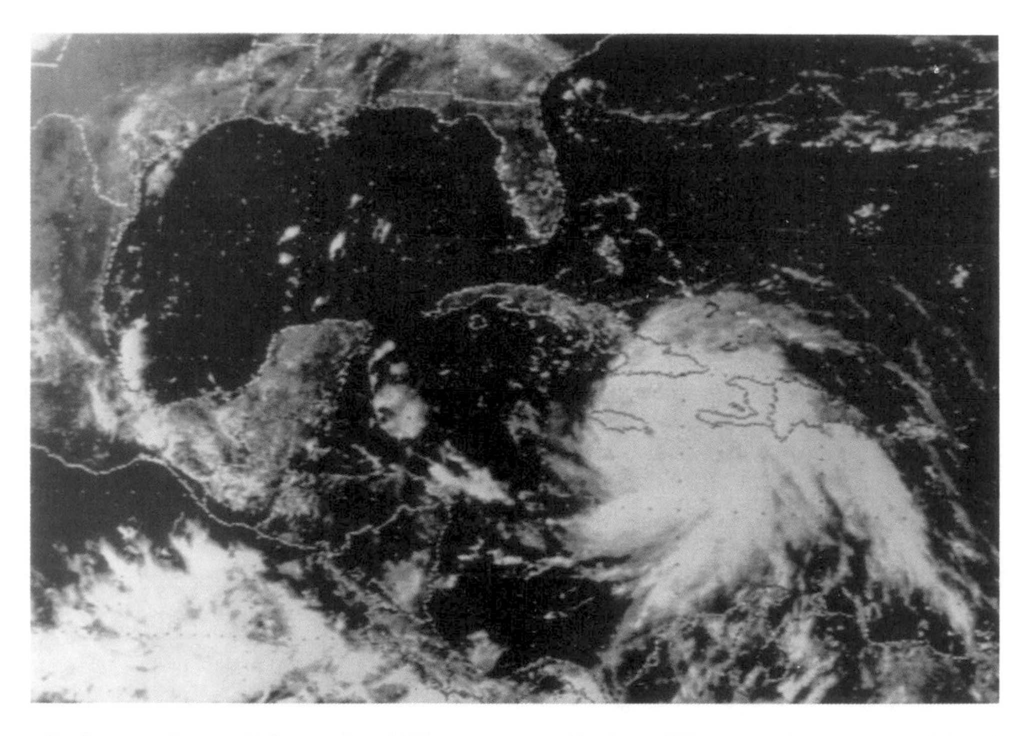

In this satellite image, taken on 11 September 1988, one can see Hurricane Gilbert touching the coast of Jamaica. The hurricane's eye was located on the island's coast near Kingston. It subsequently tracked westwards towards Mexico.

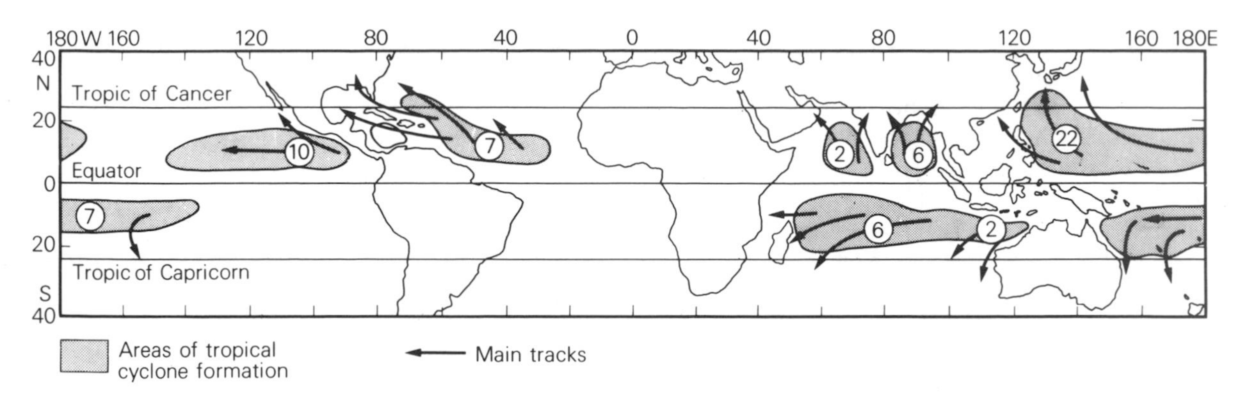

Figure 7.5 The distribution, frequency and movement of tropical cyclones (hurricanes) in relation to sea-surface temperature. For each region the percentage contribution of that region to the global total of hurricanes is shown.

Tropical cyclones tend to occur mainly within this zone (figure 7.5), and when they leave it or pass over land their violence is rapidly dissipated. They also tend to occur only in areas beyond about 5° from the Equator because closer than that to the Equator the Coriolis parameter approaches zero. The main cyclone activity in the Northern Hemisphere is in late summer and autumn during the time of the equatorial trough's northern displacement.

Early theories of cyclone development held that convection cells generated a sudden and enormous

Plate 7.1 Cyclone damage in Swaziland, southern Africa. In 1984, Cyclone Domoina deposited as much as 900 mm of rainfall in just two days. The River Usutu washed away the main railway line from Swaziland to South Africa.

release of latent heat that provided energy for the storm, but it is now believed that cyclones develop from pre-existing disturbances such as easterly waves. Most disturbances, however, do not lead to such cyclones, and it seems that the presence of an anticyclone in the upper troposphere is another necessary precondition. This is essential for high-level outflow, which in turn allows the development of very low pressure and high wind speeds near the surface.

Generally, tropical cyclones have a life span of about one week, and in that time they can wreak havoc because of the high winds that they bring, the high rainfall levels that they can produce (values of over 2000 mm per day have been recorded in the Philippines), and because the very low atmospheric pressures with which they are associated can cause sea levels to be unusually high. The width or diameter of the destructive winds of tropical cyclones is variable. The average-sized hurricane probably produces hurricane winds (winds of more than 119 $km\,h^{-1}$) over about 160 km, but more extreme values than this are known. The 'Labor Day Hurricane' that struck the Florida Keys in the United States in 1935 probably achieved wind speeds of 320–400 $km\,h^{-1}$.

As a result of the high tidal surges, high rainfall levels and high wind velocities with which they are associated, hurricanes are one of the most serious natural hazards in low latitudes (plate 7.1), particularly in low-lying coastal areas (window 7.2) (see colour plate 9). A Bay of Bengal cyclone in 1970 killed somewhere between 300 000 and 1 000 000 people in Bangladesh and India – people who lived on the overcrowded and storm-prone Ganga Delta. In Japan, cyclones or typhoons are the most dangerous of the natural hazards with which the population has to contend (see table 7.3), more serious even than earthquakes. Although Japan itself lies outside the tropics, many of these storms are derived from lower latitudes.

7.6 Tropical Rainforest

In the wet tropics (figure 7.6) the dominant form of vegetation is the tropical rainforest – the most luxuriant of all vegetation types (see colour plate 10). There are three main formations: the American formation (which has its most massive development in the Amazon Basin), the African formation (largely in the Cameroons and the Congo Basin) and the Indo-Malaysian formations of the Far East and northern Australia. Although there are differences in species between the three forma-

Window 7.2 *Cyclones in Bangladesh*

Many of the most deadly weather events in the world strike Bangladesh, a poor (per capita income $200 in 1990) and densely populated country (about 800 people per km^2). It is a nation of water, and about one-fifth of it is submerged beneath river floods in an average year. It can also be flooded by surges of sea water generated by cyclones – these types of flood can inundate over a third of the land area, for over a third of Bangladesh is less than 6 m in elevation. This is because a large portion of the country is in effect a delta built

The comparative effects of the 1991 and 1994 cyclones

Indicator	*1991*	*1994*	*Indicator*	*1991*	*1994*
Deaths	139 000	127	Places of worship damaged	3000	550
Injuries	n/a	2100	Cattle lost	1 000 000	10 000
People affected	10 800 000	390 000	Hectares of crops lost	380 000	31 000
Families affected	2 500 000	74 000	No. of cyclone shelters available	4300	500
Houses damaged	970 000	60 000	People moved to safe sites	350 000	750 000
Schools damaged	9600	150			

Source: World Disasters Report 1995 (fig. 9.1)

of sediments by two large rivers that rise in the Himalayas: the Ganges (Ganga) and the Brahmaputra. The country's coastline is also shaped like a funnel that focuses the cyclones that are generated over the warm waters of the Bay of Bengal, so that there are on average around five cyclones per year. Some of these events have catastrophic effects. The 1994 cyclone was not as devastating as that of 1991, partly because it did not produce such a devastating sea-surge, and partly because an increase in disaster preparedness allowed more people to be evacuated to new cyclone shelters.

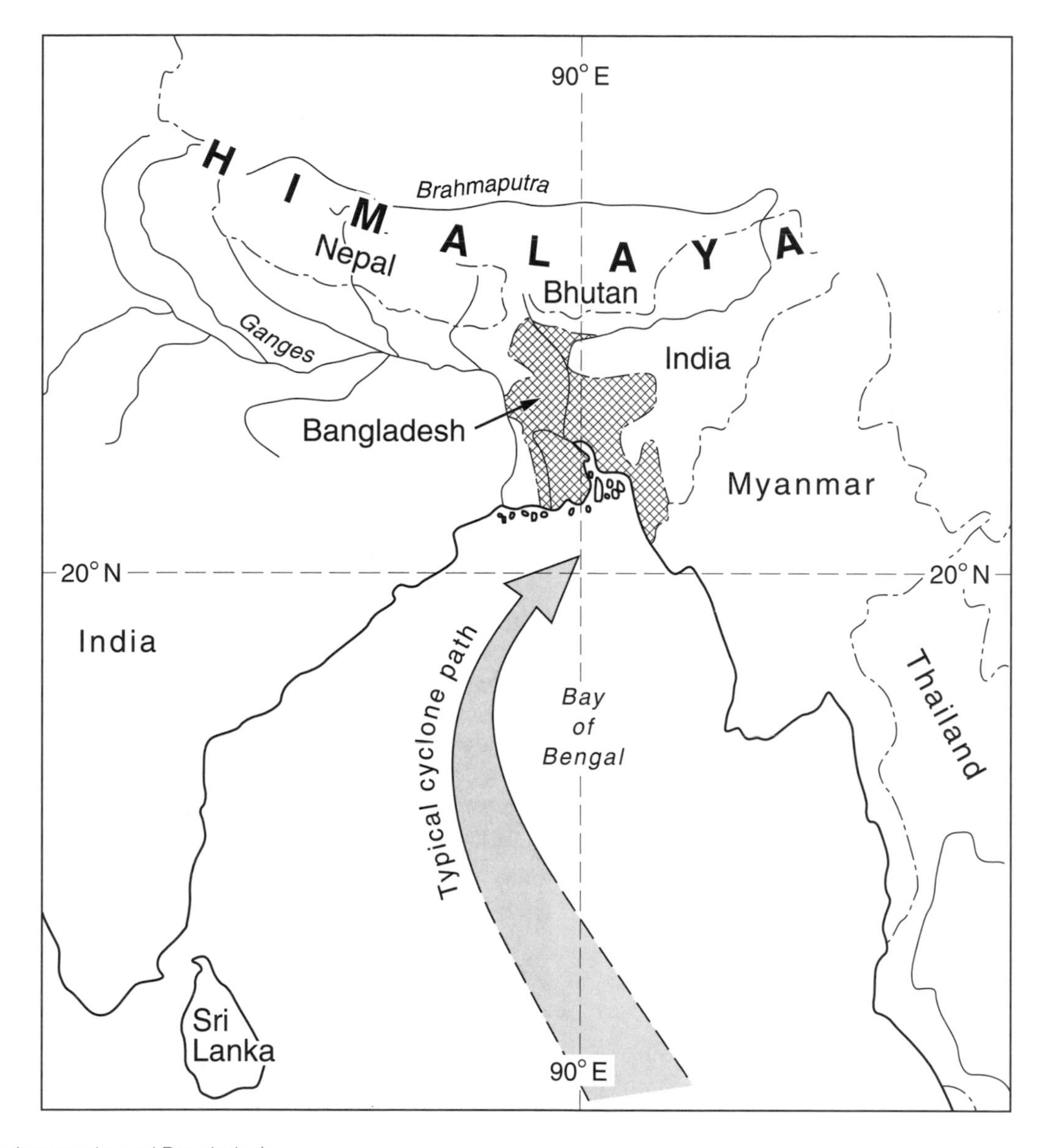

Cyclone tracks and Bangladesh.

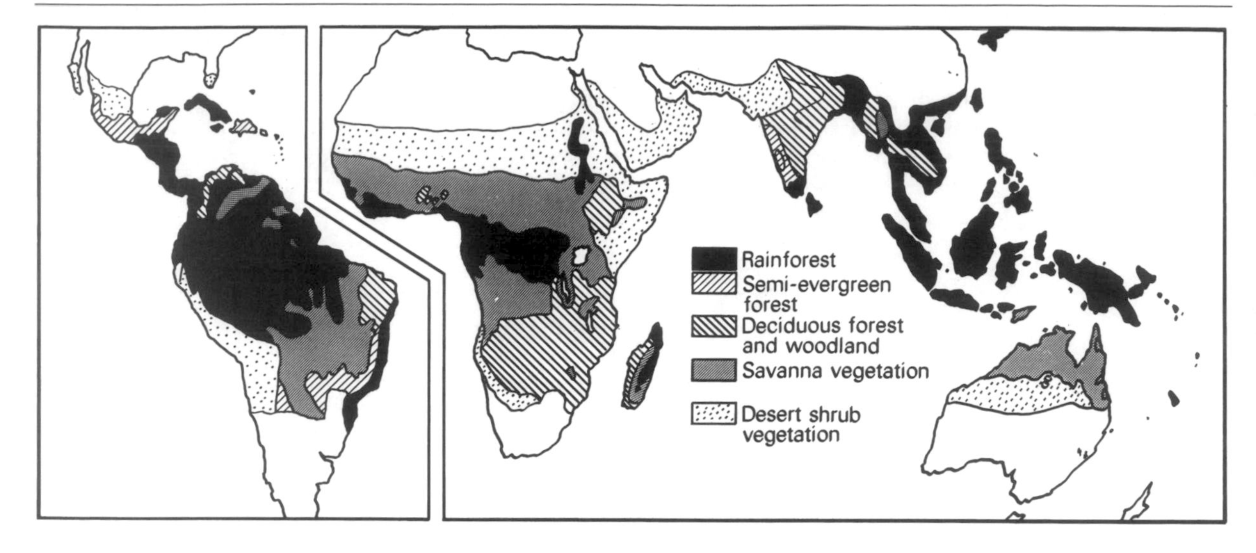

Figure 7.6 The distribution of the main types of tropical vegetation.

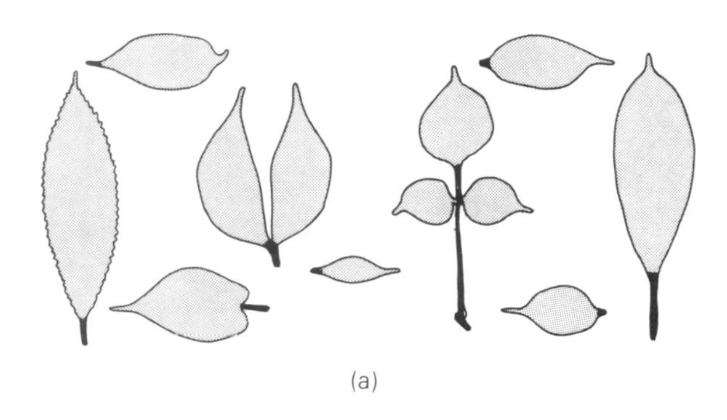

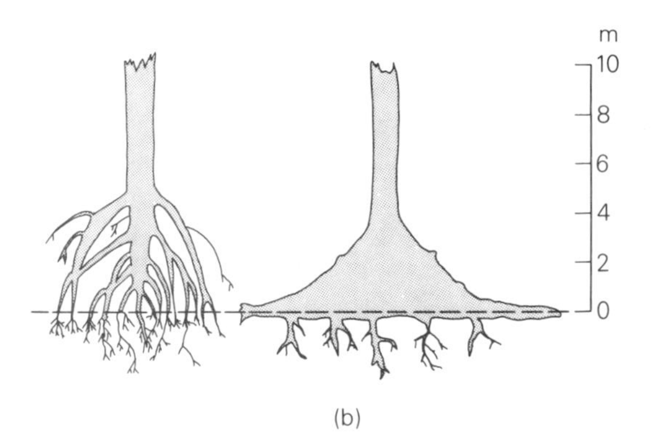

Figure 7.7 Some features of tropical plants: (a) various shapes of leaves with conspicuously differentiated 'drip-tips'; (b) stilt roots of *Uapaca* sp. (left) and heavy buttresses of *Piptadeniastrum africanum* (right).

Table 7.3 Damage caused by natural hazards in Japan, 1946–70

Phenomenon	*No. of events*	*Deaths*	*Houses destroyed*
Typhoon	59	13 745	576 378
Extra-tropical cyclone	89	8 156	65 818
Earthquake	11	5 490	113 339
Landslide	5	86	143
Hail + thunderstorm	4	28	847
Heavy snow	2	242	1 734
Volcanic eruption	1	12	12

Source: modified after data in J. Whittow, *Disasters* (Harmondsworth: Penguin, 1980)

tions, certain features of this vegetation type are fairly general:

(a) The trees are nearly all evergreens, and the vast majority cast their old leaves and grow new ones continuously and simultaneously; they are rarely leafless.
(b) Most of the trees have leathery, dark green leaves similar in shape and size to the ones that are typical in the laurel family. Particularly in the lower storeys of the forest, leaves may have 'drip tips' – devices that may have evolved for shedding heavy rain (figure 7.7a).
(c) Many trees have either buttress roots (plate

Plate 7.2 Some rainforest trees, including species of *Ficus*, display enormous buttress roots. This example is in Mauritius.

7.2) or stilt roots (figure 7.7b). These may give stability to very tall trees, but their precise function is not always clear.

(d) The trunks of many rainforest trees bear flowers and fruit – a habit called 'cauliflory'.

(e) Climbers, including lianas, are large and numerous and some are said to attain lengths of over 200 m. The forest contains large numbers of *epiphytes*. These are plants that grow attached to the trunks, branches and even the leaves of trees, shrubs and lianas. Some epiphytes send down roots to the soil and may eventually kill the tree that originally supported them – these are called 'stranglers'. In the better illuminated areas of the forest floor there are normal herbaceous plants, but in areas of great darkness some plants called *saprophytes* manage, with the help of certain fungi, to grow without the necessity for photosynthesis.

(f) Away from clearings, river banks etc., a shrub layer is ill-developed and the ground is almost bare of low-growing plants because of a shortage of light.

(g) The trees may exceed 60–90 m in height, but the taller trees rarely average more than *c.*50 m. They are in general taller than those in temperate forests, but they never reach the gigantic dimensions of the Californian redwoods or the huge *Eucalyptus* of Australia.

(h) In a European or North American forest the dominant trees belong to a few, or often only a single, species; whereas in the tropical rainforest there are seldom less than 40 species of tree over 10 cm in diameter per hectare, and sometimes over 100 species.

(i) Overall, the rainforest has a very diverse flora, especially in the Indo-Malaysian area. Malaysia has about 7900 species and 1500 genera of seed plants, whereas the British Isles, with an area about 2.3 times greater, has 1430 species and 628 genera.

A great increase in the diversity of species takes place from the Poles to the tropics. This tendency can be traced in a wide variety of groups (figure 7.8), which include molluscs, ants, lizards and birds. The pattern can be seen, moreover, in marine, freshwater and terrestrial habitats, and both in small communities and over extensive geographical regions. A single hectare of humid tropical moist forest may, for example, contain 40–100 different species of trees, while comparable areas of deciduous forest in eastern North America may have 10–30, and coniferous forests in the far north of Canada only 1–5 tree species. There are exceptions to this general trend (e.g. arid areas or islands near the tropics may have relatively few species), but overall the great richness of species in the humid tropics is one of their most important characteristics, even if there is as yet no fully acceptable explanation for this.

The vertical structure of the tropical rainforest has been described and represented by profile diagrams in many works, but there is considerable debate as to whether it consists of distinct layers or strata. The traditional view is that there are five recognisable above-ground strata:

(a) the upper tree layer of convergent trees (>25 m);
(b) the middle tree layer (10–25 m);
(c) the lower tree layer (5–10 m);
(d) the shrub layer;
(e) the herb layer.

However, we must remember that the rainforest canopy is constantly changing as individual trees pass through their life-cycles.

The biomass of the rainforest may be around 450 t ha^{-1}. The huge mass of leaves carried by the trees provides a massive litter fall (11 t ha^{-1} yr^{-1}), yet

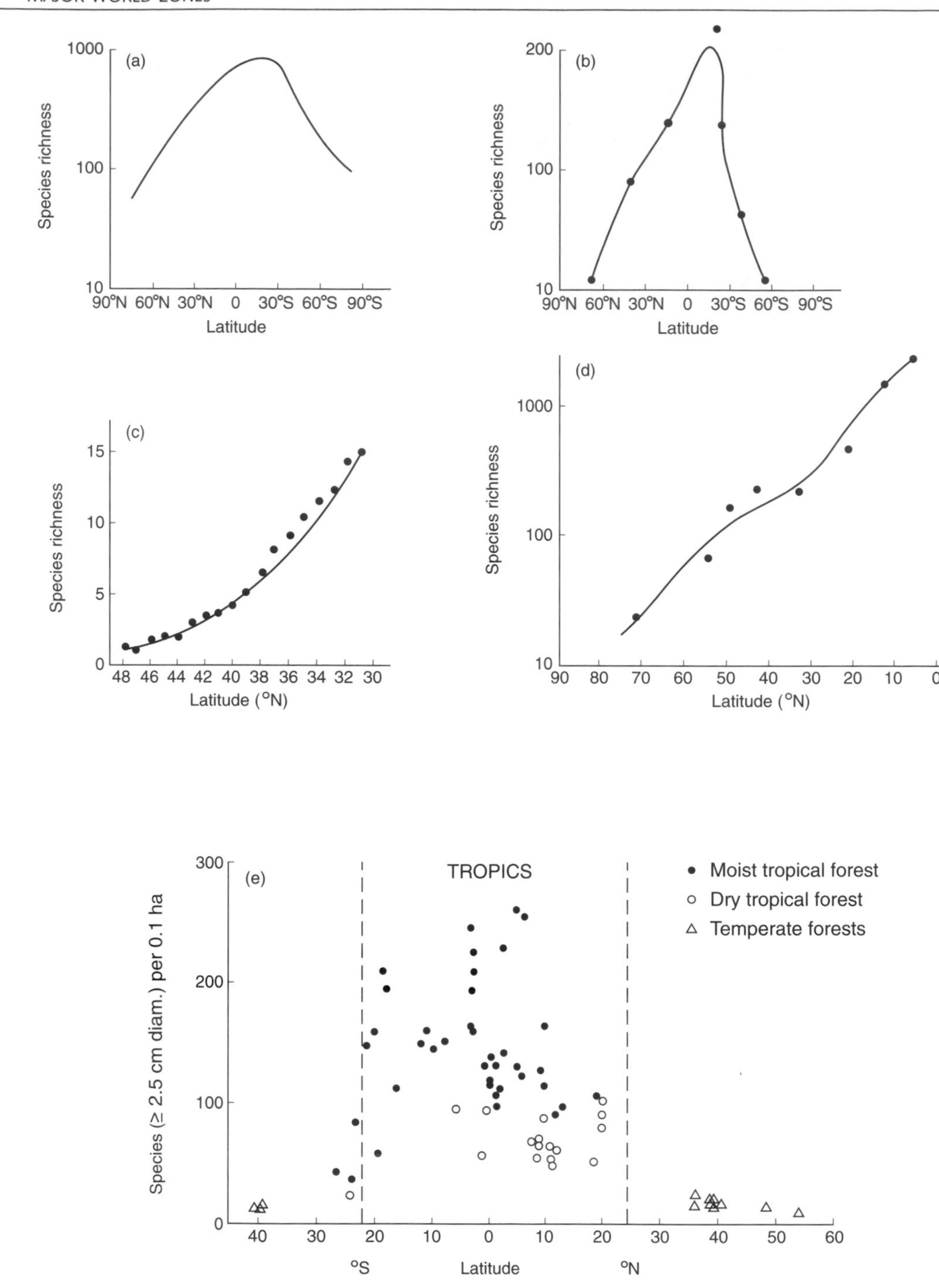

Figure 7.8 Latitudinal patterns in species richness of: (a) marine bivalve molluscs; (b) ants; (c) lizards in the United States; (d) breeding birds in North and Central America; (e) species of plants ≥ 2.5 cm diameter in standardized 0.1 ha forest samples.

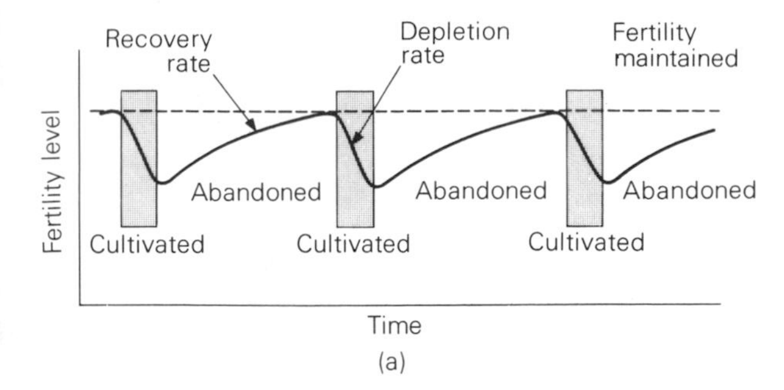

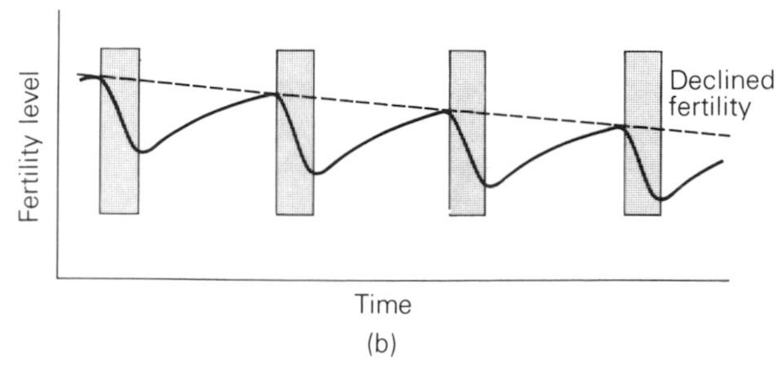

Figure 7.9 Land rotation and population density. The graphs show the relationship of soil fertility to cycles of slash-and-burn agriculture. In (a) fertility levels are maintained under the long cycles characteristic of low-density populations. In (b) fertility levels are declining under the shorter cycles characteristic of increasing human population density. Notice that in both diagrams the curves of both depletion and recovery have the same slope.

soil humus reserves and surface litter accumulations are usually small except in lowland swamps or in upland (*montane*) forests (section 8.4). This is because, in general, the litter is destroyed more rapidly than it is supplied. As humus turnover is about 1 per cent per day, there is little chance of it accumulating on the surface. The litter is rapidly decomposed with the help of fungi and other micro-organisms, and its nutrients are rapidly returned by a dense shallow root network to the trees. Thus there is a huge circulation of nutrients – but they are essentially cycled within the vegetation rather than into the soil, so that many of the soils of the rainforest, in spite of the luxuriance of the vegetation they appear to support, are actually relatively infertile.

Indeed, because so many of the nutrients are tied up in the vegetation rather than in the soil, the removal of the rainforest can lead to very serious impoverishment of the soil. Traditional farming systems involving slash-and-burn recognise this. The forest is cleared by burning and cutting, and this releases nutrients in the soil. But under conditions of rapid breakdown and rapid leaching combined with high rainfall, the fertility does not last long, and so the farmers move on, allowing the cleared area to become revegetated by secondary forest. In other words, there is an extended period of fallow. Unfortunately, as population pressures increase there is sometimes insufficient land available

Plate 7.3 Large areas of rainforest in countries like Nicaragua are being removed by slash-and-burn agriculturalists.

Plate 7.4 Areas that have been cleared for slash-and-burn agriculture in Sri Lanka typically have a few forest trees remaining, but large numbers of species may be being lost.

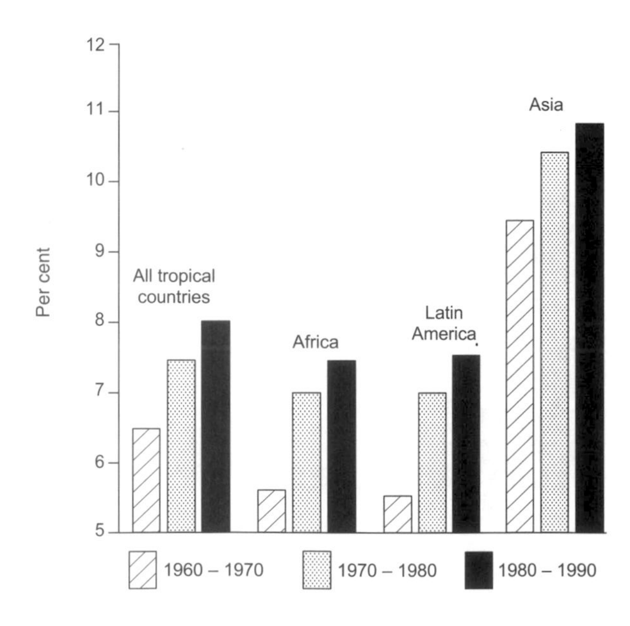

Figure 7.10 Estimated rate of tropical deforestation, 1960–90, showing the particularly high rates in Asia (modified from *World Resources, 1996–97*).

for such an extensive means of land use, and so the fallow period becomes reduced, with unfortunate long-term implications for soil fertility (figure 7.9).

Not all soils in the humid tropics, however, are so low in nutrients. On ancient, stable land surfaces, topographical stability, combined with the hot, wet climate, has led to a very deep and thorough weathering and leaching of underlying rocks. Moreover, although some underlying rocks, like quartzites, may have been low in nutrients before they were so extensively weathered, there are areas where recent geological activity – volcanism, erosion or delta and alluvium deposition – has provided areas of nutrient-rich soils. Thus some of the great deltaic and volcanic areas of south-east Asia have supported dense human agricultural populations for centuries.

Population pressure is not the only cause of the removal of the rainforest: commercial ranching, logging and the like are also making increasing inroads. Already it is estimated that something like 40 per cent of the world's rainforests have been removed, though West Africa has lost 72 per cent and south-east Asia 63.5 per cent.

Rainforests have been cleared for cultivation for some thousands of years, but the present rate of removal is especially fast (plate 7.3). A recent Food and Agriculture Organization estimate is that in the 1980s as much as 16.8 million hectares were removed per annum, compared with 9.2 million hectares in the late 1970s. In the 1990s the annual loss of rainforest may have amounted to 2 per cent of the total forest expanse. There is, however, a considerable variation in the rates of removal in different areas, with some areas under relatively modest threat (e.g. western Amazonia, and much of the Congo Basin in central Africa). But some other areas are being exploited very rapidly (e.g. the Philippines, Malaya, Thailand, Indonesia, West Africa, eastern Amazonia and Sri Lanka). Indeed, the rate of forest removal in southern Mexico and Madagascar may be as high as 10 per cent per year. Because of the richness of the flora and fauna of the rainforests this may be leading to a major spasm of biodiversity loss (plate 7.4).

Figure 7.10 shows rates of tropical deforestation for the continents, while figure 7.11 shows the spread of deforestation and fragmentation of habitat for two selected tropical locations, Sumatra and Costa Rica.

Forest cover intercepts the intense, torrential rainfall of tropical storms and protects and binds the soil. As deforestation occurs the percentage of rainfall that becomes runoff is increased. Thus rivers become choked with sediment derived from erosion and transported by runoff and are rendered more prone to floods. An indication of how rates of tropical runoff and erosion increase after vegetation removal is given in table 7.4. The five studies from different parts of Africa indicate that under forest the annual runoff in rivers as a percentage of the mean annual rainfall is a mere 0.9 per cent, whereas under crops this rises to 17.4 per cent, and under bare soil conditions to over 40 per cent. The erosion rates go up in a comparable manner. Under forest they are minimal ($0.09\,t\,ha^{-1}\,yr^{-1}$), whereas under crops they go up by 320 times to $28.8\,t\,ha^{-1}\,yr^{-1}$, and under bare soil conditions they go up 768 times to $69.1\,t\,ha^{-1}\,yr^{-1}$.

Removal of rainforest may have some more general world-wide implications (table 7.5), notably for world climate through its effect on atmospheric carbon dioxide levels. These levels in the atmosphere have an effect on the heat balance, since CO_2 is virtually transparent to incoming solar radiation but absorbs outgoing terrestrial infrared radiation – radiation that would otherwise escape to space and result in loss of heat from the lower atmosphere. Because of this 'greenhouse effect' one would expect increased CO_2 levels to lead to an increase in surface temperatures (figure 7.12). It is argued that tropical forests, because of their huge biomass and great areal extent, contain large amounts of carbon, and that if they are cut down this will be

Table 7.4 Runoff and erosion under various covers of vegetation in parts of Africa

Locality	Average annual rainfall (mm)	Slope (%)	Annual runoff (%)			Erosion ($t\,ha^{-1}yr^{-1}$)		
			A	*B*	*C*	*A*	*B*	*C*
Ouagadougou (Upper Volta)	850	0.5	2.5	2–32	40–60	0.1	0.6–0.8	10–20
Sefa (Senegal)	1300	1.2	1.0	21.2	39.5	0.2	7.3	21.3
Bouake (Ivory Coast)	1200	4.0	0.3	0.1–26	15–30	0.1	1–266	18–30
Abidjan (Ivory Coast)	2100	7.0	0.1	0.5–20	38	0.03	0.1–90	108–170
Mpwapwa (Tanzania)	*c.*570	6.0	0.4	26.0	50.4	0	78	146
Mean values			0.9	17.4	40.1	0.09	28.8	69.1

A = forest or ungrazed thicket; B = crop; C = barren soil.

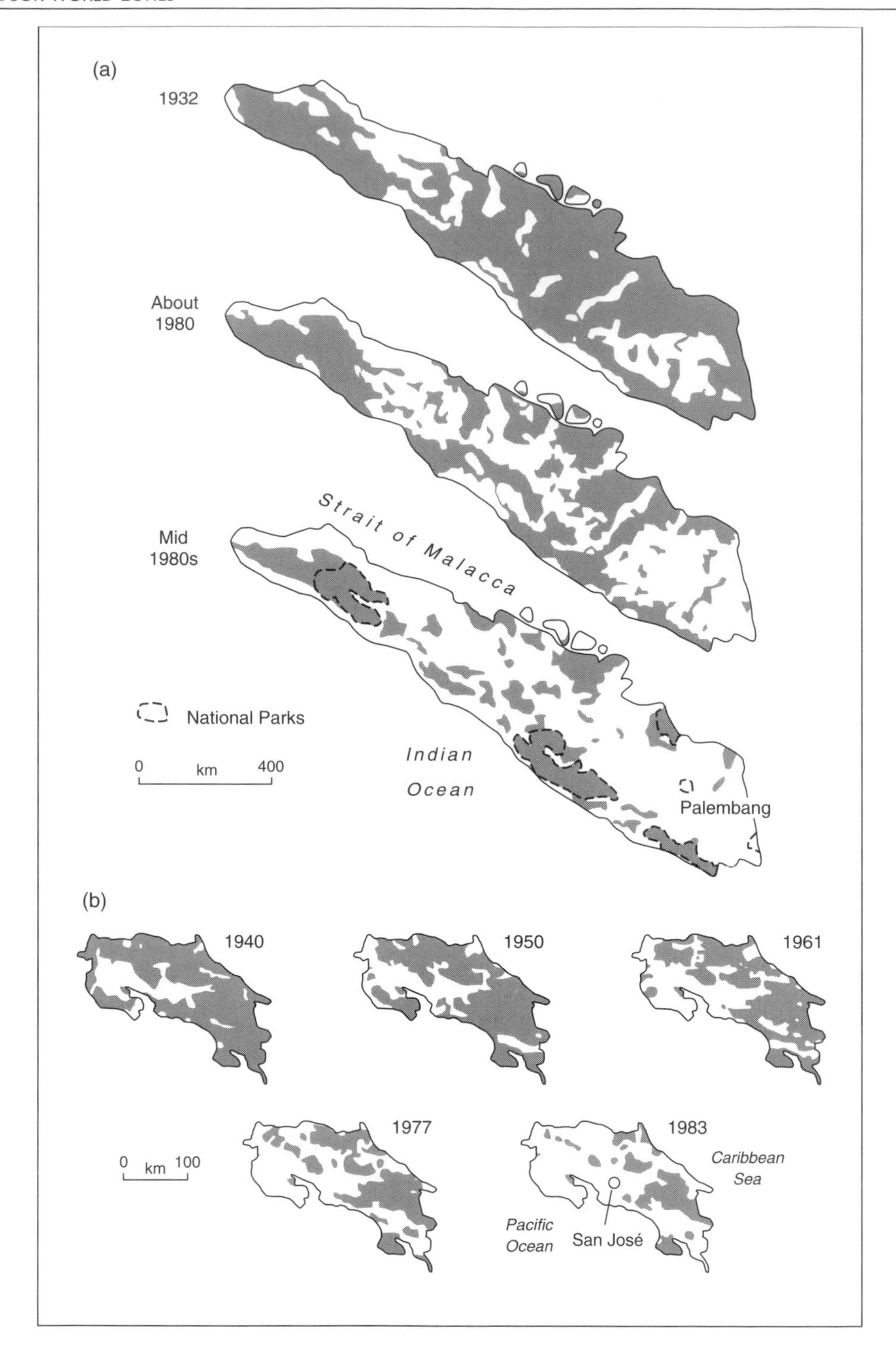

Figure 7.11 Progressive deforestation and habitat fragmentation in the rainforest environments of (a) Sumatra (south-east Asia) and (b) Costa Rica (Central America).

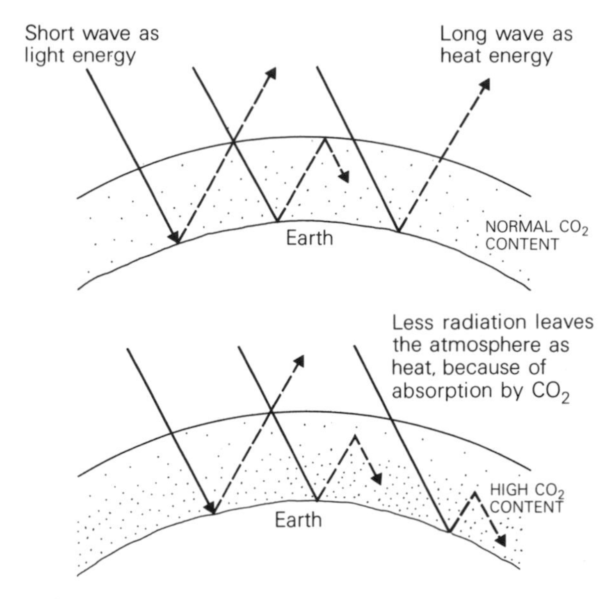

Figure 7.12 The greenhouse effect: short-wave radiation strikes the earth's surface and is transformed into long-wave radiation (heat). Since CO_2 absorbs long-wave radiation, the greater the atmospheric CO_2 content, the more heat is retained and the warmer the atmosphere becomes.

Table 7.5 Some consequences of tropical deforestation

Type of change	*Examples*
Reduced biodiversity	Species extinctions Reduced availability of tropical products Reduced capacity to breed improved crop varieties
Changes in local and regional environments	Accelerated soil erosion Reduction in soil quality (loss of nutrients, organic matter) Increased runoff and floods Increased sedimentation of rivers, estuaries, lakes etc. Possible changes in albedo, rainfall, temperature etc.
Changes in global environment	Reduction in carbon stored in the plants and soils Increase in carbon dioxide content of atmosphere leading to greenhouse effects

released in significant quantities into the atmosphere as CO_2. If this argument is correct, then deforestation, in combination with the release of CO_2 by the burning of fossil fuels such as oil and coal, could lead to an increase in world temperature levels of a few degrees in the next century or so, which in turn might cause some melting of ice caps and flooding of low-lying coastal areas because of the resulting rise in sea level.

7.7 Secondary Forest

Because of the various pressures exerted on the rainforest by human activity, increasingly large areas of the humid tropics are characterised not by true rainforest but by *secondary rainforest*. When an area of rainforest has been cleared, either for cultivation or for timber exploitation, and is then abandoned, the forest begins to regenerate; but for an extended period of years the type of forest that occurs – secondary forest – is very different in character from the virgin forest that it replaces. First, secondary forest is lower and consists of trees of smaller average dimensions than those of primary forest, although since it is comparatively rare that an area of primary forest is clear-felled or completely destroyed by fire, occasional trees much larger than the average are usually found scattered throughout the re-growth. Second, very young secondary forest is often remarkably regular and uniform in structure; and an abundance of undergrowth, unlike in primary forest, makes it laborious to penetrate. Third, secondary forest tends to be much poorer in species than primary, and is sometimes dominated by a single species, or a small number of species. Fourth, the dominant trees of secondary forest are light-demanding and intolerant of shade; most of them grow very quickly (at rates of up to 12 m in three years); and most of the trees possess efficient dispersal mechanisms (having seeds or fruits well adapted for transport by wind or animals). Fifth, because of their rapid growth the wood often has a soft texture and low density, making it less desirable for many commercial purposes than the hardwoods of the virgin forest.

Secondary forests should not, however, be dismissed as useless scrub. All but the youngest secondary forests are probably effective at preventing soil erosion, regulating water supply and maintaining water quality. They also provide a refuge for some flora and fauna.

7.8 Tropical Seasonal Forest and Savanna

As the drier season becomes more distinct, the tropical rainforest, dominated by evergreens, is gradually replaced by a forest type that is partly evergreen and partly deciduous. The tropical seasonal forest occurs in some of the monsoon lands of Asia, but is relatively rare in Africa, where savannas occur in close juxtaposition to the rainforest belt.

Between the perennially dry deserts of the subtropical high-pressure belts and the equatorial rainforests lie the world's *savannas*. They occupy about one-quarter of the world's land surface, and display great variety in form. Broadly, savannas occur in those parts of the tropics that experience a dry season of 2½ to 7½ months' duration – a season that checks plant growth. (Desert shrub vegetation occurs mainly where the number of arid months exceeds 7½.)

The savannas have been defined by T. L. Hills as 'a plant formation of tropical regions, comprising a virtually continuous ecologically dominant stratum of more or less xeromorphic (drought tolerant) plants, of which herbaceous plants, especially grasses and sedges, are frequently the principal, and occasionally the only, components, although woody plants often of the dimension of trees or palms generally occur and are present in varying densities'. Most savanna consists, therefore, of grassland with scattered trees (plate 7.5), and the biomass is intermediate between that of rainforest and that of deserts. Net primary productivity of savanna averages about $1200\,\mathrm{g\,m^{-2}\,yr^{-1}}$, compared with 2400 for rainforests and about 200 for deserts, but there is a very wide range.

Typically, the trees in a savanna area are 6–12 m in height, strongly rooted and with flattened crowns (figure 7.13). They exhibit various drought-

Plate 7.5 Savanna woodland at the end of the wet season in the Kimberley District, north-western Australia. The mix of trees and grass is one of the most striking characteristics of this biome. The area is subject to frequent burning.

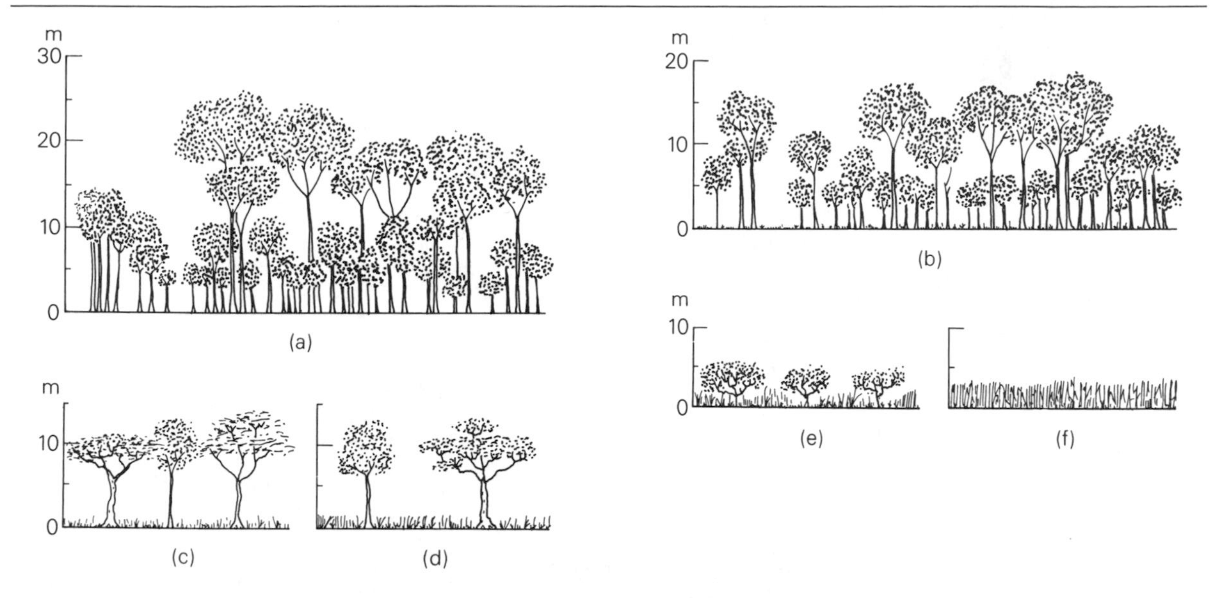

Figure 7.13 Some of the range of vegetation types found in the seasonal tropics: (a) semi-evergreen seasonal forest; (b) deciduous seasonal forest; (c) savanna woodland; (d) tree savanna; (e) shrub savanna; (f) grass savanna.

resisting features to cope with the long dry season, including partial or total seasonal loss of leaves, and some of them are fire-resistant, possessing thick bark and thick bud-scales. The tree species vary greatly (plate 7.6): in Honduras there is savanna dominated by pine trees; in Australia there are various types of eucalyptus; and in Africa there is the thorny *acacia*, various palms and the bottle-shaped baobab (*Adansonia digitata)*. The grasses are often long and reach up to 3.5 m in height, thus providing ample fuel for dry-season fires. Savannas are also characterised, where man has not decimated them, by a very diverse fauna, especially of large herbivorous mammals such as antelope, wildebeeste, zebra and giraffe, together with scavengers and predators (such as hyena and lion).

The nature and distribution of savanna owes a great deal to climate conditions, and initially scientists regarded it as a type of *climax vegetation* (see section 10.5) that occurred in those areas of the tropics in which there was an extended season of drought. However, it is now recognised that, as with many vegetation types in the world, this is a rather simplistic view, and that other factors play a role. Some workers have championed the importance of *edaphic* (soil) conditions, including poor drainage, soils that have a low water-retention capacity in the dry season, soils with a shallow profile due to the development of a laterite hardpan, and soils with a low nutrient supply (either because they are developed on a poor parent rock such as quartzite or because the soil has undergone an extended period of leaching on an old land surface). Another highly important factor is fire. The importance of this factor in maintaining and originating some savannas is suggested by the fact, already noted, that many of the savanna trees are fire-resistant. Experiments in fire control have shown how trees can re-establish themselves when burning ceases, and there are many observations of the frequency with which, for example, African herdsmen and agriculturists burn over much of tropical Africa and thereby maintain grassland. Some of the fires may be deliberately started, while others may be caused naturally by lightning, which on average strikes the land surface of the globe 100 000 times every day! In general, regular burning favours perennial grasses with underground stems that can regenerate after a fire has passed.

Plate 7.6 A savanna dominated by palms bordering the River Uruguay in Argentina.

7.9 Mangrove Swamps

One type of vegetation assemblage that is more or less restricted to within 30° of the Equator, because they are intolerant of frost, are the tidally submerged coastal woodlands called *mangrove swamps* (figure 7.14a; see colour plate 11). Mangroves are basically salt-tolerant trees that can reach 40 m in height but are generally much smaller. They occur along the borders of many tropical shores (plate 7.7), where wave action is not too intense and mud and peat are deposited, and some have stilt or prop roots to help them hold on to the shifting sediments. They may extend inland up to 60 km where there are tidal rivers, and they may also occur round the lagoons of some coral atolls.

Mangroves are most luxuriant and complex on the coasts of Malaysia and the surrounding islands, and in general mangrove vegetation in the east is taller and richer in diversity than that of western Africa and the Americas.

Mangrove swamps are important wetland environments, being among the most productive and biologically diverse ecosystems in the world. Nutrients which are released when their leaves and twigs decompose are an important food for aquatic animals, including economically important fish species. They are also valuable as a source of timber products, and their ramifying root systems act as silt traps, thereby stabilizing shorelines and helping to maintain the quality of estuarine and coastal waters.

Mangrove swamps, in spite of their great value, are being subjected to increasing human pressures, including deforestation and reclamation for agriculture and acquaculture. To give two examples, mangrove areas in the Philippines converted to fishponds have increased from about 90 000 ha in the early 1950s to over 244 000 ha in the early 1980s,

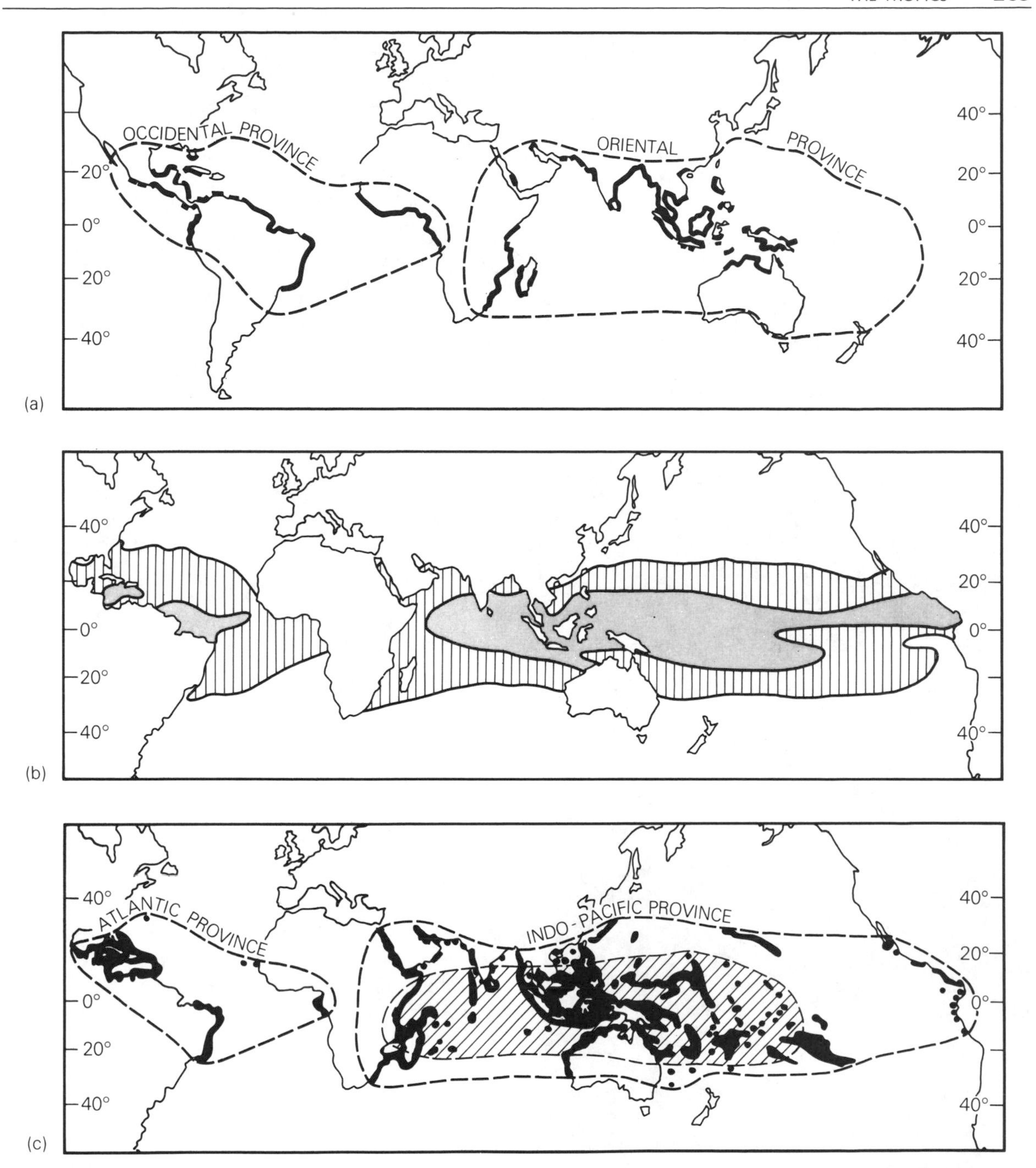

Figure 7.14 (a) The distribution of mangroves is strongly controlled by temperature conditions. Broken lines enclose the total range; thick lines denote major stretches of coast dominated by tidal woodlands. (b) Cold-month sea-surface isotherms for 20 °C and 26 °C. Since the fall in the tropical sea-surface temperatures during Pleistocene glacials was of the order of 6°, the shaded area gives an approximate idea of the reduced availability of tropical water at such times. (c) The distribution of coral reefs is also strongly related to sea temperatures. The hatched area includes the zones of most prolific reef development, and almost all oceanic atolls.

Plate 7.7 Mangrove swamp on the Island of Mahé, Seychelles, Indian Ocean. Notice the pneumatophores (breathing tubes).

Plate 7.8 Charles Darwin, the Victorian naturalist, developed one of the most important theories for coral atoll formation, based on subsidence.

while in Indonesia logging operations are claiming 200 000 ha of mangrove swamp each year.

7.10 Coral Reefs

An important environment which is more or less totally restricted to the intertropical zone is the *coral reef* (figure 7.14c). These features have fascinated scientists for over 150 years, and some of the most pertinent observations of them were made in the 1830s by Charles Darwin (plate 7.8) on the voyage of the *Beagle*. He recognised that there were three major kinds: (a) *fringing reefs*; (b) *barrier reefs*; and (c) *atolls*; and he saw that they were related to each other in a logical and gradational sequence (figure 7.15).

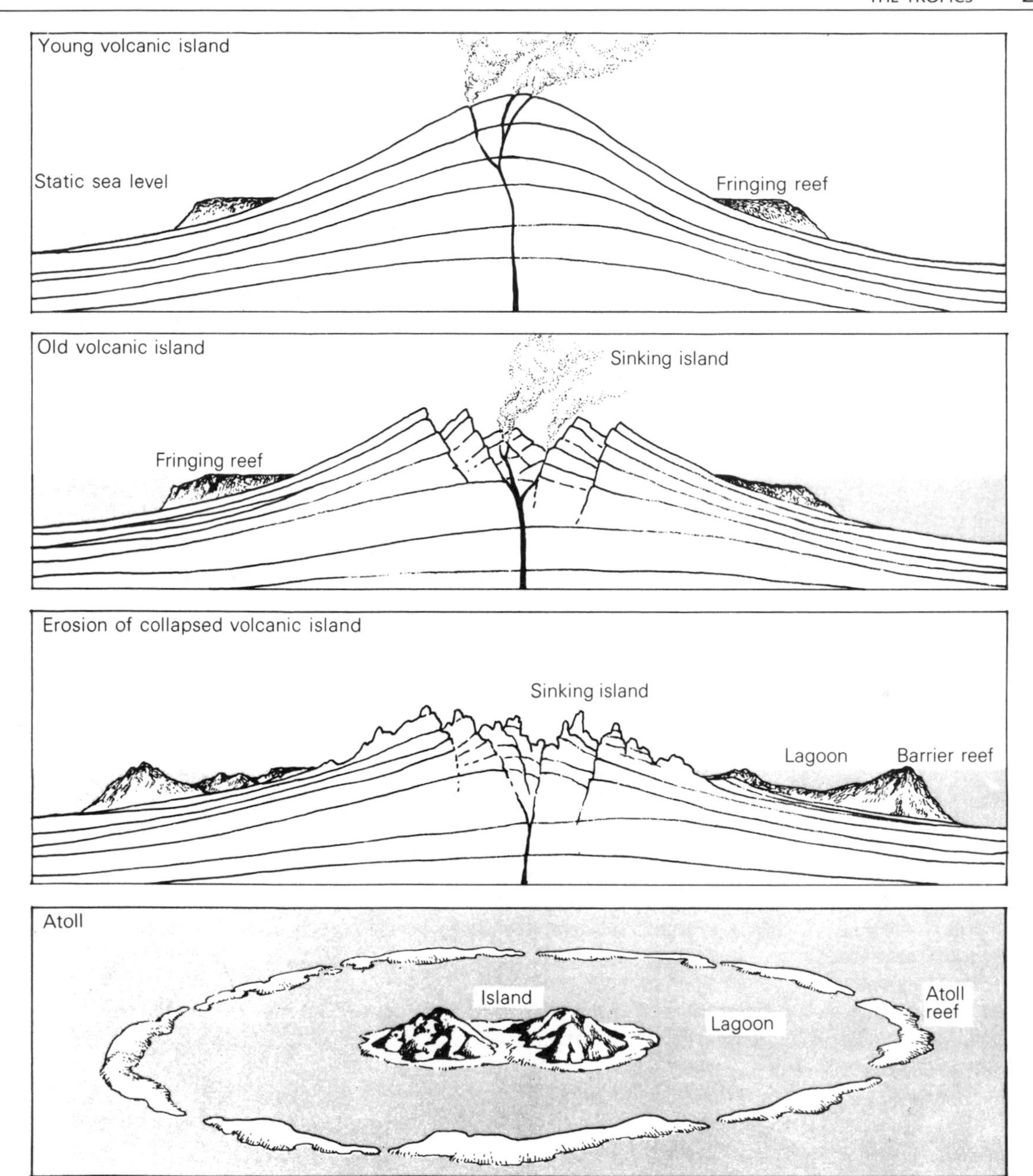

Figure 7.15 Darwin's model of coral atoll formation as a result of subsidence of a volcanic island. Note the progression from fringing reef to atoll as subsidence progresses.

A *fringing reef* is one that lies close to the shore of some continent or island. Its surface forms an uneven and rather rough platform around the coast, about the level of low water, and its outer edge slopes downwards into the sea. Between the fringing reef and the land there is sometimes a small channel or *lagoon*. When the lagoon is wide and deep and the reef lies at some distance from the shore and rises from deep water it is called a *barrier reef*. An *atoll* is a reef in the form of a ring or horseshoe with a lagoon in the centre.

Darwin's theory was that the succession from one coral reef type to another could be achieved by the upward growth of coral from a sinking platform, and that there would be a progression from a fringing reef, through the barrier reef stage until, with the disappearance through subsidence of the central island, only a reef-enclosed lagoon or atoll would survive. A long time after Darwin put forward this theory, some deep boreholes were drilled in the Pacific atolls in connection with the testing of atom bombs in the 1950s. The drill holes passed through more than 1000 m of coral before reaching the basalt substratum of the ocean floor, and indicated that the coral had been growing upward for tens of millions of years as the crust subsided at a rate of between 15 and 51 m per million years. Darwin's theory was therefore proved basically correct. There are some submarine islands (see section 1.7) called *guyots* and *seamounts*, in which subsidence associated with sea-floor spreading has been too speedy for coral growth to keep up.

Coral reefs are found where the ocean water temperature is not less than 21 °C (see figures 7.14b and c), where there is a firm substratum, and where the seawater is not rendered too dark by excessive amounts of river-borne sediment. They will not grow in very deep water, so a platform within 30–40 m of the surface is a necessary prerequisite for their development. Their physical structure is dominated by the skeletons of *corals*, which are carnivorous animals living off zooplankton. However, in

Window 7.3 The Great Barrier Reef

The Great Barrier Reef, which lies off the coast of Queensland, Australia, is by far the greatest coral formation in the world. It stretches for around 2000 km from Torres Strait to around 24° S. It has developed on a slowly foundering land surface, now submerged as a continental shelf. The seaward edge, the Outer Barrier, withstands the force of the Pacific Ocean swell in a long line of roaring surf. It is most continuous in the north, taking the form of chains of platforms which are up to 25 km long and a kilometre across. These are awash at high tide. There are some breaks in the Great Barrier Reef, for example the Grafton and Flora passages near Cairns and the Flinders and Magnetic passages near Townsville. The origin of such breaks is not entirely clear, but the Flinders Passage may be associated with the freshwaters discharged by the Burdekin River. Corals do not tolerate freshwater.

On the seaward side the Outer Barrier may rise steeply from the ocean bed some 1800 m below, while on the landward side the water depth is often only around 50 m. Coral platforms grow upward from this shallow platform, and project just above lowest tide level, when exposure periods become too long to permit further growth. Waves may deposit sand and other debris on these platforms, giving birth to low, wooded, sandy islands called cays. The shape and size of these low islands can be altered by severe summer cyclones (hurricanes).

Although the Great Barrier Reef is a massive feature which seems to withstand the power of the Pacific Ocean, many fears have been expressed about its future health. Parts of it, for example, are being nibbled away and destroyed by the crown of thorns starfish (*Acanthaster planci*), a species which is a predator on live coral. For reasons which are still the subject of debate, this starfish has greatly expanded its numbers and erosional potential in recent years. Other threats include pollution by effluents from industrial plants and tourist resorts, siltation by rivers charged with sediments derived from accelerated erosion of inland catchments, and contamination with pesticide residues from farmland.

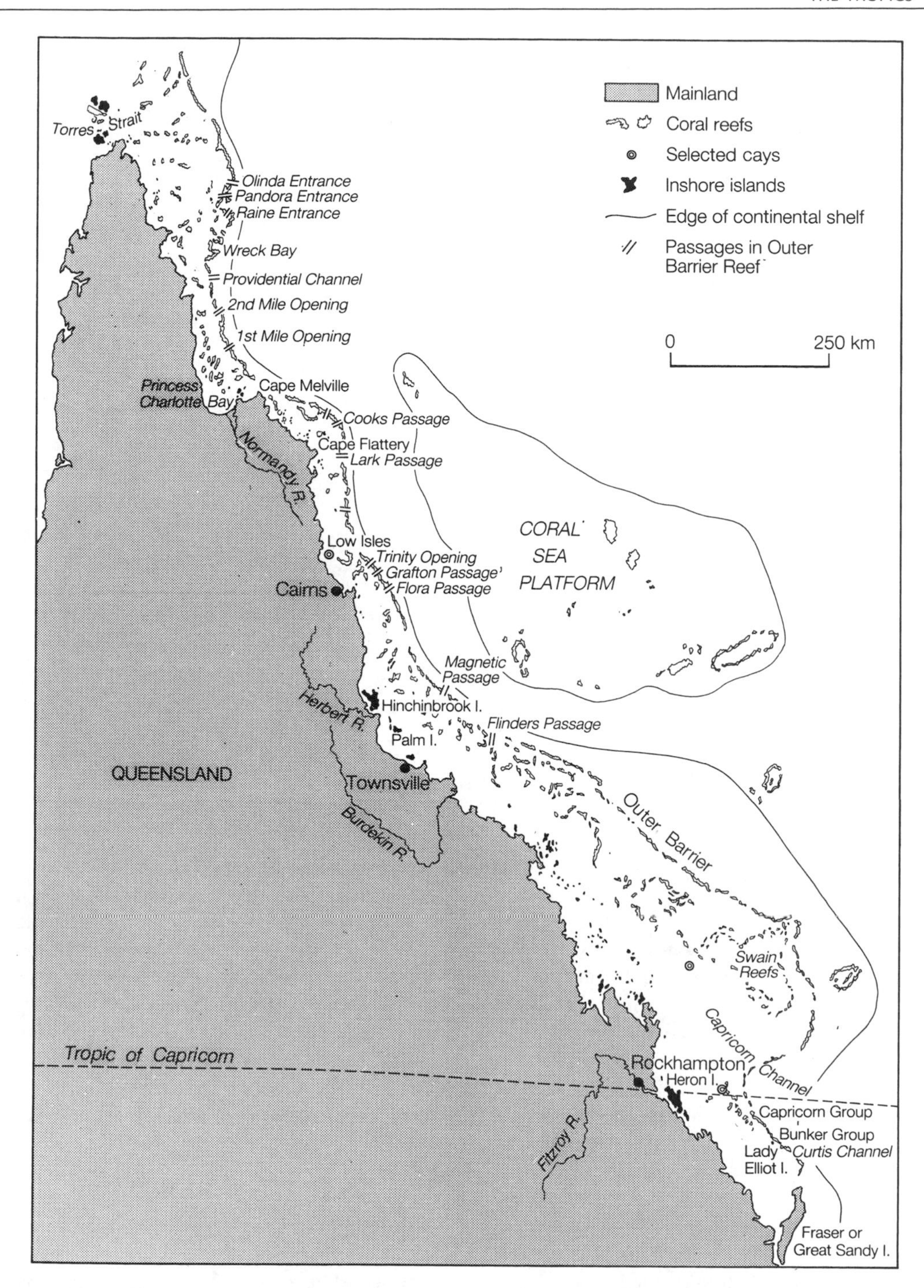

The location and form of the Great Barrier Reef, Australia.

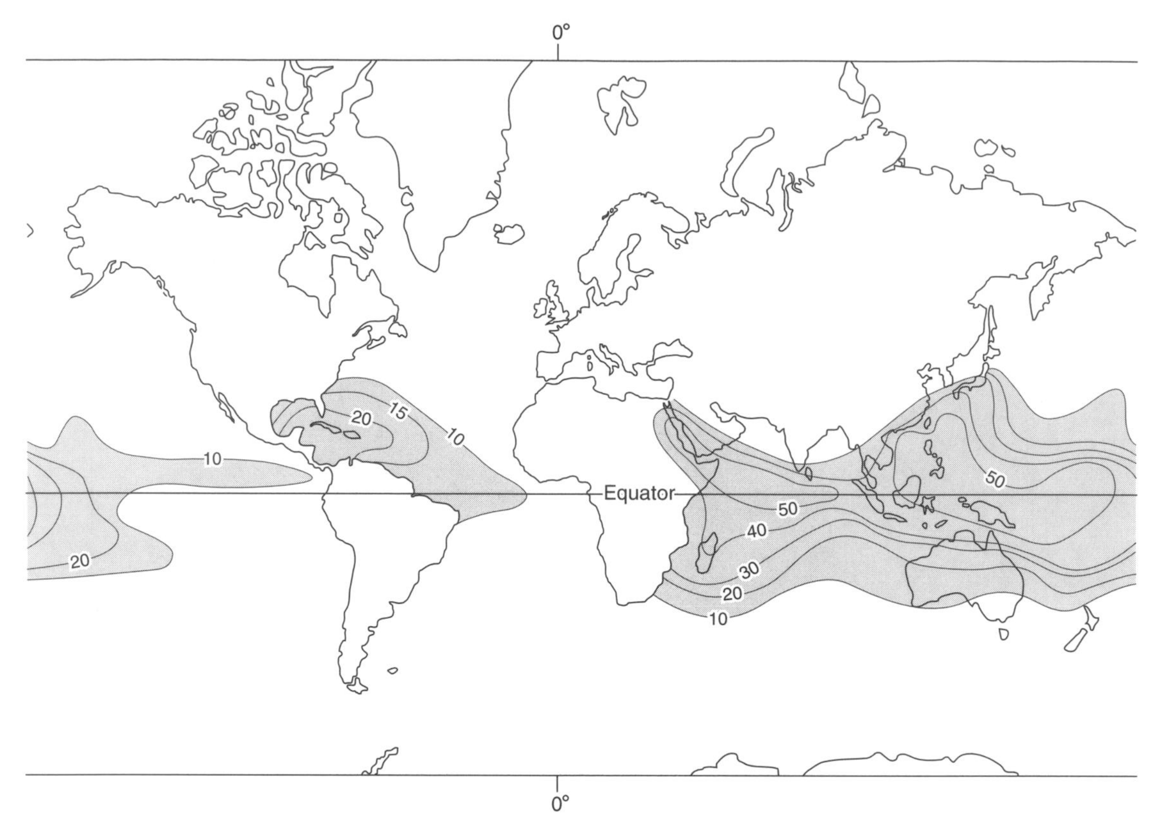

Figure 7.16 Distribution of the number of genera of reef-building corals. Note the great diversity in the Indian Ocean and western Pacific, especially close to the Equator.

addition to corals there are enormous quantities of algae, some calcareous, which help to build the reefs. The size of reefs is variable. Some atolls are very large – Kwajelein in the Marshall Islands of the South Pacific is 120 km long and as much as 24 km across – but most are very much smaller, and rise only a few metres above the water. The 2000 km complex of reefs known as the Great Barrier Reef, which forms a gigantic natural breakwater off the north-east coast of Australia, is by far the greatest coral structure on earth (window 7.3) (see colour plate 12).

Like mangrove swamps, coral reefs are extremely important habitats. Their diversity of coral genera is greatest in the warm waters of the Indian Ocean and the western Pacific (figure 7.16). Indeed, they have been called the marine version of the tropical rainforests, rivalling their terrestrial counterparts in both richness of species and biological productivity. They also have significance because they provide coastal protection, opportunities for recreation, and are potential sources of substances like drugs. They are coming under a variety of threats, of which two of the most important are dredging and the effects of increased siltation brought about by accelerated erosion from neighbouring land areas.

7.11 Weathering of Rocks

High temperatures, large amounts of rainfall and a high biomass promote the rapid decay of many rock minerals. The speed of chemical reactions, following *Van't Hoff's temperature rule,* approximately doubles for every 10 °C rise in temperature. Water

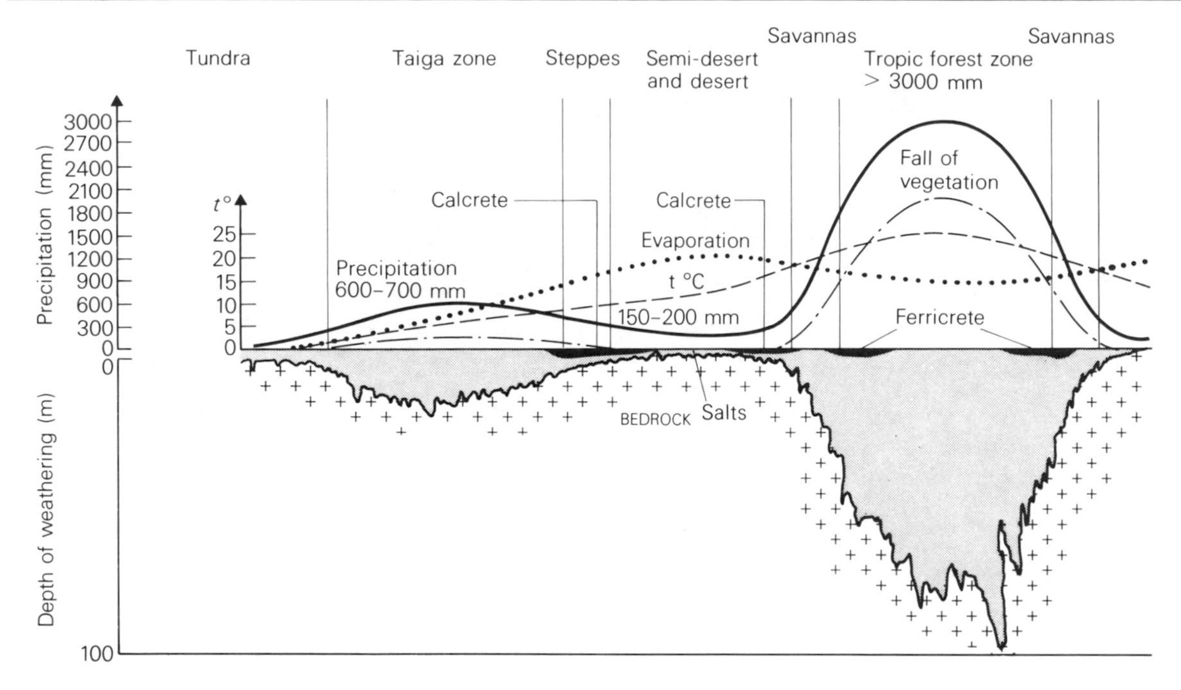

Figure 7.17 Schematic representation of the depth of weathering between the Poles and the tropics, showing the great depths of weathering that are attained in the hot, wet areas of the equatorial zone, and the moderate depths that are attained in the humid mid-latitude zone.

is the main chemical reagent in the decay of rocks, and the large quantities of organic matter decaying on and within the soil liberate organic acids and carbon dioxide gas, which play an important part in the mobilisation of some of the minerals formed during the weathering process, especially iron oxide.

The prolonged action of tropical weathering, combined with relatively slow rates of surface erosion so long as the forest cover is maintained, have led to the development over susceptible rocks of considerable thickness of weathered material called *regolith*. Such *deep weathering* (figure 7.17) is characteristic of the wet tropics, and sometimes the rocks are decomposed for tens of metres, locally exceeding 125 m. Deep weathering is especially effective where the rock is penetrated by a dense network of joints and is less effective where the joints are widely spaced. For this reason the *weathering front* (the boundary zone between the bedrock and the regolith) may be very uneven.

There are various different types of deep weathering profile, some of which are capped by an iron (*laterite*) (plate 7.9) or aluminium (*bauxite*) crust (*duricrust*). A duricrust (section 6.7) is a hard crust formation at or just below the ground surface, and of the various types found in the moister parts of the tropics probably laterites are the most extensive. They are formed because intense weathering preferentially removes silica, leaving behind the relatively insoluble residues of iron and aluminium sesquioxides. The oxides harden on exposure when vegetation is removed and produce a hard material that is resistant to weathering and erosion. Sometimes the surface is so impermeable that it has only a very limited vegetation cover. The crust often forms a distinctive capping to tabular-shaped hills.

If a change in climate or in base level causes fluvial incision into the landscape (figure 7.18) then the uneven weathering front may be exposed, leading to the exhumation of the areas of widely spaced jointing as large isolated hills called *bornhardts* or

Plate 7.9 Intense chemical weathering of basalt in Maharashtra, India, has created laterite caprocks, which lead to the development of distinctive tabular-shaped hills.

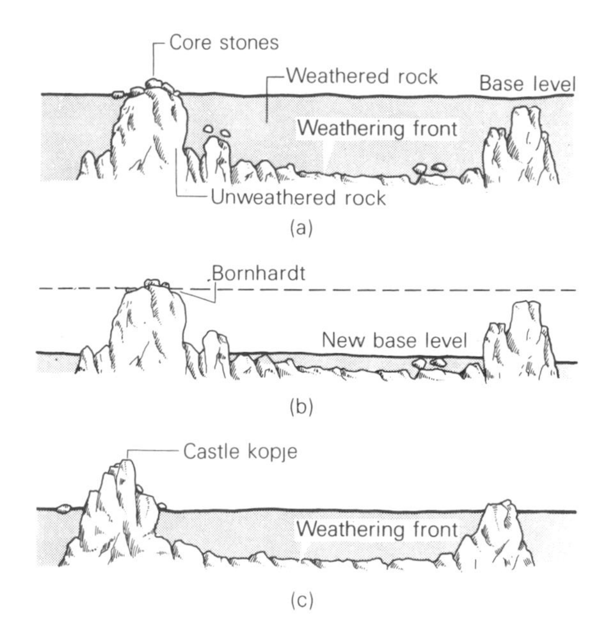

Figure 7.18 The development of tropical plains and inselbergs by a two-cycle process of deep weathering, followed by stripping. (a) Inselbergs are buried beneath weathered rock; depth of weathering varies according to the amount of jointing. (b) Weathered rock is stripped away, leaving fresh unweathered rock as inselbergs–bornhardts. (c) Deep weathering continues. Sub-aerial erosion attacks bornhardts, leading to the formation of castle kopjes.

inselbergs (plate 7.10; see colour plate 13). More closely spaced joints will produce smaller features called *castle kopjes,* and in rock where very close jointing leads to almost complete rock breakdown, the effect of stripping will be to reveal a pile of rounded boulders or core stones.

The soils of the humid tropics include *oxisols.* These soils, which are sometimes described as *ferrallitic,* are generally intensively weathered soils, which over long periods have evolved into deep, slightly acid profiles, with an accumulation of kaolinitic clays. They are rich in *sesquioxides,* rich in clays, distinctively bright in colour and highly leached of nutrients. *Ferruginous soils* and *tropical podzols* have also evolved over long periods but are less weathered and generally less deep than ferrallitic profiles. They occur in conjunction with areas of semi-evergreen, deciduous forest and in savannas with a pronounced dry season. The tropical podzols resemble their temperate counterparts but are very much deeper, and they develop most strikingly in unconsolidated quartz-rich parent materials such as alluvial sands. Other river valleys, deltas and swamps may have *alluvial hydromorphic soils* (inceptisols or entisols – see section 3.5), in which seasonal waterlogging may produce

Plate 7.10 Areas of ancient granitic and gneissic rocks in the tropics are often studded with inselbergs or bornhardts. The Sibebe inselberg in Swaziland is composed of coarse-grained granite in which there are very few joints.

gley features. In flat and low-lying portions of the seasonal tropics *vertisols* may occur. These are dark clay soils which shrink and crack in dry periods and swell in wet, and which may often be derived from the weathering of base-rich parent materials such as basalt. Clay contents are high, and *montmorillonite,* a clay mineral that expands and contracts greatly according to moisture content, is widespread. The cracks may form polygonal patterns and micro-relief called *gilgai,* and building on such soils, often described as 'expansive', may create stresses in foundations that can cause structural failure.

7.12 Slope Movements

Although the existence of deep weathering profiles indicates that in many areas the removal of material from slopes is slower than the rate of production of weathered material, this does not mean that mass movements are ineffective. The high intensity of rainfall, the high amounts of rainfall, the lack of much vegetation or humus on the forest floor and the impermeable nature of many clay-rich soils mean that there are probably considerable quantities of surface wash over tropical forest floors caused by rain splash.

The other very important process is *landsliding,* facilitated by the combination of rotted rock, clay soils, heavy rainfall and generally shallowly rooted trees. This has proved a particular hazard when cities like Rio de Janeiro, Hong Kong and Kuala Lumpur have expanded, causing building to spread on to steep slopes.

7.13 Some Problems of the Humid Tropics

Although some early European travellers marvelled at the luxuriant nature of the rainforest, believing that the tropics were fertile and merely awaited commercial exploitation, development in the humid tropics is faced with some major environmental problems:

large areas of soil infertility
lateritisation
erosion caused by intense storms
slope instability associated with deeply weathered rock
extremely enervating climate (caused by high humidity)
rapid loss of nutrients on deforestation
hurricanes
active attack on wood and other organic material by termites, fungi etc.
flooding (both natural and following deforestation) (see colour plate 14)
world climatic change (the CO_2 problem)

While it would be wrong to replace early optimism with an equal degree of pessimism, the tropics are a zone where unwise human actions can have devastating consequences. For example, some rubber plantations established in Amazonia after the First World War had a very short life. When the forest was cut down, some of the soils rapidly hardened to laterite, and others were washed away. Likewise, the attempt by the British government to establish groundnut farming in East Africa just after the Second World War foundered when the soils proved to suffer marked deterioration in their structure when cultivated by heavy machinery. They also needed heavy application of expensive fertiliser to make up for the loss of natural nutrients.

Key Terms and Concepts

biodiversity loss
coral reefs
cyclones
Darwin's theory of coral reefs
deep weathering
inselbergs
Intertropical Convergence Zone (ITCZ)
mangrove swamps
monsoon
savanna
secondary forest
tropical seasonal forest
wave disturbances

Points for Review

- What are the main characteristics of climate in the humid tropics?
- In what ways are tropical weather systems different from those in temperate latitudes?
- Discuss the character and consequences of a recent tropical cyclone of your choice.
- Why is it important to conserve tropical rainforest?
- Account for the distribution and development of savanna.
- What are the main types of coral reef?
- Assess some of the characteristics and consequences of rock weathering in the humid tropics.
- What are the main types of environmental hazards encountered in the tropics?

FURTHER READING

Aiken, S. R. and Leigh, C. H. (1992) *Vanishing Rainforests: Their Ecological Transition in Malaysia* (Oxford: Oxford University Press). A case study from a threatened area.

Corlett, R. T. (1995) 'Tropical secondary forests', *Progress in Physical Geography*, 19: 159–72. An informative review article.

Grainger, A. (1992) *Controlling Tropical Deforestation* (London: Earthscan). An up-to-date introduction with a global perspective.

Kellman, M. and Tackaberry, R. (1997) *Tropical Environments* (London: Routledge). A treatment of forests and savannas with a strong biogeographical viewpoint.

McGregor, G. R. and Nieuwolt, S. (1998) *Tropical Climatology*, 2nd edn (Chichester: Wiley). An up-to-date survey of the causes, nature and implications of tropical climates.

Park, C. C. (1992) *Tropical Rainforests* (London: Routledge). Another relatively simple introduction to many aspects of the rainforest environment.

Reading, A. J., Thompson, R. D. and Millington, A. C. (1995) *Humid Tropical Environments* (Oxford: Blackwell). An excellent geographical perspective on the moister parts of the tropics.

Richards, P. W. (1996) *The Tropical Rainforest*, 2nd edn (Cambridge: Cambridge University Press). A second edition of the classic text.

Thomas, M. F. (1994) *Geomorphology in the Tropics: A Study of Weathering and Denudation in Low Latitudes* (Chichester: Wiley). An excellent, large discussion of the geomorphology of the tropics.

Whitmore, T. C. (1990) *An Introduction to Tropical Forests* (Oxford: Oxford University Press). A most useful point of entry into the literature.

Part III

Mountain and Maritime Environments

8 Mountains

8.1 Introduction

As we saw in chapter 1, mountain ranges tend to run in long, linear belts along the margins of continents. One exceptionally long belt runs around the edge of the Pacific Ocean, while another runs from west to east along the underbelly of Eurasia. Their distribution follows closely the distributions of earthquakes, fault zones, volcanic activity, island arcs and ocean trenches. Many mountains are composed of marine sediments, which may have been metamorphosed and they are often injected with volcanic material produced during the course of mountain-building.

In addition, mountainous relief results from the operation of the various forces of weathering and erosion, including rock falls, glaciation, avalanching and river transport.

Folding, faulting and volcanism are the forces that have created mountains. The key to understanding their distribution, form and composition lies in understanding the theory of plate tectonics described in chapter 1. The purpose of this chapter is not to investigate that theme further, but rather to discuss the modifications that mountains produce in the zonal patterns of such phenomena as climate and vegetation.

Opposite Among the world's greatest mountains are those of the Western Cordillera of the Americas. Although they are primarily caused by tectonic activity, the morainic lake and large debris cones indicate that much of the variety of mountain scenery is caused by the operation of powerful weathering and erosional processes. The location is the Banff National Park, Alberta, Canada.

8.2 Mountain Climates

Fundamental to an understanding of mountain climates are the changes that occur in the atmosphere with increasing altitude, especially the decrease in temperature, air density, water vapour, carbon dioxide and impurities.

Although the sun is the ultimate source of energy, very little heating of the atmosphere takes place directly from the sun. The earth's surface receives and absorbs the short-wave radiation coming from the sun and converts it to long-wave energy, and the earth itself then becomes the radiating body. The atmosphere is therefore heated directly by the earth and not by the sun. This explains why the highest temperatures usually occur next to the earth's surface and decrease outwards and upwards. Because mountains present a smaller land area at higher elevations, they are less able to warm the surrounding air.

Furthermore, the density and composition of the air control its ability to hold heat. As one ascends a mountain the pressure of the air decreases markedly (figure 8.1b). At sea level it is generally expressed as 760 mm of mercury. By the time one reaches elevations of around 7000 m, atmospheric pressure is less than 300 mm of mercury. This is important because the ability of air to hold heat is a function of its molecular structure. At higher alti-

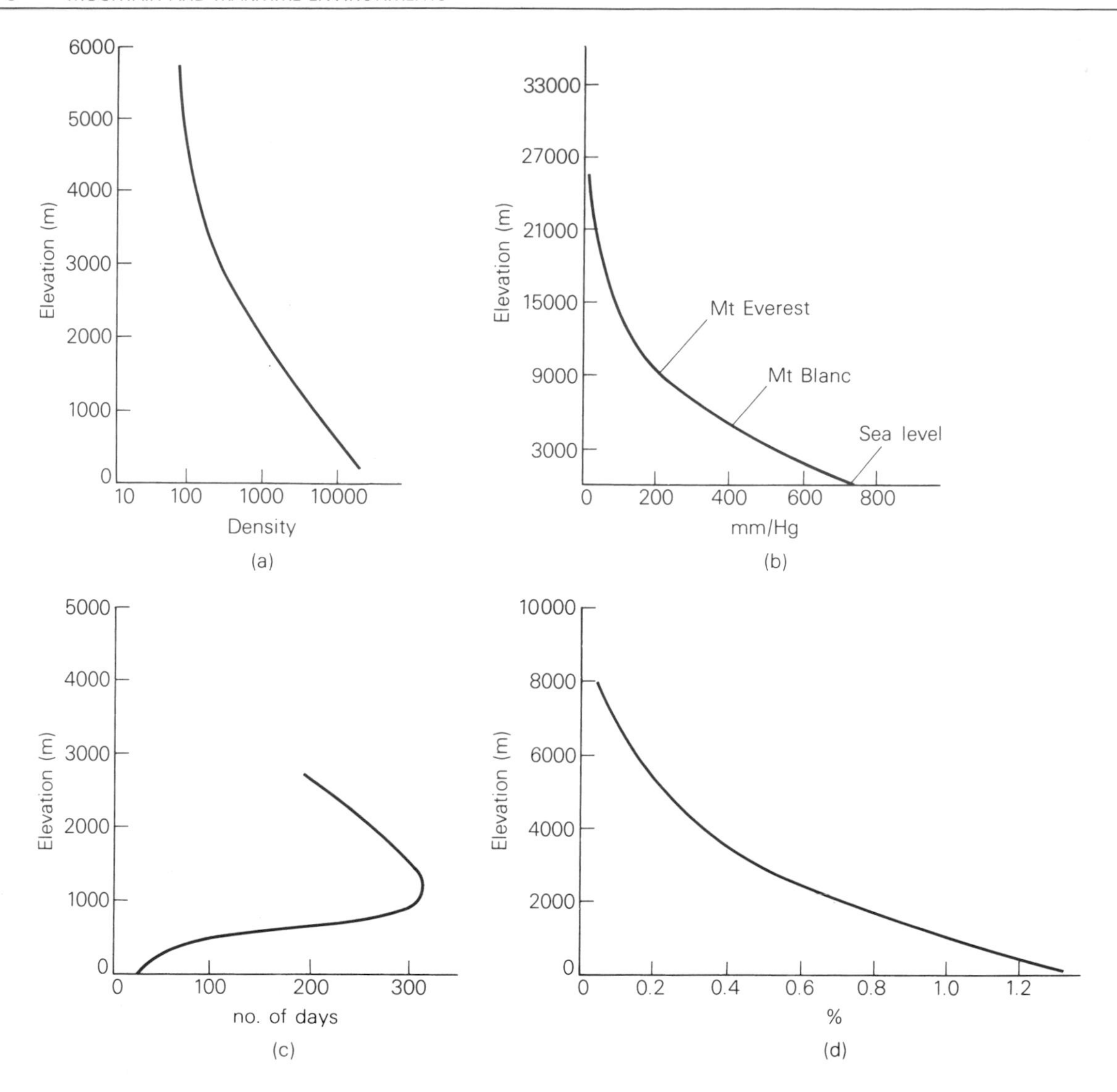

Figure 8.1 Changing climatic conditions with altitude: (a) average density of suspended particles per cubic metre; (b) atmospheric pressure; (c) number of foggy or cloudy days at different altitudes in Japan; (d) average water vapour content of air as volume percentage.

tudes, where pressures are less, molecules are spaced further apart, so there are fewer molecules in a given volume of air to receive and hold heat. Also, with altitude the air loses much of its water vapour, carbon dioxide and suspended particulate matter (figure 8.1a). These too are important in determining the ability of air to hold heat. Most water vapour, for example, is concentrated in the lower levels of the atmosphere, with half of it occurring below an elevation of 1800 m.

For these reasons a change of temperature occurs with elevation (table 8.1) and this change is called the *lapse rate* or vertical temperature gradient. The lapse rate varies according to many factors, but generally ranges from 1 to 2 °C per 300 m.

Another major influence that mountains have on

Table 8.1 Temperature conditions with elevation in the eastern Alps

Elevation (m)	Mean air temperature (°C)				Annual number of days		
	January	July	Year	Annual range	Frost free	Frost alternation	Continuous frost
200	−1.4	19.5	9.0	20.9	272	67	26
400	−2.5	18.3	8.0	20.8	267	97	31
600	−3.5	17.1	7.1	20.6	250	78	37
800	−3.9	16.0	6.4	19.9	234	91	40
1000	−3.9	14.8	5.7	18.7	226	86	53
1200	−3.9	13.6	4.9	17.5	218	84	63
1400	−4.1	12.4	4.0	16.5	211	81	73
1600	−4.9	11.2	2.8	16.1	203	78	84
1800	−6.1	9.9	1.6	16.0	190	76	99
2000	−7.1	8.7	0.4	15.8	178	73	114
2200	−8.2	7.2	−0.8	15.4	163	71	131
2400	−9.2	5.9	−2.0	15.1	146	68	151
2600	−10.3	4.6	−3.3	14.9	125	66	174
2800	−11.3	3.2	−4.5	14.5	101	64	200
3000	−12.4	1.8	−5.7	14.2	71	62	232

Source: R. Geiger (1965), *The Climate near the Ground* (Cambridge, Mass.: Harvard University Press), p. 444

climate is through the *barrier effect*. Damming of air flows can occur if mountains are sufficiently high, with the effectiveness of the damming depending upon the depth of the air mass and the elevation of the lowest valleys or passes. When the air mass is effectively dammed, the winds are usually deflected by mountains. Moreover, large mountain ranges may be the foci of anticyclonic systems (because the mountains are a centre of cold air), and the presence of such anticyclones may cause storms to detour round the mountains. Jet streams may also split to flow round mountains (such as the Tibetan plateau), but they can rejoin to the lee of the range where they often intensify and produce storms. The tornadoes and violent squall lines that form in the American Mid-West and create such havoc may also result from the great contrasts in air masses that develop in the confluence to the lee of the Rockies. Disturbances of the air by mountains can create a wave pattern much like that found in the wake of a ship. This may create zones of clear air turbulence that create problems for aeroplane pilots, or they may produce *lee waves* with beautiful lenticular (standing-wave) clouds.

Mountains also cause the forced ascent of air. When moist air blows perpendicular to a mountain range it is forced to rise; as it does so it is cooled. Eventually the dew point is reached, condensation occurs as clouds (plate 8.1), and precipitation results. For this reason, some of the rainiest places on earth are mountain slopes in the path of winds blowing off relatively warm oceans. For example, in the Pacific Ocean the precipitation around the Hawaiian Islands averages about 650 mm per annum, but at Mount Waialeale on Kauai the average annual total reaches no less than 12 344 mm! While such heavy precipitation may take place on the windward side of mountains, where the air is forced to ascend, the leeward side may receive considerably less precipitation because the air is no longer being lifted and much of the moisture has already been removed by precipitation on the windward side (figure 8.2). If the air is forced to descend on the lee side it will become heated by compression, a process called *adiabatic heating*, and this will result in clear, dry conditions. The characteristics of some local winds, including the föhn and the Chinook, are associated with this phenomenon.

The quality of solar radiation received also changes with altitude. In particular, mountains receive a greater quantity of ultraviolet light, and this is thought to have numerous harmful effects, ranging from retardation of growth in tundra plants to cancer in humans. Certainly solar energy in the ul-

Plate 8.1 Mountains create their own clouds and their own climate. In the Cape Peninsula of South Africa, Table Mountain and neighbouring peaks are often mantled in white clouds, which are locally called 'The Table Cloth'. Such clouds deposit much moisture in the form of fog drip.

traviolet spectrum is mainly responsible for the deep tans of mountain dwellers, skiers and climbers.

The amount of solar radiation that is received in mountainous areas depends very substantially on slope angle and aspect. Most mountain slopes receive fewer hours of sunshine than level surfaces, although a slope directly facing the sun may receive more energy than a level surface (especially at higher latitudes). Level surfaces in the tropics usually receive a higher solar intensity than slopes because the sun is always high in the sky at those lower latitudes. Whatever the duration and intensity of the sunlight, the effects are generally clearly evident both in the local ecology and in human activities. In the Northern Hemisphere, south-facing slopes are warmer and drier than north-facing slopes, and the number and diversity of plants and animals are greater. Timber-lines also tend to be higher on the south-facing slopes, and in the east–west valleys of the Alps in Europe most settlements are located on south-facing slopes. In spring north-facing slopes will tend to keep their snow cover longer; they therefore tend to be given over to forestry, whereas south-facing slopes are used for high-altitude pastures. The French draw a distinction between the sunny (*adret*) and shady (*ubac*) slopes.

Relief conditions can also have marked effects on temperature by creating conditions for *temperature inversions* to form. While temperatures normally decrease with elevation, during a temperature inversion the lowest temperatures occur in the valley and increase upward along the mountain slope. An explanation for this is that cold air is denser than warm air so that, as slopes cool down at night, the colder air begins to move downward, displacing the warm air in the valley. This can happen on calm, clear nights when there is rapid radiation and loss of surface heat to space, and no wind to mix and equalise the temperatures. As a consequence, enclosed valleys can become filled with a pool of cold stagnant air, which can give extremely low temperatures and severe frosts. Sensitive species of plants, whether crops or natural vegetation, may not grow in these hollows, although they may flourish at slightly higher altitudes on the slopes above.

The range of temperatures in mountains is another interesting feature of their climatology. The temperature difference between day and night and between winter and summer gradually decreases with elevation; for the higher and more isolated a mountain, the more its temperature will reflect that of the surrounding free air. Surface temperatures will vary much more than air temperatures, and even at high altitudes sunny soil surfaces may warm up in the middle of the day, though they cool down below air temperatures at night. Studies in the Karakoram Mountains in Pakistan, for example,

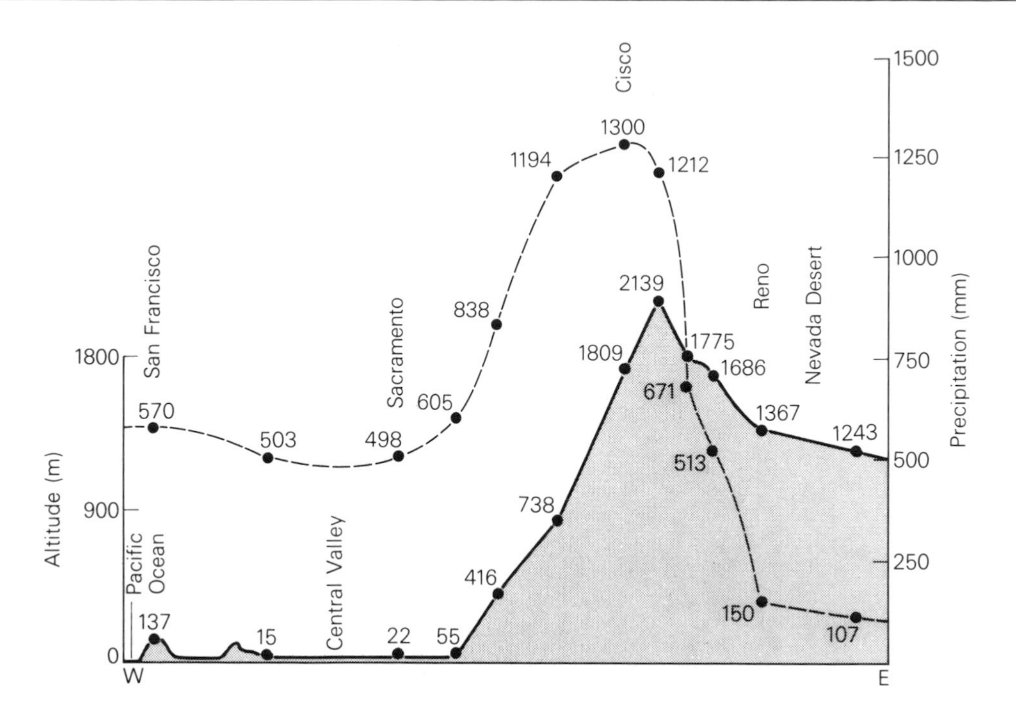

Figure 8.2 A cross-section from San Francisco on the Pacific coastline of California to Reno in the desert of Nevada, illustrating the relationships between precipitation and topography. Note the high levels of precipitation on the windward (western) side of the mountains at high altitudes, and the 'rain shadow' effect on the lee (eastern) side.

at altitudes in excess of 4000 m indicated that diurnal rock surface temperature ranges were normally between 30 and 40 °C in summer, when those in air were between 6 and 16 °C.

The moisture content of the atmosphere decreases quickly as altitude increases (figure 8.1d). At 2000 m it is only about 50 per cent of that at sea level, and at 8000 m (the height of some of the highest peaks in the Himalayas), the water vapour content of the air is less than 1 per cent of that at sea level. This decrease in water vapour with altitude might at first sight appear difficult to square with the well-known fact that precipitation may increase with altitude, but the reason for higher precipitation levels is that moist air from lower elevations has been forced upward into an area of lower temperature. In fact, at very high elevations aridity does prevail, and plants and animals may have evolved adaptations to survive the lack of moisture. Thick, corky bark and waxy leaves, rather like those of some desert plants, are common in alpine plants, while mountain sheep and goats, and beasts like the ibex and the chamois, are all able to live for prolonged periods on little moisture.

The presence of clouds in mountainous areas may increase precipitation levels. Water droplets from clouds may be caught by trees as *fog drip*. The tiny fog droplets are intercepted by the leaves and branches and grow by coalescence until they become heavy enough to fall to the ground. Many tropical and subtropical mountains sustain so-called cloud forests through the abundance of such moisture. Along the east coast of Mexico, for example, luxuriant cloud forests occur in the Sierra Madre Oriental between 1300 and 2400 m.

There is some dispute, brought about partly by a paucity of measurements and problems of maintaining instruments at high altitudes, as to whether precipitation always decreases above a certain altitude. In the tropics, however, this does seem to be established. In the Andes and Central America the zone of maximum precipitation lies at between 900

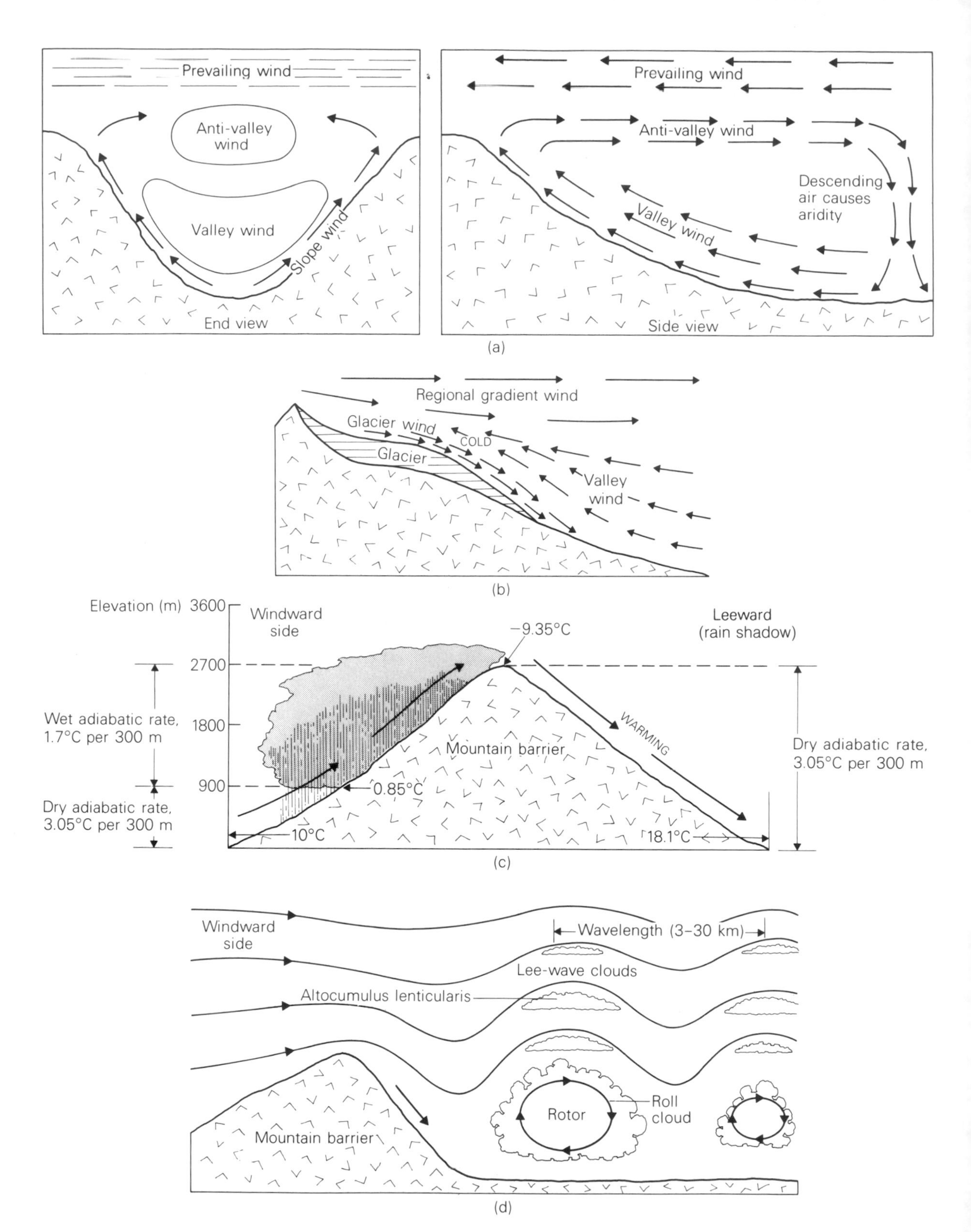

Figure 8.3 Miscellaneous types of mountain wind: (a) slope and valley winds; the view on the left is looking up valley at mid-day, while that on the right provides a vertical cross-section of the same situation viewed from the side; (b) a glacier wind; (c) föhn wind; (d) lee wave and rotors.

and 1600 m, in the Cameroon Mountains of West Africa at 1800 m, and in East Africa at 1500 m. The existence of a zone of maximum precipitation in mid-latitude mountains is being increasingly questioned; available information seems to suggest that precipitation continues to increase with altitude in middle latitudes at least up to 3000–3500 m. The difference between the low and middle latitudes in this respect may be due to the presence of the trade-wind inversion in the tropics (see section 7.1) and the higher windspeeds of mid-latitudes which cause greater orographic lifting.

Mountainous areas are some of the windiest places on earth. The winds protrude into the high atmosphere, where there is less friction to retard wind movement, and may be compressed and channelled round peaks. Wind is a major environmental stress for plants, so much alpine vegetation has a low-lying habit and often survives only in sheltered micro-habitats.

Mountains also create some of their own wind systems (figure 8.3). Winds are common that blow upslope and up valley during the day and in the reverse direction at night. The driving force of such winds is differential heating and cooling, which produces density differences between slopes and valleys and between mountains and plains. During the sunlight hours the slopes are warmed more than the air at the same elevation in the valleys, and the warm air, being less dense, moves upward as an *anabatic wind* along the slopes. At night when the air cools it moves downwards, creating a *katabatic wind* and causing a temperature inversion. An important variant on these thermal slope winds is the *glacier wind*, which arises as the air adjacent to an icy surface is cooled and moves downslope because of gravity. These may blow continuously, as the refrigeration source is always present, but they tend to reach their greatest depth and intensity at mid-afternoon, when the thermal contrast is greatest.

In addition to thermal winds, there are winds caused by barrier effects. These are generally called *föhn winds,* though in some parts of the world they have their own names (e.g. Chinook in North America, and north-wester in New Zealand). They are characterised by a rapid rise in temperature and gustiness and an extreme dryness that puts stress on plants and animals and creates a fire hazard. A typical situation in which a föhn might develop consists of a ridge of high pressure on the windward side of the mountain range and a trough of low pressure on the leeward side, creating a steep pressure gradient across the range. Air is forced up the mountain slope and condensation begins. As the air continues to rise it cools at what is called the *wet adiabatic rate* (*c*.1.7 °C per 300 m). This is a relatively slow rate of cooling because latent heat of condensation is being added to the air. On the lee side of the summit, precipitation ceases and the air starts to descend. It begins to warm as it descends, but does so at the relatively quick *dry adiabatic rate* of about 3.05 °C per 300 m. Consequently the air has the potential for arriving at the valley or plain level on the leeward side much warmer and without much of its moisture, compared with its state at the same elevation on the windward side.

A further consequence of the barrier effect on winds is the creation of *lee waves* (figure 8.3d). When wind passes over an obstacle – in this case a mountain – its normal flow is disrupted and a train of waves may be created that extend downwind for considerable distances. Such lee waves may be associated with windspeeds exceeding 160 km h^{-1} and are often marked by lenticular clouds that form at the crest of waves. Lee waves may also have *rotors,* roll-like circulations that develop to the immediate lee of mountains, usually forming beneath the wave crests.

8.3 Snow and the Snow-line

At higher elevations on mountains much of the precipitation falls as snow – snow which nourishes glaciers, snow banks and avalanches, provides the basis of winter sports, affects plants and animals, and provides water supplies to some of the world's great rivers.

On the highest peaks the snow may be permanent, in the sense that it persists on the ground throughout the year; whereas at lower levels it will be seasonal, and melt in the warmth of summer. The zone between permanent and seasonal snow is called the *snow-line*, though because it changes from year to year and from slope to slope it is often difficult to define with any precision. In general, the

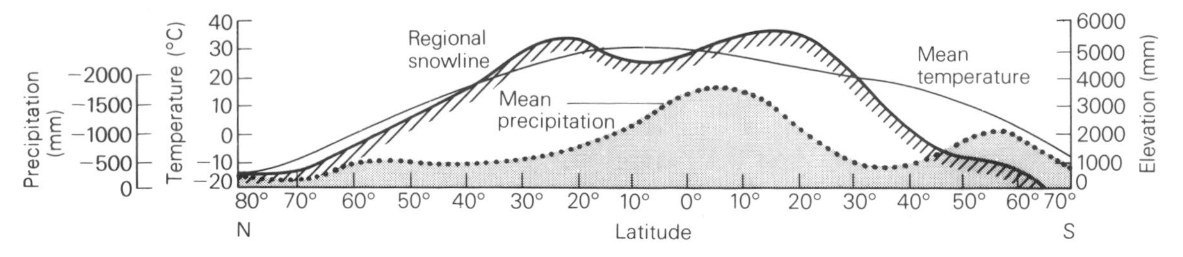

Figure 8.4 Generalised altitude of the snow-line on a transect across the world from north to south. Note the slightly lower snow-line elevation in the tropics as a result of the increased precipitation and cloudiness in these regions.

snow-line corresponds to the level where the average temperatures are 0 °C or less during the warmest month of the year, but it is also related to the amount of precipitation there is. If there is more snow, then the snow-line occurs at a lower altitude because it requires more heat to cause large quantities of snow to melt.

The regional snow-line is lowest in the high latitudes (figure 8.4), where it may occur at sea level, and highest in the tropics, where it tends to occur between 5000 and 6000 m. The highest snow-lines are found in the drier parts of the Tibetan Plateau and the Andes, where they reach levels as high as 6500 m. Indeed, the greater cloudiness and precipitation of the equatorial belt means that the snow-line may be lower near the Equator than in those areas between 20° and 30° N and S, which are under the influence of the subtropical high-pressure cells. At any given latitude, however, the snow-line is generally lowest in areas of heavy precipitation, such as coastal locations, and highest in areas of low precipitation, such as are encountered in continental interiors. Because of the nature of the prevailing winds, there is also a tendency for snow-lines to rise in elevation towards the west in the tropics and the east in middle latitudes.

8.4 Mountain Vegetation

At the end of the eighteenth century, Alexander von Humboldt, the great German scientist and geographer, visited South America and climbed some of the mountains in the tropical sector of the Andes. He recognised that both vegetation and climate changed in a vertical sequence, and he suggested that this vertical sequence from tropical rainforest at low altitudes to permanent snow at the highest elevations was a microcosm of similar zones with changing latitude from tropics to the poles. In a very broad sense there is some truth in this analogy, but in many important respects the altitudinal pattern and the latitudinal pattern are very different. In particular, the only similarity between a high tropical mountain climate (the *tierra fria* of Humboldt) and a polar climate is that of the average annual temperature. In terms of length of day, diurnal fluctuations in temperature and seasonal variability, they are totally different, and plant life reflects these differences.

The primary characteristic of mountain vegetation is the presence of sequential plant communities with increasing altitude. There is a tendency towards smaller and less elaborate plants with slower growth rates, decreased productivity, lower diversity and less inter-species competition. Forests on the lower slopes are called *submontane* or *montane forests*; the higher forest composes the *subalpine zone*; and the treeless area above is known as the *alpine zone.*

Northern Hemisphere mountain forests in middle and high latitudes are dominated by needle-leaf evergreen conifers – pine, spruce and fir. These species are closely related to those of the great belt of boreal forest (see section 5.7), which stretches south of the Arctic tundra across North America and Eurasia. On the lower and intermediate slopes of humid regions in the middle and lower-middle latitudes, broadleaf deciduous forests are the dominant vegetation type. Areas in which they occur include parts of western Europe, the eastern United States and eastern Asia. Dominant species include oak, maple, beech and chestnut. Southern Hemisphere

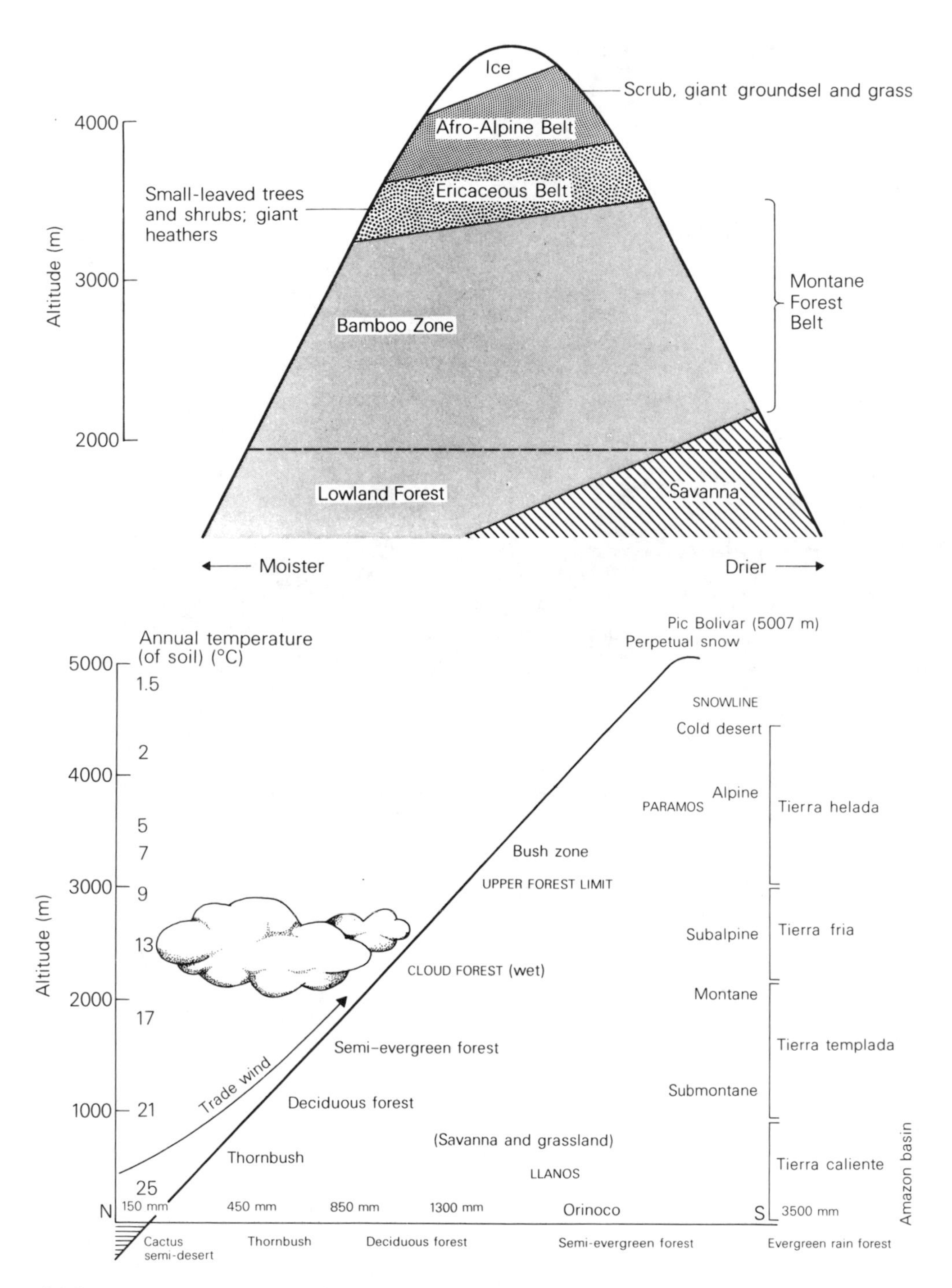

Figure 8.5 (a) Generalised representation of vegetation belts at different altitudes in the equatorial mountains of East Africa; (b) the zonation of vegetation with altitude in a north–south transect across Venezuela.

Plate 8.2 The high mountain peaks of East Africa have very distinctive vegetation. On the slopes of Mount Kenya there is a belt with various types of *Senecio*. The tall plants are the giant tree groundsel (*Senecio keniodendron*), while the bushier plants are cabbage groundsel (*Senecio brassica*). These plants occur at around 4300 m, while the peaks in the background, with the Diamond Glacier in between, lie at just below 5200 m.

mountain forests are composed primarily of broadleaf evergreen trees and may contain large numbers of tree ferns.

The generalised sequence encountered on the mountains of East Africa astride the Equator is shown in figure 8.5a. At the lowest levels, below around 2000 m, there is normal lowland forest or savanna, according to whether the area is moist or dry. Above this is the montane forest belt, distinguished by an abundance of broadleaved hardwood trees with some conifers. In some of the wetter areas the montane forest may contain bamboo (*Arundinaria alpina*), especially in the higher parts of this belt. Above the montane forest belt is the ericaceous belt, characterised by many small-leaved trees and shrubs, especially members of the Ericaceae, including giant heathers. Between that belt and the ice caps on the mountain summits is the Afro-alpine belt (plate 8.2), which includes various scrub, giant groundsel (*Senecio*) and grassland communities.

Figure 8.5b illustrates another low-latitude zonation of vegetation with altitude, based on a north–south transect across Venezuela. In that country,

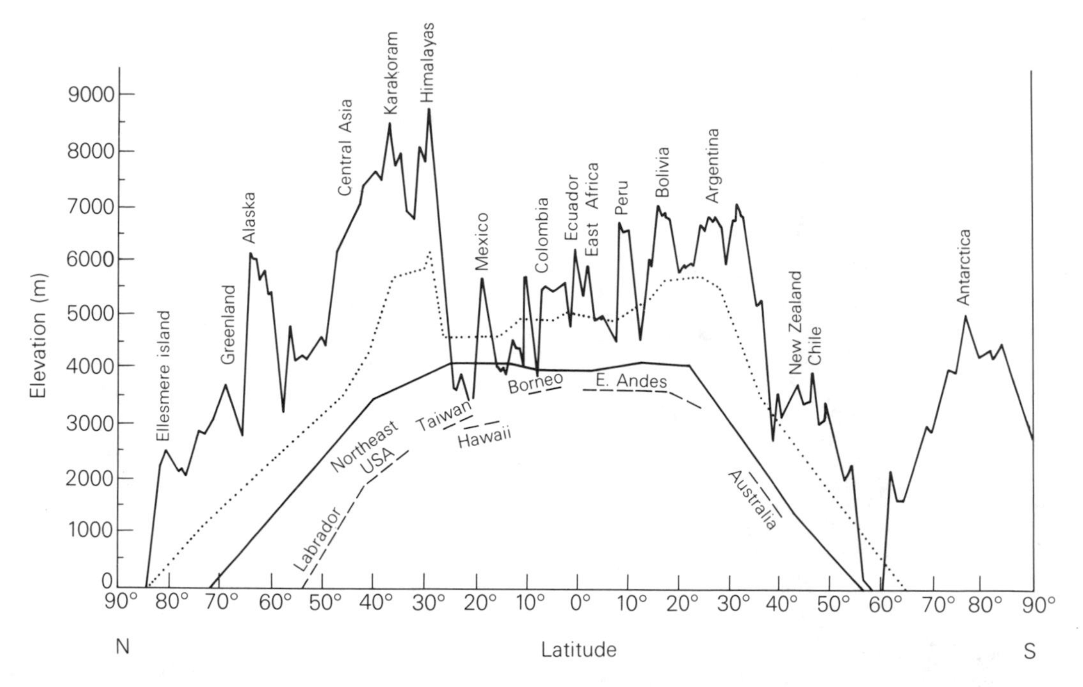

Figure 8.6 Schematic north–south view of the highest summits (top solid line) and the upper limit of vascular plants (dotted line). Timber-line altitudes are shown both for drier continental situations (lower solid line) and for the wetter marine situations (dashed line).

the vegetation communities at sea level range from semi-desert through a thorn bushland to deciduous, semi-evergreen and eventually evergreen rainforest in the Amazon Basin. With increasing altitude, the classic zones of Humboldt may be recognised: *tierra caliente, tierra templada, tierra fria* and *tierra helada.*

8.5 The Timber-line

The transition from forest to treeless tundra on a mountain slope is often a dramatic one. Within a vertical distance of just a few tens of metres, trees disappear as a life form and are replaced by low shrubs, herbs and grasses. This rapid zone of transition is called the upper *timber-line* or *tree-line*. In many semi-arid areas there is also a lower timber-line where the forest passes into steppe or desert at its lower edge, usually because of lack of moisture.

The upper timber-line, like the snow-line, is highest in the tropics and lowest in the polar regions (figure 8.6). It ranges from sea level in the polar regions to 4500 m in the dry subtropics and 3500–4000 m in the moist tropics. Timber-line trees are normally evergreens, suggesting that these have some advantage over deciduous trees in the extreme environments of the upper timber-line. There are some areas, however, where broadleaf deciduous trees form the timber-line. Species of birch (*Betula*), for example, may occur at the timber-line in parts of the Himalayas. The timber-line vegetation of tropical mountains is dominated by the Ericaceae (the heath family). They grow as evergreen shrubs and trees with characteristically tough and leathery leaves.

At the upper timber-line the trees begin to become gnarled, stunted, prostrate and deformed – characteristics to which the Germans give the name *krummholz* (plate 8.3). This is particularly true for trees in the middle and upper latitudes, which tend to attain greater heights on ridges, whereas in the tropics the trees reach their greater heights in the valleys. This is because middle and upper latitude timber-lines are strongly influenced by the dura-

Plate 8.3 At high altitudes, trees, like these *Pinus flexilis* in Colorado, USA, may have their shapes deformed by wind, an example of krummholz.

tion and depth of the snow cover. As the snow is deeper and lasts longer in the valleys, trees tend to attain greater heights on the ridges, even though they are more exposed to high-velocity winds and poor, thin soils there. In the tropics, the valleys appear to be more favourable because they are less prone to dry out, they have less frost and they have deeper soils.

There is still no universally agreed explanation for why there should be such a dramatic cessation of tree growth at the upper timber-line. Various environmental factors may play a role. Too much snow, for example, can smother trees, and avalanches and snow creep can damage or destroy them. Late-lying snow reduces the effective growing season to the point where seedlings cannot establish themselves. Wind velocity also increases with altitude and may cause serious stress for trees, as is made evident by the krummholz types. Some scientists have proposed that the presence of increasing levels of ultraviolet light with elevation may play a role, while browsing and grazing by animals like the ibex may be another contributing factor. Probably the most important environmental factor is temperature, for if the growing season is too short and temperatures are too low, tree shoots and buds cannot mature sufficiently to survive the winter months.

8.6 Alpine Tundra

Above the tree-line there is a zone that is generally called *alpine tundra*. Immediately adjacent to the timber-line, the tundra consists of a fairly complete cover of low-lying shrubs, herbs and grasses (plate 8.4), while higher up the number and diversity of species decrease until there is much bare ground with occasional mosses and lichens and some prostrate cushion plants. Some plants can even survive in favourable micro-habitats above the snow-line. The highest plants in the world occur at around 6100 m on Makalu in the Himalayas, where rocks, warmed by the sun, melt small snowdrifts.

The most striking characteristic of the plants of the alpine zone is their low growth form. This enables them to avoid the worst rigours of high winds, and permits them to make use of the higher temperatures immediately adjacent to the ground surface. In an area where low temperatures are limiting to life, the importance of the additional heat near the surface is crucial. The low growth form can also permit the plants to take advantage of the insulation provided by a winter snow cover. In the equatorial mountains the low growth form is less prevalent.

Another common characteristic of the alpine plants is that they tend to be perennials, rather than annuals. Alpine plants have to be able to make use of the limited growing season, and annuals find it difficult to germinate, flower and fruit in such a short period, whereas perennials do not have to spend time in germination and initial growth processes. The perennials also have large and extensive root systems, like many desert plants. These roots are useful during droughts; they act as reserves of food, and provide nutrients to the plant for rapid initial growth in the spring.

Plate 8.4 Above the tree-line is the alpine zone with low plants. This may provide summer pasture. In the nineteenth century glaciers almost reached to this meadow just above the village of Arolla in the Valais of Switzerland.

8.7 Mountain Hazards

Mountains have always been regarded with awe, but today their majestic scenery makes them prime country for recreation. More and more people visit them, and frequently they underestimate the hazards that the environment can pose. Mountain weather is often harsh, and at high altitudes people can suffer from severe physiological conditions. In particular they may suffer from oxygen deficiency, which often reveals itself at about 2800–3000 m as a slight breathlessness and can produce *mountain sickness*, symptoms of which include headache, dizziness, nausea, lack of appetite and insomnia. A further health hazard is a pulmonary oedema, in which fluid accumulation in the lungs impairs oxygen transfer into the blood. Wind chill applies in cold, windy mountain areas as much as it does in the polar regions (see section 4.1), and frostbite and snow blindness are additional hazards. Rapid changes of weather, and the sudden appearance of fog and cloud, frequently take climbers unawares. Exposed ridges and summits may be especially prone to lightning strikes, and elsewhere snow avalanches can be a severe hazard (see section 4.14) (windows 8.1 and 8.2). Air pollution may occur in some mountain valleys as a result of topographically caused inversions, and on steep valley sides mass movements such as rockfalls may occur with some frequency (see chapter 12). The high relief, oversteepening by glacial excavation, the presence of torrential streams at the base of slopes, the existence of rocks shattered by tectonic activity, the powerful operation of freeze–thaw processes, the existence of earthquakes and the frequent occurrence of high precipitation levels – all contribute to

Window 8.1 *Avalanches in the European Alps*

Avalanches in the European Alps cause considerable loss of life. Between 1975 and 1989, there were 1622 victims. France was the greatest sufferer, with 413 deaths (30 per year). The number of deaths varies from year to year, with a peak of 180 in 1984–5 and only 58 in 1988–9.

Skiers account for most of the fatalities, with 1259 (about 78 per cent) out of the total of 1622: 158 climbers also died. Comparatively few people (49) were killed in their houses or by avalanches hitting them while travelling on roads. Most deaths occur during the ski season. At Chamonix around 80 per cent of avalanches are recorded in January, February and March.

The hazardous nature of these mountains is evidently the result of humans choosing to expose themselves, by skiing and climbing, to unpredictable and potentially lethal events.

The Lower Arolla Glacier Ice Fall in Switzerland. In (a) one can see the area prior to the avalanche. In (b) one can see the avalanche taking place, with snow tumbling from Mont Collon down a chute on to the glacier. The avalanche is located approximately in the centre of the photo, which was taken in August 1966.

the instability of slopes in mountainous environments. Communication lines and settlements need to be located with care.

This is illustrated by a consideration of the Karakoram Highway, a strategic road which links China with Pakistan across the Karakoram Mountains at a height of 4604 m (see colour plate 15). Largely built – with great loss of life – in the 1970s, it has had to contend with a formidable series of problems. Its course has been threatened by the rapidly fluctuating positions of the glacier snouts; bridges and stretches of the road have been washed

Window 8.2 *The Chamonix Valley avalanche disaster, 1999*

On Tuesday 9 February 1999 at 2.40 pm a large mass of snow and ice broke away from a mountain – the Montagne de Peclerey – on the sides of the glacially steepened Chamonix Valley in the Alps of France. This avalanche, triggered by exceptionally heavy snowfalls (there had been over 2 m in the area since the previous Saturday), sped towards the valley floor, crossed the River Arve, and crashed into the opposite valley side, demolishing 17 chalets in the hamlets of Le Tour and Montroc. Being the height of the ski season, ten people were killed and five more were badly injured. It was the worst avalanche disaster in the valley in the twentieth century.

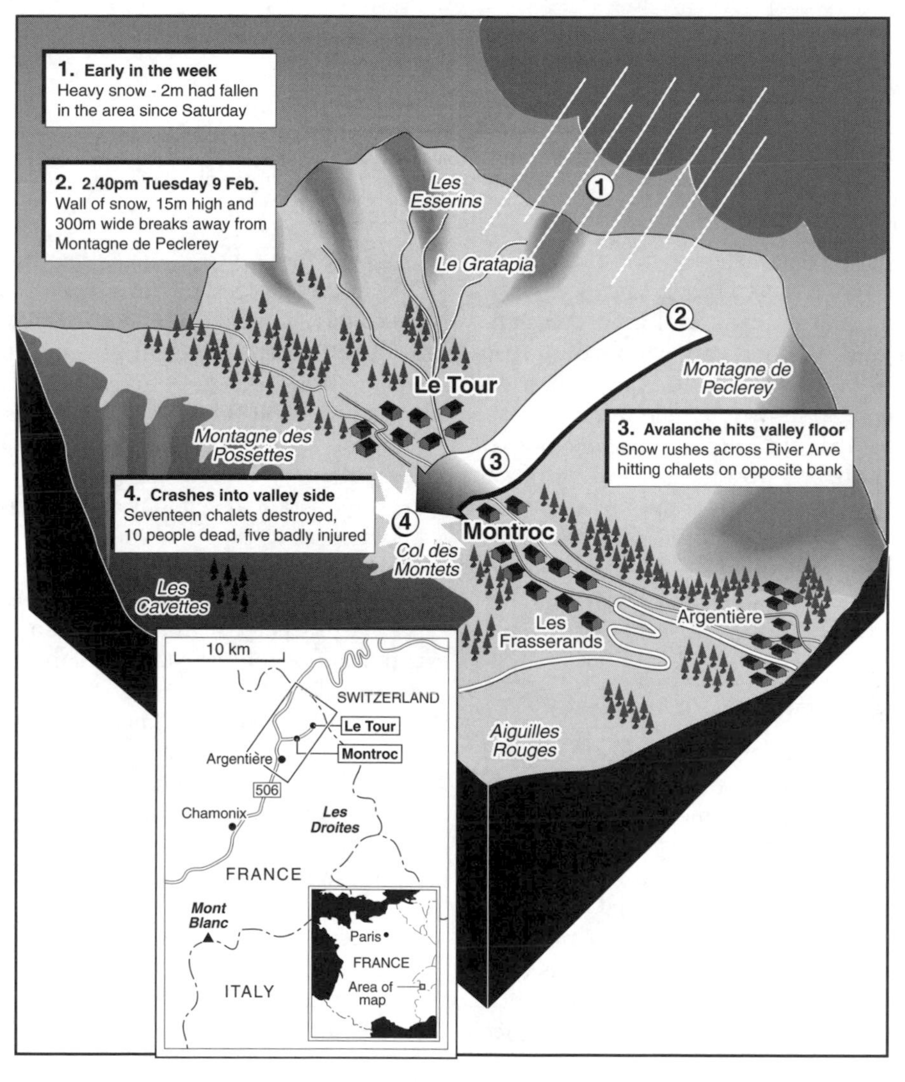

The Montroc avalanche, 1999.

Table 8.2 Distribution of major natural disasters in mountain regions 1953–1988

Region	*Earthquake*	*Volcano*	*Landslide*	*Avalanche*	*Flood*	*Storm*	*Total*
Mediterranean Basin (excl.) Turkey	21	1	4	—	13	1	40
South-west/south Asia	49	—	15	2	44	3	113
East Asia	11	2	19	2	14	10	58
South-east Asia, Australia and Oceania	21	8	7	1	9	18	64
Africa	3	2	1	—	5	2	13
Europe (excl. Mediterranean)	2	—	4	6	5	2	19
South/Central America	33	5	16	7	22	4	87
North America	14	1	4	—	6	1	26
Total	154	19	70	18	118	41	

The events are those reported in *The New York Times*, *Globe* and *Mail* (Toronto) and *The Times* of London. Thresholds of damage were at least 10 deaths and 50 injuries, or more than US$ 1 million damage and emergency assistance from outside the damage zone.
Source: K. Hewitt (1997), 'Risk and disasters in mountain lands', in B. Messerli and J. D. Ives (eds), *Mountains of the World: A Global Priority* (New York: Parthenon), pp. 371–406.

away by capricious outwash streams; in the gorge of the Hunza River it has to face great flood waves caused by the collapse of natural landslide dams across the river; dust storms are whipped up from the glacial outwash plains; in the winter there are avalanches and deep snow falls; and because of the steep, unstable slopes, the road is frequently blocked by rockfalls and debris flows. Teams of workmen labour continually to keep the road open.

On a global basis a review of major natural disasters in mountain regions has indicated that the three most serious types of event are earthquakes, floods and landslides (table 8.2).

8.8 Changing Climates

Because of the intimate relationships between altitude, climate, vegetation and the operation of geomorphological processes, mountains have been particularly exposed to the effects of past climatic changes and are likely to be affected by any future climatic changes associated with global warming.

The effects of past climatic changes on mountain ecosystems can be demonstrated by a consideration of the way in which vegetation belts shifted altitudinally during cold glacial phases on tropical mountains (figure 8.7). Detailed pollen analysis from numerous lakes and swamps in New Guinea and the Colombian Andes of South America shows that over the past 30 000 years the major vegetation zones have moved through as much as 1700 m. The boundary between the forest and the alpine zone above it was low before 30 000, shows a slight peak (of uncertain height and imprecise date) between 30 000 and 25 000, reaches an especially low point at 18 000–15 000 (more or less equivalent to the Glacial Maximum in higher latitudes), shows a steep climb as climate ameliorated between 14 000 and 9000 BP and reaches modern altitudes or slightly above them by about 7000 BP.

Similarly, we can consider the way in which alpine valley glaciers in mountainous regions have reacted in the past hundred or so years since the ending of a cold phase called the 'Little Ice Age' (see window 2.6).

Since the nineteenth century, many of the world's alpine glaciers have retreated up their valleys as a consequence of the climatic changes, especially warming, that have occurred. Studies in the changes of snout positions obtained from cartographic, photogrammetric and other data therefore permit estimates to be made of the rate at which retreat can occur. The rate has not been constant, nor the process uninterrupted. Indeed, some glaciers have shown a tendency to advance for some of the period. However, if one takes those glaciers that have shown a tendency for a fairly general retreat (table 8.3), it becomes evident that as with most geomorphological phenomena there is a wide range of values, the variability of which is probably related

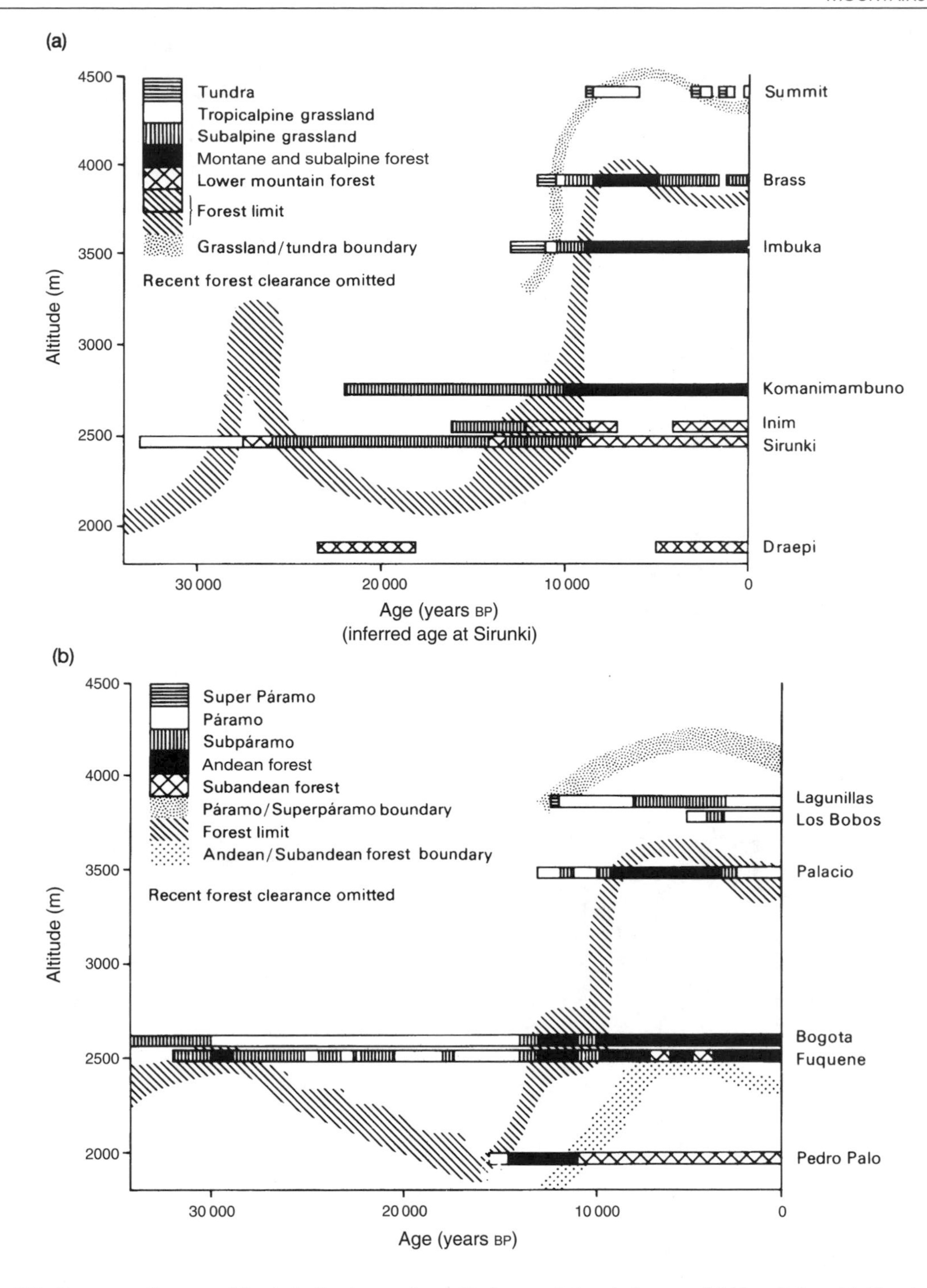

Figure 8.7 Summary diagram of the Late Quaternary low-latitude vegetational changes: (a) New Guinea highlands; (b) Colombian Andes.

Table 8.3 Retreat of glaciers in the twentieth century (metres per year)

Location	*Period*	*Rate*
Breidamerkurjökull, Iceland	1903–48	30–40
	1945–65	53–62
	1965–80	48–70
Lemon Creek, Alaska	1902–19	4.4
	1919–29	7.5
	1929–48	32.9
	1948–58	37.5
Humo Glacier, Argentina	1914–82	60.4
Franz Josef, New Zealand	1909–65	40.2
Nigardsbreen, Norway	1900–70	26.1
Austersdalbreen, Norway	1900–70	21.0
Abrekkbreen	1900–70	17.7
Brikdalbreen	1900–70	11.4
Tunsbergdalsbreen	1900–70	11.4
Argentière, Mont Blanc	1900–70	12.1
Bossons, Mont Blanc	1900–70	6.4
Oztal Group	1910–80	3.6–12.9
Grosser Aletsch	1900–80	52.5
Carstenz, New Guinea	1936–74	26.2
Rocky Mountains	1890–1974	15.2
Spitzbergen	1906–1990	51.7
Iceland	1850–1965	12.2
Norway	1850–1990	28.7
Alps	1850–1988	15.6
Central Asia	1874–1980	9.9
Irian Jaya	1936–1990	25.9
Kenya	1893–1987	4.8
New Zealand	1894–1990	25.9

to such variables as topography, slope, size, altitude, accumulation rate and ablation rate. It is also evident, however, that rates of retreat can often be very high, being of the order of 20–70 $m a^{-1}$ over extended periods of some decades in the case of the more active examples. It is therefore not unusual to find that over the past hundred or so years alpine glaciers in many areas have managed to retreat by some kilometres.

What may happen to mountain environments if further warming takes place in coming decades as a result of the enhanced greenhouse effect? Altitudinal belts are likely over time to migrate by around 160 m for each 1 °C rise in temperature (400 m if a warming of 2.5 °C takes place). Glaciers will continue to waste away as will mountain permafrost. The depth and duration of snow cover will change, thereby affecting winter sports and tourism, and the changing state of snow, ground ice and glaciers will have an impact on runoff and floods. Some plants and animals may find that their habitats shrink or disappear altogether as upward movement occurs.

■ *Key Terms and Concepts*

adiabatic heating
lapse rate
snow-line
temperature inversions
timber-line

■ *Points for Review*

- Select a mountain range of your choice and suggest how its climate differs from that of its neighbouring lowland.
- In what way does the nature of the vegetation change as you ascend a mountain?
- If you were building a mountain highway, what are the geomorphological hazards with which you might have to contend?
- How may mountain environments change in a warmer world?

FURTHER READING

Barry, R. G. (1992) *Mountain Weather and Climate*, 2nd edn (London: Routledge). A splendid survey of all aspects of mountain weather and climate.

Gerrard, A. J. (1990) *Mountain Environments* (London: Belhaven Press). A valuable point-of-entry survey of wide scope.

Messerli, B. and Ives, I. D. (eds) (1997) *Mountains of the World: A Global Priority*. (New York: Parthenon). A study of mountain environments and their sustainability, edited by two of the leading figures in mountain environment research.

Stone, P. B. (ed.) (1992) *The State of the World's Mountains: A Global Report* (London: Zed Books). A study of ecological issues in a mountain context.

9 Coasts

9.1 Coastlines

The world's coastlines, which have a length not far short of half a million kilometres, are an important habitat for man, for it has been estimated that around two-thirds of the world's population lives within a few kilometres of a coast.

Coastlines show a tremendous diversity. This is the result of the climatic conditions under which they occur, their history of sea-level change, the geological structures that lie behind them, the sediments that are available to make their beaches and the nature of the waves, currents and tides that mould them. Even within the small compass of Britain, one can contrast the sinuous inlets of the southwest, the great sea lochs of Scotland, the low glacial coastline of East Anglia, the marshes of the Thames Estuary and the imposing chalk and limestone cliffs of the south coast.

Given this great variety, it is not surprising that there have been many attempts to classify coasts on the basis of numerous different criteria. One of the best known classifications was that of the American geomorphologist D. W. Johnson, who believed that the most important division was between those coasts that had a history dominated by emergence, and those that had been characterised by submergence. He also recognised 'neutral' coasts and 'compound' coasts. Coasts of submergence would include those in which rias and fiords were important, whereas emergent coasts would be characterised by barrier beaches like those of the Baltic (see section 9.3)

An American oceanographer, F. P. Shepard, believed that a useful distinction could be drawn between what he termed 'primary' (youthful) coasts, where non-marine forces were still in evidence as having moulded the landscape, and 'secondary' coasts, where marine action was more prevalent. Primary coasts would include those where land erosion was evident (e.g. rias, fiords, drowned karst), where sub-aerial deposition could be seen (e.g. deltas – see section 15.9 – sand dunes, landslides, submerged glacial features), or where the effects of volcanism and tectonic movements could be traced. Secondary coasts would be those either where marine erosion had greatly modified the coast (e.g. wave-straightened coasts) or where marine deposition had occurred (e.g. barrier coasts, cuspate forelands, mud flats), or where the coasts had been built up by marine organisms like coral or mangrove (see chapter 7).

More recent classifications have recognised the importance of the coast's position *vis-à-vis* plate tectonic boundaries (see section 1.8) and the amount of energy supplied to different coastlines by different types of wave attack (see section 9.2). It has also become increasingly evident in recent years that virtually all coastlines have undergone a very complex series of sea-level changes during the Pleistocene and Holocene, so that any simple division into emergent or submergent coasts must be fraught with difficulty, as must any attempt to explain modern beaches and coasts without reference to their past history. The nature of world sea-level changes and the effects of ice caps have been discussed in sections 2.11 and 4.8, respectively. For the effects of tectonics, see section 11.4

9.2 Waves

Waves are crucial for an understanding of the geomorphology of beaches and coasts because waves bring most of the input of energy into the beach system. Many of the changes that one can observe result from waves of different characteristics – their height and wavelength – and the direction from which they come.

Waves obtain their energy from wind, and the wind may occur either quite near to the coast where the waves break or at a vast distance from it. This is because, once large waves have been formed by the wind, they will travel across the oceans, slowly losing height (and therefore energy), until eventually they reach shallow coastal waters and break against the land. In this way, waves generated by strong, persistent winds in the southern Pacific Ocean, for example, may extend as far as the coast of Alaska.

Before we progress further, a few terms relating to waves should be introduced. *Wavelength* is the horizontal distance separating two wave crests or two wave troughs; the *velocity* is the distance travelled by the wave in a unit period of time; the *period* of a wave is the length of time required for two crests or two troughs to pass a fixed point; and the *frequency* is the number of periods that occur within a set interval of time – say a minute.

It is important to remember, when we see a procession of waves coming in from the ocean, that it is the form of the wave that moves forward through the water and not the water itself. It is rather like the rippling motion that wind makes as it blows across a field of corn: waves follow one another across the stalks of wheat, and yet the wheat does not pile up in a heap on the far side of the field! Instead, the motion of the grain results from the nodding of the individual stalks each time a wave passes through them. Similarly, particles within a wave in the ocean do not move forward with the advancing wave itself but follow a circular orbit (figure 9.1).

If the wind that generates the waves is strong and

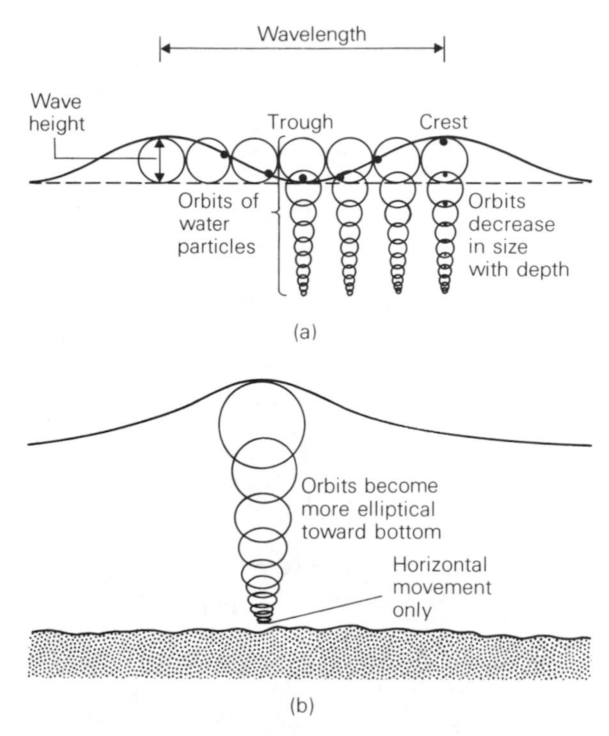

Figure 9.1 The motion of waves. (a) Wave forms are produced by the orbital motions of water particles – each water particle continues orbiting about the same position while the wave form travels. (b) Orbits of water particles become elliptical as they approach a shallow bottom; at the bottom, particles move back and forth only.

reasonably constant, the waves will increase in size as long as the wind blows, and will gain their energy from the wind. In enclosed seas the limitation on the largest size of wave that can be produced is the distance of open water in a straight line, known as the *fetch*. High winds will generate waves related to fetch by the formula:

$$H = 0.36 \sqrt{F}$$

where H is the wave height in metres, and F is the fetch in kilometres. For long fetches, H is generally held to be proportional to U^2, where U is the wind speed:

$$H = 0.0024\ U^2$$

Once again, H is the height in metres, and U is the wind speed in kilometres per hour.

Plate 9.1 The beach at Durdle Door in Dorset, southern England. It is composed of a series of cusps, and the backwash flows down the beach in definite channels between them.

There appears to be a finite limit to the size that waves can reach. In the open ocean they have occasionally been observed with a height of as much as 34 m, wavelengths of over 800 m, velocities over 120 km h^{-1} and periods of 22.5 s. As they move shorewards, however, waves change in character. They lose energy abruptly when they arrive at the coast, for as they move into shallow water they suffer friction and distortion. This process, which eventually causes the wave to break, seems to start when the depth of water is equal to about half the wavelength. The waves become shortened and their height increases relatively, until the upper part of the wave topples over. The wave breaks at the *plunge line*, where the depth of water and the wave height are approximately equal. Above this line the water rushes shorewards as *swash*, carrying sand with it. On a sloping beach it will run back down as *backwash* (plate 9.1).

The relationship between the shore profile and the waves breaking upon it determines the effect of the waves on beach materials, and hence on beach form. With small, long-wavelength waves and/or a shallow gradient shore profile, the wavefront steepens relatively gradually, and instead of breaking forcefully it spills over. As such waves move up the beach they rapidly lose volume and energy by percolating into the beach materials and thus the backwash has much less energy to remove sediments down the beach. The backwash is also not sufficiently strong, in spite of having gravity on its side, to impede the following wave-break and swash.

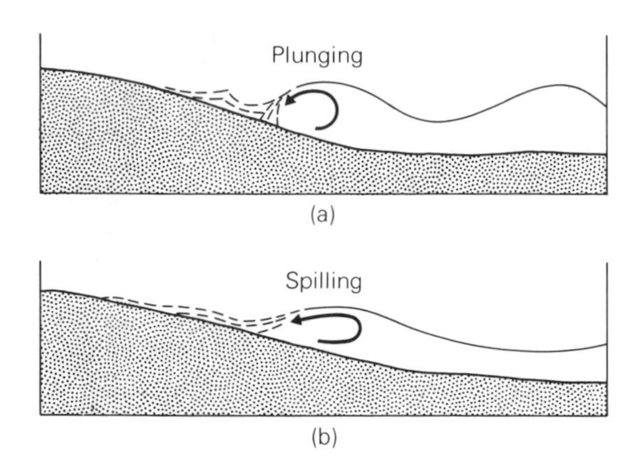

Figure 9.2 Two different types of wave which have great geomorphological significance: (a) destructive; (b) constructive.

These waves therefore tend to be constructive and to be important in building up beaches (figure 9.2).

Destructive waves, by contrast, are those with short wavelengths and high crests occurring on a more steeply sloping shore. Such waves plunge rather than spill, causing erosion and generating a powerful backwash. They comb down the beach and move sand below the tide level. In temperate areas there is often a seasonal change between winter storm wave environments, which cut, and summer low-energy waves, which result in filling.

Beaches produced by short waves tend to be rather steep, whereas beaches formed under the influence of long swells tend to be wide and gentle. The reason for this contrast is related to the volume of water put on to the beach; the large volumes associated with long waves are capable of transporting sediment over a low slope, whereas the shorter waves have a much smaller water volume and hence need a steeper gradient if they are to transport the sediment.

Another important control of beach gradient is the grain size of the material involved. In general, the steepest beaches occur on the coarsest materials. This is because percolation rates increase as sediment becomes coarser, so steeper slopes are required before swash and the greatly diminished backwash are in equilibrium.

The backwash caused by the withdrawal of water carried shoreward by wave action may return down the beach either as *undertow* (sheet flow near the seabed) or in localised *rip currents* (figure 9.3). These are threads of water about 30 m wide, which flow through the breaker line at velocities of up to 8 km h^{-1} before dispersing seaward. A light or moderate swell produces a few concentrated rips, fed by strong lateral currents in the surf zone. The currents and the rips themselves cut channels along the beach and through any sand bars that may be lying parallel to the shore.

The nature and importance of wave action varies in different environments, and R. A. Davies has classified wave environments into four main types:

(a) storm wave environments;
(b) west coast swell environments;
(c) east coast swell environments;
(d) protected sea environments.

Storm wave environments are those in which gales are frequent – namely, the regions of cyclonic frontal activity in the higher mid-latitudes of both hemispheres. In the high polar regions and the tropical zones gales are infrequent, with the notable exception of tropical cyclones (hurricanes). The waves of this environment are short, high-energy waves of varying direction.

West coast swell environments have waves that are generated at some distance, generally by temperate latitude gales associated with the great belts of westerly winds. This westerly swell is particularly in evidence along such coasts as those of Peru

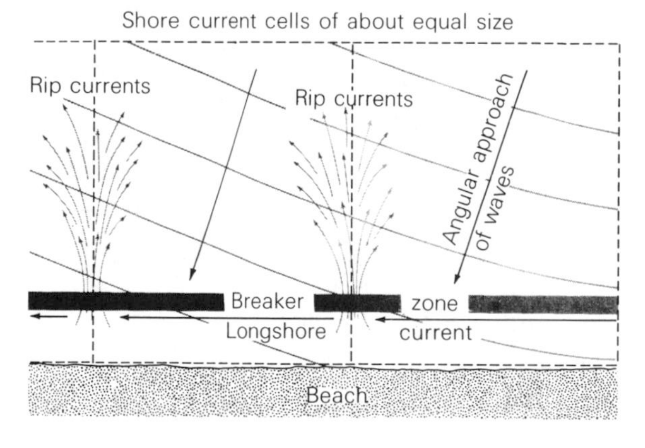

Figure 9.3 Rip currents flow out from the shore when, as a result of oblique wave attack, which forms the longshore current, the water tends to pile up; they break through the surf zone, fan out and dissipate.

and West Africa, where strong winds are few or offshore in direction so that there are not many locally produced waves of geomorphological importance. The waves are long and low and relatively consistent in frequency of occurrence and direction.

East coast swell environments are characterised by waves generated by the trade winds, and they are less subject to the regular swell generated in extra-tropical westerly wind belts. Superimposed on the swell, there may be important waves generated by the hurricanes that are such a feature of the east sides of land masses in low latitudes. With the exception of the cyclone-generated waves, mean energy levels on such coasts are low to moderate, and are generally lower than those of the west coast swell environments.

Protected sea environments are coasts of seas in which there is little penetration of oceanic swell, either because they are enclosed (e.g. the Black Sea) or because they are protected by ice cover (e.g. parts of the Arctic Ocean).

Given variations in wave type and sediment character, it is obvious that there will be a wide range of different beach forms. However, it is worth discussing the form of an idealised beach (figure 9.4). This is composed of two main elements and sundry smaller ones. The first of the main elements is the *upper beach,* which is often composed of coarse materials such as pebbles and, being coarse, tends to have a slope of up to 10°–20°. The second is the *lower beach,* which, being generally formed of sand or even mud, has a low gradient (as low as 2° or less). On many, though not all beaches, there is quite a sharp break between these two elements. The minor elements that may be superimposed on this generalised section are:

(a) the storm beach – a well-defined and semi-permanent ridge standing on the level of highest spring tides;
(b) beach ridges or *berms*, which are built up at successive levels below that of high spring tides by constructive waves;
(c) beach *cusps* (plate 9.1) – small regular embayments developed on the face of the shingle or at the junction of the shingle and sand beaches;
(d) small channels – formed in the sand by water draining from the beach at low tide;
(e) ripples – developed on the sand by wave action or tidal currents;
(f) ridges and runnels – broad and gentle rises and depressions aligned parallel to the shoreline and found to the seaward side of the sand beach.

Waves do more than merely control (with the assistance of sediment size) the gradients and pro-

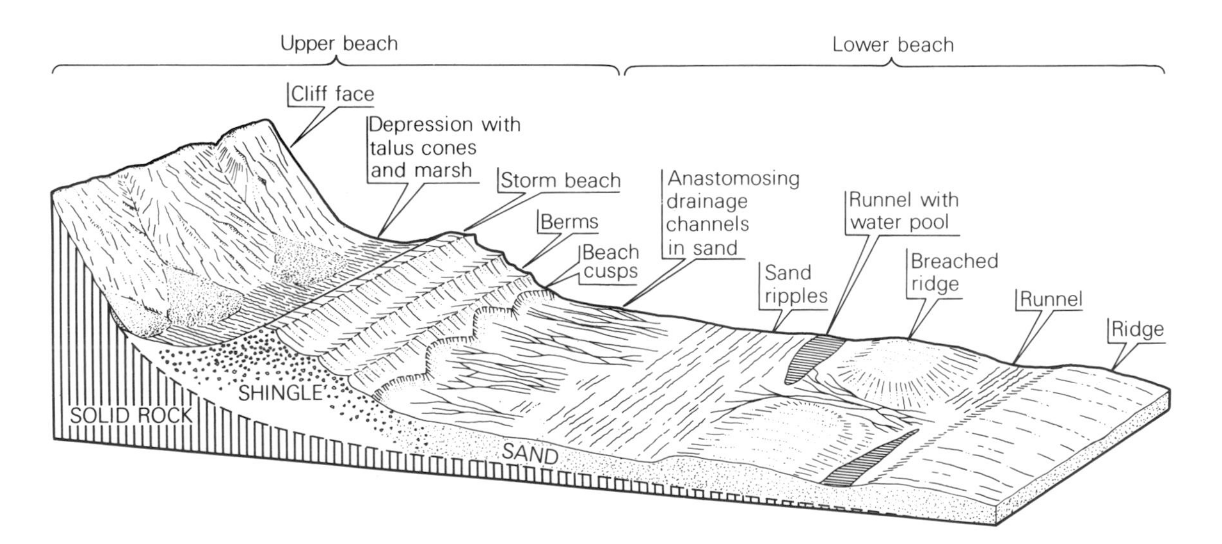

Figure 9.4 The idealised features of a sand and shingle beach.

Plate 9.2 When waves approach the coastline they are refracted as they are slowed down by the shallowing bottom. This air photograph of Start Point in Devon shows the process clearly.

files of beaches. They are also extremely important in moving material along the shore. In an idealised situation, in which waves approach a completely straight coastline with their crests parallel to that line, a given wave will break at the same instant at all points and the swash will ride up the beach at right angles to the same line. Consequently, particles of sand will move up and down the beach slope along a fixed line. In the real world, however, such simplicity is seldom found, either because the shoreline is not completely straight, or because the waves approach the coast at an oblique angle, or because the depth of the water offshore varies. As waves move towards the coast they are *refracted* (plate 9.2) as they are slowed down by the shallowing bottom. As the sections of the waves over the shallowest areas fall behind those still in deeper water the wave crests are distorted. This has an effect on the distribution of wave energy along the crest; for converging *orthogonals* (lines drawn at right angle to the wave crests) indicate a concentration of wave energy, and diverging orthogonals indicate areas of lower wave energy. Headlands will thus tend to be subjected to greater energy levels than the bays in between (figure 9.5).

Although wave refraction tends to cause wave crests to become curved in plan and to become more nearly parallel to the shoreline, the waves generally arrive in the breaker zone at an oblique approach. The swash therefore rides up the beach at an oblique angle, moving sediment with it. After the swash has spent its energy, the backwash flows down the slope of the beach under the influence of gravity, and so sediment moves back down the beach at a different angle to the shoreline from the one it went up at. As a result, over a period of time

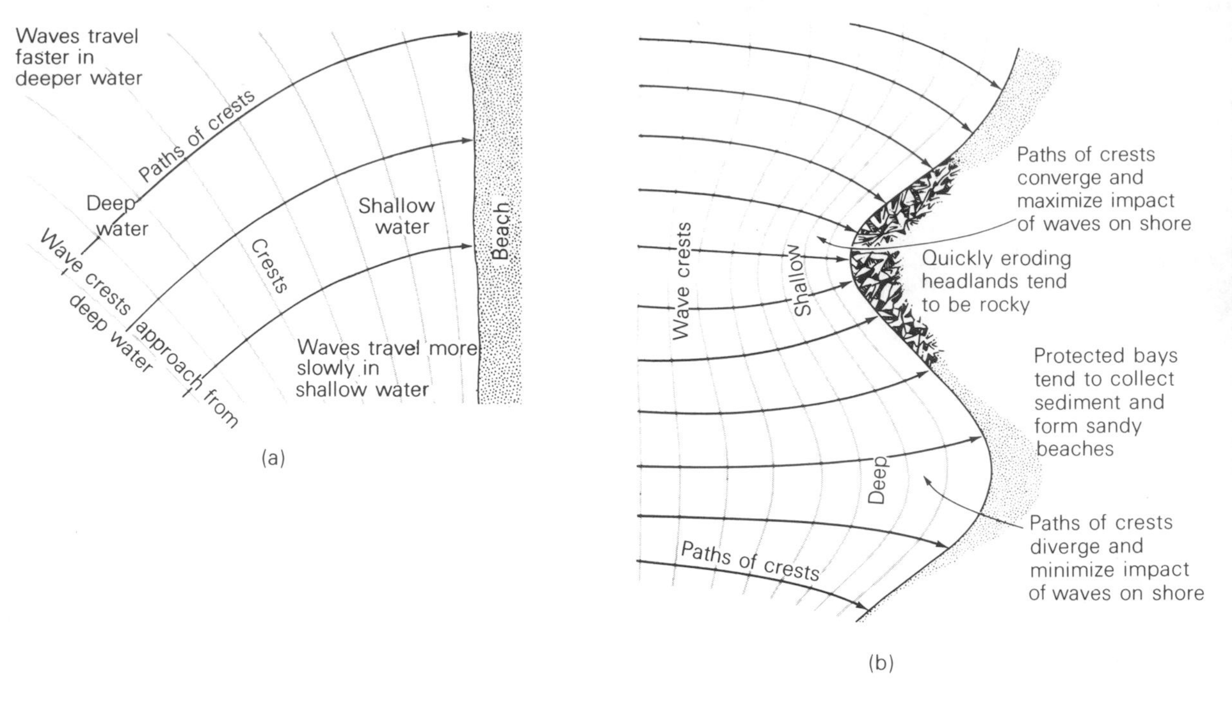

Figure 9.5 Wave refraction. (a) The part of the wave that first encounters shallow water slows in speed, while that part of the wave still in deeper water continues to move with its original speed, making the wave change angle. (b) Because of the configuration of shallow and deep water around headlands and bays in a shoreline, wave energies are concentrated at projections and dispersed at indentations.

the waves move sediment along the beach, by a process called *littoral* or *longshore drift*. Such littoral drift forms some important landforms (figure 9.6).

When a coastline possesses an embayment like an estuary or ria, littoral drift will tend to build a narrow beach across the embayment. Such a beach is called a *spit*. Because of the wave refraction in the embayment across which the spit is growing, sediment is carried around the spit end and so a landwards curvature develops. If the spit continues to grow it may evenually seal off the embayment, forming a *baymouth bar*. Another landform produced by the same process is the connection of an island to the mainland by the growth of a *tombolo*. If the littoral drift converges from opposite directions upon a given point on a shoreline, sediment may accumulate in the form of a *cuspate bar*, which may eventually expand to a *cuspate foreland.*

9.3 Barrier Beaches and Related Forms

The deposition of beach material offshore, or across the mouth of inlets or embayments, in such a way as to form barriers extending above the normal level of highest tides and partly or wholly enclosing lagoons is a widely distributed phenomenon. While some may form by the longshore growth of spits, others appear to develop offshore, especially where there is abundant sand to form the barrier, a suitably low gradient on which it can be formed and constructive waves associated with swell conditions. It has been estimated that barrier coasts make up about 13 per cent of the world's coastline. The longest stretches of barrier coastline are along the eastern coast of the United States and in the Gulf of Mexico (figure 9.7), but they are also common elsewhere, particularly on the shores of seas with a low tidal range, like the Mediterranean and the Baltic. They are less common in areas with high tidal range,

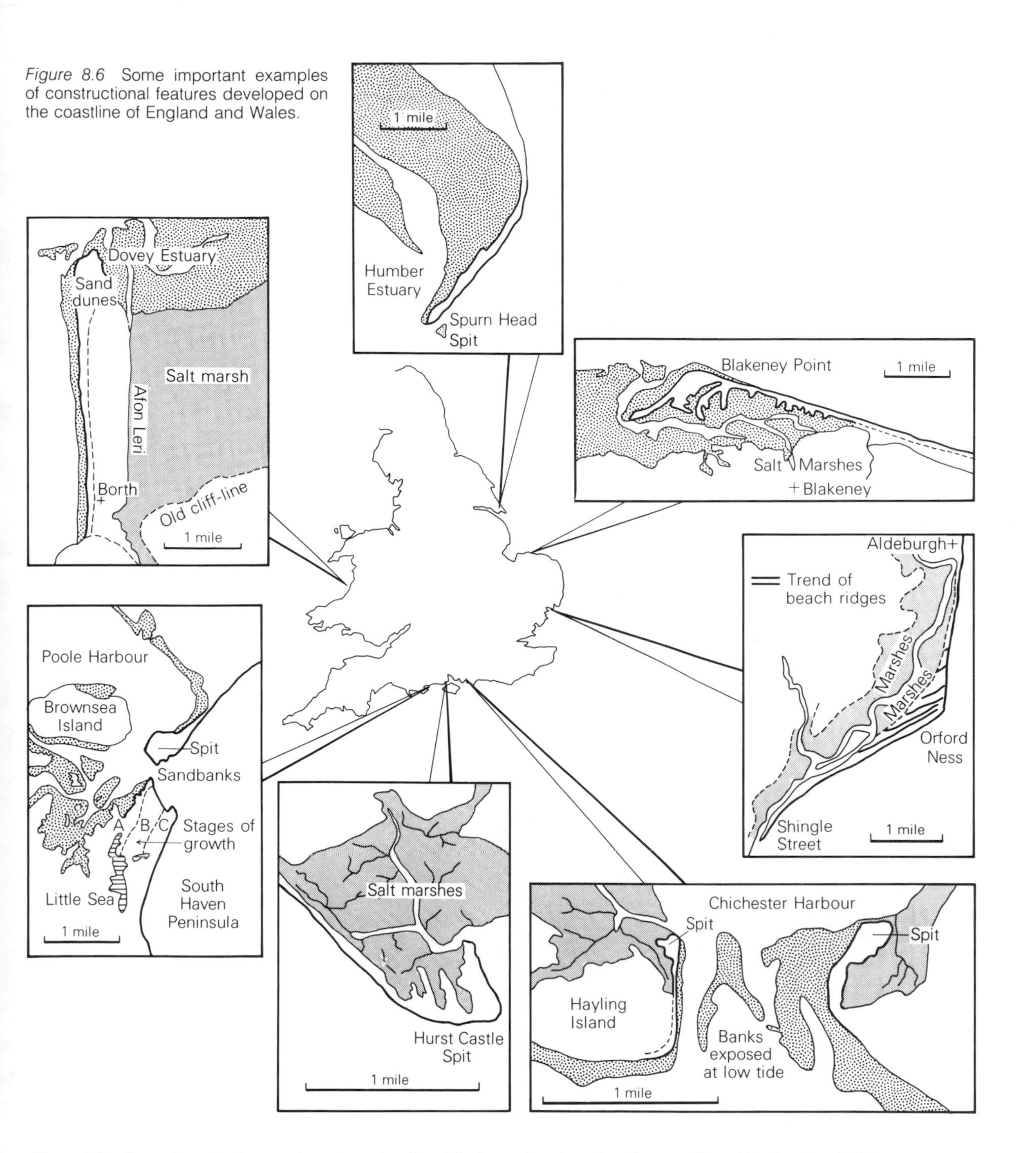

Figure 8.6 Some important examples of constructional features developed on the coastline of England and Wales.

Figure 9.6 Some important examples of constructional features developed on the coastline of England and Wales.

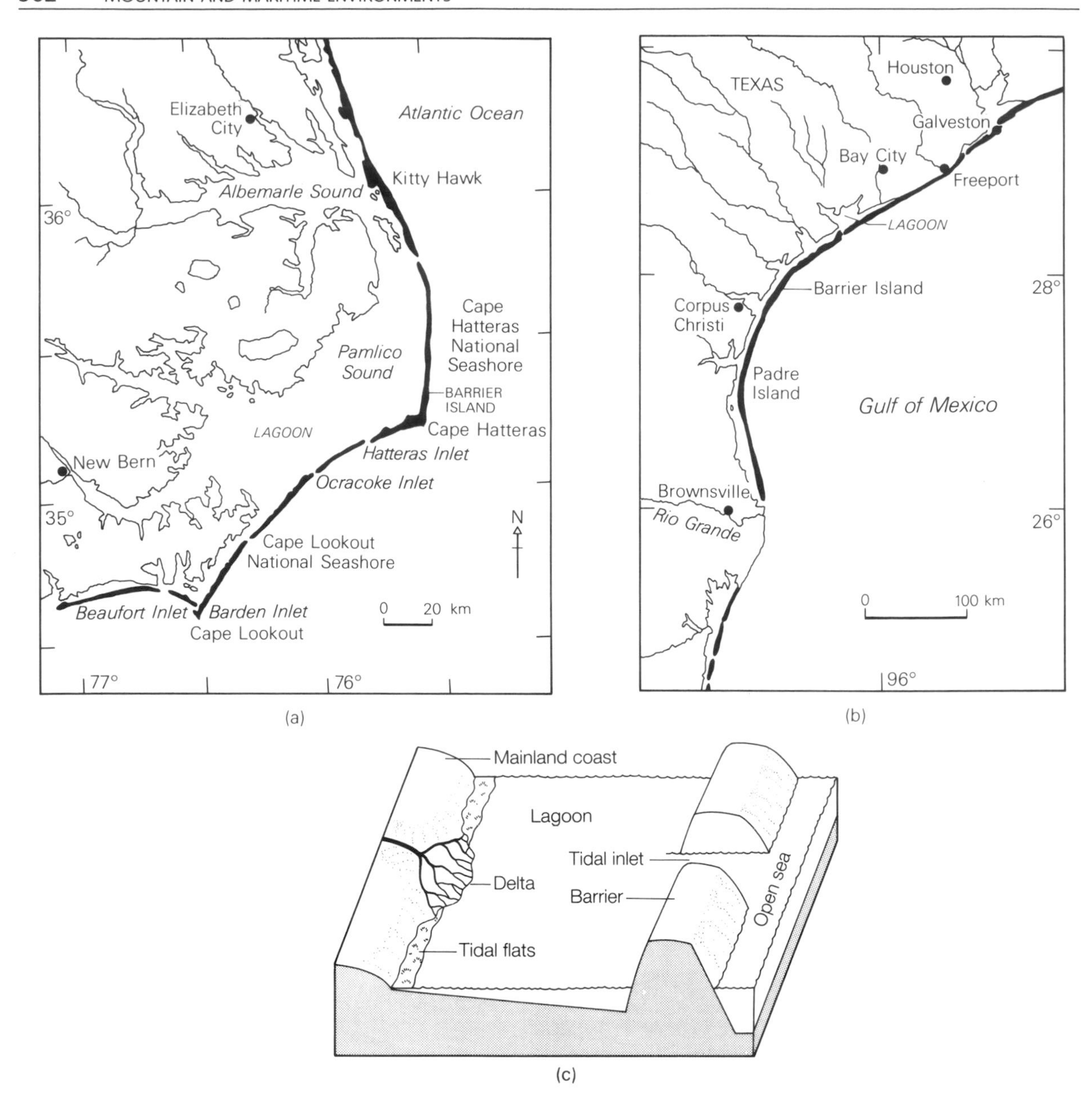

Figure 9.7 Barrier islands: (a) the outer banks of North Carolina, USA; (b) the offshore barrier island of the Texas coast; (c) a schematic representation of a barrier island, separated from the mainland by a wide lagoon, while dune ridges advance over the tidal flats.

and they become more fragmented as tidal range increases because of the action of tidal currents between the lagoon and the open sea.

The exact mechanism by which the plentiful supply of sand is thrown up into a barrier and maintained above water level is the subject of considerable debate. At one time it was thought that a fall in sea level would expose a bar of sediment that had been formed beneath the waves as a breakpoint bar, but in most parts of the world sea level has generally been rising rather than falling over the past 10 000 years or so. An alternative explanation is that they are formed as a result of rising sea level, such as has characterised the Flandrian Transgression (see section 2.11), which has partially submerged an old beach ridge capped by sand dunes. Other workers maintain that sea-level change is not a *sine qua non* for their formation, and that submerged shoals can be built up so close to sea level that eventually the action of swash takes over and constructs them into an island or bar. Once the island is formed, sand may accumulate as dunes, and so the feature can be raised above water level.

9.4 Tides

Although waves are probably the most important process whereby coastlines are moulded, tides are also extremely important, for they help to control the height range over which wave action can operate.

Tides are regular movements of the ocean water arising from the gravitational attraction of the moon and, though to a lesser extent, the sun on the earth. This attraction causes the waters to be gathered as bulges at two opposite sides of the earth, so that on an earth wholly covered with water two areas of high water would be produced. These bulges are held fixed beneath the moon as the earth spins, giving high and low tides twice a day. The sun, though much farther away, has so much more mass than the moon that it too causes tides, though they are less than half the height of moon tides. The two sets of tides are not synchronous, those related to the sun coming every 24 hours. When the earth, moon and sun line up, the combined gravitational pull of the moon and sun reinforce each other and produce very high tides – the *spring* tides. The lowest tides – *neaps* – come when the sun and the moon are at right angles to each other with respect to the earth.

The above account describes what is termed the *equilibrium tide* – one that is calculated for a uniform globe. The earth's surface, of course, is not uniform, so the heights of the actual tides are very different in different parts of the ocean. Because the oceans are of various shapes and sizes, tidal response is complicated. Their effect has been likened to connecting a great number of large and small pots and bowls in a complicated pattern and sloshing water back and forth between them.

In some constricted arms of the oceans like the Severn Estuary in Britain (maximum tidal range 13 m) and the Bay of Fundy in Canada (maximum tidal range 15–17 m), tidal ranges are high. On coasts facing the open ocean that range seldom exceeds 2 m, and in enclosed seas like the Mediterranean there is hardly any tidal range at all. In the Black Sea the range is no more than 10 cm. In the seas around the UK the complexity imposed by local factors becomes very apparent. After passing up the western coasts, the crest of a tidal wave swings around Scotland and into the North Sea, and then proceeds southwards. In passing northwards up the Irish Sea the Coriolis Force (see section 2.2) results in the tides being at least twice as high on the Welsh and English sides as on the Irish. The tides coming down the North Sea are higher in Britain than in Norway and Denmark, because the Coriolis Force drives the water towards the right. The tides running up the English Channel from the west are also forced to the right, giving the French coasts higher tides than the English. As one might imagine, where the tidal crests merge, as they do in the Straits of Dover and the southern North Sea, exceptional complexity arises.

The geomorphological consequences of tidal action are twofold: they create tidal currents, and they spread wave activity over varying height ranges.

Tidal currents are caused in two main situations. First, when there is a difference in the time of high tide between two ends of a strait, obviously a steep gradient is developed between the two ends. If the straits are narrow a strong current can develop. The second situation occurs even when there is no difference at either end of the strait, and simply results from the canalisation of the tide within a narrow strait or estuary. Velocities in tidal currents can be high (up to 8 km h^{-1} in the Straits of Messina between Sicily and mainland Italy; 17–18 km h^{-1} between Alderney, Channel Islands and the Cotentin Peninsula of France; and up to 25 km h^{-1} in the Moluccas), though the values given here are exceptional. Strong currents can cause erosion, while slower currents may be effective in moving silt and sand stirred into suspension by waves.

The tidal range of an area influences the relative

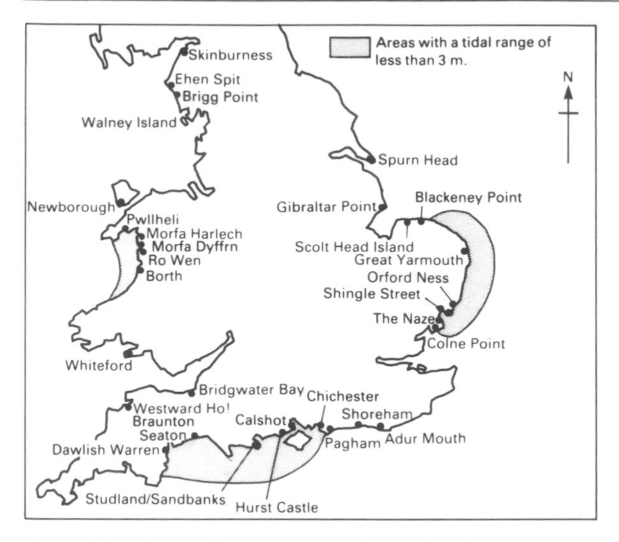

Figure 9.8 The distribution of spits in England and Wales. The main concentrations occur where the tidal range is less than about 3 m, though there are exceptions to this rule.

importance of wave activity. Thus, if the tidal range is less than about 2 m, it can probably be assumed that wave action is the dominant coastal landforming process, with the result that such features as beaches, spits and barrier islands are prominent. For example, if one looks at the English and Welsh coasts there are only three areas which experience tides with a range that is below *c.*3 m: the east Norfolk coast, the southern coast from Start Point to the Isle of Wight and part of the Welsh coast. This distribution coincides with the occurrence of spits along the coast (see figure 9.8).

9.5 Storm Surges

Although waves and tides are the two main processes that mould coastlines, storm surges or storm tides may be of considerable importance in certain locations. These occur when water is piled up against the coast by very strong onshore winds and waves. In this way sea level may be raised by some

Plate 9.3 One of the worst coastal flooding episodes that has taken place in Britain occurred in February 1953, when areas like Canvey Island were inundated by a storm surge. In this picture an amphibious DUKW vehicle churns up a muddy wake as it cruises down a Canvey Island street.

metres, so that wave attack is lifted to unusually high, and often dangerous, levels. Both tropical cyclone and temperate frontal storms may generate surges, and they are especially severe in gulfs or gulf-like seas which are enclosed on all sides except that from which the big waves come.

Areas that are known to be prone to them include the Gulf of Mexico, the Bay of Bengal, the Gulf of Tonkin and the southern part of the North Sea. Low-lying coasts such as The Netherlands and the Thames Estuary may be severely flooded if surges and high tides coincide, and are magnified in their influence by large flows coming down major rivers. It was this combination of circumstances that produced the serious floods around the southern North Sea in 1953 (plate 9.3) (see section 5.17).

9.6 Estuaries

Estuaries are the tidal mouths of great rivers. They are inlets of the sea reaching into river valleys as far as the upper limit of tidal rise. They are, however, very different from rivers *per se*, even though estuaries depend upon freshwater flow from upland rivers in order to maintain their characteristic processes. The two-way flow of estuarine water, the current set up by the mixing of saline and freshwater, and the continuous variations which take place in both discharge and velocity during the course of a tidal cycle, mean that their processes are highly distinctive.

There are two large-scale flows in estuaries which serve to mould them: tidal currents caused by the movement of the tides up the inlet, and *residual currents* caused by the mixing of fresh and saline waters of different densities. These two types of current determine patterns of erosion and deposition of sediments within the inlets. Tidal range determines the tidal current and residual current velocities and therefore the amount and source of sediments. *Micro-tidal estuaries* occur where tidal range is less than 2 m and so are dominated by freshwater flow upstream of the mouth and by wind-driven waves seaward of the mouth. They often contain a fluvial delta and spits and bars at the seaward margin. In *meso-tidal estuaries* (tidal range *c*.2–4 m) tidal currents are of greater importance, but because of the still somewhat modest tidal range tidal flow does not extend very far upstream. Thus most meso-tidal estuaries are relatively stubby. In the case of *macro-tidal estuaries*, tidal ranges in excess of 4 m produce a situation where tidal influences extend far inland. Such estuaries have long, linear sand bars parallel to the tidal flow, but their most important distinguishing characteristic is their trumpet-shaped flare. The Severn Estuary in Britain, the Delaware Estuary in the USA, and the Plate Estuary in Latin America are prime examples of this type.

9.7 Coastal Dunes

Inshore from the beach on many coasts there is a belt of sand dunes, though their presence and form depend on a combination of favourable circumstances. First, an abundant supply of sand of suitable grain size (generally about 0.25 mm diameter) is necessary for the dunes to build up. Rivers may be a major source of such sand, though substantial quantities may also have been provided during the Holocene rise in sea level (the Flandrian Transgression) which combed up sandy materials from the continental shelf and deposited them on beaches. Some sand may also be provided by the attrition of rock or coral debris. Second, given a supply of sand on the beach, it is necessary to have winds blowing onshore with sufficient frequency and velocity to transport the sand inland.

The overall importance of wind velocity in the evolution of large dune systems on coasts is attested by the preferential occurrence of dunes on temperate mid-latitude coasts where there is intense frontal activity. Larger dunes occur on windward coasts as compared with leeward. Dunes are relatively rare in the doldrum zone of low latitudes, though wind velocity is probably only one of the factors involved here. In very wet areas the beach may be so damp for so long that the wind velocities required to entrain sand grains would need to be very high, and therefore sand movement is infrequent. Tidal range also seems to play a role, for beaches with a larger tidal range expose a larger sand expanse to wind action. A large tidal range also tends to produce a lower-gradient beach, and this favours sand movement because threshold velocities for sand movement increase with the steepness of the slope up which the sand is being moved.

Thus we can conclude that large coastal dune systems are more likely to occur where there is active sand supply, impeded littoral transport, strong offshore winds, low precipitation and humidity, high tidal ranges and low beach face angles. There also needs to be suitable topography behind the beach on which dunes can accumulate – beaches backed by high cliffs are most unlikely to have big dunes.

If colossal quantities of sand are available, or if the climate is so arid that there is no vegetation cover inland from the beach, one may encounter some of the classic dunes that are characteristic of the world's great deserts (see section 6.10). This is the case in parts of Washington and Oregon in the western United States, where there are no plant species capable of binding the sand very effectively, and along the hyper-arid desert coasts of the Namib and Morocco, among others. In such situations one may find large fields of free-moving barchans and transverse ridges. On most coasts, however, including those of Britain, vegetation is a major influence that affects the growth and form of dunes (plate 9.4).

Plate 9.4 Coastal dunes at Braunton Burrows, Devon, England. Note the extensive cover of marram grass (*Ammophila arenaria*), a plant which colonises the sand at an early stage of dune evolution, and helps to promote stabilisation.

Coastal dune systems can be classified into five main types (figure 9.9):

(A) *Offshore island dune* systems are those developed on offshore or barrier islands; they serve to protect mudflats which lie in their lee. They are narrow, subject to overwash from time to time and may form an age series extending in one direction along the coast as at Blakeney, Norfolk.

(B) *Prograding ness dunes* form from an open coast where there is an abundant supply of sand at an accumulation point (ness) receiving sand by longshore drift from two directions at once. These conditions pertain on parts of the eastern coast (e.g. at Winterton Ness in Norfolk), where the prevailing wind blows offshore and is in opposition to the dominant wind.

(C) *Spit dunes* form on sandy promontories at the mouths of estuaries, and often form a fan-like series of dune ridges and intervening slacks, with the 'handle' of fan tied to the mainland. Examples include Whiteford Burrows, Glamorgan, and the Studland Dunes in Dorset.

(D) *Bay dunes*, the commonest type, accumulate in bays developed along indented coastline like those of the south-west peninsula and Pembrokeshire.

(E) *Hindshore dunes* are found on extensive sandy coasts where the prevailing wind is also the dominant one. Large dunes are driven inland as great arcs or ridges, such as those of Braunton Burrows in Devon and Newborough Warren in Anglesey.

Parallel to many beaches there is a *frontal dune* or foredune. This is formed where sand is trapped on litter deposited by high tides or on vegetation. Dunes are initiated when sand accumulates round, or is colonised by, certain species of grass that tolerate sandy and salty conditions. In Britain the two main species are sea couch grass (*Agropyron junceiforme*) and marram grass (*Ammophila arenaria*). These grasses have extensive root systems, which

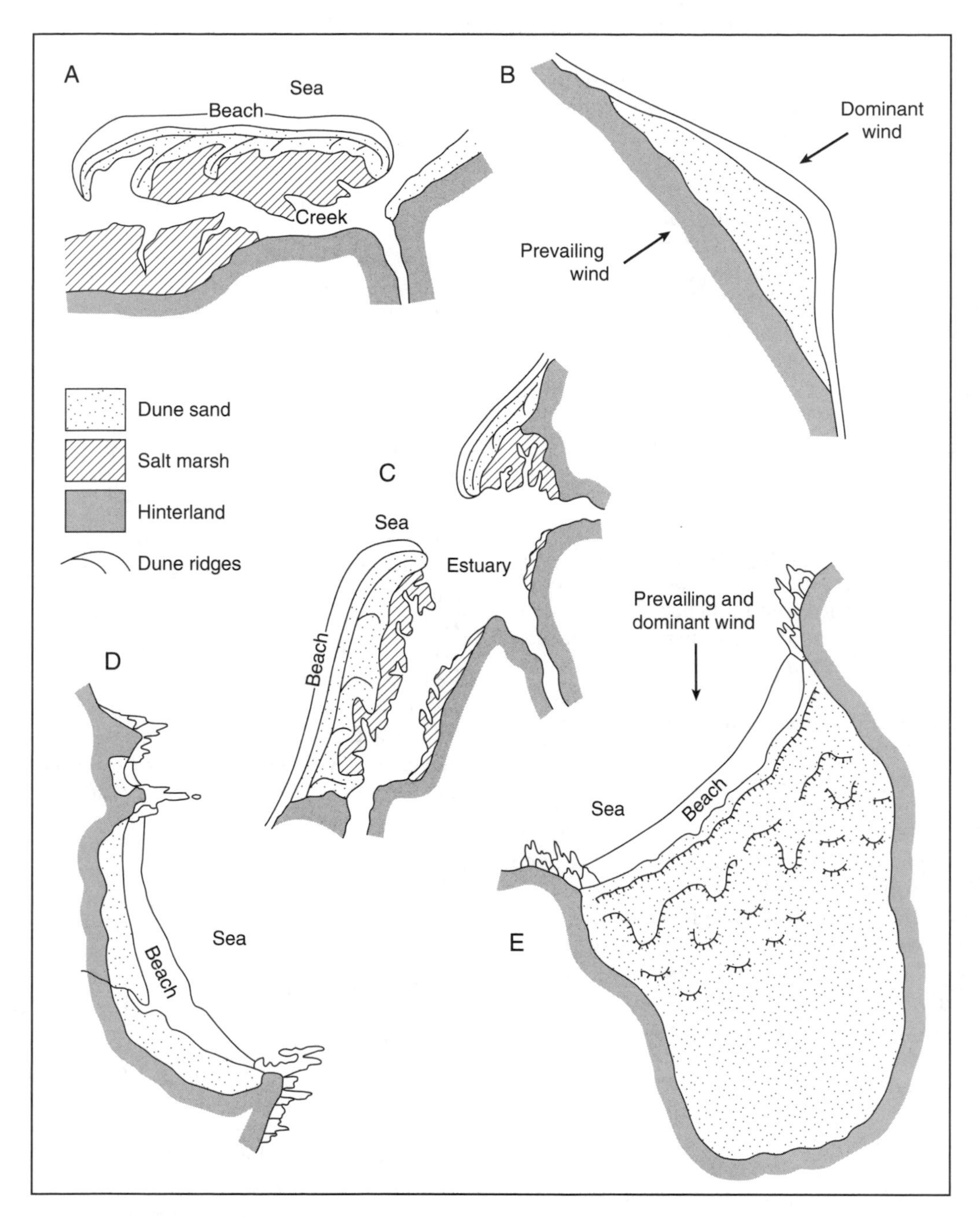

Figure 9.9 Different types of sand dune in England and Wales: (A) offshore island dunes (e.g. Scolt Head Island, Norfolk); (B) prograding ness dunes (e.g. Winterton Ness, Norfolk); (C) spit dunes (e.g. South Haven peninsula, Dorset; Whiteford Burrows, Glamorgan); (D) bay dunes (e.g. Oxwich Bay, Gower); (E) hindshore dunes (e.g. Braunton Burrows, Devon; Newborough Warren, Anglesey).

hold the sand in place, and they trap further sand, with the result that individual small mounds grow and coalesce. Under favourable conditions the dunes will grow steadily in height, for these vigorous roots may grow fast enough to allow 50–60 cm of sand to accumulate in one year, and will give a continuous dune-line backing the sandy beach. If the dunes are breached by the sea because of a storm, or if the vegetation cover is disturbed by man or animals such as the rabbit, the exposed sand may be blown beyond the ridge, leaving behind a large gap or corridor in the foredune. A *blowout* has occurred, and sand may accumulate further inland of the gap as a hairpin-shaped dune called a *parabolic*. In this way sand may move inland.

In an area where the coast is building out, new foredunes may develop seawards of the original ridge. When this happens the inner dunes will start to decay because they are cut off from their sand supply. They become degraded by rain, stabilised by an increasingly dense vegetation cover (including plants like heather, gorse and birch), and weathered so that a marked soil profile may develop. Such dunes are called *remanié* dunes. The low-lying areas between successive ridges are called *slacks*, which may be damp areas with their own plant assemblages. In some parts of the lower and middle latitudes, where the sands can be highly calcareous, the dunes may become *lithified* (cemented) as they get older to give a material that is sometimes hard enough to be used as a building stone. Such material is called *aeolianate*, and it has been extensively quarried around the Mediterranean Basin to build structures as different as Roman temples (as in Libya) or modern package tour hotels (as in Mallorca).

Sand dunes are important; they are a widespread type of landform, and they provide valuable habitats for a great diversity of interesting plants and animals. They are also a very good form of coastal defence. Finally, they are very beautiful. If they are abused by human activity they may march inland, overwhelming farmlands and buildings, or they may be degraded or breached. If, on the other hand, they are protected from overgrazing, deforestation, trampling and mining, they can prove an effective barrier against wave attack on low-lying coasts.

9.8 Salt Marshes

In our discussion of the formation of coastal dunes we have seen the important role that vegetation plays. The same is true of another major coastal environmental type – the salt marsh (plate 9.5).

In areas sheltered from wave action, as in estuaries or in the lee of spits and barriers, coastal accretion takes the form of mud flats which are subject to tidal flooding. The accretion is promoted by vegetation that is adapted to periodic inundation by salt water, whether it be the salt marsh vegetation of temperate regions or the mangrove swamp of low latitudes (see section 7.9). Sediment carried into marsh vegetation communities by the rising tide is filtered out by the vegetation and retained as the tide ebbs. In this way the level of the land is gradually built up, and vegetated marshland encroaches across the estuary or tidal flat. The rate of accretion is relatively slow at first because the vegetation cover is sporadic, but as the process goes on and the level of the land is raised, the vegetation cover becomes denser, and is therefore more efficient as a filter and stabiliser. Towards the end of the process the height of the marsh is such that the frequency of tidal inundation is reduced, and so the rate of accretion declines. Accretion tends to be uneven, so most marshes have well-developed creek systems and small basins, called *pans*.

The sequence of vegetation colonisation as the accretion process goes on varies from area to area, but it does provide a good example of the early stages of a phenomenon called *vegetation succession* (see section 10.5). In southern England, in the early stages bare mud flats accumulate patches of green algae such as *Enteromorpha*. This allows some sediment accumulation, which permits the colonisation of muddy areas by some salt-tolerant plants (*halophytes*) such as *Salicornia* and *Suaeda maritima*. These plants are only annuals and so are of limited effectiveness for trapping silt for much of the year. None the less, they do in time permit sufficient accretion to take place so that other plants can come in that, although less tolerant of inundation and salinity, provide a denser, higher and longer-lasting vegetation cover. These include *Aster tripolium*, *Limonium vulgare* and *Puccinellia maritima*. Trapping of sediment by the often fleshy

Plate 9.5 Large areas of salt marsh have developed behind the coastal barrier of Scolt Head Island in Norfolk, England. In the foreground there are old shingle ridges formed by the growth of the barrier along the coast. Note the intricate creek system which will be flooded at high tide.

foliage is now greatly accelerated, and the marsh may grow at rates approaching $1\,\mathrm{cm\,yr^{-1}}$. Creeks become established, and they are lined by a plant that likes relatively well-drained conditions on their banks, namely *Halimione portulacoides*. This plant is effective at trapping the silt and so may tend to stabilise the creeks and to cause the development of slight levées along their margins. Towards the end of the marsh-building phase plants such as *Juncus maritimus* become dominant, and when the marsh is sufficiently high for inundation to be rare, freshwater sedges and reeds appear.

In Britain the speed of marsh accretion has been greatly accelerated in some areas by the introduction of a plant called *Spartina*. This has been a particularly effective salt marsh pioneer species since the vigorous hybrid *Spartina townsendii* originated in Southampton Water as a cross between native British and American species in the 1870s. This perennial, tough spiky grass is very effective at trap-

Plate 9.6 Cliffs are moulded both by direct marine influences and by sub-aerial processes. These cliffs in North Devon, England, appear to be largely inactive at present with respect to marine attack, and may be largely formed by frost weathering and soil creep.

ping silt, and rates of accretion of 1–5 cm yr^{-1} may occur. It is very useful as a means of reclaiming land from the sea or for stabilising mud flats, but unfortunately it also reduces the beauty and variety of the normal, natural marshlands.

9.9 Coast Erosion

The sea can be a potent agent of erosion. The destructive impact of waves against obstacles in their path can be substantial, and it has been calculated that the average pressure exerted by Atlantic waves in winter is not much under 10 000 kg m^{-2}, while in great storms it may exceed 30 000 kg m^{-2}. Thus cliffs and man-made structures like sea walls are subjected to shocks of enormous intensity. Furthermore, water driven into cracks in the rock may compress the air that is already there, creating an explosive blast. This combination of bombardment and blasting is augmented by the impact of sediment thrown up by the waves. In areas of susceptible rocks these mechanical processes may be assisted by weathering achieved by solution and salt-laden spray. Furthermore, cliffs exposed to wave attack are at the same time exposed to various sub-aerial processes of frost weathering and mass movements (plate 9.6), and the sea is very effective, through the agency of longshore drift and tidal currents, at removing the debris supplied to the base of the cliff by these processes.

Table 9.1 Selected rates of cliff erosion for particular localities

Lithology	*Location*	*Rate (m per 100 yr)*
Glacial drift	Holderness, E. England	175
Pleistocene deposits	Pakefield, E. Anglia	300
Volcanic ash	Krakatau, E. Indies	3000
London clay	Isle of Sheppey	300
Pleistocene deposits	Dunwich, E. Anglia	400
Glacial drift	North Yorkshire	30
Chalk	Isle of Thanet, Kent	30
Chalk	Sussex	50
Lias shales	North Yorkshire	9

Erosion is most rapid in areas of friable, uncemented sediments and rocks. Glacial drifts, clays, Tertiary sands and gravels and volcanic ash are the sorts of materials that can be eroded very quickly; some rates of cliff retreat to which they are subjected are listed in tables 9.1 and 9.2. In more resistant rocks, rates of cliff recession will be less, but even then, if wave action is marked and longshore drift active, the rates are measurable. The form of cliffs in resistant rocks will tend to be more complex than those in softer rocks because wave action will eat out joints, fractures and faults. Caves, stacks, stumps, arches, blow-holes and enlarged joints (called *geos*) are the sorts of features that may form (see figure 9.10). The joint pattern is also of

Table 9.2 Typical ranges of rates of cliff recession in different material types

Material	*Rate (m per year)*
Granitic rocks	10^{-3}
Limestone	10^{-3}–10^{-2}
Flysch and slate	10^{-2}
Chalk and Tertiary sediment rocks	10^{-1}–10^{0}
Quaternary deposits	10^{0}–10^{1}
Unconsolidated volcanic ejecta	10^{1}

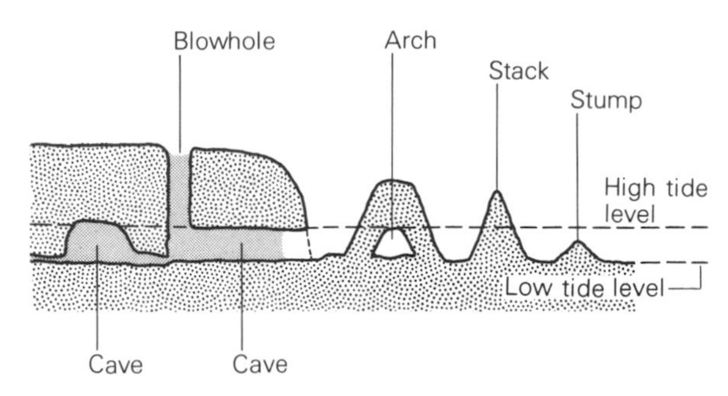

Figure 9.10 The formation of caves, blow-holes, arches, stacks and stumps by marine erosion.

considerable importance in controlling the nature of the cliff profile. If the beds dip fairly steeply towards the sea, there is a tendency for blocks of rocks to break off at the joint planes, usually at right angles to the bedding, so that the cliff profile tends to be dominated by the dip of the beds. On the other hand, where the beds have an almost vertical or horizontal dip (plate 9.7), joint blocks cannot so readily break off and slip down the bedding planes: the result is a tendency for cliffs to be more nearly vertical.

Cliff forms will be more complex where rocks of differing character are exposed in the cliff face. If resistant rocks overlie rather weak rock, the preferential attack of the sea on the weak rock will cause

Plate 9.7 The cliffs on the west coast of the Isle of Portland, southern England. The upper part of the cliff is composed of well-jointed Portland limestone. This is the main cliff former. Under the limestone is a thick bed of Kimmeridge clay. The juxtaposition of limestone and clay causes instability, which explains the presence of landslipped debris between the cliffs and the sea.

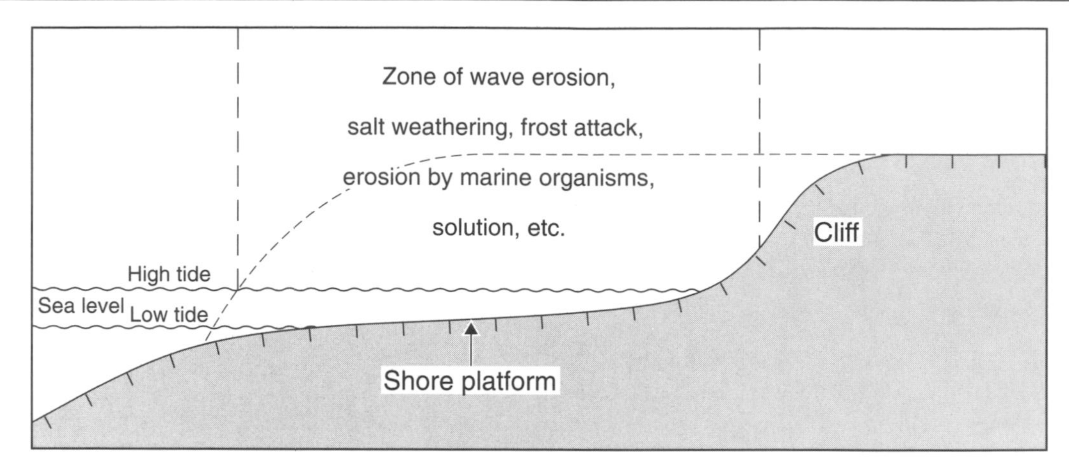

Figure 9.11 Processes that act upon a shore platform.

the undermining of the resistant upper beds, and major landslides may occur. A good example of this in England is the way in which chalk has slipped over the weak Gault clay in Folkestone Warren, Kent.

As cliffs are worn back, a *shore platform* (figure 9.11) is left in front, the upper part of which is visible as a rocky foreshore at low tide (see colour plate 16). The platform is caused partly by the abrasion of sand and shingle sweeping across it as the tide goes in and out, but in limestones, solution, grazing and boring by marine organisms, salt weathering, frost attack, wetting and drying and other types of weathering may contribute to its development. In Britain spectacular examples of shore platforms, some a few hundred metres wide, occur on either side of the Bristol Channel, where they are associated with Liassic and Carboniferous limestones, but other fine examples can be found at low tide at the foot of some of the great chalk cliffs of the south coast of England, as near Beachy Head. There is probably some limit to how wide these features can become, because as the wave platform becomes broader and broader an increasing proportion of wave energy will be expended by passage through the shallow water overlying the platform, and attack on the cliff base will become less intense.

Rock structures and character are also highly important in controlling the plan of a coastline. Where tectonic processes have created folding at right angles to the trend of the coast, a series of bays and headlands develops. South-west Ireland shows this type of *Atlantic coastline*. If, on the other hand, the general trend of the fold structures is parallel to the coast, as in Croatia, a *Dalmatian coastline* is produced, with elongated islands and inlets running parallel to the coast.

9.10 The Human Impact on the Coastline

Because of the concentration of so many settlements and human activities on the coastline, the pressures being placed on coastal environments are often acute, and the consequences of excessive erosion will be serious. We have already seen the importance of a plant that was introduced by humans, *Spartina townsendii*, in altering the nature of salt marshes, and we have also seen that sand dunes can easily be affected by deforestation, trampling and other pressures so that their effectiveness as a coastal barrier is greatly reduced.

While most areas are subject to some degree of natural erosion and accretion, the balance can be upset by human activity in a variety of different ways. In general, the changes are the unexpected and unwelcome results of various economic activities, though perversely in some areas coastal erosion has been accelerated as a result of man's attempts to reduce it.

One of the best forms of coastal defence is a good beach. Thus, if material is removed from a beach

Plate 9.8 The ruined village of Hallsands in Devon, which was attacked by the sea when its beach was depleted by shingle removal for the construction of Plymouth dockyards in 1887.

to provide minerals of economic worth or aggregates for construction purposes, accelerated *cliff retreat* may take place. The classic example of this process was the mining of 66 000 tonnes of shingle from the beach at Hallsands in Devon in southern England in 1887 to provide materials for the construction of naval dockyards at Plymouth. The shingle proved to be undergoing very little natural replenishment, and in consequence the shore level was reduced by about 4 m, and the loss of protective shingle soon resulted in cliff erosion, which amounted to 6 m between 1907 and 1957. As a consequence the little village of Hallsands was cruelly attacked by waves and is now mostly abandoned and in ruins (plate 9.8).

Another common cause of beach and cliff erosion at one point is coast protection at another (figure 9.12). Engineers frequently try, by constructing groynes, to create a broad beach to protect a cliff from erosion. Groynes will accelerate accumulation, but they mean that sediment is not transported, as it was in the past, further along the shoreline by longshore drift. Thus protection by the groynes may cause serious beach depletion and erosion further along the coast.

Piers or breakwaters can have similar effects to groynes. This has occurred at various points along the British coast: erosion at Seaford in Sussex resulted from the Newhaven breakwater; erosion at Lowestoft resulted from the pier at Gorleston; and erosion at West Bay in Dorset followed the construction of a harbour mouth jetty (plate 9.9).

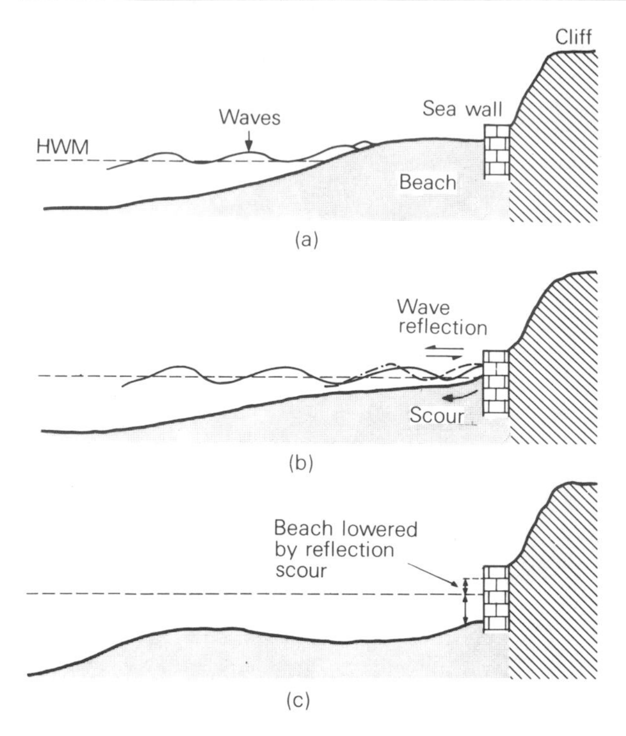

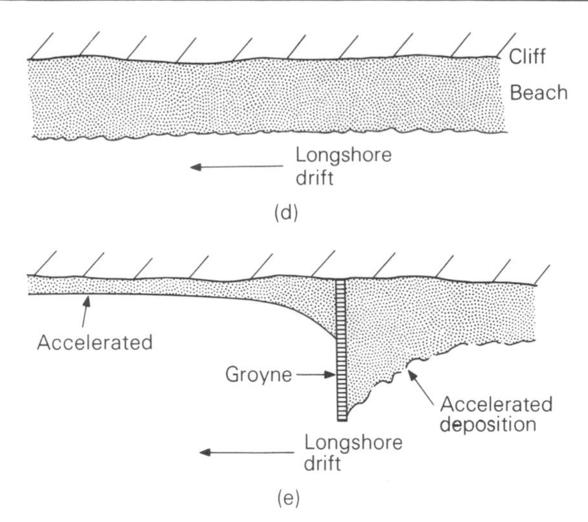

Figure 9.12 Human influences on coast erosion. (a)–(c) Sea walls and erosion: a broad high beach prevents storm waves from breaking against the sea wall and will persist, or erode only slowly; but where waves are reflected by the wall, scour is accelerated, and the beach is quickly removed and lowered. (d)–(e) The effects of groyne construction on sedimentation on a beach.

One of the reasons why sediment mining from beaches and the construction of groynes, piers and breakwaters can be so serious is that in many parts of the world beaches are no longer being so actively replenished as they were in the past. There is now abundant evidence to indicate that much of the reservoir of sand and shingle that creates beaches is in some respects a relict feature. Much of it was deposited on the continental shelf during the maximum of the last glaciation (around 18 000 years BP), when sea level was about 120–140 m below its present level. It was transported shorewards and incorporated in present-day beaches during the phase of rapidly rising post-glacial sea levels that characterised the Flandrian Transgression until about 6000 years BP. Since that time world sea levels have been relatively stable, and as a consequence much less material is being added to beaches and shingle complexes.

In some parts of the world the sediments brought to the coastline by rivers are an important source of materials for beach replenishment, for they become incorporated into beaches by longshore drift. Thus any change in the sediment load of such rivers may result in a change in the sediment budget of neighbouring beaches. On the one hand, accelerated soil erosion in their catchments may cause beach accretion and silting of estuaries, while on the other building of dams may trap river sediments so that coastal erosion occurs (figure 9.13).

In other parts of the world, vegetation modification by humans creates the potential for increased erosion. In low latitudes some small islands are, under natural conditions, covered by a dense thicket of vegetation that acts as a baffle against waves and as a means of trapping coral blocks and shingle thrown up by cyclonic storms. However, on many islands the natural vegetation has been replaced by coconut plantations: these have an open structure, shallow roots and little or no undergrowth, so

Plate 9.9 At West Bay, Dorset, the construction of a breakwater for the harbour has caused changes in the configuration of the beach. This sequence of photos shows the position of the beach in 1860, 1900 and 1976. While accretion is taking place in the foreground, severe coastal retreat has taken place on the far side of the breakwater, and coast protection has had to be adopted. Note how the beach in front of the cliff has narrowed.

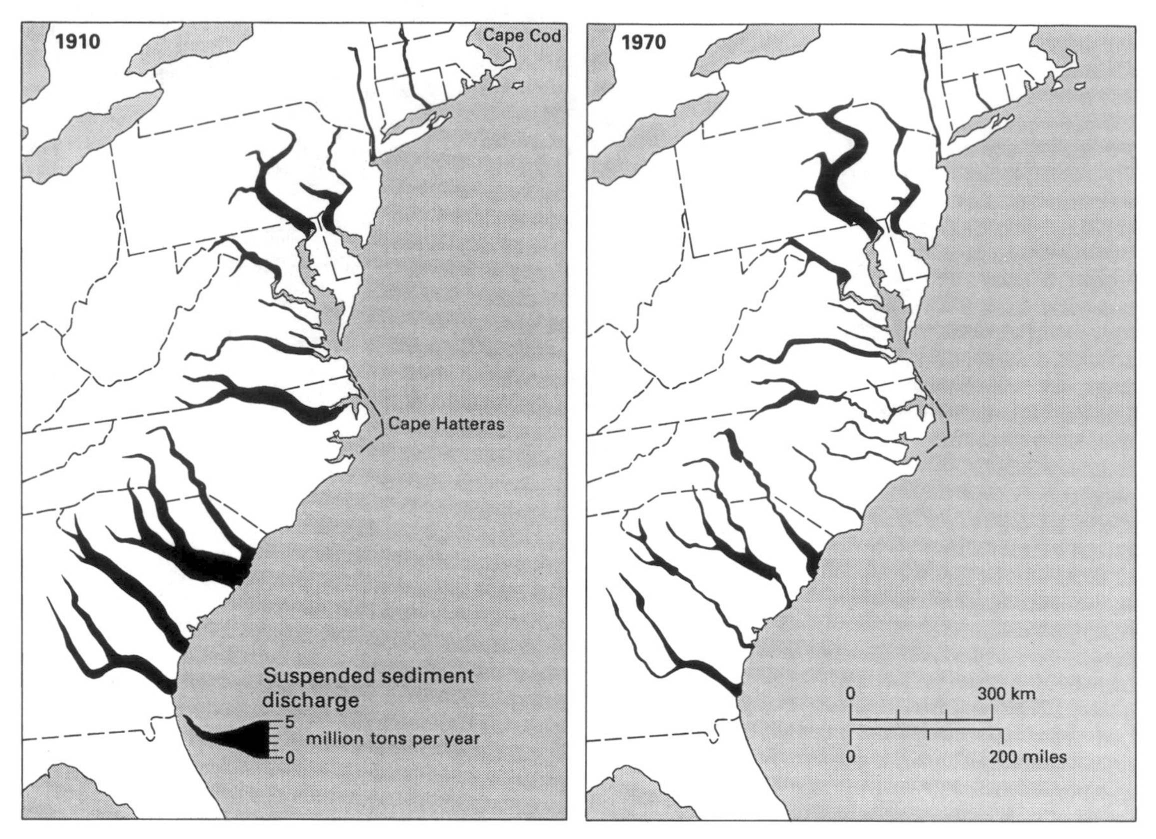

Figure 9.13 The decline in suspended sediment discharge to the eastern seaboard of the United States between 1910 and 1970 as a result of soil conservation measures, dam construction and land-use changes.

they are impotent in the face of hurricane attack. Observations during Hurricane Hattie in Belize, Central America, showed that, where the natural vegetation had been replaced by coconuts, the height of the small sandy islands (*cays*) was reduced by around 2 m; whereas, where natural sedi-ments remained, banking of storm sediments against the vegetation thicket led to net vertical increases in height of over 1 m.

In recent years a new human threat to world shorelines has emerged: accelerated sea-level rise (window 9.1). Over the past century or so, world sea levels appear to have risen by around 15 cm, and this has coincided with a general warming in global temperatures. The reason for this is that increased temperatures cause the world's ice sheets and glaciers to melt more quickly, and to deliver more water into the oceans. In addition, as the temperature of ocean water increases it also expands in volume (the so-called 'steric effect'). The human role in this relates to the probability that the burning of fossil fuels and the removal of large areas of forest in the humid tropics and elsewhere have caused the amount of carbon dioxide in the atmosphere to increase. As explained in section 7.6, increased carbon-dioxide levels in the atmosphere could lead to an increase in atmospheric temperatures as a result of the 'greenhouse effect'.

Window 9.1 ***The 'greenhouse effect' and sea-level rise***

As a result of the burning of fossil fuels and other human activities, many scientists postulate that global temperatures will rise over coming decades. Many predictions suggest that the earth will warm up by several degrees celsius over the next 40–50 years. One of the most serious consequences of a global warming of this magnitude might be a world-wide rise in sea level as a result of two distinct processes: the thermal expansion of the upper layers of ocean water; and the melting of alpine glaciers, permafrost and ice caps. However, the degree of rise that may occur is the subject of uncertainty, largely because scientists are unclear how the Antarctic ice cap will respond to warming. Might it become unstable and surge, releasing large amounts of ice into the world's oceans, and thereby causing a rapid rise in sea level? Alternatively, at the other extreme, might global warming lead to greater snowfall in Antarctica, and thus to a growth of the ice cap and a potential drop in sea level?

There is some consensus that sea level might rise by somewhere between 0.5 m and 1.0 m by the year 2100. These figures may not at first sight appear large, but in sensitive areas (e.g. low-lying coasts, deltas, marshes, wetlands, coral atolls) the effects would be marked. It has been calculated, for example, that were sea level to rise by just 1 m over the next century, up to 15 per cent of Egypt's arable land would be lost as a result of inundation, and 16 per cent of the population would have to be relocated. Some of the world's great conurbations would also be threatened, including London, Rotterdam, Tokyo, Bangkok, Miami and Calcutta, while some low-lying island states based on atolls, such as the Maldives of the Indian Ocean, might suffer near-complete inundation. The effects of such inundation would be compounded in those areas where the land is subsiding as a result of local tectonic movements, isostatic adjustments or the abstraction of oil and groundwater.

Accelerated coastal submergence will have a variety of effects, of which accelerated erosion may be the most significant. On erodible sand coasts, some calculations based on the so-called 'Bruun Rule' suggest that the coastline may erode by the order of 1 m for every 1 cm rise in sea level.

There are various ways in which humans may react to a sea-level rise. They can move away, create planning controls to limit future development in susceptible areas, build protective structures (sea walls, dykes etc.) or encourage the health of natural barriers (dunes, marshes and beaches).

■ *Key Terms and Concepts*

accelerated coast erosion
accretion
barrier beach
coastal classification
constructive waves
destructive waves
estuary classification
refraction
salt marsh
shore platform
storm surges
swell environments
tidal currents
tidal range
wave motion

Points for Review

- Attempt a classification of global coastline types.
- What role do (a) waves, (b) currents and (c) tides play in moulding coastlines?
- What types of coastline do you think might be most susceptible to erosion?
- What role does vegetation play in the development and form of coastal environments?
- How do humans accelerate rates of coastal retreat?

FURTHER READING

Bird, E. C. F. (1984) *Coasts* (Oxford: Blackwell). A simple, clear, introductory text.

Brunsden, D. and Goudie, A. S. (1997) *Classic Coastal Landforms of West Dorset* (Sheffield: Geographical Association). A booklet on the nature and evolution of one of Europe's most striking coastlines (part 1).

Carter, R. W. G. (1988) *Coastal Environments* (London: Academic Press). A comprehensive, advanced text.

Goudie, A. S. and Brunsden, D. (1997) *Classic Coastal Landforms of East Dorset* (Sheffield: Geographical Association). A booklet on the nature and evolution of one of Europe's most striking coastlines (part 2).

Hanson, J. D. (1988) *Coasts* (Cambridge: Cambridge University Press). An introductory-level text.

Pethick, J. (1984) *Introduction to Coastal Geomorphology* (London: Edward Arnold). An overview by a leading coastal geomorphologist.

Viles, H. and Spencer, T. (1995) *Coastal Problems: Geomorphology, Ecology and Society at the Coast* (London: Edward Arnold). A book that brings together material vital to any attempt to understand and manage our coasts.

Part IV

Landscapes and Ecosystems

10 Plants and Animals

10.1 Ecology and Ecosystems

The study of the interactions between organisms (plants and animals) and their environment is called *ecology*, and the term *ecosystem* is used to describe any unit that involves the interactions, in a given area, of organisms with the physical environment, so that a flow of energy leads to an exchange of materials between living and non-living parts within the system. An ecosystem can be of any size, from a cowpat or a goldfish bowl to the Amazon rainforest or the world as a whole. Scientists who study the various relationships between the different components of an ecosystem are called *ecologists*.

The key features of an ecosystem are the transmissions of energy through its different components, and the source of this energy can be traced back to solar radiation. This provides both the energy of the wind and rain and the energy used by plants in *photosynthesis*, a process that leads to the creation of plant tissue, the basic food for the animal population. The decaying plant and animal remains are incorporated into the soil, and the nutrients they release are taken up again by plants.

Let us start with a consideration of green plants. These contain chlorophyll, which can absorb energy derived from the sun and convert it into molecules containing carbon (i.e. organic molecules). This process, which is called photosynthesis, is the source of energy in living things and thus in all ecosystems. It converts radiant energy into chemical energy (although some lower organisms can convert energy in the absence of light), and fixes that energy in the form of organic compounds that do not spontaneously break down and therefore can be stored until needed by the plant.

Opposite Large herbivores, in this case elephants in the Kidepo Valley National Park, Uganda, are an important component of savanna ecosystems, modifying vegetation through their trampling, pushing, feeding and nutrient recycling.

For photosynthesis to take place, light, water, carbon dioxide and heat are required. Basically, the elements of carbon dioxide and water are combined with light energy, which is absorbed by the chlorophyll of plant leaves. Molecular processes act on this combination to produce oxygen and organic compounds in the form of glucose and carbohydrates. The oxygen and water vapour are released into the atmosphere as respiration through the stomata of plants. Plants grow when photosynthesis produces plant materials (glucose and carbohydrates) in excess of the rate of utilisation of these materials by the biochemical processes of respiration. In other words, some of the sugar produced by photosynthesis is used as an energy source by the plants for respiration; this energy is degraded from the highly chemical form to a highly dispersed form as heat, which cannot be recycled into chemical energy but must be radiated out to the atmosphere and then to space and so is lost to the ecosystem.

The rate of organic material output by the plant

cover is referred to as *production*, and it is equivalent to net photosynthesis.

Net photosynthesis = total photosynthesis − respiration.

Production is measured in terms of the weight of organic matter added to a ground area of unit size per unit time (e.g. $g m^{-2} d^{-1}$). The rate of production depends basically on heat, light, carbon dioxide and water. Under *the principle of limiting factors,* the highest factor that can be obtained is controlled by the factor that is in shortest supply. So, for example, without enough heat to maintain a temperature above 10 °C, photosynthesis will be minimal no matter how much light, water and carbon dioxide are available. Of the four essential factors, carbon dioxide is the least variable throughout the world, but the other three are highly variable over the planet, and in many regions are also variable from season to season.

Of all the major climatic regions, it is the low latitudes that receive the greatest combined total of heat, moisture and light. This is also the zone in which they are least variable. With such ample and dependable supplies of energy, it is understandable that plant productivity should be higher in the tropics than in any other major region. By contrast, in polar regions heat is insufficient to produce appreciable levels of biochemical activity for plant growth; light is severely limited in winter; and for much of the year water is frozen and so is largely unavailable for plant use. The same sort of trends can be seen in the seas (see section 10.3). The most productive environments are the tidal marshes, estuaries and coral reefs of tropical coastlines. In their shallow waters, light, heat and moisture are all present in profusion. By contrast, in the colder, darker waters offshore, production is very much less (say, $125 g m^{-1} yr^{-1}$ compared with $2500 g m^{-1} yr^{-1}$).

The fate of energy can be followed through a consideration of a simple energy transfer model, called a *food chain* (figure 10.1). Each stage in the chain is called a *trophic level*. Plants are the first in the chain and are called *producers*. Normally, some of the plant material continues as storage in perennial plants, some is eaten by plant-eating animals (*herbivores*) and some dies to form a litter on the soil surface. These herbivorous animals are the second trophic level and are called *primary consumers*. They in turn are eaten by the carnivorous animals (*secondary consumers*) of the third trophic level. Plants and animals that die are chemically decomposed by soil fauna and flora (*decomposer organisms*), which help to recycle mineral nutrients into the system (see colour plate 17). At each trophic level a conversion to heat takes place, which means that less energy becomes biomass at the succeeding trophic level. No organism can convert the food it takes into an equal amount of stored energy.

Let us consider a simple plant–herbivore–carnivore food chain consisting of grass plants, mice and snakes. The mice obtain approximately 10 per cent

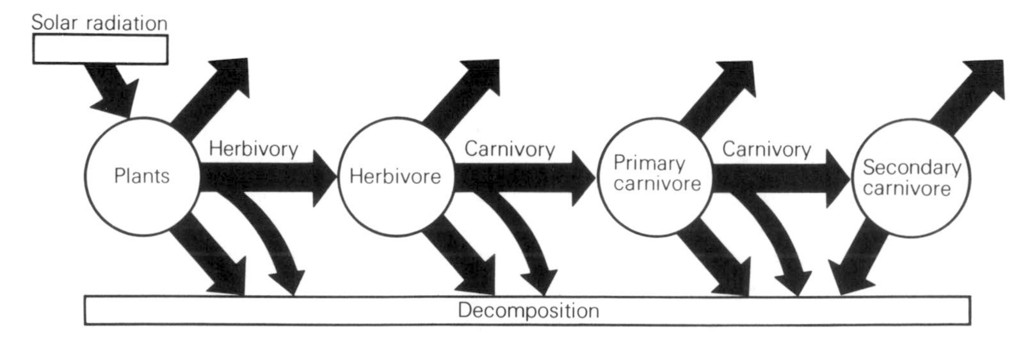

Figure 10.1 Energy flow through a food chain. At each stage some energy is available as food for the next trophic level, some is lost as excretory products, some as decay of dead organisms and some as respiration. Thus the quantity of energy decreases down the chain away from the plants; the number of secondary carnivores in a given area is therefore likely to be much smaller than that of plants. Much of the energy 'lost' to the chain ends up with the decomposer organisms, which in the terrestrial ecosystems are in the top few centimetres of the soil.

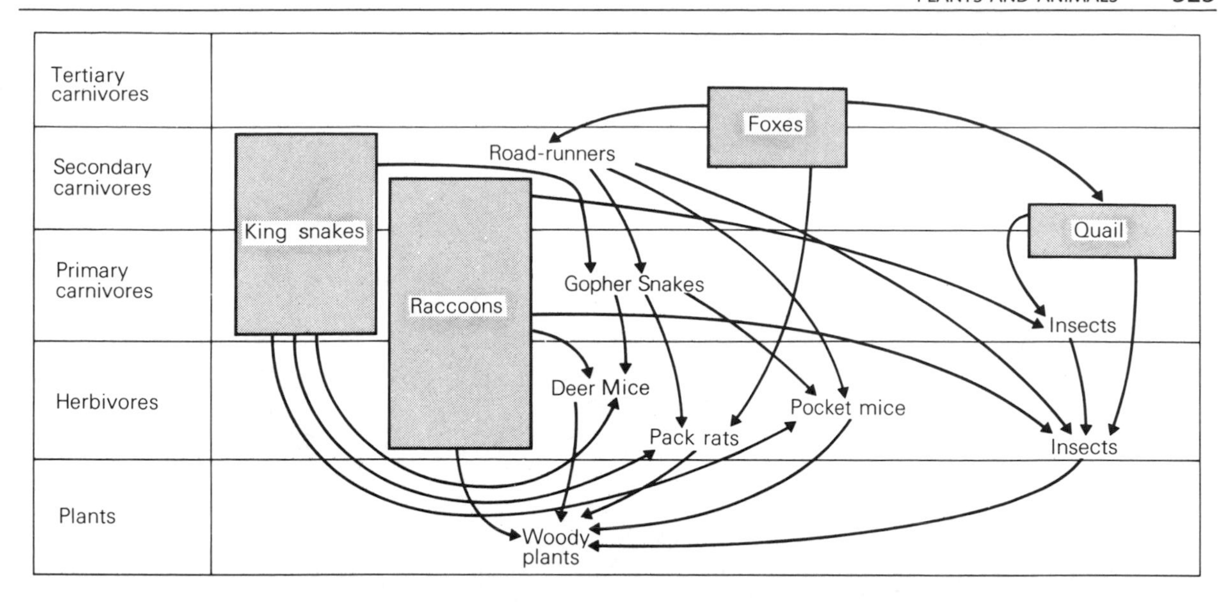

Figure 10.2 A simplified food web, showing some of the more complex linkages in an actual ecosystem – the chaparral scrub of California. The various trophic levels demarcate the animals of different groups and it is noticeable that not all animals are confined to one trophic level – raccoons, for instance, are both herbivores and carnivores. However, they are all ultimately dependent on the plants.

of the energy absorbed by the grass plants, and the snakes that feed on the mice obtain approximately 10 per cent of the energy absorbed by the mice. Thus the snake receives only 1 per cent or so of the energy originally absorbed by the plants. The fraction of the original energy available to a succeeding carnivore stage is still less. This explains why most food chains are limited to four or five trophic levels, and why the animals at the end of the food chain, for example lions, have to roam over large areas to obtain their food, because one small area cannot support many of them.

A food chain is seldom as simple as the foregoing discussion might imply. It is more realistic to think in terms of a *food web* (see figure 10.2), for the relationships between different organisms are often complex, and there are usually various different species of plants and animals at each trophic level, each animal having its own feeding patterns.

However, to understand how ecosystems function it is necessary to consider the elements of the environment other than energy; for life is sustained by a number of chemical elements that enter ecosystems via the plants. Plants are fed by various mineral nutrients that come from the atmosphere and the rocks, and many of the important elements circulate between living organisms and non-living (abiotic) pools of various scales. These movements are generally cyclic, involving the use and re-use of nutrients, and are termed *biogeochemical cycles*.

Eighteen essential nutrients are required by plants to fulfil their growth requirements. These can be divided into the macro-nutrients (e.g. carbon (window 10.1), nitrogen, oxygen, phosphorus, potassium, sulphur and magnesium) and micro-nutrients (manganese, iron, silica, sodium and chlorine, and various trace elements such as zinc, molybdenum, boron and copper).

The general nature of a nutrient cycle is shown in figure 10.3. Nutrient inputs come from the weathering of rocks. Silica, aluminium, manganese, potassium and sodium may be made available as nutrients in this way. Another major source of nutrient inputs is the atmosphere, whether in precipitation or by biological processes. Rainfall is never entirely pure water, for it contains various impurities, including salt spray and windblown dust. As far as biological processes are concerned, animals

Window 10.1 The carbon cycle

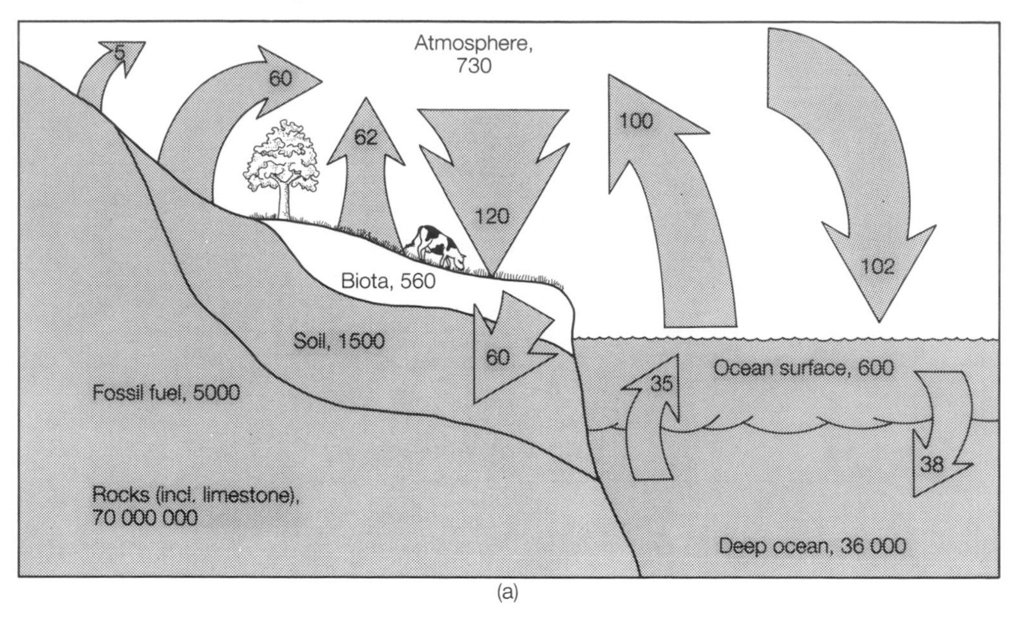

(a)

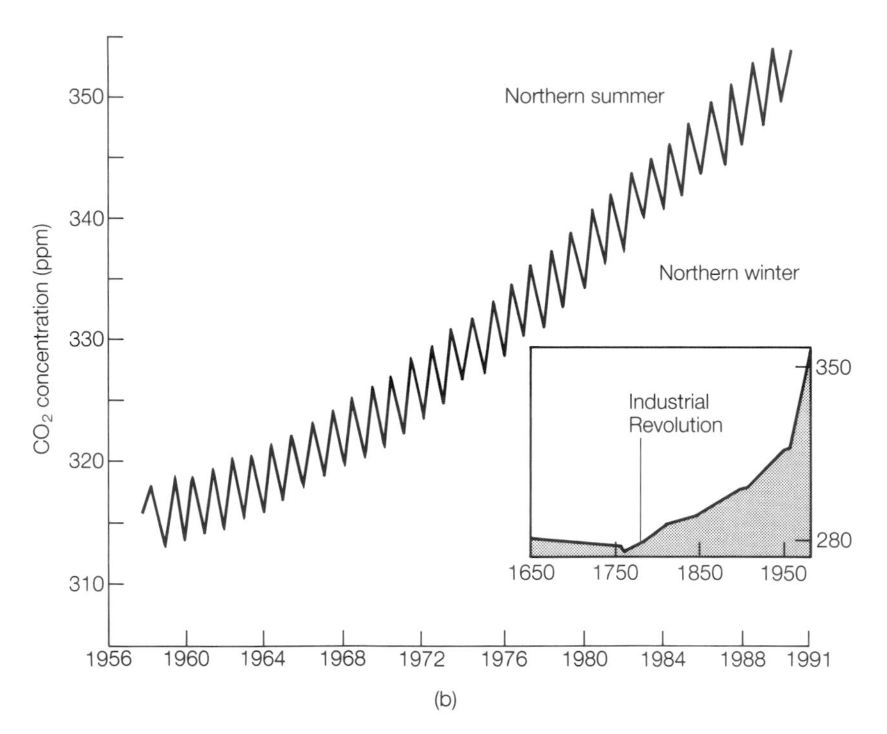

(b)

The global carbon cycle. (a) The main pools of carbon (in billions of tonnes): the arrows show how much it moves each year between these pools. (b) The increasing CO_2 concentrations in the atmosphere since the Industrial Revolution (see inset) and over the past several decades. Annual fluctuations are mostly due to seasonal changes in photosynthesis on land in the Northern Hemisphere.

Although it makes up less than 1 per cent of our planet, carbon is the basis for life on earth. Carbon compounds make up plants, animals and micro-organisms, and provide much of our food and our energy resources. The gas carbon dioxide (CO_2), which makes up only 0.03 per cent of the atmosphere, helps to control the level of radiation received from the sun and helps to make the planet sufficiently warm for life to evolve. This gas is also the source of much carbon for organisms, as well as a product of burnt or decomposing organic matter. Indeed, plants are the prime movers in the global carbon cycle, for through the process called photosynthesis they convert carbon dioxide into leaves, stems, trunks and roots, and these in turn enter the food chain as the plants are consumed by animals.

The carbon cycle consists of three main pools that store carbon for different amounts of time. The *biological pool* has the shortest storage time related to growth, death and decomposition of organisms over periods of days and years. It thus exchanges rapidly with the atmosphere and the surface layers of the ocean. Intermediate in storage time is the *soil carbon pool*, which is relatively stable and does not exchange so rapidly as biological material. World-wide, soil carbon may amount to as much as 1500 billion tonnes, which is greater than that held in the atmosphere, the biota or the surface layers of the oceans (see figure). The pool with the largest residence time is the *geochemical pool*, which consists of two components: rock (including limestone – $CaCO_3$ – and fossil fuels) and deep ocean water. Both components are huge in comparison with the biological and soil carbon pools.

Humans are now interfering with the natural rate of exchange between these pools and thereby adding large amounts of carbon, in the form of carbon dioxide and methane, to the atmosphere. Carbon is being unlocked from the geochemical pool by the burning of fossil fuels (coal, oil and gas), carbon from the soil pool is oxidised and transferred to the atmosphere as cultivation takes place, and the destruction of forests and grasslands is releasing more carbon dioxide into the atmosphere than would happen naturally by decomposition.

The result of this is that whereas the carbon-dioxide level of the atmosphere before the Industrial Revolution and human population explosion was about 270 ppm (parts per million), it has now climbed to over 360 ppm, and may reach double its natural background level (between, say, 500 and 600 ppm) by the middle of the twenty-first century. Many scientists believe that the presence of increased levels of atmospheric carbon dioxide will promote global warming – the greenhouse effect.

obtain oxygen for respiration from the open atmosphere, and plants absorb nitrogen and carbon dioxide through their leaves. Bacteria and algae may fix nitrogen in the soil. Other nutrient inputs include those brought into an area by animals that have either migrated naturally or have been introduced by man, and fertiliser applied by farmers.

Nutrient outputs are achieved in a variety of ways. Soil erosion by wind and water can lead to the export of large amounts of nutrient material, especially if fertile topsoils are involved. Leaching is also a significant cause of nutrient loss, for water percolating through the soil carries with it nutrients in solution, which may ultimately be carried into streams and rivers. Other losses may take place in the gaseous form by diffusion of gases from soil pores into the atmosphere or because of transpiration by plants. Increasingly, nutrient losses may result from human activities, including harvesting.

The energy within an ecosystem is more or less continuously renewable, but the same is not the case for the nutrients involved in the biogeochemical cycles. Although atmospheric inputs may be quite important in some areas, and for some elements, most nutrients are derived from the weathering of rock and the creation of soil by the action of organisms. Such transformations are rather slow, with the result that, unless nutrients are constantly recycled within the ecosystem by the return of decaying plant and animal remains to the soil, there is a decline in fertility. In a mature, stable woodland, for example, there will be some loss of nutrients by erosion and leaching, but these will be more or less balanced by release of fresh minerals through weath-

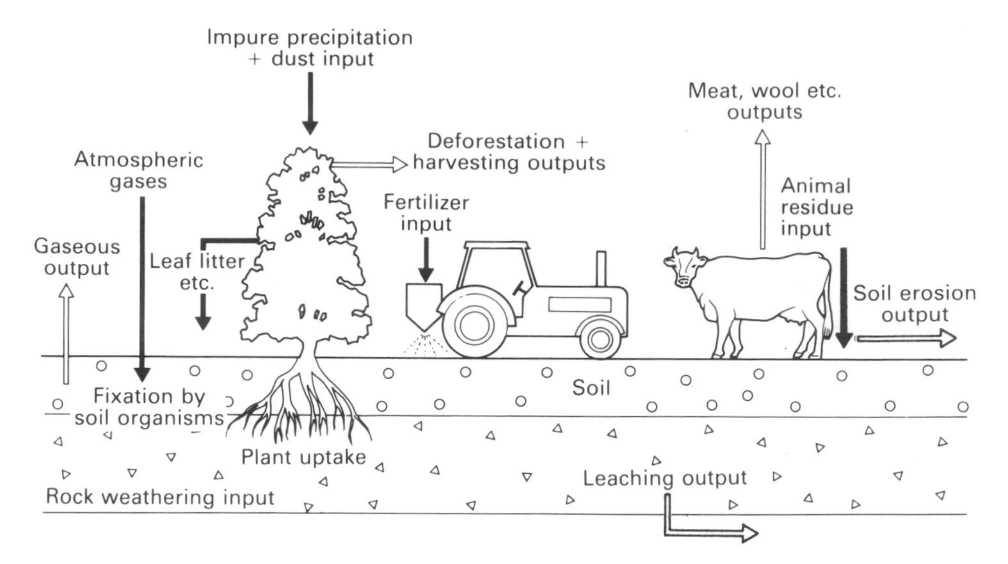

Figure 10.3 A simple model of biogeochemical cycling within an ecosystem.

ering and by additions from the atmosphere. In the great equatorial rainforests (see section 7.6) most of the nutrients are actually stored in the vegetation layer; many of the soils are relatively infertile, and under natural conditions there is very little loss of nutrients to drainage water. As a consequence, if deforestation takes place the great store of nutrients will be depleted, and the whole ecosystem will become greatly depleted in nutrients.

10.2 Plants and their Habitats

The place where an organism or community of organisms lives is called its *habitat*. A habitat possesses various environmental conditions or factors that affect the growth of plants. In order to cope with these conditions, plants may have made certain physiological and morphological adaptations by selective evolution over time. Such adaptations are beneficial in that they permit the plant to live in a particular environment, but a plant, by being so specialised that it can withstand the factors in one environment only, may lose its adaptability to changing factors. Over-specialisation can be the first step to extinction. Although we will now discuss some of the environmental factors in habitats individually, it should be remembered that factors are normally closely intermeshed and interrelated. For example, climate may affect other factors – soils, vegetation, the incidence of fire, the nature of landforms and so on. Moreover, although all factors are interrelated in a complex way, some are more independent than others. In particular, climate and geological factors tend to be independent, whereas soil, vegetation and animals tend to be more dependent on them.

Water is a most important factor in habitat, for water carries nutrients, is a raw material of photosynthesis, is essential for chemical reactions within a plant, makes up a large proportion of the plant mass and helps to maintain a relatively equable soil climate because it can absorb much heat with relatively little temperature change. According to their responses to the water environment, plants are classified into three main groups:

(a) *Xerophytes* (plate 10.1) These are plants that can withstand water shortages for long periods, either because the climate is dry or because the plants live on surfaces, like bare rock, that have limited water-storing capacity. Xerophytes are of four main types:

(i) annuals or ephemerals, which get through a drought in seed form;

Plate 10.1 Plants that can withstand water shortages for long periods are termed xerophytes. Two of the most characteristic xerophytes of the deserts of the south-west of the United States are the yucca (left) and the ocotilla (right).

(ii) plants that tap groundwater by sending down deep roots (such plants are called *phreatophytes*);
(iii) succulents, which store water in their stems or leaves (e.g. cacti);
(iv) plants that withstand drought by having a great variety of structural adaptations, such as large-spreading root systems to tap available soil water, bulbs and tubers to store water and nutrients, woody structures to withstand wilting etc.

(b) *Mesophytes* These are plants that cannot inhabit water or saturated soil, yet cannot survive a prolonged water deficiency. Most plants of temperate regions come into this category.

(c) *Hydrophytes* These are plants that cannot withstand drought. Many of them, like mangroves and water lilies, have their roots permanently in water; others, like ferns of the tropical rainforests, live in a permanently humid micro-climate.

Another major climatic factor is *temperature*. Different plants have different optimum temperatures for their growth, and different temperatures at which growth begins. Some plants cannot withstand freezing, some require daily changes of temperature, some cannot withstand great heat and some require a cold season in which they can lie dormant. Likewise, *light* is an important factor, for light is required for photosynthesis. Some plants have developed notable adaptations that enable them to seek out light (e.g. climbing plants), but if conditions are dark, as in the lower layers of the rainforest, vegetation growth will be limited, and the ground may remain bare until a break in the canopy improves lighting conditions. *Wind* is another climatic factor – it affects plants by increasing transpiration, by mechanical damage, and by scattering pollen and seeds. Deformed krummholz trees are a feature of the timber-line on mountains (see section 8.5), and some very exposed situations, such as high-latitude islands, may have only low-growing cushion plants and dwarf shrubs. This is the main reason why islands like the Falklands have such limited tree cover.

Another major group of factors in habitats are those associated with soil conditions. These are called *edaphic* conditions. Plainly, soils with limited nutrients, a high salt content, very shallow depth, waterlogging or impermeable and impenetra-

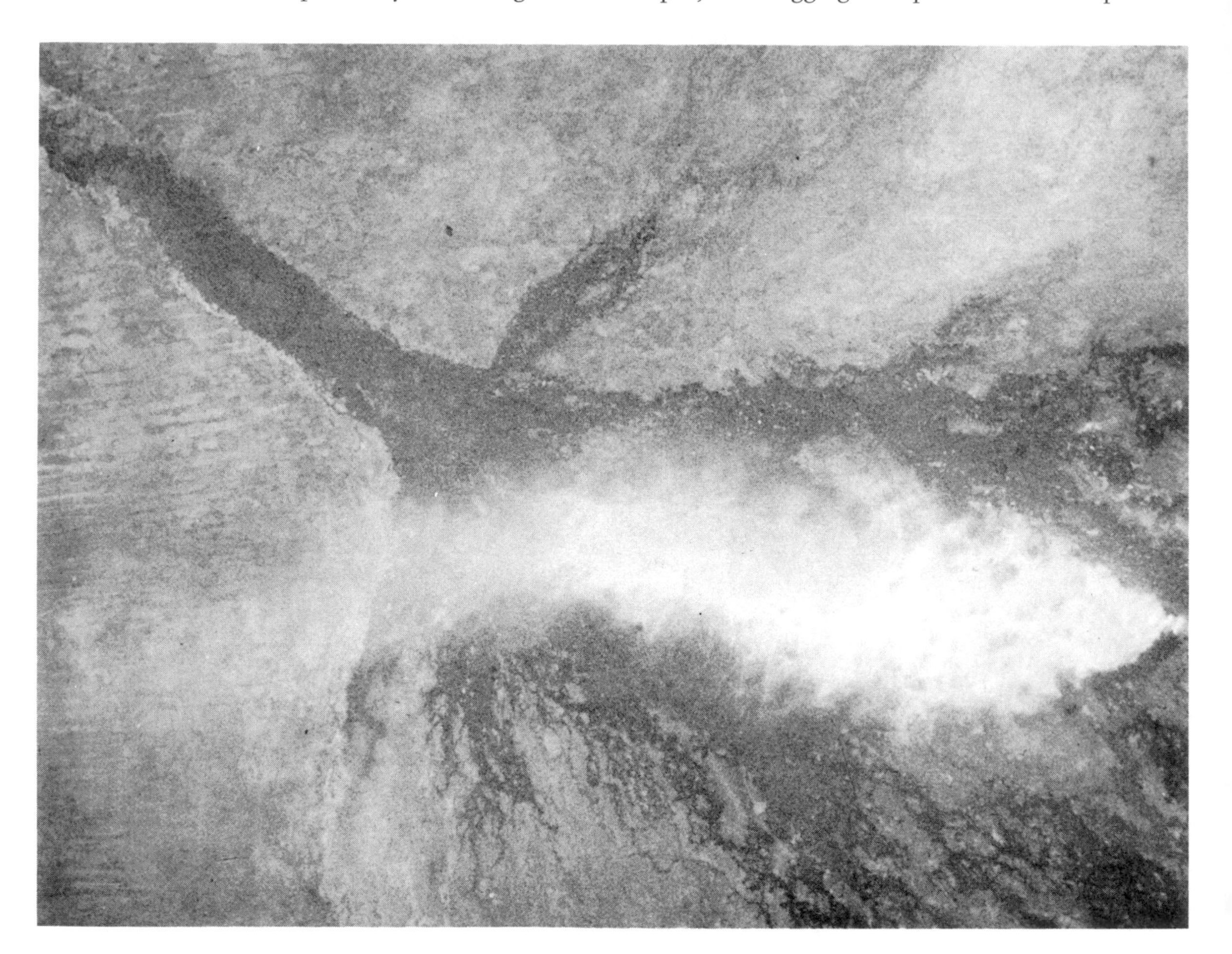

Plate 10.2 Fire is a major control of certain vegetation types and is a habitat factor of some importance. In savanna areas fires may be of regular occurrence and sometimes huge, as is shown by this photograph of the Okavango Delta in Botswana by a Space Shuttle crew. The white plume is a belt of smoke crossing the delta from east to west. The stripes on the left of the photo are fossil dunes (compare plate 2.3).

ble layers (like some laterites or podzols) will impose certain limitations on plant growth. For example, in many desert areas the presence of salt in the soil (see section 6.7) requires that, if they are to survive, plants will need to make adaptations that will enable them to overcome the effects of the salt. Such plants are called *halophytes*. Likewise, in some tropical areas (see section 7.8) it has been shown that, whereas the normal type of vegetation in an area is rainforest, there are patches of grassland (savanna) which occur because the trees cannot grow in soils that either are very highly leached of nutrients (as on old erosion surfaces underlain by chemically rather inert rocks like quartzites) or have a laterite horizon which limits root growth.

A factor of some importance which is rather difficult to place into any other category is *fire* (plate 10.2). It is a factor that is plainly more prevalent in certain environments than others – e.g. where there is a long dry season or where lightning is common – but it can also be important where fires are deliberately started (see section 5.9). In environments with a marked dry season, some shrubs have seeds that remain dormant until their hard pods have been cracked open by fire and seedlings can emerge. Cork oaks, some pines and many savanna trees have barks that protect them from all but the hottest fires. Some of the great vegetation types – e.g. savanna grassland, prairie and maquis – are a response to fire, while others require fire to open up the forest to enable new seedlings or new species to spread.

Finally there are *biotic* factors, for we must remember that no plant is independent of the activity of other organisms. The decomposers in the soil and litter layers are vital in nutrient cycling; insects cause pollination; birds and mammals disperse seeds; the grazing of herbivorous animals influences the plants on which they live; and one plant will compete with another for light, moisture and nutrients. Some plants are entirely dependent on others for providing support (e.g. lianas) or for providing their nutrients (e.g. parasites like the mistletoe).

10.3 Life in the Oceans

We can investigate some of these themes further by a brief consideration of life in the oceans. Marine life is diverse, and the range of size alone is very great: from microscopic one-celled plants to seaweeds over 30 m long, and from microscopic bacteria to the great blue whale, the largest animal ever to live on earth. None the less, species numbers are much greater on land than in the sea – some millions of species to the sea's hundreds of thousands. This is primarily because of the greater variety of habitats and environments on land.

Marine organisms occupy two major habitats: the *benthic* realm (the sea bottom), and the *pelagic* realm (the water above). Benthic organisms may live either *on* the bottom (*epibionts*) or *in* the bottom (*endobionts*). The *sessile* (immobile) types may, as in the case of barnacles, corals etc., cement themselves to a hard substrate. Other epibionts, such as seaweed, attach themselves to the substrate by root-like structures. Reclining epibionts, such as mature oysters, simply lie unattached on the bottom. Others crawl along the seabed or burrow beneath the surface.

Pelagic organisms may be classified into *nekton*, which actively swim and can make headway against normal currents, and *plankton*, which cannot swim against a normal current.

As on land, the organisms can be regarded in terms of their position in a food chain. At the first trophic level are the *phytoplankton*. The second trophic level constitutes the *herbivorous zooplankton*, and then at higher levels there are miscellaneous carnivores ranging from carnivorous zooplankton through plankton-eating fish and fish-eating fish:

Bacteria and biogenic detritus

↓

Phytoplankton

↓

Herbivorous zooplankton

↓

Carnivorous zooplankton

↓

Fish (*planktivores*)

↓

Fish (*piscivores*)

Table 10.1 Productivity and biomass of continents and oceans

	Continents	*Oceans*	*Difference*
Area (10^6 km^2)	149	361	×0.4
Mean net primary productivity per unit area (g m^{-2} yr^{-1})	773	152	×5.1
World net primary production (10^9 t yr^{-1})	115	55	×2.1
Biomass per unit area (kg m^{-2})	12.3	0.01	×1230
Total world biomass (10^9 t)	1837	3.6	×471

Source: modified from data in R. H. Whittaker (1975) *Communities and Ecosystems* (London: Collier-MacMillan), table 5.2

The production of these various organisms is surprisingly low when one considers the enormous volume of the world's oceans. As we have already seen, photosynthesis is powered by sunlight, and without photosynthesis production is of necessity very limited. In the oceans sunlight intensity diminishes very rapidly with depth. On average, at a depth of 10 m only 10 per cent of the light entering the sea is available for photosynthesis; at 100 m, only 1 per cent. In turbid coastal waters these values may be reduced even further.

Another highly important control of productivity is temperature. Over much of the ocean, temperatures lie below the temperature optimum for many species. Ocean water, especially at depth, is cold. Only 8 per cent of ocean water is warmer than 10 °C, and more than one-half is colder than 2.3 °C.

Although taken as a whole the oceans are unproductive with respect to their enormous volumes, there are some areas of high productivity – especially the continental shelves and areas in the open ocean where upwelling enriches the surface water in nutrients. The continental shelves are productive because, being shallow, they are warm and sunlit. Moreover, their productivity is enhanced where they receive nutrient-rich river discharge from the land. Areas of upwelling are productive because nutrients are continuously being brought up from depth to replenish stocks. This process goes on most effectively in those areas in mid-latitudes, such as on the west sides of continents, California, the coasts of Chile and Peru, south-west and north-west Africa, parts of north-west Australia, and also on the coast of Somaliland, and southern Arabia. Because of this great productivity, particularly of small pelagic organisms, these areas can provide food to support large fish stocks. The key nutrient in such locations appears to be phosphorus, and it is a peculiarity that phosphates are more soluble in cold water than warmer water, so that it occurs in larger quantities where cold upwelling waters prevail. At a more local scale, estuaries and coral reefs may be amongst the most highly productive of all ecosystems, being equal with that of tropical rainforest, with a mean net primary production of 1500 and 2500 respectively (g m^{-1} yr^{-1}).

Overall, however, the oceans are biological deserts, and the strong contrast between continental areas and marine areas of the earth's surface is brought out by table 10.1. The mean productivity and biomass are strikingly different. Thus although oceans cover about two-thirds of the world's surface, they account for only about one-third of its production.

10.4 Community

Within a given landscape, it is often apparent that the plants and animals are not distributed independently of one another, but that similar combinations are often found living together in habitats that are characterised by similar combinations of environmental conditions. This is termed the *community*. Three types of interaction between the different members of the community can be identified. The first is *competition*, in which individual plants or animals are competing for a scarce environmental resource. Competition for light is particularly important; in the absence of any limiting factors, the most successful plant species are those that grow the fastest or the tallest, or that produce dense shade or large accumulations of litter. The second of these types of interaction is where species are *complementary* to one another, co-existing by occupying different niches rather than demanding the same environmental resources. This is well shown in the stratification of vegetation in a forest, where the plants occupying the lower strata are capable of surviving in the shade cast by the taller species. The third category of interaction is called *dependence*. Here a species can occur only through the presence of others, which provide either physical support or nutrients.

Within a plant community some species are called *dominants*. These are plants that occupy more space, absorb more nutrients and contribute more biomass than the rest. They are almost always the tallest plants in the community, and as such exert considerable influence on the environment within the community – by affecting the light regime beneath their canopy, by creating areas of protection from exposure, and by utilising a larger proportion of available nutrients. If one compares an oak forest with a beech forest one can see this significance. Oak tree leaves are so arranged that light can reach the lower layers of the woods, thereby permitting the development of rich shrub and herb vegetation. By contrast, beech leaves are so arranged that little light can penetrate, so that the ground layer is relatively impoverished with regard to species and cover. There is thus a tendency for beech to be more dominant in a beech wood than oak in an oak wood.

10.5 Succession

The habitats of plants change through time, and the plants change with them. Plants, in fact, are able themselves to change various aspects of the environments they inhabit. *Succession* occurs because altered environmental conditions favour certain species which therefore can compete more successfully than before for nutrients, light, space and so on. As a result, populations of well-adapted species replace earlier ones now less well equipped to compete in the altered conditions. Thus, during the course of succession, individual species populations come and go, giving rise to a gradual progressive change in the community (figure 10.4).

The process of succession may start with bare ground. This might be produced by a mud flat being formed on a coast, by a sand dune developing behind a beach, by a glacier uncovering a rock surface as it melts, by a volcanic cone erupting, by a farmer clearing some land, or by fire destroying the previous vegetation cover. Over time, such bare ground becomes colonised by vegetation, and the vegetation in turn alters the nature of the environment. Eventually, in the mature stage of the succes-

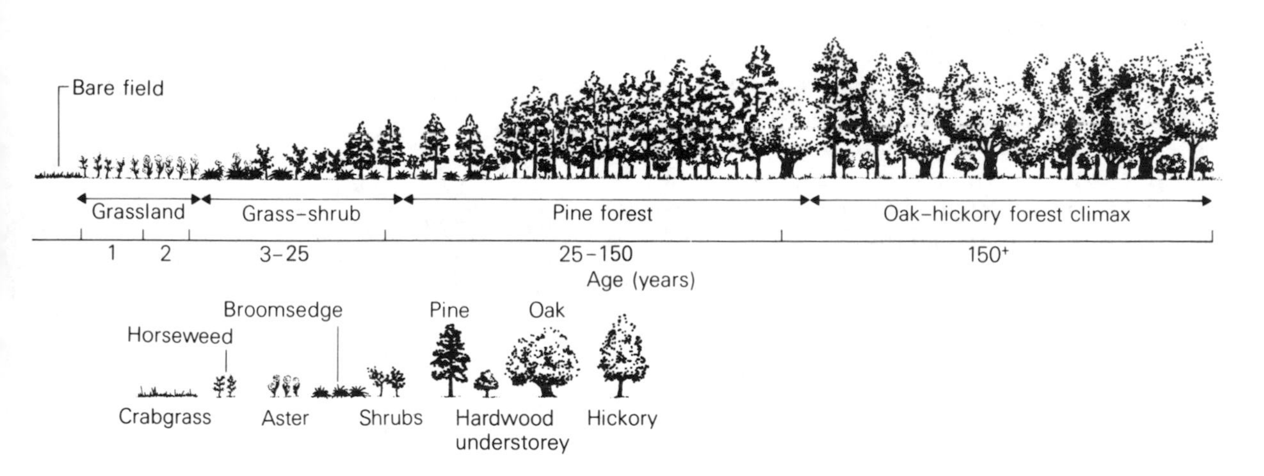

Figure 10.4 Succession from open land to a mid-latitude deciduous forest. Note how in the initial stages of the succession low, fast-growing grasses and shrubs dominate. Each stage alters the soil and micro-climate, enabling other species to establish themselves and become dominant. The final stage consists of high trees that overshade and force out some of the low shrubs of earlier stages.

sion process, the ecosystem becomes stable and relatively unchanging, and this stage is referred to as *climax*. Where this climax lasts for hundreds or thousands of years and appears to be in equilibrium with environmental conditions it is sometimes called the *climatic climax*. The climax community represents a steady state between plant cover and the physical environment. The whole group of plant communities that successively occupy the same site from the pioneer to the final stage of climax is called a *sere*, and the relatively transitory communities that occupy the site at a given time are *seral stages*. Seres can be classified according to environmental conditions: *hydroseres* are initiated in fresh water; *haloseres* are initiated in saline water; dry places give rise to *xeroseres*, of which *lithoseres*, are initiated on bare rock and *psammoseres* on dry sand.

Let us consider an example of succession. A glacier retreats and exposes a small lake. Initially only aquatic plants, such as water lilies and algae, and semi-aquatic plants, such as reeds and rushes, inhabit the water. As the plants grow and die their remains are deposited on the bottom, and after many years a thick organic layer may develop. In the shallow water near the shore, where plant productivity is highest, the organic layer gradually builds up to the water surface, where it provides new habitats for certain types of mosses, grasses and shrubs. These plants in turn both add organic matter to the habitat and help to stabilise the surface. Meanwhile inflowing streams are probably adding sediment to the organic mass. Little by little the community of plants encroaches on the open water and adds more and more organic matter. When the pond has completely filled, the aquatic and semi-aquatic plants no longer have anywhere to live because their original environment has been so greatly changed. Eventually the types of tree characteristic of that particular climatic zone colonise what had once been pond.

Another glacial example (this time from Glacier Bay, Alaska) will serve to demonstrate further the nature of succession. As the glacier melts and retreats, morainic material is deposited adjacent to the snout of the glacier. These morainic ridges can be dated, and so offer a means of studying how vegetation composition changes through time. On the youngest ridges, in close proximity to the glacier, lichens, mosses and herbs colonise the bare moraine, and may be joined by some small trees like *Salix* and *Populus deltoides*. In the next stage alder (*Alnus*) becomes more important, and this, with some of the herbs from the previous stage (e.g. *Dryas*), increases the soil nitrogen content through nitrogen-fixing activities. This increase in soil nitrogen, together with a reduction of pH brought about by decomposition of the alder leaves, creates conditions suitable for spruce (*Picea*) and hemlock, which make up the final woodland stage.

Other simple examples of succession described in this book include those on coastal sand dunes (see section 9.7) and those on salt marshes (see section 9.8).

A number of trends or progressive developments underlies most successional processes and sequences:

(a) There is a progressive development of the soil, with increasing depth, increasing organic content and the differentiation of soil horizons towards the mature soil of the climax community.
(b) The height of plants increases and strata become more evident.
(c) Productivity increases, as does biomass.
(d) Species diversity increases from simple communities of early succession to the richer communities of later succession.
(e) As height and density of above-ground plant cover increases, the micro-climate within the community is increasingly determined by the characteristics of the community itself.
(f) Populations of different species rise and fall and replace one another, and the rate of this replacement tends to slow through the course of succession as smaller and shorter-lived species are replaced by large and longer-lived ones.
(g) The final community is usually more stable than earlier communities and there is a very tight cycling of nutrients.

The rate of succession varies according to how favourable environmental factors are. If a tundra area is denuded of its natural vegetation cover by human activity or natural agency it will take a very considerable time for full successional development to take place, whereas in an area in the humid trop-

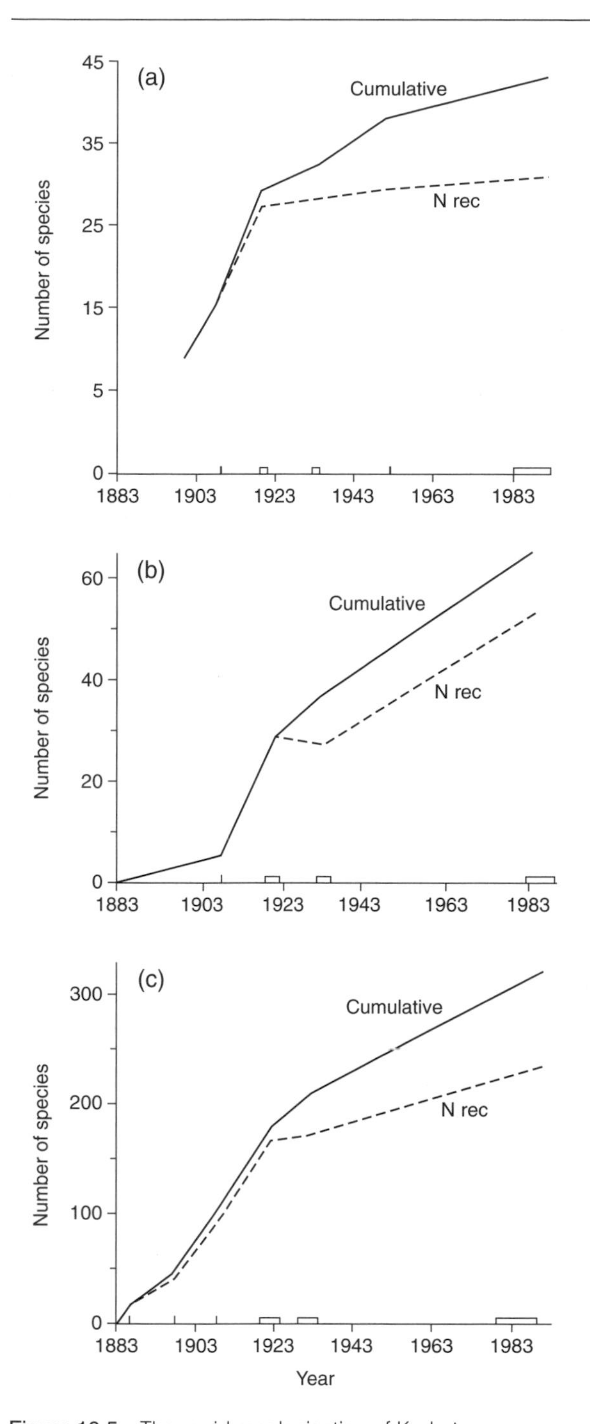

Figure 10.5 The rapid recolonisation of Krakatau (Indonesia) since the volcanic eruption of 1883. (a) 'Resident' land birds; (b) butterflies; (c) plants (N rec is the number recorded at each survey).

ics the various stages of succession may occur with great rapidity. For example, when the Indonesian volcano Krakatau exploded in 1883, all life on the island was killed and smothered with a layer of fresh ash some 30 m in thickness. Yet only 50 years later, a rich and maturing tropical rainforest with over 250 component plant species and many epiphytes had re-established itself (figure 10.5), together with 720 insect species and a few reptiles.

When he first proposed the concept of succession in 1916, F. E. Clements argued that for any major regional climatic zone only one type of *climax vegetation* could be expected, even though this might be derived from a wide variety of pioneer sites (dunes, marsh, pond etc.). It is evident, however, that there may be some *arresting factors* (topographical, edaphic, biotic etc.) that may halt the succession process and give rise to *subclimaxes*, in which plants are held in an apparently stable situation by non-climatic controls. Human activity may be one such arresting factor, and if its effects are permanent then a man-controlled *plagioclimax* community may be created (as in the case of many British heather moorlands). Some plant communities that were originally thought to be climatic climax types – savannas and mid-latitude grasslands, for instance – are now recognised as being partly the result of human activities, with fire playing a major role. In general, it is now believed that, instead of having just one climatic zone (the *monoclimax* idea), a whole species of possible stable communities might develop according to the relative importance of the different types of environmental factors (the *polyclimax* idea).

10.6 Dispersal and Migration of Plants

In order for a newly exposed area of land to be colonised as a first stage in the process of succession, there needs to be a means whereby plants can enter the area. This brings us to the question of the ability of plants to disperse and migrate. *Dispersal* involves dissemination from the parent plant and distribution to a new place; *migration* involves successful dispersal and establishment.

Many plants are adapted to the use of wind as an agent of dispersal. Trees like the sycamore, and flowers like the dandelion, have seeds that are specially adapted to wind transport, and many light seeds and spores are spread very effectively by this mechanism. Other plants may be distributed by water. Some mangroves have seedlings that float easily, for example, while coconuts have fibrous outer husks which favour their transport in the sea. Animals and humans also play a role in dispersal. Seeds may be carried adhering to the fur, hair, skin or feathers of beasts by hooks or sticky surfaces, or they may be carried internally. Since the development of long-distance transport, plant dispersal has been accelerated by man – sometimes deliberately and sometimes accidentally. Finally, dispersal may be achieved by mechanical means. Some plants have seed pods that explosively scatter their contents in the wind, while other plants send out runners and suckers over quite long distances.

In spite of the range of dispersal mechanisms that are available, plants are not evenly distributed over the face of the earth, for there are certain *barriers to dispersal*. These include sheets of water, deserts, high mountains, forests and other environmental factors. It is a characteristic of remoteness that the flora and fauna will be impoverished.

10.7 Dispersal and Migration of Animals

As we saw in chapter 3, the world can be divided into certain major regions or realms which have distinct assemblages of animals. Australia, for example, has a very different fauna from Asia, and has certain distinctive characteristics like the presence of marsupials. Even quite small areas of land, like islands or mountains, may have species that are found only in those particular environments and nowhere else. Such unique species are described as being *endemic*. From this, it is evident that animals are faced by barriers to dispersal in much the same way that plants are, permitting different areas to acquire different species by the process of evolution.

Once again, *isolation* is an important consideration. Britain's fauna is impoverished in comparison with that of mainland Europe because Pleistocene glaciation drove out many types of animals and the postglacial rise in sea level cut the country off before they could come back again. Being still further removed from the mainland, Ireland's fauna is even more impoverished, and Ireland today lacks certain beasts that are encountered in England and Wales: these include the poisonous adder, the mole, the common shrew, the weasel, the dormouse, the brown hare, the yellow-necked field mouse and the English meadow mouse.

The importance of isolation is still more important in oceanic islands distant from major land masses (table 10.2). New Zealand, for instance, had a very limited fauna before certain species were introduced by Europeans over the past two centuries (see window 3.2). The only mammals were two

Table 10.2 Higher plant species richness and endemism of selected islands

Island or archipelago	*Species number*	*Endemics*	*% endemic*
Borneo	20 000–25 000	6000–7500	30
New Guinea	15 000–20 000	10 500–16 000	70–80
Madagascar	8 000–10 000	500–800	68.4
Cuba	6514	3229	49.6
Japan	5372	2000	37.2
Jamaica	3308	906	27.4
New Caledonia	3094	2480	80.2
New Zealand	2371	1942	81.9
Seychelles	1640	250	15.2
Fiji	1628	812	49.9
Mauritius, including Réunion	878	329	37.5
Cook Islands	284	3	1.1
St Helena	74	59	79.7

These figures should be treated cautiously as different values for some islands can be found; in some cases even within a single publication.
Source: R. J. Whittaker (1998) *Island Biogeography* (Oxford: Oxford University Press), table 3.1

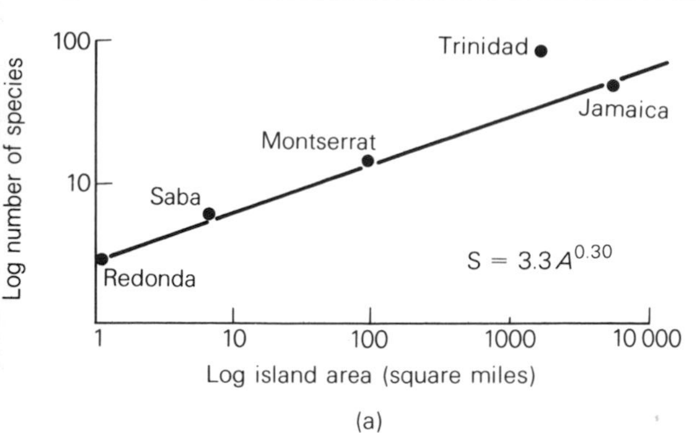

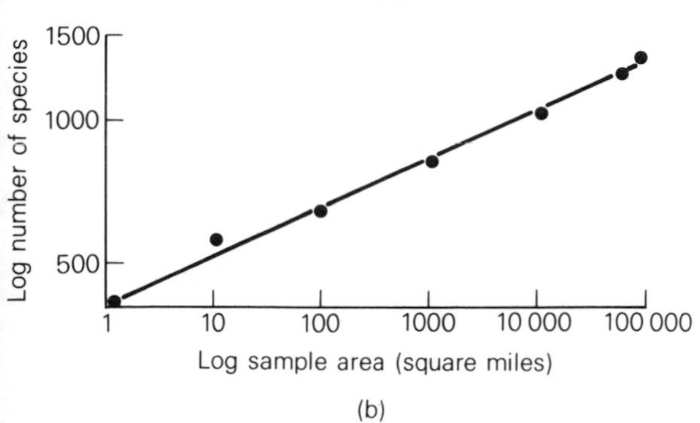

Figure 10.6 Two examples of the relationship between the area of an ecosystem and the number of species that it possesses: (a) the number of amphibian and reptile species living on West Indian islands of various sizes; Trinidad, joined to South America because of low sea levels as recently as 10 000 years BP, lies well above the area-species curve of the other islands; (b) the area-species curve for the number of species of flowering plants found in sample areas of England.

families of bats, though there were various flightless birds, including the kiwi, and the moa (now extinct). The latter were probably descended from flighted birds, but in the absence of predators they took to the ground.

10.8 The Importance of Area

In discussing factors that affect the nature of plant and animal geography we have so far said very little about one extremely important factor: the areal size of the ecosystem. A substantial body of work has now shown that there is a relationship between the number of species in an area and the size of that area (figure 10.6). Throughout the world larger islands support more species than smaller ones. Generally speaking, if the number of species in a given area is plotted against island area (both on log scales), a linear relationship results:

$$\log S = \log C + Z \log A$$

where S is the number of species, C is a constant giving the number of species when A has a value of 1, A is the area of the island, and Z is the slope of the regression line relating S and A. The value of C shows considerable variation from area to area and for different types of organisms, but Z is remarkably constant, with most values falling in the range 0.24–0.34.

The reason for this extremely widespread and important relationship is the subject of some controversy. Some scientists believe that as an area increases so does the range of topographical relief and habitat diversity. Other scientists argue that differences in area alone may be responsible for differences in species numbers. It is probable that habitat diversity is an important factor affecting species diversity, and that habitat diversity does tend to increase with island size, but there is also evidence that area in itself has some significance.

An explanation for the importance of area has been put forward in the influential *Equilibrium*

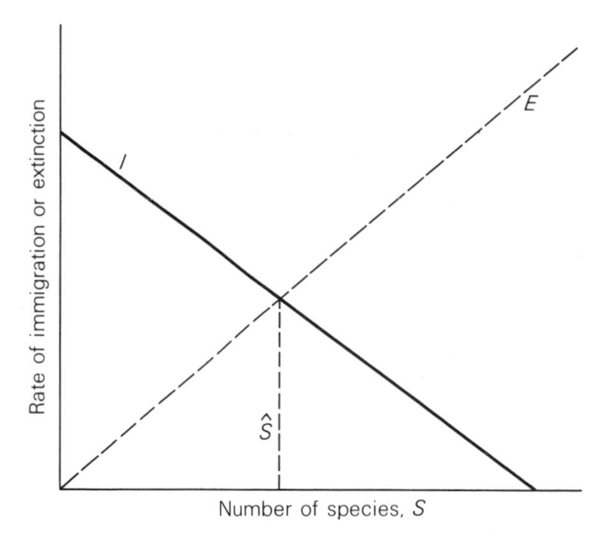

Figure 10.7 A diagrammatic representation of the MacArthur and Wilson model showing an island's extinction curve E and immigration curve I as functions of the number of species (S) on the island. The equilibrium number of species for the island is $\hat{S}$, where the curves intersect.

Theory of R. H. MacArthur and E. O. Wilson. This is illustrated in figure 10.7. The model envisages the number of species on an island as a dynamic equilibrium resulting from a continuing immigration of new species to the island and the extinction of species already there. As the number of species present on the island increases, fewer of the immigrants are new, so the immigration rate of new species falls, reaching zero when all mainland species are present. Furthermore, as the number of species increases, the rate of extinction increases for two reasons. First, there is an increasing number of species to become extinct, and, second, each species may be rarer (because of competition) when more species are present, so each may have an increased chance of extinction. At the point where the immigration and extinction curves cross, the two rates are equal, with new species arriving at the same rate as the old ones are becoming extinct.

Let us see how this explains the importance of area. Imagine two islands at the same distance from the mainland but one much larger than the other. Since they are equally accessible, they will enjoy similar immigration rates. However, the smaller island will usually support smaller populations of any species than the larger one, and small populations are more likely to become extinct than large. Thus the smaller island will have a higher rate of extinction and will hold fewer species at equilibrium.

The relationship between species diversity and the area of an island is a concept that can be applied to isolated habitats on the continents: lakes, desert oases, sphagnum bogs, patches of woodland, mountain summits and so on. Similar relationships to those found for true oceanic islands have been found for habitat 'islands' on the continents. This has its own intrinsic biogeographical interest, but it is of interest also in terms of the design of nature reserves, for many such reserves may be small patches of a particular habitat preserved in what is otherwise an inhospitable environment. They too may be 'islands'. This implies very clearly that, since area is an important consideration, nature reserves should in general be as large as we can make them. A large reserve will support more species at equilibrium by allowing the existence of larger populations with lower extinction rates.

10.9 Human Impact on Plants and Animals

As *Homo sapiens* has diffused over the face of the earth, as the world population has increased, and technological competence has developed, people have become more and more capable of having a significant impact on plants and animals. Our impacts are legion:

(a) *Genetic change* We have changed the genetics of plants and animals so that they are of greater use to us, and so that in many cases they cannot survive without human help. This is the process of *domestication*. The effect of domestication is to replace the processes of natural selection by those of human selection. We have been domesticating plants and animals for at least 10 000 years.

(b) *Ecosystem change* We have altered habitats (e.g. by irrigation, forestry, ploughing, hunting) and modified the numbers of species, their distribution and their population sizes.

(c) *Protection, preservation and conservation* We have deliberately tried to counter the effects of (b) by creating reserves for animals and by protecting their habitats.

10.10 Domestication

In the past 10 000 years or so, we have sought to replace the processes of natural selection by human selection of species, with the aim of changing the genetics of the organism so that the new characteristics (which are regarded as desirable) are passed down to succeeding generations. This process is called *domestication*. Our agricultural crops and our main domestic animals are the consequence of this form of environmental control, and different crops and animals have been domesticated at different times and in different places (figure 10.8). One of the most active phases of domestication occurred in the Near East (modern Palestine, Iraq and Iran) after about 11 000 BP. From that area came such useful species as cattle, sheep, goats, barley, wheat, oats and rye. Rice and chicken came from south-east Asia, and maize and various types of beans from Central America.

Domestication has been a most important stage in our cultural evolution, enabling us to produce

17 A large termite mound in the Kimberley district of tropical north-western Australia. Termites consume large quantities of organic material and are very important in leading to its decomposition and recycling through the ecosystem.

18 In southern Africa, the native fynbos heathland, one of the most attractive and diverse habitats, is being disrupted by the explosive spread of introduced exotic plants (in this case *Acacia cyclops*) from Australia and elsewhere. Ecological invasions are one of the most striking manifestations of the human impact on fauna and flora.

19 In recent years the importance of wetlands has become increasingly clear. They are often immensely productive and varied habitats for many types of wildlife. Their nature can vary greatly with the seasons, as shown by Mrazek Pond in the Everglades of Florida, USA: left shows the dry season and right shows the wet season.

20 A dark dyke cutting discordantly through lighter-coloured granites in the desert terrain of southern Jordan.

21 A large debris slide, Black Ven, on the coast of Dorset, southern England. Clay-rich Lias rocks, saturated with water, slide and flow into the English Channel. The coast is retreating rapidly, and the slides threaten property and pose a hazard to tourists and field parties.

22 A pediment near Grunau in southern Namibia. Notice the abrupt break of slope between the granite hills and the plain. The pediment is a gently sloping rock-cut feature with only a limited cover of debris.

23 A podzol soil from the North York Moors, northern England. Note the upper organic layer, the bleached layer, the zone of iron accumulation and the underlying parent material.

24 Weathering in Cambrian/Ordovician sandstones in southern Jordan. The outer surface of the sandstone has become case hardened, but the weakened sandstone behind is being weathered by salt crystallisation and other processes to give a series of cavernous features or niches, called *tafoni*.

25 Deeply weathered granite enclosing core stones near Chapman's Peak in the Cape, South Africa. Tors and inselbergs may be formed when such cores of more resistant rock are exposed as the weathered material (regolith) is stripped away.

26 The floodplain of a river in Swaziland, southern Africa. The yellow material is a series of sandy point bars that were deposited when the meandering river was in flood.

27 A series of terraces near Arthur's Pass, South Island, New Zealand. They probably result from a combination of rapid tectonic uplift of the snow-covered Southern Alps and changes in riverload resulting from the waxing and waning of glaciers in the catchment area.

28 Amongst those rivers which carry a large sediment load, those draining the heavily cultivated loess lands of China have pre-eminence. Rivers like the Yellow River (Huang He) gain their name from the large amount of loess-derived material which they transport.

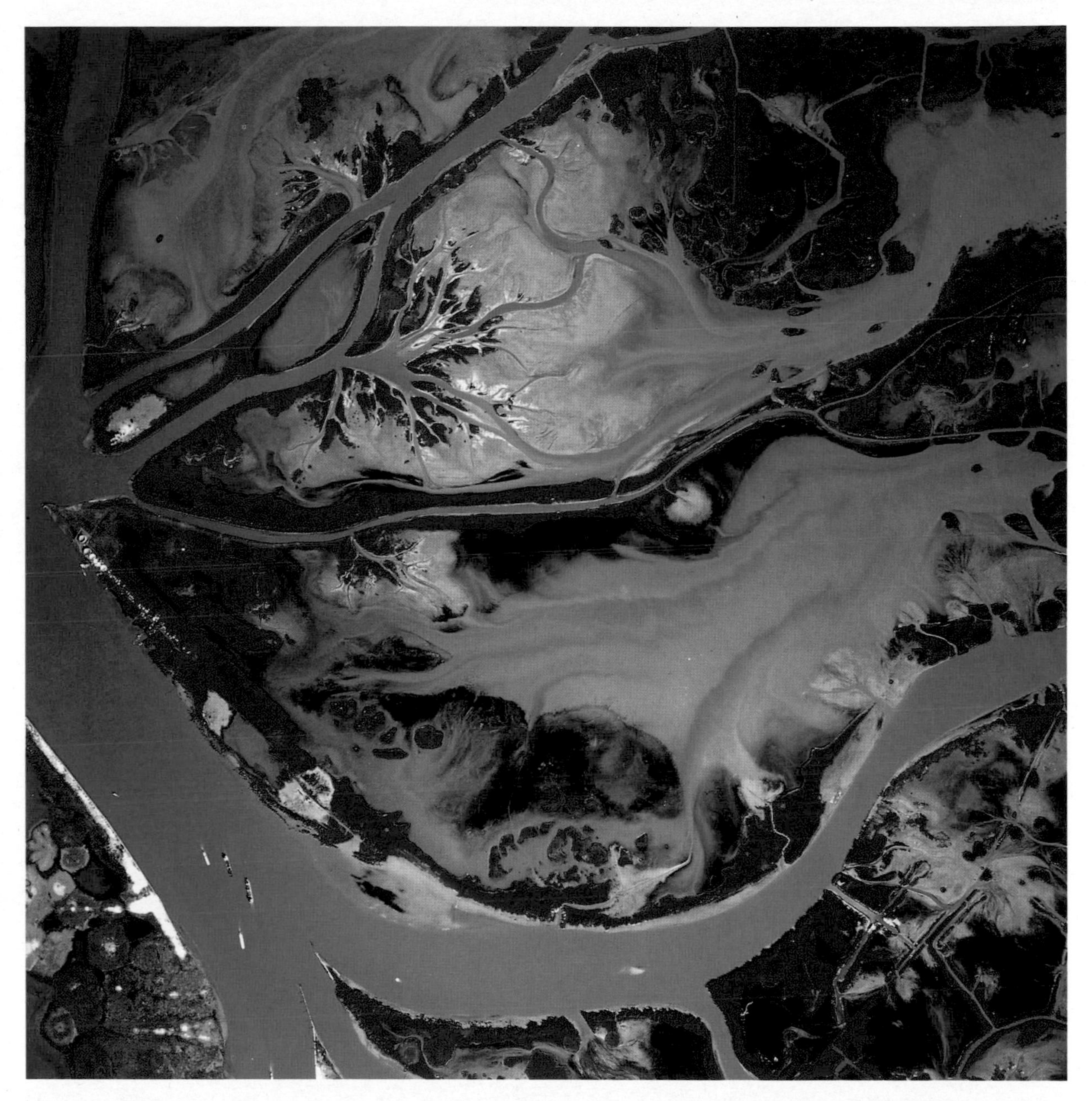

29 Small subdeltas in the vicinity of the Head of Passes, Mississippi River Delta, USA (shown by ER2 false colour infrared photography from 62 000 feet, 13 December 1989). Note the development of levées along the major channels and the delicate nature of the components of this environment. The lobes have formed as a result of subsidence in the areas behind the natural levées. The main river at this point is about 1200 metres wide.

30 A severely damaged bridge over the Usutu River at Sidvokodvo, Swaziland, southern Africa, illustrates the power of cyclone-generated floods in tropical environments. During the flood the river reached a height of some 7 metres above the bridge parapet. Vegetation was largely scoured out of the river channel and large fig trees transported by the flood contributed to the damage that was caused.

31 Large cities and industrial areas produce substantial quantities of atmospheric pollutants, such as sulphur dioxide and ozone. These pollutants can affect plant growth. This picture shows a smog over Cape Town, South Africa.

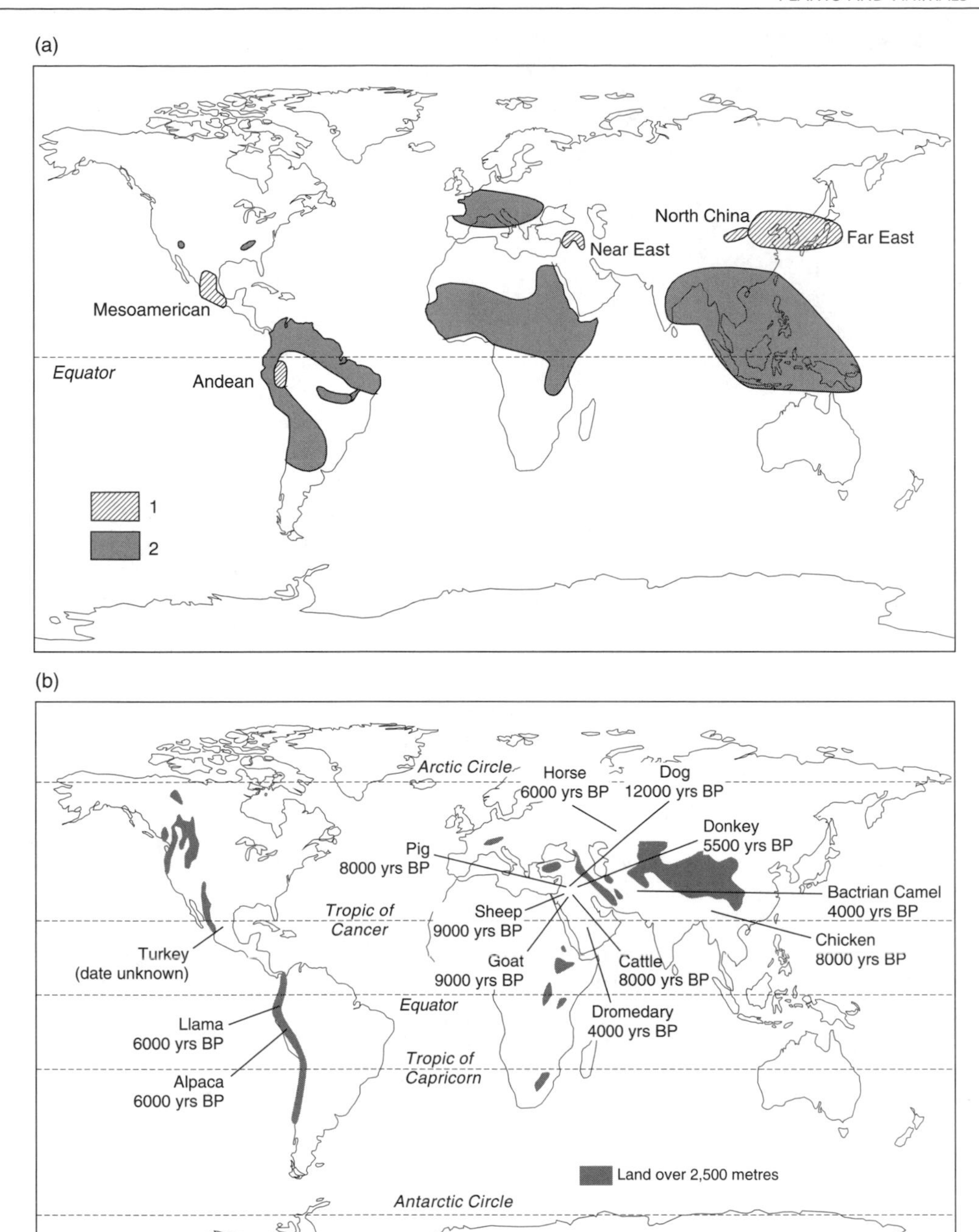

Figure 10.8 (a) Major areas of domestication of plants identified by various workers. The hatched areas (1) are the prime centres in which a number of plants were domesticated and which then diffused outwards to various neighbouring regions. The dark areas (2) are broader regions in which plant domestication occurred widely and which may have received their first domesticated plants from prime centres. (b) The places of origin, with approximate dates, for the most common domesticated animals.

Plate 10.3 A dwarf cow with enormous horns – an example of the diversity of forms that can be developed through domestication.

much larger volumes of food from a much smaller area than was possible for most hunters and gatherers. This in turn has permitted the great population densities that we have today. The breeding of plants and animals has also had major consequences of biogeographical significance. First, because of this breeding, the domesticated plant or animal is often reproductively isolated from its wild ancestors and so becomes effectively a new species. Second, domestic varieties show marked changes in shape and size; domesticated maize, for example, has a very large husk compared with its wild predecessors. Third, domesticated species often show very considerable variety (plate 10.3). Consider, for instance, the extraordinary range in the characteristics of dogs, from the diminutive chihuahua to the spotted dalmatian, and from the shivering whippet to the robust mastiff.

10.11 Introductions

Another powerful way in which we have altered ecosystems is through the introduction of plants and animals. We have crossed the oceans and other barriers and taken plants and animals with us (figure 10.9). Sometimes the introductions have been deliberate, supplying people with food, sport, ornament or a means of combating local pests and predators. Frequently, however, the introductions have been accidental, as when seeds have been carried in ballast, or when rats have jumped ship in a foreign port.

Sometimes these introductions, whether deliberate or accidental, have found their new habitat so congenial, or free from competition, that they have undergone a major population explosion, have eventually become pests and have dramatically altered the environment (window 10.2). In Britain, for example, many elm trees died in the 1970s because of the accidental introduction of the Dutch elm disease fungus which arrived at certain ports, notably Avonmouth and those in the Thames Estuary. Ocean islands have been particularly vulnerable. Their relatively simple ecosystems appear to be unstable, and introduced species often find that the comparable lack of competition enables them to broaden their ecological range more than they were able to do on the continents. Moreover, be-

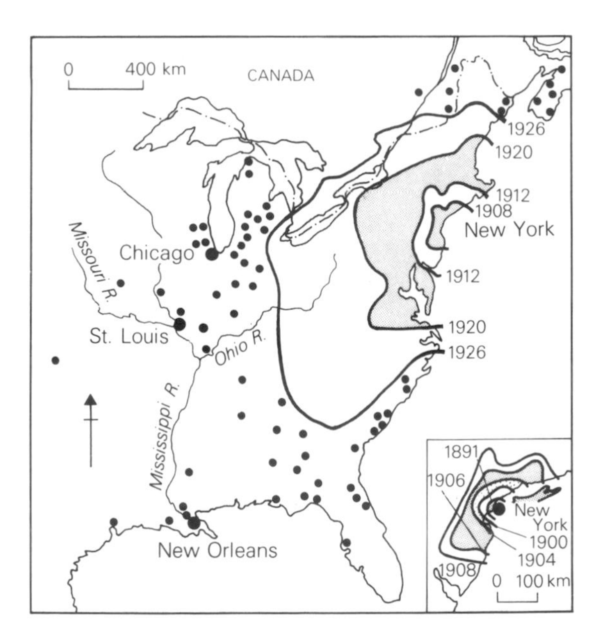

Figure 10.9 The spread of the European starling (*Sturnus vulgaris*) in North America following its release in 1891 in New York. Outside the 1926 isochrone, the dots are instances of early phases of colonisation. Less than a century after its release, the starling had become well established in most states and in Canada.

Window 10.2 *The introduction of the Nile perch to Lake Victoria*

Lake Victoria, the largest freshwater lake in the world, has a very rich fish fauna. These include over 300 species of cichlids, of which more than 90 per cent are endemic to the lake. In the 1960s, to improve the possibilities for commercial fishing, the Nile perch (*Lates*), a large beast, was introduced to Lake Victoria with the aim of converting the large biomass of small, unpalatable cichlids to a suitable table fish. The Nile perch is a voracious predator and the cichlids are sensitive to heavy predation because of their low reproductive potential. This is because they produce a limited number of large eggs, which they brood in the mouth. Their populations have crashed and over 200 species, about two-thirds of the total, have already disappeared or are threatened by eradication. Half a century ago cichlids made up over 99 per cent of the lake's fish biomass; today they are less than 1 per cent. The Nile perch are now feeding on other species, like freshwater prawns, and some of the larger perch have become cannibalistic. The whole ecosystem is being transformed and it remains to be seen whether the commercially successful Nile perch fisheries will be sustainable in the longer term. Many of the cichlids that the perch fed on were algae-eaters: with them gone, the dead, decaying algae suck oxygen from the waters of the lake.

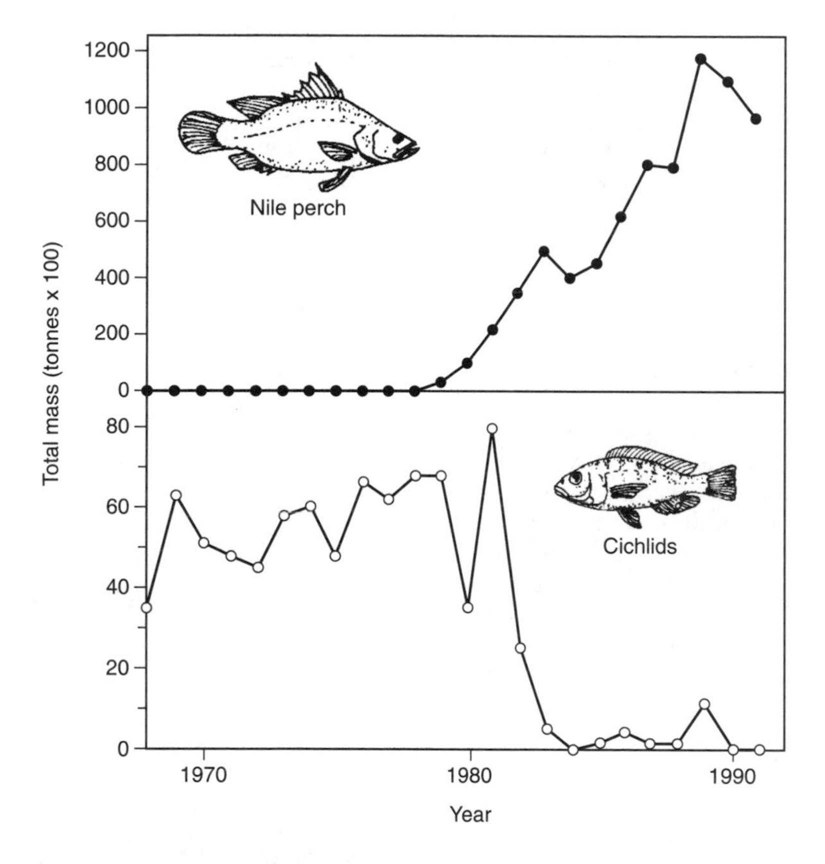

Commercial landings of Nile perch (*Lates*) and cichlids in the Kenyan part of Lake Victoria since the late 1960s.

cause many of the natural species inhabiting remote islands were selected primarily by their ability for dispersal, they were not necessarily dominant or even highly successful in their original continental setting. Therefore introduced species may prove more vigorous and effective than the indigenous ones. A clear illustration of this comes from the atoll of Laysan. Rabbits and hares were introduced there in 1903 in the hope of establishing a meat cannery. The number of native species on the island at that time was 24; by 1923 it had fallen to four. At that point all the rabbits and hares were systematically exterminated to prevent the island from being turned into a desert by overgrazing. By 1961 there were 16 species on the island.

However, it is not only simple ecosystems that are prone to disturbance by invaders. In South Africa, for example, there is an immensely species-rich and beautiful natural heath vegetation called fynbos, which occurs in the coastal regions of the Western Cape (see colour plate 18). Over large areas it has been displaced by various types of Australian trees of the *Acacia* family, which, freed from the organisms which control them in their native continent, have spread explosively in their new homeland (see window 3.3).

There are also many examples of the deliberate introduction of an organism to check the explosive invasion of a particular plant. One of the most spectacular examples of this involves the history of the prickly pear (*Opuntia*) in Australia. This plant came from the Americas and was introduced some time prior to 1839. By 1900, 4 million hectares had been invaded by it, and by 1925 more than 24 million hectares. Sometimes the density of the cactus was such that other more useful plants were excluded. To combat this menace one of *Opuntia's* natural enemies, a South American moth, *Cactoblastus*, was introduced to remarkable effect: by 1940 something like 95 per cent of the prickly pear had been eradicated.

Another introduction to Australia was the water buffalo. They were deliberately introduced into northern Australia, bringing their own blood-sucking fly with them. This fly breeds in cattle dung. Australia's native dung beetles, accustomed only to the sheep-like pellets of the grazing marsupials (the natural fauna of Australia before the arrival of Europeans) could not tackle the large dung pats of the buffalo. Thus unconsumed pats abounded and the flies were able to breed undisturbed until African dung beetles were introduced to compete with the flies.

10.12 Pollution

One consequence of human activity that appears to be having an increasingly widespread effect on plants and animals is pollution (plate 10.4). Some of the *air pollutants* that have been released into the atmosphere have had detrimental impacts on plants; for example, sulphur dioxide, one of the pollutants resulting from the burning of fossil fuels, is toxic to plants. Experiments with growing lettuces in different parts of Leeds, northern England, as early as 1913 showed that the size and

Plate 10.4 The great metal smelter at Sudbury in Canada has one of the largest chimneys in the world and produces polluting fumes that extend over large areas. Trees in the immediate vicinity are badly affected.

quality of lettuce leaves were very closely related to the amount of sulphate in the air in different parts of the city. Likewise, lichens are very sensitive to air pollutants and are rarely found in central areas of cities like Bonn, Helsinki, Stockholm, Paris and London. Overall, it has been calculated that over one-third of England and Wales, extending in a belt from the London area to Birmingham and broadening out to include the industrial Midlands, most of Lancashire and West Yorkshire and part of Tyneside, has lost nearly all of its epiphytic lichen flora, largely because of sulphur-dioxide pollution.

Another form of air pollution that affects plant growth is *photochemical smog* (see section 16.3). In California, Ponderosa pines in the San Bernadino Mountains, as much as 120–130 km to the east of Los Angeles, have been extensively damaged by smog coming from that metropolis. The gas that seems to be the culprit is ozone; in normal, clean air it occurs in low concentrations of only 4 parts per hundred million (pphm), but during smog conditions in Los Angeles it can reach concentrations of 70 pphm. In summer months in Britain ozone concentrations produced by photochemical reactions can reach 17 pphm. Fumigation experiments in the laboratory suggest that damage to plants can occur if ozone concentrations stay above 5–6 pphm for any extended period – a figure not very much above the natural background levels.

One of the problems of air pollution is that it can have repercussions over very wide areas. If we return to the example of sulphur dioxide, this gas is pumped into the atmosphere in great quantities by manufacturing plants and power stations in industrial areas, and it can create a phenomenon called *acid rain* (window 10.3). Normally, water in the atmosphere will have an acidity that is relatively small (pH of *c.*5.7). However, recently much stronger acids have been observed in snow and rain in the north-eastern United States and in Scotland,

Window 10.3 Acid precipitation

One consequence of acid rain is that buildings can be subjected to accelerated weathering. In this example, the limestone of which the great cathedral at Lincoln is made is being attacked by sulphate-rich rain, some of which may be derived from large power stations downwind on the River Trent.

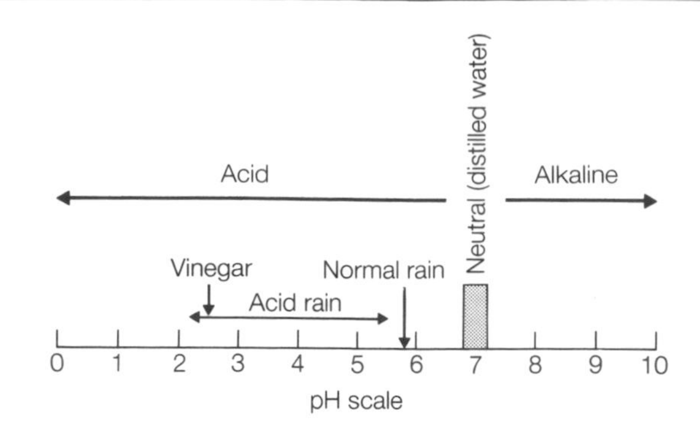

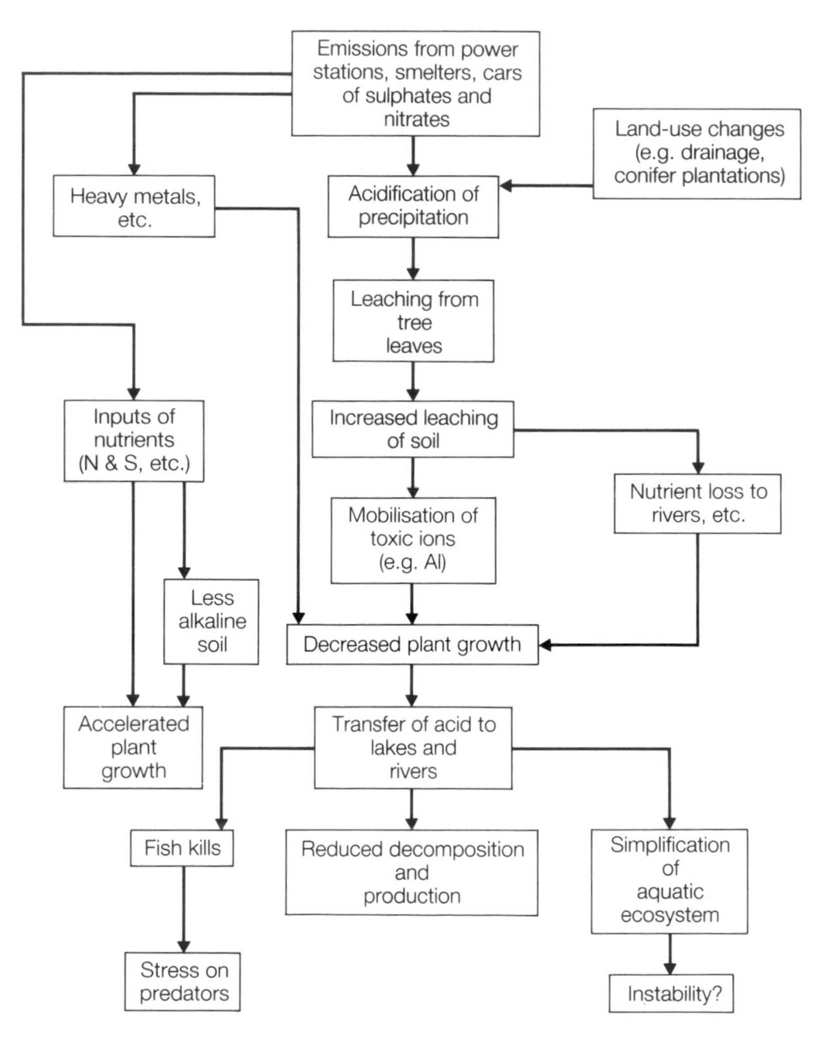

Pathways and effects of acid precipitation through different components of the ecosystem, showing some of the adverse and beneficial consequences.

Distilled water has a pH of 7, and is said to be neutral. Most rain, under natural conditions, would have a pH of just under 6, and would thus be slightly acid. This is because rain absorbs some carbon dioxide from the atmosphere, which produces a weak solution of carbonic acid. However, in some parts of the world rainfall is more acid than this and has a pH that is much lower. Large parts of Scandinavia, for example, have rain with pH values of just over 4. This is more serious than it seems at first glance because a change in pH of one unit means a tenfold change in acidity.

Areas of highly acidic precipitation are the result of the burning of fossil fuels (coal, oil and gas) which release two gases – sulphur dioxide (SO_2) and nitrogen oxide (NO_x) – which react in the atmosphere with water to produce acid solutions. It has been estimated that in Britain about 70 per cent of the acidity of rainfall comes from sulphur products and about 30 per cent from nitrogen products.

Areas downwind of major industrial areas suffer particularly from acidification of precipitation, especially those with plentiful snow and rain, and that have rocks and soils with few of the alkaline minerals (lime being a notable example) which can help to neutralise acid rain when it reaches the ground surface. Thus areas underlain by ancient igneous rocks like granite may be more vulnerable than limestone areas. Parts of Scandinavia possess many of these properties and so suffer seriously from the various ecological consequences of acid rain and snow. Among the consequences of acid precipitation are damage to trees, fish deaths in rivers and lakes, and dissolving of limestone buildings.

Research suggests that some of the damage caused by acid rain may not be directly because of the lowered pH of water and rain. One important indirect effect of acid rain is the liberation from rock and soils of various metals that can be harmful to plants and soil biota. Such toxic metals include aluminium, cadmium, copper and zinc.

What can be done to reduce acid rain? Sulphate emissions can be reduced by burning less fossil fuel, switching to lower-sulphur fuels (e.g. gas rather than coal), removing sulphur from coal before combustion, burning coal in the presence of neutralising limestone, and removing sulphurous gases from power station emissions by a process called 'flue gas desulphurisation'. Nitrates, many of which come from power stations, can be removed in similar ways, while those that are produced by cars can be reduced by modifications to engines and exhaust systems.

with pH values as low as 2.1–2.4 (the latter the acidic equivalent of vinegar). In much of the north-eastern United States and western Europe the sulphate pumped into the atmosphere causes average annual rainfall values to be around pH 4. As a result, the rain feeding many rivers and lakes is much more acidic than it used to be, and these waters have become more acid. Many fish and other organisms inhabiting freshwaters are sensitive to this acidity, so the ecology of many lakes has changed considerably.

In addition, high levels of soil and rain acidity appear to be injurious to trees, and acid rain probably contributes to the widely observed damage that has occurred to the forests of western and central Europe and various other parts of the world. Europe's forests exhibited signs of injury, especially to conifers: yellowing of needles, casting off of older needles, and damage to the fine roots through which these trees take up their nutrients (window 10.4).

Another example of the widespread implications of pollution is provided by the role of certain *pesticides* related to DDT. These were introduced on a world-wide basis after the Second World War and proved highly effective in the control of insects like malarial mosquitoes. However, evidence soon accumulated to suggest that DDT was persistent, capable of wide dispersal, and prone to becoming concentrated at high levels in certain animals at the top trophic levels in a food chain or web. For example, a lake might be sprayed to kill gnats, introducing a small concentration of DDT into the lake water. Bacteria and algae might accumulate some of the pesticide in their tissues, and would then be eaten by fish, which would have slightly higher levels of DDT, and so on, until the fish-eating birds at

Window 10.4 Forest decline

Forest decline is an environmental issue that attained considerable prominence in the 1980s. The common symptoms of this phenomenon (modified from *World Resources 1986*, table 12.1) are:

1. *Growth-decreasing symptoms*
   Discoloration and loss of needles and leaves
   Loss of feeder-root biomass (especially conifers)
   Decreased annual increment (width of growth rings)
   Premature ageing of older needles in conifers
   Increased susceptibility to secondary root and foliar pathogens
   Death of herbaceous vegetation beneath affected trees
   Prodigious production of lichens on affected trees
   Death of affected trees
2. *Abnormal growth symptoms*
   Active shedding of needles and leaves while still green with no indication of disease
   Shedding of whole green shoots, especially in spruce
   Altered branching habit
   Altered morphology of leaves
3. *Water-stress symptoms*
   Altered water balance
   Increased incidence of wet wood disease

Forest decline is a major environmental issue in many parts of the world. Although the causes of the phenomenon are complex and varied, it is believed that this damaged forest in Maine, USA, is the result of acid rain.

The decline is widespread in much of Europe and is particularly severe in Poland and the Czech republic.

Many hypotheses have been put forward to explain this dieback: poor forest management practices, ageing of stands, climatic change, severe climatic events (such as the severe droughts in Britain during 1976), nutrient deficiency, viruses, fungal pathogens and pest infestation. However, particular attention is being paid to the role of pollution, either by gaseous pollutants (sulphur dioxide, nitrous oxide or ozone), acid deposition on leaves and needles, soil acidification and associated aluminium toxicity problems and excess leaching of nutrients (for example, magnesium), over-fertilisation by deposited nitrogen, and trace metal or synthetic organic compound (e.g. pesticide, herbicide) accumulation as a result of atmospheric deposition.

The arguments for and against each of these possible factors have been expertly reviewed by Innes (1987), who believes that in all probability most cases of forest decline are the result of the cumulative effects of a number of stresses. He draws a distinction between the predisposing, inciting and contributing stresses (1987: 25):

> Predisposing stresses are those that operate over long time scales such as climatic change and changes in soil properties. They place the tree under permanent stress and may weaken its ability to resist other forms of stress. Inciting stresses are those such as drought, frost and short-term pollution episodes, that operate over short time scales. A fully healthy tree would probably have been able to cope with these, but the presence of predisposing stresses interferes with the tree's mechanism of natural recovery. Contributing stresses appear in weakened plants and are frequently classed as secondary factors. They include attack by some insect pests and root fungi. It is probable that all three types of stress are involved in the decline of trees.

As with many environmental problems, interpretation of forest decline is bedevilled by a paucity of long-term data and detailed surveys. Given that forest condition oscillates from year to year in response to variability in climatic stress (e.g. drought, frost, wind throw) it is dangerous to infer long-term trends from short-term data. There may also be differences in causation in different areas. Thus while widespread forest death in eastern Europe may result from high concentrations of sulphur dioxide combined with extreme winter stress, this is a much less likely explanation in Britain, where sulphur dioxide concentrations have shown a marked decrease in recent years. Indeed, in Britain the direct effects of gaseous pollutants appear to be very limited.

Source

J. Innes (1987) 'Air pollution and forestry', *Forestry Commission Bulletin*, 70.

Results of forest damage surveys in Europe: percentage of trees with >25% defoliation (all species)

	Mean of 1993/4		*Mean of 1993/4*
Austria	8	Latvia	33
Belarus	33	Lithuania	26
Belgium	16	Luxembourg	19
Bulgaria	26	Netherlands	11
Croatia	24	Norway	16
Czech Republic	56	Poland	52
Denmark	35	Portugal	7
Estonia	18	Romania	21
Finland	14	Slovak Republic	40
France	8	Slovenia	18
Germany	24	Spain	16
Greece	22	Switzerland	20
Hungary	21	UK	15
Italy	19		

Source: data from *Acid News*, 5 (1995), p. 7

the top trophic level would have very high concentrations indeed. Studies of birds like the osprey, the bald eagle and the peregrine falcon have shown that, as DDT levels have built up in them, so the thickness of their egg shells has been reduced, thereby impairing their success at reproduction. Some dramatic declines in the populations of such species have therefore occurred.

A final example of the effects of pollution on organisms is provided by a consideration of the process of *eutrophication*. What precisely is eutrophication? Fundamentally it is the enrichment of waters by nutrients. Among these nutrients, phosphorus and nitrogen are particularly important as they regulate the growth of aquatic plants. The process does occur naturally – for example, when lakes get older – but it can be accelerated by human activities, both by runoff from fertilised and manured agricultural land and by the discharge of domestic sewage and industrial effluents. The anthropogenically accelerated eutrophication – often called 'cultural eutrophication' – commonly leads, as in the case of the Black Sea, to excessive growth of algae, serious depletion of dissolved oxygen as algae decay after death and, in extreme cases, to an inability to support fish life. It can affect all water bodies, from streams, to lakes, to estuaries and coastal seas. Coastal and estuary waters are sometimes affected by algal foam and scum, often called 'red tides'. Some of these blooms are so toxic that consumers of seafood that has been exposed to them can be affected by diarrhoea, sometimes fatally. These blooms, produced by certain types of phytoplankton (tiny pigmented plants), can grow in such abundance that they change the colour of the seawater not only to red but also to brown or even green. They may be sufficiently toxic to kill marine animals such as fish and seals. Long-term studies at the local and regional level in many parts of the world suggest that these so-called red tides are increasing in extent and frequency as coastal pollution worsens and nutrient enrichment occurs more often.

What has happened, particularly since the Second World War, is that various human actions have speeded up the natural processes. The growth in fertiliser usage in the past five decades has been increasingly rapid. In spite of the increasing costs of energy supplies and hydrocarbons (from which many of the fertilisers are derived) in the 1970s, world fertiliser production has continued to rise inexorably, and fertiliser-derived nitrates reach groundwater and rivers. For example, the mean annual nitrate concentration of the River Thames, which provides most of London's water supply, increased from around 11 mg per litre in 1928 to 35 mg per litre in the 1980s.

A topic of considerable recent importance is the effect that human activities may be having on the layer of ozone which screens all life on earth from the harmful effects of the sun's ultraviolet radiation. Ozone (O_3) is a naturally occurring form of oxygen which consists of three oxygen atoms rather than two. It exists throughout the atmosphere in very low concentrations, never exceeding about one molecule in every 100 000 present. It is especially abundant in the stratosphere between 10 and 40 km above the ground. The 'ozone layer' contains about 90 per cent of atmospheric ozone and is important because it provides a thin veil which absorbs ultraviolet (UV) radiation from the sun. Indeed, the ozone layer prevents about 97 per cent of UV-B light from reaching the earth's surface. Too much ultraviolet radiation can damage plants, including the phytoplankton that live in the oceans. Such phytoplankton are crucial in that they form the base of the food chain in the oceans. In humans, it can cause skin cancers; it may also cause eye cataracts and damage the body's immune system. Thus it is clear that any reduction in the thickness of the ozone layer is worrying.

In the 1980s, satellite observations, ground measurements and readings from instruments on balloons and in aircraft began to suggest that the ozone layer was becoming thinner, especially over the Antarctic. More recent measurements have indicated that the ozone layer is also thinning over America and northern Europe. Here, ozone decreased on average by around 3 per cent in the 1980s. In the 1970s concern was expressed about possible damage to the ozone layer by high-flying supersonic aircraft such as military jets or Concorde. However, current concern among scientists is focused on a range of manufactured gases of recent origin. These include chlorofluorocarbons (CFCs) and halons. These gases have been extremely useful in many ways – for example, as refrigerants, for extinguishing fires, for making foams and plastics, and for use in aerosol spray cans. This is because they have some valuable properties: they are stable, non-flammable and non-toxic. Unfortunately, their stability means that they can persist a long time in the atmosphere and can thus reach the ozone layer without being destroyed. Once they are in the ozone layer, UV radiation from the sun starts to break them down. This sets off a chain of chemical reactions in which reactive chlorine atoms are released. These act as a *catalyst* causing ozone (O_3) to be converted into oxygen (O_2).

Global production of CFC gases increased greatly during the 1960s, 1970s and 1980s, from around 180 million kg per year in 1960 to nearly 1100 million kg per year in 1990. However, in response to the thinning of the ozone layer, many governments signed an international agreement called the Montreal Protocol in 1987. This pledged them to a rapid phasing out of CFCs and halons. Production has since dropped substantially. However, because of their stability, these gases will persist in the atmosphere for decades or even centuries to come. Even with the most stringent controls that are now being considered, it will be the middle of the twenty-first century before the chlorine content of the atmosphere falls below the level that triggered the formation of the Antarctic 'ozone hole' (see below) in the first place.

The most drastic decline in stratospheric ozone has been over Antarctica. This has led to the formation of the 'ozone hole' which expanded to an area of 24 million sq km during September–October 1992 and again in the same months of 1993. Record low ozone levels of less than 100 ozone units were registered during a few days in October 1993. These compare with values from years before the ozone hole (1957–78) of 330–250 units.

The destruction of ozone is greatest over the Antarctic because of the unique weather conditions during the long, dark winter of the south polar regions. Strong winds circulate in a great vortex above the Antarctic, essentially isolating the polar stratosphere from the rest of the atmosphere. Under very cold conditions, with temperatures below –80 °C, ice clouds form, called polar stratospheric clouds. These provide ideal conditions for the transformation of chlorine (derived from the breakdown of CFCs and halons) into potentially reactive compounds. When sunlight returns in the spring months, UV radiation from the sun triggers the reaction between these chlorine compounds and ozone, thereby leading to ozone destruction.

No such clear ozone hole develops over the Arctic because the more complex arrangement of land and sea here leads to a less well-developed vortex system of winds. In addition, the winter stratosphere at the North Pole tends to be warmer than its southern counterpart. None the less, ozone depletion does seem to have occurred, producing an ozone 'crater' rather than a hole.

10.13 Habitat Change

Pollution is but one of the ways in which human beings have so changed the environment of other organisms that their distribution and well-being have been affected. Of very great importance have been the changes in the nature of the habitats in which organisms live, and the area that is covered by such organisms.

Habitat destruction can take many forms, and they have been listed thus in the *World Conservation Strategy*:

(a) Replacement of the entire habitat by settlements and other human constructions.
(b) Replacement of the entire habitat by cropland, grazing land, plantations etc.
(c) Replacement of the entire habitat by mines and quarries.
(d) The effects of dams, which drown certain habitats, block spawning migrations, and change water conditions (temperature etc.).
(e) Drainage, channelisation and flood control.
(f) Chemical, nutrient and solid waste pollution.
(g) Over-exploitation of water, causing, for example, lakes to dry up.
(h) Removal of materials, such as vegetation, soil etc.
(i) Dredging and dumping.
(j) Overgrazing and overbrowsing by domestic stock.
(k) Erosion and siltation.
(l) The introduction of invasive alien plants and animals.

Losses to the heritage of nature

Habitat

Lowland herb-rich hay meadows: 95% now lacking significant wildlife interest and only 3% left undamaged by agricultural intensification.

Lowland grasslands of sheep walks. On chalk and Jurassic limestone: 80% loss, largely by conversion to arable or improved grassland (mainly since 1940), but some scrubbed over through lack of grazing.

Lowland heaths on acidic soils: 40% loss, largely by conversion to arable or improved grassland, afforestation and building; some scrubbed over through lack of grazing.

Limestone pavements in northern England: 45% damaged or destroyed, largely by removal of weathered surfaces for sale as rockery stone, and only 3% left completely undamaged.

Ancient lowland woods composed of native, broad-leaved trees: 30–50% loss, by conversion to conifer plantation or grubbing out to provide more farmland.

Lowland fens, valley and basin mires: 50% loss or significant damage through drainage operations, reclamation for agriculture and chemical enrichment of drainage water.

Lowland raised mires: 60% loss or significant damage through afforestation, peat-winning, reclamation for agriculture or repeated burning.

Upland grasslands, heaths and blanket bogs: 30% loss or significant damage through coniferous afforestation, hill land improvement and reclamation, burning and over-grazing.

Species

The large blue butterfly became extinct in 1979, but ten more species are vulnerable or even seriously endangered, and out of a total British list of 55 resident breeding species of butterfly another 13 have declined and contracted in range substantially since 1960.

Three or four of our 43 species of dragonflies have become extinct since 1953, six are vulnerable or endangered, and five have decreased substantially.

Four of 12 reptiles and amphibians are endangered.

At least 36 breeding species of bird have shown appreciable long-term decline during the last 35 years as a result of habitat loss or deterioration, 30 in the lowlands and six in the uplands.

The otter has become rare or has disappeared in many parts of England and Wales.

Bats in general have decreased and several of our 15 species, notably the greater horseshoe and mouse-eared bats are at risk of extinction. Others such as Bechstein's, Leisler's and the barbastelle are rare, and even the pipistrelle is no longer common. Problems are food supply, destruction of breeding and hibernation roosts and pollution.

There have been local increases and extensions of range in some vertebrate populations (e.g. wild cat and pine marten), but the overall balance of change is on the debit side.

Figure 10.10 Summary of the current habitat and species change in Britain.

The effects of habitat change can be both beneficial and detrimental according to the species involved. For example, the spread of open country as a result of agricultural practice in Britain encouraged the spread of the rabbit and its success was further increased as a result of the decline of its natural predators (such as hawks and foxes) brought about by game-guarding landlords. By the 1950s there were more than 60–100 million rabbits in England and it was only the introduction of a South American virus, *Myxomatosis*, that has kept their numbers under control.

For other species, however, agricultural practices have caused adverse changes in their habitat. One of the most important habitat changes in Britain has been the removal of many of the hedgerows that are such a characteristic part of the landscape. They have been removed at a very fast rate to create larger fields (which are more efficient for large farm machines); to provide more usable space for crops and animals; because as farms become amalgamated many boundary hedges become redundant; and because, as many farms become more specialised, hedges are less necessary to divide farms up for different uses. Their removal is highly detrimental to many types of nesting bird. Figure 10.10 illustrates some of the major habitat changes that have been taking place in Britain in recent decades.

10.14 Extinction

Extinction is a process in which human activity plays a role, but it is not a process caused only by man, for as an examination of the fossil record shows, extinction is a normal feature of evolution. What we have done is to accelerate the rate of extinction by a variety of acts: predation, as with hunting of animals, collecting of eggs and plants etc.; by introducing alien animals and plants which compete with native species; by altering the food supply as well as by removing for human consumption the food of beasts which are higher up the food chain; by polluting air, water or soil with toxic substances; by introducing disease into the environment, either

Plate 10.5 One of the most famous examples of the process of extinction caused by human activities is provided by the hapless dodo. This appealing bird disappeared from Mauritius *c.*1680.

deliberately as with *Myxomatosis* which decimated rabbit stocks in Britain and Europe, or accidentally, as with Dutch elm disease fungus which has severely depleted the elm trees of much of eastern North America and England; and by changing or removing various types of natural habitat with which particular plants or animals are inextricably entwined for their existence.

Some species are more vulnerable than others to human pressures because of certain inherent characteristics which they possess. Some species require large areas to maintain themselves. This is particularly true of the large carnivores, which need expansive hunting territories to find enough prey. Slow reproducers, often large mammals or birds, represent another vulnerable group: their recovery from a natural catastrophe or human exploitation is very slow. The Californian condor, a bird now down to fewer than ten individuals in the wild, does not breed until at least six years of age, lays only one egg, and fails to nest at all in some years. A pair of these birds on average requires some 10–15 years to replace themselves. Species that forage or breed in large groups are also susceptible to rapid extermination, as are species that have become very highly specialised. For example, island species (plate 10.5), because they may have evolved in isolation and have no developed defences against certain types of predator, may be vulnerable if such predators are introduced. Other species are highly specialised in the sense that they have evolved in unique types of habitat or require highly specialised food supplies. If those habitats or supplies are removed, the species become extinct.

There is some evidence that a major wave of extinction occurred when man was still predominantly a hunter and a gatherer. The date varies according to the level of stone tool technology achieved in different parts of the world, and according to when *Homo sapiens* arrived in an area. In North America, for example, many major animals disappeared rather quickly round about 11 000 BP, and this has been attributed to the southward movement of tribes with sophisticated hunting techniques who had migrated across what is now the Bering Strait towards the end of Pleistocene times. Climatic changes may also have played a role, but this can-

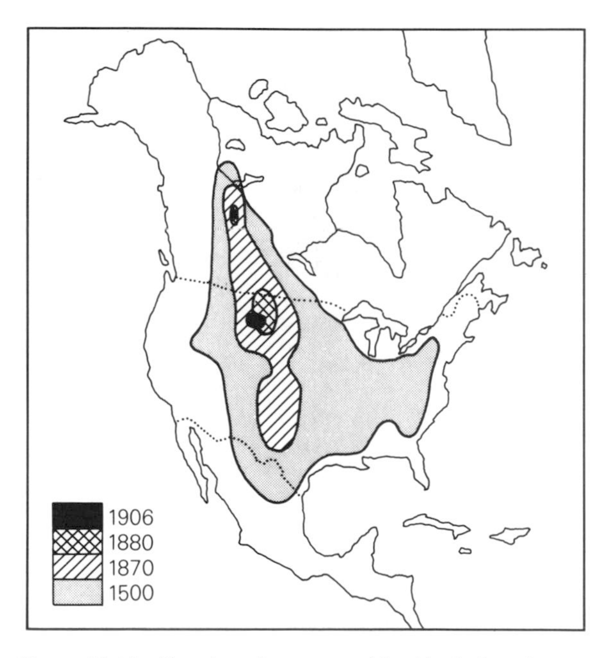

Figure 10.11 The changing range of the North American bison (*Bison bison*) under the influence of hunting pressures, especially in the nineteenth century.

not be the explanation for the rapid extinctions of fauna that have taken place when man has arrived at islands like Madagascar and those of the Pacific.

Before a state of complete extinction is attained, the geographical spread of particular organisms may become severely reduced. This is illustrated for the North American bison (*Bison bison*) or buffalo (figure 10.11). Prior to European colonisation this mammal was an essential element of the environment, thriving on rich summer herbage in the Prairies, and moving into forest or woodland in winter for warmth. Its range was huge, and the animals relied on their strength, size and numbers for protection against predators like wolves. In the nineteenth century hunters brought the once great herds to the point of extinction, but fortunately there were a few small areas where they were able to survive, including the Badlands of South Dakota.

Extinctions are still proceeding at a fast rate (figure 10.12) in spite of conservation efforts aimed at stemming the loss. Possibly one of the most fundamental ways in which humans are causing extinction is by reducing the area of natural habitat available to a species. Even wildlife reserves tend to be small 'islands' in an inhospitable sea of artificially modified vegetation or urban sprawl. We know from many of the classic studies of true island biogeography that the number of species living at a particular location is related to area; islands support fewer species than do similar areas of mainland, and small islands have fewer species than do large ones. Thus it may well follow that if humans destroy the greater part of a vast belt of natural forest, leaving just a small reserve, initially it will be 'supersaturated' with species, containing more than is appropriate to its area when at equilibrium. Since the population sizes of the species living in the forest will now be greatly reduced, the extinction rate will increase and the number of species will decline towards equilibrium.

Reduction in area leads to reduction in numbers, and this in turn can lead to genetic impoverishment through inbreeding. The effect on reproductive performance appears to be particularly marked. In-

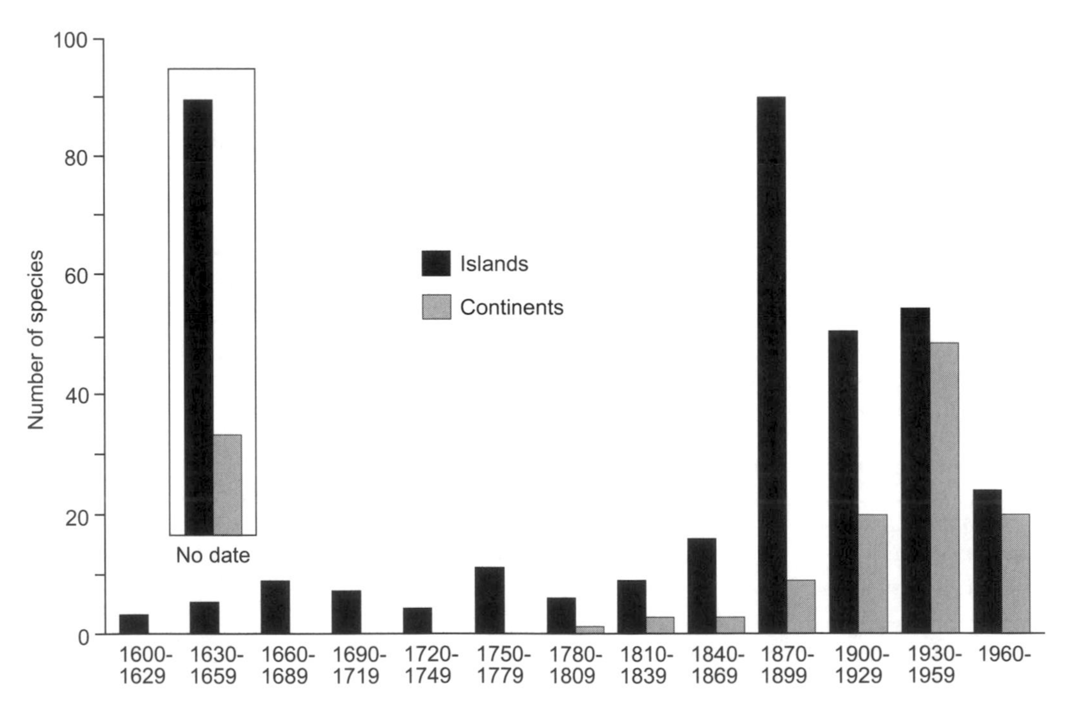

Figure 10.12 Time series of extinctions of species of molluscs, birds and mammals from islands and continents since about AD 1600.

breeding degeneration is, however, not the only effect of small population size for, in the longer term, the depletion of genetic variance is more serious since it reduces the capacity for adaptive change. Space is therefore an important consideration, especially for those animals that require large expanses of territory. For example, the population density of the wolf is about one per 20 km^2. The significance of this is apparent when one realises that most nature reserves are small: 93 per cent of the world's national parks and reserves have an area less than 5000 km^2, and 78 per cent less than 1000 km^2.

Equally, range loss, the shrinking of the geographical area in which a given species is found, often marks the start of a downward spiral towards extinction. Such a contraction in range results from habitat loss or from such processes as hunting and capture. Particular concern has been expressed in this context about the pressure on primates, notably in south-east Asia. Of 44 species, 33 have lost at least half of their natural range in the region. In two cases, those of the Javan leaf monkey and the Javan and grey gibbons, the loss of range is no less than 96 per cent. Recent figures produced by the International Union for the Conservation of Nature and the United Nations Environment Programme for wildlife habitat loss show the severity of the problem. In the Indomalayan countries 68 per cent of the original wildlife habitat has been lost, and the comparable figure for tropical Africa is 65 per cent. In these regions, only Brunei and Zambia have lost less than 30 per cent of their original habitat, while at the other end of the spectrum, Bangladesh, the most densely populated large country in the world, has suffered a loss of 94 per cent.

Although habitat change and destruction is clearly a major cause of extinction in the modern era, a remarkably important cause is the introduction of competitive species. When new species are deliberately or accidentally introduced to an area, they can cause the extermination of local fauna by preying on them or out-competing them for food and space. As we have seen elsewhere, island species have proved to be especially vulnerable. The

Window 10.5 *Biodiversity*

Biodiversity has five main aspects:

(a) The distribution of different kinds of ecosystems, which comprise communities of plant and animal species and the surrounding environment, and which are valuable not only for the species they contain, but also in their own right.
(b) The total number of species in a region or area.
(c) The number of endemic species (species whose distribution is confined to one particular location) in an area.
(d) The genetic diversity of an individual species.
(e) The sub-populations of an individual species, that is, the different groups which represent its genetic diversity.

Biodiversity has recently become a major environmental issue. With environments being degraded at an accelerated rate, much diversity is being irretrievably lost through the destruction of natural habitats. At the same time, science is discovering new uses for biological diversity.

The fundamental concern is the finality of the loss of biodiversity. Once a species has gone it cannot be brought back. The dodo (a bird) is dead and gone, and will never be seen again.

The earth's genes, species and ecosystems have evolved over a period of 3000 million years. They form the basis for human survival on the planet. However, human activities are now leading to a rapid loss of many of the components of biodiversity. Human self-interest argues that this process should be stemmed, for ecosystems play a major role in the global climate, are a source of useful products, preserve genetic strains which crop breeders use to improve cultivated varieties of plants, and conserve the soil.

Window 10.6 *Wetlands – a habitat under threat*

The term 'wetlands' is relatively new, and incorporates many other terms, such as marsh, swamp, bog and fen. Whatever name is given to them, wetlands are distinguished because they are areas where there is an interplay between land and water and a sharing of the characteristics of both. They lie at the junction between dry, terrestrial ecosystems and permanently wet, aquatic ones. Their soils are formed and conditioned by waterlogging; their plants are adapted to wet conditions, playing a role in trapping silts and – because they decay slowly – often causing the formation of peat; and much of their animal life is adapted either to dwelling in deep water (fish and shellfish) or to moving seasonally into the wetland (e.g. waterfowl).

Wetlands are areally important, for although they occur as rather small patches (the Everglades of Florida are a notably large exception), they cover around 6 per cent of the earth's surface (just a little less than the tropical rainforest). However, they are also important because they have many functions, values and benefits. Coastal wetlands make up about one-quarter of the total and inland wetlands the remainder.

The benefits of preserving wetlands are several: they can reduce flood risk by temporarily storing runoff water; coastal marshes absorb wave energy and reduce erosion, thereby buffering the shore against storms; they also trap sediment, and may contribute water to the groundwater table; they can trap pollutants and remove toxic residues; and they may provide fuel, fish, food and fibre. However, one of the most important reasons for protecting them is that they are immensely productive ecosystems that provide very important habitats for all sorts of wildlife. Indeed, in terms of net primary productivity it has been calculated that, although they only cover around 6 per cent of the earth's surface, they contribute about 24 per cent of primary productivity. In other words, their productivity is out of all proportion to the area they cover. Equally, wetlands provide habitats for a wide range of plants and animals. For example, they act as breeding grounds and areas of wintering for migratory birds.

In spite of their importance, wetlands are under threat from a whole range of human activities. Various estimates suggest, for example, that in the United States around one-half of the natural wetland area has been lost. The prime cause of this, amounting to over three-quarters of the total, is draining and conversion of the land for agriculture or other forms of development, while most coastal wetlands in the USA are being lost to dredging, canals, and marina, port and urban expansion. Other wetlands are also suffering a great loss of area, including the mangrove swamps of the Philippines, the inland deltas and swamps of Africa and the Sundarbans of the Ganges–Brahmaputra delta.

World Conservation Monitoring Centre, in its analysis of the known causes of animal extinctions since 1600 AD, believe that 39 per cent were caused by species introductions, 36 per cent by habitat destruction, 23 per cent by hunting and 2 per cent by other reasons (*World Resources 1994–5*, p. 149).

There are certainly some particularly important environments in terms of their diversity of life forms (biodiversity – window 10.5). Such biodiversity 'hot-spots' need to be made priorities for conservation. They include coral reefs, tropical forests (which support well over half the planet's species on only about 6 per cent of its land area), and some of the Mediterranean climate ecosystems (including the extraordinarily diverse fynbos shrublands of the Cape region of South Africa, described in window 3.3). Some environments are crucial because their loss would have consequences elsewhere. This applies, for example, to wetlands which provide habitats for migratory birds and produce the nutrients for many fisheries (window 10.6) (see colour plate 19).

10.15 Conservation

As a result of all the pressures that we have been placing on plants and animals, there has arisen a desire to protect some species and some ecosystems from alteration, and to preserve them for posterity. This desire has a long history but was given a great

Plate 10.6 In 1864 George Perkins Marsh wrote a book called *Man and Nature*, in which he put forward ideas that form the basis of the conservation ethic.

boost by the work of George Perkins Marsh (plate 10.6) in the late nineteenth century.

The motives for the protection of species and landscapes are many:

(a) *Ethical* It is asserted that wild species have a right to co-exist with us on our planet, and that we have no right to exterminate them.
(b) *Scientific* It is asserted that we know very little about our environment, and that we should learn all we can before we destroy it.
(c) *Aesthetic* Plants and animals are beautiful, and so enrich our lives.
(d) *The need to maintain genetic diversity* By protecting species we maintain the species diversity upon which future plant and animal breeding work will depend. Once genes have been lost they cannot be replaced.
(e) *Environmental stability* It is argued that in general the more diverse an ecosystem is, the more checks and balances there are to maintain stability. Thus environments that have been greatly simplified by human activity may be inherently unstable.
(f) *Recreational* Preserved habitats have great recreational value, and in the case of game reserves they may well have economic value as well (e.g. the safari industry in East Africa).
(g) *Economic* Many of the species in the world are still little known, and it is possible that they are great storehouses of food crops, medicinal plants etc., which, when knowledge improves, will be a useful economic resource. At present we derive some 85 per cent of our food from no more than 20 species of plants, and there are doubtless many more plants that would be of great value to human welfare if we understood their properties.

■ *Key Terms and Concepts*

acid precipitation
benthic
biogeochemical cycles
climax
community
conservation
dispersal
domestication
ecology
ecosystems
edaphic conditions
endemic
equilibrium theory of island biogeography
eutrophication
forest decline
habitat
limiting factors
migration
nutrient cycle
pelagic

photochemical smog
photosynthesis
sere
succession
trophic level

Points for Review

- Describe the main components of a biogeochemical cycle of your choice.
- What do you understand by trophic levels? Give examples from an ecosystem with which you are familiar.
- Why are most parts of the oceans so unproductive whereas fewer parts are highly productive?
- What are the main characteristics of the succession concept?
- Why are oceanic islands such special ecosystems?
- Why should one be careful about introducing a new plant or animal species to an ecosystem where it does not naturally occur?
- What do you consider to be the main causes of plant and animal extinctions at the present time?
- How would you convince a sceptic that species richness should be conserved?

FURTHER READING

Begon, M., Harper, J. L. and Townsend, C. R. (1996) *Ecology: Individuals, Populations and Communities*, 3rd edn (Oxford: Blackwell Scientific). A first-class and wide-ranging review of many aspects of ecosystems.

Brown, J. H. and Lamolino, M. V. (1998) *Biogeography*, 2nd edn (Sunderland, Mass.: Sinauer). An excellent, large, comprehensive text.

Colinvaux, P. (1986) *Ecology* (New York: Wiley). One of the standard texts on ecology.

Cox, C. B. and Moore, P. D. (2000) *Biogeography: An Ecological and Evolutionary Approach*, 6th edn (Oxford: Blackwell Scientific). A standard text for many years.

Dickinson, G. and Murphy, K. (1998) *Ecosystems* (London: Routledge). A useful new introductory-level book.

Odum, E. P. (1997) *Ecology: A Bridge between Science and Society*, 2nd edn (Sunderland, Mass.: Sinauer). A very sound volume from a well-known stable.

Tivy, J. (1993) *Biogeography*, 3rd edn (Harlow: Longman). A very useful introduction to biogeography by a geographer.

Whittaker, R. J. (1998) *Island Biogeography* (Oxford: Oxford University Press). A study of the importance and fascination of island ecosystems.

11 Tectonic Features

11.1 Introduction

Tectonic activity, associated with movement of the plates that make up the earth's crust, is a most important part of the environment; for not only does it affect landscape evolution, but it also has certain implications in the shape of major natural hazards. We live in an active world, and areas that are affected by tectonic processes undergo variable rates of uplift or subsidence. It is useful to have some conception of the speed at which such vertical movements are occurring.

Regarding *uplift*, the *old shield areas* tend to be relatively highly stable and have rates of crustal movement that are generally less than 1 mm in each 1000 years. *Ancient mountain belts*, associated with earlier phases of plate collision and orogenic activity, may have slightly higher rates (up to 5 m in each 1000 years); *active block-faulted regions*, such as the basin and range province of the western United States, may have higher rates still (up to 10 m in each 1000 years); but the highest rates of all (up to 20 m in each 1000 years) occur in the *major orogenic belts*, those undergoing current mountain-building. Even this maximum rate may not appear very fast, but it is important to remember that, if that rate persisted over a period of a million years, and if no erosion occurred in the meantime, the mountain range thereby created would be very much higher than any in existence today. A more general rate of uplift for an active mountain belt is probably around 3–5 m per 1000 years, and allowing for a rate of erosion of about 0.5–2 m per 1000 years, it may be reasonable to postulate that an 8000 m mountain could form in around 2–8 million years.

The main areas of *subsidence* are probably the deltaic basins of some of the great rivers like the Rhine and the Mississippi, where the weight of sediments progressively depresses the crust. The Netherlands, for example, currently appear to be subsiding at a rate of up to 2.5 m in each 1000 years, thereby contributing to the flooding problems of that area.

Let us now move to a consideration of one of the most important classes of tectonic features: volcanoes (plate 11.1).

11.2 Volcanoes

As we saw in chapter 1, volcanoes are not scattered randomly around the world. Indeed, one remarkable fact is that almost all the volcanoes in the world occur within a couple of hundred kilometres of the sea, and that only a few active volcanoes occur in the centre of continents. Furthermore, if one plots their distribution on a world map, many of them appear to occur as a series of distinctive narrow chains. It is therefore evident that volcanoes occur in certain preferred locations, most of which can be related to the positions, types and movements of the world's major tectonic plates:

(a) *along the mid-ocean ridges*, where sea-floor

Plate 11.1 Remnants of volcanoes (background) and lava flow, on Fuerteventura, Canary Islands.

spreading is taking place and where upwelling magma creates lithospheric crust on a constructive plate margin;

(b) *in island arcs* that are created by the subduction of oceanic crust, as in the Pacific Ocean;

(c) *in recent orogenic belts*, where at least one continental area is in collision with a subduction zone (e.g. the Andes);

(d) *at isolated places within the boundaries of plates*, where rising plumes of material within the mantle create 'hot spots' (as within the border of the Pacific Plate) and 'blow torch' their way through the crust;

(e) *along some of the great continental rift valleys and fractures*, notably in East Africa.

According to the type of volcano, there are differences in the composition of lava. Lava produced by upward movement of mantle material is *basaltic* and occurs in locations (a), (d) and (e). Lava that is produced by the process of subduction, either in island arcs or in association with Andean-type orogeny, is described as *andesitic*. Andesites, named after the Andes, where they make up almost all the major volcanoes, are very different from basalts because during the process of subduction a great deal of other material is added from the mantle and the continental crust. This affects the type of eruptions and volcanoes with which they are associated. Basalt is normally very free-flowing, with something like the consistency of syrup. Thus it forms bubbling lava lakes or fast-moving rivers of molten rock. Andesite, by contrast, is much more viscous and moves more slowly, forming thicker, but generally less extensive lava sheets. Moreover, andesite, being viscous, tends to trap gas until the pressure is sufficient to blow the lava apart, and so andesite eruptions are often accompanied by a great deal of explosive activity, which shoots lava fragments (called *pyroclastic* material) into the air.

Pyroclastic materials

The word 'pyroclastic' means fire-broken. Pyroclastic materials are quite simply those materials that have been ejected from volcanoes in a solid but fragmentary form. The most straightforward types of pyroclastic deposits are made up of fragments that have been shot into the air and fallen back down again as what are called *tephra* (pyroclastic fall deposits). They occur in a variety of sizes including *ash* (very fine material of less than 4 mm across), *lapilli* (little stones between 4 and 32 mm across), and *bombs* and *blocks* (which are much larger).

There is, however, another type of pyroclastic material – that produced by *pyroclastic flows*. The fragments occur as mudflows or as avalanches down the side of a volcano, or they may roll away as a very hot, fast-moving cloud called *nuée ardente*. The heat, the gas and the blast associated with their movement make them extremely dangerous. Some material produced by the frothing-over of a volcano is called *ignimbrite*.

Types of volcanic eruption and volcanic landscape forms

The form that a volcanic eruption takes can be highly variable, and may be classified on the shape of the vent, on the nature of the eruption itself and on the nature of the deposits that it produces (plate 11.2).

Plate 11.2 Organ pipe lavas in the northern Czech Republic.

With regard to the nature of the vent, by far the most important division is between central and fissure eruptions. *Central vent eruptions* result when lava and other material are ejected from a hole in the ground, fed by a single pipe-like supply channel extending deep down below ground. Ejected material piles up around the vent as the eruption continues, producing a heap of material that is called a *volcano*. Central vent eruptions occur in all situations where volcanic activity can occur. This is in contrast to *fissure eruptions*, which occur only in those places where the earth's crust is being subjected to tensional forces, trying to pull it apart. Where this happens a long narrow series of deep

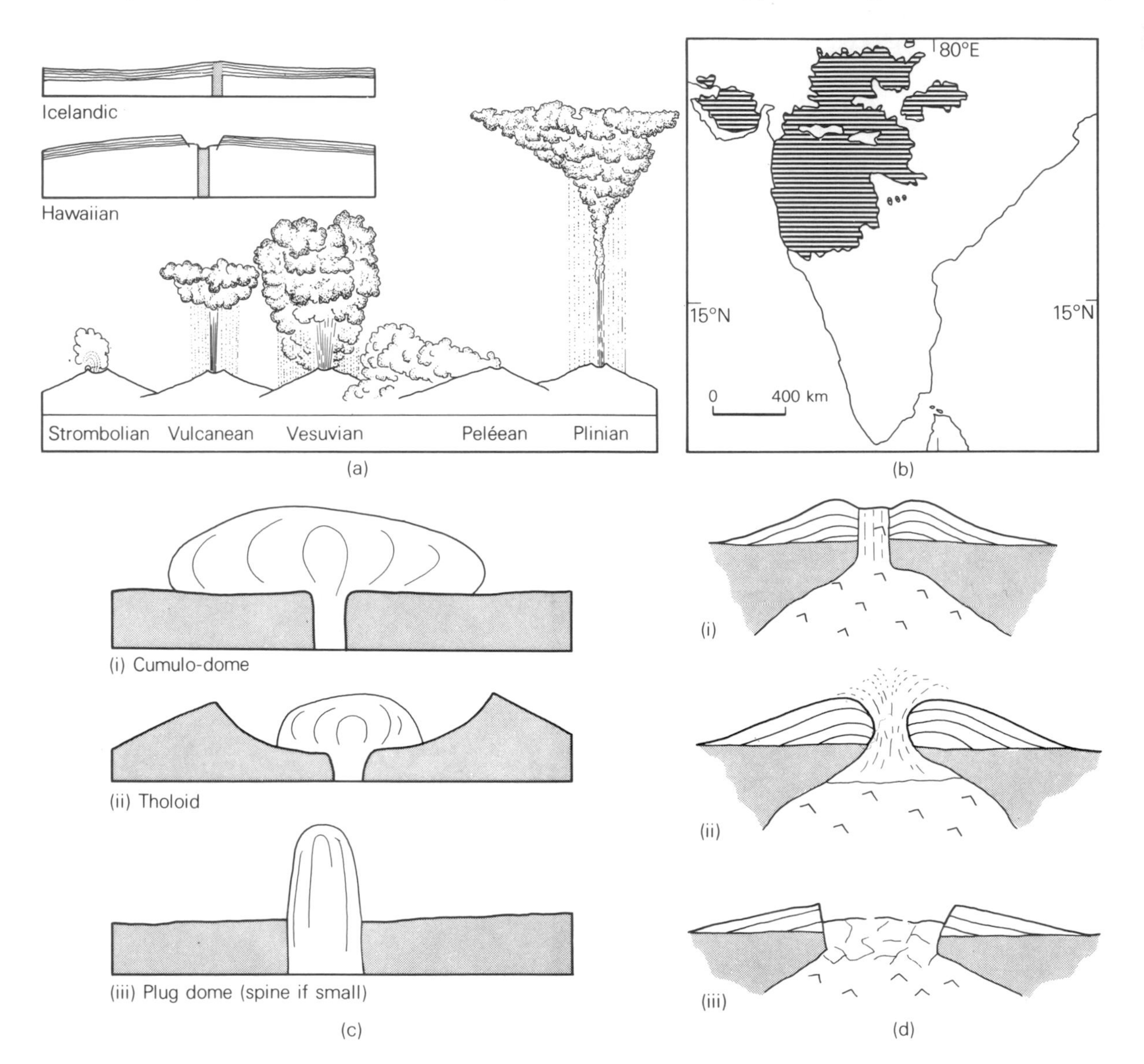

Figure 11.1 Some major volcanic landforms: (a) types of eruption; (b) a lava plateau – the Deccan of India; (c) acid lava (viscous) extrusion forms; (d) the stages in the formation of a caldera by collapse: (i) initial volcano; (ii) explosion; (iii) collapse.

cracks may form through which magma forces its way. Mid-ocean ridges, like that which Iceland is astride, are favourite spots for this kind of activity. If the material does not reach the surface a dyke is formed. If it does reach the surface then a fissure eruption takes place and basalt lavas pour out over the surface. The volumes involved can be enormous in the case of *basaltic flood eruptions*.

It is not very easy to make a hard and fast line between these two main types. Volcanic cones may occur in lines, and some of the basalt flowing from volcanic cones may emerge from fissures. Lava does not always flow out of the central vent on a volcano and can be erupted from a vent or fissure on its flanks.

Traditionally, vulcanologists have grouped eruptions into a sequence of progressively more violently explosive types (see figure 11.1a). *Hawaiian eruptions* are the mildest of them all, and tend to be typified by basalt flowing with great ease from a central vent. Slightly more explosive are *Strombolian eruptions*, named after a small volcanic island between Sicily and Italy. In this type, gas escapes spasmodically, producing small explosions. In the case of *Vesuvian eruptions*, named after Mount Vesuvius, near Naples, there are longish periods of inactivity during which gas pressure builds up behind the lavas that clog the vent. The blockage is removed by a substantial explosion, or series of explosions, during which large quantities of pyroclastic ash are blown out. When very large quantities of material are ejected by the gas blast there is said to be a *Plinian eruption*, and when the material ejected is in the form of one of those dangerous clouds known as nuées ardentes the eruption is said to be *Peléean*.

The landforms produced by volcanic activity are the end result of two opposing forces – constructive and destructive. The constructive forces are those involving the deposition of lava and the ejection of pyroclastic material. The destructive forces may be the result either of the normal processes of erosion found in most environments (such as wind, water and mass movement) or of the volcano's own explosive activity. Since most volcanoes have long lives, they may, during their existence, go through several phases of construction, erosion and explosion.

Basalt plateaux are formed by the eruption of fluid basalts through fissures (window 11.1) rather than central vents. The voluminous flows may pile up one on top of another, drowning earlier relief, to produce flat plateaux like the Deccan of India (figure 11.1b and plate 11.3) or the Columbia River plateau in North America. When they become dissected by erosion, as in the case of the Drakensbergs in South Africa, the seaward side of the Deccan behind Bombay or parts of Ethiopia, thick piles of basalt lavas give rise to characteristic 'stepped' topography. There are some remnants of an ancient basalt plateau in Britain: northern Ireland, Skye, Mull, Rhum and many of the small islands of the Inner Hebrides contain fragments of a plateau that, about 60 million years ago, covered much of the area north of the Irish and west of the Scottish coasts which is now below the sea, and which was once united with similar remnants in the Faroes, Iceland and Greenland. They are the result of eruptions associated with the plate movements that led to the opening up of the North Atlantic.

If, rather than being erupted through a fissure, the lava erupts through a single central vent and flows easily in the fashion of the *Hawaiian* type of volcano, then a broad, convex-swelling volcano is produced, usually with a relatively small sunken crater or *caldera* on the crest of the swelling. These are called *shield volcanoes*. The slopes are gentle and are made up of many layers of lavas, but they can form large volcanoes. Mauna Kea and Mauna Loa in Hawaii both reach over 4000 m above sea level; if one allows for the fact that part of them is below sea level, they are the highest mountains on earth – over 9000 m high.

The classic shape for a volcano is *conical*, with smooth, symmetrical profiles. Lava (plate 11.4) and pyroclastic material occur in interleaved layers and such volcanoes are usually, though not always, made of the relatively viscous andesitic lavas.

Among the destructive forces that mould volcanoes are the eruptive, explosive and subsidence forces that create craters. *Eruption craters* are the holes at the top of a volcanic cone, and when activity ceases the crater will become partially filled with debris. *Explosion craters* are produced when surface water percolating downwards comes into contact with hot magma and is converted into steam. The resulting explosion blasts up to the surface,

Window 11.1 Lanzarote

Lanzarote, one of the Canary Islands off the north-west coast of Africa, is a popular holiday resort for Europeans seeking winter sun. Between 1 September 1730 and 16 April 1736, however, the island was the scene of a major eruption. One-quarter of its area was given an entirely new landscape and farms and villages were buried. The activity took place along a belt of fissures some 4 km wide and stretching over a distance of 18 km. Over 30 cones of cinder were produced, as well as extensive flows of jagged lava. Much of this spectacular scenery is now incorporated in the Timanfaya National Park, where fumaroles are still active.

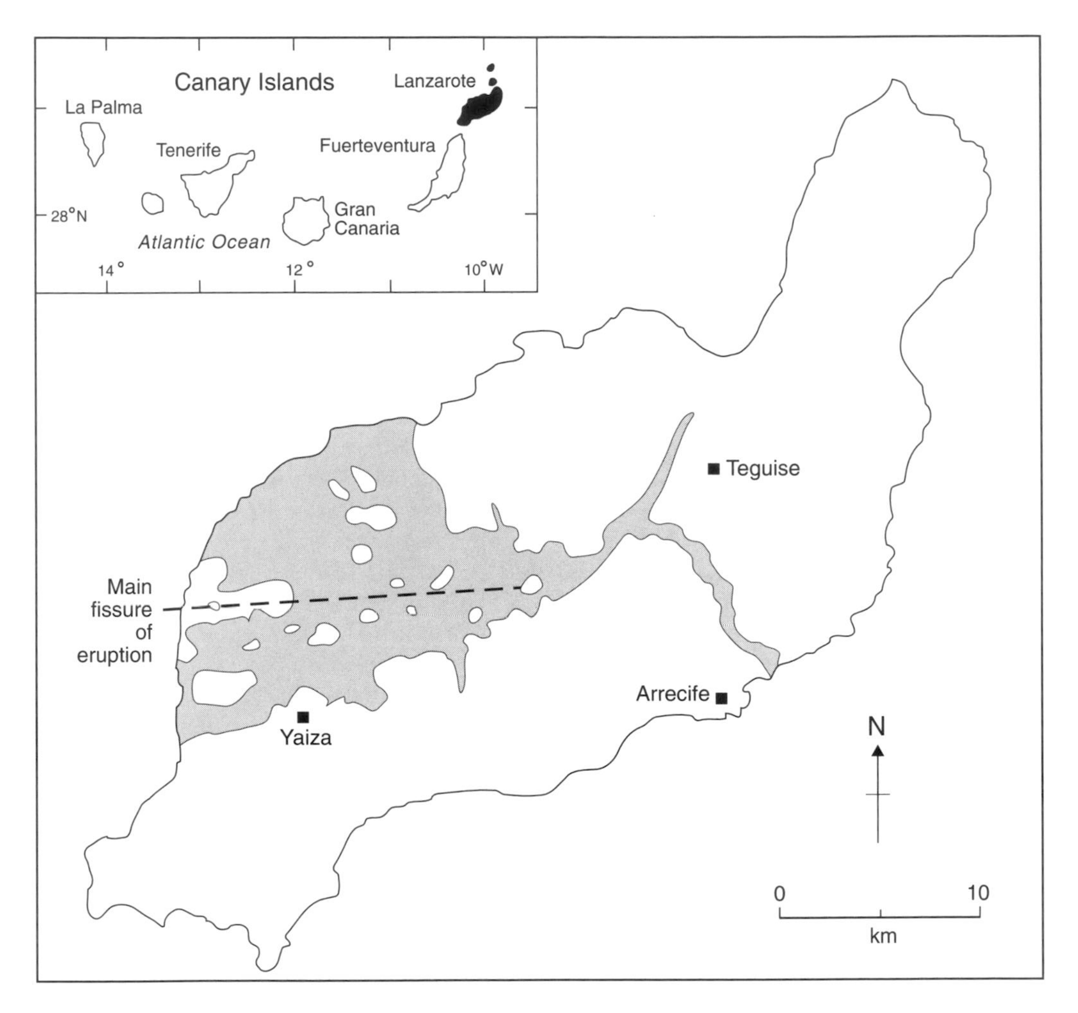

The area covered by the eruption between 1730 and 1736 in Lanzarote, Canary Islands, shown by stipple.

Plate 11.3 The Deccan Plateau in central India is composed of thick layers of lava, and its margins are marked by stepped escarpments produced by erosion of beds of variable resistance.

Plate 11.4 A lava flow in the Galápagos Islands. The example is composed of pahoehoe, a type which produces festooned, ropey surface structures.

blowing out a large hole in the ground in the process. Such craters are usually simple, circular depressions surrounded by low rims of ejected debris, and they may become occupied by a lake. Such lake-filled craters are often called *maars*. *Subsidence craters* may be either small, in which case they are called *pit craters* or *collapse craters*, or greater than *c.*1 km in diameter, in which case they are called *calderas*. The latter are usually circular in outline, with broad bottoms and steep to precipitous walls (figure 11.1d). The largest known are some tens of kilometres across. Explosive activity may also play some role in their development.

The shape of volcanoes and associated landforms is also affected by erosive processes (figure 11.2). On a classic conical cone, for example (figure 11.2b), gullies will cut down and will continue to develop until they are only separated from each other by sharp ridges. In this state the cone looks rather like a partly opened umbrella, and the pattern of alternative V-shaped ridges and gullies is known as *parasol ribbing*. As the erosive processes

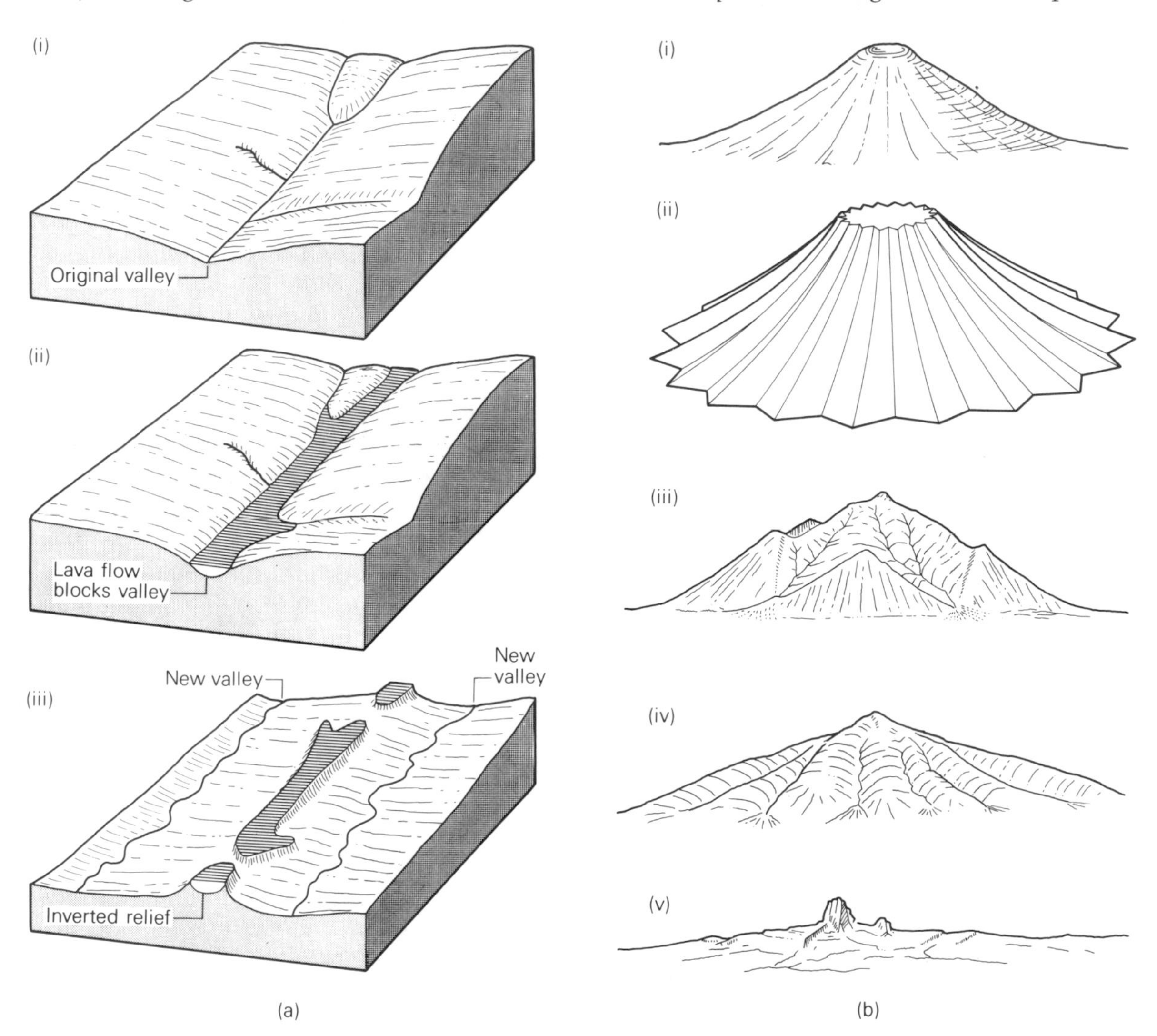

Figure 11.2 Post-eruption modifications of volcanic landscapes: (a) stages in relief inversion following a lava flow; (b) stages of the erosion of a volcano: (i) intact cone; (ii) parasol ribbing; (iii) planeze stage; (iv) residual volcano; (v) volcanic skeleton.

Plate 11.5 A volcanic plug at Le Puy in the Massif Central, France, exposed by differential erosion of volcanic deposits.

proceed, however, the cone will gradually become reduced to a rather shapeless hill – a process that may take some tens of millions of years. Pyroclastic deposits may also be intricately dissected by gullies to give badlands topography. Erosion will tend to etch out the weaker layers in a volcanic sequence, in some cases producing caverns; whereas elsewhere some resistant parts of the original volcano, such as the necks or feeder pipes of central-vent-type volcanoes, may stand up after most of the surrounding material has been eroded. The Puys of the Massif Central in France (plate 11.5) result from such differential erosion, as does Arthur's Seat in Edinburgh, the site of a large Carboniferous volcano.

If the lava that flows from a volcano is harder than the rocks that make up the valleys into which it flows, then relief inversion may occur (figure 11.2a).

Extremely viscous lavas of the acidic type, if they are not afflicted by an explosion, give rise to a number of distinctive landforms. The features formed include *cumulo-domes, tholoids* and *plug domes*, the last being produced when the lava is so viscous that it moves up like a piston, producing a roughly cylindrical body (figure 11.1c).

The permeability of lava flows may give a partial explanation of why valleys in basalts tend to be broad, flat-floored and steep-sided. They have a U-shaped cross-profile reminiscent in many ways of a glacial valley, and they also tend to possess steep amphitheatric heads.

Volcanoes as hazards

Volcanoes pose many dangers (table 11.1). First, lava can flow at considerable speeds (as much as 48 km per hour) and cover considerable distances. The longest recorded flow of historic times was on Hawaii, when a flow from Mauna Loa ran downslope for 53 km before reaching the sea. Fissure eruptions can cover even wider areas.

Falls of tephra also pose problems. Heavy ash falls will destroy vegetation (including crops), cause buildings to collapse, disrupt drainage, create suffocation, provide erodible material for subsequent mudflows and, if hot, cause fires. Pyroclastic flows, such as nuées ardentes, are especially serious. In 1902 Mont Pelée exploded on the Isle of Martinique in the Caribbean, and there were only two survivors out of a population of over 20 000. Most of them were killed by the nuées ardentes, which travelled at speeds of 33 m s^{-1} and had temperatures of as much as 1200 °C. More recently there has been an eruption on the island of Montserrat (window 11.2).

If there is heavy rainfall during the course of an eruption and the rain mixes with volcanic ash on the steep slopes of a volcanic cone, then a great mudflow or *lahar* may result (see window 11.4). The most disastrous mudflow of this type occurred on the Kelut volcano in Java in 1919, when 5500 people were killed. Other destructive mudflows are

Table 11.1 Deaths resulting from some of the most severe volcanic eruptions in the twentieth century

Year	*Locality*	*Deaths*
1985	Nevada del Ruiz, Colombia	25 000
1902	Mt Pelée, Martinique	20 000
1919	Kelut, Indonesia	5 500
1951	Mt Lamington, Papua	3 000
1982	El Chichon, Mexico	2 000
1902	La Soufrière, St Vincent	1 680
1963	Mt Agung, Bali, Indonesia	1 500
1911	Taal, Philippines	1 335
1965	Taal, Philippines	500
1991	Mount Pinatubo, Philippines	350
1980	Mt St Helens, Washington, USA	70

Window 11.2 Montserrat

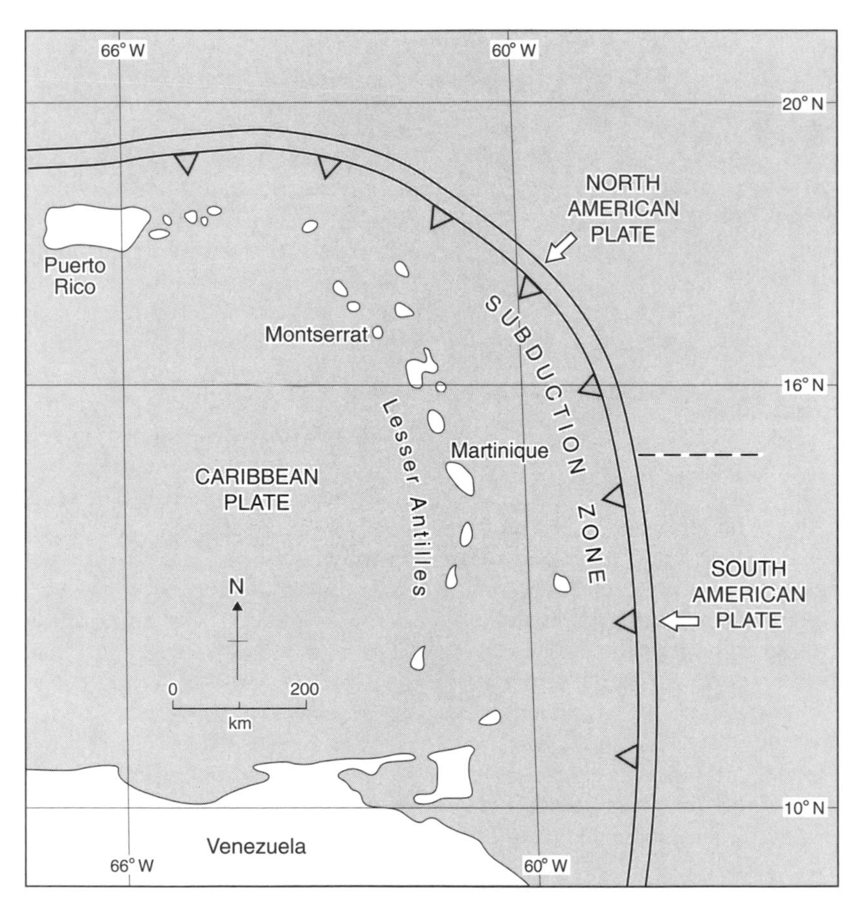

The tectonic situation of Montserrat.

Montserrat is a small island in the West Indies and it is part of an island arc that was formed as a result of subduction of the North and South American plates under the less dense Caribbean plate. Thus, the magma which rises beneath the Soufrière Hills volcano on Montserrat (and other islands in the Lesser Antilles chain) is re-melted material that once comprised the North and South American plates. This magma periodically forces its way up through weaknesses in the crust to give a volcanic eruption.

The first eruption of the Soufrière Hills volcano in recorded history began in July 1995. In June 1997 large pyroclastic flows killed 19 people and the island's only airport had to be evacuated. In August of that year the island's capital, Plymouth, was largely destroyed, but its inhabitants had already been evacuated. In effect, by the end of 1997, nearly two-thirds of the island had been claimed by the volcano and the southern end of the island had been almost entirely abandoned. Around 7000 of the island's pre-eruption population of 11 000 have left the island.

The Montserrat eruption was not the first to afflict the inhabitants of the Lesser Antilles. In 1902 the Mont Pelée volcano killed 20 000 people when white hot clouds of gas and tephra (called *nuées ardentes*) rolled down the side of the volcano and into the town of St Pierre.

Source

J. Horrocks (1998) 'Death and destruction on Montserrat', *The Geography Review*, 11 (4): 18–22.

caused by the rupturing of crater lakes.

How serious are all these hazards? C. D. Ollier (1988) sums up the situation in a rather surprising way:

In the past 500 years volcanoes have probably killed, directly or indirectly, over 200,000 people, of whom half died in the eruptions on Tamboro, Krakatoa, and Mount Pelée. This toll is in fact very low when compared with those of earthquakes, floods, wars, or road accidents.

On the credit side volcanoes provide fertile land, energy and materials for industry, and a livelihood for many people in the tourist trade. Even when activity has long ceased, most volcanic regions retain a great natural beauty, and frequently display most spectacular scenery. Such areas are often tourist attractions, and a large number of National Parks are located on volcanic centres. On balance volcanoes do more good than harm.

The scale of change that can be wrought by one volcanic eruption can be gauged by his account of the Mount St Helens eruption (plate 11.6) that took place in 1980 (from Costa and Baker, 1981):

The northward blast of the lateral eruption reduced the mountain's elevation by 396 m . . . ash was blown 19 km into the atmosphere. Waves of superheated dust and gas knocked down millions of trees. Mudflows filled lakes, rivers and streams, destroying fish, roads and bridges . . . Two earthquakes of magnitude 5.0 were recorded during the eruption. The blast left a crater 1.6 km across and 900 to 1500 m deep . . . An estimated $4.2 \times 10^9 m^3$ of debris had been blown out of the volcano by the start of summer . . . the ash fall closed schools, factories, stores, offices, airports, and highways, and idled 370,000 workers in Washington State . . . Damages from the eruption are estimated to be $2.0 billion . . . Shipping on the Columbia River . . . was halted by the Coast Guard because of sediment and log jams. The harbor of Portland was reduced in depth from 12 to 4 m in places by sedimentation . . . .

Prediction of volcanic eruptions

Observations of many volcanoes have led to the discovery of a number of phenomena that may be taken as warning signs of impending eruptions. It needs to be said, however, that these predictions can seldom be given with complete certainty.

The commonest means of eruption prediction is the measurement of earth tremors by seismic methods. As the ascent of magma causes tremors, there is usually a marked increase in the number and violence of local tremors with a focus at shallow depth just before an eruption. Another method is to undertake tilt measurements because before an eruption a volcano is liable to swell as a result of lava pushing up inside it. Tiltmeters measure very accurately the tilt of the ground, and any change in tilt may be regarded as a warning. Likewise, the tem-

Plate 11.6 The eruption of Mount St Helens in 1980 produced severe damage in the north-western parts of the United States, and waves of superheated dust and gas knocked down millions of trees.

perature of crater lakes, hot springs and *fumaroles* often shows a sharp increase before eruption, so that constant or regular readings of water temperatures may give some warning. The composition of gases erupted from craters or fumaroles may vary before eruption, so gas monitoring can be employed for predictive purposes. In addition, movement of lava at depth may cause changes in local gravity, and magnetic field measurements of these effects may give some way of predicting impending eruptions.

Once a volcano has made up its mind that it is going to erupt there is precious little that can be done to stop it. However, there are some ways of making sure that the effects are as small as possible. For instance, lava tends to flow down pre-existing drainage lines or depressions on a slope so these areas should be avoided. Given that lava flows relatively slowly this is often possible. The speed and direction of lava flows may also be influenced by bombing them, building diversion walls and spraying them with water. However, because mudflows and nuées ardentes travel so much faster than lava flows, and are so much less dense, they are much less easy to cope with.

Some environmental consequences of volcanic eruptions

Although the direct consequences of volcanic eruptions, such as ash fall, gas clouds, debris flows, explosions and the flow of lava, have serious consequences and may create serious hazards, there are a number of other consequences that are of potential significance.

A rather specialised consequence is the existence of *jökulhlaups* (see window 4.6). When a volcano beneath an ice cap erupts, large amounts of ice above it will melt. But if there is insufficient heat to melt through the entire thickness of the ice cap, the ice sheet immediately over the volcano may collapse, forming an immense hole, and the water melted by the eruption will be ponded up. The ice-dammed water may sometimes burst out from the ice cap, causing enormous quantities of water and sediment to gush across the outwash plains around the ice cap. Individual flows have been known to exceed $400\,000\,m^3\,s^{-1}$, and a boulder $400\,m^3$ in size was carried 14 km by such a flow in Iceland in 1918.

Table 11.2 Some notable twentieth-century tsunamis

Date	*Zone of origin*	*Height of wave (m)*	*Damaged area*	*Additional comments*
1908	Strait of Messina	5	Messina, Reggio, Calabria	>8000 deaths
1923	Kamchatka	6	Waiakea, Hawaii	
1925	Pacific Ocean	11	Zihuatanejo, Mexico	
1932	Pacific Ocean	10	Cuyutlán-San Blas, Mexico	10–75 deaths (estimated)
1933	Honshu	>20	Japan	3000 deaths
1946	Aleutian Islands, Alaska	17	Wainaku, Hawaii	173 killed; 163 injured
1957	Aleutian Islands	16	Hawaii	Heavy damage; 61 killed; 282 injured
1960	Chile	>10	Hawaii	
1964	Alaska	8.5	Crescent City, California	119 deaths; 200 injured; $104 million in damage
1975	Hawaii Island	8	Hilo, Hawaii	
1976	West Pacific	5	Philippines	3000 drowned
1983	Hokkaido	6–14	Japan	100 deaths

Source: from D. Alexander (1993) *Natural Disasters* (London: UCL Press), table 2.7

Window 11.3 *Tsunamis*

On 17 July 1998, an earthquake with a magnitude of 7.1 took place under the sea off New Guinea. This produced a fearsome tsunami that swept on shore and killed more than 2200 villagers. Waves were up to 15 m high. Most tsunamis affect the Pacific Ocean, and 86 per cent of them are the products of undersea earthquakes around the Pacific Rim, where powerful collisions of tectonic plates form highly seismic

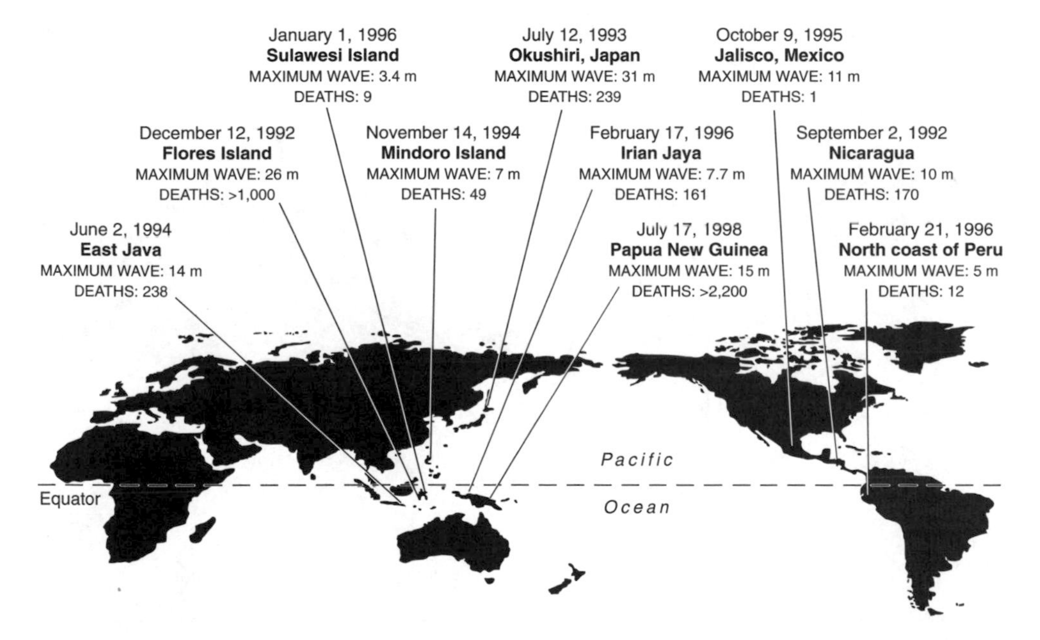

In the 1990s, at least 4000 people were killed by tsunamis which struck at coastal communities around the Pacific Ocean.

subduction zones. In all, destructive tsunamis claimed more than 4000 lives during the 1990s. In the past, however, individual events have caused even greater mortality than this. A tsunami associated with the catastrophic eruption of Krakatau in Indonesia in 1883 killed more than 36000 people. The New Guinea tsunami was not especially great in terms of the wave it generated. The maximum wave height during the 1993 Okushiri event in Japan was 31 m, while the East Aleutian Island event of 1946 had a maximum wave height of 35 m. Methods for predicting tsunamis are being developed, and coastal communities need inundation maps that identify far in advance what areas are likely to be flooded so that they can lay out evacuation routes.

Source

F.I. Gonzáles (1999) 'Tsunami!', *Scientific American*, 280 (5): 44–55.

Window 11.4 ***Mount Pinatubo, 1991***

Mount Pinatubo is one of a chain of volcanoes in the Philippines that makes up the Luzon volcanic arc. On 14 June 1991 there was a massive eruption, but this had been preceded for several months by earthquakes and emissions from vents. Some 300 people lost their lives during the eruption, but hundreds of thousands of people were successfully evacuated. The eruption coincided with a tropical storm (Typhoon Yunya) which meant that large amounts of volcanic debris were converted to a lethal slurry that generated debris flows called *lahars*. These forced the evacuation of 200000 people from the region. The eruption was the world's largest known eruption in more than half a century and about ten times more magma was involved than had occurred in the famous 1980 eruption of Mount St Helens in the USA. Mount Pinatubo also expelled large amounts of dust and sulphur dioxide. These probably formed the densest veil in the Northern Hemisphere since the 1883 eruption of Krakatau and caused a cooling of the world climate for several years.

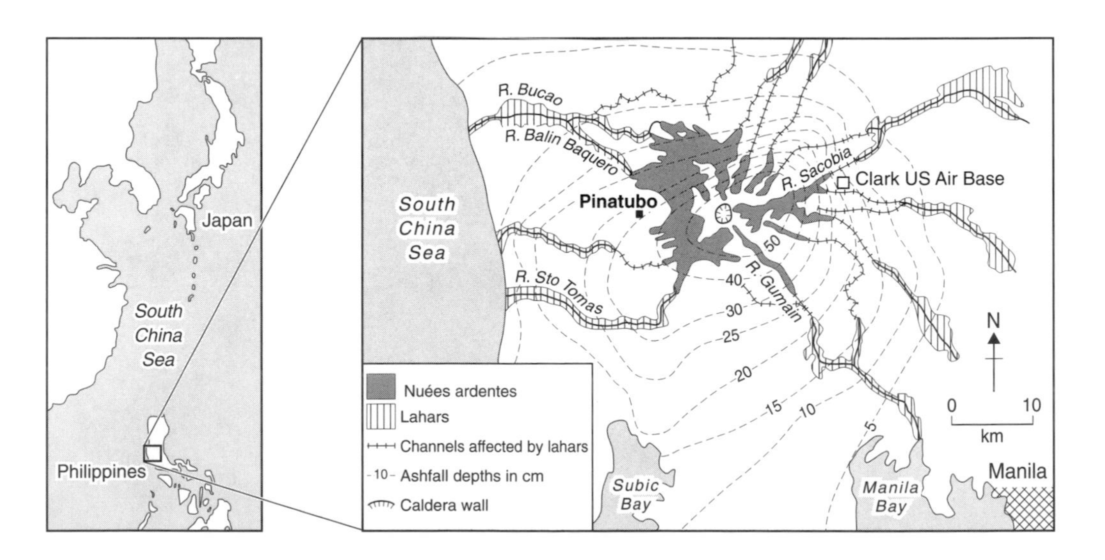

The areas affected by nuées ardentes, lahars and ash from Pinatubo in June 1991.

(For comparison, it is worth noting that the Amazon, which normally carries more discharge than any other river has an estimated flow of about 200 000 $m^3 s^{-1}$.)

A no less spectacular effect of some volcanic eruptions is the production of some 'tidal waves' called *tsunamis* (table 11.2) (window 11.3). The explosion of the volcano Krakatau in the East Indies in 1883 was so violent that great waves ravaged the shores of Java and Sumatra, killing over 36 000 people. The biggest wave is estimated to have been no less than 15 m in height at the shore, and where it washed up on to the land its momentum carried it ever higher, to as much as 30 m! Some tsunamis can result from earthquake activity without volcanic eruptions taking place.

Volcanic eruptions also have an effect on world climate over various time scales. This is because some explosive eruptions can produce a dense dust veil in the lower stratosphere (see window 11.4 on Mount Pinatubo) which shuts out some of the incoming radiation from the sun, resulting in cooling. The ash emissions of Krakatau were injected at a height of 32 km and produced a global decrease of solar radiation received at the earth's surface by 10–20 per cent. A series of cold, wet years followed in Europe; and many of the coldest and wettest summers in British history, including 1695, 1725, the 1760s, the 1840s, 1879, the 1880s, 1903 and 1912, occurred in conjunction with times of high volcanic dust inputs into the stratosphere and upper atmosphere. Moreover, the period of more or less worldwide warming of the 1920s, 1930s and 1940s coincides with a period when there were no major explosive eruptions in the Northern Hemisphere, suggesting that the absence of a volcanic dust pall in those decades was one factor in the warming process. Going back further, a study of the Byrd ice core that has been drilled through the Antarctic ice cap has produced evidence of particularly heavy and frequent volcanic dust falls at 20 000–16 000 years BP – the same time as the phase of maximum cold of the last glaciation of the Pleistocene.

11.3 Intrusive Igneous Rocks

Not all igneous rocks are extrusive in the sense that volcanic lavas are. Many are said to be *intrusive* because, rather than being poured out on to the earth's surface, they have cooled and solidified within the earth's crust. Intrusive rocks occur in a great variety of forms. Some form enormous bodies as much as 1600 × 160 km in size, while others may form a small dyke only a few centimetres wide. Some are relatively fine-grained while others are fairly coarse. Some have a very dense network of joints while others may have very few. Some are *concordant* with the structures into which they are intruded, while others cut across the structures and are said to be *discordant*. Bearing these sorts of considerations in mind, one can classify igneous rock masses as follows (see figure 11.3):

Major
- Concordant: lopoliths
- Discordant: batholiths with associated bosses and stocks, diapirs

Minor
- Concordant: sills, laccoliths, bysmaliths, phacoliths
- Discordant: dykes, ring complexes including cone sheets, ring dykes and cauldron subsidence

These different types of intrusion have different effects on relief. *Lopoliths* are intrusive complexes which have a saucer-like form and are basically concordant to the structures of the rocks into which they are intruded. They are of enormous size (the Bushveld igneous complex in South Africa covers some 55 000 km^2), and are often layered so that there is a tendency towards the development of outward-facing scarps with variations in the resistance of the various layers. *Batholiths* are also large; they are nearly always formed of granite, they are common, and they are exposed at the surface by erosion of the rocks into which they were intruded discordantly. These rocks are often altered (*metamorphosed*) by their contact with the magma that made up the batholith. This zone of alteration is called a *metamorphic aureole*. Examples of batholiths include the granite masses of the West Country in England and the uplands of Brittany. They are particularly widespread in the old Precambrian shield areas, as in Zimbabwe. Batholith exposures are often characterised by such landforms as tors and inselbergs, though their development depends very much on the nature of the jointing within the granite.

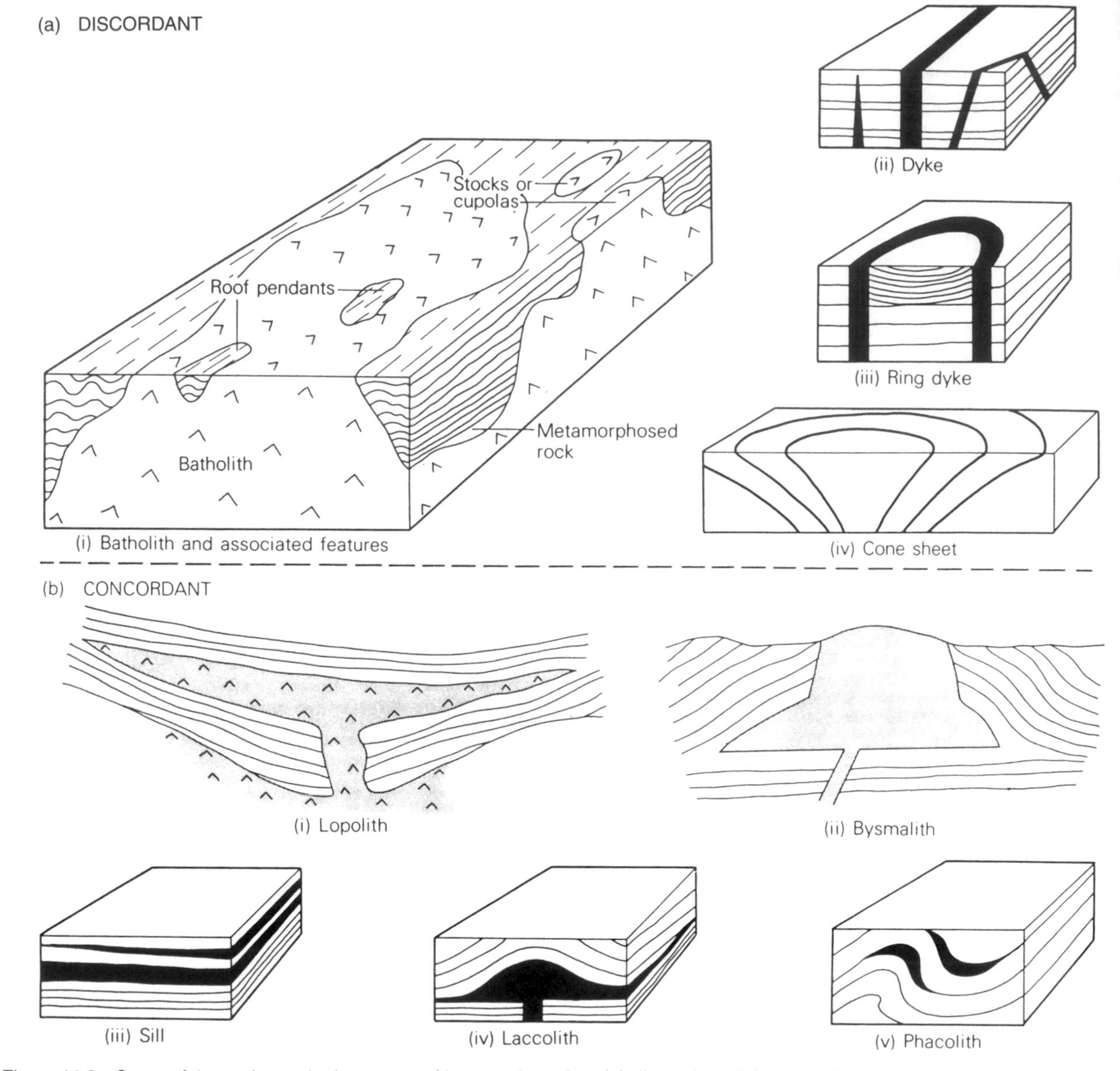

Figure 11.3 Some of the major and minor types of igneous intrusion: (a) discordant; (b) concordant.

Sills are minor intrusions approximately concordant with the bedding planes of the rocks; they may be of any thickness and extend over large areas. In Britain the most striking example is the Great Whin Sill of northern England. It is up to 70 m thick and forms a natural rampart, part of which was used by the Romans in locating Hadrian's Wall. One of Britain's finest waterfalls, High Force, occurs where the River Tees plunges over its resistant outcrop.

Rather more complex in form (see figure 11.3) are *laccoliths, bysmaliths* and *phacoliths*. If, as in the case of the classic Henry Mountains of the United States, the sediments surrounding the intrusions are removed, then almost perfect exhumation may reveal some striking relief forms. *Dykes* are thin, long discordant intrusions that may have a variety of relief effects according to how they weather and erode in comparison to the rock into which they are intruded. Where they are more resistant than the surrounding rocks they project; where they are weaker than the surrounding rocks they form depressions; and where they are harder they may stand higher than the surrounding rocks but be surpassed in resistance and elevation by the baked margins of the country rocks. Dykes may occur in various configurations, thereby creating patterns in the landscape (see colour plate 20).

11.4 Earthquakes

Another major demonstration of the power of tectonic forces is provided by the existence of earthquakes (plate 11.7). These are motions of the ground, ranging from a faint tremor to a wild motion capable of shaking buildings apart and causing gaping fissures and steps to open up in the ground. An earthquake can be likened to a form of wave motion in which energy is transmitted through the surface layer of the earth in widening circles from a point of sudden release – the *earthquake focus*. Like the ripples produced when a pebble is thrown into a pond, these *seismic waves* travel outwards in all directions, gradually losing energy.

Earthquakes are the results of earth movements, many of which are associated with the boundaries of the major plates. If stress is placed upon a section of rock by forces operating in opposite directions, the rock initially deforms because it is to some extent elastic. If the stress is maintained long enough, this strain may exceed the elastic strength of the rock, so that fracture takes place along a line called a *fault* (see section 11.6). At the moment of fracture the rock regains its original shape, but in a new position. The quaking and shaking takes place during the sudden movement of the rock back to its original shape, after the stress has been released. The place at the ground surface above the point where the stress release occurred is called the *epicentre*.

Plate 11.7 Earthquakes are a major type of natural hazard, ranking third (after floods and cyclones) in terms of the total loss of life that they cause. In December 1972 the city of Managua in Nicaragua was severely damaged.

There have been various attempts to define the degree of destructiveness (*intensity*) of earthquakes. Two of the most used scales are the Mercalli scale of intensity and the Richter scale of magnitude. An earthquake's intensity depends upon such things as magnitude, distance from epicentre, acceleration, duration, amplitude of waves, type of ground, water table and the nature and type of constructions affected. Its determination is based on interviews and observations, and is rated on a scale of I to XII (table 11.3). Magnitude is a more objective instrumentally determined measure of the amount of energy released by an earthquake. The Richter scale is logarithmic, and the numbers are proportional to the amplitude of seismic waves measured 100 km from the earthquake's epicentre.

Table 11.3 Scale of earthquake intensity with approximately corresponding magnitude

	Mercalli intensity	*Description of characteristic effects*	*Richter magnitude corresponding to highest intensity reached*
I	*Instrumental*	Detected only by seismographs	3.5
II	*Feeble*	Noticed only by sensitive people	to
III	*Slight*	Like the vibration due to a passing lorry; felt by people at rest, especially on upper floors	4.2
IV	*Moderate*	Felt by people while walking; rocking of loose objects, including standing vehicles	4.3
V	*Rather strong*	Felt generally; most sleepers are awakened and bells ring	to 4.8
VI	*Strong*	Trees sway and all suspended objects swing; damage by overturning and falling of loose objects	4.9–5.4
VII	*Very strong*	General alarm; walls crack; plaster falls	5.5–6.1
VIII	*Destructive*	Car drivers seriously disturbed; masonry fissured; chimneys fall; poorly constructed buildings damaged	6.2
IX	*Ruinous*	Some houses collapse where ground begins to crack, and pipes break open	to 6.9
X	*Disastrous*	Ground cracks badly; many buildings destroyed and railway lines bent; landslides of steep slopes	7–7.3
XI	*Very disastrous*	Few buildings remain standing; bridges destroyed; all services (railways, pipes and cables) out of action; great landslides and floods	7.4–8.1
XII	*Catastrophic*	Total destruction; objects thrown into the air; ground rises and falls in waves	>8.1 (maximum known, 8.9)

Table 11.4 Loss of life by disaster type, 1947–1980

	Loss of life	*% of total*
Hurricane	498 516	40.8
Earthquake	450 048	36.8
Flood	194 435	15.9
Severe storm	22 977	1.9
Snowfall and extreme cold	13 197	1.1
Volcanic eruption	9 430	0.8
Tornado	7 648	0.6
Heatwave	7 470	0.6
Landslide	5 493	0.4
Avalanche	5 025	0.4
Tsunami	4 526	0.4
Fog	3 550	0.3
Total	1 222 315	100

Source: from data in D. Alexander (1993) *Natural Disasters* (London: UCL Press), table 1.1

On a global basis, earthquakes are one of the most serious of natural hazards (tables 11.4 and 11.5). Although it is difficult to collate totally reliable data, for the period from 1947 to 1980 earthquakes were only marginally less serious than hurricanes in terms of the loss of life they caused. They caused almost 37 per cent of all deaths produced by natural disasters. Individual earthquake events have the potential to inflict the greatest loss of life and property of any geological hazard. Loss of life in an extreme earthquake can be more than ten times that of extreme volcanic eruptions. They also tend to be more sudden and less predictable than volcanic eruptions.

The shocks produced by earthquakes can create *slope instability*. One of the most horrendous examples of this was the tragedy that afflicted the town of Yungay in Peru in May 1970 (window 11.5).

One of the most important consequences of earthquake activity is that changes in the level of the ground surface may take place, thereby creating

Table 11.5 Deaths resulting from some severe earthquakes of the twentieth century

Year	*Location*	*Magnitude (Richter scale)*	*Deaths (max. estimate)*
1905	Kangra, India	8.6	20 000
1907	Afghanistan	8.1	12 000
1908	Messina, Italy	7.5	200 000
1915	Avezzano, Italy	7.5	30 000
1917	South Java	?	15 000
1918	South-east China	7.3	10 000
1920	Kansu, China	8.5	200 000
1923	Tokyo–Yokohama, Japan	8.3	163 000
1927	Nansham, China	8.0–8.3	180 000
1933	North-central China	7.4	10 000
1934	Bihar, Nepal	8.4	10 700
1935	Quetta, Pakistan	7.5–7.6	60 000
1939	Chillan, Chile	8.3	40 000
1939	Erzincan, Turkey	8.0	32 700
1948	Kagi, Formosa	7.3	19 800
1960	Agadir, Morocco	5.6–5.9	14 000
1962	Buyin-Zara, Iran	7.3	14 000
1968	Dash-i-Bayaz, Iran	7.3–7.8	18 000
1970	Chimbote, Peru	7.8–7.9	67 000
1974	West-central China	6.8	20 000
1975	Haicheng, China	7.3–7.4	10 000
1976	Guatemala	7.5	23 000
1976	Tangshan, China	7.8–8.1	750 000
1978	Tabas, Iran	7.7–7.8	25 000
1985	Michoacan, Mexico	8.1	22 000
1988	Armenia (USSR)	6.9	25 000
1990	Gilan, Iran	7.4	50 000
1995	Kobe, Japan	6.9	5 100

fault traces, changes in sea levels and so on (plate 11.8). An example of this is provided by the effects of an earthquake that took place about 1550 years ago and affected a considerable part of Crete, Karpathos, Rhodes and the south coast of Turkey near Alanya (figure 11.4). It was at its most intense in western Crete, where some 9 m of uplift have taken place. The Roman harbour at Falasarna was abandoned as it was raised over 6 m above sea level. Likewise, the Alaska earthquake of 1964 produced ground displacement of 10–15 m, and displacements of the order of a few metres are not uncommon. Some seismic movements are more gradual, and occur by a process called *seismic creep*. Although

Window 11.5 *The Mount Huascaran disaster, 1970*

Severe shaking induced by earthquakes can cause landslides and rock and snow avalanches in mountainous areas. These are major causes of loss of life and property during earthquake events. One of the greatest landslide disasters ever recorded occurred when an offshore earthquake under the Pacific Ocean triggered a massive avalanche of rock and snow from the overhanging face of the Nevados Huascaran Mountain. Huascaran, at 6654 m the highest peak in the Andes of Peru, generated in 1970 a turbulent flow of mud and boulders (estimated to have a volume of $50–100 \times 10^6 m^3$) that moved as a wave up to 30 m high and at an average speed of 270–360 km h^{-1}. It buried the towns of Yungay and Ranrahirca, killing at least 18 000 people.

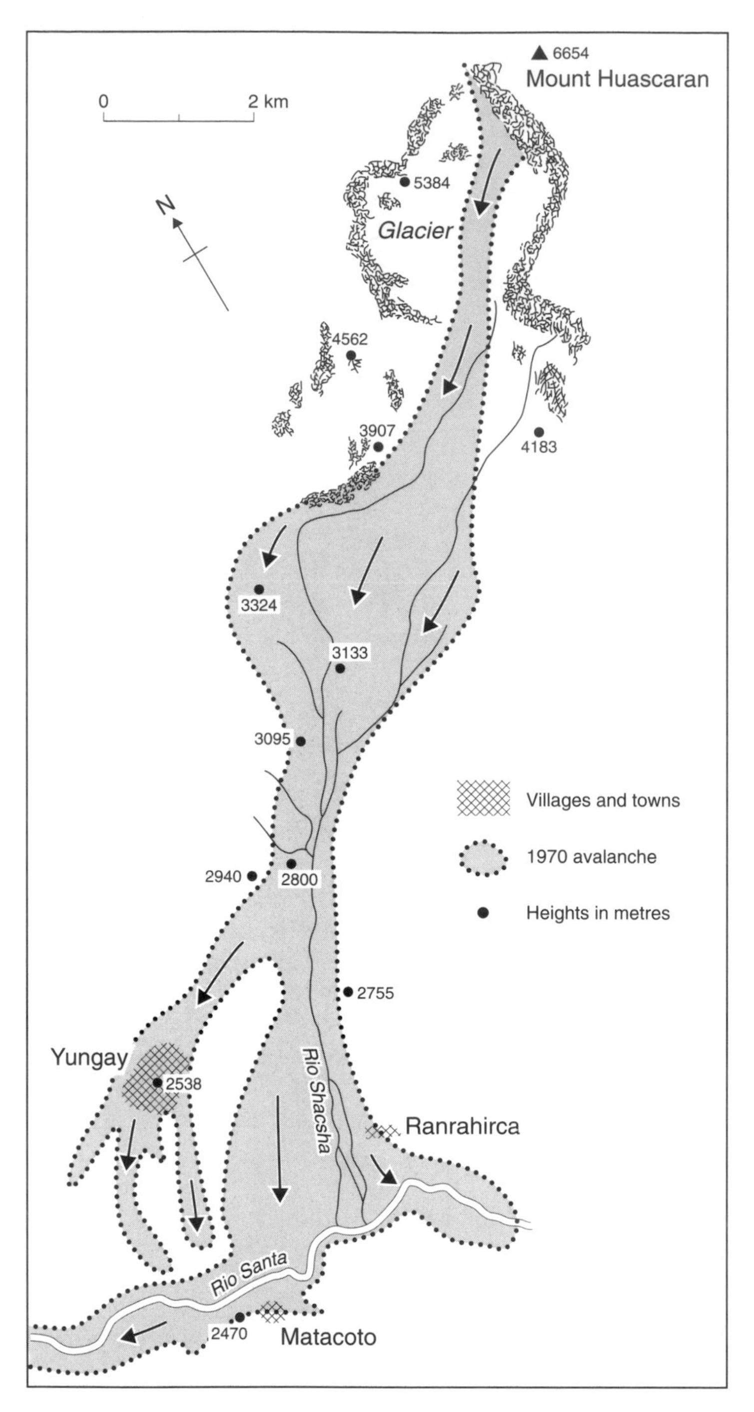

The 1970 Huascaran avalanche.

Plate 11.8 Seismic activity in Turkey is of frequent occurrence, as the dreadful 1999 earthquake demonstrated. In southern Turkey this Lycian tomb and other structures have been flooded as a result of subsidence associated with seismic events.

their effects are less immediately drastic, detailed surveys with precise levelling instruments in parts of California, where there are major faults like the Garlock and the San Andreas, reveal movements of 5–13 m in 1000 years.

The principal danger in an earthquake is construction failure, and the extent of *earthquake vibration damage* is controlled partly by the characteristics of the ground upon which the buildings are constructed. In unconsolidated surface materials, such as alluvium, artificial fill or swamp deposits, earthquake vibrations last longer and the wave amplitudes are greater, causing more shaking than on bedrock. Thus, the greatest damage during the San Francisco earthquake of 1906 occurred in the areas of susceptible superficial materials, whereas buildings on bedrock received relatively little damage. Vibration, however, is not the only problem. When the ground shakes and vibrates during an earthquake, unconsolidated sediment becomes more compact, for the material assumes a more dense arrangement by particles moving together and decreasing in volume. As a consequence ground subsidence occurs, and this too can damage buildings. The ground locally subsided 0.6 m by compaction of sediment in the 1906 San Francisco earthquake and by 1.8 m in the 1964 Alaska earthquake. The consequence of settlement will be especially severe if building foundations are built across deposits of varying physical properties, which would cause one part of the foundations to settle more than another.

In areas where unconsolidated material has a water table close to the surface, ground shaking may cause the sediment to be transformed from a solid to a liquid by a process called *liquefaction*. The liq-

(a)

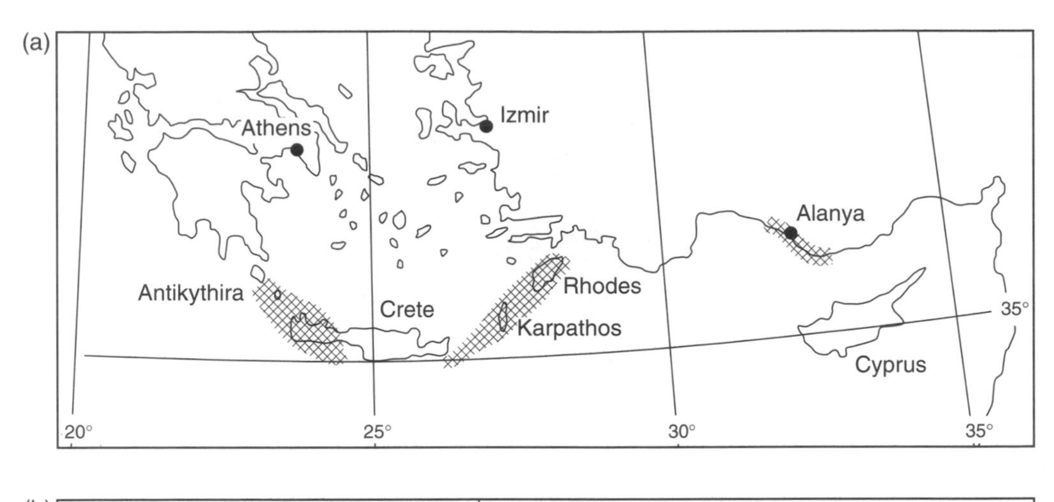

(b)

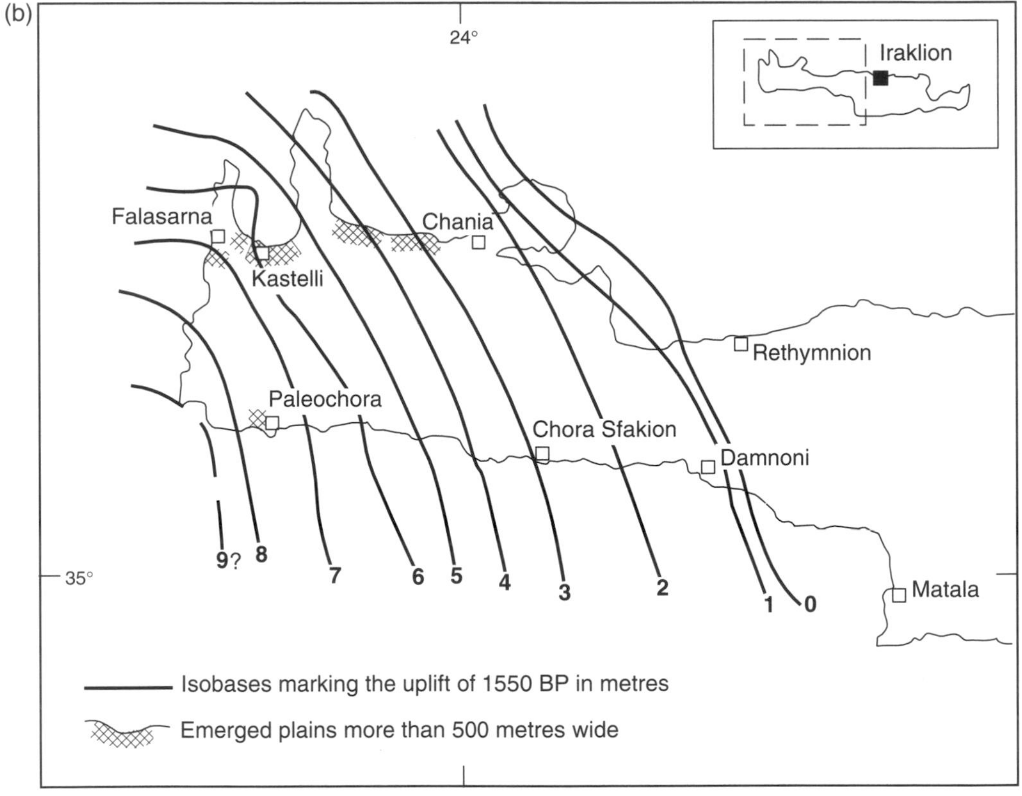

Figure 11.4 The uplift achieved by an earthquake 1550 years ago: (a) the stippled area shows the regions with appreciable uplift as a result of the event; (b) map of western Crete showing the differential uplift that occurred.

uefied sediments may slide or flow, causing further damage, especially on slopes. In many earthquakes fire is a major hazard, as stoves are overturned, gas mains are fractured and so on.

Tsunamis, which – as we have already noted – sometimes result from great volcanic eruptions like that of Krakatau, can also result from earthquakes and can cause great loss of life. The amount of damage they cause is related partly to the intensity of the earthquake that produced them but also to the nature of the coastline, and its distance from the epicentre. Broadly, gently sloping continental shelves allow tsunamis to build to great heights, whereas deep water extending close to the shore minimises bottom friction and wave size. Narrow V-shaped bays and harbours that concentrate the wave energy are thought to provide conditions favouring maximum wave heights. Tsunamis are especially common in the Pacific, averaging more than two per year, but most are small and not destructive. They are more rare in the Atlantic because it is not ringed by active plate boundaries, but major tsunamis were caused by the great Lisbon earthquake in 1755 (plate 11.9); they killed 25 000 people in the eastern Atlantic and caused 3.5–4.5 m waves as far away as the West Indies.

11.5 Man-made Seismic Activity

Although we can do relatively little to predict or stop earthquakes, an increasing amount of evidence suggests that human activity is capable of triggering off seismic events, some of which can have unfortunate consequences.

Plate 11.9 As this contemporary print demonstrates, the great Lisbon earthquake of 1755 generated devastating tsunamis or tidal waves. Their effects were felt as far away as the West Indies.

Plate 11.10 An earthquake produced this fault which disrupted the regular rows of garlic in a California field.

In Denver, Colorado, in the early 1960s, nerve gas waste was disposed of in a well at a great depth in the hope of avoiding contamination of useful groundwater supplies. The waste was pumped in at high pressures and triggered off a series of earthquakes, the timing of which corresponded very closely to the timing of waste disposal into the well. It appears that the increased fluid pressure reduced the frictional force across the contact surface of a fault, thereby allowing slippage to occur, which created the earthquakes. The significance of the 'accident' has now been verified experimentally at an oil field in Colorado, where variations in seismicity have been produced by deliberately controlled variations in the fluid pressure in a zone that is seismically active.

However, perhaps the most important seismicity induced by human activity results from the impoundment of great volumes of water in reservoirs behind dams. With the ever-increasing number and size of reservoirs, this threat increases. Detailed monitoring has shown that earthquake clusters occur in the vicinity of some dams after their reservoirs have been filled, whereas prior to construction seismic activity was less clustered and less frequent. It has even been found that the seismic activity varies as the level of water in the reservoir changes with the seasons. Some of these earthquakes can eventually damage the dam, thereby causing flooding and loss of life.

Miscellaneous other human activities appear to affect seismic activity. In the gold-mining area of South Africa, for example, and also in the coal-mining area of Staffordshire, England, excavation and

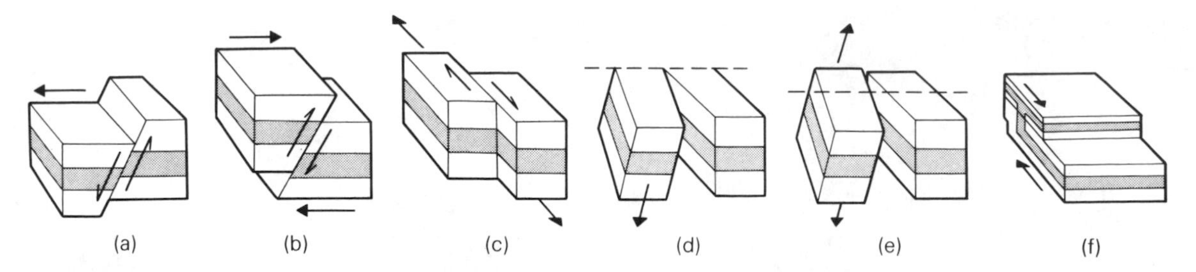

Figure 11.5 Different types of fault: (a) a normal fault produced by stretching; (b) a reversed fault caused by compression; (c) a strike-slip (transcurrent) fault produced by shearing; (d) a hinge fault produced by slumping; (e) a rotational fault produced by twisting; (f) a low-angle overthrust fault.

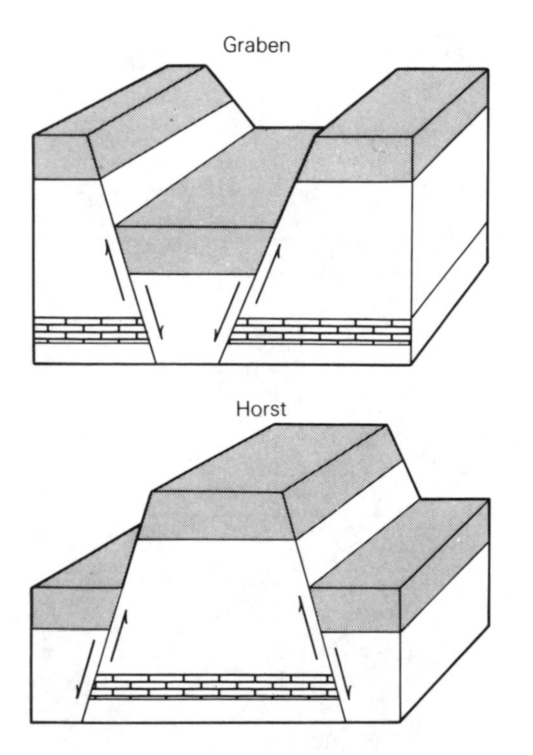

Figure 11.6 The development of graben and horst. When faulting causes one block to be lower than those on either side, a graben is formed, and where one is raised a horst is formed.

blasting associated with underground mining have produced small tremors. There are also cases where seismicity and faulting can be attributed to fluid extraction, as in the oil fields of Texas and California and the gas fields of the Po Valley in Italy.

11.6 Faults

A major consequence of tectonic activity is the formation of *faults*. These are breaks in the surface rocks of the earth's crust as a result of sudden yielding under unequal stresses. Faulting is accompanied by a displacement along a fault plane (plate 11.10), and since faults are often of great horizontal extent they can be followed along a fault line for many kilometres. In some places clearly recognisable layers of rocks can be seen offset on opposite sides of a fault, and the amount of displacement can be measured.

Faults take various forms that can be classified according to the angle of inclination and the relative direction of displacement (figure 11.5). A *normal fault* has a steep or nearly vertical plane, and movement is predominantly in a vertical direction, with one side being raised (or upthrown) relative to the other (which is downthrown). In a *reverse fault* the beds on one side of a fault are thrust over those on the other side as a result of compressive forces. With *transcurrent faults*, which are also called tear, strike-slip and wrench faults, the displacement is horizontal along the line of the fault. The Great Glen in Scotland is such a fault, where the horizontal displacement is about 100 km. Elsewhere faulting can be associated with slumping and twisting, while a *low-angle overthrust fault* involves movement that is predominantly in a horizontal direction, with the fault plane also in a more or less horizontal position. One slice of rock slides over the adjacent ground surface.

In general, faults occur in groups. Regions that are divided by a series of normal faults into rela-

Plate 11.11 Contorted strata in Jurassic sedimentary rocks at Stair Hole, Lulworth, Dorset, England. Folding has created anticlinal and synclinal structures.

tively elevated or depressed blocks are said to be *block-faulted*. Upstanding fault blocks, bounded by striking scarps, are called *horsts*, and examples of these in Europe include the Vosges, the Black Forest and the Harz Mountains (figure 11.6). Fault blocks depressed below their surroundings are called *fault troughs*, and long fault troughs between approximately parallel fault scarps are known as *rift valleys* or *graben*. Between the horsts of the Black Forest and the Vosges the Rhine flows through such a graben. An example of a very large graben produced by continental rifting is the Red Sea, separating the Arabian Peninsula from Africa.

Faults and fractures influence relief in three main ways: by the tectonic displacements that they create (see section 11.1), by the modification of the

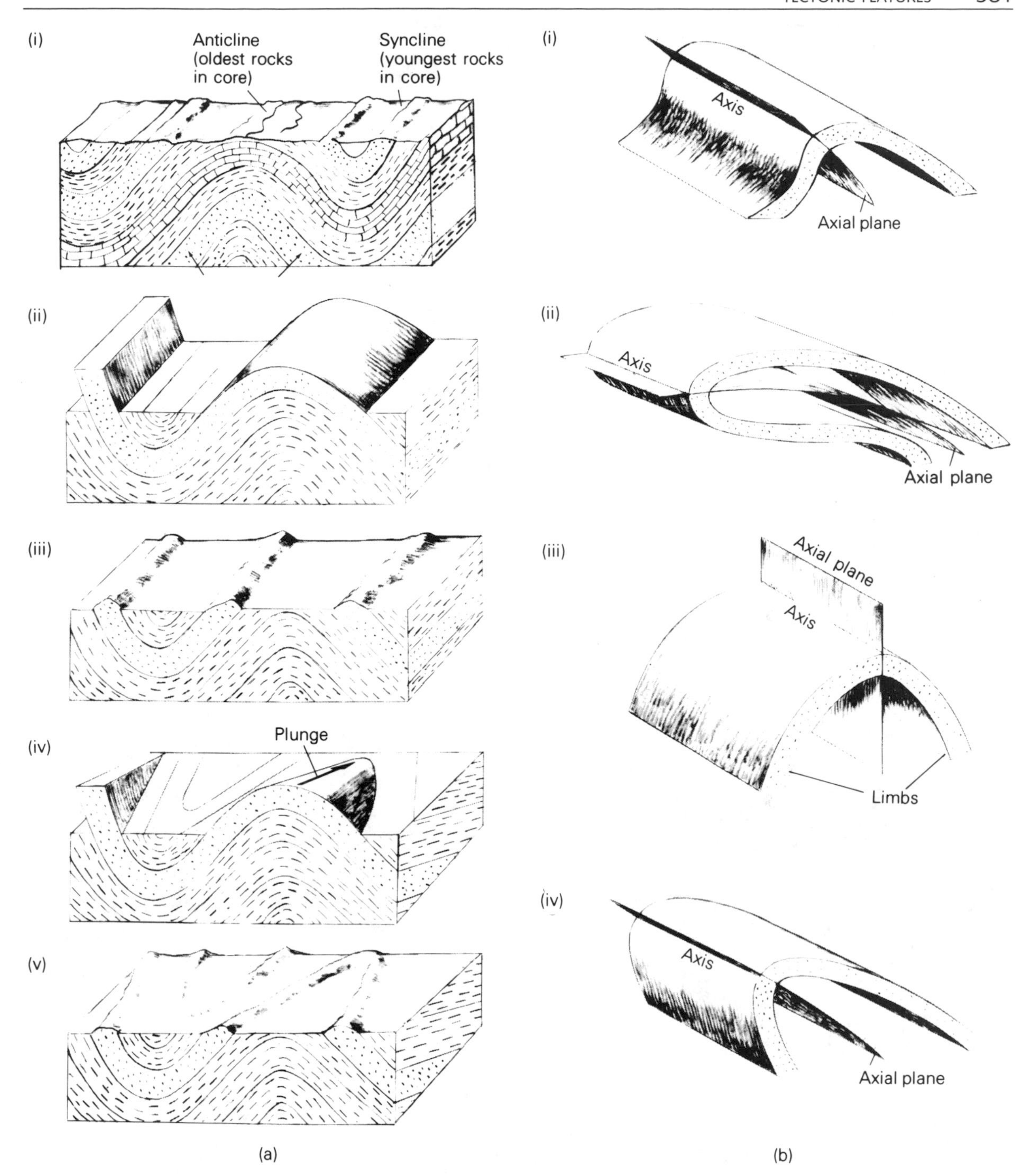

Figure 11.7 Different types of folds. (a) An open symmetrical anticline and syncline: (i) because of erosion the oldest rocks will be found in the core of an anticline and the youngest in the core of the syncline; (ii) and (iii) symmetrical folds before and after erosion; (iv) and (v) symmetrical plunging folds before and after erosion. (b) Different types of anticlinal folds showing in each case the relation of the axial plane and the axis to the geometry of the fold: (i) asymmetrical fold; (ii) recumbent fold; (iii) symmetrical fold; (iv) overturned fold.

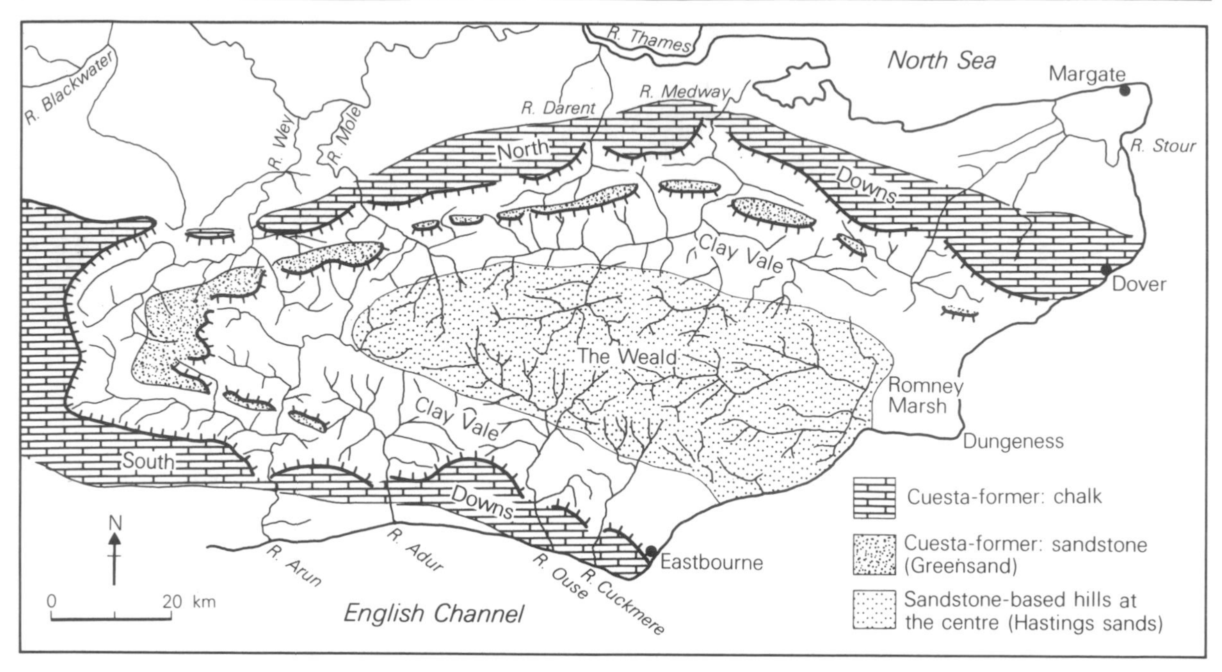

Figure 11.8 Drainage pattern, rock type and relief in the Weald of south-east England.

Plate 11.12 A great escarpment formed in gently dipping sandstones in the Brecon Beacons, south Wales.

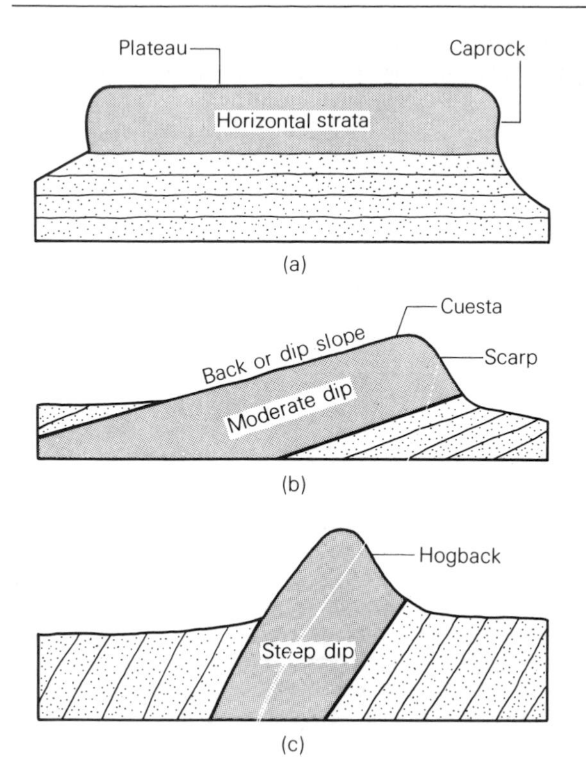

Figure 11.9 Three different ways in which landforms result from differences in the dip of strata of tilted sedimentary rocks: (a) a plateau in horizontal strata; (b) a cuesta in gently dipping strata; (c) a hogback ridge in more deeply dipping strata.

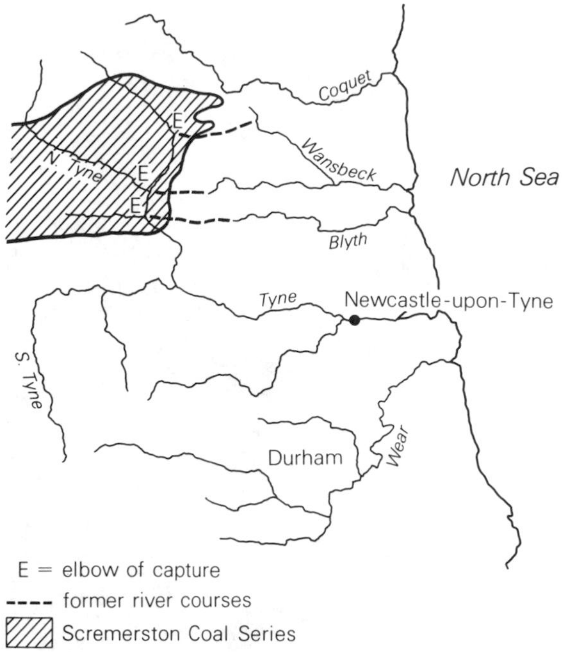

E = elbow of capture
---- former river courses
Scremerston Coal Series

Figure 11.10 The courses of some rivers in northern England which have been affected by river capture. The North Tyne, by cutting back along the weak Scremerston Coal Series, has beheaded the Blyth and the Wansbeck. River capture is also called 'stream piracy'.

rocks in their vicinity (crush zones, breccias and mineral veins), and by the juxtaposition of rocks of differing weathering qualities along fault lines. Among the landforms produced are fault scarps with distinctive facets, rivers with disrupted courses, cave systems with marked changes of course and many alluvial fans.

11.7 Folding

One way in which the rocks of the earth's crust respond to stresses set up by tectonic activity is to deform by crumpling – a process called *folding* (plate 11.11). Folds vary in size from a few millimetres to hundreds of kilometres across.

Upfolds or arches of layered rock are called *anticlines*, and downfolds or troughs are called *synclines*. A steplike bend in an otherwise gently dipping or horizontal bed is a *monocline*. Anticlines that lack a well-defined elongation and plunge from a point nearly equal in all directions are called *domes*, and the corresponding synclinal structures are called *basins*. As figure 11.7 shows, there are various subdivisions of the main types of fold caused by differences in asymmetry, angles and so on.

Because folding creates patterns in rocks of different character, it is a process that has an important effect on landscape. During the building of the Alps in Europe, for example, some of the Cretaceous sedimentary rocks of southern England were deformed. In the Weald (figure 11.8) a great dome was formed on which were superimposed some smaller anticlines, giving a corrugated effect. This was subsequently eroded to produce a landscape

Plate 11.13 The Avon Gorge in Bristol, England, is an example of a river that cuts across geological structures, either because of superimposition or because of antecedence.

dominated by escarpments and vales, according to the different resistances of the rocks exposed as the dome was dissected by rivers.

The precise form of the escarpments will depend in part on the angle at which the rocks lie after they have been deformed. Moderately dipping strata tend to form a typical *cuesta scarpland* like that of the Cotswolds in the West Midlands of England, a *cuesta* being a ridge or upland with a steep scarp slope and a gentle back or dip slope. Horizontal strata will tend to produce plateau-like relief (plate 11.12), whereas if the dip is very steep the result is a ridge with deep slopes on both sides, called a *hogback* (figure 11.9).

During the early stages of erosion anticlines are identical with mountains or ridges, and synclines with valleys, but frequently the anticlines may be rendered less resistant to erosion because of the presence of many joints caused in the process of flexuring. As time passes, therefore, relief inversion can sometimes occur so that the anticline becomes a valley and the syncline a hill.

Another way in which streams may gradually adjust to structures is by *river capture*. In areas of variable lithology, streams may be able to expand their networks preferentially in areas of less resistant rocks. This is evident in the north-east of England, where the North Tyne (figure 11.10) has cut back through the relatively unresistant Scremerston

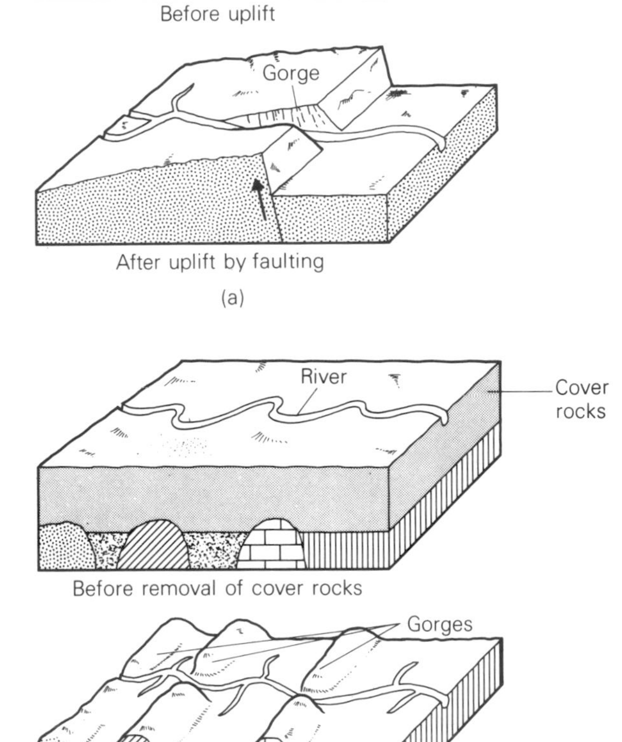

Figure 11.11 Examples of the way in which rivers may become unadjusted to structures: (a) antecedence; (b) superimposition.

Coal Series rocks, beheading the Wansbeck and Blythe Rivers. A distinct change in direction occurs at the point of capture, creating what is called the *elbow of capture*.

Although in general drainage lines become adapted to the arrangement of rocks of different resistance caused by folding, there are cases where there appears to be a striking lack of adjustment (plate 11.13). Such maladjustment is brought about in two main ways: by the development of structures across drainage lines, and by the lowering of drainage lines on to a new set of structures (figure 11.11).

The first of these situations results when a fold rises across the line of a stream, but sufficiently slowly that the stream, by continuing to cut down, can persist across the axis of folding. Drainage that cuts across structures in this way is termed *antecedent*. One of the most spectacular examples of this is provided from the great fold mountains of the Himalayas. The Indus, the Ganga and the Brahmaputra persist in their southward escape from the rising Himalayas as the Indian and Eurasian plates collide at a rate of 5 cm per year, even though the Brahmaputra in particular flows for 1300 km along the direction of the mountain axis.

The second of these two mechanisms produces drainage that is said to be *superimposed*. Such drainage arises if a stream network, however well adapted to the structure within which it developed, is let down on to a completely different set of structures. It is then thrown completely out of adjustment. This is what happens when streams cut through a cover of folded sedimentary rocks on to some older rocks beneath that have a markedly different arrangement of structures.

■ Key Terms and Concepts

antecedence
anticline
earthquake intensity
earthquake magnitude
fault
fold
lahar
natural hazard
pyroclastic deposit
superimposition
syncline
tephra
tsunami

■ Points for Review

- In what ways are volcanoes hazardous?
- Account for the world distribution of earthquakes.
- What steps can be taken to reduce the effects of volcanic and earthquake hazards?
- What influence do faults and folds have on landscape?
- Why do some rivers cut across geological structures?
- Attempt to classify the main types of volcano.

FURTHER READING

Abbott, P. L. (1996) *Natural Disasters* (Dubuqua, Iowa: WmC. Brown). Disasters from an American viewpoint.

Alexander, D. (1993) *Natural Disasters* (London: UCL Press). A comprehensive review of the physical, technological and social components of natural disasters.

Chester, D. (1993) *Volcanoes and Society* (London: Edward Arnold). A survey of volcanoes and an assessment of how society responds to and makes use of areas with significant volcanic risk.

Costa, J. E. and Baker, V. R. (1981) *Surficial Geology: Building with the Earth* (New York: John Wiley). An American text that covers aspects of geology that impact upon humans.

Dolan, C. (1994) *Hazard Geography*, 2nd edn (Melbourne: Longman). Hazards from an Australian perspective.

Francis, P. (1993) *Volcanoes: A Planetary Perspective* (Oxford: Clarendon Press). An authoritative survey by an expert researcher.

Ollier, C. D. (1988) *Volcanoes* (Oxford: Blackwell). A useful text.

Oppenheimer, C. (1996) 'Volcanism', *Geography*, 81: 65–81. A review article on volcanoes in an accessible journal.

Scarth, A. (1994) *Volcanoes* (London: UCL Press). A particularly readable survey.

Smith, K. (1992) *Environmental Hazards* (London: Routledge). A review of many types of hazard, how to assess risk and how to reduce disaster.

Sparks, B. W. (1971) *Rocks and Relief* (London: Longman). A traditional text that links landforms to the rocks they are made of.

12 Slopes

12.1 Mass Movements

Mass movement is a term that covers the movement of material on slopes under the influence of gravity without the benefit of a contributing force such as flowing water, wind or ice. In reality, however, there is a continuum between a river, where there is a great dominance of water over debris, and rockfalls, where there is a great abundance of debris and very little water (plate 12.1). Because we are dealing with a continuum, and because some mass movement events may include more than one type of mass movement, any classification is bound to be arbitrary to a certain extent. However, a very useful classification is that of Carson and Kirkby (1972), in which a triangular diagram (figure 12.1) can be used to classify the main types of movement according to mechanism (heave, slide and flow) and moisture content.

Plate 12.1. A debris fall taking place in glacial debris in the Karakoram Mountains, Pakistan, 1980. The jeep tracks give an idea of scale. The debris is falling into the Hunza River, thereby contributing to its enormous load.

Slides

The simplest type of movement is a fall (figure 12.2a). *Falls* are generated on the steepest hill-slope faces and occur where there are discontinuities such as joints or bedding planes along which weathering can occur. The debris falls from the cliff face through the air and accumulates at the base as a cone of talus (scree). Weathering, especially frost action, and heavy snow and rainfall can cause the fall to occur. Related to falls are *topples* (figure 12.2b), which are forward movements about some pivotal point.

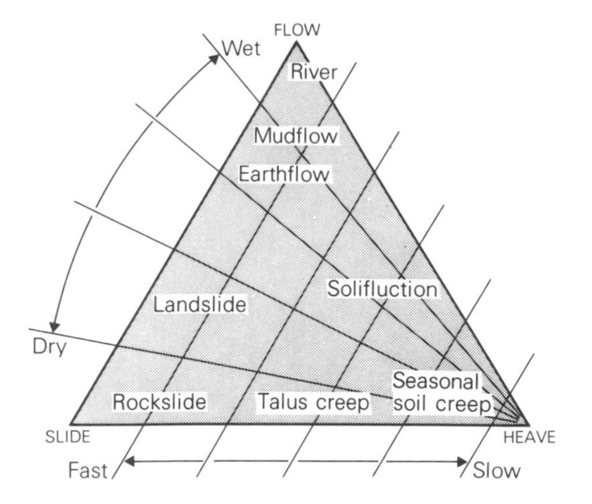

Figure 12.1 A classification of mass movement processes of slopes.

Landslides (plate 12.2) and *rockslides* involve the movement of material as a mass over a discrete surface or plane, and there are two main types. *Rotational slides* (figure 12.2d) have a curved surface of rupture along which sliding occurs and they produce slumps by their backward rotational movement. *Translational slides* (figure 12.2c) have a relatively flat, planar surface of movement along one or several surfaces (see colour plate 21).

Flows

The distinction between debris slides and debris flows is usually difficult to make. Frequently the saturation of the soil on a slope by a cloudburst will initiate failure as a debris slide; however, as motion increases the water-soaked mass will begin to flow.

Flow movements (figure 12.2e) occur in a wide variety of forms. Materials involved vary from huge blocks of rock to clay; the water content ranges from fairly dry to very wet; and velocities range from little more than creep to hundreds of metres per second (figure 12.3). *Debris flows* (plate 12.3) involve the relatively rapid movement of fairly coarse material. They are characteristic of mountain fronts in arid regions, for the mountain slopes

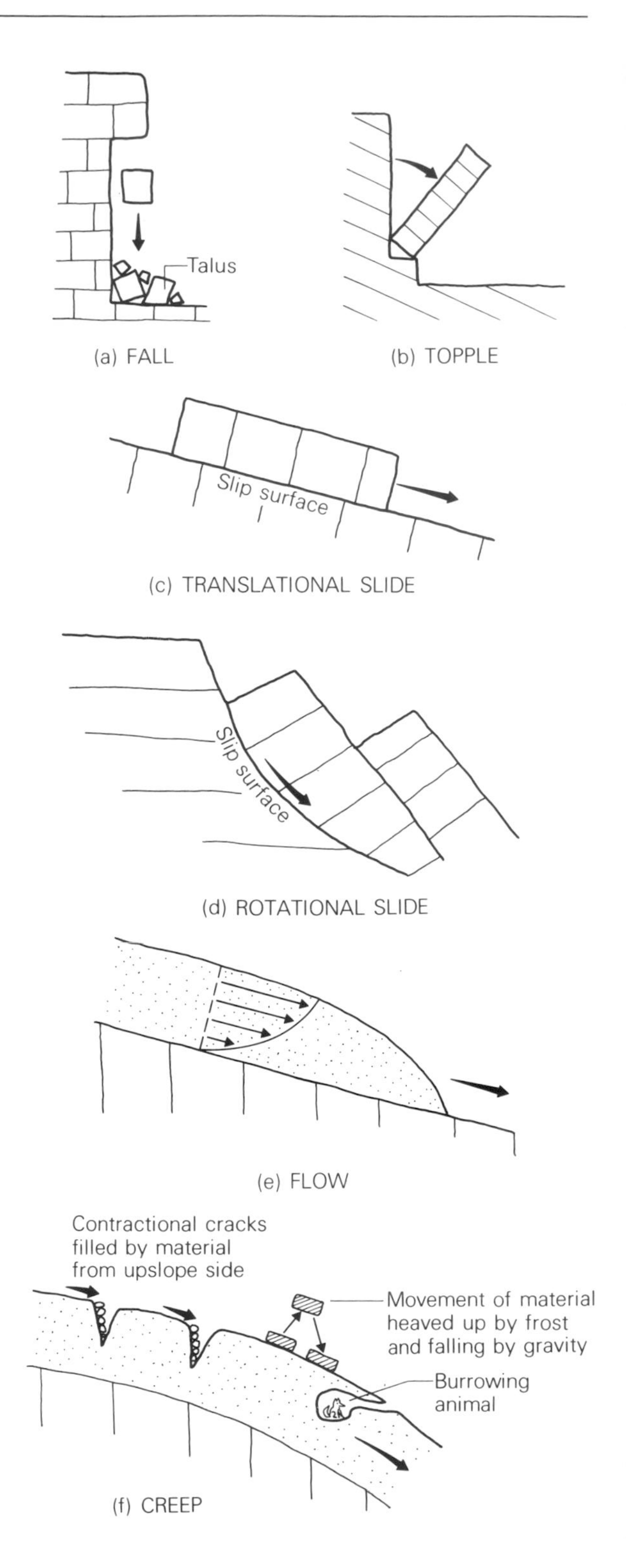

Figure 12.2 Examples of some of the major types of mass movement.

Plate 12.2 The Alport Castles in the Peak District of England are a major landslide complex. The masses in the left of the picture have slipped down from the escarpment to the right. The cause of the instability is the presence of well-jointed millstone grit overlying weak and impermeable shales.

are often covered with a debris layer that is unstabilised by vegetation, so that when a severe storm occurs the water mobilises the debris into a slurry; movement then resembles the flow of wet concrete and occurs as a series of pulsations or surges. Flow can take place for quite long distances into the plains, and helps to form alluvial fans. It tends to take place along pre-existing channels. Flowslide flows characteristically move at velocities of 3–10 m s^{-1}; they carry blocks of rock that are often 2–8 m across, have water contents of around 20 per cent, and may flow for many kilometres. They can carry away cars, tractors and houses, and they can rapidly and easily block culverts and bridges on roads and railways.

Solifluction, which literally means soil flowage, occurs widely in periglacial areas, though it is by no means restricted to them. It comprises two main processes: the flow of water-soaked debris, and the creep of surface material by freeze–thaw action. The water-soaked debris occurs as a result of thawing of permafrost or seasonally frozen ground in the active layer (see section 4.9), assisted by snow-melt. In addition, the presence of permafrost just below the surface creates a relatively impermeable layer that helps to maintain water saturation in the near-surface material. Creep is caused by freezing and thawing, which causes changes in the volume of the surface layers and so makes them potentially unstable for the soil flow processes to operate. Solifluction is also favoured by the presence of silty soils, which can hold a great deal of moisture, by a moderate slope, and by a limited amount of binding vegetation.

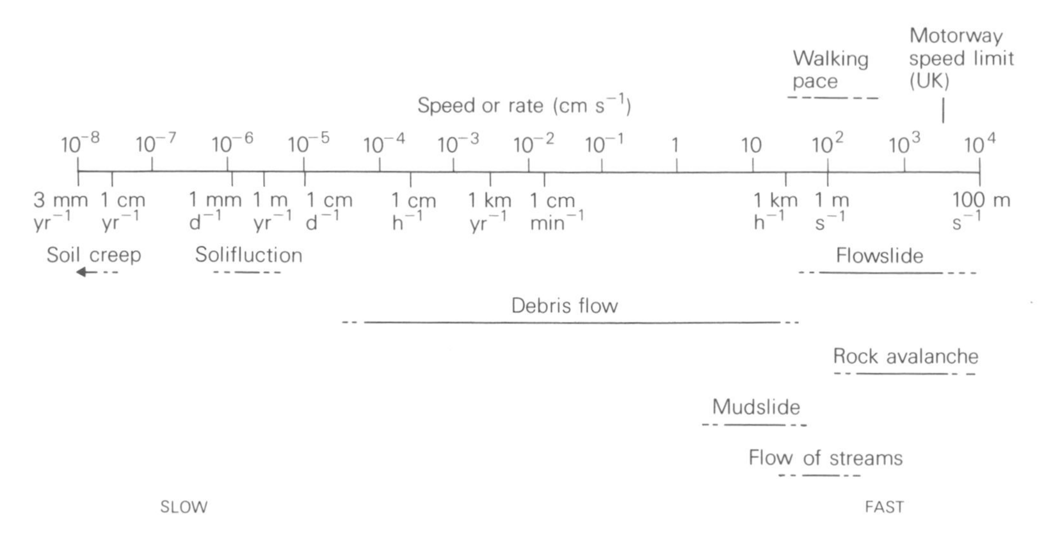

Figure 12.3 The rate of movement of various mass movement processes.

Plate 12.3 Debris flows in the Karakoram Mountains, Pakistan, are generated on steep, unvegetated scree slopes. Debris has crossed irrigated fields and plunged into the Hunza River.

Soil creep itself is a general term, applied to the slow downhill movement of soil and related material such as talus (figure 12.2f). A number of processes, each capable of producing only very slight movements, combine to cause it. Rainwash and raindrop impact may move small particles downslope, and may remove fine material from the downslope side of large stones, thus facilitating the movement of the stones. The swaying of vegetation and the treading of animals, both wild and domestic, also tend to give a slight downslope movement to weathered material on a slope. Ploughing may also contribute to such movement. However, the two prime causes of soil creep are expansion and contraction, caused either by seasonal wetting and drying, or by seasonal and diurnal freeze–thaw processes. In the case of freeze–thaw, for example, frost heaving, arising from the growth of ice crystals and needle particles, lifts those particles up at right angles to the slope, but they fall back perpendicularly under the influence of gravity. Equally, as a soil dries out a crack may form as a result of contraction. When it rains again this crack will fill with material from upslope and so a net movement of material downslope will have taken place.

Various techniques have now been developed to measure rates of soil creep. On moderately steep slopes in humid temperate areas soil layers commonly creep downhill at a mean rate of about 1–2 $mm\,yr^{-1}$. In the moist tropics the rates may be 3–6 $mm\,yr^{-1}$, and in semi-arid regions with cold winters they may be 5–10 $mm\,yr^{-1}$. Thus, although this is rather an unspectacular process in comparison with some of the more violent types of mass movement, over a period of years soil creep may achieve a considerable amount of geomorphological work.

Plate 12.4 A huge landslide which developed along the Pacific Coast Highway, Los Angeles County, California, USA. Note the landslide scar at its rear and the way in which the toe of the slide extends across the highway on to the beach.

Table 12.1 Factors involved in slope failure

Factors leading to an increase in shear stress

Removal of lateral or underlying support
- Undercutting by water (e.g. river, waves) or glacier ice
- Weathering of weaker strata at the toe of the slope
- Washing out of granular material by seepage erosion
- *Man-made cuts and excavations, draining of lakes or reservoirs*

Increased disturbing forces
- Natural accumulations of water, snow, talus
- *Man-made pressures (e.g. stockpiles of ore, tip-heaps, rubbish dumps and buildings)*

Transitory earth stresses
- Earthquakes
- *Continual passing of heavy traffic*

Increased internal pressure
- Build-up of pore-water pressures (e.g. in joints and cracks, especially in the tension crack zone at the rear of the slide)

Factors leading to a decrease in shearing resistance

Materials
- Beds that decrease in shear strength if water content increases (clays, shale, mica, schist, talc, serpentine) (e.g. *when local water table is artificially increased in height by reservoir construction*) or as a result of stress release (vertical and/or horizontal) following slope formation
- Low internal cohesion (e.g. consolidated clays, sands, porous organic matter)
- In bedrock: faults, bedding planes, joints, foliation in schists, cleavage, brecciated zones and pre-existing shears

Weathering changes
- Weathering reduces effective cohesion, and to a lesser extent the angle of shearing resistance
- Absorption of water leading to changes in the fabric of clays (e.g. loss of bonds between particles or the formation of fissures)

Pore-water pressure increase
- High groundwater table as a result of increased precipitation *or as a result of human interference (e.g dam construction)*

Human influences are italicised
Source: modified from R. U. Cooke and J. C. Doornkamp (1974) *Geomorphology in Environmental Management* (Oxford: Clarendon Press), p. 131

12.2 Slope Instability

Many types of mass movement occur sporadically, and it is often their sporadic behaviour that creates hazards (plate 12.4). There is seldom only one cause for such instability; usually it results from a sequence of events that ends with downhill movement. The causes, however, can be divided into those that are disturbing forces and those that result from a weakening in the strength of the material upon which the disturbing forces operate. The sort of factors that lead to increases and decreases in the power of the disturbing forces are listed in table 12.1.

The presence of water in slope materials is an important cause of their stability or instability, for water creates various forces in the soil (see figure 12.4). If the soil's pore spaces are not completely filled with water (i.e. if the soil is unsaturated), then a suction force is exerted which tends to draw the soil grains closer together. This suction is caused by a process called *capillary tension* – the force that enables sand castles to be made of wet sand but not of dry sand. However, if the soil pore spaces are completely filled with water, then the soil is said to be saturated. In this state the water exerts a pressure within the pore spaces that tends to produce forces that push the grains apart. Therefore, when *pore-water pressures* reach high levels in slope materials these materials may become unstable. Such

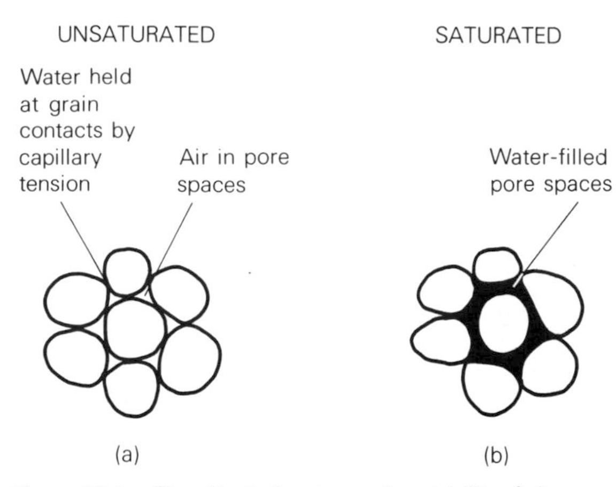

Figure 12.4 The effect of water on the stability of slope materials. In the case of unsaturated grains suction forces draw the grains together; in the case of saturated grains high pore-water pressures tend to push grains apart.

conditions could be caused by extremely heavy rainfalls, the blocking of drains, the seepage of irrigation water, or slope saturation arising from the impounding of a reservoir behind a dam.

A few examples (as, for example, in window 12.1) will illustrate why slope instability occurs and can be hazardous. In 1963, in the Italian Alps, a huge rockslide ($240 \times 10^6\,m^3$) fell into the large reservoir impounded behind the Vaiont Dam. The slide was 2 km long, 1.6 km wide and in parts 150 m thick. As it collapsed into the reservoir it displaced so much water that the dam was overtopped and a flood wave of 70 m high poured down the valley from the dam, killing as many as 3000 people. There were certain factors that made this slope potentially unstable: the rocks in the area are alternating bands of clays and limestones which have been heavily

Window 12.1 ***The Abbotsford landslide, Dunedin, New Zealand, 1979***

Dunedin is a city in the south-east of the South Island of New Zealand. One of its suburbs is called Abbotsford. On 8 August 1979, 7 ha of hillside began to slide downslope at a speed of 3 m per minute into part of this suburb. The developments of cracks in the area in the previous few months had alerted the inhabitants to the risk of a slope failure taking place, so that nobody was injured. However, 69 homes were destroyed by the event and 200 residents had to leave the area permanently. The Abbotsford landslide is instructive because it shows the number of different factors that can be involved in generating a landslide:

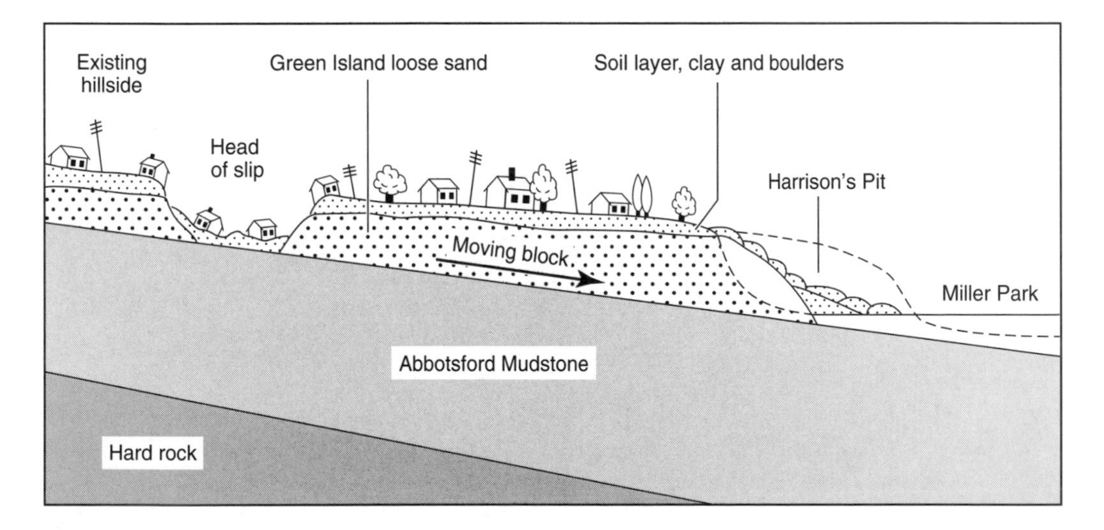

The setting of the Abbotsford mudslide, New Zealand.

1 The suburb was built on loose sand (the Green Island loose sand).
2 The sand overlies a mudstone which provides a surface over which movement can occur.
3 The base of the slope had been disturbed by the quarrying of material from Harrison's Pit for roadworks.
4 There was heavy snow and rain before the event to build up pore water pressures in the slope.

Source
C. Dolan (1994) *Hazard Geography*, 2nd edn (Melbourne: Longman), pp. 193–4.

folded, so that they have a steep dip and rapid changes in their permeability to seeping water; the valley has been greatly steepened, first by glacial scour and then by post-glacial fluvial incision; and because of the severe tectonic stresses on the area and the opening of joints associated with the erosion caused by glaciers, the rocks were deeply split up by joints. What actually triggered the rockslide was the water in the reservoir, for this led to the saturation of the lower part of the potentially unstable slope, increased pore-water pressure in the rock, and thereby reduced its strength below a critical threshold.

Some of the most severe examples of accelerated slope instability occur on coasts. A classic example is provided by Folkestone Warren, southern England, where a large landslide complex developed where a thick bed of chalk overlies impermeable Gault clay. The complex lies to the east of Folkestone Harbour and the direction of littoral drift of sediment is from the west. Thus the construction of large harbour breakwaters at Folkestone in the nineteenth century blocked the eastward movement of coastal shingle so that the Warren became depleted in beach material. This made it more susceptible to undercutting and oversteepening by wave action. A series of failures occurred, with a major slip in 1915 disrupting the Folkestone–Dover railway line. Remedial action, such as weighting and protecting the toe of the slip with masses of concrete, has been employed to make the area less unstable.

In parts of Canada and Scandinavia there is a type of landslide caused by the failure of materials called *quick clays*. These clays are highly deceptive, for in their natural undisturbed state they seem to be strong, and are often seen to be supporting steep stream banks. However, for reasons that are not fully understood, they sometimes change into a liquid state of negligible strength. Failures come without warning and may be disastrous. One theory, which may well apply to the Canadian

Table 12.2 Landslides associated with earthquakes

Year	*Location*	*Magnitude*	*No. of landslides*	*Vol. of landslide material (m^3)*
1976	Guatemala	7.6	≅50 000	1.16×10^8
1989	Loma Prieta, USA	7.0	≅1 500	7.5×10^7
1994	Northridge, USA	6.7	>11 000	
1974	Izu-Oshima Kinkai, Japan	6.7	>51	
1983	Kaoiki, Hawaii	6.7	≅300	
1983	Coalinga, California, USA	6.5	9 389	1.9×10^6
1980	Mammoth Lakes, California, USA	6.2	5 250	1.2×10^7
1980	Mt Diablo, California, USA	5.8	103	
1986	San Salvador, El Salvador	5.4	≅400	3.78×10^5
1957	Daly City, California, USA	5.3	23	6.7×10^4

Source: D. K. Keefer (1999) 'Earthquake induced landslides and their effect on alluvial fans', *Journal of Sedimentary Research*, 68: 84–104

examples, is that during the Ice Age some fine-grained sediments were deposited in marine coastal areas. The clays were coagulated by the electrolytes in the sea water and settled quickly. Some of the chemical constituents from the seawater, such as sodium (Na^+), helped to bind the clays together and give them strength. However, post-glacial isostatic uplift elevated these deposits high above present sea levels, and subsequently fresh rainwater leached the sodium 'glue' from the clays, leaving them with a very unstable structure. Upon shaking, this structure collapses, expelling the interstitial water, liquefying the clay and allowing it to flow on gentle slopes. The shaking may be caused by earthquakes. Thus we have a combination of factors that make certain parts of the St Lawrence Valley area prone to slope failures: glacial action has provided the right material, post-glacial uplift has provided the right setting, and earthquakes provide the trigger.

Indeed, one important trigger of landslides in tectonically active areas is the shaking effect of earthquakes. As table 12.2 indicates, there are many records of earthquakes generating large numbers of landslides in affected areas, and of them leading to the movement of large volumes of material. An extreme example of this is the magnitude 7.6 earthquake that struck Guatemala in 1976, generating around 50 000 landslides.

The incidence of debris flows and mudflows can be accelerated by human activity. The removal of vegetation, perhaps by fire or in the process of converting forested land to agricultural use, can expose potentially unstable materials to the effects of massive storms. However, catastrophic rainfalls can lead to debris flows in steep, humid terrain even without human intervention.

Whatever the exact combination of forces that causes them, major mass movements can cause great loss of life (see table 12.3), and they play an important role in moulding the forms of slopes.

12.3 Other Processes Operating on Slopes

The various forms of mass movement described above are not the only ones that operate on slopes and mould their form. Weathering is extremely important, since it attacks bare rock faces, creates a layer of weathered material (regolith) and affects the strength of the slope by forming materials like clays. Solution is also extremely important in removing material from slopes.

Table 12.3 Major mass movement disasters of the twentieth century

Year	*Location*	*Nature*	*Deaths*
1920	Kansu, China	Earthquake-triggered flow and slide in loess	200 000
1936	Loen, Norway	Rockfall into fiord, causing tsunami-like flood-wave	73
1941	Huaraz, Peru	Avalanche and mudflow	7 000
1956	Santos, Brazil	Landslides	100+
1959	Montana, USA	Earthquake-triggered landslide	26
1962	Mt Huascarán, Mexico	Ice avalanche and mudflow	*c.*4 000
1963	Vaiont Dam, Italy	Landslide-created flood	*c.*3 000
1966	Rio de Janeiro, Brazil	Landslides in shanty towns	279
1966	Hong Kong	Landslides on steep tropical slopes after heavy rain	64
1966	Aberfan, Wales	Debris flow from colliery waste tip	144
1970	Mt Huascarán, Mexico	Earthquake-triggered ice and rockfall and debris flow	25 000
1971	Quebec, Canada	Landslides in quickclay	31
1971	Romania	Mudslide in mining village	45
1972	West Virginia, USA	Landslide and mudflow in mining tip waste	400
1974	Mayunmarca, Peru	Rockslide and debris flow	450
1976	Pahire-Phedi, Nepal	Landslide	150
1976	Hong Kong	Landslide	22
1978	Myoko Kogen Machi, Japan	Mudflow, overwhelming ski resort	12
1985	Mameyes, Puerto Rico	?	129

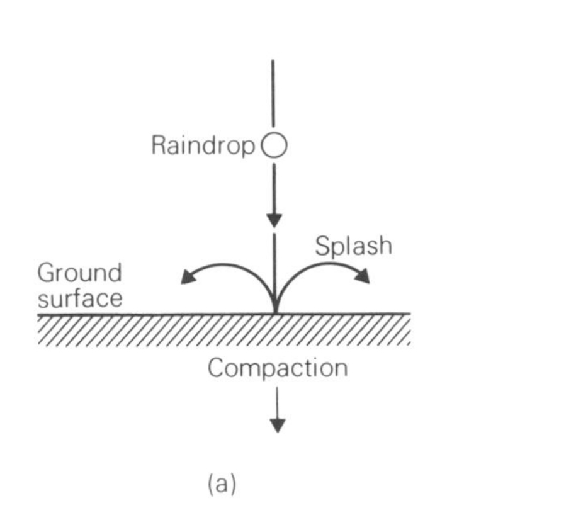

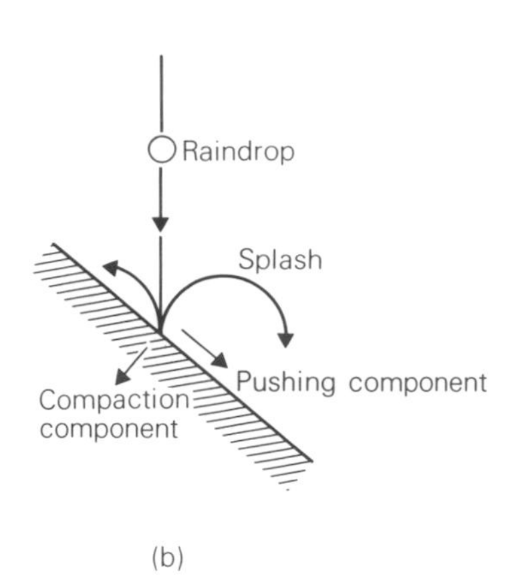

Figure 12.5 The effect of rainsplash on slopes. (a) On a flat surface the net effect of splashing and pushing of grains is equal in all directions and no net erosion takes place. (b) On a slope, the splash trajectories are much longer downslope than they are upslope and grains will be pushed downslope more easily; as a result, erosion increases with slope angle.

A third highly important process is *surface wash* (figure 12.5). This is the downslope transport of regolith across the ground surface through the action of water. There are two distinct processes involved: the impact of raindrops on the ground surface (*raindrop impact*), and the flow of water across it (*surface flow*). Surface wash probably has its greatest effect in areas with limited vegetation cover, poorly structured soils and occasional storms of high energy. For these reasons it may attain its optimum impact in semi-arid areas, though it also operates effectively in tropical rainforests, where there is often only very limited litter on the ground surface, and where torrential tropical storms manage to produce large raindrops that fall from driptips of the very tall trees (section 7.6).

There are also some subsurface transport processes that may be of importance. In soils with the right sort of structure, water may move laterally as *throughflow* (see section 14.6), and in doing so it may transport fine particles through the regolith by a process called *lateral eluviation*. Lateral movement of soil water may also take place in concentrated lines of seepage or flow, and tunnelling or piping may occur by mechanical corrasion. Sometimes such tunnelling causes ground subsidence, and may also initiate the formation of gullies which create gashes in many eroded hillsides.

Slopes are going to be further affected by such processes as river erosion at their bases; and by the removal of weathered material by fluvial, aeolian and other processes.

12.4 Slope Forms

It is essentially the balance between the power of all these different processes and the nature of the material on which they operate that explain the evolution of slope forms. Three main types of slope occur. The first is a *denudation slope*, which is one on which ground loss is occurring. The second is a *transportation slope*, which is one that undergoes neither ground loss nor ground gain because at each point the material brought down from upslope is equal to that carried away downslope. The third type of slope is an *accumulation slope*, one on which ground gain is occurring.

In the case of denudation slopes, there is an important distinction to be made between those that are controlled by weathering and those that are controlled by erosion, or *removal*. On the former, the potential rate at which weathered material can be removed from the slope exceeds the rate of weathering; hence weathered material is removed shortly after it is formed, and the rate at which denudation occurs is thus limited by the rate of weathering. Control by removal occurs where the potential rate at which weathering produces debris exceeds the rate at which the debris can be removed; ground loss is then dependent on the efficiency of the removal process. In the case of a weathering-controlled slope, the relative resistance of different rocks to weathering controls the slope form, whereas in the case of a removal-controlled slope, the slope form is very much influenced by the nature of the regolith that is produced.

Strictly speaking, these terms should be applied only to a point on a slope, for slopes need not be subject to the same type of control throughout their length. A typical slope profile may well possess, for example, a weathering-controlled cliff, a basal concavity, and a talus cone where accumulation is dominant. Some of the different forms of debris accumulation that occur on slopes are illustrated in figure 12.6, and include debris cones (plate 12.5), alluvial fans (see also section 6.11) and rock avalanche tongues.

There are basically three fundamental slope forms that may exist in various combinations on any one slope profile (figure 12.7). A *straight slope* is a section of a profile on which the mean angle does not change with distance; a cliff or *free face* is a special form of straight slope. A *convexity* is a section of a profile on which slope angle increases downslope, while a *concavity* is a section of a profile on which slope angle decreases downslope.

Let us now consider the circumstances that might produce each of these fundamental types. A *straight slope* may develop in a number of ways. For example, if a slope is undercut by a river or by very active weathering at its base, it may be oversteepened and then fail; the resulting slope will then rest at an angle that is related to the strength of the material and the water that it contains. In general, straight slopes, or the straight part of slopes, are to be expected where rapid types of mass movement operate as a major slope-forming process. For example, if weathering attacks a cliff face, boulders will be detached and these will accumulate at their angle of repose (generally, 32–38°) as rockfall screes. Scree slopes tend to have a straight upper segment and a concave lower segment. The straightness of the cliff itself may be controlled by joints in the bedrock.

Straight slopes are controlled by threshold conditions; when the threshold is reached as a result of attack from below, or because of weathering attack on joints, or because of a build-up in pore-water pressure, the slope adjusts rapidly to an angle that is related to its material properties. By contrast, on curved slopes there is usually a slow evolution towards a characteristic form involving a variety of processes.

Convexities tend to occur on the upper portions of slopes, though in some humid tropical areas they may exceed 90 per cent of total slope profile length, and in locations where stream undercutting occurs convexities may also occur at the base of the slope. Rainsplash and soil creep are the two processes most responsible for the development of the upper convexity, though weathering will tend to round off

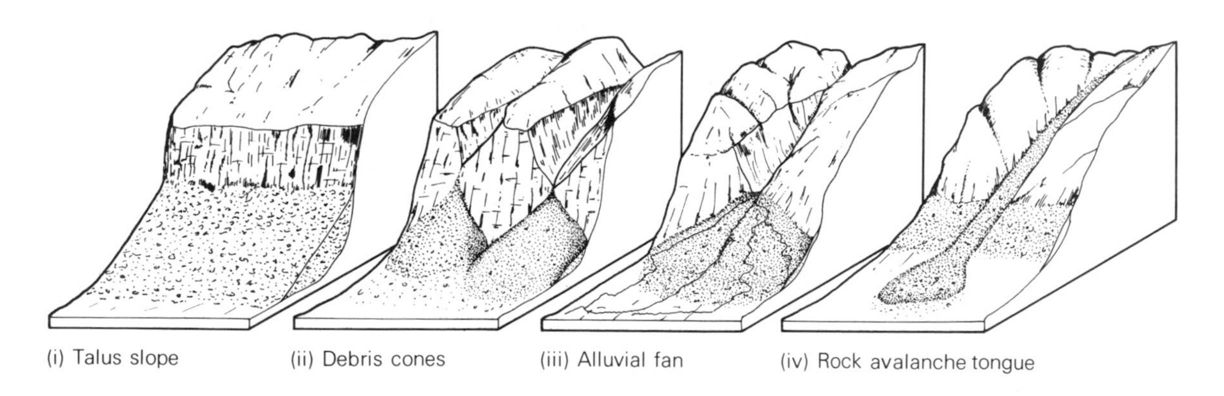

Figure 12.6 The four main forms of debris accumulation that may occur below rock slopes.

Plate 12.5 Enormous debris cones produced by severe frost weathering in the Karakoram Mountains, Pakistan.

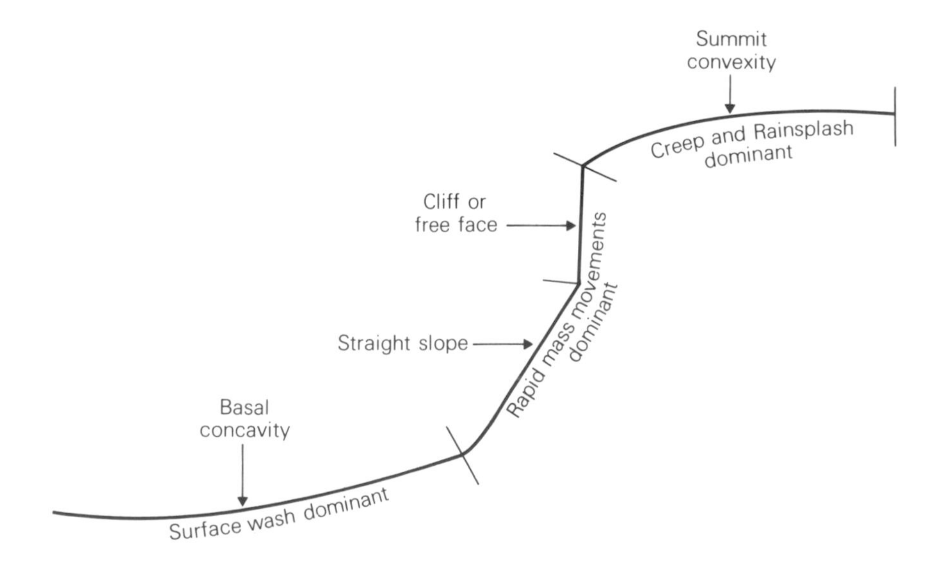

Figure 12.7 The three fundamental slope forms (straight, convex and concave) shown in an idealised four-segment profile, with some suggestions for the dominant types of process acting on each segment type.

Plate 12.6 One of the most important geomorphologists of all time was G. K. Gilbert. Working in the late nineteenth century and the first few years of the twentieth century, he stressed the importance of slope processes in moulding the landscape.

exposed right-angles of bare rock, and on slopes where a resistant caprock overlies weak clays rounding may occur as a result of a process called *cambering*. The mechanism whereby soil creep can produce rounding was postulated by G. K. Gilbert (plate 12.6) in 1909 (figure 12.8). The two parallel lines represent the position of the ground surface at two points in time. If one assumes uniform soil depth, and no change of soil depth through time at any one point, the volume of soil between A and B will have passed point B and the volume between point A and C will have passed point C for this amount of lowering to have occurred. Thus the volume of soil passing a position progressively further down the slope increases commensurately with position from the drainage divide. If the transport rates for the creep processes are proportional to slope angle, then clearly the slope angle must also increase with distance from the divide, in order for those volumes of sediment to have been transported from the slope. This is one of the explanations for slope-top convexity.

Slope concavity, which tends to occur towards the base of a profile, but which may be a very lengthy part of the profile on pediment slopes in arid regions and elsewhere, tends to occur where rapid downcutting is not taking place at the base of the slope. It also tends to be associated with surface wash processes, and therefore may be most common on slopes composed of soils with low infiltration capacities. The normal explanation for the association of surface wash with slope concavity is much the same as the explanation given for most river long profiles being concave (see section 15.4): namely, as one moves downslope, discharge tends to increase because the contributing area to the process also increases. As discharge increases, so the transporting ability of the flow grows greater, and as flow depth increases velocity can be maintained on lower slope angles. Moreover, because of abrasion and sorting processes, particle size decreases downslope and any given flow can carry a greater load of fine material than of coarse material. This combination of factors and tendencies combines to produce an excess transporting capacity in the flow so that slope can be reduced while the power to transport load is maintained. The steepest concavities occur in the coarsest materials, and the gentlest concavities in the finest.

12.5 Slope Development through Time

One of the most controversial subjects in geomorphology is the question of the way in which slope form evolves through time. Basically there have been three

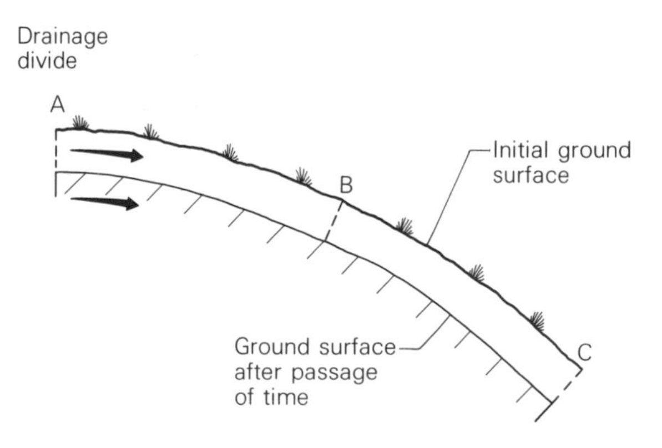

Figure 12.8 G. K. Gilbert's model of slope-top convexity through soil creep processes. For explanation see text.

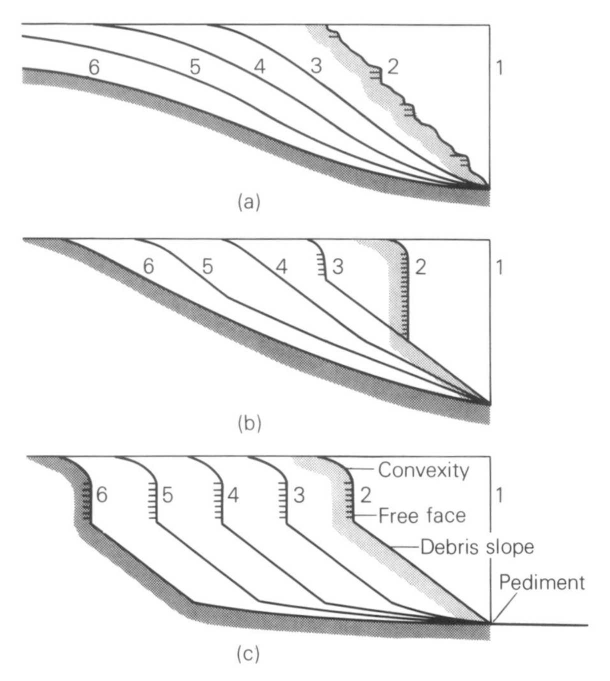

Figure 12.9 Three models of slope evolution: (a) slope decline; (b) slope replacement; (c) parallel retreat. 1 represents the initial cliff form, and 2–6 represent subsequent stages in development of the slopes.

main theories of slope evolution: slope decline (an idea associated with the name of W. M. Davis), slope replacement (an idea associated with the name of W. Penck) and parallel retreat (an idea associated with L. C. King). These three main ideas are illustrated in figure 12.9.

In the case of *slope decline* the steepest part of the slope progressively decreases in angle, accompanied by the development of a convexity in the upper portion and a concavity in the lower portion of the slope. In the case of *slope replacement* the maximum angle decreases through replacement from below by gentler slopes, causing the greater part of the profile to become occupied by the concavity, which may be either smoothly curved or segmented. In the case of *parallel retreat* the maximum angle remains constant, the absolute lengths of all parts of the slope except the basal concavity remain constant, and the concavity increases in length.

The ideas of W. M. Davis (plate 12.7) on slope decline (figure 12.9a) were first formulated towards the end of the nineteenth century, and they were presented in his *Cycle of Normal Erosion*. In this extraordinarily influential model, Davis envisaged that a mass of land would be quickly uplifted from beneath the sea by tectonic movements. He made the assumption that, because the rate of uplift was so rapid, the processes of denudation were able to act almost from the start on what was, in effect, a stable mass. In the stage of youth, under a humid environment a system of rivers would quickly develop, and they would cut down rapidly, producing steep valley side slopes. However, as time progressed the rate of deepening of the V-shaped valleys would decrease, for the rivers would have lowered their channels nearer and nearer to the *base level of erosion* (which in most cases is the level of the sea into which the rivers flow). In this stage of maturity, with the retardation of down cutting, lateral erosion would become more significant. Weathering, soil creep and rainwash would gradually cause the retreat of the valley sides and the lowering of the areas between the valleys, called the

Plate 12.7 A great contemporary of G. K. Gilbert was the American geomorphologist W. M. Davis, who produced a model of slope evolution called the cycle of normal erosion, in which slope angles were thought to decline through time.

interfluves. By the end of the mature stage Davis envisaged that slope angles in general would have been considerably reduced by the process of divide wasting, and smoothly curving slope profiles, with no major breaks, would dominate the landscape. A decrease in the vertical height separating interfluve summits and valley floors would occur. In the third stage – that of old age – erosion processes would have slowed down, for the reduction of river gradients would have led to a decline in stream energy, and slopes would have been degraded by slope processes. Rivers would, however, still broaden their valleys by meandering, and broad flood plains would develop. By the end of this stage, the land surface would assume the form of a very gentle undulating plain, which Davis called a *peneplain*. This would stand only a little above the base level of erosion.

The cycle of erosion postulated by L. C. King (figure 12.9c) was different from that of Davis, possibly because, whereas Davis initially formulated his ideas in the humid temperate landscapes of Appalachia in the United States, King formulated his ideas in South Africa, where there are broad

Plate 12.8 The slopes of Monument Valley, Arizona, illustrate how slope retreat through time can lead to the development of steep-sided residuals.

plains above which stand some steep-sided island-like hills, called inselbergs. King's examination of the African landscape led him to conclude that it was made up of two basic elements: gently concave lower slopes, called *pediments*, and steep slopes, or *scarps*, bounding upland blocks (see colour plate 22). The former have angles that generally range from 0.25° to 7°, while the latter tend to have angles of 15–30°. King believed that the landscape evolved as a result of the extension of pediments (*pedimentation*) and by the back-wearing or parallel retreat of scarps. In its stage of youth the cycle is initiated by uplift of a previously formed plain. Rivers cut down rapidly towards base level, exploiting joints, and cliff-like slopes are very important. Once these have reached a stable angle of repose related to the characteristics of the material of which they are made, King envisages them retreating, parallel to each other, away from the drainage lines, leaving continually expanding pediments between the slopes and the drainage lines. The residuals of these *pediplains* will become progressively smaller in area, until they are left as small, steep-sided, isolated hills (plate 12.8).

Unlike Davis or King, W. Penck (plate 12.9) worked in areas such as the Alps and the Andes, where tectonic processes were extremely active, and as a result he believed that the nature of uplift was vital, as it controlled the rate of river erosion, which in turn affected the evolution of valley side slopes. In his theory of slope replacement, he rejected the simple Davisian assumption of rapid uplift followed by a period of stability, and attempted to establish a relationship between the nature of a slope and the tectonic history of a region: convex slopes were formed in periods of accelerated uplift, straight slopes in periods of constant uplift, and concave slopes in periods of decreasing uplift or stability. He saw the rate of retreat of a slope as being determined by gradient. Thus a steep slope would undergo more rapid weathering and debris removal than a gentle slope, on which the more slowly moving weathered layer would tend to persist and thereby cushion the rock from atmospheric agencies. As a result, a steep slope beneath a gentle slope will weather back so quickly that in due course the gentle upper slope will be destroyed. In addition, rapid retreat will destroy the steeper slopes on either side of a drainage divide, and these will be replaced by gentler slopes developing at their bases (figure 12.9b).

Plate 12.9 The German geomorphologist Walther Penck believed that slope evolution was more complex than had been envisaged by Davis, and stressed how the nature of tectonic processes would affect the changing form of slope profiles through time.

No one of these three models of slope evolution is universally correct. They are all ideal models that simplify the complexity of the real world, and it is unlikely that the retreat of any natural slope corresponds to any one of them absolutely. The extent to which it does correspond depends to a large extent on the structure of an area, and on the comparative rates of operation of different slope processes as influenced by climate. Indeed, under certain circumstances slopes may evolve in ways other than the three discussed above. In some rainforest areas, for example, slopes are dominated by long convexities and may even tend to become steepened through time.

Because of the great length of time that it takes for most slopes to evolve, their evolution can scarcely be checked in one lifetime. However, if one is able to locate landforms of similar material quality under the same climatic conditions, but which are known to be different ages, then it may be possible to see which slope evolution model is most appropriate in any set of environmental conditions. For example, geomorphologists have used clifflines that have been abandoned by the sea at different times, lake shorelines exposed as a lake progressively dries up, moraines that have been deposited by a gradually retreating glacier, and slag heaps of known initial form but of different age.

■ *Key Terms and Concepts*

cycle of normal erosion
falls
flows
parallel retreat
pediplain
peneplain
pore-water pressure
quickclay
rainsplash
slides
slope concavity
slope convexity
solifluction

■ *Points for Review*

- What are the main types of mass movements on slopes?
- In what ways might humans induce landslides?
- How do slope forms change through time?
- What processes might cause (a) slope convexity and (b) slope concavity?
- What are the main differences between peneplains and pediplains?

FURTHER READING

Anderson, M. G. and Brookes, S. M. (eds) (1996) *Advances in Hillslope Processes* (Chichester: Wiley). A massive, two-volume collection of edited papers.

Carson, M. A. and Kirkby, M. (1972) *Hillslope Form and Process* (Cambridge: Cambridge University Press). A classic text.

Crozier, M. J. (1986) *Landslides: Causes, Consequences and Environment* (London: Croom Helm). An excellent introduction by a leading expert from New Zealand.

Dikau, R., Brunsden, D., Schrott, L. and Ibsen, M-L. (1996) *Landslide Recognition* (Chichester: Wiley). A European compilation.

Parsons, A. J. (1988) *Hillslope Form* (London: Routledge). An intermediate-level text.

Selby, M. J. and Hodder, A. P. W. (1993) *Hillslope Materials and Processes*, 2nd edn (Oxford: Oxford University Press). A thorough textbook.

Young, A. and Young, D. (1990) *Slope Development*, 2nd edn (Basingstoke, Macmillan). A simple introduction.

13 Soils and Weathering

13.1 Factors of Soil Formation

Soil is the support of vegetable life, and thus plays the most fundamental role in providing sustenance for all living things. It can be defined as 'a natural body of animal, mineral and organic constituents differentiated into horizons that comprise a profile of variable depth which differs from the material below in morphology, physical make-up, chemical properties and composition, and biological characteristics.' The study of soils is called *pedology*.

The decay of rock by weathering processes plays a fundamental role in the formation of most soils. But soil is far more complex than simply decayed rock: it is the product of a wide range of different factors.

In the 1940s a pedologist called H. Jenny attempted to summarise the various factors involved in soil formation by the following expression:

$$S = f(cl, o, r, p, t, \dots)$$

Where S denotes any soil property, cl the regional climate, o the biota (comprising plants and animals), r the topography, p the parent material, t the time (or period) of soil formation, and the dots represent additional, unspecified factors. These factors are often closely interrelated. Regional climate, for example, will affect both the nature of the biota and the form of the topography, and certain types of parent material may be a feature of certain climatic zones (e.g. boulder clay in glacial regions, and aeolian sand in desert areas).

Climate

Because the nature of weathering, of erosion and of organic activity are so closely controlled by climatic conditions, climate is a vital factor in soil formation. As we saw in chapter 2, world soil types show a broad correspondence to world climatic zones. There are considerable differences between tropical soils (see section 7.11), arid zone soils (see section 6.7) and tundra soils (see section 4.13)

Biota

Soil is more than simply weathered rock because its organic component is also important. The vegetation cover of an area provides partially decayed organic matter, called *humus*, and a cycling of plant nutrients takes place between plants and soils (see section 10.1), aided by the role of 'decomposing' micro-organisms such as bacteria, actinomycetes and fungi. Some micro-organisms, including worms and termites, mix up the soil and thereby affect its character. Charles Darwin observed, for example, that the English garden worm can turn over 6.5 tonnes of soil per hectare per year.

Increasingly, human activity is becoming an important component of the biotic factor in soil development. Indeed, as table 13.1 shows, humans can have both beneficial and detrimental effects on all five of the classic factors of soil formation. Of particular importance are their effects on soil salin-

Table 13.1 Suggested effects of the influence of man on five classic factors of soil formation

Factors of soil formation	*Type of effect*	*Nature of effect*
Climate	Beneficial	Adding water by irrigation; rainmaking by seeding clouds; removing water by drainage; diverting winds etc.
	Detrimental	Subjecting soil to excessive insolation, to extended frost action, to wind etc.
Biota	Beneficial	Introducing and controlling populations of plants and animals; adding organic matter including 'nightsoil'; loosening soil by ploughing to admit more oxygen; fallowing; removing pathogenic organisms as by controlled burning
	Detrimental	Removing plants and animals; reducing organic matter content of soil through burning, ploughing, overgrazing, harvesting etc.; adding or fostering patho genic organisms; adding radioactive substances
Topography	Beneficial	Checking erosion through roughening, land forming and structure building; raising land level by accumulation of material; land levelling
	Detrimental	Causing subsidence by drainage of wetlands and by mining; accelerated erosion; excavating
Parent material	Beneficial	Adding material fertilisers; accumulating shells and bones; accumulating ash locally; removing excess amounts of substances such as salts
	Detrimental	Removing, through harvest, more plant and animal nutrients than are replaced; adding materials in amounts toxic to plants or animals; altering soil constituents in a way to depress plant growth
Time	Beneficial	Rejuvenating the soil through adding of fresh plant material or through exposure of local parent material by soil erosion; reclaiming land from under water
	Detrimental	Degrading the soil by accelerated removal of nutrients from soil and of vegetation cover; burying the soil under solid fill or water

Source: modified from O. W. Bidwell and F. D. Hole, (1965) 'Man as a factor of soil formation', *Soil Science*, 99: 65–75

ity (section 6.7), laterite formation (section 7.11) and soil erosion (section 13.6).

Topography (relief)

Relief conditions in an area are the most significant control of soil formation and character, especially at the local scale. Slope steepness, for example, affects soil moisture conditions and the rate of mass movements and erosive processes. Thus, if a slope is steep the runoff is rapid, erosion removes the soil as fast as it forms, little water enters the soil and the soil profiles are thin and poorly developed. On more level terrain, runoff is inhibited, and more water enters the terrain to weather minerals and to redistribute the soluble components of the soil profile. Erosion is less rapid, so that thicker profiles may occur, and waterlogging may be an influence on soil form. Thus soils and landforms may be related in a topo-sequence or *catena*. The catena concept (figure 13.1) was orginally put forward by a scientist called G. Milne, who worked in East Africa in the 1930s. He originally defined a catena as: 'a unit of mapping convenience . . . a grouping of soils which while they fall wide apart in the natural system of classification on account of fundamental and morphological differences, are yet linked in their occurrence by conditions of topography and are repeated in the same relationships to each other wherever the same conditions are met with'.

Parent material

When a soil forms directly from underlying rock or sediment, the soil minerals and type may bear a direct relationship to the original material. Weathering processes may mean that the balance of ions in the soil is different from that in the parent material, but none the less the nature of such *residual soils* owes much to the nature of the parent ma-

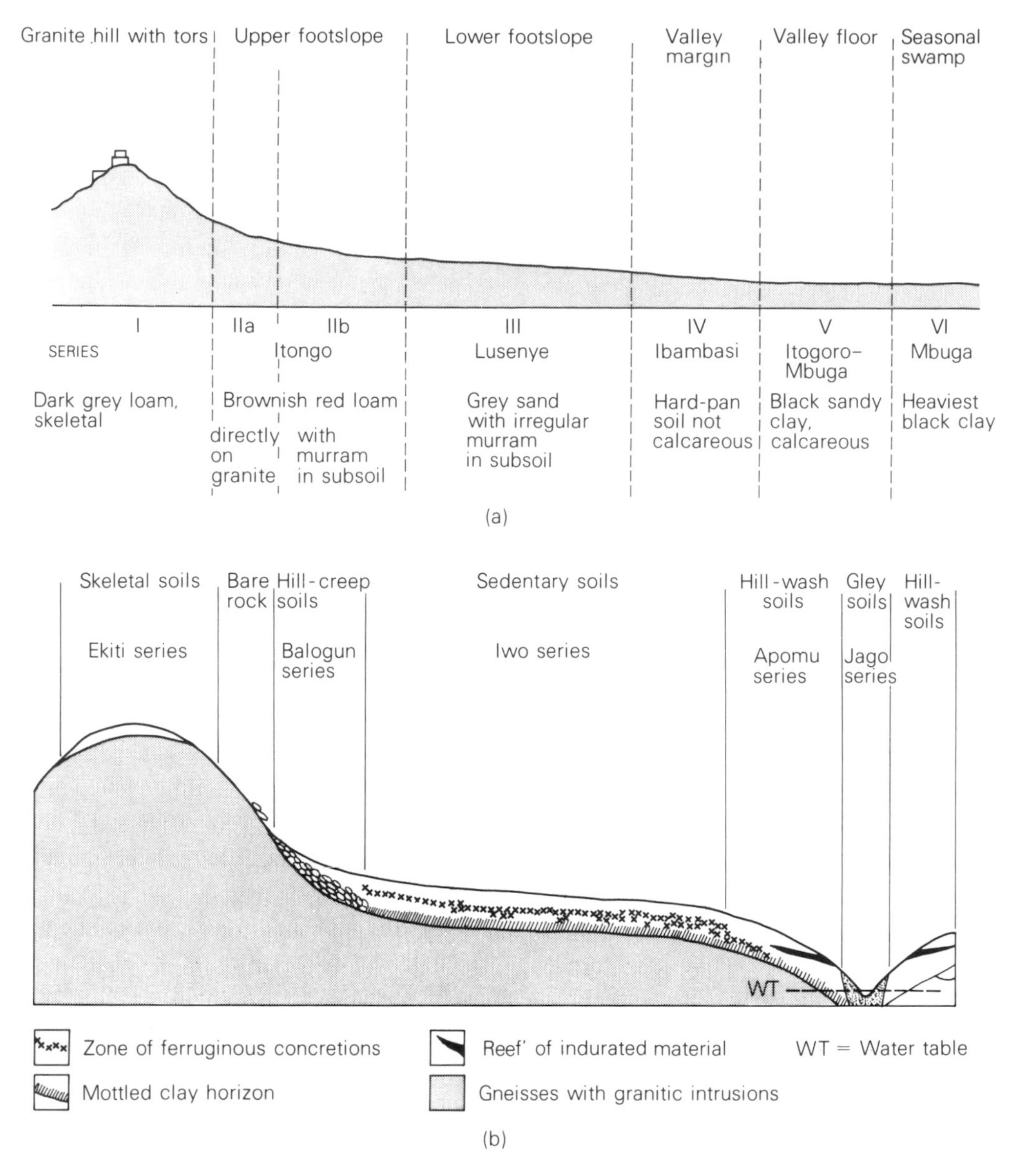

Figure 13.1 Examples of two catenas in Africa: (a) the Ukiriguru catena from Tanzania as proposed by Milne; (b) the Iwo catena, western Nigeria.

terial. However, in a second category of soil, known as *transported soil*, the soil may be largely independent of the underlying solid rock because the parent material has been transported from somewhere else. Soils derived from glacial drift, alluvium or windblown dust (loess) come into this category.

Time

Whether a soil is deep or shallow, it requires a period of time to form. One would expect soils that have developed on old, stable, flat land surfaces, such as some of the great pediplains of Africa, to be very highly weathered. By contrast, soils developed on recently formed glacial moraine or sand dunes might be expected to show only very limited development.

Together these five factors account for the character of the four main constituents that make up a soil: mineral matter, organic matter, air and water. The mineral matter includes all those minerals, such as clays, weathered from the parent material as well as those formed in the soil by recombination from substances in the soil solution. The organic matter (which includes material like humus) is derived mostly from decaying vegetation which is decomposed with the assistance of the many different forms of animal life that live in the soil. Air and water occupy the spaces between the structures of the soil. If the soil is well drained its pores and fissures will be open to the penetration of the atmosphere; if it is saturated with water most of the air will be driven out, and the soil is then said to be *anaerobic*.

The mineral portion of the soil is derived from the parent material by the various weathering processes described later in the chapter. Different soils will have very different textures according to the relative proportions of materials of different sizes. At the one extreme there are soils composed largely of coarse rock fragments (these are often termed *skeletal soils*), while at the other there are soils formed of very fine materials, called clays, which have a particle size of less than 0.002 mm. Particle size is an important consideration because it affects the erodibility of the soil, its tillability, its moisture and its warmth. Sandy soils, for example, will tend to be freely drained and easy to cultivate, but on the other hand in a dry year they will suffer from drought and may be susceptible to wind erosion.

Clay minerals are probably the most important mineral constituents of the soil, and they result from chemical weathering of parent material. They consist of minute platy-structured mineral fragments which can be identified only by sophisticated microscopic and chemical methods. They are members of a group of minerals that are characterised by a layered, crystalline structure, and they are built from layers of silica and aluminium atoms with their attendant oxygen atoms arranged like a sandwich. Three major groups of aluminosilicate clay minerals may be distinguished: the *kaolinite*, the *montmorillonite* (or smectite) and the *hydrous mica groups*. They differ in crystal structure, surface electrical charges and resulting mechanical properties, though intermediate forms frequently occur. Most weathered rock material contains a mixed assemblage of the different clay minerals, and it is difficult to generalise about the factors that produce one particular assemblage rather than another. In the early stages of weathering, parent material type may be important, with granite, for example, tending to alter to kaolinite and basic igneous rocks to montmoril-lonite. But other factors also play a role. Acid soil conditions are conducive to the production of kaolinite, while alkaline conditions favour the evolution of montmorillonite. In the United States kaolinite is particularly abundant in the warm, wet south-eastern parts of the country, the hydrous mica group becomes increasingly prominent in the cooler north-east, while montmorillonite reaches its fullest development in the drier west.

Clay minerals may make the soil poorly drained, cold and hard when dry. They are also prone to shrink and swell according to their moisture content, which may cause problems for buildings constructed upon them. On the other hand, they tend to retain moisture longer in dry years and, being cohesive, tend to be resistant to wind erosion.

Additional minerals are formed in the soil itself, for weathering releases elements from the parent material's minerals, and these may precipitate elsewhere in the soil profile. Lime concretions may occur in dry areas, for example, while iron and manganese concretions may develop in poorly drained soils.

Soil organic matter also can take on a variety of forms: peat, mull, moder and mor. Under wet, ill-drained anaerobic conditions organic accumulation may be in the form of layers of more than 40 cm thick. This is *peat*. *Mor* tends to develop under acidic conditions associated with heathland or coniferous plant communities, in which the role of soil fauna is limited. *Mull* forms in freely drained, base-rich soils with good aeration. Plant growth is good and provides much litter, but there is also a rich soil fauna, including earthworms. The organic matter is broken down and well mixed into the mineral components of the soil. *Moder* is a form of organic matter intermediate between mor and mull because it has a higher soil fauna level.

Organic matter decomposes to form humus, and this humus, together with clay minerals, plays an important role in absorbing ions released by chemical weathering of rock or dead tissue. The clay–humus complex acts as an invaluable reservoir of soil nutrients and serves as a buffer against rapid loss of nutrients by leaching away in soil water. The nature and amount of clay and humus in a soil are therefore highly significant in terms of fertility.

13.2 Soil Profiles

The vertical succession down through the soil is called a *profile*, and recognition of the differences between the profiles of different soils is the basis of schemes of soil classification. The soil profile is separated into various bands, which result from the balance between inputs and outputs into the soil system, and the redistribution of, and chemical changes in, the soil constituents (see colour plate 23). These bands are called *horizons*, and some of the major processes involved are as follows:

(a) *Organic accumulation:* takes place mainly at the ground surface owing to accumulation of decaying vegetable matter.
(b) *Eluviation:* the mechanical translocation of clay or other fine particles down the profile.
(c) *Leaching:* the removal of material from a horizon in solution, and its movement down (and possibly out of) the profile.
(d) *Illuviation:* the accumulation in the lower part of the profile of material eluviated (washed down) from above.
(e) *Precipitation:* the formation of solid matter, sometimes in the form of nodules or concretions, in the subsoil from solutions leached from above or derived by capillary rise from groundwater.
(f) *Cheluviation:* the downward movement of material, akin to leaching, but under the influence of chelating agents.
(g) *Organic sorting:* the separation of material, usually of differing grain sizes, by organic activity (e.g. the churning action of worms and termites).

The *inputs* into the soil system include:

(a) Water, dust and aerosols from the atmosphere.
(b) Gases from the atmosphere and from the respiration of soil fauna.
(c) Organic matter from decaying vegetation and animals.
(d) Excretions from plant roots.
(e) Chemicals (nutrients) from weathered plant materials.

The *outputs* from the soil system include:

(a) Losses of soil material by soil erosion and soil creep.
(b) Nutrient losses into water percolating through the soil.
(c) Nutrients taken up by plants growing in the soil.
(d) Evaporation.

The soil system is not only influenced by simple inputs and outputs: some *recycling* may occur. For example, some of the chemicals or plant nutrients lost as an output (output c) may well return in leaf litter at a later time (input c). Man may modify the nature of these inputs and outputs in a number of ways; for example, by adding artificial nutrients in the form of fertilisers, by accelerating or slowing the rates at which soil erosion takes place (see section 13.6), by changing the amount of water movement in the soil through drainage activities, or by modifying the nature of the vegetation cover.

Traditionally, three soil horizons were recognised. The uppermost, the A horizon, consists of thoroughly weathered rock material and humus. It is a band that has been strongly affected by the washing out, or leaching, of soluble materials and by the downward dispersion, or eluviation, of clay particles. Beneath this is the B horizon, which consists of less thoroughly weathered parent material, a minor proportion of organic matter and materials carried down, or eluviated, from above. The C horizon consists dominantly of still less weathered parent material, affected little or not at all by organic action, eluviation or illuviation.

This three-fold division is extremely useful, but rather oversimplified. For this reason various schemes of elaboration have been developed. Each uses number or letter systems in addition to A, B and C. For instance, the surface layer of plant litter at the top of the A horizon can be subclassified as L, F or H according to the dominance of litter proper (L), fermented organic material (F) or humus (H). Part of the A horizon can be subclassified Aa, Ah or Ae according to whether it is bleached (a = *a*shy), *h*umus-rich or *e*luviated. Likewise, if the B horizon has an *ir*onpan caused by the accumulation of iron leached down from above, this can be labelled Bir. If the B or C horizons have an accumulation of *cal*cium carbonate, they can be labelled Bca or Cca. If the soil is subjected to waterlogging in the lower part of the profile, a *gley horizon* results: this is the product of an oxygen shortage, which produces grey coloured iron compounds. A well-developed gley horizon (depicted by G) is uniformly pale, but where the soil dries out from time to time the pale grey colouration becomes mottled with yellows, reds and purples. Such moderate gleying is connoted by a small g.

13.3 Types of Weathering

Most rocks and sediments in the top few metres of the earth's crust are exposed to physical, chemical and biological conditions much different from those prevailing at the time they were formed (see colour plate 24). Because of the interaction of these conditions, the rock or sediment gradually changes into soil-like material. These changes are collectively called *weathering*, but two main types are recognised:

(a) *Processes of disintegration* (physical or mechanical weathering)
 - (i) Crystallisation
   - Salt weathering (by crystallisation, hydration and thermal expansion)
   - Frost weathering
 - (ii) Temperature changes
   - Insolation weathering (heating and cooling)
   - Fire
   - Expansion of dirt in cracks
 - (iii) Wetting and drying (especially of shales)
 - (iv) Pressure release by erosion of overburden
 - (v) Organic processes

(b) *Processes of decomposition* (chemical weathering)
 - (i) Hydration and hydrolysis
 - (ii) Oxidation and reduction
 - (iii) Solution and carbonation
 - (iv) Chelation
 - (v) Biological–chemical changes (organic weathering)

Mechanical weathering involves the breakdown or *disintegration* of rock without any substantial degree of chemical change taking place in the minerals that make up the rock mass. *Chemical weathering* involves the *decomposition* or decay of such minerals. In most parts of the world both types of weathering tend to operate together, though in differing proportions, and the one may accelerate the other. For example, the mechanical disintegration of a rock will greatly increase the surface area that is then exposed to chemical attack.

13.4 Mechanical Weathering

Mechanical weathering can be brought about by a variety of processes: freeze–thaw action (section 4.13), heating and cooling (insolation) (section 6.8), the growth of salt crystals (section 6.8), the prising effect of tree roots and unloading or pressure release.

This last process is the reason why, in many parts of the world, the rock close to the surface is cut by joints that more or less parallel the surface. The term *exfoliation* (plate 13.1) is used for the sheeting that explains this onion-layered appearance of many rock outcrops. The cause is the upward expansion of rock as an overlying or confining burden is removed, as for instance by intense glacial erosion (section 4.6). At depth the rock is under high confining pressure equivalent to the weight of the overlying mass; as erosion removes the overlying rock – say, a sedimentary cover from above a batholith – the remaining rock can expand, usually either upward or towards valley walls. The release of pressure results in joints that are oriented at right angles to the direction of the release – hence they usually parallel the land surface. When men dig mines or excavate quarries they can trigger the same effect as erosion, and the pressure release may occur as violent rock bursts.

Plate 13.1 As a consequence of pressure release resulting from the erosion of overlying material, this granite near Kyle in Zimbabwe is being broken up into a series of sheets which parallel the land surface. Such sheeting is called exfoliation.

13.5 Chemical Weathering

Solution

Solution is one of the simplest chemical weathering processes to visualise, for some rocks may literally dissolve, rather like sugar in a drink or salt in a saucepan of water. Limestone, which is composed of calcium carbonate, is especially prone to weathering in this way, for it is readily attacked by carbonic acid. This acid is produced by the union of water and carbon dioxide, the latter coming either from the atmosphere or, in larger concentrations, from the soil:

$$\underset{\text{water}}{H_2O} + \underset{\text{carbon dioxide}}{CO_2} \rightarrow \underset{\text{carbonic acid}}{H_2CO_3}$$

Thus the formula which expresses the dissolution of calcium carbonate in the presence of water and carbon dioxide can be written as follows:

$$\underset{\text{calcium carbonate}}{CaCO_3} + \underset{\text{carbon dioxide}}{CO_2} + \underset{\text{water}}{H_2O} \rightarrow \underset{\text{calcium in solution}}{\boxed{Ca^{2+}}} + \underset{\text{hydrogen carbonate in solution}}{\boxed{2HCO_3^-}}$$

Water enriched in CO_2 attacks many silicate minerals. Olivine, for example, which is found in some igneous rocks, can be dissolved almost completely in a sequence of reactions represented in a very simplified form in the formula:

$$\underset{\text{olivine}}{MgSiO_4} + \underset{\text{water}}{2H_2O} + \underset{\text{carbon dioxide}}{4CO_2}$$

$$\rightarrow \underset{\text{magnesium bicarbonate}}{2MgC(CHO_3)_2} + \underset{\text{soluble silica}}{SiO_2}$$

Sometimes (plate 13.2), granite rocks can show solutional forms, as indicated by the development of flutes in the Seychelles.

Hydrolysis

The common rock-forming minerals weather by a process called hydrolysis, a class of chemical reaction with water involving the action of H and OH ions. This produces as a by-product a substance that is very different in character from the minerals of which the rock was initially composed. For instance, the weathering of feldspar (an aluminium silicate mineral that is found in igneous rocks like granite) can lead to the formation of clays, as shown by the following equation:

$$\underset{\text{orthoclase feldspar}}{K_2O.Al_2O_3.6SiO_2} + \underset{\text{water}}{11H_2O}$$

Plate 13.2 Solutional flutes developed on granite on the island of Praslin in the Seychelles, Indian Ocean.

$$\rightarrow \underset{\text{kaolinite}}{Al_2O_3.2SiO_2.2H_2O} + \underset{\text{silicic acid in solution}}{4H_4SiO}$$

$$+ \underset{\text{hydroxyl}}{2OH^-} + \underset{\text{potassium in solution}}{2K^+}$$

Kaolinite is one type of clay mineral. The silicic acid and the potassium may go into solution and be leached away, leaving only the clay mineral behind.

Hydration

Hydration occurs when minerals incorporate water into their molecular structure. This includes some of the minerals commonly found in igneous rocks, together with some of the constituents of sedimentary rocks, as well as many of the silicate clay minerals. Hydration often causes swelling, and it is believed to be a major cause of the crumbling of coarse-grained igneous rocks which are disrupted by the progressive expansion of their hydrated minerals.

Oxidation

Rusting is a familiar process to most of us. Iron objects exposed to moisture are soon affected. Most rocks contain iron-bearing minerals, and these change in the presence of moisture and oxygen. This process is responsible for the first visible signs of chemical weathering in many rocks – their discoloration to red and yellowish-brown colours.

Mineral composition and weathering

The mineral composition of a rock is an important control of the rate at which weathering may occur, for some materials are more susceptible to chemical weathering than others. For example, igneous rocks show a greater variety in the minerals they contain, and it is possible to rank them in order according to their susceptibility following the *Goldich system* (table 13.2). Granite is composed primarily of quartz, orthoclase, muscovite and biotite, which are four of the least susceptible minerals in the series. Thus granite would tend to be

Table 13.2 The Goldich system

	Dark-coloured minerals	*Light-coloured minerals*	
Least stable	Olivine		Most susceptible
		Lime plagioclase	
	Augite		
		Lime-soda plagioclase	
	Hornblende	Soda-lime plagioclase	
		Soda plagioclase	
	Biotite		
		Orthoclase	
		Muscovite	
Most stable		Quartz	Least susceptible

Plate 13.3 Deeply weathered granite on the Cape Peninsula, South Africa. Note the cores of intact rock surrounded by weathered material (regolith). The cores occur where jointing is sparse, while the regolith has been developed in zones where the joints are close together.

less prone to chemical attack, and to be more resistant than an igneous rock such as gabbro, which is composed primarily of two more susceptible minerals, augite and lime-soda plagioclase. Mineral composition is not the only factor influencing the rate of rock decay – factors such as crystal size and the way in which the crystals interlock may be of equal importance.

Chemical composition is also important in affecting the weathering of sedimentary rocks. These have four common cement types: silica, calcium carbonate, iron oxides and clay. If silica is the cement, the rock will tend to be relatively resistant. Thus, silica-cemented quartz sandstone is one of the most durable rock types, for neither the cement nor the cemented grains succumb easily to chemical attack. On the other hand, calcium carbonate cement may be prone to solutional processes, and an iron oxide cement to oxidation processes, both of which may weaken the rock. Clay cements tend to be especially weak, and physical weathering processes, such as freeze–thaw, may cause rapid weathering of the rock.

The importance of joints in chemical weathering

Water is the prime control of the intensity of chemical weathering of all the types discussed above. Thus the presence of joints and other discontinuities in a rock mass is highly important, since they help to control the rate at which water can move within the rock, and also control the surface area for the weathering processes to operate on. Other things being equal, a densely jointed rock will tend to weather more quickly than a rock with just a few

Window 13.1 *The karst of south China*

Probably the most spectacular and extensive karst topography on earth is located in southern China. It is a common subject in Chinese literature, art and ceramics, which is not surprising when one considers that no less than 1.2 million km^2 of China are composed of exposed karstic surfaces.

Why is karst so well developed in south China? A variety of factors contribute. First, the area is quite humid, with mean annual precipitation of around 2000 mm. This suggests that there is plenty of water to achieve solutional alteration of the bedrock. Second, the area has undergone slow uplift through the Cenozoic era and this has exposed broad plateaux of gently dipping to horizontal carbonate strata. It has also permitted drainage to become incised, thereby providing plenty of available relief, a necessity if spectacular landforms are to develop. Third, and perhaps most importantly, the limestones of the area, which are from Late Precambrian to Triassic in age, form huge thicknesses (over 3000 m) of virtually uninterrupted, massive, crystalline beds in which extensive and striking features can develop.

There are two main types of classic karst in the area, to which the Chinese give the names *fenglin* and *fengcong*. Fenglin consists of isolated towers rising above flat plains and is akin to the tower karst of Western geomorphologists. These towers are steep sided, and often exceed 150 m in height. Fencong is akin to what Western workers call cone or cockpit karst, with clusters of steep-sided hills rising from intervening depressions, sinkholes and valleys. Individual hills may vary in height from 30 to 500 m.

Why should there be two types of karstic form? This is a difficult question, and there are several different possibilities. Part of the explanation may be geological, in that the steepest towers develop on especially massive, dense and gently bedded limestones. The cones and depressions seem to occur preferentially in areas that have been subjected to greater rates of uplift by tectonic forces, and which thus have greater potential

The karst of China, shown here near Yangshuo, Guangxi, is one of the most striking rock-controlled landscapes on earth. Particularly characteristic of the area are large towers.

for downcutting by drainage. This is the case near Guilin, where the greatest uplift has been in the east and fencong has developed, while in the west of the area the uplift has been less and fenglin has developed. An alternative theory, not generally supported by Chinese workers, is that the towers are older than the cones, and that through time the cone landscape is transformed into a tower landscape, with isolated remnants rising up above ever-broadening plains.

The Chinese karst lands are of great importance to the Chinese people. In particular, in common with many limestone areas, they possess large reserves of groundwater, and their striking morphology has considerable tourist appeal.

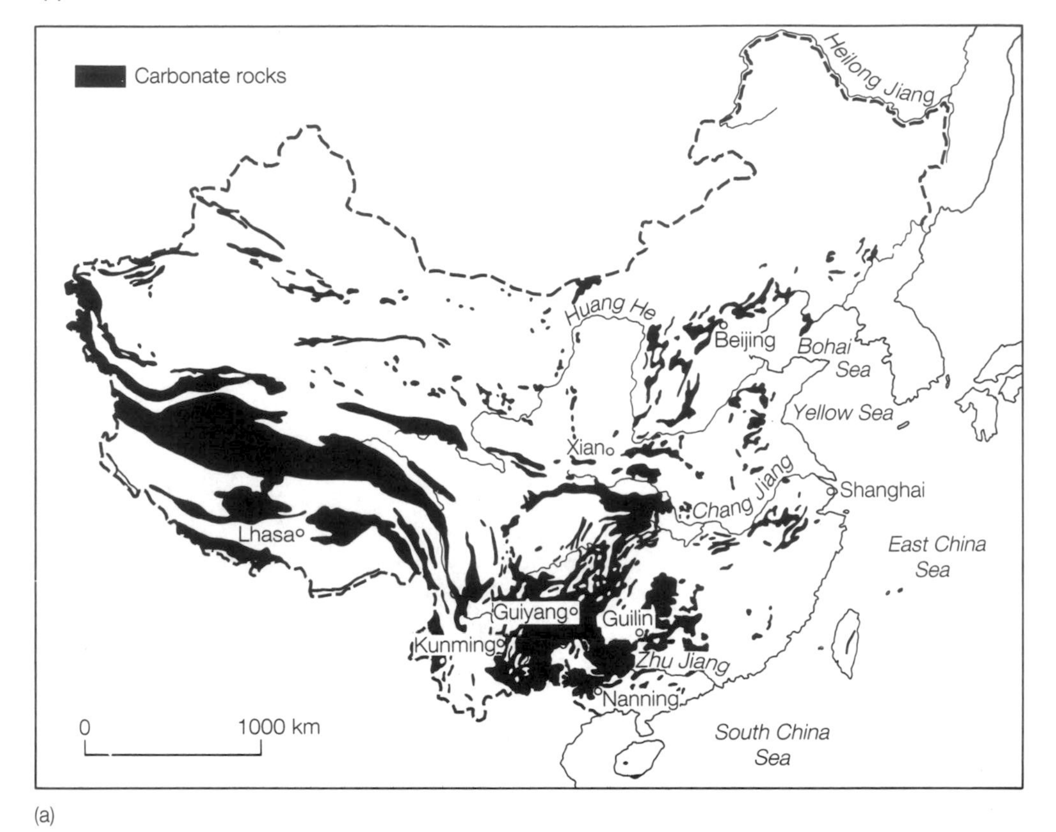

(a)

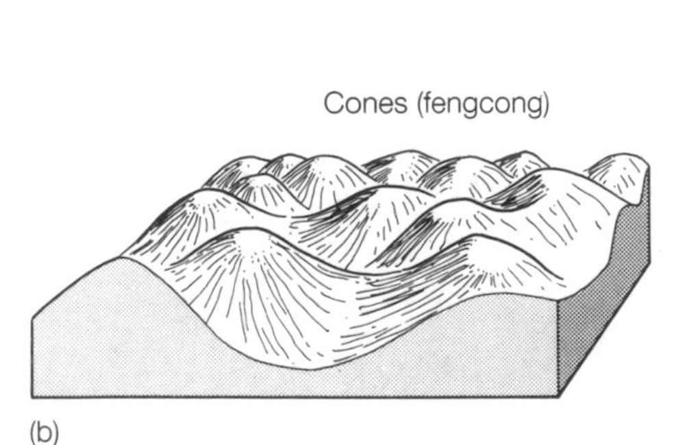

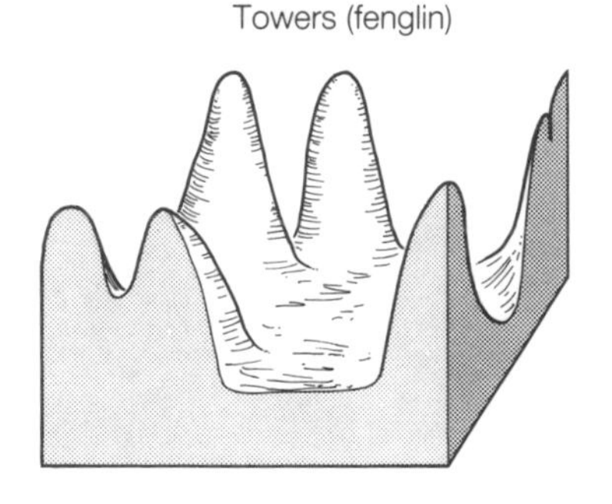

(b)

The karst landscapes of China: (a) the distribution of carbonate rocks, primary limestones; (b) the shapes of cones and towers.

Plate 13.4 Lapiés (karren) produced by limestone solution in Mallorca.

joints. This is thought to be particularly important when considering the development of landforms in igneous rocks, including tors and inselbergs. For example, a major thesis of tor formation is that deep chemical weathering penetrates a rock mass unevenly according to the spacing of joints. As weathering proceeds, zones of relatively sound rock remain where the joints are widely spaced (plate 13.3); whereas a thick regolith, containing much clay produced by the chemical weathering of feldspar minerals, occurs where the joints are closest together (see colour plate 25). Subsequent erosional stripping will remove the regolith, exposing the sounder rock as a series of upstanding tors and piles of boulders (core-stones). The formation of tors is explained at greater length in section 5.16.

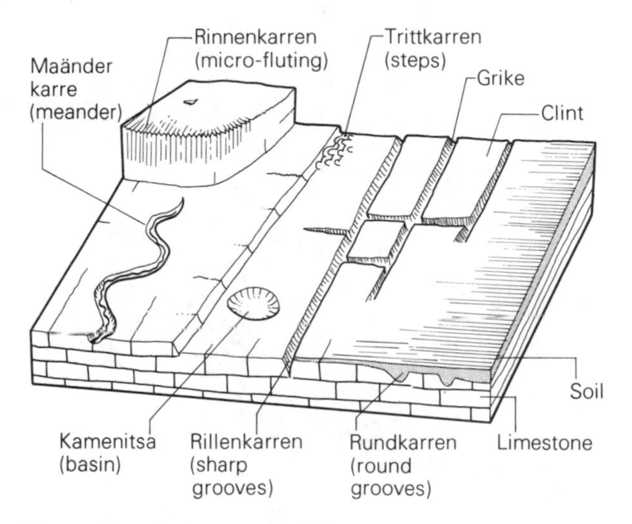

Figure 13.2 Surface solutional forms from limestone areas. Some of the minor solutional forms (karren and lapiés) that develop on limestone surfaces. Rinnenkarren are small-scale flutes that occur on steep surfaces, rillenkarren are sharp-edged rills that occur on bare surfaces, rundkarren are rounded runnels that tend to occur beneath a soil cover, meanders develop on slightly sloping surfaces, while step karren resemble miniature cirques cut into a sloping limestone surface. In northern England, where such features are widespread on glacially stripped limestone pavements, the flat areas between the grooves are called clints, while the deep grooves that break them up are called grikes.

Organic weathering

The action of carbonic acid is probably supplemented in many cases by the action of other acids, especially those derived from the decomposition of vegetable matter, which are termed *humic acids*. Moreover, bacterial action and the respiration of plant roots tend to raise carbon-dioxide levels in the soil atmosphere, and thereby help to accelerate the solutional processes. Bacteria can also contribute to a process called *reduction*, for some of them obtain part of their oxygen requirement by reducing iron from the ferric to the ferrous form. Some of these ferrous compounds tend to be markedly more soluble in water than the original ferric ones, so that they can be relatively easily mobilised and removed from the soil. *Chelation* is another important weathering reaction. The word 'chelate' means 'clawlike', and refers to the tight chemical bonds that hydromolecules may impose on metallic cations. The process occurs because plant roots are surrounded by a concentration of hydrogen ions which can exchange with the cations in adjacent minerals; the metallic cations are then absorbed into the plant. Through this process, otherwise relatively insoluble elements such as aluminium can be mobilised.

The weathering of limestones

Limestones are a type of rock in which the effects of chemical weathering, especially solutional attack, are particularly evident. Some of the world's most striking limestone landscapes are in China (window 13.1). But limestones are extremely variable in their composition, hardness, jointing and so on, so that their response to solutional modification is very varied. Thus the Cretaceous chalk of southern England is a relatively soft, pure limestone with many joints and bedding planes, and it results in a very different scenery from some of the harder, denser, more crystalline, widely jointed Carboniferous limestones that form some of the classic limestone terrain of areas such as South Wales, the Mendips and the Ingleborough district of Yorkshire. Areas where the limestone is thick, massive and extensive, and where the water table is at depth, may develop

Plate 13.5 In limestone areas that have been glaciated bare limestone surfaces, called pavements, occur. Near Ingleborough in north-west England the pavement has been weathered by solutional processes to give clints and grikes.

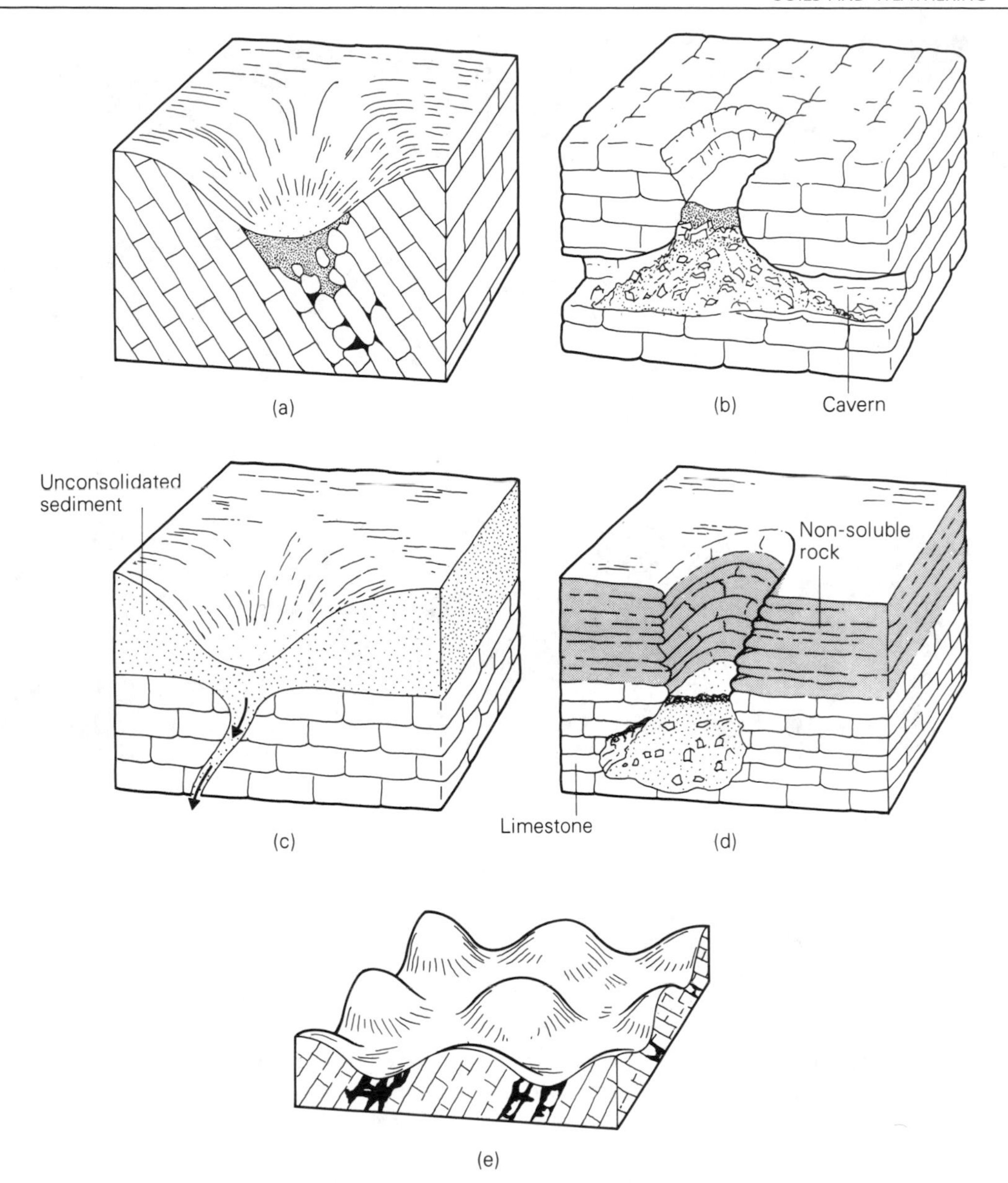

Figure 13.3 Five different types of limestone solution hollow (doline): (a) solution doline produced by widening of joints; (b) collapse doline produced by a cave roof falling in; (c) subsidence produced by removal of unconsolidated sediment down a fissure or joint; (d) subjacent karst collapse doline; (e) cockpits (intersecting star-shaped dolines), typical of tropical areas.

characteristic solutional features which in extreme cases produce a type of landscape called *karst*.

Within an area of karst relief it is possible to distinguish three main types of relief features:

(a) *Surface solutional forms* (plate 13.4) These are termed *clints* and *grikes* in northern England, *lapiés* in France and *karren* in Germany. They tend to be small features which are given a bewildering range of names; some characteristic forms are illustrated in figure 13.2. They are often especially well developed on

Plate 13.6 In the tropics, as this air photograph of Jamaica shows, one of the characteristic forms of karst consists of a succession of cone-like hills alternating with enclosed polygonal depressions – cockpits. The roads and buildings give an indication of scale.

limestone surfaces that have been stripped of their cover of soil or superficial debris by glacial action. Such surfaces are termed *limestone pavements* (see section 4.6) (plate 13.5).

(b) *Closed depressions and individual hills* In most karstic areas the ground surface is pocked by a series of closed depressions of variable size and steepness. Some of the largest forms, called *poljes*, may owe at least part of their form to faulting rather than solution, but the smaller types of depression, called *dolines*, tend to be the result either of solutional enlargement of joints or of the collapse of the ground surface into a cavern produced by underground solution (figure 13.3). Surface drainage may disappear down such holes (termed *swallow holes* in many British limestone areas), as is the case with the famous example of Gaping Ghyll in Yorkshire. In the tropics the landscape may be composed of a succession of cone-like hills alternating with enclosed polygonal depressions or 'cockpits' (plate 13.6). Alternatively, landscapes may be characterised by wide plains from which tall, isolated hills, called *towers* (plate 13.7), arise.

(c) *Underground features, including caves and caverns* Much of the water in karstic areas flows underground in cave passages (plate 13.8) and along joints. Surface drainage tends, therefore, to be relatively ill developed. The caves are formed by the solutional enlargement of major joints in the limestone, aided by the corrosional activity of the streams that may flow in them. Some caves, especially those marked by a tube-like cross-section, may develop beneath the water table (*vadose caves*). In the past, when climatic conditions were different, or before weathering processes had ren-

Plate 13.7 In China, and some other parts of the low latitudes, a typical limestone landscape consists of tower karst, in which tall, isolated hills rise above wide plains.

dered the rock permeable by solutional widening of joints, surface drainage may have occurred in limestone regions; these are represented in the landscape today by dry valley systems (see section 5.15).

Rates of chemical denudation

By measuring the amount of water being discharged from an area of known size by springs or streams, and by determining the amount of soluble material present in the water, it is possible over a period of years to make an estimate of the rate at which limestone weathering takes place. Some early studies, notably by a French geomorphologist called J. Corbel, indicated that perhaps some of the highest rates occurred in cold areas (possibly because carbon dioxide is more soluble in water at low temperatures, thereby producing waters rendered more acidic with carbonic acid). Other investigators, however, have argued that, as there is more carbon dioxide and humic acid generated in the soils of the wet tropical areas, and as the rate of chemical reaction increases with temperature, rates of limestone denudation should be highest in the low-latitude humid areas. However, as figure 13.4 shows, the main control of the rate of limestone denudation appears to be not temperature so much as the availability of water to cause solution to occur. Thus, the highest rates of limestone denudation appear to be in the wettest areas. Cold, dry tundra areas such as the far north of Canada have rates similar to hot areas like the margins of the Sahara (both having rates of less than 5 mm of vertical lowering per 1000 years), while cold wet areas such as western Norway may have rates similar to the rainforest clad slopes of the East Indies (both having rates of several hundreds of millimetres of vertical lowering per 1000 years).

A global attempt to assess the factors involved in affecting rates of chemical denudation has been made by M. Meybeck in the *American Journal of Science* (1987). He suggests (figure 13.5a) that a major control is the amount of runoff in the region, but that a secondary control in terms of silica transport is temperature, with hot regions tending to have about five times more silica removal than cold regions. He also shows (figure 13.5b) that for a given amount of runoff, sedimentary rocks provide the largest amounts of dissolved material for stream removal, that volcanic rocks have an intermediate position and that the lowest rates of removal of dissolved material occur in areas underlain by metamorphic and plutonic (igneous) rocks.

13.6 Accelerated Soil Erosion

Although it becomes clear from this consideration of rates of weathering that, given time, soil and weathering profiles will develop to an advanced stage, especially on susceptible rock types, this cannot happen if the rates of erosion exceed those of weathering. Soils are a resource that generally form slowly, whereas erosion can occur with great rapidity, especially when human activity accelerates its operation. It has been estimated that in the United States soil erosion on agricultural land operates at a rate of about 30 $t\,ha^{-1}\,yr^{-1}$, which is approximately eight times more quickly than topsoil is formed.

Plate 13.8 Limestone areas may have numerous underground caves and caverns. Some of the largest and most spectacular examples occur in the Mulu area of Sarawak.

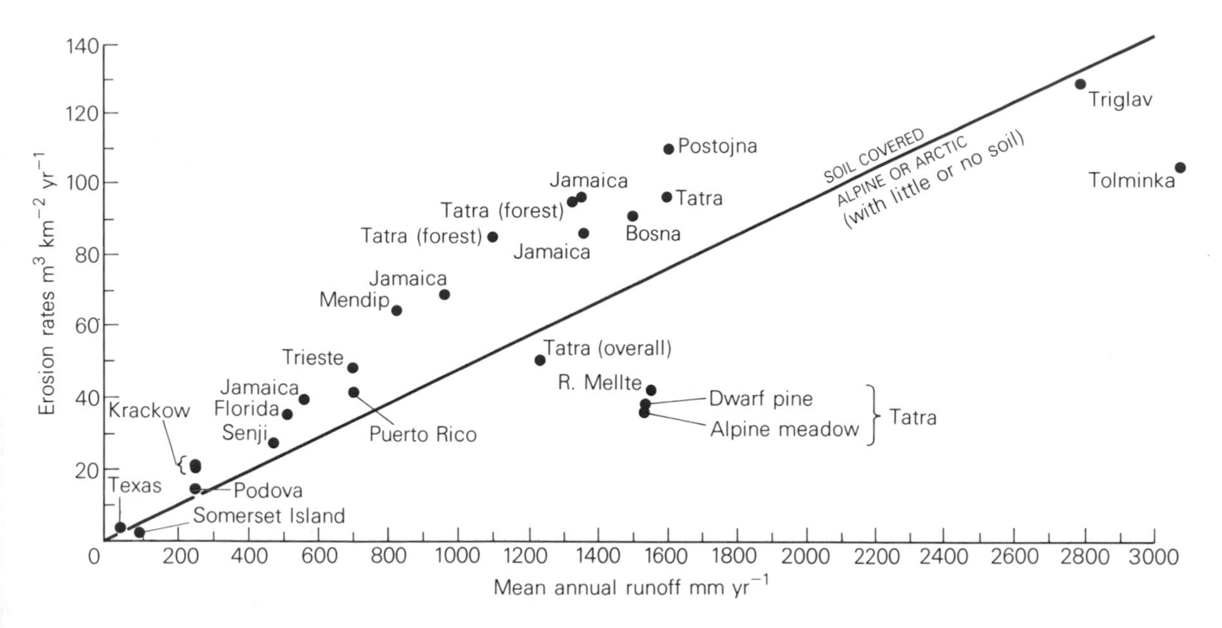

Figure 13.4 The relationship between mean annual rate of solution of limestone and mean annual runoff.

Plate 13.9 This bizarre landscape, which has a relief of about 4 m, is an example of severe soil erosion caused by gully erosion in Swaziland.

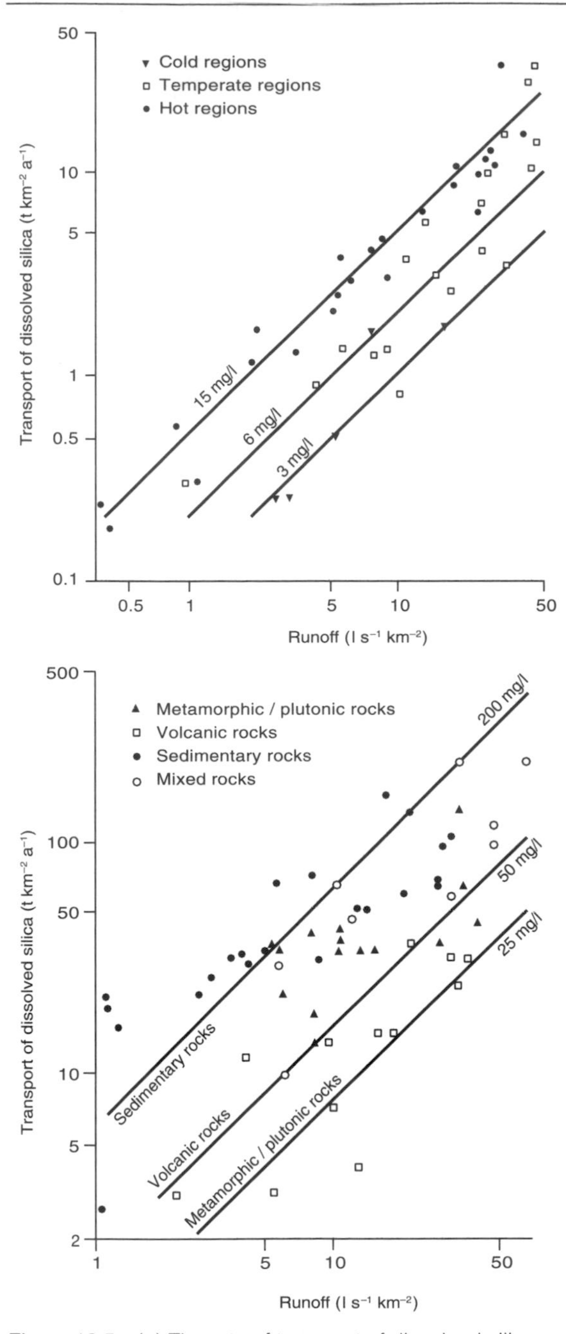

Figure 13.5 (a) The rate of transport of dissolved silica plotted against runoff (specific discharge) for cold, temperate and hot regions. (b) The rate of transport of all major ions plus dissolved silica plotted against runoff (specific discharge) for various major drainage basins underlain by sedimentary, volcanic and metamorphic/plutonic rocks.

Concern about accelerated erosion focuses on two main categories of impact. The first of these relates to the threat it poses to our ability to grow crops and to feed the world's population. Soil erosion reduces soil depth and often means that the most fertile, humus- and nutrient-rich portion of the soil profile is lost. The second category of impact is what is termed 'off-farm impact'. This includes accelerated siltation of reservoirs, rivers and drainage ditches; and eutrophication of water bodies by the transport of nutrients attached to soil particles, damage to property by soil-laden water and debris flows.

The main cause of *accelerated erosion* (plate 13.9) is the replacement of natural forest cover by agricultural land. Forest protects the underlying soil from the direct effects of rainfall; runoff is greatly reduced, tree roots bind the soil, and the litter layer protects the ground from rainsplash. The rates of erosion that will result when forest is removed will be particularly high if the ground is left bare, whereas under crops the increase will be less marked. Furthermore, the method of ploughing, the time of planting, the nature of the crop and the size of fields will have an influence on the severity of the erosion.

Soil erosion resulting from deforestation and agricultural practice is especially serious in tropical and semi-arid areas (plate 13.10) (see section 6.15), but as table 13.3 shows, there are many studies which demonstrate the large increases in river suspended loads that follow on from various types of land cover change and surface disturbance. In the examples given, rates of sediment yield range from 4 to 310 times.

This is brought out further when one considers the data presented in table 13.4 which compares annual rates of soil erosion under natural conditions with a cultivated crop and when the soil is left bare. The difference is substantial.

There is some evidence that soil erosion is becoming a more serious problem in parts of Britain, in spite of the fact that the country's rainfall is much less intense, and so less erosive, than in many parts of the world. The following practices may have caused this state of affairs:

- Ploughing on steep slopes that were formerly under grass in order to increase the area of ar-

Plate 13.10 Deforestation and overgrazing near St Michael's Mission in Zimbabwe has led to the development of this gully system in a once flat valley floor. Note the exposure of the tree roots.

Table 13.3 Evidence of increases in suspended sediment transfer following catchment disturbance

Location	*Land use change*	*Increase in sediment flux*
Monitored catchment studies		
Westland, New Zealand	Clearfelling	× 8
Severn and Wye Rivers, Wales, UK	Clearfelling	× 8
Northern England, UK	Afforestation: ditching and ploughing	× 100
Texas, USA	Forest clearance and cultivation	× 310
Lake and reservoir catchment studies		
Lake Sacnao, Mexico	Forest clearance and Mayan urban expansion	× 35
Lake Ipea, Papua New Guinea	Land use intensification: introduction of sweet potato	× 10
Seeswood Pool, Warwicks, UK	Agricultural intensification	× 4.5
Old Mill Reservoir, Devon, UK	Increased grazing pressure	× 4.5
Dayat er Roumi, Morocco	Land drainage	× 45
Llyn Geirionydd, Wales, UK	Mining activity	× 4
Krageholmssjön, south Sweden	Modern agricultural practices	× 4.5

Source: J. Woodward and I. Foster (1997) 'Erosion and suspended sediment transfer in river catchments', *Geography*, 82: 353–76

Table 13.4 Annual rates of erosion in selected countries (t/ha)

	Natural	*Cultivated*	*Bare soil*
Australia	0.0–64	0.1–150	44–87
Belgium	0.1–0.5	3–30	7–82
China	0.1–2	150–200	280–360
Ethiopia	1–5	8–42	5–70
India	0.5–5	0.3–40	10–185
Ivory Coast	0.03–0.2	0.1–90	10–750
Nigeria	0.5–1	0.1–35	3–150
UK	0.1–0.5	0.1–20	10–200
USA	0.03–3	5–170	4–9

Source: R. P. C. Morgan (1995) *Soil Erosion and Conservation*, 2nd edn (Harlow: Longman), table 1.1

able cultivation.

- Use of larger and heavier agricultural machinery, which tends to increase soil compaction.
- Use of more powerful machinery which permits cultivation in the direction of maximum slope rather than along contours. Rills often develop along the wheel ruts ('wheelings') left by tractors and farm implements, and along drill lines (plate 13.11).
- Use of powered harrows in seedbed preparation and the rolling of fields after drilling.
- Removal of hedgerows and the associated increase in field size. Larger fields cause an increase in slope length and thus a higher risk of erosion.
- Declining levels of organic matter resulting from intensive cultivation and reliance on chemical fertilizers, which in turn lead to reduced aggregate stability.
- Widespread introduction of autumn-sown cereals to replace spring-sown cereals. Because of their longer growing season, autumn-sown cereals produce greater yields and are therefore more profitable. The change means that seedbeds with fine tilth and little vegetation cover are exposed throughout the period of winter rainfall.

Measurements in the English East Midlands near Bedford indicate that, even there, rates of soil loss under bare soil conditions on steep slopes can reach 17.7 t ha^{-1} yr^{-1}, compared with 0.7 under grass and virtually nothing under woodland (see table 13.5). This erosion is caused primarily by processes like rainsplash, but elsewhere in eastern England wind is also an active process. Ever since the 1920s dust storms have been recorded in the Fenlands, the Breckland, Lincolnshire and Humberside. Another type of severe soil erosion in Britain is peat erosion. Deposits of blanket peat in upland areas like the Pennines are heavily degraded, producing pool and hummock topography, areas of bare peat and incised gullies. River waters draining the moors are rendered brown and turbid by the eroded peat par-

Table 13.5 Annual rates of soil loss under different land-use types in eastern England (t ha^{-1})

Plot	*Splash*	*Overland flow*	*Rill*	*Total*
Bare soil				
Top slope	0.33	6.67	0.10	7.10
Mid-slope	0.82	16.48	0.39	17.69
Lower slope	0.62	14.34	0.06	15.02
Cultivated soil				
Top slope	0.60	1.11	—	1.71
Mid-slope	0.43	7.78	—	8.21
Lower slope	0.37	3.01	—	3.38
Grass				
Top slope	0.09	0.09	—	0.18
Mid-slope	0.09	0.57	—	0.68
Lower slope	0.12	0.05	—	0.17
Woodland				
Top slope	—	—	—	0.00
Mid-slope	—	0.012	—	0.012
Lower slope	—	0.008	—	0.008

Source: from P. C. Morgan (1977) 'Soil erosion in the United Kingdom: field studies in the Silsoe area, 1973–75', *National College of Agricultural Engineering, Occasional Paper 4*

Plate 13.11 Soil erosion along wheel tracks on a sandy soil near Studland in Dorset, southern England.

ticles they contain. Some of the peat erosion may be an essential but natural process, for the high water content and low cohesion of undrained peat make the bogs inherently unstable, especially if a severe rainstorm occurs. However, there is evidence to suggest that the natural rate has been accelerated in the past 200–300 years as a result of a variety of human pressures: heavy sheep grazing, burning, peat cutting, military manoeuvres, the incision of pack horse trails, increased hiking and rambling, and severe air pollution from the factories of northern England.

Because of the adverse effects of soil erosion, a whole array of techniques has now been widely adopted to conserve soil resources in the face of water erosion:

1 Revegetation
   (a) deliberate planting;
   (b) suppression of fire, grazing, etc., to allow regeneration.
2 Measures to stop stream bank erosion (e.g stone banks and rip-rap).
3 Measures to stop gully enlargement:
   (a) maintaining cover at critical times of year;
   (b) rotation of crops;
   (c) growing cover crops;
   (d) agroforestry.
4 Slope runoff control:
   (a) terracing;
   (b) deep tillage and application of humus;
   (c) digging transverse hillside ditches to interrupt runoff;
   (d) contour ploughing;
   (e) preservation of vegetation strips (to limit field width).
5 Prevention of erosion from point sources such as roads and feedlots:
   (a) intelligent geomorphic location of roads, feedlots etc.;
   (b) channelling of drainage water to non-susceptible areas;
   (c) covering of banks, cuttings etc. with vegetation.

The implementation of such methods is vital if this basic, but only slowly renewable, resource is to remain productive to feed the world's rapidly growing human population.

■ *Key Terms and Concepts*

accelerated erosion
biological weathering
catena
chelation
chemical denudation
chemical weathering
clay minerals
doline
exfoliation
factors of soil formation
Goldich system
horizon
hydrolysis
karst
oxidation
physical weathering
profile
soil conservation
solution

Points for Review

- What do you consider to be the main factors that determine soil type?
- Describe the main types of physical (mechanical) weathering.
- What is solution and what landforms does it produce?
- Why is it important to consider organic processes in the study of soils and weathering?
- What factors control rates of chemical denudation?
- Why is soil erosion important and what can be done to control it?

FURTHER READING

Bland, W. and Rolls, D. (1998) *Weathering* (London: Arnold). An advanced overview.

Boardman, J., Foster, I. D. L. and Dearing, J. A (eds) (1990) *Soil Erosion on Agricultural Land* (Chichester: Wiley). An edited series of advanced research papers providing some useful case studies.

Brady, N. C. and Weil, R. R. (1996) *The Nature and Properties of Soils*, 11th edn (London: Prentice-Hall). The classic text on soils.

Bridges, E. M. (1997) *World Soils*, 3rd edn (Cambridge: Cambridge University Press). An excellent introductory world survey.

Gillieson, D. (1996) *Caves* (Oxford: Blackwell). A clear and comprehensive study of underground geomorphology.

Hudson, N. (1971) *Soil Conservation* (London: Batsford). A general introductory-level textbook.

Jennings, J. N. (1985) *Karst Geomorphology* (Oxford: Blackwell). A masterful summary of limestone landform and karstic processes.

Morgan, R. P. C. (1995) *Soil Erosion and Conservation*, 2nd edn (Harlow: Longman). A general introduction that is especially strong on methods of controlling erosion.

Pimental, D. (ed.) (1993) *World Soil Erosion and Conservation* (Cambridge: Cambridge University Press). A series of advanced, edited papers that look at soil erosion in a regional context.

14 The Hydrological Cycle

14.1 Introduction

Water occurs in three states: as crystalline ice (the solid state) as water (the liquid state), and as water vapour (the gaseous state). Water may pass from the solid to the liquid state by melting, and from the liquid to the solid state by freezing. It may pass from the gaseous to the liquid state by condensation, and from the liquid to the gaseous state by evaporation. If temperatures are below freezing point, molecules can pass from the gaseous to the solid state by sublimation.

When water changes state there are exchanges of heat energy that have considerable meteorological importance. When water evaporates, for example, *sensible heat*, which we can feel and measure by thermometer, passes into a hidden form held by the water vapour and known as the *latent heat of vaporisation*. This change results in a drop in temperature of the remaining liquid, as when evaporation of sweat causes cooling of the skin. Likewise, when the reverse process of condensation occurs an equal amount of energy is released as sensible heat and the temperature rises correspondingly. Similarly, when water freezes heat energy is released, whereas melting absorbs an equal quantity of heat, called the *latent heat of fusion*.

Water is present in the air in varying amounts; the measure of it is referred to as *humidity*. The humidity varies according to the nature of the surface that the air is in contact with and according to the temperature. At any specific temperature the quantity of moisture that can be held by the air has a definite limit, with warm air being able to hold more than cold. The proportion of water vapour present at a given temperature relative to the maximum quantity that could be present is the *relative humidity*. If a parcel of air is warmed, therefore, its relative humidity would fall, and if it were cooled the relative humidity would rise. If the cooling proceeds beyond a critical point – the *dew point* – the air becomes saturated and the excess vapour condenses into a liquid or solid state, producing clouds (window 14.1).

Although relative humidity is a very important concept, and gives an important indication of the state of water vapour in the air, it is only a statement of the quantity of water vapour present compared with a saturation quantity. The actual quantity of moisture present is known as the *specific humidity*; this is defined as the mass of water vapour in grams contained in a kilogram of air. Warm air, like that at the Equator, has a very high potential specific humidity, whereas cold air has a very low potential specific humidity. For this reason warm air can provide a very large amount of precipitation in comparison with cold air.

14.2 Precipitation

Understanding the different states of water gives us the key to understanding the nature of rainfall and other types of precipitation such as fog, snow and hail; for only where large masses of air experience

Window 14.1 Clouds

Clouds form when the dew point is reached, with some of the water vapour in the air condensing into minute water droplets or ice crystals. These, being extremely small and light, float into the air, and occur as either layers or as heaps. The most extensive *layer clouds* (stratus) form along fronts and near lows and troughs where converging air streams in the lower atmosphere force air to rise. These are rather featureless clouds and tend to produce overcast skies. For example, *nimbostratus* is a grey, dark, heavy opaque rain cloud. Heap or *convective clouds* tend to be less extensive, but have considerable vertical development; they are formed by pockets of warm, unstable air rising. The most common type is the *cumulus* cloud. The third type of cloud, *cirrus*, is high, white and thin, and consists of delicate veil-like patches of feathery appearance.

There are ten main cloud genera:

(a) *Cirrus (Ci)* Small clouds in the form of white, sometimes silky, patches or bands. Often occur as feathery filaments with their ends swept into hooks by wind shear.
(b) *Cirrostratus (Cs)* Transparent white veil: smooth and uniform or fibrous. Commonly produces haloes.
(c) *Cirrocumulus (Cc)* Thin layer, which is either continuous or in patches that are made up of small elements. Commonly forms a 'mackerel sky'.
(d) *Altostratus (As)* Grey, featureless layer cloud that can be fibrous or uniform. The sun may penetrate weakly, but this type of cloud does not produce optical phenomena.
(e) *Altocumulus (Ac)* Very variable in form – can be continuous or patchy. As it is usually waved or in rolls, lumps and laminae, it is better known as *lenticular* or *crenellate*.
(f) *Nimbostratus (Ns)* A grey, dark, heavy, opaque raincloud.
(g) *Stratus (St)* Forms a grey uniform layer that may be continuous or patchy and often produces rain, or snow.

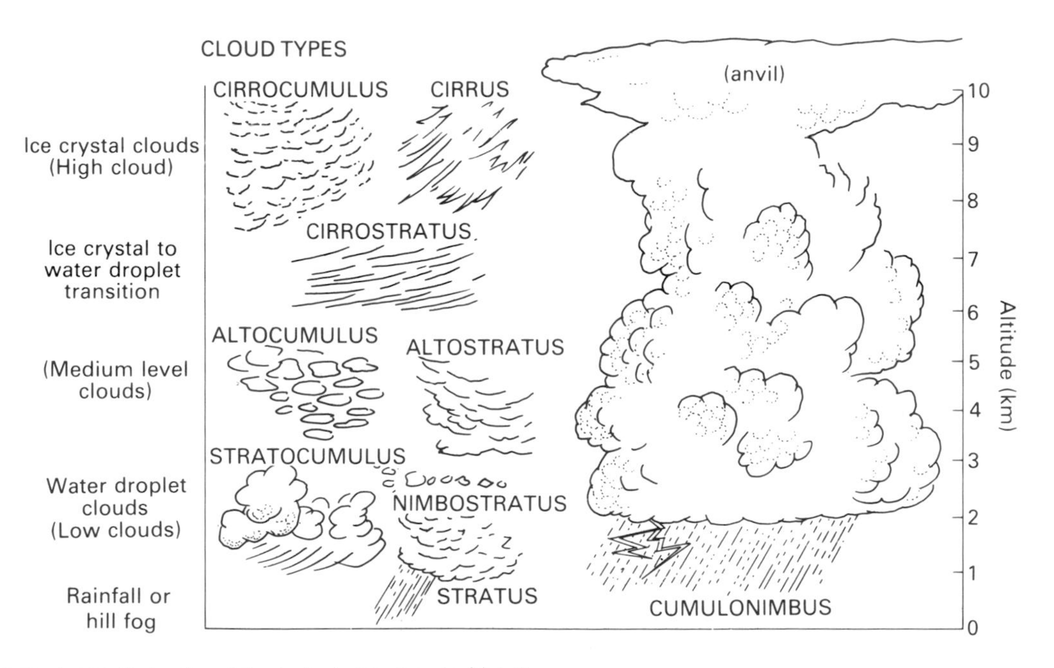

The major types of cloud and the typical altitudes at which they occur.

(h) *Stratocumulus (Sc)* Grey or white layer with dark areas. Usually in rolls, undulations and rounded masses but not fibrous. Elements often in a regular pattern.
(i) *Cumulus (Cu)* Separate dense white clouds with a well-defined form and strong vertical development. Flat base, upper parts brilliant white and cauliflower-like.
(j) *Cumulonimbus (Cb)* Extreme vertical development of a cumulus cloud. It has a huge tower, is dark at the base and is often associated with precipitation and thunder. Top smooth, occasionally fibrous and pulled out laterally.

A cumulonimbus cloud, illustrating its great vertical development and an upper 'anvil' of fibrous clouds.

a steady drop in temperature to below the dew point can precipitation occur in appreciable amounts. Such a cooling normally results from a parcel of air rising to higher altitudes.

Rising air experiences a drop in temperature even though no heat energy is lost to the outside. This is a consequence of the decrease in air pressure that occurs at higher altitudes, which permits rising air to expand. The expansion means that individual molecules of gas are more widely diffused and do not move so fast, so that the sensible temperature of the expanding gas is lowered. Parcels of air that descend undergo the opposite, with compression leading to warming.

The rate at which the air cools with altitude is termed the *adiabatic lapse rate*. For dry air it is about 10 °C per 1000 m of vertical rise. However, if the air contains moisture and uplift takes place, the cooling produces condensation when the dew point is passed, and this condensation releases latent heat, so that as a result the adiabatic lapse rate is reduced to around 3–6 °C per 1000 m. This lower rate is termed the *wet adiabatic lapse rate*.

During the rapid ascent of a parcel of air in the

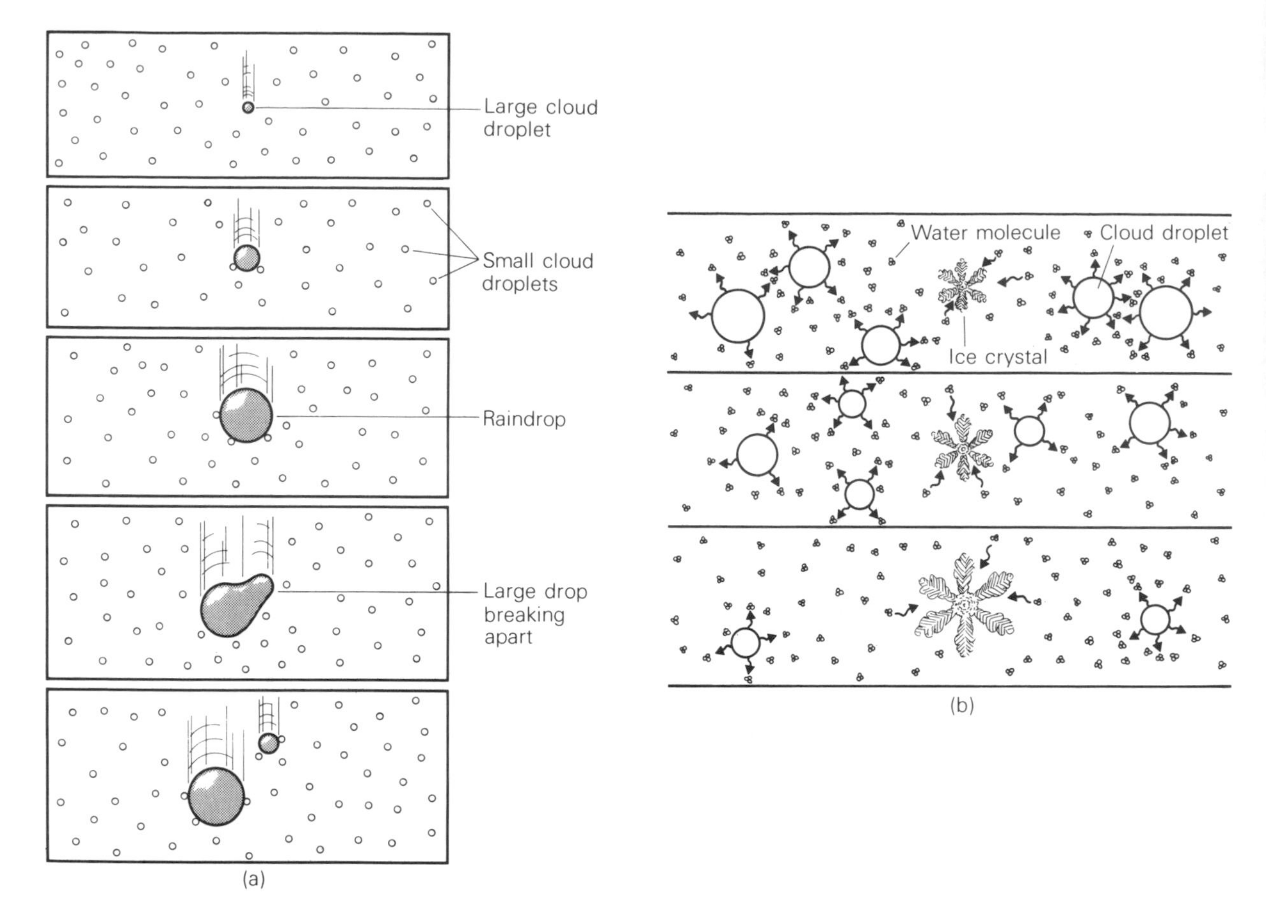

Figure 14.1 Two mechanisms for rain formation. (a) The collision–coalescence process: because large cloud droplets fall more rapidly than smaller droplets, they are able to sweep up the smaller ones in their path and grow. (b) The Bergeron–Findeisen process: ice particles grow at the expense of cloud droplets until they are large enough to fall. Note that the size of these particles has been greatly exaggerated in this diagram.

saturated state, clouds form. The clouds contain water particles that may progressively coalesce to form first drizzle and then rain. In the warm clouds of the low latitudes rain forms directly by liquid condensation and droplet coalescence.

The *collision–coalescence process* was proposed by an American physicist called I. Langmuir, and operates as follows. Cloud droplets are mostly very light and remain suspended in the air, but a few have sufficient weight to start falling slowly through the cloud. As they fall so they collide with other droplets and thus grow even bigger. The deeper the cloud, the bigger the drops grow and the faster they fall (figure 14.1a).

In the middle and high latitudes, however, another mechanism operates, called the *Bergeron–Findeisen* process (figure 14.1b). This theory is based on the observation that many clouds extend so high into the atmosphere that the temperature may be as low as –20 to –40 °C. Clouds at such levels are made up of a mixture of ice crystals, supercooled water droplets and water vapour. At these low temperatures, if air is saturated with respect to water then it is supersaturated with respect to ice. This means that, in terms of the ice crystals, there is too much vapour in the air. Thus some of the vapour freezes on to the ice, so that crystals become enlarged. But this process reduces the amount of water vapour in the air, making it unsaturated with respect to water. Some droplets therefore evaporate and redress the balance, and by this process there is a continuous transfer of moisture through vapour to ice, and the ice particles grow rapidly. Such particles eventually become large enough to fall as snow, and as they move to lower altitudes they may be converted into rain. This theory forms the basis of many rain-making experiments, where the growth of large ice crystals is stimulated by introducing ice or similar crystals into clouds of supercooled water droplets.

The next question to ponder is what causes a parcel of air to rise. There would appear to be two main mechanisms: the spontaneous rise of warm air, and the forced rise of moist air.

Spontaneous rise is associated with convection, a form of atmospheric motion consisting of strong updrafts taking place within a convection cell. A major cause of such updrafts is the heating of a parcel of air by heat radiating from the earth's surface. The warm air produced is less dense than the surrounding air and thus rises. As the air rises it is cooled adiabatically so that eventually it is cooled beyond its dew point and condensation occurs. Heat is then liberated by the condensation process, and this serves to fuel the convection cell further. Unequal ground heating under conditions of high temperature and humidity is thus the sort of circumstance that favours the production of the unstable air that produces this type of convectional rain.

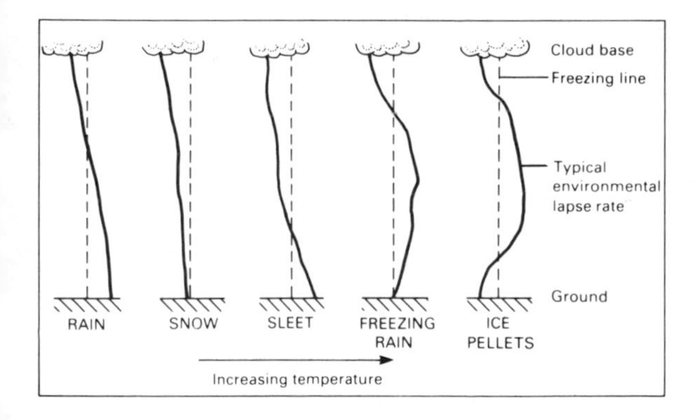

Figure 14.2 The type of precipitation that reaches the ground depends on the temperature structure of the air layer between the cloud base and the ground. In the situation illustrated, precipitation falls from a mixed cloud. The resultant precipitation type depends on the relation between the environmental lapse rate and the freezing temperature. In all cases typical profiles are shown and temperature increases from left to right.

Precipitation can also be caused by *forced ascent* of moist air. For example, when prevailing winds encounter a mountain barrier, the air is forced to rise to surmount the range. This type of relief-produced precipitation is termed *orographic*. A layer of dense cold air may operate in rather a similar way to a mountain range. Being dense it will tend to remain close to the ground, acting as a barrier to the movement of the lighter warm air. The warm air may therefore be forced to rise over the cold air, as is the case with frontal precipitation of the type found in mid-latitudes (see section 5.2).

In addition to rain there are other types of precipitation. As figure 14.2 indicates, the nature of the precipitation that actually reaches the ground depends on the conditions that exist between the cloud and the ground. One of these is *fog*, a substance made up of tiny droplets of liquid water that forms when air is cooled below its dew point so that some of the invisible water vapour it contains condenses to form small water droplets. It differs from cloud only in that, whereas clouds form when air rises and is cooled, fog is formed when air cools near the ground surface. *Radiation fog* develops when the relatively warm air is cooled by contact with a cool land surface as on clear nights when calm conditions exist. If warm air flows over an area of cold seas, as for example where uplift takes place along the west coasts of some land masses (e.g. California and Namibia), then *sea fogs* may develop. In coastal Arctic regions *steam fog* is produced when cold air comes into contact with warm water. In areas where fogs are frequent and there are trees on which the droplets may impinge, *fog water* may be a significant component of the hydrological cycle.

Another important type of precipitation is *snow*. The conditions necessary for snow to fall are exactly the same as for rain except that temperatures at ground level are lower. Once again, the fundamental requirement is an upward movement of air produced by an area of low pressure, by the convergence of air streams or by the physical barrier of a range of mountains. As warm air holds more moisture than cold, the heaviest snowfalls tend to occur when the temperature is close to freezing rather than at much colder temperatures. The snowiest place in the world for which we have records is a place called Paradise in the Mount Rainier area of the north-west United States, where in the winter of 1971–2 a total of 31 m of snowfall was recorded.

Hail, consisting of spherical lumps of ice from 5 mm in diameter upwards, forms in vigorous cumulonimbus clouds in which the upcurrents are strong enough to carry the weight of the stones as they grow. The hailstones originate as pellets of snow or frozen raindrops which rise and fall in the violent air currents of a storm cloud, growing by accretion as cloud droplets freeze on to them. Hail is relatively rare in cold areas because convectional activity is not sufficiently severe, and also in very warm areas because any stones that do form are likely to melt away before reaching the ground. It therefore tends to be a feature of areas like the central plains of North America and of subtropical areas like South Africa. The record weight for a hailstone is of a specimen weighing 766 g and having a diameter of 44 cm which fell in Kansas in 1970. Hailstorms can cause great damage, especially to crops; the damage caused each year in the United States exceeds $500 million.

14.3 Interception

Under most conditions, rain and snow do not fall directly on the ground surface but are *intercepted* by the branches and foliage of plants (figure 14.3). They can then take three possible routes to the ground surface: (a) dropping off the plant leaves (a process known as *throughfall*), (b) undergoing secondary interception by plants at lower levels, or (c) flowing down stems and trunks (a process termed *stemflow*). But not all intercepted moisture finds

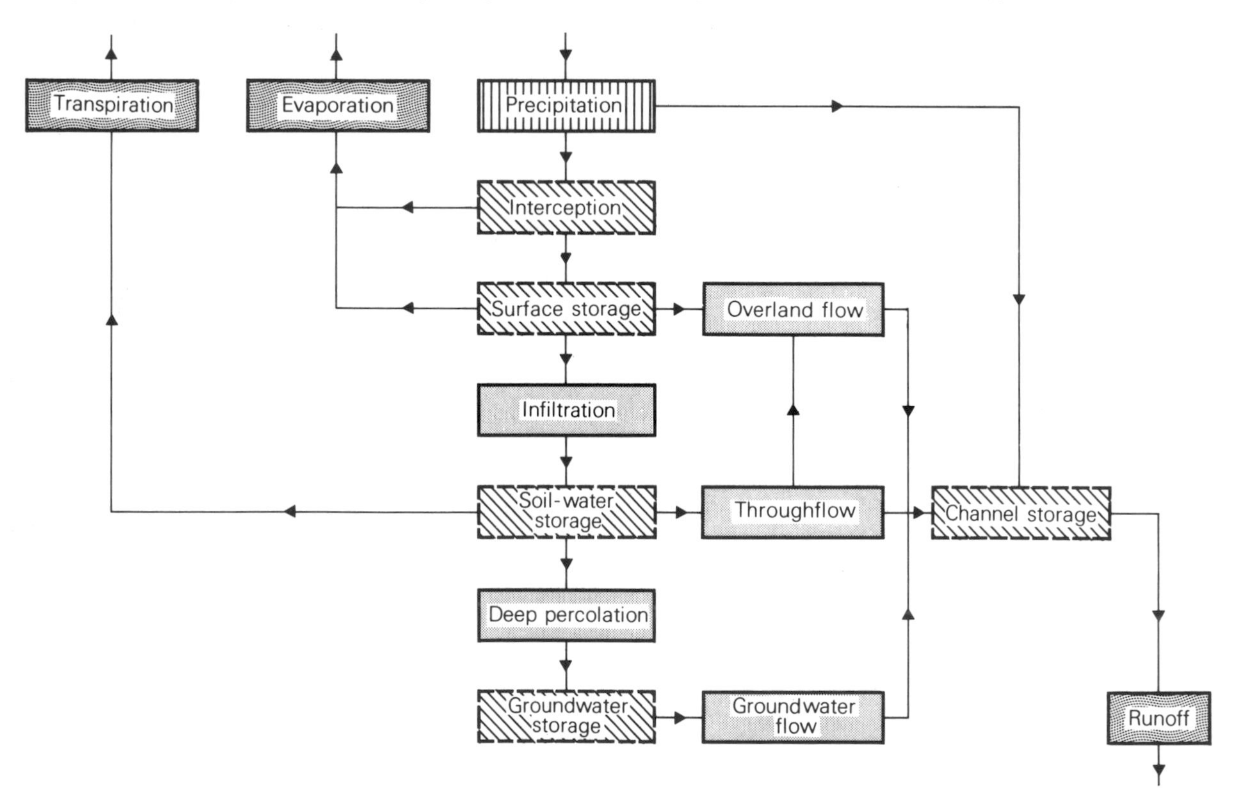

Figure 14.3 The drainage-basin cycle showing the main inputs, stores (bounded by a broken line), transmission processes and outputs.

its way to the ground. Some is lost from plant leaves by evaporation back to the atmosphere. This amount will vary greatly with the nature of the vegetation and the nature of the rainfall. For light falls of rain it can be substantial (as much as 60 per cent), but for heavier storms it tends to even out at about 15 per cent.

In winter deciduous trees are capable of relatively little interception, as they have lost their leaves. This does not apply to coniferous trees, and even in summer there is evidence that coniferous trees intercept more than deciduous because the drops appear to find it easier to cling to individual needles. Some leaves of trees in the equatorial forests may have drip-tips that tend to shed raindrops easily. In general, however, tropical evergreen forests, with their closed crown covers and several layers of vegetation, are quite effective at preventing intercepted rainfall from reaching the ground and are thought to have interception losses of about 30 per cent. An average figure for the interception loss under a grass cover is about 20 per cent of the total precipitation.

14.4 Evapotranspiration

When one compares the amount of water that comes into a drainage basin in rainfall with the amount that leaves the basin as river discharge, it becomes apparent that there is a very substantial loss within the system. If we take the case of south-east England, the stream output from an area is only about 30 per cent of the input, and in desert areas the loss is even greater. This output difference between the input and the output values is predominantly because of *evapotranspiration*.

There are two components to evapotranspiration. One of these is *evaporation*, which is the direct loss of water to the atmosphere from water, soil and other surfaces as a result of physical processes. The other is *transpiration*, which is the biological process of evaporation of water from the leaves of plants through openings known as stomata. Transpiration rates vary markedly during the growing season and with the nature and amount of the vegetation cover. Evaporation rates vary markedly according to temperature, wind speeds, atmospheric turbulence, humidity, number of sunshine hours and so on. Climatologists distinguish between *potential evapotranspiration* (E_p) (the amount of water loss that would occur if sufficient water were always available for the needs of vegetation that covers the area), and *actual evapotranspiration* (E_a) (the real quantity of water that is lost given that sometimes there is no moisture available).

If we take an idealized model (figure 14.4a), in

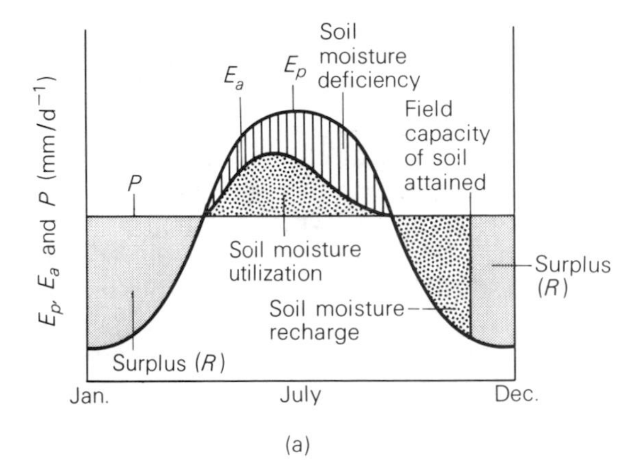

(a)

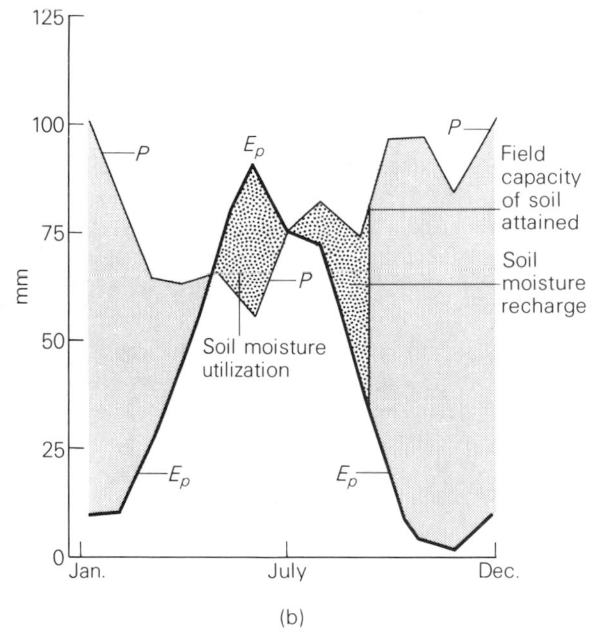

(b)

Figure 14.4 Soil moisture budgets: (a) an idealised representation of the annual soil moisture budget for a mid-latitude location in the Northern Hemisphere (for explanation see text); (b) an annual soil–water-balance diagram for Huddersfield Oakes, Yorkshire, England.

which we assume that precipitation inputs are uniformly distributed throughout the year, we can see the importance of the difference between E_p and E_a; for, while the precipitation inputs are uniformly distributed throughout the year, E_p shows a strong annual cycle because of higher summer temperatures and rapid vegetation growth. During the period when precipitation (P) is greater than E_p there is a water surplus, and so runoff (R) can be generated. However, when, as in the summer months, E_p exceeds P, then moisture is lost from the soil as plants draw upon soil moisture. Under these conditions of *soil moisture deficiency*, E_a is less than E_p. Once the period of soil moisture deficiency has ended because P once again exceeds E_p, there follows a period of soil moisture recharge in which soil moisture is restored to the field capacity – the amount of water that is held when a soil has first been saturated and then allowed to drain under gravity until no more water moves downward. When this recharge of soil moisture is complete a period of water surplus will exist and R can once again occur. This idealised model compares reasonably well with a real water-balance diagram for a station in Yorkshire, England (figure 14.4b).

14.5 Infiltration

Once it has passed through the vegetation layer, rainfall hits the soil surface. Depending on the nature of the soil, some of it will run off the surface, and some of it will enter the soil by a process called *infiltration*. There is a maximum rate at which the soil in a given condition can absorb water, and this upper limit is called the *infiltration capacity* of the soil. If rainfall intensity is less than this capacity, then the infiltration rate will be equal to the rainfall rate. If rainfall intensity exceeds the ability of the soil to absorb moisture, infiltration will occur at the capacity rate; the excess rainfall over infiltration will cause water to accumulate on the ground surface, and *runoff* will commence.

Water that enters the soil by infiltration has to move through small cracks and pore spaces in the soil. Hence its rate of movement is low – lower than the rate at which water will run off the surface. The soil therefore plays a major part in determining the volume of storm runoff, its timing and its peak rate flow.

Water is drawn into the soil by forces of gravity and capillary attraction. The rate of infiltration declines rapidly during the early part of a storm and reaches an approximately constant value after one or two hours of rain. There are several factors that account for this. First, the filling of fine soil pores with water reduces capillary forces drawing water into pores and fills the storage potential of the soil. Second, as they are moistened, soil clay particles swell and reduce the size of pores. Third, the impact of raindrops during rainsplash breaks up soil aggregates, splashing fine particles over the surface and washing them into pores where they impede the entry of water.

Many factors influence the nature of the infiltration capacity curve over time. Long, intense rainfall packs down the loose soil surface, disperses fine soil particles and causes them to plug soil pores. It also causes clay minerals to swell. Soil properties are also vitally important, for coarse-textured soils with a sandy texture have large pores down which water can easily drain, while the exceedingly fine pores in clay retard drainage. Likewise, if the soil particles are held together as aggregates by organic matter, and have a good friable crumb structure, then that will permit rapid infiltration and drainage. The depth of the soil profile and its initial moisture content are important determinants of how much water can be stored in the soil before it is saturated and overland flow can occur. Similarly, vegetation plays a role. Vegetation and litter help to prevent soil from packing by raindrop impact and provide the necessary humus for binding soil particles together in open aggregates. Soil fauna that live on the organic material assist in this process by churning together the mineral particles and the organic material.

The infiltration capacity of a surface can be changed by human activity. Overgrazing by domestic animals may trample and puddle the soil so that its structure is destroyed; the removal of forests will reduce the protection against rainsplash and will decrease the soil's organic content; and the building of cities may lead to the extension of large areas of impermeable concrete, tarmac and roof tiles over the ground surface (see section 16.4). The first

Table 14.1 The influence of land use on infiltration rates

A qualitative ranking of the influence of land use on the minimum rate of infiltration

Highest infiltration	Woods, good
	Meadows
	Woods, fair
	Pasture, good
	Woods, poor
	Pastures, fair
	Small grain crops with good rotation
	Small grain crops with poor rotation
	Legumes after row crops
	Pasture, poor
	Row crops with good rotation
	Row crops with poor rotation (one-quarter or less in hay or sod)
Lowest infiltration	Fallow

Rates of infiltration on grazed and ungrazed lands in the United States (mm h^{-1})

State	*Ungrazed*	*Heavily grazed*
Montana	34.4	10.2
Oklahoma	222.3	62.2
Colorado	62.2	25.4
Montana	147.3	58.4
Wyoming	34.3	24.2
Louisiana	45.7	17.8
Kansas	33.0	20.3
Arizona	40.6	30.5
Mean	77.5	31.1

part of table 14.1 gives a qualitative impression of how different the infiltration rates are on heavily grazed land compared with ungrazed land: on the heavily grazed land the mean rate is only 31.1 mm h^{-1} compared with 77.5 on ungrazed land.

14.6 Surface Runoff

Traditionally, it has been assumed that surface runoff is the result of rainfall intensities being in excess of the infiltration capacity of the soil, and indeed many observations have been made, during heavy storms, of water standing on or flowing over the ground surface of many river basins. In 1935 an American hydrologist, R. E. Horton, presented a highly influential *model of overland flow* (figure 14.5a). It predicts that prolonged rain falling on the slopes of a drainage basin possessing relatively uniform infiltration capacity will, if the rainfall intensity is greater than the infiltration capacity, produce overland flow over all the basin more or less simultaneously. Horton believed that such overland flow was the prime contribution to the rapid rise of river flow levels during storms, and that it was a major cause of slope erosion. He maintained that, according to the intensity of the storm and the nature of the surface materials, there would be a critical distance downslope from the drainage divide where the depth of overland flow would become sufficient to generate a shear stress component to

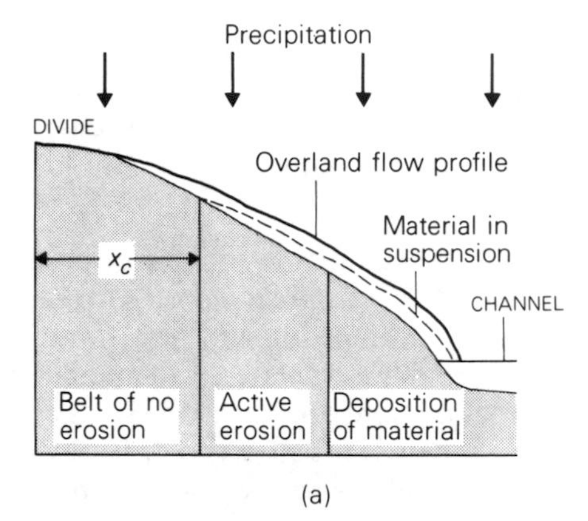

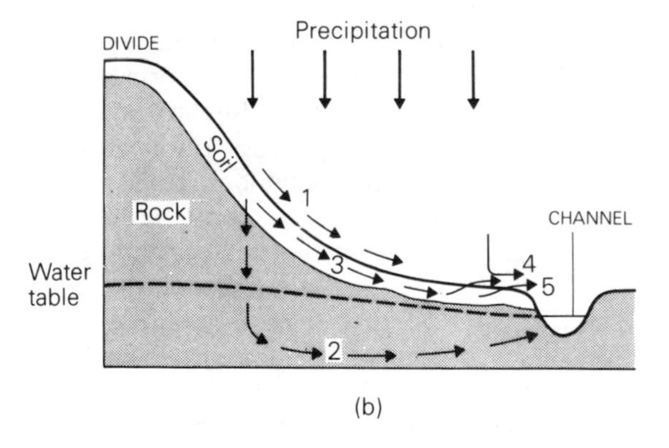

Figure 14.5 Water movement into channels from hillslopes: (a) the classic model of R. E. Horton; (b) the more complex throughflow and saturation overland flow model: 1, infiltration excess Horton-type overland flow; 2, groundwater flow; 3, shallow subsurface stormflow; 4, saturated overland flow composed of direct precipitation; 5, return flow.

entrain the surface soil particles and to erode rills. Above that critical line Horton believed there was a zone called *the belt of no erosion* where slope-top convexity was the norm, associated with the effects of rainsplash and soil creep (see section 12.4). Below that line there would be active surface erosion by overland flow, and slopes would thus tend towards concavity.

The sequence of events that takes place in the Horton scheme can thus be summarised as follows:

(a) Rainfall intensity exceeds the soil's infiltration capacity.
(b) A thin water layer forms on the surface and downslope surface flow is initiated.
(c) The flowing water accumulates in surface depressions.
(d) When full, these depressions begin to overflow.
(e) Overland flow enters small rills, which coalesce to form rivulets, which discharge into channels and produce a rapid rise in stream level.

Hortonian overland flow occurs in areas with high-intensity rains, in areas devoid of vegetation and in areas with thin soils of low-infiltration capacity. Semi-arid areas (plate 14.1) and cultivated fields in areas of high rainfall are the sorts of places where the process can be observed. It can also be seen where the soil has been compacted by vehicles or animals.

However, two pieces of evidence suggest that Horton's model in its simple form may not have very wide applicability. First, on some slopes where surface water has been observed in storms, measured infiltration rates are very high. Rainfall intensities in the United Kingdom, for example, rarely exceed 20 mm in an hour for any prolonged period, even in flood-causing storms, and yet the infiltration capacities of the soils may exceed 200 mm h^{-1}. Second, it has been noted that runoff is not generated from all over the catchment in the way envisaged in Horton's model, but that it is restricted to certain specific parts of the basin.

The explanation for these two pieces of evidence lies in the consideration of a process called throughflow, for it has now become evident that the downslope movement of water within the soil layers is more important than overland flow. Most soils occur on sloping land and so soil water movement will be towards the base of the slope. The vertical component of this movement is called *percolation*, and the horizontal component is called *throughflow*. Unless there is a storm of extremely high intensity, or unless the soil is very hard, all the effective rainfall will infiltrate into the soil, filling up empty soil pore spaces and establishing a zone of higher water content near the surface. The leading edge of the newly penetrating water is called the *wetting front*: this separates a zone of high permeability above from one with drier soil and a low permeability below. If the storm is of sufficiently long duration, the wetting front eventually reaches the base of the soil profile and all storage is filled. Thereafter, water moves downslope through throughflow. However, soil adjacent to the stream chan-

Plate 14.1 In desert areas, Hortonian overland flow can occur. Silty soils have low infiltration capacities, and when a storm occurs little wadis on the island of Bahrain can rapidly fill with storm runoff.

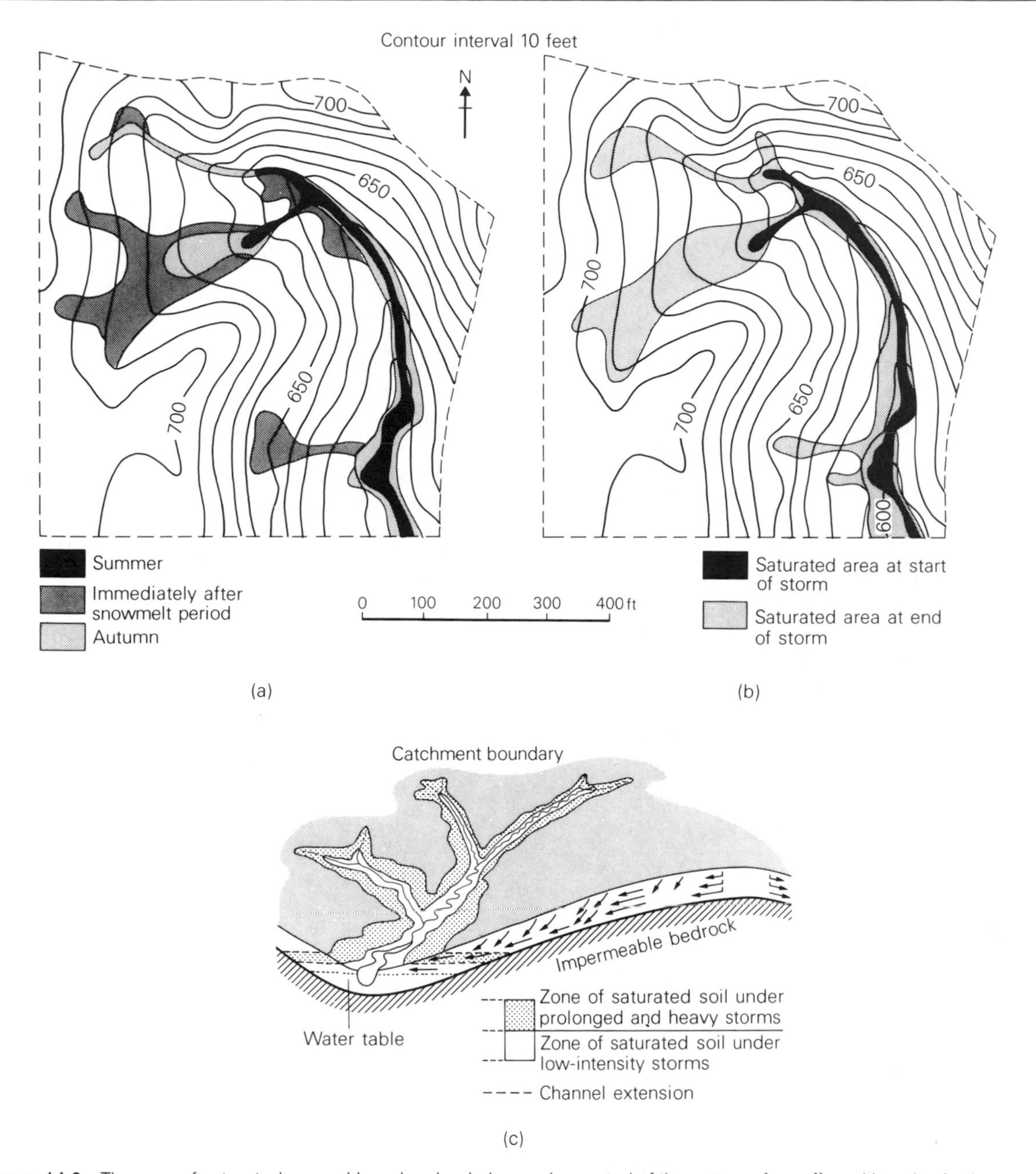

Figure 14.6 The area of saturated ground in a river basin is a major control of the nature of runoff, and it varies both seasonally and during the course of one event: (a) the seasonal variation of the pre-storm saturated area in a small steep-sided, well-drained hillside with a narrow valley floor in Vermont, USA; (b) the saturated areas as they expand during a single rainstorm of 46 mm; (c) the effect of a rainstorm on a catchment is to cause stream expansion, and the extension of the saturated zone adjacent to stream channels. The soil thickness has been exaggerated for clarity.

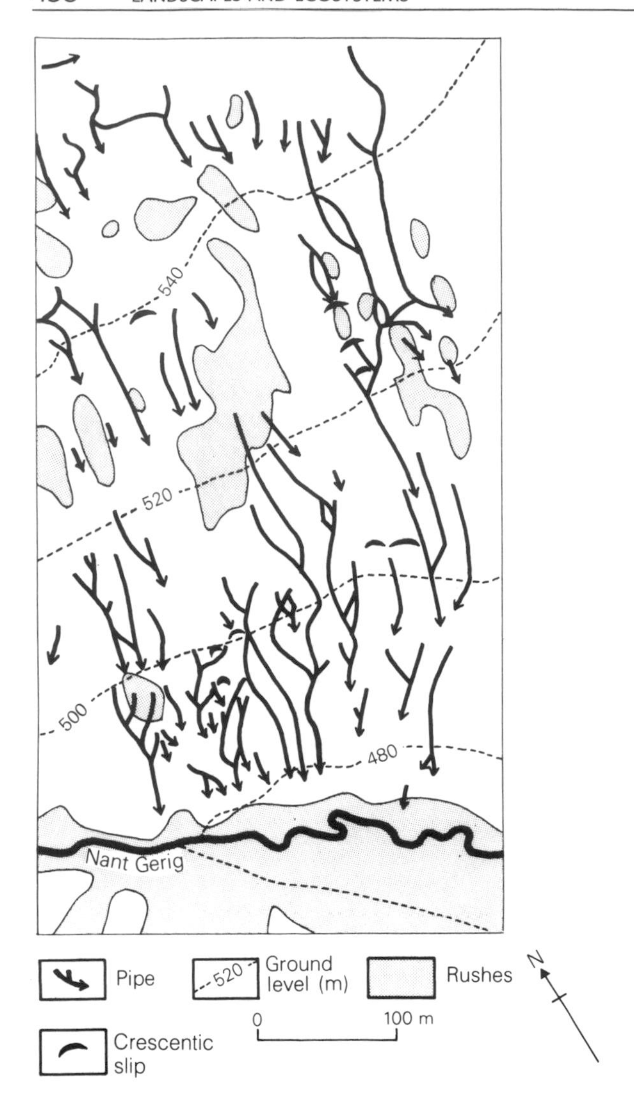

Figure 14.7 Pipe network in the lower Nant Gerig Valley, central Wales.

nels, or soil at the base of a slope, may already be saturated and will expand as throughflow water reaches it (figure 14.6), and water falling on this saturated area will not be able to infiltrate into the soil so overland flow will occur. This will cause river level to rise and will be a major cause of any rapid rise in stream level, but some water will enter the sides of river channels directly from throughflow. The lateral movement of throughflow in the soil may produce small pipes or tunnels (figure

Table 14.2 Estimated flow velocity of various hydrological processes

Flow type	*Velocity range* ($m\,h^{-1}$)
Open channel	300–10 000
Overland flow	50–500
Pipeflow	50–500
Soil matrix throughflow	0.005–0.3
Groundwater flow	
Sandstone	0.001–10
Shale	0.00000001–1
Jointed limestone	0–500

Source: adapted from D. Weyman (1975) *Runoff Processes and Streamflow Modelling* (Oxford: Oxford University Press)

14.7), and these can transmit water downslope to channels at some speed, thereby contributing to a rapid rise in stream levels during a storm. Water that flows slowly through the soil matrix rather than in pipes travels much more slowly (table 14.2).

The sequence of events that takes place in the throughflow and saturation overland flow model can be summarised thus (see figure 14.5b):

(a) Early in the storm there is a small wedge of saturated ground near the stream channel (point 1 in the figure). The saturated area in the soil is close to the surface and throughflow or subsurface storm flow (SSSF) (point 3) has just begun.
(b) As the storm proceeds the saturated area intersects the ground surface over the lower part of the slope and water seeps over the surface as something called return flow (RF) (point 5).
(c) This type of overland flow is augmented by direct precipitation (DPS) on to the saturated area and on to the water surface in the channel (point 4).
(d) As the storm goes on the saturated wedge migrates upslope, so that the contributing area for runoff in the basin increases.
(e) When the storm ceases there is a very sudden drop in DPS, and RF also decreases very rapidly. However, SSSF continues for a long time, with flow through the soil matrix being rather slow and that in pipes rather faster.

The importance of throughflow processes in small British catchments can be established by empirical

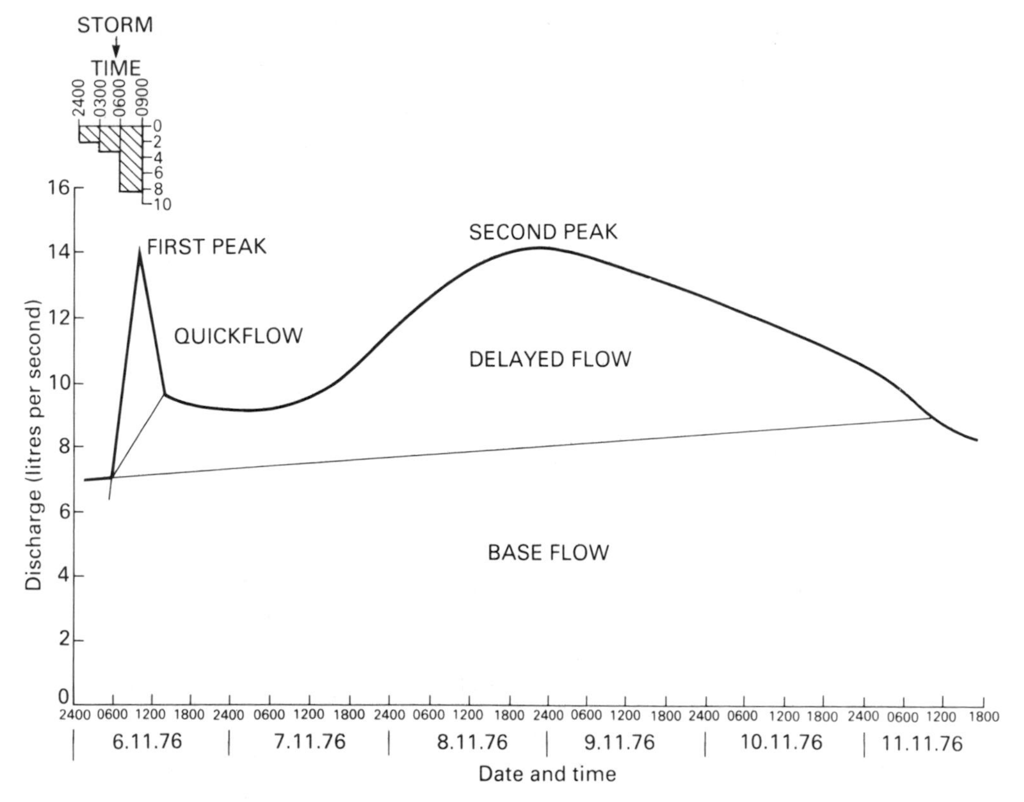

Figure 14.8 A stream hydrograph (i.e. a plot of discharge against time) for a small drainage basin in the Mendip Hills, southwest England. The bedrock consists of impermeable Devonian Old Red Sandstone, and this is overlaid by deep, permeable brown earth soils. Note the two peaks in the hydrograph.

studies of stream hydrographs. In various studies it has been found that following a period of high-intensity rainfall there are two distinct peaks in the hydrograph (figure 14.8). The first seems to occur fairly shortly after the peak of the storm, sometimes with a lag of only an hour, and reflects rapid runoff from rain that has fallen directly on the channel, on roads or on highly compacted surfaces. The second hydrograph peak occurs some hours after the first, and is related to slower throughflow processes, sometimes assisted by the presence of tile drains.

14.7 Groundwater

If one excludes the water that is locked up in the polar ice caps and in glaciers, some 97 per cent of all freshwater that is found on the planet is stored underground. This is known as groundwater. It is found in the pores that exist in sediments such as sand and gravel, and in the fractures that are found in rocks like limestone and sandstone.

Groundwater is a very important resource. Some individual countries, such as Barbados, The Netherlands and Denmark depend almost entirely on groundwater. More than one-third of water use in the UK and France is supplied from this source, while the USA is approximately 50 per cent dependent on groundwater supplies.

Precipitation that succeeds in passing from the soil layer downwards into the underlying bedrock (figure 14.5b, point 2) will at some point meet a zone of saturation – the *groundwater zone*. The top of this zone is generally called the *water table*. Normally the shape of the water table is a subdued image of the surface topography. Groundwater tends to flow towards a low point, and this is often represented by the river channel.

The classic model of the rate of groundwater motion is *Darcy's law*, named after a mid-nineteenth-century engineer from Dijon in France. Darcy said that, for a given aquifer, the rate of water flow from one place to another is directly proportional to the drop in vertical elevation between the two places and inversely proportional to the horizontal distance the water travels. The relationship between the horizontal and vertical distances can be seen to be akin to the angle of an inclined plane; it is, in fact, the force of gravity that makes water run 'downhill' underground, just as a ball rolls with varying speeds down planes of different slopes. Darcy also found that the greater the permeability of the rock, the faster the flow.

It is important to note, however, that there is a difference between porosity and permeability. The former is a measure of how much water can be stored in a rock or sediment, whereas permeability measures the properties of a rock or sediment which determine how fast water can flow through it.

Porosity results from the open texture, coarse-grained constituents and loose cementation of rocks, and open pores may provide *primary permeability*. *Secondary permeability* (or *perviousness*) is the result of the presence of joints, cracks and fissures through which water can flow. Typical groundwater velocities lie in the range 1.0–0.001 $m\,d^{-1}$. Coarse-grained sedimentary rocks (gravel, conglomerate, sandstone) show the highest permeabilities, and fine-grained sedimentaries (e.g. clays) or igneous rocks (including granite) the lowest. For unconsolidated sediments, the coarser varieties (sand and gravels) are more permeable than the finer-grained varieties (silts and clays), even though they are in general less porous. The reasons for this are twofold. The angular and platy shape of many clay particles means that they have a tendency to interlock and thus isolate the spaces between them. This inhibits water movement through the sediment. More importantly, the smaller grain sizes in silt and clay result in a greater surface area of particles relative to volume which means that water tends to be held in the pores by surface tension effects.

Some rocks may have a very high secondary permeability along joints and bedding planes which have either been enlarged by solution (as in the case of limestones) or along cooling joints (as in the case of basaltic lavas). Groundwater tends to seep out where a rock of high permeability (the *aquifer*) is

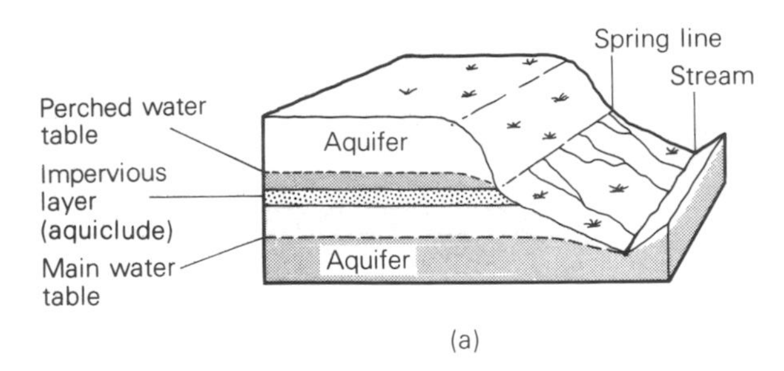

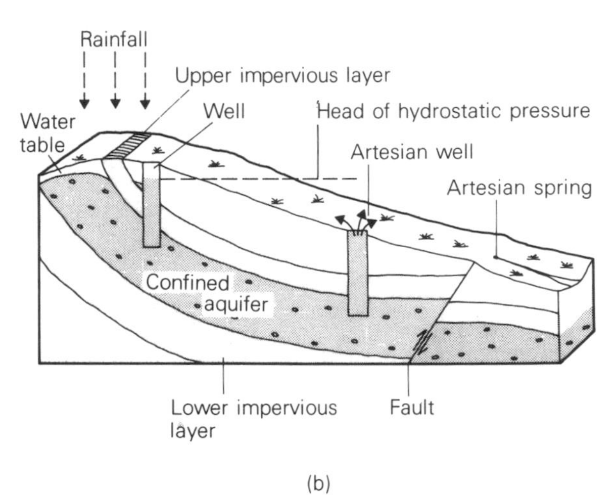

Figure 14.9 (a) A perched water table. In an area of gently dipping sedimentary rocks some layers will have high porosities and permeabilities (e.g. sandstones) while others (e.g. shale) will not. The sandstone could act as an underground storage reservoir for groundwater – an *aquifer*. By contrast, a shale bed, by dint of its lower permeability, would virtually preclude groundwater flow. Such a bed is called an *aquiclude*. In this example, a thin impermeable layer has blocked downward percolation of water to the main water table, and where a perched water table meets the valley side a seep or spring can occur. (b) An artesian spring and well. The water table in the confined aquifer lies near the top of the dipping layers. The upper well drilled through the top impervious layer is not artesian, since the head of hydrostatic pressure is not sufficient to force water to the surface. The top of the artesian well, lower down the slope, lies below the level of the head of the hydrostatic pressure, and so water can flow out at the surface. The artesian spring rises where a fault has occurred in the strata, which provides a route along which water, under sufficient hydrostatic pressure, can move.

underlain by less permeable material; for example, if a limestone overlies a clay, groundwater may be discharged through springs.

Groundwater occurs in two main forms which are related to geological structure and topography. When it is not overlain by relatively impermeable materials it is said to be unconfined. In many places a zone of unconfined water may exist near the surface of the ground where downward movement of percolating water is impeded by the presence of an underlying impermeable stratum called an *aquiclude*. A body of water that is isolated in this way from other groundwater is termed a *perched water table* (figure 14.9a).

The second main mode of groundwater occurrence is in the *confined* state. Such water is also called *artesian*. Artesian water (plate 14.2) is confined beneath a relatively impermeable stratum such that, if a well penetrates the confined zone, water will rise into the well to an elevation above the land surface (figure 14.9b).

Plate 14.2 One of the most famous artesian basins is that of Australia. This is an artesian bore at Moree in New South Wales.

The low velocities and high-storage capacities of some groundwater systems, especially in rocks like sandstones, mean that the outflow of groundwater at springs and into the river channel can maintain river flows over long periods without rainfall. Groundwater flow is therefore a major component, along with some of the slower forms of deep throughflow, of what is called *base flow*. Some groundwater is extremely old and may be derived from rain that fell as long as 30 000–40 000 years ago. For example, some of the large groundwater reserves that occur in the Nubian sandstone and other aquifers under the Sahara, may have accumulated when rainfall levels were higher during pluvial episodes. It is therefore essentially fossil water, so that if it is over-exploited it may not be replenished.

Groundwater resources are coming under increasing human pressures. One result is that many aquifers, especially in arid areas, are being over-exploited (plate 14.3), with water being abstracted from them at unsustainable rates. New technologies, such as centre-pivot systems, are spreading widely (plate 14.4). Over-abstraction causes a number of serious problems (window 14.2). Yields from wells and boreholes may decline, which ultimately increases the cost of pumping and thus the price of supply. Moreover, as the water level falls land subsidence may occur. In coastal areas saltwater may intrude into aquifers to replace the exploited freshwater. Groundwater may also be polluted by seepage from sewers, landfill sites, mines and from agricultural land that has been treated with sludge, fertiliser or herbicides.

However, in some industrial areas, recent reductions in industrial activity have led to less groundwater being taken out of the ground. As a consequence, groundwater levels in such areas have begun to rise, a trend assisted by considerable leakage from ancient, deteriorating pipe and sewer systems. This is already happening in British cities including London, Liverpool and Birmingham. In London, because of a 46 per cent reduction in groundwater abstraction, the water table in the Chalk and Tertiary beds has risen by as much as 20 metres. Such a rise has numerous implications, both good and bad:

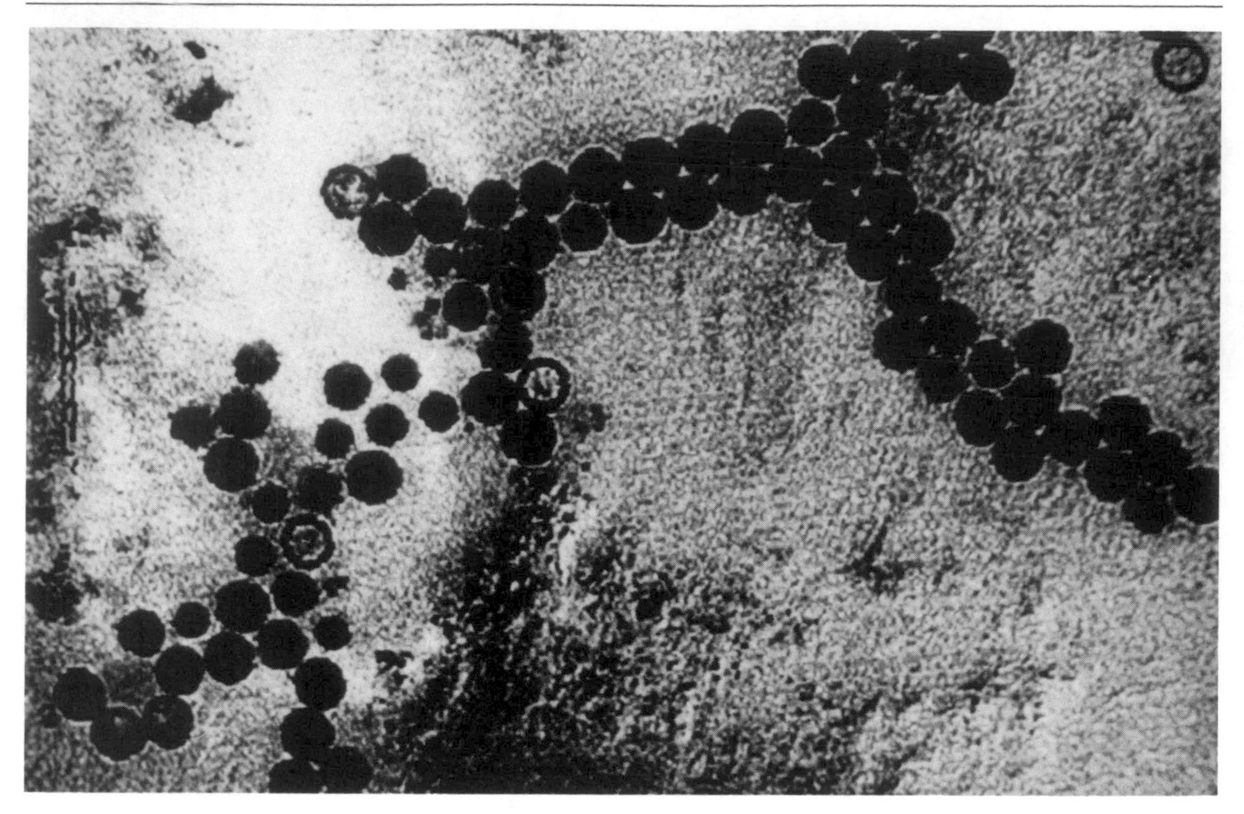

Plate 14.3 Groundwater is a major resource in many arid areas. This satellite image shows some fields in Libya which are irrigated by pumping groundwater up from great depth and then distributing it over the ground by means of a rotating sprinkler arm – the centre-pivot irrigation technique.

Window 14.2 ***Groundwater depletion in Saudi Arabia***

Most of Saudi Arabia is desert, so that climatic conditions are not favourable for rapid large-scale recharge of aquifers. Also, much of the groundwater that lies beneath the desert is a fossil resource, created during more humid conditions – pluvials – that existed in the Late Pleistocene, between 15 000 and 30 000 years ago. In spite of these inherently unfavourable circumstances, Saudi Arabia's demand for water is growing inexorably as its economy develops. In 1980 the annual demand was 2.4 billion cubic metres (bcm). By 1990 it had reached 12 bcm (a five-fold increase in just a decade), and it is expected to reach 20 bcm by 2010. Only a very small part of the demand can be met from desalination plants or surface runoff; over three-quarters of the supply is obtained from predominantly non-renewable groundwater resources. The *downdraw* on aquifers is thus enormous. It has been calculated that by 2010 the deep aquifers will contain 42 per cent less water than in 1985. Much of the water is used ineffectively and inefficiently in the agricultural sector to irrigate crops that could easily be grown in more humid regions and then imported.

Saudi Arabia is not alone in its voracious appetite for groundwater. In many parts of the world such problems have grown with increasing population levels and consumption demands, together with the adoption of new exploitation techniques (for example, the replacement of irrigation methods involving animal or human power by electric and diesel pumps).

Plate 14.4 Centre-pivot irrigation technology, employed here in the High Plains of the United States, can produce large crops in otherwise dry areas, but excessive pumping can cause rapid groundwater depletion.

- increase in spring and river flows;
- re-emergence of flow from 'dry springs';
- surface water flooding;
- pollution of surface waters and spread of underground pollution;
- flooding of basements;
- increased leakage into tunnels;
- reduction in stability of slopes and retaining walls;
- reduction in bearing capacity of foundations and piles;
- increased *hydrostatic uplift* and swelling pressures on foundations and structures;
- swelling of clays as they absorb water;
- chemical attack on building foundations.

14.8 Streamflow and the Hydrograph

If one measures the amount of runoff flowing in a stream continuously, one can then plot a graph of flow as a function of time. This is called a discharge *hydrograph*, and it expresses the sequence of relationships that occur between runoff and the other components of the basin water balance, together with their adjustments to the physical characteristics of the basin.

When a storm occurs the flow rises relatively rapidly; this portion of the hydrograph is called the *rising limb*. A point is reached, called the 'peak' or 'crest', when the quantity of water passing the gauging position has reached a maximum. Thereafter a 'recession' occurs, and the shape of the *recession limb* is controlled by the amount of water stored in the basin and the way in which it is held in the soil and the bedrock. Further rain during the recession can cause two or more peaks to occur, but if there is no further rain, discharge decreases until the extra water in storage arising from the recent rain has been depleted and the flow now approaches its original volume. This original volume, which may depend on a contribution from groundwater, is the *base flow* of the hydrograph.

The shape and size of the hydrograph are controlled by two sets of factors: those that are permanent and those that can be classed as transient. In general, the permanent factors are those involving the character of the basin, while the transient ones are associated with climate and related features (see figure 14.10). It should be remembered, however, that many of the factors are interdependent rather than independent, and that virtually all controls are related to one another in some way.

Let us then consider some of the more important controls on the shape of the flood peak shown on a

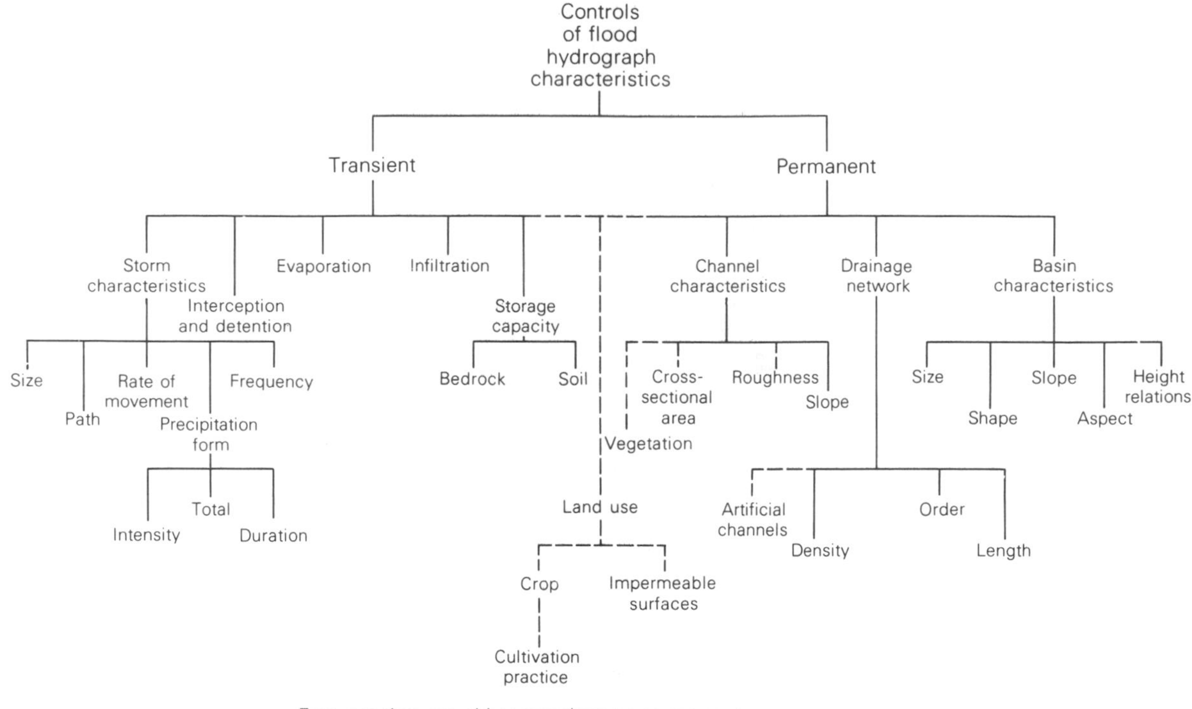

Figure 14.10 Controls of flood hydrograph characteristics.

stream hydrograph. First, other things being equal, basin area affects the size of the flood peak because this controls the amount of rainfall. Large basins receive more rain than small basins, and so have larger flood runoff discharges. However, this may not be a simple relationship; large basins, for example, tend not to have such steep slopes as smaller ones, and slope angle may control the nature of surface runoff rates.

Shape of basin is another important factor (figure 14.11a). In general, a round basin will generate a more peaked flood hydrograph than a long thin one with the same area because when a storm occurs over the basin, the distances from the edges of the round basin to the gauging station are broadly comparable, whereas in a thin basin the water from the far end will take longer to reach the gauging station than will the water from nearer to the gauging station. For the same sorts of reason basins of very complex form may have rather complex hydrograph shapes.

Basin slope is an obvious control of peak discharge (figure 14.11b); for, other things being equal, channel flow will be faster down a steep slope than down a gentle slope, and the same probably applies to throughflow in pipes and to surface runoff.

Several characteristics of the flood hydrograph hinge on the efficiency of a basin's drainage network (figure 14.11c). For example, given that channel flow is quicker than overland flow or throughflow, the less distance water has to travel to reach a channel the faster it will reach the gauging station. Thus if there is a dense network of stream channels, the rising limb of the hydrograph may be steep (figure 14. 11d).

The geology of a catchment area will also exert a fundamental influence on runoff. Especially important are the types of rock into which the basin has been eroded, and the main structural features of the area. Rock type affects the nature of the soil,

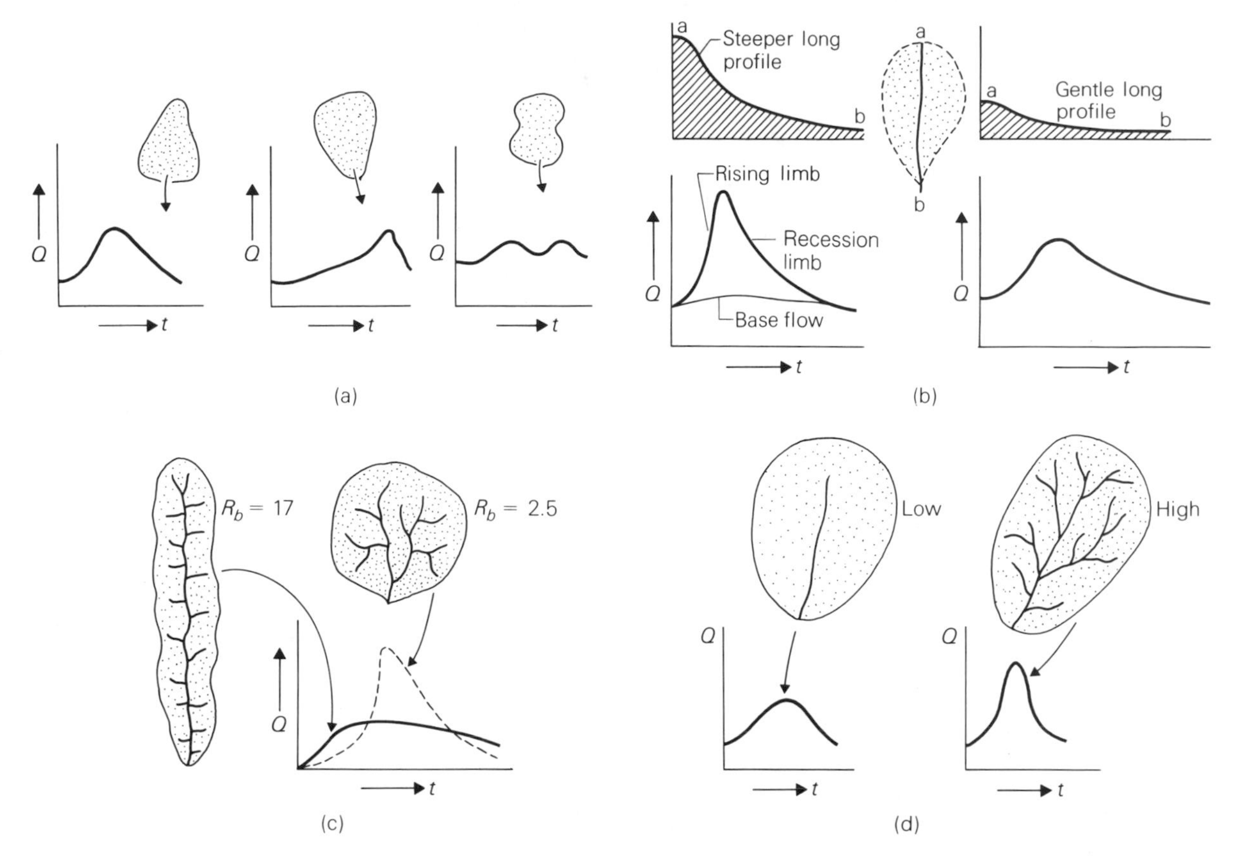

Figure 14.11 Diagrams to illustrate probable hydrograph shape in relation to some permanent characteristics of the drainage basin (*Q* is discharge, and *t* is time): (a) the shape of the basin; (b) the nature and steepness of the stream long profile; (c) the network characteristics (the long thin basin has a high bifurcation ratio (*Rb*) in comparison with the rounder basin); (d) the drainage density.

the texture of the stream network, the nature of groundwater and its speed of movement. Rock structure is largely important as a factor guiding the movement of groundwater towards the streams. Thus it is probable that the time lag between rainfall and groundwater runoff peaks will be smaller in the case of a synclinal catchment, where the rocks dip towards the channel, than in the case of a catchment with horizontally bedded strata (figure 14.12).

Soil type, through its influence on infiltration rates and the speed of throughflow, is plainly very important. So is vegetation cover; the amount of vegetation affects the transpiration rate, the amount of interception, the ease with which water can flow across a slope, the soil structure and so on. It is through our influence on vegetation and soil that we may greatly modify the natural runoff characteristics of a catchment.

With regard to the more transient factors, meteorological conditions are of great significance. For example, if the precipitation in a catchment falls as snow rather than rain, the hydrograph will be modified. A blanket of snow has a storage effect, so for some parts of the year its presence will lead to relatively low discharges given the amount of precipitation, whereas in the season of spring meltwater will be released very rapidly into streams. Moreover, if a blanket of snow is present and a

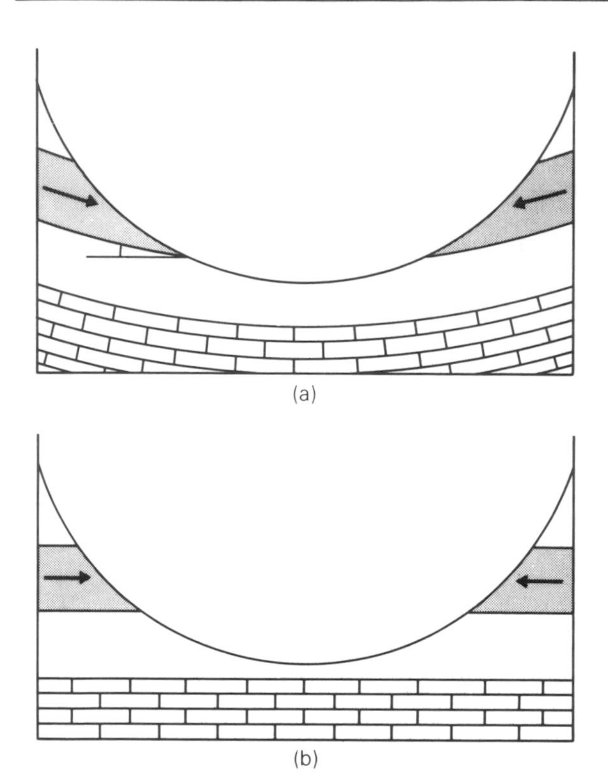

Figure 14.12 Groundwater will tend to move more speedily in synclinal strata (a) than in horizontal bedded strata (b).

rainstorm occurs, the snow may absorb the rain, only to release it when melting occurs. The intensity of rainfall is one of the most important factors in determining the proportions of the rainfall that go to surface and groundwater runoff. Thus heavy rain may exceed the infiltration capacity of the soil, so that it runs off rapidly, while rain falling at lower intensities will be largely absorbed by the soil, so that its addition to streamflow will probably be much delayed. The duration of rainfall is also important, for the infiltration capacity of a soil often decreases through time, while the size of the saturated wedge may increase in time. Therefore, the longer the rain continues, the smaller the infiltration capacity will become and the larger the contributing portion of the basin.

Land-use changes

That land-use changes could affect the nature of streamflow has been known for centuries. Studies of the torrents of the French and Austrian Alps in the late eighteenth and early nineteenth centuries showed that the increase in flooding was the result of wholesale felling of forests. The first experimental study in which planned land-use change was executed to enable observation of the effects on streamflow began at Wagon Wheel Gap, Colorado, in 1910. Here streamflows from two similar watersheds of about 80 ha each were compared for eight years. One valley was then clear-felled and the records were continued. After the clear-felling the annual water yield was increased by 17 per cent above that predicted from the flows of the unchanged control valley.

As regeneration of vegetation takes place after forest has been cut or burned, so streamflow tends to revert to normal, though the process may take some decades. This is illustrated in figure 14.13, which shows the dramatic effects produced on the Coweeta catchments in North Carolina by two spasms of clear-felling together with the gradual return to normality in between. The substitution of one forest type for another may also have effects on streamflow. This can be exemplified from the same Coweeta catchments. Two experimental catchments were converted from a mature decidu-

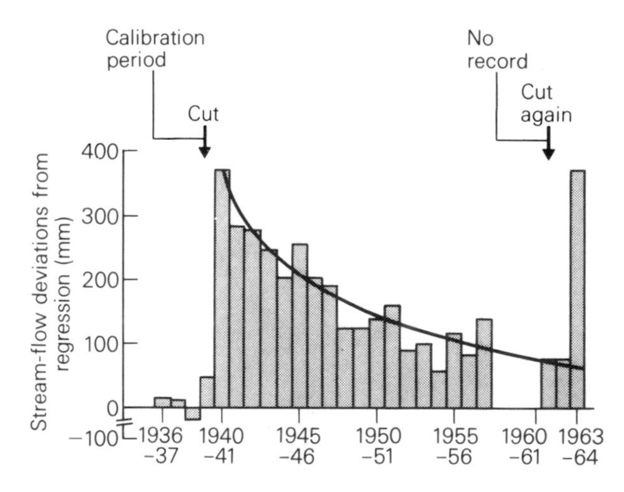

Figure 14.13 The increase of water yield after clear-felling a forest: a unique confirmation from the Coweeta catchment in North Carolina.

ous hardwood forest cover to a cover of white pine (*Pinus strobus*). Fifteen years after the conversion took place annual streamflow was found to be reduced by about 20 per cent. The reason for this notable change is that the interception and subsequent evaporation of rainfall is greater for pine than it is for hardwoods during the dormant season (window 14.3).

Changes in river bank vegetation may have a particularly strong influence on riverflow. In the south-western United States, for example, many streams are lined by the salt cedar (*Tamarix pentandra*). The roots of the shrub abstract large quantities of water from the water table or the capillary fringe just above. It was introduced by man to help combat bank erosion, but it spread unexpectedly explosively, and in the Upper Rio Grande Valley in New Mexico it came to consume approximately 45 per cent of the area's total available water.

Window 14.3 ***Forests and runoff***

There are many reasons why the removal of a forest cover and its replacement with pasture, crops or bare ground have such important effects on streamflow. A mature forest probably intercepts a higher proportion of rainfall, tends to reduce rates of overland flow, and promotes soils with a higher *infiltration* capacity and better general structure. All these factors will tend to produce both a reduction in overall runoff levels and less extreme flood peaks, though this is not invariably the case.

Reforestation of abandoned farmlands reverses the effects of deforestation: increased interception of rainfall and higher levels of evapotranspiration can cause a decline in water yield to rivers. This can cause problems for human activities.

Reviews of catchment experiments from many parts of the world have pointed to two conclusions:

- Pine and eucalypt forest types cause an average change of 40 mm in annual flow for a 10 per cent change in cover with respect to grasslands; that is, a 10 per cent increase in forest cover on grassland will decrease annual flow by 40 mm, and a 10 per cent decrease in cover will increase annual flow by the same amount.
- The equivalent effect on annual flow of a 10 per cent change in cover of deciduous hardwood or scrub is 10–25 mm: that is, if 10 per cent of a grassland catchment is converted to hardwood trees or scrub vegetation, the annual runoff will decrease by 10–25 mm.

The increase in annual flow that results from tree or scrub removal tends to be most marked in two particular environments: those with very high rainfall and those with very low rainfall. In the former, evaporation from forest will tend to be higher than that from other land covers because of high levels of rainfall interception. In the latter, evaporation from forest is likely to be higher than from other land covers because forests, composed of trees that have deep root systems, are better able to make use of soil and groundwater reserves.

Having discussed changes in annual flows, now let us turn to a consideration of how forest removal influences low-season flows and flood peaks. The higher losses from forests in wet seasons from rainfall interception and increased losses in dry seasons from transpiration (because of the deep root systems of trees) both tend to increase soil-moisture deficits in dry seasons compared to those under other land uses. On the other hand, in forests at high altitudes, where there is a lot of water deposition on to trees from clouds, this may provide a significant component of the dry-season flows into rivers and also increase runoff. The same applies in areas with high-intensity storms where high-intensity rainfall may lead to high levels of surface runoff. The higher infiltration rates under indigenous forest compared with other land uses may help soils and their below-ground aquifers to recharge themselves. In steeply sloped areas forests may have the additional benefits of reducing landslips and preserving the soil aquifer which may be the source of dry-season flows. Both these effects of afforestation may therefore benefit stream flows in the low season.

When it comes to flood peaks there is still a great deal of controversy as to how important forest cover is with respect to the largest types of event. Some authors suggest that management practices associated with forestry (e.g. the building of roads, culverts and drainage ditches) or subsequent activities (e.g. grazing) which promote the flood, by causing compaction of the soil and reducing its infiltration capacity, increase this type of hazard.

Summary of the effects of some land-use changes on stream runoff

Land-use change	*Hydrological component affected*	*Principal hydrological process involved*
Afforestation (deforestation has the opposite effects in general)	Annual flow	Increased interception Increased transpiration in dry periods
	Seasonal flow	Increased interception and increased dry-period transpiration reduce dry-season flow Drainage improvements associated with planting may increase dry-season flows Cloud water (mist and fog) deposition on trees will augment dry-season flows
	Floods	Interception reduces floods by removing a portion of the storm rainfall, and allowing soil moisture storage to increase
Agricultural intensification	Water quantity	Alteration of transpiration rates affects runoff Timing of storm runoff altered through land drainage
Draining wetlands	Seasonal flow	Lowering of water table may induce soil-moisture stress, reduce transpiration and increase dry-season flows Initial dewatering on drainage will increase dry-season flows
	Annual flow	Initial dewatering on drainage will increase annual flow Afforestation after drainage will reduce annual flow
	Floods	Drainage method, soil type and channel improvement will all affect flood response

At present, *field drainage* is spreading to wide areas of Britain and Europe to increase crop productivity. Ditches are dug and tile drains are installed. This process may well affect runoff, though the exact nature and direction of the effect has been subject to some debate. On the one hand, the installation of ditches increases the drainage density in any area, which should facilitate rapid runoff for the reason explained in section 14.6. On the other hand, effective drainage may greatly reduce the area of saturated land in a catchment, thereby reducing the amount of runoff generated by returning flow and saturation overland flow.

14.9 Annual River Regimes

In the above section we have seen why the flow of streams varies in response to individual precipitation events. It is also, however, possible to consider why the annual regimes of rivers vary in different parts of the world (figure 14.14).

Some rivers have *simple regimes* because the climate provides one major peak of flow. One such environment is the glacial environment, where maximum flow occurs in the summer when glacial ablation is at a maximum. Low flows occur in winter because most of the winter precipitation is in the form of snow. Another simple regime is that of some tropical rivers, which are influenced by

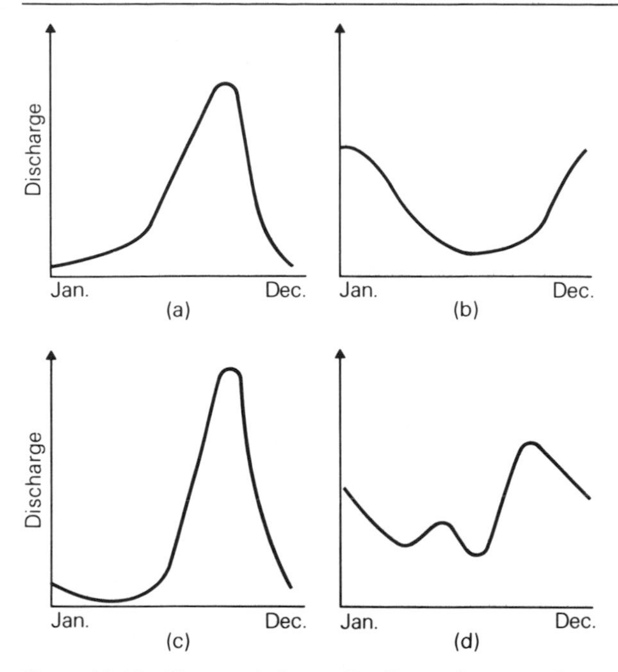

Figure 14.14 Characteristic runoff regimes of some major world zones in the Northern Hemisphere: (a) a regime dominated by snow and ice melt, as would occur in an area of high altitude or in high latitudes; (b) the mid-latitude oceanic regime with low summer discharge levels caused by high rates of evapotranspiration; (c) the tropical monsoonal type of regime dominated by high rainfall inputs during the summer and a long winter dry season; (d) the complex regime of an equatorial river, with rainfall throughout the year, but with two clear peaks.

one relatively short summer wet season. The Blue Nile is such an example. In areas with a Mediterranean climate the regime may also be simple but will be the reverse of that just described, for most of the precipitation occurs in the winter months, so that there is very little flow in the hot, dry summers.

Other rivers, however, have more *complex regimes*. For example, in some Mediterranean areas there is an autumn maximum associated with the start of the rainy season, and then a spring maximum associated with snowmelt from the higher areas of the catchment. In some equatorial regions there are two peaks associated with two peaks of rainfall. Another cause of complex regimes is a river being so large that it has tributaries from a variety of different climatic zones. The Nile, for example, crosses several climatic zones – the equatorial lands of central Africa, the monsoonal lands of Ethiopia, the extreme desert of the Sahara and the Mediterranean climate of the Egyptian coastal zone. Furthermore, the timing of the peaks of flow in the Nile are affected by the role that the East African lakes play in regulating the flow of the White Nile, and the role that the great Sudd swamps play in delaying the progress of the flood through the Sudan.

14.10 Lakes

As has been suggested, in many parts of the world, such as East Africa, the hydrological cycle is complicated by the presence of lakes, which act as stores within the system which may regulate peaks in discharge. They may also provide surfaces from which substantial evaporation can occur. Some lakes provide the base level for a drainage basin instead of the sea. This is particularly the case in some arid areas where large depressions like Death Valley (California), the Dead Sea and the Caspian Sea lie below sea level.

The basins occupied by lakes show not only a wide range of dimensions but also a wide range of origins (window 14.4). Some of these origins are listed in table 14.3.

Window 14.4 *The Great Lakes of North America*

The Great Lakes of North America, which lie astride the border of Canada and the USA, consists of five main water bodies – Superior, Michigan, Huron, Erie and Ontario – which have a combined area of almost 0.25 million km^2. However, about 10 000 years ago, as the great Laurentide ice sheet waned, there were even greater bodies of water to the south of the ice front, and the largest of these, called Lake Agassiz, attained at its maximum an area of 350 000 km^2.

Why did such large freshwater bodies develop? The answer lies in the history of ice sheet development during the Quaternary, and in the various factors that have combined to give the basins in which the lakes are located. One of these is erosion, for the ice sheet and its various lobes scoured and deepened areas of relatively erodible rock, especially shales. A second major factor is the presence of a large structural and topographic basin in central Canada, part of which now contains Hudson Bay. Thus, whenever a large ice sheet has accumulated across the high latitudes of central Canada, drainage towards the Hudson Bay has been impeded, and so lakes have tended to form. The presence of a great ice sheet has also caused substantial isostatic warping, and this also has served to influence the direction and pattern of drainage. Plainly, however, a crucial influence has been the position of the ice front with respect to the watershed between southward-flowing drainage (which flows into the Mississippi system) and the northward- and eastward-flowing system (which flows towards Hudson Bay and the Atlantic).

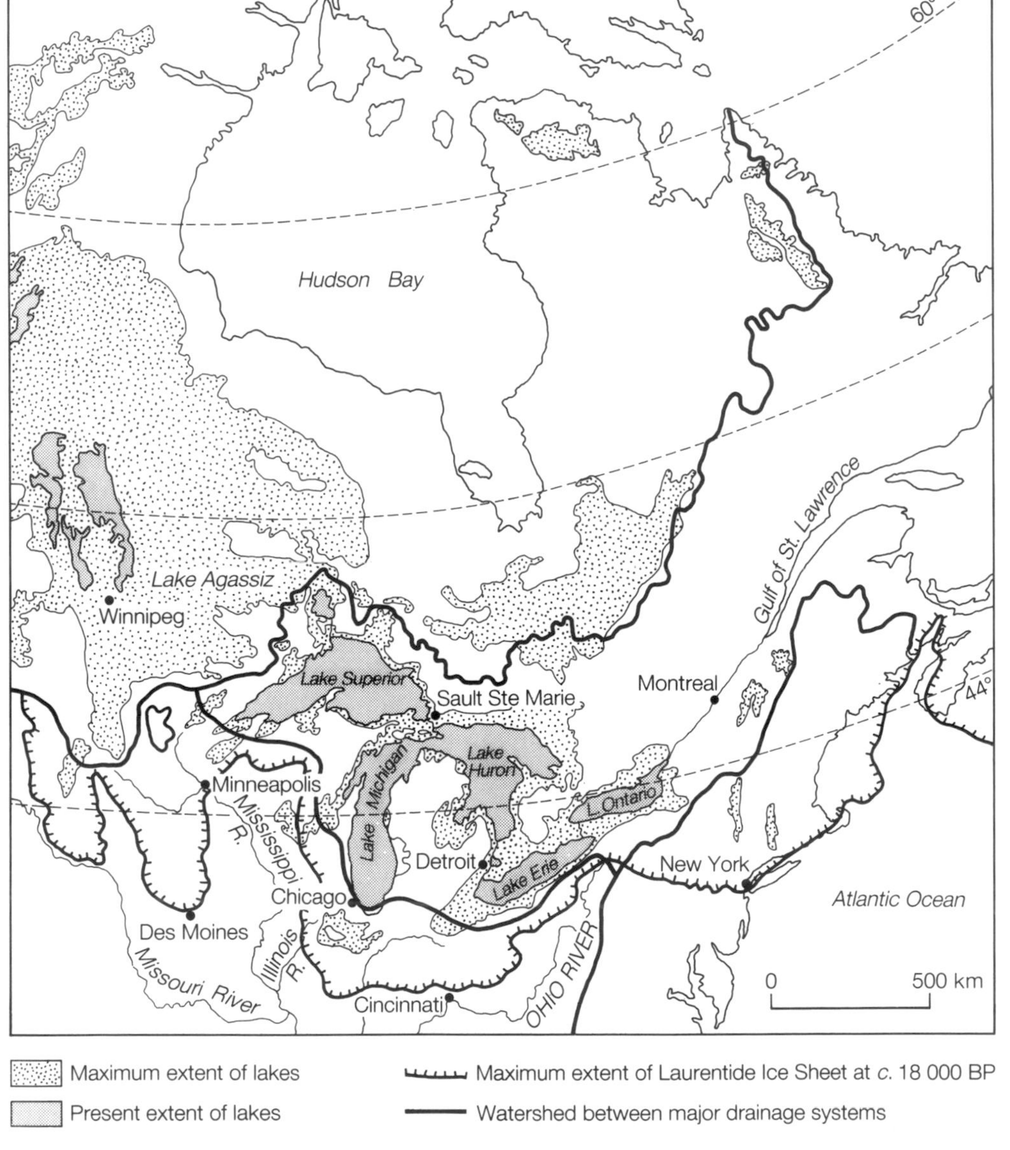

The extent of glaciation in eastern North America at around 18 000 years ago, and the extent of lakes in relation to the major drainage divides.

The Niagara Falls between Lake Erie and Lake Ontario. During the ice advances and retreats of the Quaternary the drainage systems of large parts of North America were transformed. A crucial factor was the position of the ice front with respect to the southward-flowing Mississippi system and the north-eastwards-flowing St Lawrence system, of which the Niagara Falls are a part.

Table 14.3 The origins of lake basins

Mechanism	*Example*
Tectonic activity	Rift valley lakes in East Africa
Volcanic	Valleys blocked by lava flows, or crater and caldera lakes (e.g. the Eifel region, Germany)
Mass movements	Valleys blocked by landslides
Solution of limestone and other susceptible rocks	Many of the lakes of Florida, the Breckland meres of East Anglia and the flashes of the Cheshire salt lakes
Fluvial	Oxbow lakes etc.
Glacial action	Basins created by glacial excavation or blocked by morainic debris (e.g. the tarns of the English Lake District or the Italian lakes)
Shoreline processes	Lakes formed behind barrier beaches etc.
Wind action	Basins created by deflational activity or by the blocking of river systems by dune movement (e.g. the pans of South Africa, the playas of the High Plains in the USA and the lakes in the Sandhills of Nebraska)
Organic processes	Basins dammed by such processes as beaver activity
Meteorite showers	Impact craters (e.g. Lonar Lake, India)
Man-made	Reservoirs

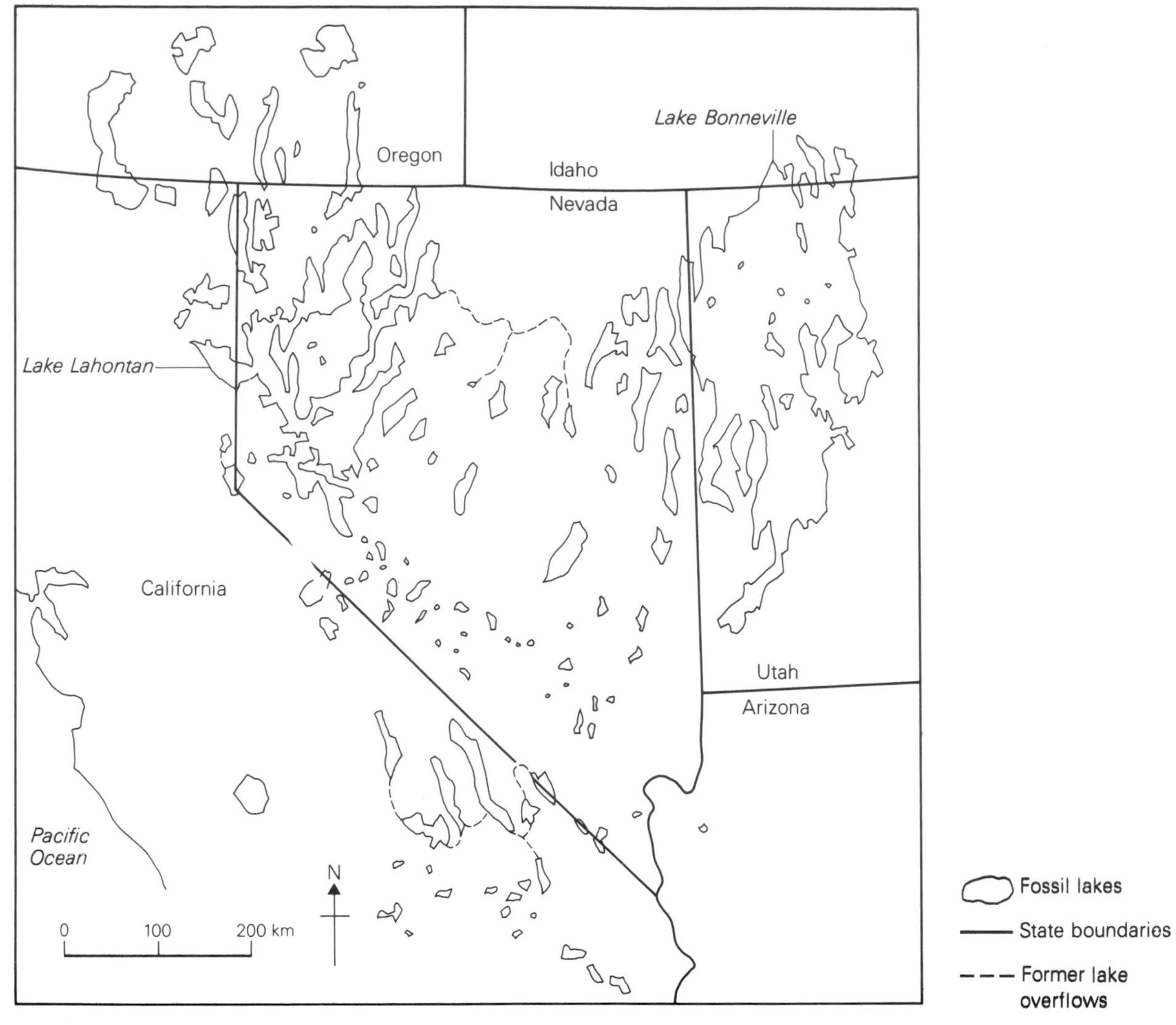

Figure 14.15 The great pluvial lake basins of the south-west USA. Compare this map with a map of present-day lakes in the area from your atlas.

Most lake basins are transient features of the landscape. Excessive sedimentation will fill in the basins, or erosion of their outlets will gradually cause them to be drained. They also respond very greatly to climate change, and many of the great basins of the south-western United States, which are now dry or saline, were, in pluvial epochs, full of relatively fresh water (figure 14.15). Some of these, which are today closed sumps, were then linked into a series of interconnecting basins. Lakes have also responded to more recent climatic changes as the water budgets have responded to changes in rainfall inputs and evapotranspirational outputs (which are largely controlled by temperature).

The Great Salt Lake in Utah has an especially impressive record of fluctuating levels dating back to the middle of the nineteenth century. As figure 14.16 shows, the lake rose from an elevation of about 1280 m in 1851 to a peak of around 1283 m in 1873. Thereafter it declined very markedly, reaching the lowest recorded level in 1963. Since then it has risen from around 1278 m in elevation back to

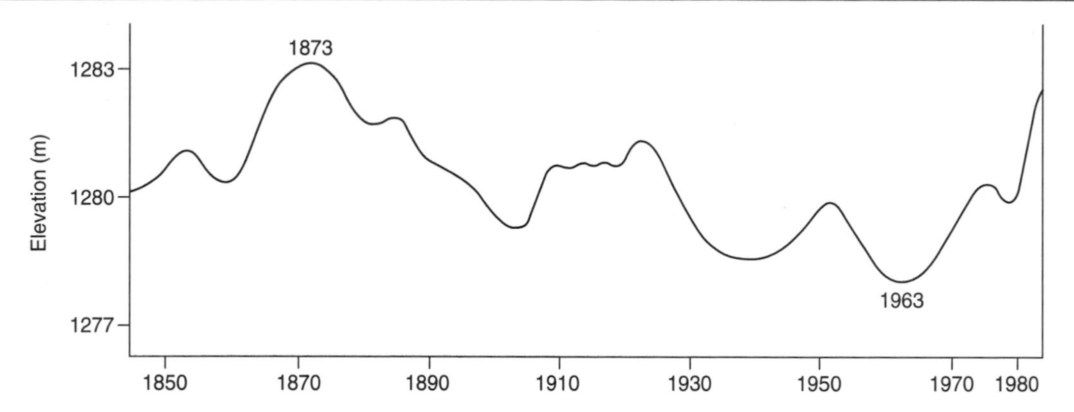

Figure 14.16 The changing level of the Great Salt Lake, USA, 1851–1984.

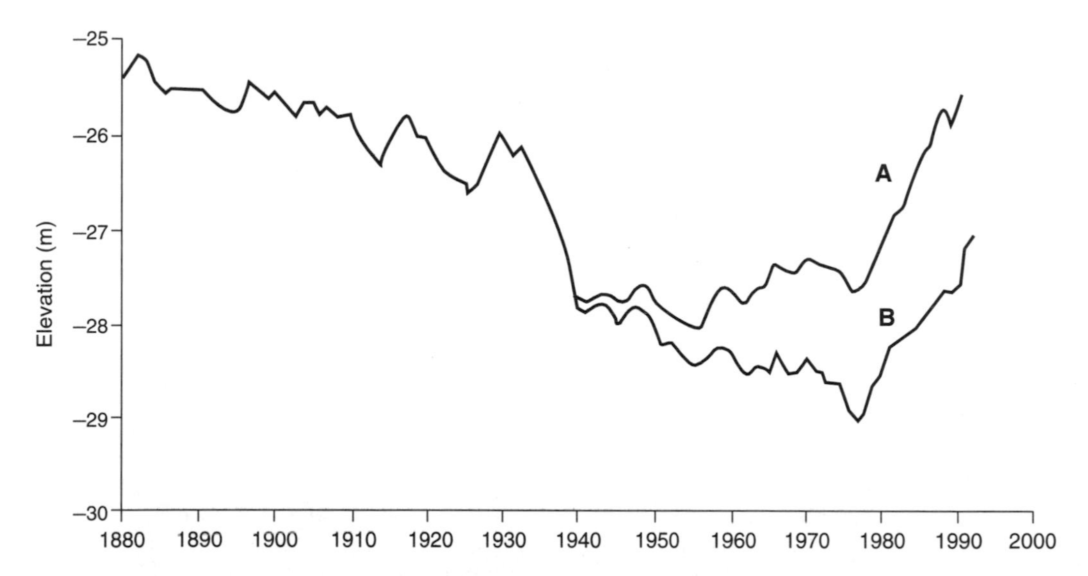

Figure 14.17 Annual fluctuations in the level of the Caspian Sea for the period 1880–1993. Curve (A) shows the changes in level which would have occurred but for anthropogenic influences; while curve (B) shows the actual observed levels.

the sort of elevation achieved in the 1870s. There has thus been a total fluctuation of this great water body of the order of 5 m over that time period. The latest rise is a consequence of a run of very wet years since the mid-1970s.

Humans have a range of impacts on lake basins, which include eutrophication because of the addition of nutrients to their catchments, acidification of acid precipitation or land-use changes, contamination of chemical pollutants and heavy metals, explosive invasion by exotic introduced organisms, siltation because of land-use and land-cover changes, and changes in their water balance because of such processes as inter-basin water transfers or changes in land use.

Even the world's largest lake, the Caspian, has been modified by human activity. The most important change was the fall of 3 m in its level between 1929 and the late 1970s (see figure 14.17). This decline was undoubtedly partly the product of climatic change, for winter precipitation in the northern Volga Basin, the chief flow-generating area of the Caspian, was generally below normal for the period because of a reduction in the number of moist cyclones penetrating into the Volga Basin from the Atlantic. None the less, human actions have contributed to this fall, particularly since the early 1950s because of reservoir formation, irrigation, municipal and industrial withdrawals and agricultural practices.

Since the 1970s, the Caspian has seen a restoration in its levels caused by a decrease in the difference between evaporation and precipitation over its catchment. But for anthropogenic effects its level would have returned to pre-1930 levels.

Perhaps the most severe change to a major inland sea is that taking place in the Aral Sea of the CIS (see window 6.5). Since 1960 the Aral Sea has lost more than 40 per cent of its area and about 60 per cent of its volume, and its level has fallen by more than 14 m. This has lowered the artesian water table over a band of 80–170 km in width, has exposed 24 000 km^2 of former lakebed to desiccation, and has created salty surfaces from which salts are deflated to be transported in dust storms to the detriment of soil quality. The mineral content of what remains has increased almost three-fold over the same period. It is probably the most dire ecological tragedy to have affected the CIS, and, as with the Caspian's decline, much of the blame rests on excessive use of water which would otherwise replenish the sea.

■ *Key Terms and Concepts*

aquifer
Bergeron–Findeisen process
collision–coalescence process
Darcy's law
evapotranspiration
hydrograph
infiltration
lapse rate
overland flow
throughflow

■ *Points for Review*

- Why does it rain?
- What do you understand by 'evaporation' and 'potential evapotranspiration'?
- Under what circumstances does runoff occur on slopes?
- Where would you expect to find the best groundwater resources?
- What causes some streams to have highly peaked hydrographs?
- In what ways do human activities affect the amounts of water in (a) rivers and (b) lakes?

FURTHER READING

Downing, R. A. and Wilkinson, W. B. (eds) (1991) *Applied Groundwater Hydrology: A British Perspective* (Oxford: Clarendon Press). An advanced textbook on all aspects of groundwater in the British context.

Gordon, N. D., McMahon, T. A. and Finlayson, B. L. (1992) *Stream Hydrology: An Introduction for Ecologists* (Chichester: Wiley). A wide-ranging survey of hydrology and its importance.

Jones, J. A. A. (1997) *Global Hydrology: Processes, Resources and Environmental Management* (Harlow: Longman). A modern view of hydrology from a global perspective.

Newson, M. D. (1994) *Hydrology and the River Environment* (Oxford: Clarendon Press). An introductory text.

Price, M. (1996) *Introducing Groundwater*, 2nd edn (London: Chapman and Hall). A useful introduction to underground water.

Shaw, E. H. (1994) *Hydrology in Practice* (London: Chapman and Hall). A good survey with an engineering perspective.

Summer, G. (1996) 'Precipitation weather', *Geography*, 81 (4): 327–45. Two review articles in a very important and accessible journal, specifically aimed at students and their teachers.

Ward, R. C. and Robinson, M. (1990) *Principles of Hydrology*, 3rd edn (London: McGraw Hill). The standard text.

Wilby, R. L. (1997) *Contemporary Hydrology* (Chichester: Wiley). A modern examination of the environmental significance of water.

15 Rivers

15.1 Introduction

Rivers are a most important component both of our environment and of the landscape, as sources of water, transporters of sediment, causes of erosion and flooding and routeways for navigation. In this chapter we will first of all examine the form of river basins, the nature of river long profiles and the main features of channel cross-sectional forms and patterns. We will then consider certain particular river environments – floodplains, terraces and deltas – before turning to the two aspects of river basins of special significance for people: the transport of sediment and the problem of floods. We will conclude with a brief discussion of human impacts on river systems.

15.2 Morphometry of Drainage Basins

Morphometry is the measurement of shape, and it is one of the most important types of quantification in hydrology and geomorphology. The great virtue of measuring shape, as opposed to simply describing it in words, is the precision that can be gained; one can compare and contrast different river basins, undertake correlations between different variables in the drainage basin and identify basins that are different from the norm. This was appreciated by one of the founders of modern morphometric studies in geomorphology, R. E. Horton.

One of the first aims of fluvial morphometry was to establish a hierarchy of streams ranked according to *order*. Many systems have been developed, but that of one of Horton's followers, A. N. Strahler, is objective and straightforward. According to Strahler's system, fingertip tributaries at the head of a stream system are designated as first-order streams, two first-order streams join to form a second-order stream, two second-order streams form a third-order stream, and so on (figure 15.1a). It takes at least two streams of any given order to form a stream of the next higher order. Using stream-ordering systems of this type, it became apparent to Horton that there was an orderly arrangement of streams of different orders within a drainage basin, and this led him to propose a group of *laws of drainage composition*. He noted, for example, that in most basins the number of streams of different orders decreases with increasing order in a regular way. Thus if one plots a logarithm of the number of streams of a given order against the order, the points lie on a straight line. This is known as the *law of stream numbers* (figure 15.1b). Similarly, if the logarithms of the mean lengths of the stream segments of different orders are plotted against stream orders, the result is usually a more or less straight line. This is the *law of stream lengths*. The *law of basin areas* follows the same general pattern. In a drainage network the mean basin areas of the orders approximate to a direct geometric sequence in which the first term is the average length of a first-order basin.

In addition to stream ordering, various aspects of drainage network form have been found to be

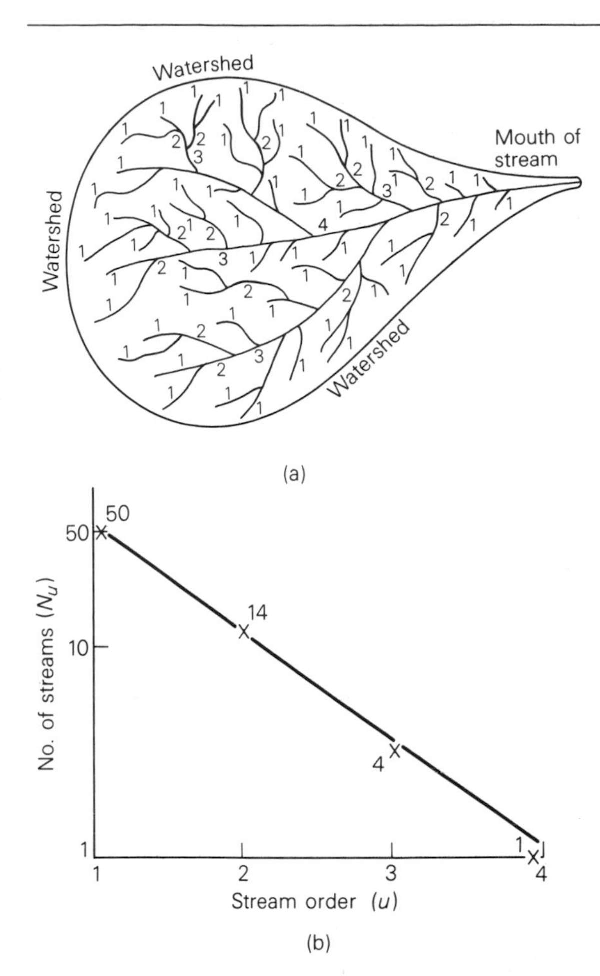

Figure 15.1 Drainage basin morphometry: (a) the Strahler system of stream ordering applied to a small drainage basin; (b) the relationship between the numbers of streams of different orders as determined for the basin above.

useful. For example, *drainage density*, which is a measure of the length of stream channel per unit area of basin, gives a useful quantitative measure of landscape dissection. This measurement shows a very great range in different areas, and although some of this may be due to the differences in portrayal of streams on maps of different scales, much of it is real, and can be accounted for by varying conditions in different catchments. On highly permeable surfaces with very little potential for the generation of runoff, drainage densities may be less than 1 km km^{-2}. At the other extreme, in areas of extreme erosion on impermeable but weak rocks – badlands – densities of over 700 km km^{-2} have been reported. In Britain most drainage densities are around 2–8 km km^{-2}, but there is considerable variation according to climatic condition and rock type. In addition to the nature of the rocks and soils in a basin, precipitation characteristics are an important control of drainage density, as they affect the extent of the vegetation cover, and thus the erodibility of the ground surface. Figure 15.2 (b) and (c) suggests that the highest drainage densities are in areas where effective precipitation is low.

15.3 Drainage Basin Patterns

Although drainage density is one of the significant determinants of the nature of a river basin, it is also important to remember that the pattern of the drainage network may also be subject to considerable variety. Some characteristic examples are shown in figure 15.3.

The least regular patterns are called *deranged* and are found in areas where the drainage network has been modified by a process like glacial erosion and deposition. A look at a map of large parts of Finland will demonstrate this type. Most drainage patterns are said to be *dendritic*, and such systems are characteristic of well-adjusted rivers that have been developed on relatively uniform materials. Departures from this 'normal' type occur where there is strict structural control. For example, *annular* patterns may develop on domes or anticlines, *trellised* patterns may develop where a gently sloping surface cuts across rocks of different type, and *rectangular* patterns may evolve where faults or joints guide stream channels. Slope may be another important control. For instance, on very steep slopes (as for example, on the side of an escarpment, the side of a mine spoil tip or the side of a deep tidal creek) *parallel* patterns may develop. Finally, although most drainage systems flow into the oceans or into each other, *endoreic* or *centripetal* systems develop in certain arid areas because of such processes as deflation and dune blocking.

It also needs to be appreciated that drainage patterns may be influenced by their long-term evolu-

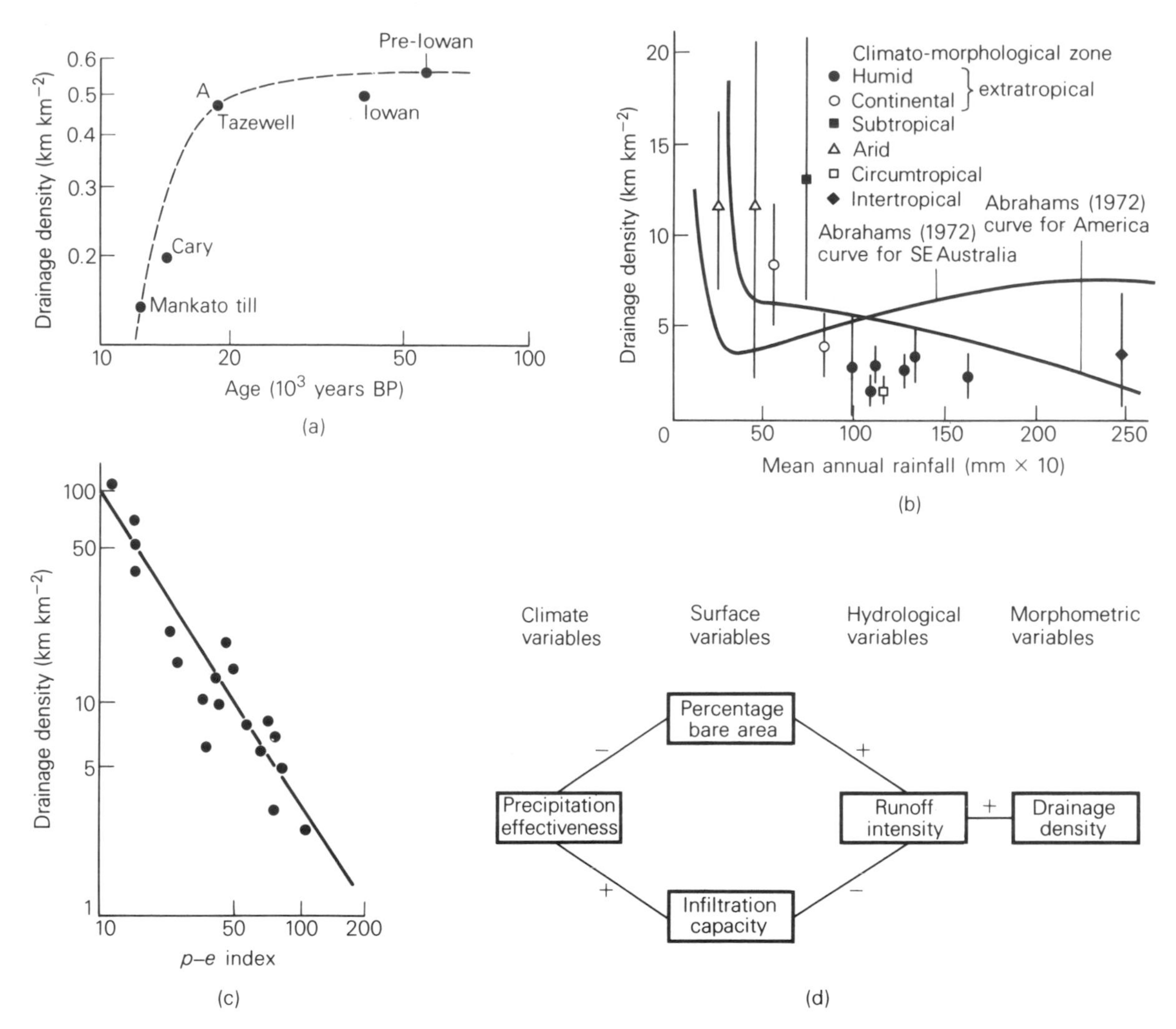

Figure 15.2 Drainage density and its controls. (a) Drainage density is greater on surfaces of greater age, as inferred from measurements on tills of varying age from North America, but an equilibrium value appears to be attained after about 20 000 years. (b) Global variation of mean annual precipitation, with each line representing the range of values represented by a particular data source. (c) Drainage density as a function of precipitation effectiveness, derived by subtracting evapotranspiration losses (*e*) from precipitation (*p*). (d) A correlation structure relating climate, surface, hydrological and morphometric variables.

tionary history. This is particularly the case where there appears to be a striking lack of correspondence between drainage pattern and geological structures. As we saw in section 11.7, antecedence and superimposition may be the cause of such anomalies.

15.4 Concave Long Profiles

The great majority of rivers have concave long profiles (figure 15.4), and through time there seems to be a trend towards a removal of irregularities and the adoption of a smooth long profile with a gradient that becomes progressively less as one moves downstream. There are various factors that, work-

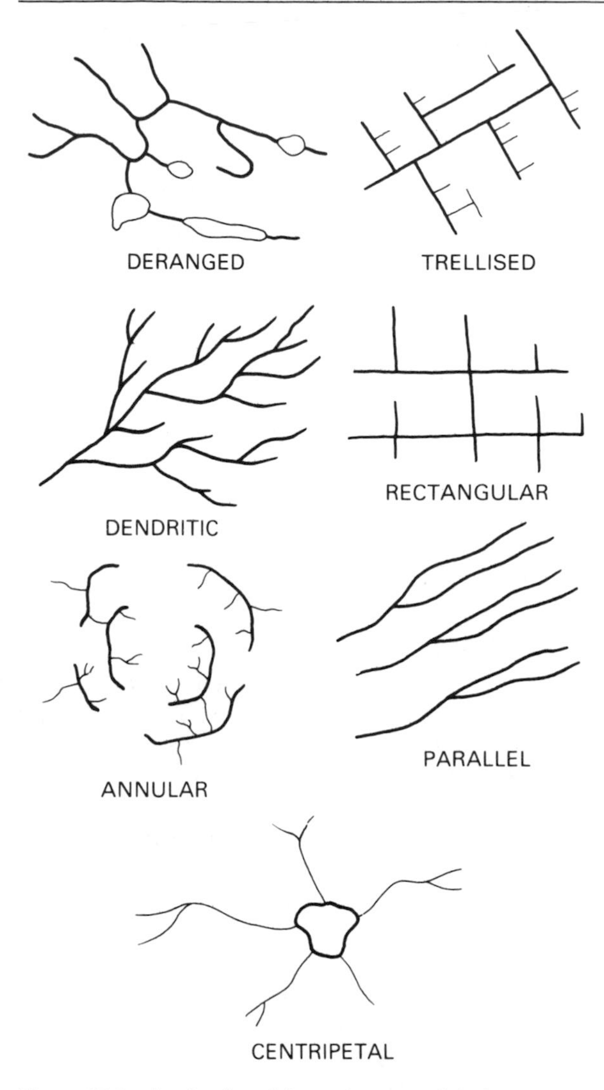

Figure 15.3 A selection of the main types of drainage pattern.

ing in combination, serve to explain this general tendency.

First, the *calibre* of a stream's load tends to decrease, partly because it is the finest materials that can be moved furthest, and partly because material becomes ground down by abrasion or attrition the further it travels. The coarse material of the upper tracts of a stream gradually gives way to finer material downstream, and the river can transport such finer material over a progressively gentler slope. In its upper tract, it can be argued, the gradient is steep in order to maintain the velocity and competence to move the coarse debris that is supplied to it there, while downstream the river can flatten its gradient and still transport the load.

Second, although the total amount of load to be transported increases in a downstream direction, the river discharge itself increases steadily as its

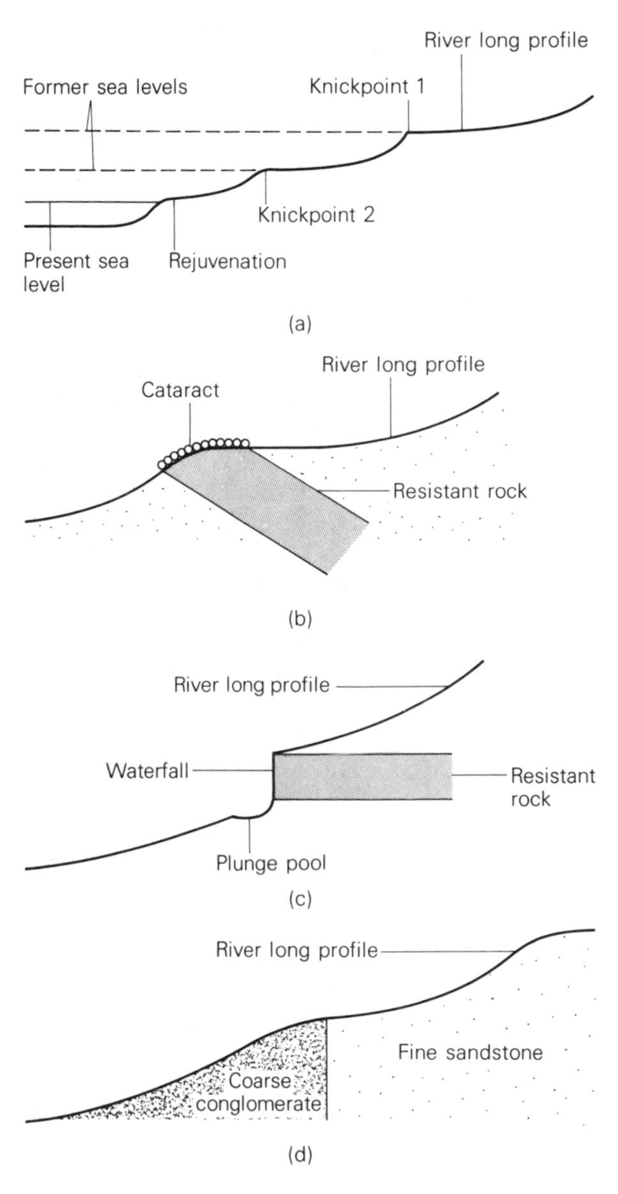

Figure 15.4 Various reasons why river long profiles may show irregularities. For explanation see text.

catchment area is increased by the entrance of more and more tributaries. The cross-sectional area of the channel therefore grows larger, and the *hydraulic radius* (the area of water per unit of channel–water contact) will also increase. Thus the river becomes more efficient, and as a result has greater 'spare' energy to devote to load transport. Therefore, although the overall amount of load to be carried increases downstream, it can be transported by an ever more efficient stream across gentler and gentler slopes towards the mouth. Moreover, contrary to what one might expect if one looked at the turbulent streams of the upper tract in comparison with the smoother streams of the lower tract, the mean velocity of streams tends either to remain constant or to increase slightly downstream. This is because the roughness of the bed decreases downstream as the bed material becomes finer, and once again the increasing depth of water makes the channel less prone to the effects of bed roughness.

The smooth concave curve of a river's long profile may not always exist (figure 15.4). For example, although there may be a general tendency for the grain size of bed material in a stream to decrease downstream, if the stream cuts across a particularly resistant rock in its lower course it may be loaded with some coarse debris, so that its long profile then steepens (figure 15.4d). The same resistant bedrock may cause a waterfall or cataract to form by limiting stream incision (figure 15.4c). Some rivers may contain insufficient abrasive material in their bed load to be able to cut down into underlying rocks, and this has been suggested as one of the reasons why some of the great tropical rivers have long profiles with great falls and cataracts (figure 15.4b). Extreme weathering under tropical conditions has furnished the rivers with a load that is largely silt and clay size. Alternatively, some rivers may have irregularities in their long profiles because of changes in the base level to which they have to cut down. So, for example, if sea level falls (figure 15.4a), a river may cut down to adjust to the new level. This rejuvenation will, it is argued, first affect the mouth of the stream, so that the steep new lower section will undergo headward recession as a *knickpoint*.

15.5 Channel Cross-section and Hydraulic Geometry

The size and shape of a river's cross-section varies according to a large range of factors. One of the prime controls is discharge, and studies of streams by American geologists like L. B. Leopold and T. Maddock have established some of the relationships between channel width, depth, discharge and velocity levels. Width, depth and velocity all increase with increasing discharge levels, though in general width seems to change more rapidly than depth in a downstream direction.

The material of which the river's banks are composed is also highly significant. If the banks are composed of unstable, non-cohesive coarse material, the stream will tend to be wide relative to its depth; whereas banks with high contents of silt and clay are cohesive and thus tend to resist erosion, promoting the development of deeper and narrower cross-sections.

The investigation of the geometric relationships of channel cross-sections pioneered by Leopold and Maddock is known as *hydraulic geometry*. They were able to show the interrelationships, both in time and space, between width, depth, velocity, channel roughness, channel curvature and discharge.

If you plot the mean depth of a stream against its discharge for several different times at a point in the channel, you will find that there will be an increasing depth with increasing discharge. Stream width and velocity will also rise. If the data for these three variables are plotted on logarithmic graph paper, the points are found to lie close to a straight line, indicating that each tends to be a power function of discharge. One may therefore write three equations:

$$w = aQ^b$$

$$d = cQ^f$$

$$v = kQ^m$$

where Q is discharge, w is width, d is mean depth and v is mean velocity. The other terms, a, b, c, f, k and m, are empirical constants.

It is important to understand channel shape in order to be able to identify the controls of water

Table 15.1 Estimation of Manning's channel roughness coefficient

		Manning's coefficient values for bed-material size		
Bed profile	*Vegetation (tree roots, aquatic weeds etc.)*	*Sand and gravel*	*Coarse gravels*	*Boulders*
Uniform	None	0.020	0.030	0.050
Undulating[a]	None	0.030	0.040	0.055
Uniform	Some	0.040	0.050	0.060
Undulating[a]	Some	0.050	0.060	0.070
Highly irregular	None	0.055	0.070	0.080
Highly irregular	Extensive	0.080	0.090	0.100

[a] Undulating = pools and riffles well developed
Source: Modified after G. E. Petts (1983) *Rivers* (London: Butterworth), table 3.4

flow in channels. The rate at which water flows in any particular channel is influenced by three main factors: channel shape in cross-section, channel slope and the roughness of the channel's bed and sides. Channel shape is best described by a term we have already mentioned – the hydraulic radius, the area of the channel cross-section divided by the length of its wetted perimeter. Where the river has a large wetted perimeter in relation to its cross-sectional area, there will be much friction and its velocity will be relatively reduced. None the less, the larger the cross-sectional area of the channel, the greater the flow that it will be able to carry. The roughness of the materials that make up the channel walls will depend on their shape and size and the presence of obstructions within or protruding into the flow, such as fallen trees, aquatic plants or roots. With a very rough bed there will be considerable resistance to flow. However, roughness is very difficult to quantify, though one measure of it, called Manning's roughness coefficient, may be estimated visually (table 15.1).

These three factors are related to the mean velocity of flow in metres per second by the Manning equation:

$$\bar{v} = \frac{R^{0.67}\, S^{0.5}}{n}$$

Where $\bar{v}$ is the mean velocity of flow in m s^{-1}, S is the channel slope in m m^{-1}, R is the hydraulic radius in m and n is the boundary roughness.

15.6 Channel Patterns

The patterns that river channels take fall into one of three categories: braided, when a channel divides into several parts; meandering, when the channel twists; and, least frequently, straight. Any one river may take each of these three patterns at various points along its channel, and sometimes it may be difficult to separate one pattern from another. A meandering river may, for example, have straight and braided reaches. Definitions are also likely to be arbitrary, though generally a river is said to be straight if the reach is straight for at least ten times the width of the channel along that reach; and it is said to be meandering if the distance from a point A along its channel to a point B further down the channel is equal to, or more than, 1.5 times the distance from A to B measured along the valley. The ratio of channel distance to valley distance is known as the *sinuosity ratio*.

Rivers are seldom absolutely straight. Moreover, many *straight channels* prove to be unstable, both in the laboratory and in nature. If we take the deepest points along a stream channel and join them in a line, we have marked what is called the *thalweg* of the channel. It has frequently been noticed that, although a stream may have straight banks, the thalweg winds its way from side to side within the channel. In addition, the profile of a straight stretch shows a series of *pools* and *riffles* (deeps and shallows). These tend to be spaced about five to six bed widths apart, and once a pool-and-riffle sequence has been established, many channels develop a side-

Plate 15.1 An aerial view of the braided course of the Rakaia River, Canterbury, New Zealand. Note the wide, shallow bed choked with sand bars, and the multi-thread channel.

to-side swing. Alternate pools migrate to opposite sides, and riffles remain at the crossings between bends. There thus seems to be a propensity for water to flow in a sinuous course. A straight reach may therefore be abnormal and unstable, and can, for example, result from the river flowing along a feature like a fracture zone or a fault, which controls its natural tendency to wander in a wave-like manner.

Braided streams (plate 15.1) have received less attention than meandering channels, but have none the less been recognised in a wide variety of different environments, including semi-arid areas of low relief which receive their discharge from mountain areas, glacial outwash plains, periglacial areas underlain by permafrost and highland areas in all climates. The distinctive features of braided channels include a wide, shallow bed choked with sand bars, together with rapid shifts of sand banks and channels. In general, they appear to develop in coarser deposits than those that support meanders, and are initiated as short submerged bars which are pointed downstream. Once initiated, these accrete rapidly as finer material is trapped, allowing downstream extension; and their existence, by reducing channel width, can encourage bank erosion. The sorts of environmental conditions that favour braiding are the provision of large quantities of coarse bedload (as by glacial outwash (plate 15.2) or on alluvial fans), the presence of non-cohesive banks, rapidly fluctuating discharges and steep slopes. This last factor is made evident by a consideration of figure 15.5. It is apparent that, immediately channel gradient exceeds a critical value compared with channel discharge ($S = 0.012Q^{-0.44}$), shown by the regression line, the channel changes from a meandering to a braided one.

Plate 15.2 Braided rivers are a common feature of areas with glacial outwash. They tend to occur, as in this example from the Valais of Switzerland, where there are steep slopes, a high sediment load and variable discharge.

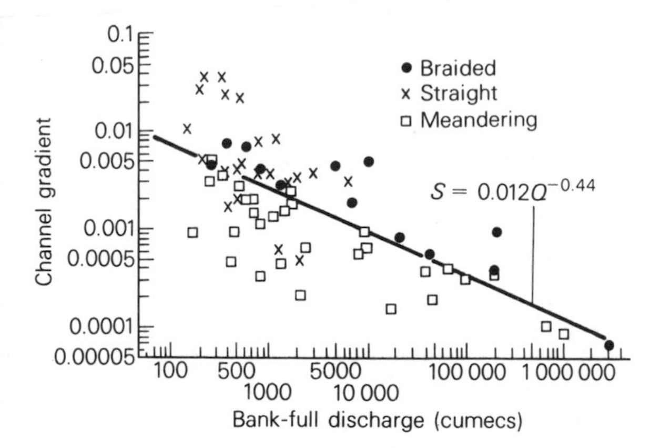

Figure 15.5 The relationship of discharge to gradient in straight, braided and meandering rivers.

The problem of why *meanders* develop is a difficult one. The old and simple idea that chance initial irregularities in the configuration of the land or the occurrence of hard rock outcrops – either of which could cause a deflection of the stream that subsequently would grow into a shapely, well-developed meander – could be the explanation is quite inadequate. Laboratory and field observations show that the reverse tends to be the case – the best meanders are developed in homogeneous material like alluvium, in which they are free to move both downstream and to the side. The sideways movement is accompanied by erosion along the outside of the bend, where the current is strongest, and deposition of a curved bar on the inside of the bend, called a *point bar* (see colour plate 26),where the current is weakest. As the meanders move sideways and downstream, so do the point bars, building up an accumulation of sand and silt that covers the part of the valley floor over which the channel has migrated. This, as we shall see, assists in the development of floodplains. As meanders move, they may progress unevenly so that loops become so close to each other that the river cuts through the land in between, shortening its course and leaving the abandoned bend behind as an *oxbow lake* (figure 15.6).

The proportions of meanders do not vary much. It is unusual to find meander wavelength less than eight times or more than twelve times bed width. The reason for this is that very tight bends (with a wavelength : width ratio of around 5) and very open bends (with a wavelength : width ratio of, say, 20) would increase resistance to flow round the bend.

Human activity has caused changes in the nature of river channel patterns, both deliberately and accidentally. For example, for navigation and flood control many river channels have been deliberately straightened. Removing meanders helps to reduce flooding as the shortened course that the river then follows increases both the gradient and the velocity so that water runs away more quickly. One of the most impressive examples of this change is the Mississippi River, where 16 cutoffs have reduced the length of the river between Memphis and Baton Rouge by 270 km over a distance of 600 km. Non-deliberate channel changes may result from accidental side-effects of human activity on the

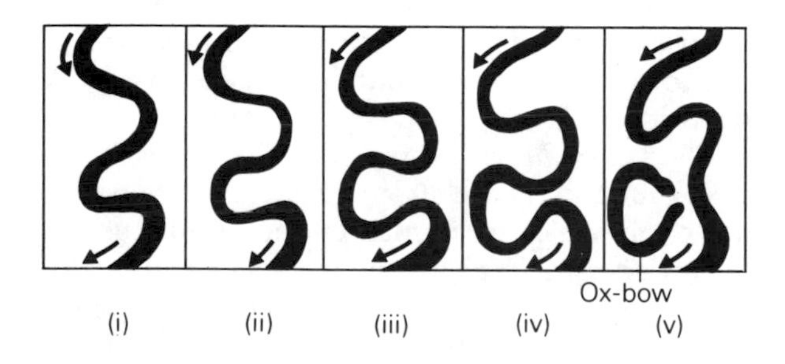

Figure 15.6 Successive stages in the development of meanders, showing the formation of an oxbow lake by a cut-off.

Table 15.2 Accidental channel changes caused by man

Phenomenon	*Cause*
Channel incision	'Clear-water erosion' below dams caused by sediment removal
Channel aggradation	Reduction in peak flows below dams: addition of sediment to streams by mining, agriculture etc.
Channel enlargement	Increase in discharge level produced by urbanisation
Channel diminution	Discharge decrease following water abstraction or flood control; trapping and stabilising of sediment by artificially introduced plants

Plate 15.3 The river Cuckmere in Sussex, England, has a floodplain which is traversed by an especially fine series of meanders.

amount of sediment and discharge in a channel (table 15.2).

15.7 Floodplains

The area of relatively flat land that borders the river channel and is inundated at the time of high river flow is described as the *floodplain* (plate 15.3). The features of floodplains are two main types: those due to lateral accretion, and those due to overbank deposition (figure 15.7).

The *lateral accretion* materials are deposited as rivers cut and fill and meander laterally. Point-bar accretion, which we encountered in our discussion of meanders, is perhaps the main agent of lateral accretion in some areas, though accumulation may also take place in oxbow lakes and the like. Studies in America suggest that possibly 80–90 per cent of floodplain deposits consist of lateral accretion and channel deposits. However, when rivers overtop their banks, as normally happens every one to three years, various overbank deposits can form. Of these, some of the most important are natural *levées*.

During floods, when a river overflows its banks the velocity of flow of sediment-laden water rapidly decreases as it spreads out over the floodplain. Along the strip where the decrease is particularly rapid, i.e. along the river banks, much coarse sediment is deposited; lesser amounts of fine sediment are distributed more widely over the plain. In this way successive floods build up ridges on both sides of a river channel to form features called natural levées. These confine the river within its banks between flood stages, but some levées have been built up so many metres above the surrounding plain that the plain may be lower than the river surface. If a river then breaks through its levée and water escapes, a series of distributary channels may deposit areas of coarse sediments, called *crevasse splays*, over the finer-grained materials of the backswamp areas.

15.8 Terraces

If for any reason a floodplain is cut into by the river that formed it, then alluvial terraces, or in some cases, rock-cut terraces, will be formed (see colour plate 27). Most rivers have terrace remnants along their courses (plate 15.4), and sometimes they may occur at a whole range of different heights, and may be either at the same height on either side of the valley, in which case they are termed 'paired', or at different heights, in which case they are said to be 'unpaired' (figure 15.8a).

Further complexity in terrace type can be introduced by the nature of their development (figure 15.8b), according to whether the river has cut down into one age of floodplain fill or into several.

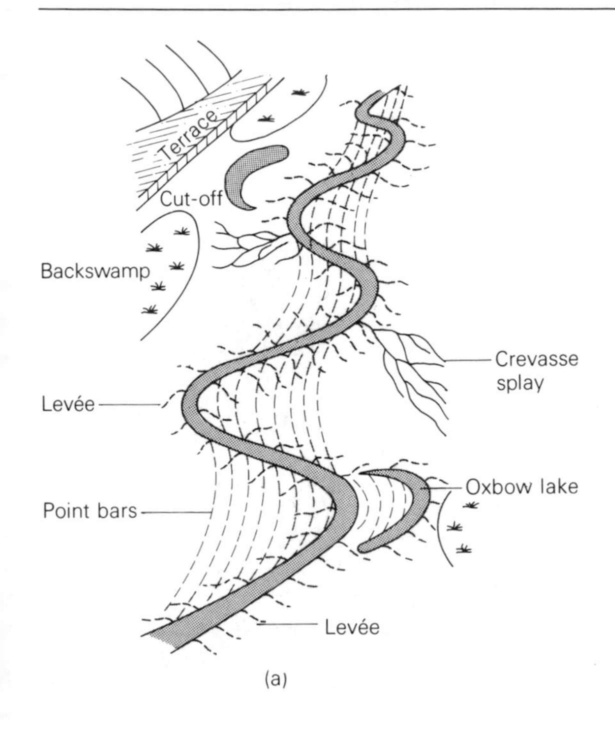

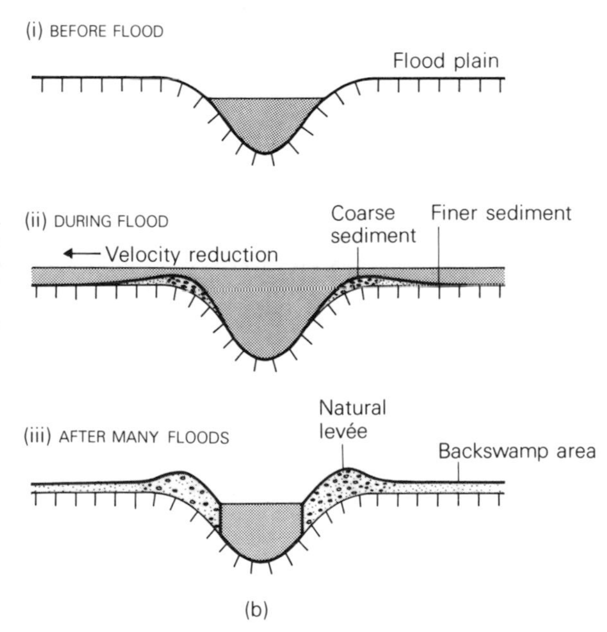

Figure 15.7 River floodplains: (a) some of the major characteristics of a floodplain indicated diagrammatically; (b) the formation of natural levées by river floods. As a river flood stage overflows its banks, it rapidly decreases in velocity away from the channel and drops most of its load, with the coarser material close by and the finer silts and clays across the floodplain.

The major question is, why does the river sometimes cut down and sometimes cease to cut down and begin to build up a floodplain? As with so many geomorphological features, there is a wide range of possible explanations, including changes in base level, climate, land use and vegetation cover (table 15.3).

In the case of the terraces of the upper Thames in England, of which there are four, they all appear to have developed in the Pleistocene; and studies of the sub-Arctic mollusca and pollen they contain, together with the presence of fossil ice-wedge casts (indicative of permafrost) in their rather short-travelled gravels, suggest that the majority of these terraces were caused by the rivers of the Cotswolds being heavily laden with coarse debris provided during cold phases by intense frost weathering and solifluction. During warmer interglacial conditions the rivers had very little load because of the well-vegetated nature of the terrain and the limited operation of mass movement processes, and so the rivers tended to cut down. In its lower course, however, the Thames may have been influenced much more by the rises and falls of sea level in response to the waxing and waning of the great ice sheets of the Pleistocene. When sea levels are low, rivers will tend to cut down into their floodplains, whereas when they are high they will tend to aggrade (figure 15.9).

15.9 Deltas

When a river enters a lake or the sea, its velocity rapidly diminishes and it deposits its load of sediment. If conditions are favourable this sediment will build up a distinctively shaped depositional landform called a *delta* (see colour plate 29). The simplest deltas are formed by streams entering freshwater lakes. Because the density of the entering river water is the same as that of the surrounding lake water, the river's current mixes in all directions in a cone-like pattern and rapidly slows to a halt. The coarsest material is inevitably dropped first, followed by medium and fine sediments further out. Where the lakebed slopes away at a greater angle than the floor of the river channel, the coarse material builds up a kind of depositional platform (figure 15.10a). *Foreset beds*, inclined downcurrent from the delta front, are

Plate 15.4 Terraces on the edge of the Canterbury Plains, South Island of New Zealand. The Southern Alps are in the background. Changes in sediment and discharge loadings associated with glaciation have contributed to their development as has tectonic activity.

Table 15.3 Some possible explanations of why rivers sometimes cut down (incise) and why they sometimes build up (aggrade)

Incision	*Aggradation*
Lack of load to deposit	Too much sediment for the river to cope with
Increased vegetation cover	
Decrease in mass movements	
Decrease in frost weathering	More erosion, weathering and mass movements on slopes
Absence of inputs from wind	
No input of glacial drift	
Sediment trapped by dam	
Change in base level	Change in base level
Land rises owing to tectonics, isostatic adjustment etc.	Land sinks or sea-level rises
Sea-level falls owing to subsidence, water being ponded up in ice caps etc.	
Change in velocity of flow (increase)	Decrease in velocity of flow
Change of slope	
Change in climate	
Change in amount of discharge	Change in amount of discharge

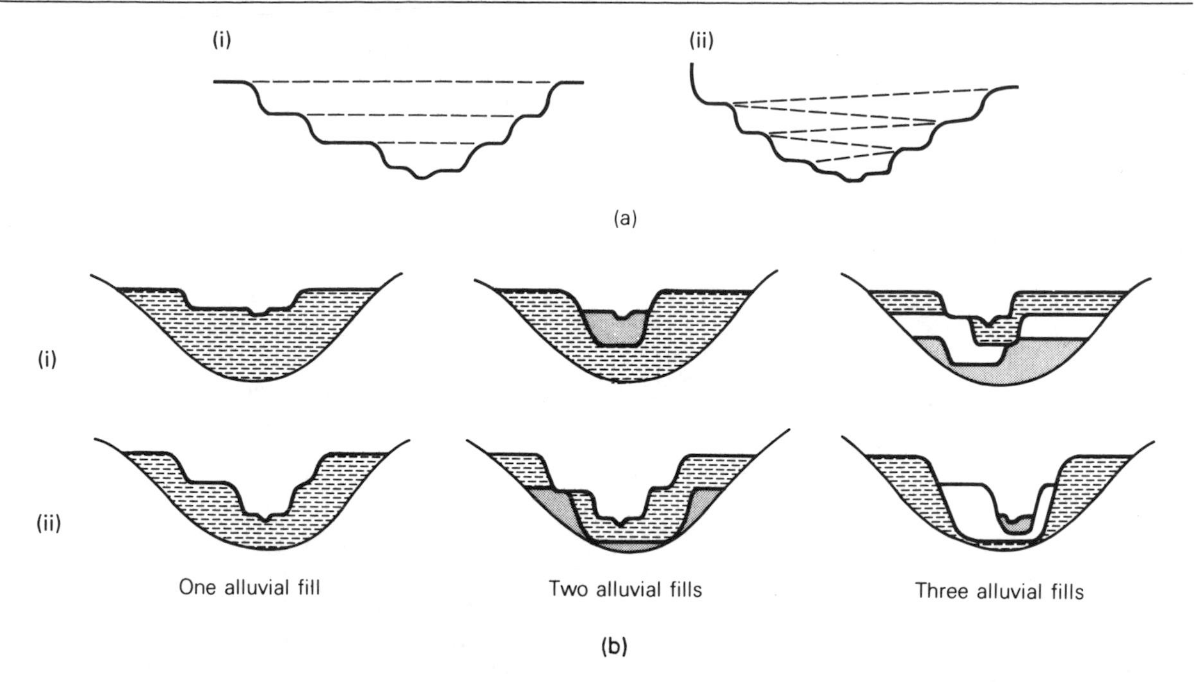

Figure 15.8 River terraces. (a) Paired (i) and unpaired (ii) terraces. (b) Valley cross-sections showing some possible terrace and alluvial fill combinations: (i) one terrace; (ii) two terraces.

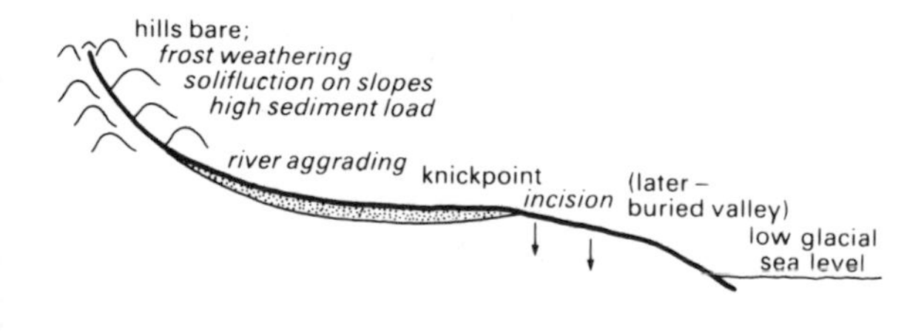

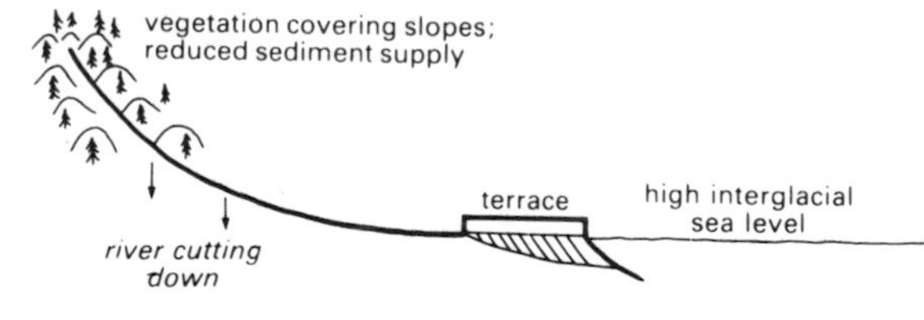

Figure 15.9 A model of terrace formation in relation to the sea-level and climatic changes of the Pleistocene.

covered by thin, horizontal *topset beds* and are preceded on the lake floor by thin, horizontal *bottomset beds*. Deltas formed in the sea are, broadly speaking, much the same, except that there is a difference in the density of seawater and freshwater; the lighter river tends to 'float' on the denser seawater, and consequently vertical mixing is more limited than in a lake. A lower mixing rate means that the current dissipates more slowly, so that the river sediments spread out along a much longer distance. As a consequence, the slope of foreset beds on large marine deltas may be only a few degrees. If the river branches where it enters the sea to form distributaries, the distributaries will build outwards as long sand fingers, called *bar-finger sands*; whereas the finer muds and silts get washed into the still water in between the distributaries, so that they gradually fill up with mud and swamp deposits.

The form of deltas is extremely varied (figure 15.11), and much of this variation can be attributed to the varying balance between stream discharge and wave power. As coastal processes become progressively dominant, the delta becomes progressively less irregular at the shoreline, and may,

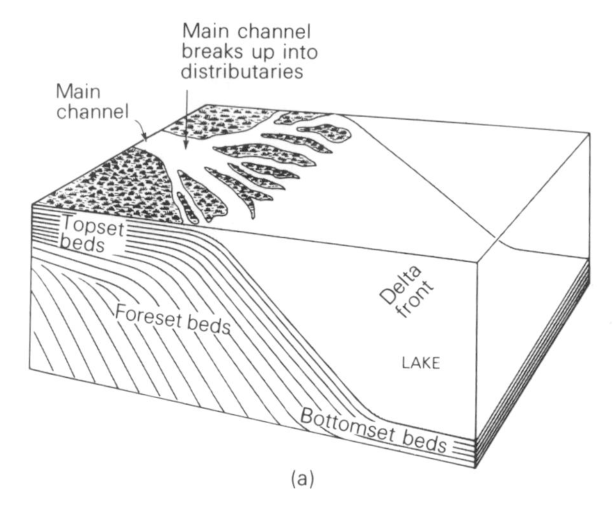

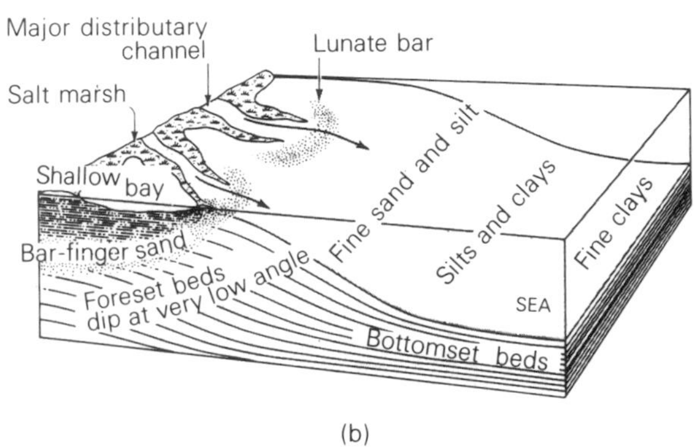

Figure 15.10 River deltas: (a) a typical freshwater delta, formed in a lake, and characterised by well-defined topset, foreset and bottomset beds, of the type described from pluvial Lake Bonneville by G. K. Gilbert. Note that the slope of the delta front may be fairly steep (up to 25°); (b) a typical marine delta. The foreset beds are fine-grained and are deposited at very low angles (generally less than 5°); crescent-shaped sand accumulations called lunate bars form at the distributary mouths, where the current velocity suddenly decreases.

as with the Senegal Delta on the exposed west coast of Africa, create only a slight protuberance. At the other extreme, if the discharge of the river is high, if its sediment load is considerable and if it occurs in a relatively sheltered environment, then a complex bird's foot form of delta, like that of the Mississippi, may form (window 15.1). Such *bird's foot deltas* also tend to develop where the river's load is predominantly fine-grained, whereas the *arcuate* form (of which the Niger is a classic example) tends to be deposited by streams that carry coarser material as bedload.

Not all rivers have deltas, for conditions may not always be favourable to their development. Among conditions favouring deltaic accumulation are:

(a) a river with a large sediment load, for it can add more sediment at its mouth than can be removed by waves and currents. Rivers draining rapidly eroding mountains (e.g. the Indus) or highly erodible sediments like loess (e.g. some of the great Chinese rivers) have large deltas, other factors being equal;
(b) a river with a large discharge, for plainly a small river entering a large ocean will make relatively little impact, especially if waves and currents disperse its sediment load;
(c) reasonably shallow water offshore, so that the sediments can accumulate, rather than disappearing into a submarine canyon or an ocean trench;
(d) a relatively sheltered coast, on which wave and current action is inhibited. The development of deltas is more likely in seas such as the Gulf of Mexico, the Mediterranean and those inland from the island arcs of the western side of the Pacific, than on exposed coasts;
(e) a low tidal range: this factor reduces tidal scour, though it has to be noted that some deltas do occur in areas with quite substantial tidal range (e.g. the Ganga and Irrawaddy).

15.10 Stream Transport

One of the main reasons why streams are of such geomorphological importance is that they carry a sediment load, transporting material from one part of a basin and depositing it elsewhere. The sediment load can be divided into three main types (figure 15.12): dissolved load, suspended load and bedload. *Dissolved load* is that portion of the stream's load that is in soluble form; it is derived primarily from the chemical weathering of rocks within its catchment and, normally to a lesser extent, from atmospheric inputs – rainwater is never completely pure. In areas of soluble rocks, such as limestone, the dissolved load may be the most important component of a river's load.

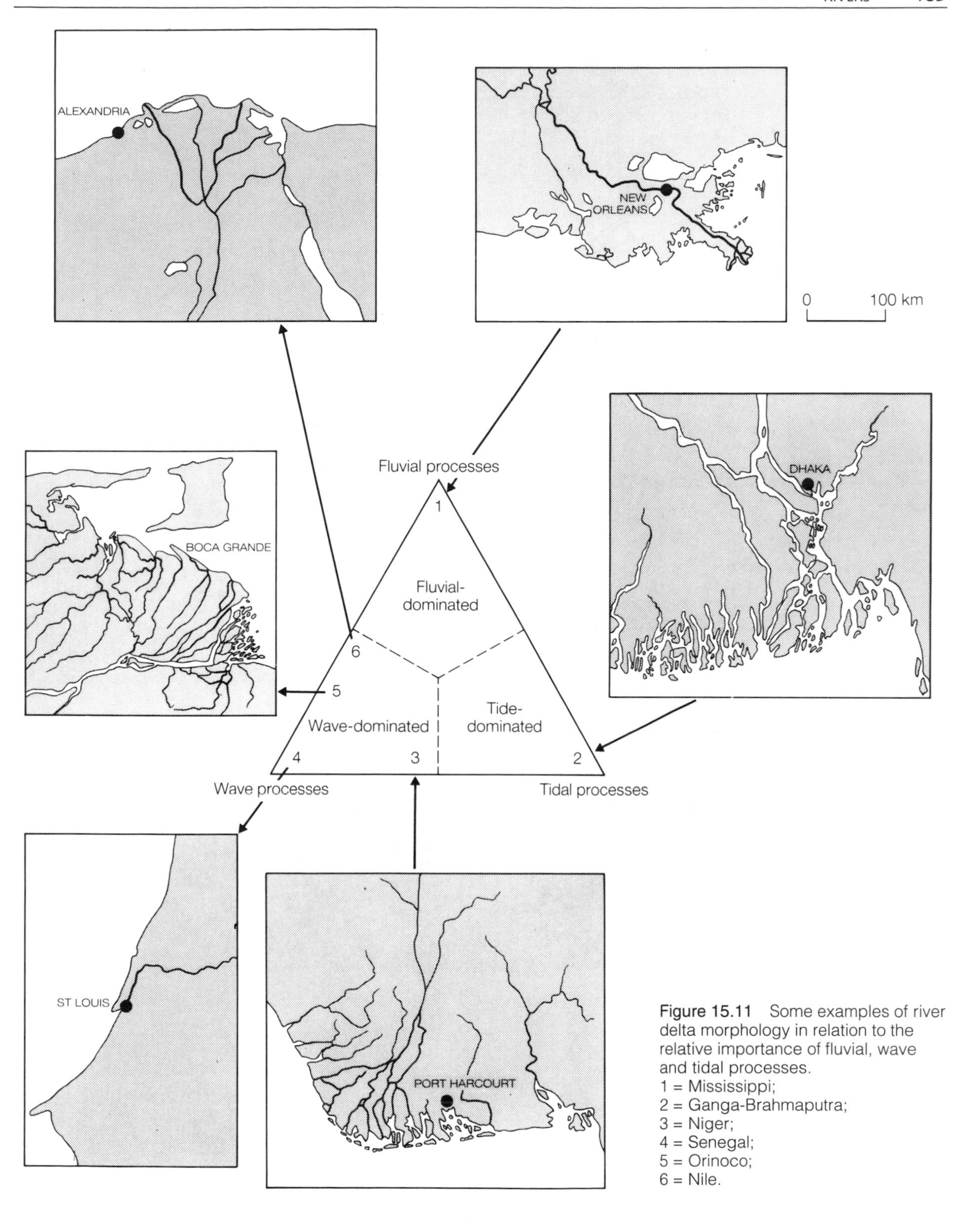

Figure 15.11 Some examples of river delta morphology in relation to the relative importance of fluvial, wave and tidal processes.
1 = Mississippi;
2 = Ganga-Brahmaputra;
3 = Niger;
4 = Senegal;
5 = Orinoco;
6 = Nile.

Window 15.1 *The Mississippi Delta*

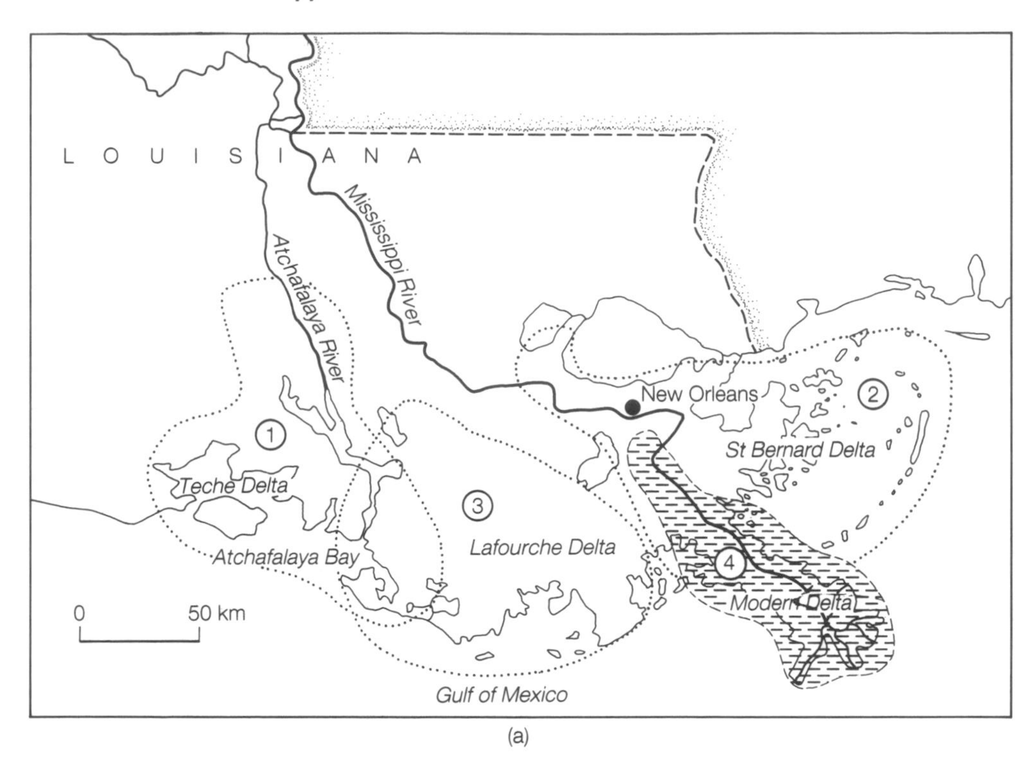

(a)

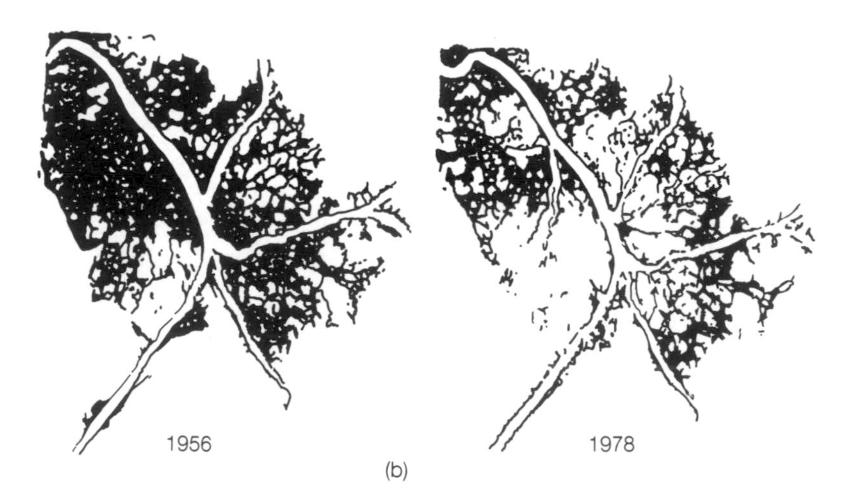

(b)

The Mississippi Delta. (a) The evolution of the delta over the past 5000 years: the river is now showing a tendency to revert to the Atchafalaya River. The numbers 1–4 show the successive cycles of delta growth until the present modern delta (4). (b) The changing outline of the modern bird's foot delta, showing the loss of land between 1956 and 1978.

The Mississippi of the southern USA is one of the world's greatest rivers; only the Amazon drains a greater area. Its mouth is formed of an intricate and delicate delta, also one of the world's biggest. Its bird's foot form is distinctive in shape, and is said to be 'river-dominated'. The river has carried more sediment than can be carried away by waves, tides and currents. However, the delta and its highly important wetland habitats are rapidly shrinking.

Many processes probably contribute to this loss, including rising sea levels, subsidence produced by groundwater abstraction, changes in the sites of sediment deposition as the delta evolved, catastrophic storms, and human interference with river flow and sediment transport.

In the case of the Mississippi Delta two of these processes have been accorded particular prominence. First, as part of the natural process of their evolution, deltas change their courses to find shorter, steeper paths to the ocean. This process is called avulsion. In effect, as deltas get bigger they become unstable, and the river shifts its course. Very little sediment then accumulates along the abandoned channel and the area starts to sink and decay. The Mississippi seems to shift its path to the Gulf of Mexico in a cyclic fashion round about every 1000 years, and at present it is attempting to send more and more of its flow and sediment via the Atchafalaya arm, a course that it had previously largely abandoned 3800 years ago. This means that the modern bird's foot delta between New Orleans and the Gulf is becoming starved.

Second, channelisation of the Mississippi and the construction of great levées have had a profound effect by increasing the velocity of the river, reducing the amount of overbank deposition onto swamps and marshes, and changing the salinity conditions of marshland plants. These changes were wrought by man to improve navigation and to control flooding, but they have upset the natural sediment balance. As a consequence the area suffers from the effects of subsidence. The reason for this is that the growth of the delta depends on a balance between the amount and the type of sediment that the river transports, and upon the rate at which the sediment becomes compacted. The Mississippi carries mostly clay-rich mud: when this clay settles out, up to 80 per cent of its initial volume is water; and the water is gradually squeezed out by the weight of new sediment deposited on top and this makes the delta sink through time. As long as the river floods bring in more sediment, this can offset the settling and subsidence of earlier deposits. The human intervention through channelisation and levée construction changes that balance.

The Mississippi River has been transformed over large parts of its lower course by various types of human manipulation. Of particular significance is the construction of great levées, as shown here between La Place and New Orleans.

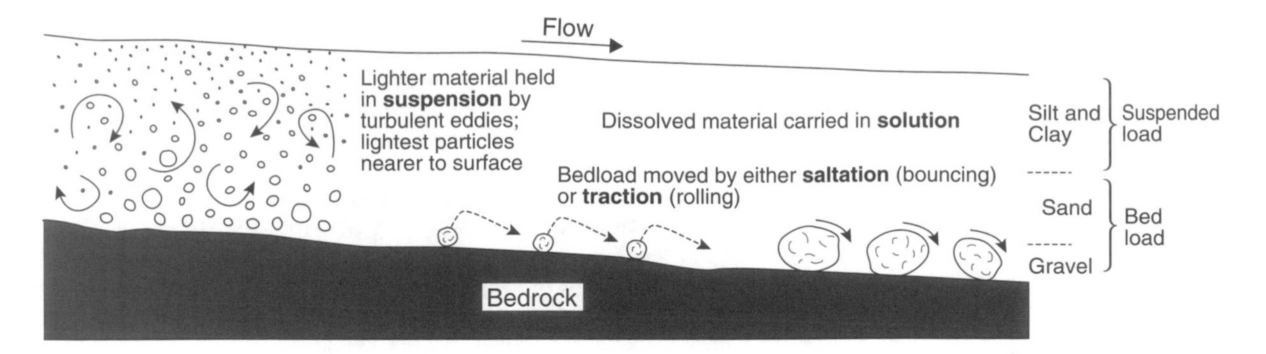

Figure 15.12 The three main means of stream transport: suspension, solution and bedload.

Plate 15.5 Traditionally, the most difficult component of a river's load to measure is bedload. However, some devices have been developed for this purpose, including the Helley–Smith Sampler.

Elsewhere the most important contribution comes from the *suspended load* (see colour plate 28). Fine particles of clay and silt are held up in the water by the upward elements of flow in turbulent eddies in the stream; as fine particles settle only very slowly once they are in suspension, they can travel long distances in this state.

The third component of a river's load is the *bedload* (plate 15.5). This moves along the channel floor by rolling or sliding and with an occasional jump (called *saltation*). Most of the bedload is composed of material that is too coarse to be lifted into suspension, such as gravel, cobbles and boulders. Coarse material needs high velocities to move it, and at low flow many streams do not have the competence to move much bedload. *Competence*, the maximum particle weight that can be moved at a particular velocity, tends to vary according to the sixth power of this velocity. In other words, if the velocity of a stream is increased by a factor of 4, the weight of individual boulders it can move is increased something like 4^6 or 4096 times. This explains to a considerable degree how rivers can achieve a great deal of their work during rare events of high magnitude.

A related concept is that of *capacity*. It has been estimated that the capacity of a stream (that is, its ability to move total bedload) varies according to the third power of the velocity; in other words, if stream velocity is doubled, capacity is increased 2^3

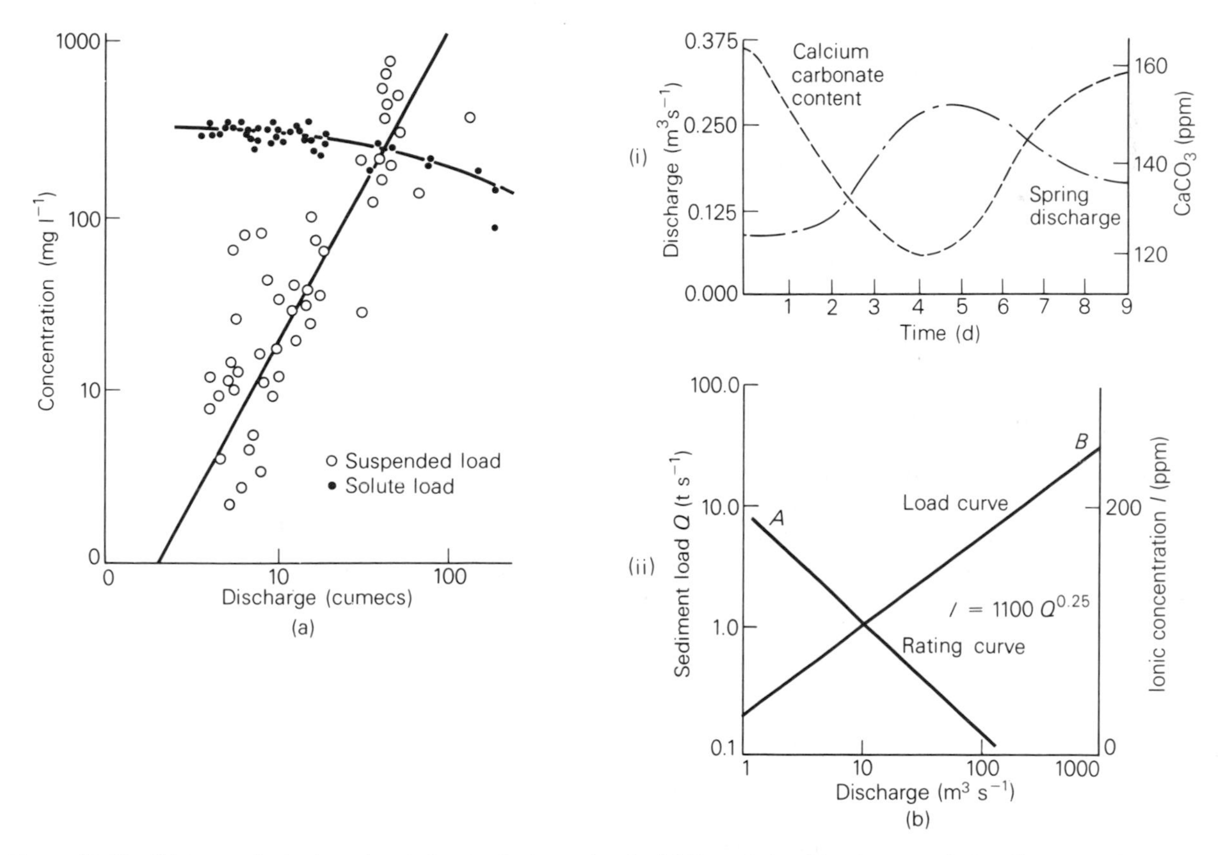

Figure 15.13 Water quality in rivers in relation to discharge levels. (a) The relationship of suspended sediments and solutes to discharge for the River Avon in Wiltshire, England. (b) Dissolved load in relation to discharge: (i) calcium carbonate concentration through a flood hydrograph for a Mendip limestone spring, England; (ii) an example of a solute rating curve for the Mekong River, south-east Asia.

or 8 times. Capacity depends upon stream gradient, discharge and the calibre of the load.

Suspended sediment concentrations tend to increase markedly with discharge levels at any point on a stream. If one plots discharge against suspended load carried in tons per day (see figure 15.13a) on log–log paper the result is a straight line relationship. In the example given, for a discharge of 4 cubic metres per second (cumecs) the suspended sediment concentration is around 4 $mg\,l^{-1}$, while at 100 cumecs the suspended sediment concentration has shot up to 1000 $mg\,l^{-1}$. Given that the flow has increased by 25 times and the suspended sediment concentration by 250 times, the quantity of material carried as suspended sediment load at bank-full conditions in that particular channel (corresponding to *c.*100 cumecs) is over 6000 times greater than at low flow!

Unlike suspended sediment concentrations, *solute concentrations* tend to decline with increasing discharge (figure 15.13b), probably because at lower flows most of the river's discharge is derived from water that has been in contact with rock and soil for an extended period and has thereby picked up ions released by chemical weathering. The decrease may be attributed to the dilution of the groundwater and the deep throughflow water by near-surface flow and overland flow during storms. However, although dissolved load contents may be reduced in their concentrations as discharge increases, in many rivers it has been found that something like a ten-fold increase in discharge causes only a halving of the solute concentrations. Consequently, solute load, which is the product of discharge and concentration, still continues to increase with discharge.

When one considers rivers in general, it is difficult to obtain any precise information on the relative importance of these three modes of material transport in rivers. A particularly severe problem is to assess the role of *bedload transport*, for this is especially difficult to estimate. For lowland rivers it is unusual for the bedload to be more than about 10 per cent of the total load, and it is generally much less than this. For solutes the picture is clearer, and on a global basis they seem to make up about 38 per cent of the total load of the world's rivers. Thus, if one wants a rough rule of thumb, suspended load, solute load and bedload occur in the ratio 5:4:1. It should be pointed out, however, that for individual streams the ratios may be widely different from this. For example, an area with soluble rocks and a soil cover well protected by vegetation may have a higher proportion of its total load made up by solutes, whereas an area of highly erodible sediments with a limited vegetation cover and much disturbance of the ground surface by human activities like ploughing will have a relatively high proportion of its total load made up by suspended sediment. In Britain, the mean ratio of suspended sediment to solute load based on a study of 35 catch-

Table 15.4 Sediment yields

(a) *Suspended sediment yields for selected major river basins*

River	*Country*	*Annual sediment yield ($t\,km^{-2}\,yr^{-1}$)*
Ganga	Bangladesh	1568
Yangtze	China	549
Indus	Pakistan	510
Mekong	Thailand	486
Colorado	USA	424
Missouri	USA	178
Mississippi	USA	109
Amazon	Brazil	67
Nile	Egypt	39
Danube	Moldavia	27
Congo	Congo	18
Rhine	Netherlands	3.5
St Lawrence	Canada	3.1

(b) *Estimates of long-term sediment yield obtained from selected surveys of reservoir sedimentation in Britain*

River	*Reservoir*	*Annual sediment yield ($t\,km^{-2}\,yr^{-1}$)*
Rede	Catcleugh	43.1
Bradgate	Cropston	45.6
Loxley	Strines	49.7
N. Esk	N. Esk	26.0
N. Tyne	Hopes	25.0
Churret	Deep Hayes	6.7
Wyre	Abbeystead	34.8
Mean		33.0

Source: modified from data of D. E. Walling and B. J. Webb in J. Lewin (ed.), *British Rivers* (London: Allen and Unwin, 1981), figure 5.12

ments is 1.09, indicating that these two modes of transport are more or less equal.

15.11 Rates of Fluvial Denudation

By long-term monitoring of the discharge levels of rivers, together with their loads of dissolved, suspended and bedload materials, it is possible to calculate the amount of material that is being removed per unit area of the land's surface in tonnes per square kilometre per year (table 15.4). This figure can also be expressed as an overall rate of surface lowering in millimetres per 1000 years. Such figures indicate the rate at which rivers are denuding the face of the earth. Sometimes rates of denudation can be calculated by other means, such as by examining the rates at which sediments accumulate in lakes or man-made reservoirs of known age (table 15.4). Data are now available for a large number of catchments, and this has encouraged many scientists to attempt to ascertain how the rates may be controlled by different environmental factors. However, the situation is complex, for there are a large number of factors that might promote a high rate of denudation in a particular river basin:

(a) limited vegetation cover to protect the ground surface from rainsplash etc.;
(b) intense rainfall;
(c) highly seasonal rainfall (e.g. dry tropical, monsoonal, 'Mediterranean');
(d) rapid mass movements and weathering (to feed the river with sediment);
(e) rapid tectonic uplift;
(f) large store of erodible materials (e.g. loess, unconsolidated glacial outwash);
(g) the presence of glaciers in the catchment;
(h) steep slopes and long slopes;
(i) disturbance of the ground surface (e.g. by grazing animals, ploughing, construction etc.);
(j) lack of conservation measures.

When we try to establish zones where rates are highest, it becomes evident that over the years there has been a wide range of opinions about this (table 15.5). The most detailed analysis of available data has been made by D. Walling and A. H. A. Kleo and can be taken as the best available information (figure 15.14). They recognise three climatic zones where rates may be especially high: the seasonal climatic zones of the 'Mediterranean', the monsoonal and the semi-arid areas. The zones with the

Table 15.5 Different interpretations of where maximum fluvial denudation rates occur

Investigators	*Zone*
Corbel	Mountainous relief (×2 to ×5 that of plainlands) Glaciated catchments
Fournier	Tropical monsoon and savanna regions
Strakhov	Tropical mountainous regions
Langbein and Schumm	Semi-arid regions
Walling and Kleo	Seasonal climates ('Mediterranean', semi-arid, tropical monsoonal)

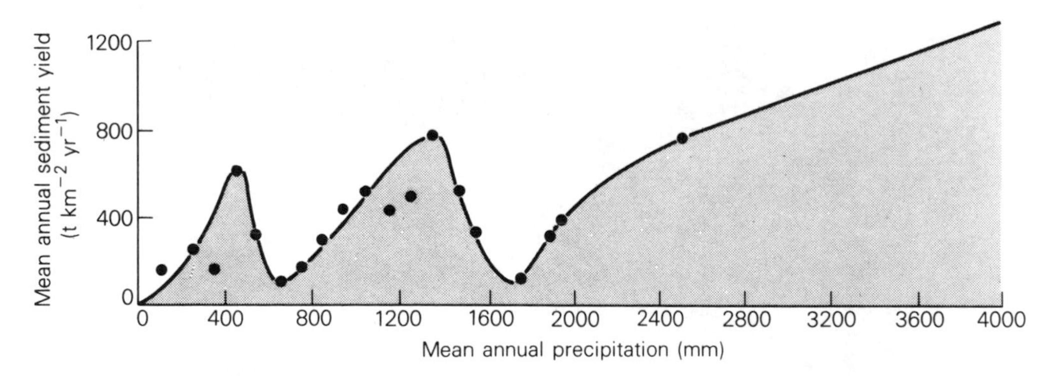

Figure 15.14 Relationship between mean annual sediment yield and mean annual precipitation.
Source: D. E. Walling and A. H. A. Kleo (1979) 'Sediment yield of rivers in areas of low precipitation: a global view', *International Association of Scientific Hydrology Publication*, 128: 479–93.

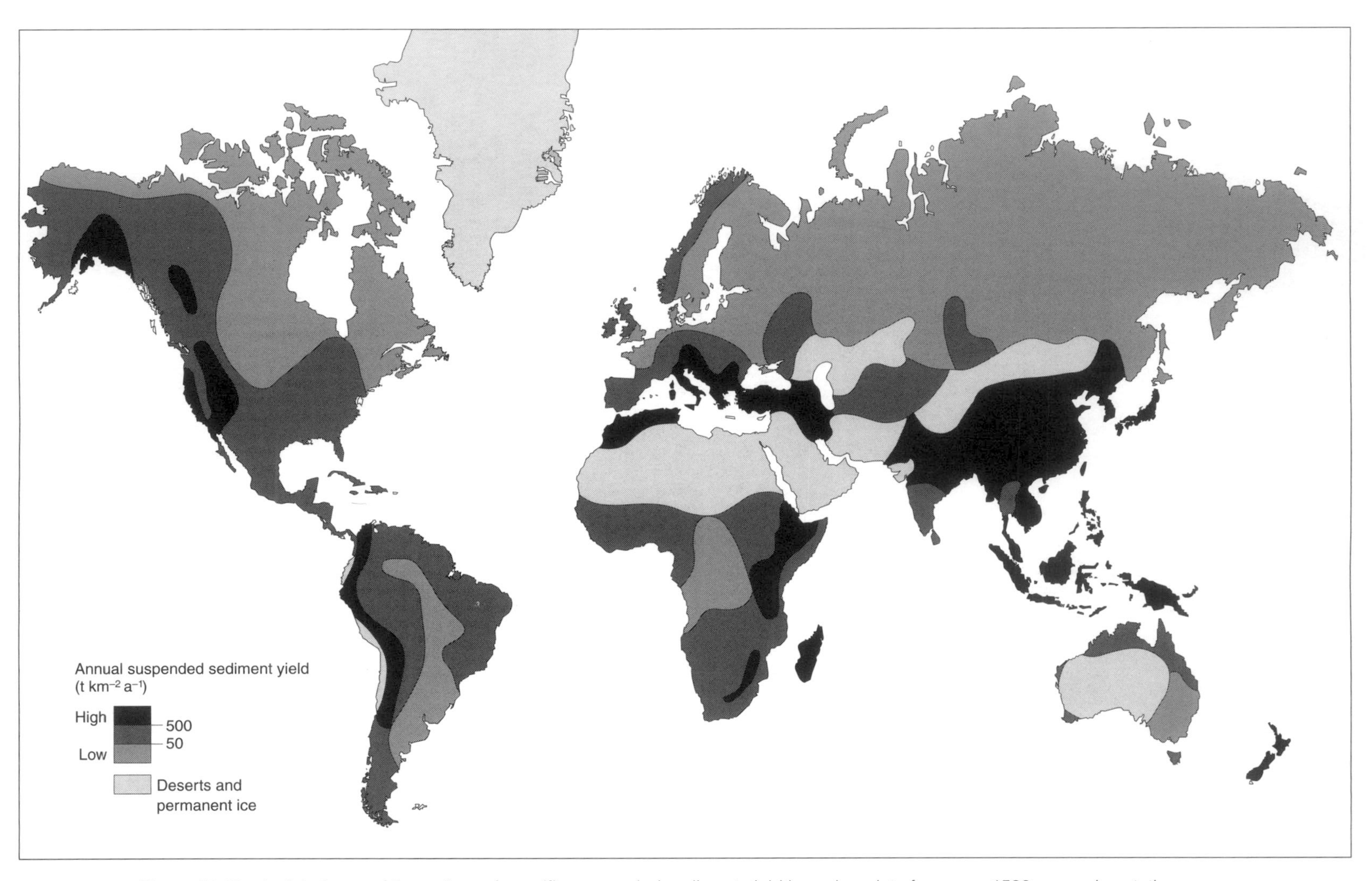

Figure 15.15 A global map of the pattern of specific suspended sediment yield based on data from over 1500 measuring stations.

highest rates have values of between 600 and 1000 $t\,km^{-2}\,yr^{-1}$, whereas the mean value for British rivers tends to be between 10 and 200 $t\,km^{-2}\,yr^{-1}$.

Walling's data on rates of denudation are plotted in figure 15.15. In addition to the impact of climate, it demonstrates that many of the zones with high suspended sediment yields are associated with mountainous terrain where relief is high and there is current tectonic activity. Notice the high rates of denudation down the western cordillera of the Americas, around the Mediterranean basin, in the highlands of East Africa, in the mountains of the Middle East, and in south and south-east Asia.

15.12 Floods

As far as people are concerned, one of the most important aspects of the study of rivers is that connected with the main hazard that they pose – floods (plate 15.6). Floods can be defined as the stage or height of water above some given datum such as the banks of the normal channel. In this sense, a flood occurs whenever a river overflows its banks. To the inhabitants of a floodplain, however, a flood occurs whenever water rises sufficiently so that life and property are damaged or threatened. There is a good deal of evidence that this threat is becoming increasingly serious in many parts of the world, for a number of reasons:

(a) Climatic changes
  (i) More frequent occurrence of storms of high intensity, perhaps associated with El Niño events.
  (ii) Heavier snowfalls
(b) Changes in land use
  (i) Urbanisation
  (ii) Deforestation of basin slopes
  (iii) Drainage of soils
(c) Ground subsidence
  (i) Natural seismic activity
  (ii) Man-made subsidence (arising from mining, oil abstraction, draining of peat, pumping of groundwater)

Plate 15.6 The remains of a new high-level road bridge across the Usutu River in Swaziland, following a flood generated by Cyclone Domoina in 1984. In one spell, some areas of the catchment received as much as 900 mm of rainfall.

Table 15.6 US flood deaths and damage, 1969–1989

Year	*Deaths*	*Property loss ($ million)*
1969	297	903
1970	135	225
1971	74	288
1972	540	3449
1973	105	859
1974	121	576
1975	114	1051
1976	187	1000
1977	212	1393
1978	120	1000
1979	100	4000
1980	97	1500
1981	90	1000
1982	155	3500
1983	200	4100
1984	126	4000
1985	304	3000
1986	80	4000
1987	82	1490
1988	29	114
1989	81	415
Total	3249	$37 863 million

Average deaths per year = 155
Average property loss per year = $1803 million

Source: Bureau of the Census, *Statistical Abstract of the United States*, 1990

(iii) Increasing population levels in flood prone areas
(iv) Spread of infrastructure on to new flood-plain areas.

Floods are a serious problem in the United States of America (table 15.6) where the average number of deaths from this cause is 155 per year and the average annual property loss per year amounts to $1803 million.

The causes of floods are legion and include:

- excessive levels of precipitation
- melting snow
- failure of man-made dams
- failure of natural and man-made levées
- rupture of a glacial lake or release of subglacial water
- ponding up of rivers behind natural dams caused by glacial advance, mass movements etc.

In addition to *causes*, we must also remember that there may be various flood-intensifying conditions, for example the degree of urbanisation, lack of vegetation cover, presence of impermeable soils, existence of steep slopes and so on. And as we saw in section 14.8, some basins, because of their various transient and permanent characteristics (figure 14.10), are prone to having a rather peaked flood hydrograph with a very steep rising limb.

The most common type of river flood is that caused by excessive levels of precipitation (window 15.2) – a phenomenon normally associated with depressions, hurricanes, thunderstorms and other low-pressure systems (see colour plate 30). This was the cause of the flood of August 1952 which caused disaster in Lynmouth, on the margins of Exmoor in north Devon. During the first two weeks of the month the area had already received the total average precipitation for August, so the moorland soils were saturated when there followed a torrential downpour of as much as 386 mm in 36 hours. Slopes in the area are high (plate 15.7), river gradients are steep, and bridges and debris jams caused temporary dams that eventually broke. These factors aggravated the effects of high rainfall levels. In all, 34 people were killed and 200 000 tons of boulders were moved into the little town.

A second major category of river floods is that associated with snowmelt. If over the winter months snow falls in large quantities and then melts, particularly if it melts very quickly, rivers may be rapidly swollen as the thaw takes place. Another important contributory factor relates to the state of the ground during the melting phase, for if the ground is frozen to a great depth it will be rendered impermeable so that there will be little percolation of the meltwater to groundwater. An example of snowmelt floods in the UK is provided by the events of 1947, when the spring thaw caused by the movement of a warm air mass from the Atlantic ended one of the worst winters that has been experienced in living memory. Flooding was extremely widespread right across the country, from Wales and the West Country to Yorkshire and the East Anglian Fens.

Of the other causes of river floods, reference can only be made to a few examples. If dams fail (perhaps because they are badly made or are ruptured by an earthquake) the water from the reservoir may

Window 15.2 *The 1993 Mississippi floods*

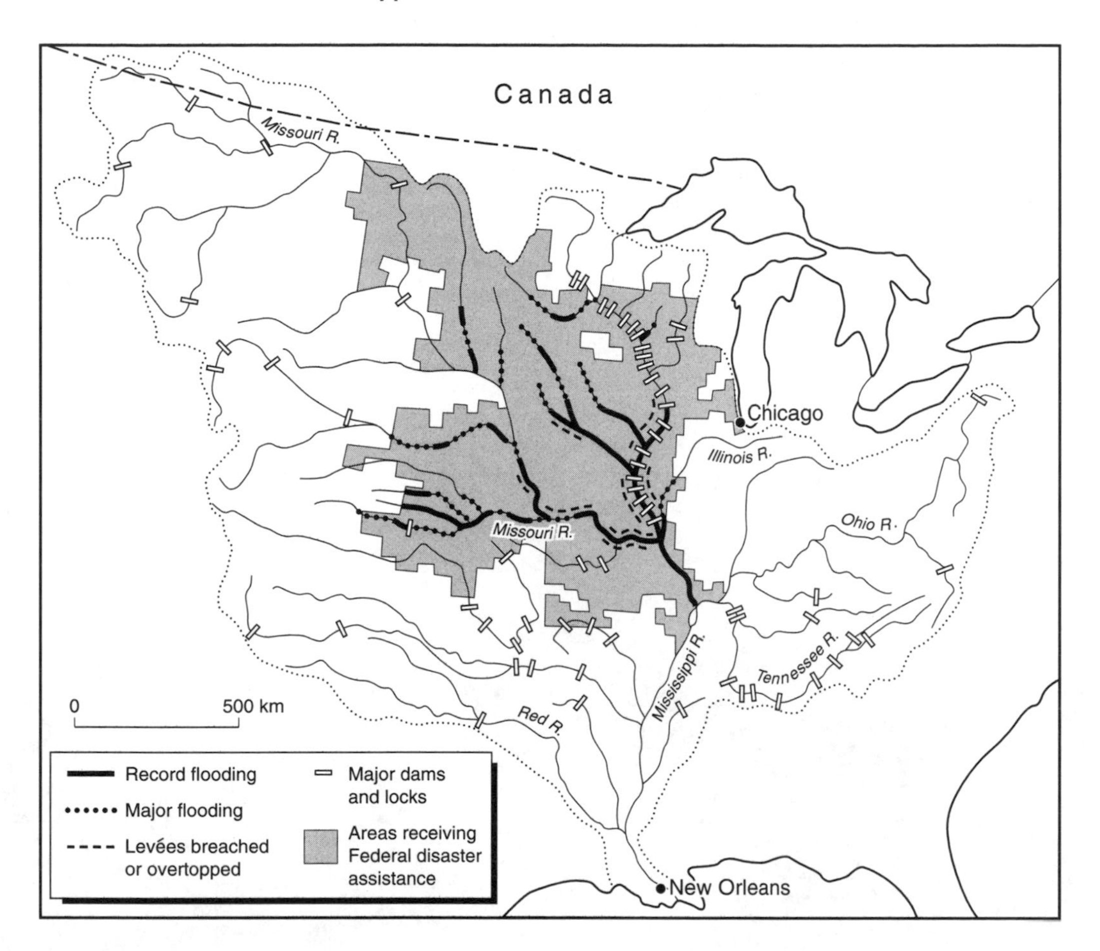

The 1993 Mississippi Basin floods.

The great floods of July 1993 in the Mississippi Basin were the worst since records began in 1895. Estimated losses amounted to $15–20 billion, 50 people were killed, 37 000 people had to be evacuated and over 30 million hectares of farmland were flooded.

The 1993 floods followed up to eight months of unusually high rainfall, with excessive winter snowpacks in the Rockies and persistent high rainfall throughout the spring and early summer. The final trigger for the floods was an unusually strong quasi-stationary jet stream running north-east across the upper part of the Mississippi Basin with a strong blocking anticyclone centred over the eastern USA for more than five weeks. This caused a persistent southerly airstream that carried warm, moist air up the Mississippi Basin from the Gulf of Mexico. This deposited large amounts of rain. While the heavy rain falling on a saturated basin was an extreme natural event, the question has been raised as to whether it was exacerbated by human actions such as the construction of embankments (levées) which reduce natural overbank storage on floodplains, the straightening of channels, the draining of wetlands and the building of settlements on floodplains that had supposedly been protected.

Source

J. A. A. Jones (1997) *Global Hydrology* (Harlow: Longman), pp. 233–7

Top ten Mississippi River floods, St Louis, 1883–1993

		Discharges (in cubic ft/sec)			*Discharges (in cubic ft/sec)*
1993	August	1 030 000	1909	July	860 600
1903	June	1 019 000	1973	April	852 000
1892	May	926 500	1908	June	850 000
1927	April	889 300	1944	April	844 000
1883	June	862 800	1943	May	840 000

Source: Illinois State Water Survey Miscellaneous Publication 151 (1994), Champaign, Ill.

Plate 15.7 One contributing factor to the great Lynmouth flood of August 1952 was the very deeply dissected and steeply sloping terrain of the Lyn catchments.

Table 15.7 Some US dam failures

Date	*Place*	*Deaths*
16 May 1874	Connecticut River, near Williamsburg, Massachusetts	143
31 May 1889	Little Conemaugh River, Johnstown, Pennsylvania	over 2100
27 Jan. 1916	Otay River, San Diego	22
12 Mar. 1928	St Francis dam, north of Los Angeles	420
14 Dec. 1963	Baldwin Hill, Los Angeles	5
26 Feb. 1972	Buffalo Creek, West Virginia	118
5 June 1976	Teton River, Idaho	14

cause inundation in the valley downstream (table 15.7). A damburst flood resulted in over 2100 drownings in Pennsylvania in 1889, and the rupture of the Bradfield Reservoir near Sheffield in Yorkshire in 1864 caused 250 deaths. Likewise, if a river breaks through its natural levées or man-made embankments along its sides, wide areas of the floodplain may be inundated. In some mountainous areas major landslides, glacier surges or mudflows may cause rivers to be blocked behind a natural dam. The water level will rise up behind this dam, and when it eventually ruptures, a flood-wave may pass downstream. The classic area for such behaviour is in the Karakoram Mountains in Pakistan, where the great tributaries of the mighty Indus River rise in the snow-clad hills of some of the steepest terrain on earth.

15.13 The Human Impact on Rivers

Humans have done much to alter rivers. This is partly because in recent decades demand for freshwater has increased rapidly. Global water use has more than tripled since 1950, and now stands at 4340 km^3 per year – equivalent to eight times the annual flow of the Mississippi River.

Some of the changes made to rivers have been deliberate, as with the construction of dams. Large dams are capable of causing almost total regulation of the streams they impound, and so are effective at controlling floods and droughts. However, dams have a whole series of environmental consequences that may or may not be anticipated. A particularly important consequence of impounding a reservoir behind a dam is the reduction in the sediment load of the river downstream. One can illustrate this by looking at the amount of sediment going down the Mississippi and Missouri rivers over the period 1938–82 (figure 15.16). Because of the construction of many dams, downstream sediment loads were reduced by about half over that period. Sediment retention is also well illustrated by the Nile, following the construction of the Aswan High Dam (table 15.8). Until its construction, the late summer and autumn period of high flow was characterised by high silt concentrations, but since it has been finished this silt load is rendered lower throughout the year and the seasonal peak is removed. The Nile now transports about 8 per cent of its natural load below the Aswan High Dam.

Such sediment removal has various possible consequences, including a reduction in flood-deposition nutrients on fields, less nutrients for fish in the sea, accelerated coastal erosion (because the river no longer brings material down which could build up beaches), and accelerated riverbed erosion (since less sediment is available to cause bed aggradation). This last process is often called 'clear water erosion'. One of the reasons why the fragile eastern seaboard of America is eroding quickly is the reduction of the sediment coming down the rivers that feed it.

Another major change that has occurred along many rivers is the construction of embankments, and the alteration of river channel geometry by a process called *channelisation*. These can speed up flood peaks, stop water spilling over on to floodplains, and change the habitats of stream fauna. Likewise, transfers of water from one river system to another – inter-basin water transfers – can have detrimental effects on the rivers that are deprived of some of their flow, and can cause the desiccation of the lakes that they feed. This has been the problem with the Aral Sea in central Asia (see window 6.5).

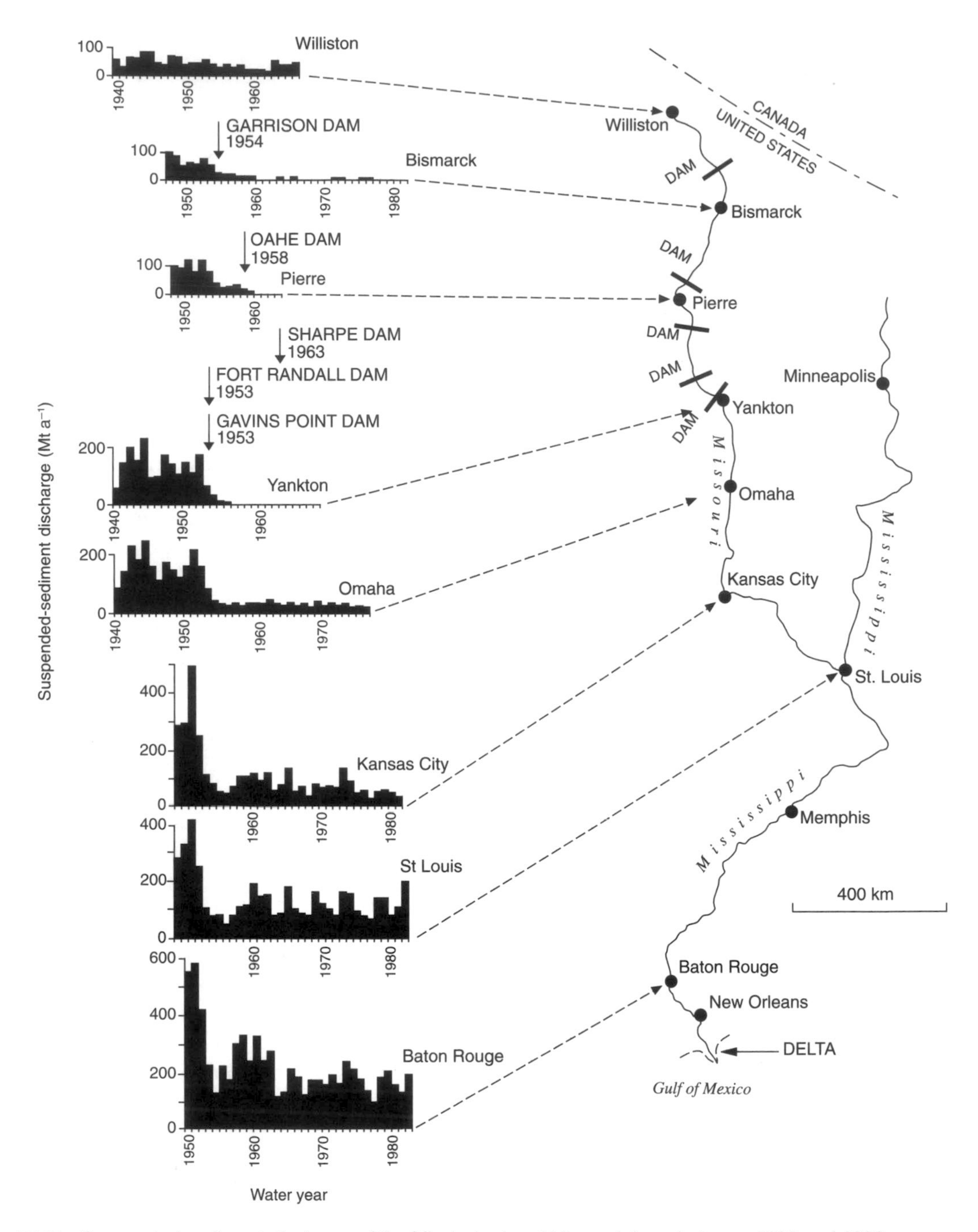

Figure 15.16 Suspended sediment discharge of the Mississippi and Missouri rivers between 1938 and 1982.

Table 15.8 Silt concentrations (in parts per million) in the Nile at Gaafra before and after construction of the Aswan High Dam

	Before construction[a]	*After construction*	*Ratio*
January	64	44	1.5
February	50	47	1.1
March	45	45	1.0
April	42	50	0.8
May	43	51	0.8
June	85	49	1.7
July	674	48	14.0
August	2702	45	60.0
September	2422	41	59.1
October	925	43	21.5
November	124	48	2.58
December	77	47	1.63

[a] Averages for the period 1958–63.
Source: A. A. Abul-Atta (1978) *Egypt and the Nile after the Construction of the High Aswan Dam* (Cairo: Ministry of Irrigation and Land Reclamation), p. 199

Human activities can affect river flow and flood characteristics. Urbanisation, for example, may affect flood runoff because urbanisation produces extended impermeable surfaces of bitumen, tarmac, tiles and concrete, and the construction of sewers and storm drains accelerates runoff (see also section 16.4). Likewise, as we saw in section 14.8, deforestation can also lead to marked changes in river flow, as can land drainage.

In recent decades human activity has had an increased influence on the quantity and quality of materials carried in rivers. For example, it is estimated that about 500 million tonnes of dissolved salts reach the oceans each year as a result of our activities. These inputs have increased by more than 30 per cent the natural values for sodium, chloride and sulphate, and have created an overall global augmentation of river solute loads of about 12 per cent. Some of the inputs are the products of industrial and municipal waste production (e.g. sewage), others are the result of the application of fertilisers to soils and pesticides and herbicides to crops, while others are the result of land-use changes. For example, the cutting or burning of a forest may release a great deal of nutrients into rivers.

Of even greater significance is the addition of suspended material to rivers as a consequence of soil erosion. Urbanisation (and especially construction activity), mining, overgrazing, deforestation and ploughing have caused severe soil erosion (see sections 13.6 and 16.6) and have thereby introduced increasingly large amounts of sediments into rivers. Such sediment makes rivers turbid, to the detriment of fauna and flora, and can also cause deposition downstream. For example, in a classic case study in the Sierra Nevada Mountains of California, the America geomorphologist G. K. Gilbert demonstrated the serious effects that gold mining could have. During the gold rush days in the second half of the nineteenth century gold was recovered by a process called hydraulic mining, which led to the addition of great quantities of sediments into the river valleys. This in itself raised their bed levels, changed their channel configurations and thereby caused flooding of lands that had previously been immune. Of even greater significance was the fact that the rivers transported increased quantities of debris into the estuarine waters of San Francisco Bay, causing extensive shoaling.

■ *Key Terms and Concepts*

- bedload
- braided streams
- capacity
- channelisation
- clear water erosion
- competence
- delta
- drainage density
- floodplain
- Horton's laws
- hydraulic geometry
- hydraulic radius
- inter-basin water transfer
- long profile
- meanders
- morphometry
- rates of denudation

■ *Points for Review*

- Why do some rivers have high drainage densities and others have low?
- Why do most rivers have concave long profiles?
- Why do some rivers braid and others meander?
- Why do some rivers display terraces?
- Where might you expect rivers to have the highest loads of (a) dissolved solids and (b) suspended material?
- By reference to a recent flood event of your choice illustrate the conditions that gave rise to it.
- In what ways can humans modify the form of river channels?
- Using local topographical maps, compare various morphometric indices between two small streams.

FURTHER READING

Gregory, K. J. (ed.) (1997) *Fluvial Geomorphology of Great Britain* (London: Chapman and Hall). A large compendium of some of the key fluvial landform sites in Great Britain.

Knighton, D. (1998) *Fluvial Forms and Processes: A New Perspective*, 2nd edn (London: Arnold). A standard text.

Lewin, J. (ed.) (1981) *British Rivers* (London: Allen and Unwin). An edited review of many aspects of British rivers.

Smith, D. I. and Stopp, P. (1978) *The River Basin* (Cambridge: Cambridge University Press). A valuable, short introduction.

Smith, K. and Ward, R. C. (1998) *Floods: Physical Processes and Human Impacts*. (Chichester: Wiley). An excellent discussion of floods.

Thornes, J. (1979) *River Channels* (London: Macmillan). A very concise introduction.

16 Cities

16.1 Introduction

The world is becoming increasingly urbanised. In 1980 there were 35 cities with populations of over 4 million; by 2025, 135 cities will probably have reached this size. Over the period 1950–90 the total population of the world's cities increased tenfold, and is now not far short of 3 billion. Cities thus contain around half the world's population. Urban populations are concentrated into a relatively small area; for example, only 3.4 per cent of land in the USA is urbanised. This makes the urban impact upon the environment even more intense.

The impact of urban areas on the environment and ecology can be devastating. Problems have been felt for a long time in many countries where industrial cities developed early. In many less-developed countries, huge expansion in population has occurred relatively recently, leading to burgeoning environmental problems.

What impacts do cities have on the environment? And how do these affect ecology? Cities do all of the following:

- produce a major demand for natural resources in the surrounding area;
- obliterate the natural hydrological system on the site of the city;
- reduce biomass and alter the species composition on the site of the city;
- produce waste products which can alter the environment in and around the city;
- create new land through reclamation and landfill.

Together, these impacts make up the 'ecological footprint' of a city; that is, the environment is affected by pollution, resource extraction, development and transport caused by the city itself. Cities demand raw materials such as timber, coal and oil: these must be extracted from the surrounding area or be transported into the city. They also require agricultural products, energy and labour. As the various parts of the world become increasingly interconnected, the ecological footprints of major cities become bigger and bigger. This means that a vast proportion of the earth's surface is being sucked into the urban system one way or another. On the sites of cities, the entire enterprise of urbanisation leads to drastic changes in geomorphology, climate, hydrology and ecology.

16.2 The Climate of Cities

Large cities create their own climatic conditions (figure 16.1). In terms of their effect on climate, urban areas have been likened to volcanoes, deserts and irregular forests: they generate dust, they give out heat, they are relatively lacking in vegetation cover, and they have large structures that affect the frictional characteristics of the ground surface. The changes they bring about in the natural climatic conditions of an area can be significant; some of the most important effects are listed in table 16.1.

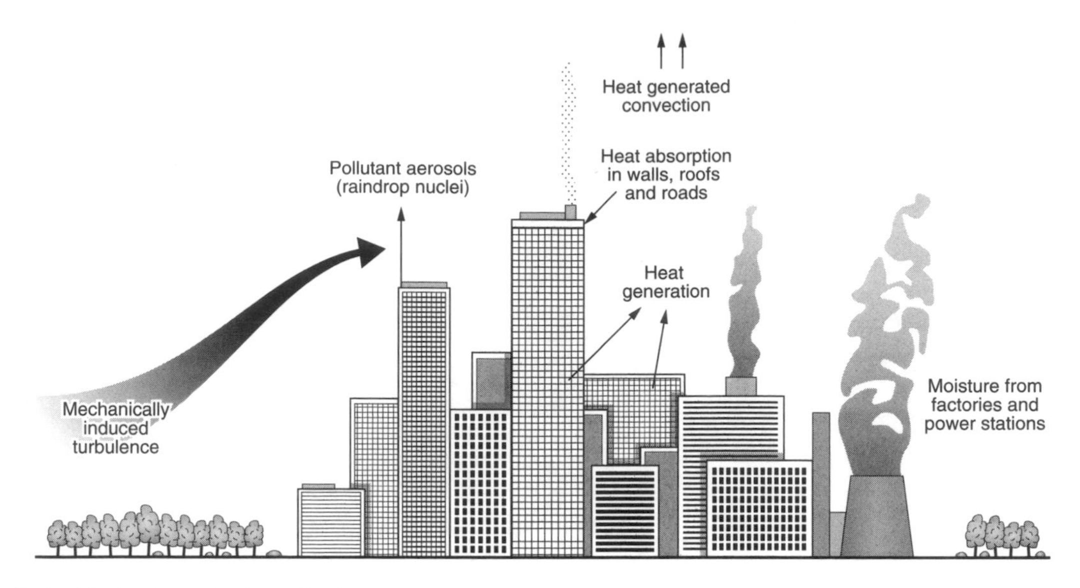

Figure 16.1 Mechanisms of urban climates.

In comparison with the rural areas across which they sprawl, city surfaces absorb appreciably more solar radiation because a higher proportion of the reflected radiation is retained by the high walls and dark-coloured roofs and roads of city streets. The concreted and tarmacked city surfaces have great thermal capacity and good conductivity, so that heat tends to be stored during the day and released at night. By contrast, the plant cover of rural areas act very much like an insulating blanket, so rural areas tend to experience relatively lower temperatures by day and night, an effect that is enhanced by the evaporation and transpiration that take place. A further thermal change is achieved because cities produce a great deal of artificial heat as a result of the activities of industrial, commercial and domestic energy users. Climatologists therefore talk about cities as being *urban heat islands* (window 16.1).

Generally speaking, the highest temperature anomalies between town and country occur in the densely built-up area near the city centre, and decrease markedly at the city perimeter. Observations in some large Canadian cities suggest temperature gradients of as much as 4° per kilometre between the urban centre and the urban edge. Temperature differences also tend to be greatest during the night. If we again consider the metropolis as a heat 'island', protruding distinctly out of the cool 'sea' of the surrounding landscape, we can say that the rural/urban boundary exhibits a steep temperature gradient or 'cliff' to the island, while much of the rest of the urban area appears as a 'plateau' of warm air with a steady but weaker horizontal gradient of increasing temperature towards the city centre. The urban core may be a 'peak' where the urban maximum temperature is found. The difference between this value (u) and the background rural temperature (r) defines the *urban heat island intensity* (ΔT_{u-r}).

The average annual urban–rural temperature differences for various large cities are listed in table 16.2, and it can be seen that the values range from 0.6 to 1.3 °C. In general, larger cities have a greater urban heat island intensity, with ΔT_{u-r} being proportional to the log of the city's population. None the less, sizeable nocturnal temperature differences have been found even in relatively small cities. Factors such as building density and the nature of the economic activities may be at least as important as city size, and cities that occur in areas with very high wind velocities will tend to have smaller temperature differences than those that occur in areas of lesser wind velocity. Strong winds tend to lessen the heat build-up of the urban heat island.

Table 16.1 Average changes in climatic elements caused by cities

Element	*Parameter*	*Urban compared with rural (– less; + more)*
Radiation	On horizontal surface	–15%
	Ultraviolet	–30% (winter); –5% (summer)
Temperature	Annual mean	+0.7 deg.
	Winter maximum	+1.5 deg.
	Length of freeze-free season	+2–3 weeks (possible)
Wind speed	Annual mean	–20 to –30%
	Extreme gusts	–10 to –20%
	Frequency of calms	+5 to +20%
Relative humidity	Annual mean	–6%
	Seasonal mean	–2% (winter); –8% (summer)
Cloudiness	Cloud frequency + amount	+5 to +10%
	Fogs	+100% (winter); +30% (summer)
Precipitation	Amounts	+5 to +10%
	Days	+10%
	Snow-days	–14%

Phenomenon	*Consequence*
Heat production (the heat island)	Rainfall + Temperature +
Retention of reflected solar radiation by high walls, dark coloured roofs, and roads	Temperature +
Surface roughness increase	Wind – Eddying +
Dust increase (the dust dome)	Fog + Rainfall + (?)

Source: after H. Landsberg in J. F. Griffiths (1976) *Applied Climatology: An Introduction* (Oxford: Oxford University Press), p. 108

Table 16.2 Annual mean urban–rural temperature differences of cities

City	*Temperature difference (°C)*
Chicago	0.6
Washington, DC	0.6
Los Angeles	0.7
Paris	0.7
Moscow	0.7
Philadelphia	0.8
Berlin	1.0
New York City	1.1
London	1.3

Source: from data in J. T. Peterson, in T. R. Detwyler (ed.) (1971) *Man's Impact on Environment* (New York: McGraw Hill), table 11.2, p. 136 and other sources

The existence of the urban heat island has a number of implications. Snow may lie for a shorter period; frosts may be less frequent; more air conditioning may be necessary in summer; temperatures may prove oppressive for humans; some animals will be attracted to the thermally more favourable urban habitat; and plants may bud and bloom earlier than they do in the countryside.

The urban heat island effects are relatively straightforward to comprehend: it is much harder to measure and explain the urban-industrial effects on clouds, rain, snowfall and associated weather hazards. However, certain changes can be anticipated. For example, the thermal increase of the city would tend to promote upward movement of air; there might also be increased vertical movements caused by the presence of high buildings; pollution from industries might provide nuclei for cloud and raindrop formation; and industries might emit considerable quantities of water vapour.

Table 16.3 illustrates the difference in summer rainfall, thunderstorms and hailstorms between various rural and urban areas in the United States. These data indicate that rainfall increased by 9–27 per cent, that the incidence of thunderstorms increased by 10–42 per cent and that hailstorms increased by 67–430 per cent.

An interesting example of the effects of a major conurbation is provided by the London area. It is believed that in this city the mechanical effect of the buildings has been an important factor affecting weather conditions, by creating an obstacle to

Window 16.1 *The implications of some urban heat islands*

As cities grow, so does their heat island effect. In Columbia, Maryland, USA, for example, when the town had only 1000 inhabitants in 1968, the maximum temperature difference between residential areas and the surrounding countryside was only 1 °C. By 1974, when it had grown to a town with a little over 20000 inhabitants, the maximum heat island effect had grown to 7 °C.

Thus the annual average temperatures over the hearts of great cities can be substantially higher than those over the surrounding countryside. This is clear from the temperature map of Paris. The outlying weather stations have mean annual temperatures of 10.6–10.9 °C, whereas in central Paris the value is 12.3 °C, about 1.5 °C higher. These values have all been reduced to a uniform elevation of 50 m above sea level to correct for possible orographic effects.

Urban climates are often characterised by different precipitation characteristics from rural areas. For example, it is remarkable that there tends to be more rain in Paris during the week than at weekends. There is a gradual increase in average rainfall from Monday to Friday (when factories are producing more heat and aerosols), then a sharp drop for Saturday and Sunday. The weekend average for May to October was 1.47 mm, whereas the workday average was 1.93 mm, a decrease of 24 per cent for the weekend.

In winter months the consequences of urban heat islands can be particularly significant in cold regions. For example, the average date of the last freezing temperatures at the end of winter in Washington, DC in the USA is about three weeks earlier than in the surrounding rural areas. In autumn the city has, on average, the first freezing temperature on about 3 November, whereas in the outlying suburb 0 °C will usually be observed about two weeks earlier. Thus, in all, the frost-free season will be about 35 days longer than it is in the countryside. Similar figures have been obtained for some other great cities. Data for Moscow, Russia, suggest an increase of around 30 days without freezing, while those for Munich in Germany suggest an increase that may be as great as 61 days.

In the summer months the urban heat island effect can lead to an increasing demand for air conditioning, and because the energy requirements of air conditioning are greater than those of heating, the savings in winter heating bills are more than offset. Moreover, air conditioning can aggravate the heat island effect because air conditioning plant discharges heat to the outside air, where it mixes with air that has already been warmed up by the hot air forming adjacent to sunlit walls and pavements.

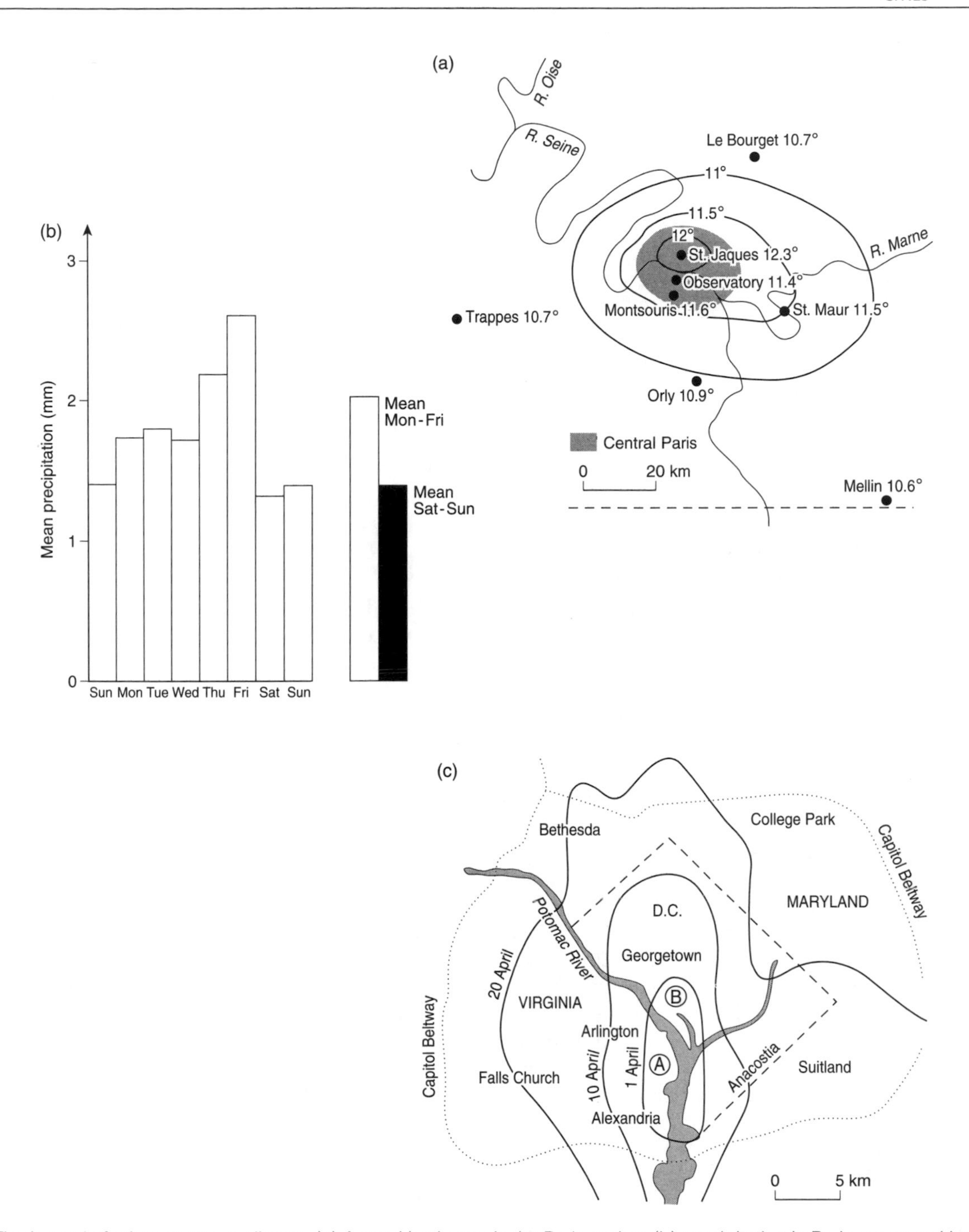

The impact of urban areas on climate. (a) Annual isotherms in the Paris region; (b) precipitation in Paris, averaged by day of week; (c) average annual date of last freezing temperature in spring in Washington, DC (A, International Airport; B, White House).

Table 16.3 Areas of maximum increases (urban–rural difference) in summer rainfall and severe weather events for eight American cities

	Rainfall		*Thunderstorms*		*Hailstorms*	
City	*Percentage*	*Location*[a]	*Percentage*	*Location*[a]	*Percentage*	*Location*[a]
St Louis	+15	B	+25	B	+276	C
Chicago	+17	C	+38	A, B, C	+246	C
Cleveland	+27	C	+42	A, B	+90	C
Indianapolis	0	—	0	—	0	—
Washington, DC	+9	C	+36	A, B	+67	B
Houston	+9	A	+10	A, B	+430	B
New Orleans	+10	A	+27	A	+350	A, B
Tulsa	0	—	0	—	0	—

[a] A, within city perimeter; B, 8–24 km downwind; C, 24–64 km downwind.
Source: after S. A. Changnon (1973) 'Atmospheric alterations from man-made biosphere changes', in W. R. D. Sewell (ed.) *Modifying the Weather: A Social Assessment* (Victoria, BC: University of Victoria Press), figure 1.5, p. 144

air flow and by causing frictional convergence of airflow. A long-term analysis of thunderstorm records in south-east England is highly suggestive, indicating the higher frequencies of thunderstorms over the conurbation compared with elsewhere. The similarity between the conurbation and the area in the south-east having the highest frequency of thunderstorms is striking. Moreover, the historical analysis of weather records for the capital shows a steadily increasing number of thunderstorms as the city has grown.

One of the most celebrated examples of the effect of a city and industrial complex on local precipitation is provided by the Chicago area of the American Mid-West. At La Porte, Indiana, some 48 km downwind of a large industrial complex between Chicago and Gary, the amount of precipitation and the number of days with thunderstorms and hail have increased markedly since 1925. There has been a 30–40 per cent increase in precipitation, and this seems to parallel the curve of increase in atmospheric pollution associated with the increasing production of the Chicago iron and steel complex. Moreover, peaks in steel production have been seen to be associated with highs in the La Porte precipitation curve.

Cities also affect wind velocities. As we have already seen, buildings of the high-rise type exert a powerful frictional drag on air moving over and around them. This creates turbulence, with characteristically rapid temporal and spatial changes in both speed and direction, though the average speed of the winds tends to be lower in built-up areas than in rural areas. The overall annual reduction of wind speed in central London is about 6 per cent but for the higher-velocity winds the reduction is more than doubled. The heat island effect also affects wind characteristics. In some English cities, such as Leicester and London, observations have shown that, on calm, clear nights when the urban heat island effect is at a maximum, there is a surface inflow of cool air towards the zones of highest temperatures. These so-called 'country breezes' have low velocities and become quickly decelerated by intense surface friction in the suburban areas.

16.3 Air Pollution and its Problems

Even 600 or 700 years ago, the growth of cities such as London was beginning to reduce the quality of the air. In medieval London the pollution from coal-burning was regarded as such a serious matter that a commission was set up to investigate the problem in 1285, and by 1306 offenders were being punished by *grievous ransoms*, which probably included fines and confiscations. Lime furnaces appear to have been a particular cause of offence. In the modern era the concentration of large numbers of people, factories, power stations and cars means that large amounts of pollutants may be emitted into urban atmospheres. If weather conditions permit, the level of pollution may build up. The nature of the pollutants (table 16.4) has changed as technologies have changed. For example, in the early

Table 16.4 Major urban pollutants

Type	*Some consequences*
Suspended particulate matter (characteristically 0.1–2.5 μm in diameter)	Fog, respiratory problems, carcinogens, soiling of buildings
Sulphur dioxide (SO_2)	Respiratory problems, can cause asthma attacks. Damage to plants and lichens, corrosion of buildings and materials, production of haze and acid rain
Photochemical oxidants: ozone and peroxyactyl nitrate (PAN)	Headaches, eye irritation, coughs, chest discomfort, damage to materials (e.g. rubber), damage to crops and natural vegetation, smog
Oxides of nitrogen (NO_x)	Photochemical reactions, accelerated weathering of buildings, respiratory problems, production of acid rain and haze
Carbon monoxide (CO)	Heart problems, headaches, fatigue etc.
Toxic metals: lead	Poisoning, reduced educational attainments and increased behavioural difficulties in children
Toxic chemicals: dioxins etc.	Poisoning, cancers etc.

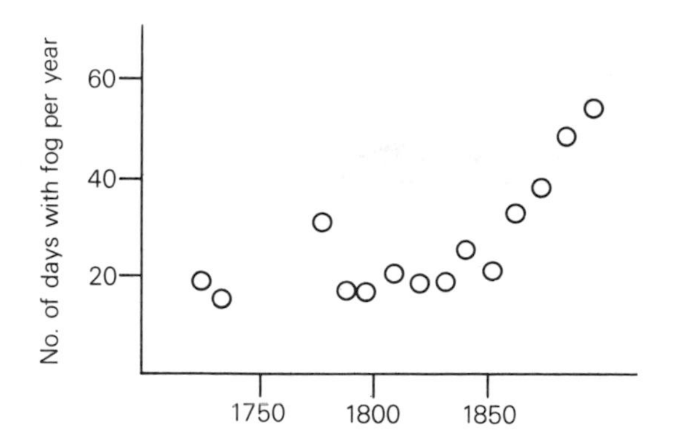

Figure 16.2 Number of fogs in London per year based on decadal means.

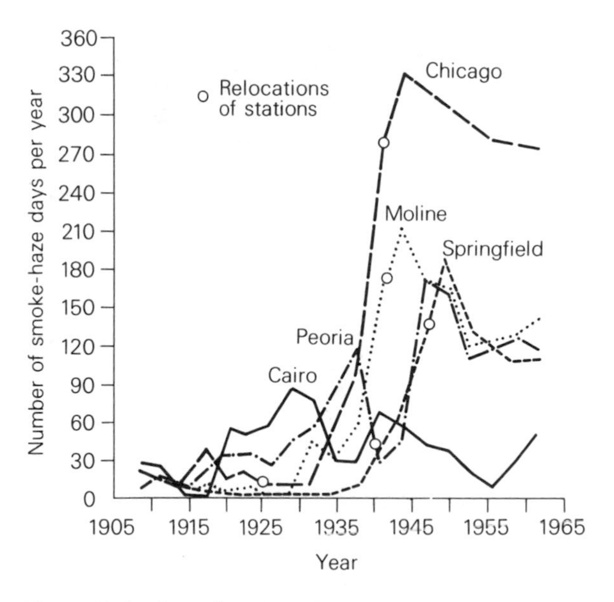

Figure 16.3 Annual number of days with smoke-haze conditions in Illinois cities as based on a three-year moving average.

phases of the Industrial Revolution in Britain the prime cause of air pollution in cities may have been the burning of coal, whereas now it may be vehicular emissions. Different cities may have very different levels of pollution, depending on factors such as the level of technology, size, wealth and anti-pollution legislation. Differences may also arise because of local topographical and climatic conditions (see colour plate 31).

With the rapid increase in the size of urban areas and their degree of industrialisation in the past two centuries, the pollution problem has accelerated in importance. This is made apparent by a consideration of the data for the number of days with fog in London (figure 16.2). Prior to 1750 there would appear to have been about 18–20 days of fog each year; by the end of the nineteenth century this figure had risen to over 50 days. A comparable situation is evident when one considers some of the data for cities in the Illinois industrial area in the United States (figure 16.3). Chicago, the biggest city in the group, was the first one to show a sizeable increase in the number of days with smoke-haze conditions. Moderately industrialised cities like Moline, Peoria and Springfield also show sizeable increases beginning in the 1930s; but Cairo, a non-industrial city with minimal twentieth-century growth, has shown no marked upward trend in smoke-haze conditions.

However, smoke control legislation, such as the British Clean Air Acts, can lead to a very considerable reduction in particle amounts, and in London, as a result of legislation, emissions of smoke had by 1970 been reduced to one-tenth of what they were in 1956. This legislation has also succeeded in increasing the number of hours of sunshine recorded in city centres. In Manchester, over the same period, the number of hours of sunshine recorded

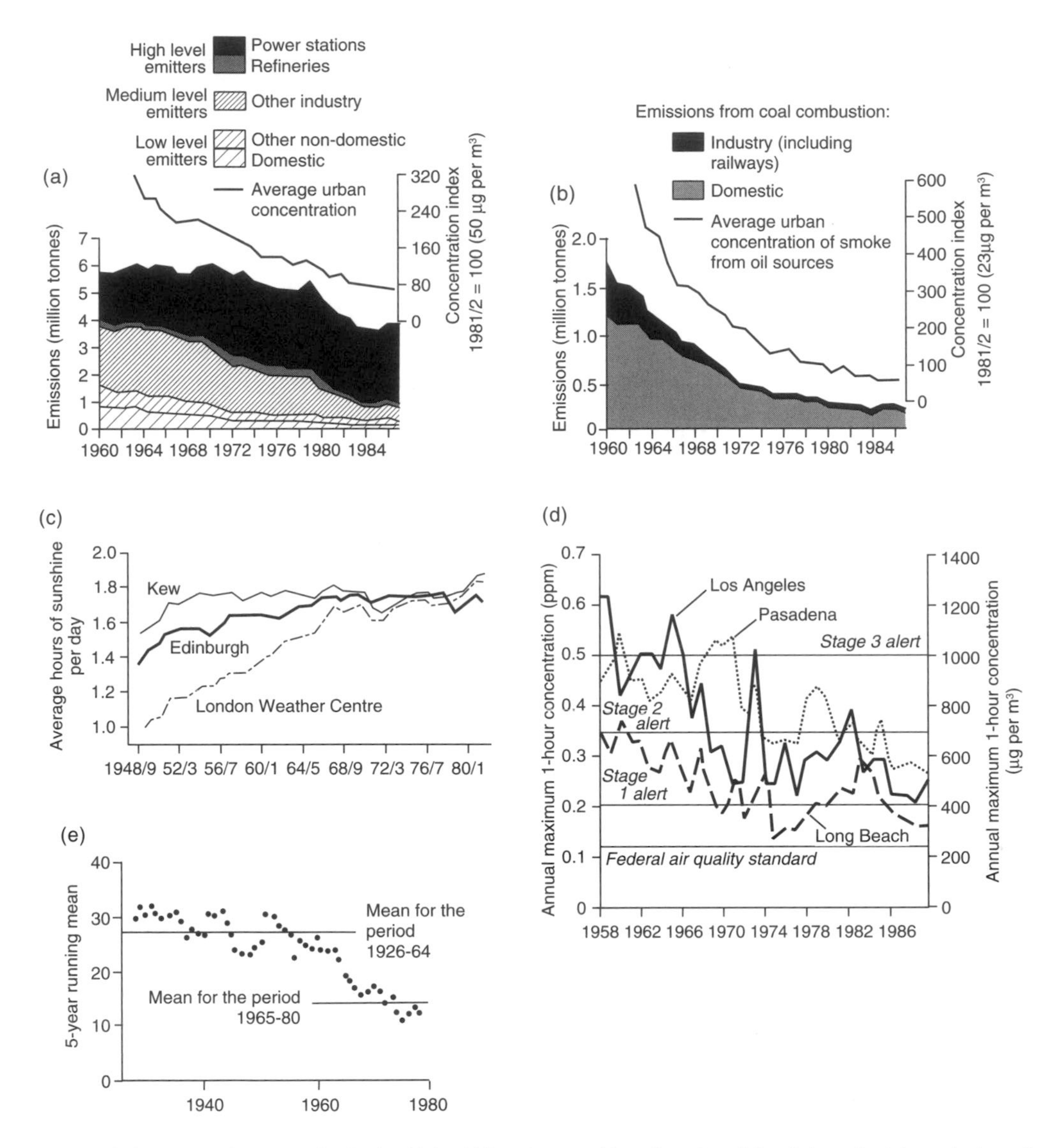

Figure 16.4 Trends in atmospheric quality in the United Kingdom and Los Angeles: (a) sulphur dioxide emissions from fuel combustion and average concentration in the UK; (b) smoke emissions from coal combustion and average urban concentration of oil smoke in the UK; (c) increase in winter sunshine (10-year moving average) for London and Edinburgh city centres and for Kew, outer London, UK; (d) annual maximum hourly ozone concentration at selected sites in the Los Angeles basin, 1958–89; (e) annual fog frequency at 0900 GMT in Oxford, central England, 1926–80.

in the winter months of November–January increased from around 70 to about 110.

The sulphurous yellow 'pea-souper' fogs or smogs produced by coal-burning are only one of the types of fog that occur in urban areas. There is a second widespread type called *photochemical smog*. The name originates from the fact that most of the less desirable properties of such fog result from the chemical reactions induced by sunlight. This smog appears 'cleaner' than fog produced by the burning of coal in the sense that it does not contain the very large particles of soot. However, the eye irritation and damage to plant leaves that it produces still make it unpleasant. Photochemical smog occurs particularly where there is large-scale combustion of petroleum products, as in cities dominated by the motor car, such as Los Angeles. A portion of the exhaust emissions is converted into harmful substances like ozone by the effects of sunlight. However, such smogs are fortunately not universal. Because sunlight is a crucial factor in their development, they are most common in the tropics or during seasons of strong sunlight. Their especial notoriety in Los Angeles is due to a meteorological setting dominated for long spells by subtropical anticyclones with weak winds, clear skies and a subsidence inversion,

Window 16.2 Urbanisation and hydrology

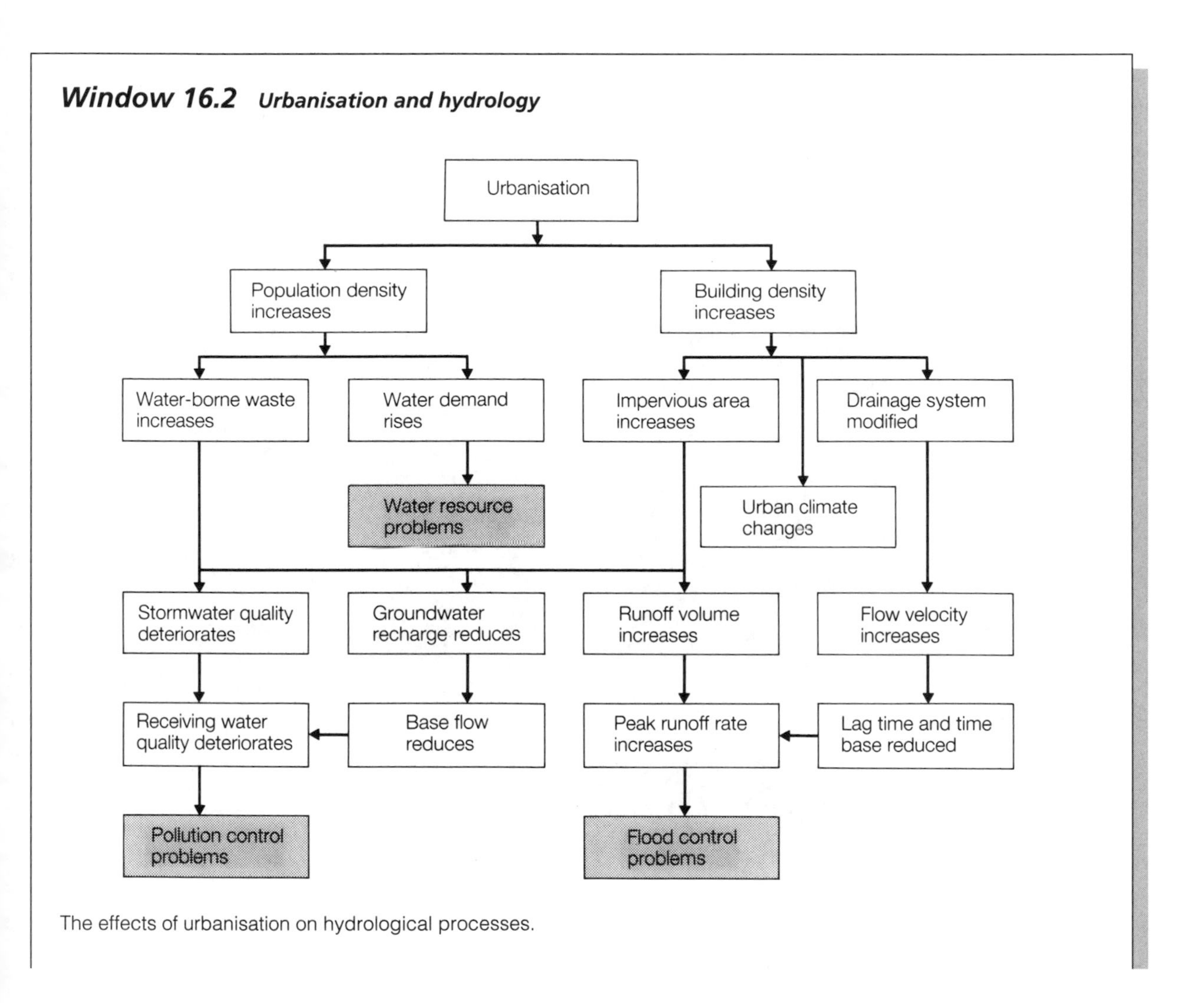

The effects of urbanisation on hydrological processes.

There are five main ways in which the process of urbanisation alters the functioning of the hydrological cycle:

(1) The replacement of vegetated soils with impermeable surfaces:
- reduces the amount of storage on the ground surface and within the soil itself, thereby increasing the proportion of precipitation that runs off into channels
- increases the speed at which overland flow occurs
- decreases the evapotranspiration losses
- reduces the degree of percolation to groundwater because large areas of the surface are impermeable

(2) The installation of storm water sewers increases the drainage density which in turn:
- reduces the distance that overland flow has to move over the surface before a channel is reached
- increases the speed of water flow because sewers have a smoother, more efficient shape than natural channels
- reduces the amount of water that can be stored within the channel system because storm water sewers are engineered to remove water as quickly as possible

(3) The process of construction work:
- clears the surface of vegetation, so facilitating overland flow
- leads to disturbance and churning of the soil surface, which tends to make it more erodible

(4) The encroachment of embankments, roads etc. on to the river channel:
- reduces the width of stream channels and the degree of storage within them, thereby increasing the height of floods
- reduces the free discharge of flood water and so causes water to be ponded upstream

(5) The climatology of the urban area is modified, which often leads to more precipitation, especially in the summer, and to heavier more frequent thunderstorms.

The construction of a massive storm drain in Dhaka, Bangladesh, illustrates one major way in which humans can transform urban hydrology.

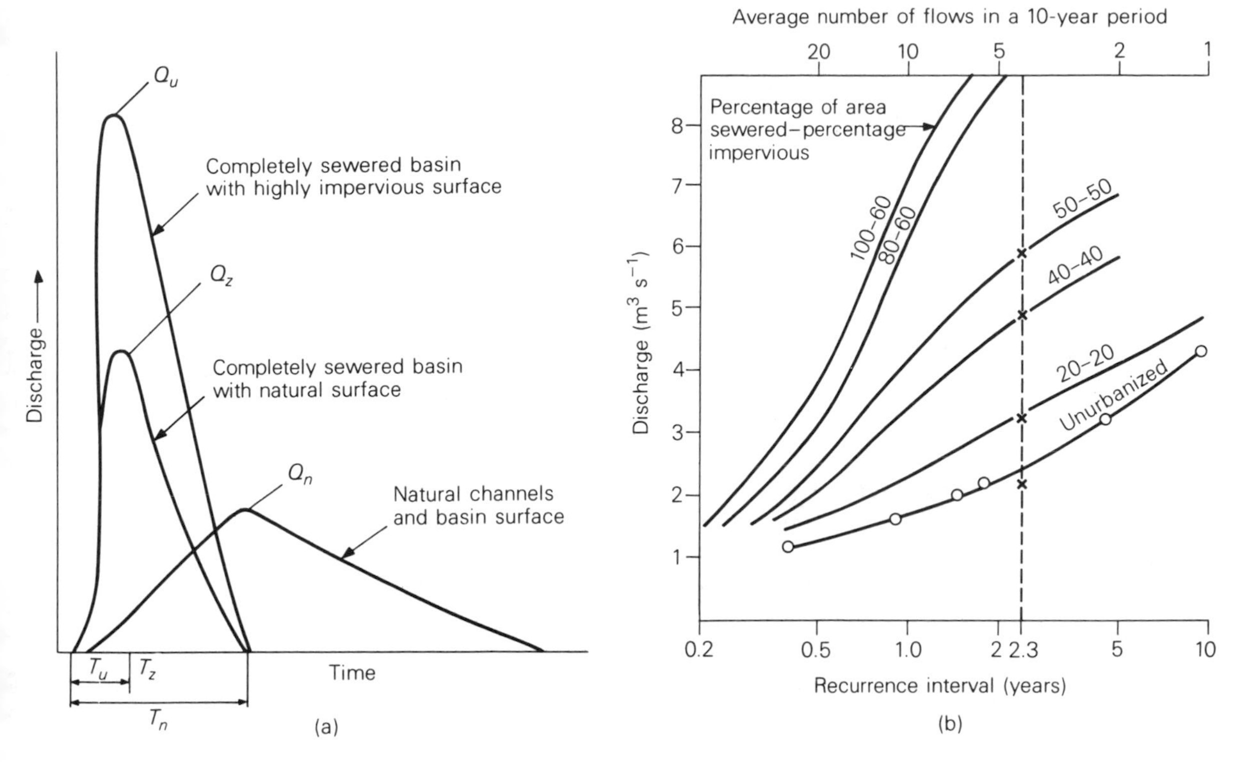

Figure 16.5 The effect of cities on floods. (a) The effect of urban development on flood hydrographs. Peak discharges (Q) are higher and occur sooner after runoff starts (T) in basins after they have been developed or sewered. (b) Flood frequency curves for a 1 square mile basin in various stages of urbanisation.

combined with the general topographical situation partially bounded by hills and the enormous vehicle density of that affluent city.

The recent improvements in the air quality of the UK and in the Los Angeles basin as a result of changes in legislation are shown in figure 16.4. It shows the decreasing fog frequency and increasing sunshine levels for the UK, and the steady fall that has taken place in the concentrations of carbon monoxide, nitrogen oxide and ozone since the late 1960s in Los Angeles.

By contrast, in many cities in poorer countries, pollution is increasing at present. In certain countries, heavy reliance on coal, oil and even wood for domestic cooking and heating means that their levels of sulphur dioxide and suspended particulate matter (SPM) are high and climbing. In addition, rapid economic development is bringing increased emissions from industry and motor vehicles, which are generating progressively more serious air-quality problems.

16.4 Urbanisation and River Flow

As we have already seen in connection with urban climatic modification, city surfaces are very different from rural surfaces. Once of the most important consequences of this is that city surfaces respond in a very different manner to storms, so that the runoff they produce is different in quantity and timing, with obvious implications for flooding (window 16.2).

Research in both the United States and Britain has shown that, because urbanisation leads to an extension of impermeable surfaces of tarmac, tiles and concrete, the infiltration capacity of the sur-

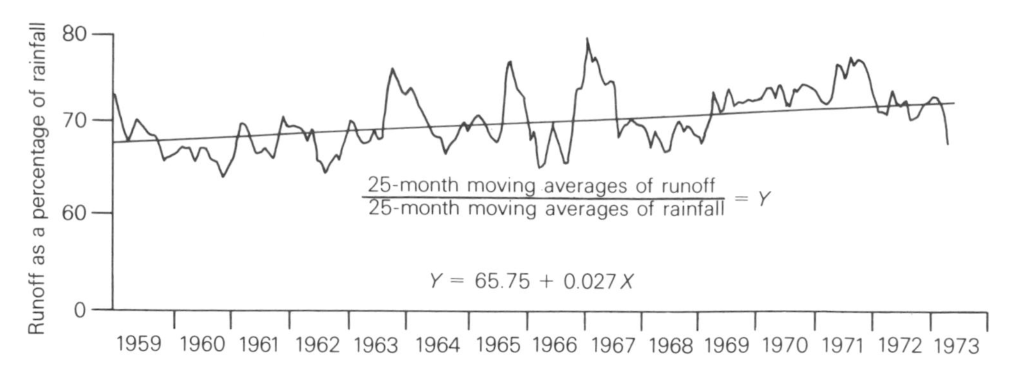

Figure 16.6 Trends in 25-month moving averages of runoff as a percentage of 25-month moving averages of rainfall for the River Tame, Lee Marston, England, 1959–74. The trend reflects increasing amounts of imported runoff.

face is reduced. This means that there is a tendency for flood runoff to increase in comparison with rural sites (figure 16.5). Furthermore, the installation of sewers and storm drains further accelerates runoff; in general, the greater the area that is sewered, the greater the discharge for a particular storm intensity. Some additional runoff may be generated in urban areas because low vegetation densities mean that losses from evapotranspiration are limited.

However, in many cases the effect of urbanisation is greater for small floods, and as the size of the flood and its recurrence interval increase, so the effect of urbanisation diminishes. A probable explanation for this is that, during a severe and prolonged storm, a non-urbanised catchment may become so saturated and its channel network so extended that it will begin to behave hydrologically as if it were an impervious catchment with a dense surface-water drain network. Hence, under these conditions a rural catchment produces floods of a type and size similar to those of its urban counterpart. Moreover, in an urban catchment it seems probable that at very high discharges some throttling of flow may take place in surface-water drains and sewers, thereby damping down the peak of the flood. None the less, a study in Jackson, Mississippi, found that the 50-year flood rate for an urbanised catchment was three times higher than for a rural one.

River flow in urban areas may be modified in one further way. Cities import much water for industrial and municipal purposes, and such *inter-basin water transfers* are having an increasing impact on river regimes. Under dry weather flow conditions, streamflow in urban catchments in the Midlands and south-east England is now mainly imported water discharged from factories and so on. This is illustrated by the River Tame (figure 16.6), which contains about 90 per cent effluent in its 95 per cent duration flow. This proportion is increasing as more water is imported.

16.5 Thermal Pollution of River Water

The pollution of water by altering its temperature is termed *thermal pollution*. As many river fauna are directly affected by temperature, this particular environmental impact has some significance. In industrial countries probably the main source of thermal pollution is from power stations, which release large quantities of condenser cooling water into rivers. Water discharged from power stations has been heated on average by some 6–9 °C, but usually has a temperature of less than 30 °C.

The extent to which water affects river temperatures depends very much on the state of flow. For example, below the Ironbridge power station in England, the River Severn undergoes a temperature increase of only 0.5 °C during floods, compared with an 8 °C increase at times of low flow. As electricity generation expands, this problem will also expand, although as power stations become more efficient the amount of thermal pollution per unit of electricity produced will probably decline.

Thermal pollution may also follow directly from

urbanisation. This results from various sources: changes in the temperature regimes of streams brought about by reservoirs; changes produced by the urban heat island effect; changes in the configuration of urban channels (e.g. their width : depth ratio); changes in the degree of shading of the channel, either by covering it over or by removing the natural vegetation cover; changes in the volume of storm runoff; and changes in the groundwater contribution. Studies in New York City have shown that the basic effect of the city on water temperature was to raise it in summer by as much as 5–8 °C and to lower it in winter by as much as 1.5–3 °C.

16.6 Soil Erosion and Sediment Yield Associated with Construction and Urbanisation

In the process of building a city the ground surface becomes greatly disturbed as the vegetation is removed, ditches are dug, excavations and fillings are made and construction vehicles churn up the surface. This can cause very high rates of erosion and sediment yield to occur, rates that in the course of a year may be equivalent to many decades of natural or even agricultural erosion. Studies in Maryland, in the United States, showed that, during the construction phases of urban development, sediment yields reached 55 000 $t\,km^{-2}\,yr^{-1}$, while in the same area rates under forest were around 80–200 $t\,km^{-2}\,yr^{-1}$. Likewise, studies in Devon have shown that suspended sediment concentrations in streams draining construction areas were ×2 to ×10 (occasionally up to ×1000) those from undisturbed areas, while studies in Virginia showed that the rates during construction were ×10 those from agricultural land, ×200 those from grassland and ×2000 those from forest in the same area.

However, construction does not go on for ever, and once the disturbance ceases, roads are surfaced and gardens and lawns are created, the rates of erosion go down and may be of the same order as those under natural or pre-agricultural conditions.

16.7 Animals in Cities

The growth of cities has often been detrimental to wildlife, but there are many examples of animals that have benefited from, and adapted to, the

Plate 16.1 Many wild animals are now adapting in various ways to city life. This fox patrols the streets of Oxford during the night hours, seeking appropriate food in the form of garbage or domestic animals.

changes wrought by humans, to the extent that they have become very closely linked to them (plate 16.1). Such animals are called *synanthropes*. Pigeons and sparrows, for instance, now form permanent and numerous populations in almost all of the world's large cities; human food supplies are the food supplies for many synanthropic rodents such

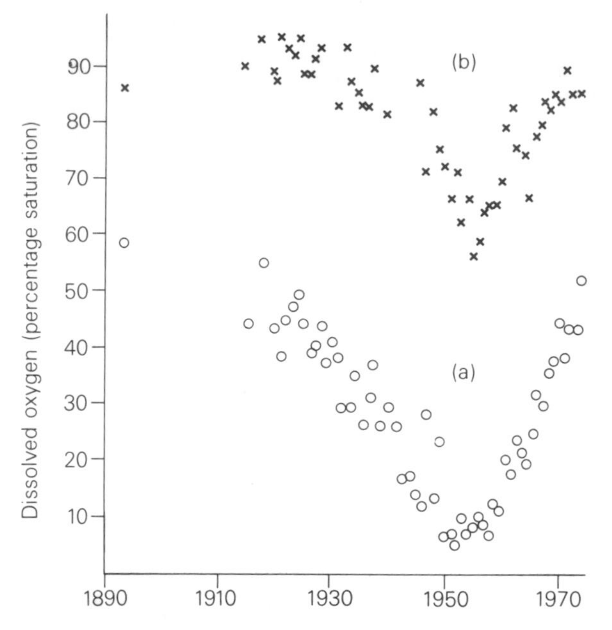

Figure 16.7 The average dissolved-oxygen content of the River Thames in England at half-tide in the July–September quarter since 1890: (a) 79 km below Teddington weir; (b) 95 km below Teddington weir.

as rats and mice; a once shy forest bird, the blackbird, has in the course of a few generations become a regular and bold inhabitant of many gardens; and the grey squirrel in many places now occurs much more frequently in parks than in forests. The marked increase in many species of gulls over recent decades in temperate regions is largely attributable to the increasing availability of food scraps on urban refuse tips. It may seem incongruous that the niches of scavenging birds that exploit rubbish tips should have been filled by seabirds, but the gulls have proved to be ideal replacements for the kites which humans have removed, as they have the same ability to watch out for likely food sources from aloft and then to hover and plunge when they spy suitable morsels for sustenance.

Animals that can tolerate disturbance, are adaptable, utilise patches of open or woodland-edge habitat, creep about inside buildings, tap human food surreptitiously, avoid recognisable competition with man or attract human appreciation and esteem may increase in the urban milieu. For these sorts of reasons the north-eastern megalopolis of the United States hosts thriving populations of squirrels, rabbits, racoons, skunks and opossums, while some African cities are now frequently blessed with the scavenging attentions of hyenas.

Many species have suffered from the air and water pollution associated with cities. As industrialisation and municipal waste production proceeded, for example, the quality of the River Thames declined to such an extent that it contained virtually no fish (figure 16.7). The amount of dissolved oxygen in its waters declined dramatically until about 1950, when, because of more efficient controls on effluent discharge, the downward trend in water quality has been reversed, and many fish, long absent from the Thames in London, are now returning.

■ *Key Terms and Concepts*

ecological footprint
photochemical smog
synanthropes
thermal pollution
urban heat island

■ *Points for Review*

- What do you understand by the 'ecological footprint' of a city?
- Why can cities have higher temperatures and more rain than the countryside?
- How do levels of air pollution vary between different cities?
- How do cities increase the risk of floods?
- Why might the biodiversity of cities be greater than that of the neighbouring countryside?

FURTHER READING

Douglas, I. (1983) *The Urban Environment* (London: Edward Arnold). A pioneer work.

Kirby, C. (1996) 'Urban air pollution'. *Geography*, 80: 375–92. A useful review in a very accessible journal.

Landsberg, H. E. (1981) *The Urban Climate* (New York: Academic Press). The classic study.

Oke, T. J. (1987) *Boundary Layer Climates*, 2nd edn (London: Routledge). A thorough review of local-scale climates which includes an authoritative study of urban climates.

Part V

Conclusion

17 Nature, Humans and the Environment

17.1 Introduction

The purpose of this final chapter is to try to extract some of the key themes that have arisen throughout the book and are fundamental to an understanding of the field of physical geography. They can be summarised as follows:

(a) The environment of humans is complex, and to understand the evolution and character of an area of landscape it is necessary to recognise that many factors are involved.
(b) These factors should not be seen as discrete entities, for they interact with each other.
(c) The importance of different factors may vary according to the scale of the investigation.
(d) Humans are an important factor in the environment – they both modify it through their actions, and have their actions modified by the environment.
(e) The environment is always changing. This is partly because of human activity, but there are also natural environmental changes.
(f) It is often difficult to ascertain whether any particular change is the result of purely natural changes, or whether humans have contributed.
(g) The problems of explanation are heightened by the fact that different processes can lead to similar end-forms – this is the *principle of equifinality*.
(h) Physical geographers, because of their subject matter, attitudes and approaches, undertake studies that are of direct relevance to humans.
(i) We are moving into an era when it is possible to talk of global environmental change.

Opposite In 1988 many seals died in the North Sea (this photograph was taken on the beach near King's Lynn, Norfolk, in September 1988); there was considerable discussion as to whether pollution had lowered the animals' resistance to a viral infection.

17.2 Environmental Complexity

The factors that can influence the nature of an area or cause a change in the environment are numerous. In general, therefore, it is not appropriate to seek single-cause explanations for phenomena. For example, there is a prima facie case for believing that a major vegetation type such as savanna is unlikely to be the result of merely one influence, such as climate; as we saw in section 7.8, other influences may also play a role, including soil quality, fires, environmental history and topographical position. Likewise, as was demonstrated in chapter 13, the nature of a soil profile is influenced by numerous factors, including the regional climate, the biota, the topography, the nature of the parent material, the period over which soil formation has taken place and human activities.

17.3 Environmental Interactions

Having recognised that the environment is complex, the next stage in understanding is to appreciate that the factors with which one is concerned do not exist as discrete entities, but interact with each other in a variety of ways. The whole concept of *systems* involves a consideration of inputs, throughputs and outputs interacting until some sort of equilibrium state is reached. Likewise, the study of *ecosystems* involves an investigation of the relationships between organisms and their environments – relationships that may involve all sorts of feedback loops. Within ecosystems, the *cycling* of materials is important (see chapter 10), as is the transference of energy. Organisms interact with each other through such processes as competition and dependence (section 10.4). Not only do they modify the environment, but as we saw in the concept of succession (section 10.5), this very act of modification may so change the nature of the environment that certain species may benefit while others will be disadvantaged.

17.4 The Importance of Scale

The relative importance of these many different and interacting processes varies according to the scale with which one is concerned. Scale has been an important consideration in this book, from the global scale to the local scale exemplified by environments like cities and river catchments. For example, at a global scale the main influence on the pattern of soils will be climate, with the great soil orders (section 3.5), such as aridisols, alfisols and oxisols, broadly corresponding to major climatic belts. On the other hand, within a local area such as a river valley climate may be fairly uniform; there, the main influence on the nature of soils will be relief (section 13.1), with soils and landforms being related in a toposequence or catena. Likewise, the broad pattern of a vegetation type like the savanna may be accounted for by the presence of a seasonal water deficit in the tropics (see section 7.8), but within this broad zone in which climate is a predisposing factor for savanna development, the precise character, distribution and boundaries may be caused by the presence of such local factors as human settlements or lateritic hardpans.

17.5 The Influence of Environment on Humans

Although the power of humans to change their environment is increasing, owing to both exponential population growth and rapid technological change, their activities cannot be seen as being immune to the influence of their physical environment. As we saw in chapter 5, even a country such as Britain, which has a highly urbanised population with a considerable degree of technological skill, has to face the effects of a wide range of environmental hazards. In some environments the hazards are much greater; this has been stressed for periglacial (section 4.14), arid (section 6.16), humid tropical (section 7.13), mountainous (section 8.7) and coastal (section 9.10) situations. As population levels continue to rise over the next decades, placing greater pressure on marginal areas and on limited resources, the significance of some of these hazards and problems may increase. The basic knowledge acquired by physical geographers and those in related fields about the operation of environmental systems may help to reduce these threats.

17.6 The Human Impact

Although during much of the nineteenth century and the first half of the twentieth many geographers were concerned with *environmental determinism* – the effect that the environment had on the actions and character of people and their societies – it has become more apparent in recent decades that humans themselves are an extremely important environmental influence. Among the important consequences of human activities that have been described in this book are:

The dispersal of animal species (sections 3.4 and 10.11)
Thermokast development in periglacial areas (section 4.11)
Origins of mid-latitude grasslands (section 5.9)
Development of Mediterranean scrubs (section 5.10)
Land subsidence (sections 5.17 and 6.13)
Increases in runoff and sediment yield (section 6.11)
Groundwater depletion (section 6.13)
Sediment trapping by dams (section 6.14)
Expansion of deserts (desertification) (section 6.15)

Secondary rainforest formation (section 7.7)
Savanna origins (section 7.8)
Tropical soil infertility and erosion (section 7.13)
The 'greenhouse effect' caused by high CO_2 levels (section 7.13)
Introduction of *Spartina* to British salt marshes (section 9.8)
Sea-level change (section 9.10)
Coast erosion acceleration (section 9.10)
Domestication of plants and animals (section 10.10)
Atmospheric pollution (acid rain etc.) (sections 10.12 and 16.3)
Depletion of the ozone layer (section 10.12)
Eutrophication of waters (section 10.12)
Habitat removal (section 10.13)
Extinction of animals (section 10.14)
Conservation of species (section 10.15)
Initiation of mass movements on slopes (section 12.2)
Soil modification (section 13.1)
Soil erosion (sections 13.6 and 16.6)
Infiltration rates in soil (section 14.5)
Streamflow (section 14.8)
River channel changes (section 15.6)
River sediment type (section 15.10)
River flooding (sections 15.12 and 16.4)
Urban heat islands (section 16.2)

From this list it is apparent that the range of human activities and their environmental effects are substantial. Some of the effects may be *deliberate,* such as the modification of stream discharge by the construction of a dam or the reduction of coastal erosion by the construction of a groyne, but some may be *inadvertent,* such as the salinisation of an irrigation system, or the accelerated erosion that a groyne may cause down the coast from where it was emplaced.

17.7 The Ever-changing Environment

It is remarkable, but partly uncoincidental, that at the same time that scientists have become concerned with anthropogenic impacts on the environment, they have also become increasingly aware of the frequency, magnitude and consequences of natural environmental changes (table 17.1) at a whole range of time scales from relatively short-lived events like ENSO phenomena and events at the decade and century scales (e.g. the Little Ice Age), through the major fluctuations of the Holocene and the Younger Dryas, and the cyclic events of the Pleistocene to the longer-term causes of the Cenozoic climate decline. Much of the reason for this concern arises from the development in the past four decades of new technologies for dating and for environmental reconstruction, including the coring of the ocean floors, lakes and ice sheets. We know that all environments, including the humid tropics and the dead hearts of deserts, have been affected by climatic change, and those climatic changes can be of very abrupt onset. It is no longer possible to see most components of the environment as being in equilibrium with some supposedly stable present-day climate. The likelihood is otherwise.

Table 17.1 Examples of environmental change

Example	*Section or Window*
Cretaceous–Tertiary boundary	Window 1.9
El Niño	Section 2.8 Window 2.5
Little Ice Age	Window 2.6
Quaternary	Section 2.9 Window 5.3 Section 8.8
Medieval warm epoch	Section 2.9
Cenozoic climatic decline	Section 2.9
Pluvials	Window 2.8 Section 6.5
Ice Age in Britain	Window 2.7
Dust Bowl	Section 6.5 Window 6.2
Lake fluctuations	Section 14.10

Equally, since the 1960s, global tectonics have become a central concern in the earth sciences and from that has arisen a whole series of major research themes that include not only the development and global pattern of such phenomena as earthquakes, volcanoes and mountain ranges but also the development of more meso-scale features such as erosion surfaces, escarpments on passive margins, atolls and deltas. It is no longer possible to explain the development of the British landscape without reference to the opening of the North Atlantic, the igneous activity of the early Tertiary, the rifting and subsidence of the North Sea and its margins and even the collision of the African and Eurasian plates. The study of neotectonics is fertile ground for understanding many geomorphological phenomena. However, global tectonics are also fundamental to understanding long-term climatic evolution, particularly because of the uplift of the Himalayas, Tibetan Plateau and the Western Cordillera of the Americas, and major patterns in biogeography.

Other highly important changes in our environment result from the passage of time. Important concepts here include the ideas of long-term slope evolution by workers like W. M. Davis (section 12.5), and the ideas of plant succession in ecology (section 10.5). Thus, to understand the environment it is important to appreciate the significance of both time and history. In particular, we must remember that the present may be atypical, that in the present landscape there may be many relict features (see, for example, sections 5.13–5.16), and that we must undertake long-term monitoring of natural phenomena to comprehend their range of variability through time.

17.8 Humans or Nature?

In many cases of environmental change it is not possible to state, without risk of contradiction, that it is people rather than nature that are responsible (plate 17.1). Most systems are complex and humans are but one component of them, so that many of our actions can lead to end-products that are intrinsically similar to those that may be produced by natural forces (table 17.2). It is one example of

Plate 17.1 Many parts of the world's coastlines are suffering from erosion. Is this a natural phenomenon or is it caused by human actions? Will the rate of erosion increase with a rise in sea level caused by global warming? Here, the main railway line along the Spanish coast near Barcelona is under attack.

Table 17.2 Man or nature? Some examples

	Some possible causes	
Nature of the change	*Anthropogenic*	*Natural*
Desertification in a semi-arid area	Overgrazing, wood collection etc.	Drought and climatic change
Gully development in a valley bottom	Runoff from a new road, removal of protective vegetation etc.	Change in climate or base level
Increasing coast erosion	Effect of groynes up the coast, or of a reduced sediment supply because of the damming of rivers	More frequent storms or an increase in the height of sea level
Greater river flood intensity	Removal of natural vegetation, urbanisation, installation of field drains	Higher-intensity rainfall
Early twentieth-century climatic warming	The CO_2-generated 'greenhouse effect'	Changes in solar emission and the amount of volcanic dust in the atmosphere

the equifinality problem which we shall investigate in the next section. The determination of cause is often a ticklish problem, given the intricate interdependence of different components of ecosystems, the varying relaxation times that different ecosystem components may have when subjected to a new impulse, and the frequency and complexity of environmental changes. Obviously, the problem does not apply to the same extent to changes that have been brought about deliberately and knowingly by man, but it does apply to the many cases in which we may have initiated change inadvertently and non-deliberately.

We have been living on earth and modifying it to different degrees for several millions of years, so it is problematical to reconstruct the environment before our intervention. We seldom have any completely clear baseline against which to measure anthropogenic changes. Moreover, even without our intervention, the environment would be in a perpetual state of flux at a great many different time scales. In addition, there are spatial and temporal discontinuities between cause and effect. For example, erosion in one locality may lead to deposition in another, while destruction of one small but key element of an animal's habitat may lead to population declines throughout its range. Likewise, in a time context, a considerable interval may elapse before the full implications of an activity are apparent. For instance, changes in a soil profile caused by human-induced soil erosion may lead to changes in the vegetation cover, which in turn may trigger changes in the quality of stream water, which may eventually modify the habitat of fishes in a lake. Likewise, it is conceivable that erosion in a mountainous area will deposit silt in a river that will eventually reach the sea and therefore modify the sediment budget of a beach, setting off a phase of coastal progradation that may in turn modify the drainage conditions of a coastal marsh area. The ultimate impact will be removed a considerable distance from the initial cause, both in terms of space and time. Primary impacts give rise to a myriad of successive repercussions throughout the ecosystems which it may be impractical to trace and monitor.

17.9 Equifinality

The problems of explanation in physical geography caused by the number of factors involved and their interaction, by the difficulties of scale, by the frequency of change and by the problem of deciphering the role of humans as against that of nature are heightened by the fact that different processes can lead to similar end-forms – the problem of equifinality. When seeking an explanation for a particular phenomenon it is important to remember that, although certain phenomena appear to be broadly similar in type, their form may be an inadequate guide to their origin. One should not be dogmatic as to the origins of many natural phenomena. Space has not allowed the development of this theme to its full extent, but the arguments that were put forward in section 5.15 for the development of dry valleys and mis-fit streams provide a striking illustration of this problem. Many landforms, including stone pavements (section 6.7), desert crusts (section 6.7), pediments (section 6.12), linear dunes (section 6.10), inselbergs (section 7.11) and patterned ground (section 4.12), have been explained in a variety of different ways, and such explanations have not been mutually exclusive. More than one may be correct. When conducting one's own field investigations, therefore, it is necessary to adopt the principle of *multiple working hypotheses*, seeking to formulate and test as many explanations as possible.

17.10 The Relevance and Application of Physical Geography

Humans cannot be seen in isolation from their environment, and thus the study of our environment, and our interaction with it, has intrinsic importance. Physical geography is just such a study, and it should not be seen as divorced from human geography. The understanding of the environment and its complexity, the appreciation that it changes frequently and rapidly, the recognition that the environment both influences and is influenced by people and the concern with certain major problems facing the human race (pollution, conservation, extinction, erosion etc.) ensure that physical geography is an important field of study that has great relevance and application to human problems.

A consideration of the role of the *applied geomorphologist* may illustrate these points further, and demonstrate some of the skills of the physical geographer. These include:

(a) mapping landforms, and describing what they are like;
(b) using landforms as indicators of other distributions (such as soils);
(c) identifying change through (i) monitoring and (ii) historical analysis;
(d) analysing the causes of change and hazards;
(e) assessing what happens when humans intervene in the system accidentally;
(f) deciding what to do to prevent or control the undesirable consequences of such changes.

In the first instance the applied geomorphologist will attempt to map the distribution of phenomena that have been identified. Both the identification and the mapping are skills, and they depend partly on having 'an eye for the country'. So, for example, if a road is to be built through a desert area, a map of landforms will help to determine the best alignment (e.g. the avoidance of shifting sand dunes, areas prone to flash floods, and locations where

Table 17.3 Examples of natural hazards

Example	*Section or window*
Avalanches	Section 4.14
	Window 8.1
	Window 8.2
Biological	Section 5.17
Coast erosion	Section 9.9
Desertification	Window 6.3
Droughts	Section 5.17
Dust storms	Window 6.2
Earthquakes	Section 11.4
	Window 11.5
Floods (coastal)	Section 5.17
Floods (river)	Section 5.17
	Window 5.5
	Section 15.12
	Window 15.2
	Section 16.4
Fog	Section 5.17
Glacier surges	Window 4.2
Hurricanes	Window 7.1
	Window 7.2
Jökulhlaups	Window 4.6
Slope instability	Section 12.2
Soil erosion	Section 13.6
Storms	Section 5.17
Thermokarst	Section 4.11
Tornadoes	Section 5.17
Tsunamis	Window 11.3
Volcanic	Section 11.2
	Window 11.1
	Window 11.2
	Window 11.4

Plate 17.2 The railway line to Walvis Bay in Namibia suffers from the hazard of sand encroachment and dune movement. Geomorphologists can advise on the selection of the best route for roads and railways in desert areas.

salt weathering may well be extremely detrimental to a road's foundations). It will also help in determining suitable materials for the construction of the road; for certain landforms, especially depositional ones, may be composed of materials with particularly desirable or undesirable properties. Thus landforms can be both resources and hazards. Indeed, the general description of a particular terrain type may provide the engineer with some important information. For example, an alluvial fan (see section 6.11) is a depositional surface across which streams are prone to wander, where there may be alternations of cutting and filling, where the cause of deposition is often a mudflow charged with coarse debris and possessing considerable transportational power when confronted with a man-made structure. All these characteristics – instability, cut-and-fill, coarse debris and the activities of mass movements – need to be considered if an alluvial fan environment is to be safely and successfully exploited.

Second, the applied geomorphologist may use the fact that certain types of landform are especially distinctive on air photos to map the distribution of other phenomena that are less easily determined but are closely related in their distribution to the presence of the more distinct forms. For example, using the principle of the catena concept (section 13.1), the mapping of particular slope forms may enable a map to be made showing the boundaries of the main types of soil in an area. The boundaries can then be checked by field sampling and modified where necessary.

In addition to mapping, the applied geomorphologist is concerned with processes, changes and hazards (table 17.3) in our ever-changing environment. It is important to know, for example, the rates at which geomorphological changes are taking place – how fast a coast is retreating towards a town or housing estate, how quickly a sand dune is encroaching on an irrigation canal or railway (plate 17.2), how stable the snout of a glacier near a new motorway may be, how active faults are in the vicinity of a nuclear plant, how much the sea level is changing in a flood-prone, low-lying area, and whether a landslide near the site of a new dam is active or relict. The use of cartographic sources, historical records and air photos may help in this endeavour, as will the instrumentation and direct monitoring of such processes. The increasing concern of geomorphology with such processes over the past few decades has enabled an increasingly important contribution to be made in this field.

Fourth, having identified the changes taking place in the landscape and the environment, it is necessary to be aware of the causes of these changes. As has been mentioned, the causes of change are complex, involving many human and natural factors. In order to take the appropriate action to ameliorate the threat of any particular change it is vital that the real cause or causes of change be identified.

Fifth, if the decision is made to intervene in the environment to remove or reduce the threat, then it is important to recognise that the intervention itself may have all sorts of repercussions through the environment. For example, as engineers have learnt to their cost, you cannot stop coast erosion at one point by the simple expedient of building a groyne without considering the possibility that this will increase the rate of erosion further down the coast because the sediment supply to replenish the beach will have been retained upcoast by the groyne. Each engineering scheme needs to be seen in the context of not only the individual location, but also the wider area. In many countries all major construction schemes have to be preceded by an environmental impact statement or assessment, in which all the possible ramifications of a particular design and location are considered before construction can proceed. The physical geographer is able to appreciate the complex interactions within the environment at a variety of scales and so can assist in this task.

Thus, not only does the physical geographer have the pleasure of investigating the variety and beauty of the world's landscapes, but at the same time he or she can contribute to the welfare of society.

17.11 Global Environmental Change

In the past two decades it has become apparent that the human race may be capable of causing global environmental change. There are two aspects of this: 'cumulative' global change and 'systemic' global change. The former refers to the snowballing effect of local changes, which add up to produce change on a world-wide scale or change which affects a significant part of a specific global resource (e.g. acid rain or soil erosion). Many such changes are already in operation. The latter – systemic global change – refers to changes operating at the global scale and includes, for instance, global changes in climate and atmospheric chemistry brought about by the emission of gases into the atmosphere (e.g. the enhanced greenhouse effect and stratospheric ozone depletion). In turn, systemic global changes may have many impacts on our environment. For example, the global warming that might result from the enhanced greenhouse effect could modify the whole atmospheric circulation of the globe, cause changes in rainfall patterns and hurricane activity, lead to the melting of ice, cause sea levels to rise, and displace vegetation zones altitudinally and latitudinally. The physical geography of the world could be very different at the end of the twenty-first century from what it is now, and the monitoring, prediction and management of such changes provide a major opportunity for physical geographers to make a major contribution.

Index

The index aims to complement the text in drawing out the interconnections between the various elements of the environment. Entries are included for specific locations where these illustrate particular aspects of the environment but not where they are only mentioned in passing. Readers should search for information using the most specific term possible; more general headings are included to help give an overview of various aspects of the environment. Alphabetical arrangement is word by word, ignoring 'and', 'in', 'by', etc. Page numbers in *italics* refer to figures and plates.

Index compiled by Ann Kingdom